Lecture Notes in Computer Science 9049

Commenced Publication in 1973
Founding and Former Series Editors:
Gerhard Goos, Juris Hartmanis, and Jan van Leeuwen

More information about this series at http://www.springer.com/series/7409

Matthias Renz · Cyrus Shahabi
Xiaofang Zhou · Muhammad Aamir Cheema (Eds.)

Database Systems
for Advanced Applications

20th International Conference, DASFAA 2015
Hanoi, Vietnam, April 20–23, 2015
Proceedings, Part I

 Springer

Editors
Matthias Renz
Universität München
München
Germany

Cyrus Shahabi
University of Southern California
Los Angeles
USA

Xiaofang Zhou
University of Queensland
Brisbane
Australia

Muhammad Aamir Cheema
Monash University
Clayton
Australia

ISSN 0302-9743   ISSN 1611-3349   (electronic)
Lecture Notes in Computer Science
ISBN 978-3-319-18119-6   ISBN 978-3-319-18120-2   (eBook)
DOI 10.1007/978-3-319-18120-2

Library of Congress Control Number: 2015936691

LNCS Sublibrary: SL3 – Information Systems and Applications, incl. Internet/Web, and HCI

Springer Cham Heidelberg New York Dordrecht London

Printed on acid-free paper

Springer International Publishing AG Switzerland is part of Springer Science+Business Media
(www.springer.com)

Preface

It is our great pleasure to welcome you to DASFAA 2015, the 20th edition of the International Conference on Database Systems for Advanced Applications (DASFAA 2015), which was held in Hanoi, Vietnam during April 20–23, 2015. Hanoi (Vietnamese: *Hà Nội*), the capital of Vietnam, is the second largest city in Vietnam and has collected all the essence, unique features, and diversification of Vietnamese culture. The city is preserving more than 4000 historical and cultural relics, architecture and beauty spots, in which nearly 900 relics have been nationally ranked with hundreds of pagodas, temples, architectural works, and sceneries. Handcraft streets and traditional handcraft villages remain prominent and attractive to tourists when visiting Hanoi, many of which centered around the Hoan Kiem Lake in the Old Quarter, close to the conference venue. Hanoi has recently been included on TripAdvisor's list of best destinations in the world, ranked 8th among the world's top 25 destinations.

We are delighted to offer an exciting technical program, including two keynote talks by Amr El Abbadi (University of California, Santa Barbara) and Luc Vincent (Google Inc.); one 10-year best paper award presentation; a panel session on "Big Data Search and Analysis;" a poster session with 18 papers; a demo session with 6 demonstrations; an industry session with 3 full paper presentations; 3 tutorial sessions; and of course a superb set of research papers. This year, we received 287 submissions, each of which went through a rigorous review process. That is, each paper was reviewed by at least three Program Committee members, followed by a discussion led by the meta-reviewers, and a final meta-review prepared for each paper. At the end, DASFA 2015 accepted 63 full papers (the acceptance ratio is 22%).

Two workshops were selected by the Workshop Co-chairs to be held in conjunction with DASFAA 2015. They are the Second International Workshop on Big Data Management and Service (BDMS 2015), and the Second International Workshop on Semantic Computing and Personalization (SeCoP 2015). The workshop papers are included in a separate volume of proceedings also published by Springer in its Lecture Notes in Computer Science series.

The conference received generous financial support from the Hanoi University of Science and Technology (HUST). We, the conference organizers, also received extensive help and logistic support from the DASFAA Steering Committee and the Conference Management Toolkit Support Team at Microsoft.

We are grateful to all conference organizers, Han Su (University of Queensland) and many other people, for their great effort in supporting conference organization. Special thanks also go to the DASFAA 2015 Local Organizing Committee: Tuyet-Trinh Vu, Hong-Phuong Nguyen, and Van Thu Truong, all from the Hanoi University of Science

and Technology, Vietnam. Finally, we would like to take this opportunity to thank all
the meta-reviewers, Program Committee members, and external reviewers for their expertise and help in evaluating papers, and all the authors who submitted their papers to
this conference.

February 2015   Quyet-Thang Huynh
                                                                          Qing Li
                                                                   Matthias Renz
                                                                  Cyrus Shahabi
                                                                 Xiaofang Zhou

Organization

General Co-chairs

Qing Li   City University of Hong Kong, HKSAR,
                            Hong Kong
Quyet-Thang Huynh   Hanoi University of Science and Technology,
                            Vietnam

Program Committee Co-chairs

Cyrus Shahabi   University of Southern California, USA
Matthias Renz   Ludwig-Maximilians-Universität München,
                            Germany
Xiaofang Zhou   University of Queensland, Australia

Tutorial Co-chairs

Arbee L.P. Chen   NCCU, Taiwan
Pierre Senellart   Télécom ParisTech, France

Workshops Co-chairs

An Liu   Soochow University, China
Yoshiharu Ishikawa   Nagoya University, Japan

Demo Co-chairs

Haiwei Pan   Harbin Engineering University, China
Binh Minh Nguyen   Hanoi University of Science and Technology,
                            Vietnam

Panel Co-chairs

Bin Cui   Peking University, China
Katsumi Tanaka   Kyoto University, Japan

Poster Co-chairs

Sarana Nutanong   City University of Hong Kong, China
Tieyun Qian   Wuhan University, China

Industrial/Practitioners Track Co-chairs

Mukesh Mohania   IBM, India
Khai Tran   Oracle, USA

PhD Colloquium

Khoat Than   Hanoi University of Science and Technology,
                          Vietnam
Ge Yu   Northeastern University, China
Tok Wang Ling   National University of Singapore, Singapore
Duong Nguyen Vu   John Von Neumann Institute - VNU-HCMUS,
                          Vietnam

Publication Chair

Muhammad Aamir Cheema   Monash University, Australia

Publicity Co-chairs

Yunjun Gao   Zhejiang University, China
Bao-Quoc Ho   VNU-HCMUS, Vietnam
Jianfeng Si   Institute for Infocomm Research, Singapore
Wen-Chih Peng   National Chiao Tung University, Taiwan

Local Organizing Committee

Tuyet-Trinh Vu   Hanoi University of Science and Technology,
                          Vietnam
Hong-Phuong Nguyen   Hanoi University of Science and Technology,
                          Vietnam
Van Thu Truong   Hanoi University of Science and Technology,
                          Vietnam

Steering Committee Liaison

Stephane Bressan   National University of Singapore, Singapore

Webmaster

Viet-Trung Tran   Hanoi University of Science and Technology,
                          Vietnam

Program Committees

Senior PC members

Ira Assent	Aarhus University, Denmark
Lei Chen	Hong Kong University of Science and Technology (HKUST), China
Reynold Cheng	University of Hong Kong, China
Gabriel Ghinita	University of Massachusetts Boston, USA
Panos Kalnis	King Abdullah University of Science and Technology, Saudi Arabia
Nikos Mamoulis	University of Hong Kong, China
Kyriakos Mouratidis	Singapore Management University, Singapore
Mario Nascimento	University of Alberta, Canada
Dimitris Papadias	Hong Kong University of Science and Technology (HKUST), China
Stavros Papadoupoulos	MIT, USA
Torben Bach Pedersen	Aalborg University, Denmark
Jian Pei	Simon Fraser University, Canada
Thomas Seidl	RWTH Aachen University, Germany
Timos Sellis	RMIT University, Australia
Raymond Wong	Hong Kong University of Science and Technology (HKUST), China

PC Members

Nikolaus Augsten	University of Salzburg, Austria
Spiridon Bakiras	City University of New York, USA
Zhifeng Bao	University of Tasmania, Australia
Srikanta Bedathur	IBM Research, Delhi, India
Ladjel Bellatreche	University of Poitiers, France
Boualem Benatallah	University of New South Wales, Australia
Bin Cui	Peking University, China
Athman Bouguettaya	Commonwealth Scientific and Industrial Research Organisation (CSIRO), Australia
Panagiotis Bouros	Humboldt-Universität zu Berlin, Germany
Selcuk Candan	Arizona State University, USA
Jianneng Cao	A*STAR, Singapore
Marco Casanova	Pontifical Catholic University of Rio de Janeiro, Brazil
Sharma Chakravarthy	University of Texas at Arlington, USA
Jae Chang	Chonbuk National University, Korea
Rui Chen	Hong Kong Baptist University, China
Yi Chen	New Jersey Institute of Technology, USA
James Cheng	The Chinese University of Hong Kong (CUHK), China
Gao Cong	Nanyang Technological University (NTU), Singapore

Ugur Demiryurek	University of Southern California (USC), USA
Prasad Deshpande	IBM Research, India
Gill Dobbie	University of Auckland, New Zealand
Eduard Dragut	Temple University, USA
Cristina Dutra de Aguiar Ciferri	Universidade de São Paulo, Brazil
Sameh Elnikety	Microsoft Research Redmond, USA
Tobias Emrich	Ludwig-Maximilians-Universität München, Germany
Johann Gamper	Free University of Bozen-Bolzano, Italy
Xin Gao	King Abdullah University of Science and Technology (KAUST), Saudi Arabia
Chenjuan Guo	Aarhus University, Denmark
Ralf Hartmut Güting	University of Hagen, Germany
Takahiro Hara	Osaka University, Japan
Haibo Hu	Hong Kong Baptist University, China
Yoshiharu Ishikawa	Nagoya University, Japan
Mizuho Iwaihara	Waseda University, Japan
Adam Jatowt	Kyoto University, Japan
Vana Kalogeraki	Athens University of Economy and Business, Greece
Panos Karras	Skoltech, Russia
Norio Katayama	National Institute of Informatics, Japan
Sang-Wook Kim	Hanyang University, Korea
Seon Ho Kim	University of Southern California (USC), USA
Hiroyuki Kitagawa	University of Tsukuba, Japan
Peer Kröger	Ludwig-Maximilians-Universität München, Germany
Jae-Gil Lee	Korea Advanced Institute of Science and Technology (KAIST), Korea
Wang-Chien Lee	Portland State University (PSU), USA
Sang-Goo Lee	Seoul National University, Korea
Hou Leong	University of Macau, China
Guoliang Li	Tsinghua University, China
Hui Li	Xidian University, China
Xiang Lian	University of Texas–Pan American (UTPA), USA
Lipyeow Lim	University of Hawaii, USA
Sebastian Link	University of Auckland, New Zealand
Bin Liu	NEC Laboratories, USA
Changbin Liu	AT & T, USA
Eric Lo	Hong Kong Polytechnic University, China
Jiaheng Lu	Renmin University of China, China
Qiong Luo	Hong Kong University of Science and Technology (HKUST), China
Matteo Magnani	Uppsala University, Sweden
Silviu Maniu	University of Hong Kong (HKU), China
Essam Mansour	Qatar Computing Research Institute, Qatar

Marco Mesiti	University of Milan, Italy
Yasuhiko Morimoto	Hiroshima University, Japan
Wilfred Ng	Hong Kong University of Science and Technology (HKUST), China
Makoto Onizuka	Osaka University, Japan
Balaji Palanisamy	University of Pittsburgh, USA
Stefano Paraboschi	Università degli Studi di Bergamo, Italy
Sanghyun Park	Yonsei University, Korea
Dhaval Patel	IIT Roorkee, India
Evaggelia Pitoura	University of Ioannina, Greece
Pascal Poncelet	Université Montpellier 2, France
Maya Ramanath	Indian Institute of Technology, New Delhi, India
Shazia Sadiq	University of Queensland, Australia
Sherif Sakr	University of New South Wales, Australia
Kai-Uwe Sattler	Ilmenau University of Technology, Germany
Peter Scheuermann	Northwestern University, USA
Markus Schneider	University of Florida, USA
Matthias Schubert	Ludwig-Maximilians-Universität München, Germany
Shuo Shang	China University of Petroleum, Beijing, China
Kyuseok Shim	Seoul National University, Korea
Junho Shim	Sookmyung Women's University, Korea
Shaoxu Song	Tsinghua University, China
Atsuhiro Takasu	National Institute of Informatics, Japan
Kian-Lee Tan	National University of Singapore (NUS), Singapore
Nan Tang	Qatar Computing Research Institute, Qatar
Martin Theobald	University of Antwerp, Belgium
Dimitri Theodoratos	New Jersey Institute of Technology, USA
James Thom	RMIT University, Australia
Wolf Tilo-Balke	University of Hannover, Germany
Hanghang Tong	City University of New York (CUNY), USA
Yongxin Tong	Hong Kong University of Science and Technology (HKUST), China
Kristian Torp	Aalborg University, Denmark
Goce Trajcevski	Northwestern University, USA
Vincent S. Tseng	National Cheng Kung University, Taiwan
Stratis Viglas	University of Edinburgh, UK
Wei Wang	University of New South Wales, Australia
Huayu Wu	Institute for Infocomm Research (I^2R), Singapore
Yinghui Wu	University of California, Santa Barbara (UCSB), USA
Xiaokui Xiao	Nanyang Technological University (NTU), Singapore
Xike Xie	Aalborg University, Denmark
Jianliang Xu	Hong Kong Baptist University, China
Bin Yang	Aalborg University, Denmark

Yin Yang	Advanced Digital Sciences Center, Singapore
Man-Lung Yiu	Hong Kong Polytechnic University, China
Haruo Yokota	Tokyo Institute of Technology, Japan
Jeffrey Yu	The Chinese University of Hong Kong (CUHK), China
Zhenjie Zhang	Advanced Digital Sciences Center (ADSC), Singapore
Xiuzhen Zhang	RMIT University, Australia
Kevin Zheng	University of Queensland, Australia
Wenchao Zhou	Georgetown University, USA
Bin Zhou	University of Maryland, Baltimore County, USA
Roger Zimmermann	National University of Singapore (NUS), Singapore
Lei Zou	Beijing University, China
Andreas Züfle	Ludwig-Maximilians-Universität München, Germany

External Reviewers

Yeonchan Ahn	Seoul National University, Korea
Cem Aksoy	New Jersey Institute of Technology, USA
Ibrahim Alabdulmohsin	King Abdullah University of Science and Technology, Saudi Arabia
Yoshitaka Arahori	Tokyo Institute of Technology, Japan
Zhuowei Bao	Facebook, USA
Thomas Behr	University of Hagen, Germany
Jianneng Cao	A*STAR, Singapore
Brice Chardin	LIAS/ISAE-ENSMA, France
Lei Chen	Hong Kong Baptist University, China
Jian Dai	The Chinese Academy of Sciences, China
Ananya Dass	New Jersey Institute of Technology, USA
Aggeliki Dimitriou	National Technical University of Athens, Greece
Zhaoan Dong	Renmin University of China, China
Hai Dong	RMIT University, Australia
Zoé Faget	LIAS/ISAE-ENSMA, France
Qiong Fang	Hong Kong University of Science and Technology (HKUST), China
ZiQiang Feng	Hong Kong Polytechnic University, China
Ming Gao	East China Normal University, China
Azadeh Ghari-Neiat	RMIT University, Australia
Zhian He	Hong Kong Polytechnic University, China
Yuzhen Huang	The Chinese University of Hong Kong, China
Stéphane Jean	LIAS/ISAE-ENSMA, France
Selma Khouri	LIAS/ISAE-ENSMA, France
Hanbit Lee	Seoul National University, Korea
Sang-Chul Lee	Carnegie Mellon University, USA

Feng Li	Microsoft Research, Redmond, USA
Yafei Li	Hong Kong Baptist University, China
Jinfeng Li	The Chinese University of Hong Kong, China
Xin Lin	East China Normal University, China
Yu Liu	Renmin University of China, China
Yi Lu	The Chinese University of Hong Kong, China
Nguyen Minh Luan	A*STAR, Singapore
Gerasimos Marketos	University of Piraeus, Greece
Jun Miyazaki	Tokyo Institute of Technology, Japan
Bin Mu	City University of New York, USA
Johannes Niedermayer	Ludwig-Maximilians-Universität München, Germany
Konstantinos Nikolopoulos	City University of New York, USA
Sungchan Park	Seoul National University, Korea
Youngki Park	Samsung Advanced Institute of Technology, Korea
Jianbin Qin	University of New South Wales, Australia
Kai Qin	RMIT University, Australia
Youhyun Shin	Seoul National University, Korea
Hiroaki Shiokawa	NTT Software Innovation Center, Japan
Masumi Shirakawa	Osaka University, Japan
Md. Anisuzzaman Siddique	Hiroshima University, Japan
Reza Soltanpoor	RMIT University, Australia
Yifang Sun	University of New South Wales, Australia
Erald Troja	City University of New York, USA
Fabio Valdés	University of Hagen, Germany
Jan Vosecky	Hong Kong University of Science and Technology (HKUST), China
Jim Jing-Yan Wang	King Abdullah University of Science and Technology, Saudi Arabia
Huanhuan Wu	The Chinese University of Hong Kong, China
Xiaoying Wu	Wuhan University, China
Chen Xu	Technische Universität Berlin, Germany
Jianqiu Xu	Nanjing University of Aeronautics and Astronautics, China
Takeshi Yamamuro	NTT Software Innovation Center, Japan
Da Yan	The Chinese University of Hong Kong, China
Fan Yang	The Chinese University of Hong Kong, China
Jongheum Yeon	Seoul National University, Korea
Seongwook Youn	University of Southern California, USA
Zhou Zhao	Hong Kong University of Science and Technology (HKUST), China
Xiaoling Zhou	University of New South Wales, Australia

Contents – Part I

Social Networks I

Information Integration and Data Quality

Information Retrieval and Summarization

Contents – Part II

Social Networks II

Industrial Papers

Demo

Data Mining I

Leveraging Homomorphisms and Bitmaps to Enable the Mining of Embedded Patterns from Large Data Trees

Xiaoying Wu[1]([✉]) and Dimitri Theodoratos[2]

[1] State Key Laboratory of Software Engineering, Wuhan University, Wuhan, China
xiaoying.wu@whu.edu.cn
[2] New Jersey Institute of Technology, New York, USA
dth@njit.edu

Abstract. Finding interesting tree patterns hidden in large datasets is an important research area that has many practical applications. Along the years, research has evolved from mining induced patterns to mining embedded patterns. Embedded patterns allow for discovering useful relationships which cannot be captured by induced patterns. Unfortunately, previous contributions have focused almost exclusively on mining patterns from a set of small trees. The problem of mining embedded patterns from large data trees has been neglected. This is mainly due to the complexity of this task related to the problem of unordered tree embedding test being NP-Complete. However, mining embedded patterns from large trees is important for many modern applications that arise naturally and in particular with the explosion of big data.

In this paper, we address the problem of mining unordered frequent embedded tree patterns from large trees. We propose a novel approach that exploits efficient homomorphic pattern matching algorithms to compute pattern support incrementally and avoids the costly enumeration of all pattern matchings required by previous approaches. A further originality of our approach is that matching information of already computed patterns is materialized as bitmaps. This technique not only minimizes the memory consumption but also reduces CPU costs by translating pattern evaluation to bitwise operations. An extensive experimental evaluation shows that our approach not only mines embedded patterns from real datasets up to several orders of magnitude faster than state-of-the-art tree mining algorithms applied to large data trees but also scales well empowering the extraction of patterns from large datasets where previous approaches fail.

Keywords: Tree pattern mining · Bitmap view · Holistic twig-join algorithm

The research of this author was supported by the National Natural Science Foundation of China under Grant No. 61202035 and 61272110.

M. Renz et al. (Eds.): DASFAA 2015, Part I, LNCS 9049, pp. 3–20, 2015.
DOI: 10.1007/978-3-319-18120-2_1

1 Introduction

Nowadays, huge amounts of data are represented, exported and exchanged between and within organizations in tree-structure form, e.g., XML and JSON files, RNA sequences, and software traces [14,17]. Finding interesting tree patterns that are hidden in tree datasets has many practical applications. The goal is to capture the complex relations that exist among the data entries. Because of its importance, tree mining has been the subject of extensive research [1,2,5,7,9,11,15,16,18–22, 28,29,31]. In this context, mining a large data tree, as opposed to mining multiple small trees, allows the discovery of large patterns. Large patterns are a natural result of big data and are more informative than smaller patterns [32].

Tree pattern mining has evolved from mining induced patterns to mining embedded patterns. Embedded patterns generalize induced patterns: while induced patterns involve parent-child edges, embedded patterns involve ancestor-descendant edges. As such, embedded patterns are able to extract relationships hidden (or embedded) deeply within large trees which might be missed by induced patterns [29,31].

The Problem. Unfortunately, previous contributions have focused almost exclusively on mining patterns from a set of small trees. The problem of mining *embedded patterns* from *large data trees* has been neglected. This can be explained by the increased complexity of this task due mainly to three reasons: (a) embeddings generate a larger set of candidate patterns and this substantially increases their computation time; (b) the problem of finding an unordered embedding of a tree pattern to a data tree is NP-Complete [12]. This renders the computation of the frequency of a candidate embedded pattern difficult; and (c) mining a large data tree is more complex than mining a set of small data trees. Indeed, the single large tree setting is more general than the set of small trees, since the latter can be modelled as a single large tree rooted at a virtual unlabeled node, whereas it is not feasible to split a single large tree into small subtrees and mine the subtrees to extract large frequent patterns.

As our experiments showed, a state-of-the-art method for mining frequent embedded tree patterns at a low frequency threshold crashes after 36 hours consuming 4GB of memory for a data tree of only 10K nodes.

Contribution. In this paper, we address the problem of mining embedded unordered frequent tree patterns from a single large tree. We provide a novel approach which leverages homomorphic matches of the tree patterns to the data tree and encodes the occurrences of previously generated frequent patterns as bitmaps to efficiently compute incrementally the frequency of the generated patterns. Our main contributions are as follows:

- Checking the existence of an embedding of a pattern to a data tree is NP-complete [12], but this test can be done in polynomial time for a homomorphism [13]. Homomorphisms comprise embeddings. Therefore, we design a new pattern frequency computation approach which exploits a holistic twig-join algorithm [3] to compute the homomorphisms of a pattern to the data tree in linear time on the input and output, and then filters out homomorphisms which are not embeddings with a polynomial procedure. Our approach turns out to be much more efficient than computing the embeddings directly as previous approaches do (Section 3).

- Our approach involves an incremental method that computes the embeddings of a new candidate pattern based on the embeddings of already computed frequent patterns. We encode the embeddings of previously computed frequent patterns in inverted lists, a technique which records in polynomial space a potentially exponential number of embeddings (Section 3.2).
- A further originality of our method is that the inverted sublists are materialized as bitmaps. Exploiting bitmaps not only minimizes the memory consumption but also reduces CPU costs by translating pattern frequency computation to bitwise operations (Section 3.2).
- We extend our incremental embedding computation method with a technique which produces the encoding (inverted lists) of the embeddings of a new candidate pattern without first producing the embeddings of this pattern. This technique further reduces, by a large margin, the CPU cost and memory consumption (Section 3.4).
- We run extensive experiments to evaluate the performance and scalability of our approach on real and synthetic datasets. The experimental results show that our approach mines embedded patterns up to several orders of magnitude faster than a state-of-the-art algorithm mining embedded tree patterns when applied to a large data tree. Further, our algorithm scales smoothly in terms of execution time and space consumption empowering the extraction of patterns from large datasets where previous approaches crash (Section 4).

2 Preliminaries

Trees. A *rooted labeled tree*, $T = (V, E)$, is a directed acyclic connected graph consisting of a set of nodes V, and a set of edges $E \subseteq V \times V$, satisfying the following properties: (1) there is a distinguished node called the *root* that has no incoming edges; (2) there is a unique path from the root to any other node; and (3) there is a labeling function lb mapping nodes to labels. A tree is called *ordered* if it has a predefined left-to-right ordering among the children of each node. Otherwise, it is *unordered*. The *size* of a tree is defined as the number of its nodes. In this paper, unless otherwise specified, a tree is a rooted, labeled, unordered tree.

Tree Encoding Scheme. We assume that the input data tree T is preprocessed and the position of every node is encoded following the regional encoding scheme [3]. According to this scheme, every node in T is associated with its positional representation which is a *(begin, end, level)* triple. For every label a in T, an inverted list L_a of the positional representations of the nodes with label a is produced, ordered by their *begin* field. Fig. 1(a) shows a data tree and the positional representation of its nodes. Fig. 1(b) shows the inverted lists of its

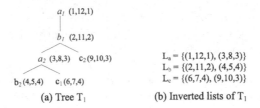

(a) Tree T_1

$L_a = \{(1,12,1), (3,8,3)\}$
$L_b = \{(2,11,2), (4,5,4)\}$
$L_c = \{(6,7,4), (9,10,3)\}$

(b) Inverted lists of T_1

Fig. 1. A tree and its inverted lists

Table 1. Occurrence information for a pattern on the tree T_1 of Fig. 1

pattern P	occu. relation OC	embbed occu. relation OC^*	occu. lists OL	emb. occu. lists OL^e	bitmaps OL^e
A B C	$\{a_1b_1c_1, a_1b_1c_2, a_1b_2c_1,$ $a_1b_2c_2, a_2b_2c_1\}$	$\{a_1b_2c_1, a_1b_2c_2,$ $a_2b_2c_1\}$	$L_A=\{a_1,a_2\}$ $L_B=\{b_1,b_2\}$ $L_C=\{c_1,c_2\}$	$L_A=\{a_1,a_2\}$ $L_B=\{b_2\}$ $L_C=\{c_1,c_2\}$	$L_A=11$ $L_B=01$ $L_C=11$

labels. In the following and depending on the context we use the same symbol T to refer interchangeably to the tree T and its set of inverted lists.

Tree Morphisms. There are two types of tree patterns: patterns whose edges represent child relationships (child edges) and patterns whose edges represent descendant relationships (descendant edges). In the literature of tree pattern mining, different types of morphisms are employed to determine if a tree pattern is included in a tree.

Given a pattern P and a tree T, a *homomorphism* from P to T is a function m mapping nodes of P to nodes of T, such that: (1) for any node $x \in P$, $lb(x) = lb(m(x))$; and (2) for any edge $(x,y) \in P$, if (x,y) is a child edge, $(m(x), m(y))$ is an edge of T, while if (x,y) is a descendant edge, $m(x)$ is an ancestor of $m(y)$ in T.

Previous contributions have constrained the homomorphisms considered for tree mining in different ways. An *isomorphism* from a pattern P with child edges to T is an injective function m mapping nodes of P to nodes of T, such that: (1) for any node x in P, $lb(x) = lb(m(x))$; and (2) (x,y) is an edge of P iff $(m(x), m(y))$ is an edge of T. If isomorphisms are considered, the mined patterns are qualified as *induced*. An *embedding* from a pattern P with descendant edges to T is an injective function m mapping nodes of P to nodes of T, such that: (1) for any node $x \in P$, $lb(x) = lb(m(x))$; and (2) (x,y) is an edge in P *iff* $m(x)$ is an ancestor of $m(y)$ in T. Patterns mined using embeddings are called *embedded* patterns. Induced patterns are a subset of embedded patterns. In this paper, we consider mining embedded patterns.

Support. We identify an *occurrence* of P on T by a tuple indexed by the nodes of P whose values are the images of the corresponding nodes in P under a homomorphism of P to T. The values in a tuple are the positional representations of nodes in T. An *embedded occurrence* of P on T is an occurrence defined by an embedding from P to T.

The set of occurrences of P under all possible homomorphisms of P to T is a relation OC whose schema is the set of nodes of P. If X is a node in P labeled by label a, the *occurrence list of X on T* is a sublist L_X of L_a containing only those nodes that occur in the column for X in OC. Let OC^e be the subset of OC containing all the embedded occurrences from P to T. The *embedded occurrence list of X on T*, denoted L_X^e, is defined similarly to L_X over OC^e instead of OC. Clearly, L_X^e is a sublist of L_X.

We define the *occurrence list set of P on T* as the set OL of the occurrence lists of the nodes of P on T; that is, $OL = \{L_X \mid X \in nodes(P)\}$. Similarly, we define the *embedded occurrence list set of P on T*, OL^e, as the set of the embedded occurrence lists of the nodes of P on T.

As an example, Table 1 shows the occurrence relations and lists as well as their embedded versions for a pattern on the tree T_1 of Fig. 1(a).

The *support* of pattern P on T is defined as the size of the embedded occurrence list of the root R of P on T, L_R^e. A pattern is *frequent* if its support is no less than a user defined threshold *minsup*. We denote by F_k the set of all frequent patterns of size k, also known as *k-patterns*.

Canonical Form. A unordered pattern may have multiple alternative isomorphic representations. In order to design efficient mining algorithms, a process is needed for minimizing the redundant generation of the isomorphic representations of the same pattern, and for efficiently checking whether two representations are isomorphic. To this end, the concept of canonical form of a tree is used by pattern mining algorithms as a representative of the corresponding pattern. A detailed study of various canonical representations of trees can be found in [6]. Our approach also employs a canonical form for tree patterns.

Problem Statement. Given a large tree T and a minimum support threshold *minsup*, our goal is to mine all frequent unordered embedded patterns.

3   Proposed Approach

As existing pattern mining approaches, our approach for mining embedded tree patterns from a large tree iterates between the candidate generation phase and the support counting phase. In the first phase, we use a systematic way to generate candidate patterns that are potentially frequent. In the second phase, we develop an efficient method to compute the support of candidate patterns.

3.1   Candidate Generation

In order to systematically generate candidate patterns, we adopt the equivalence class-based pattern generation method introduced in [29,30] outlined next.

Equivalence Class-Based Pattern Generation. Let P be a pattern of size k-1. Each node of P is identified by its *depth-first position* in the tree, determined through a depth-first traversal of P, by sequentially assigning numbers to the

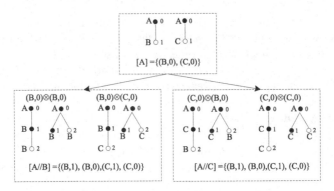

Fig. 2. An example of equivalence class expansion. The black nodes of each pattern represent the immediate prefix to all patterns in the equivalence class.

first visit of the node. The *rightmost leaf* of P, denoted *rml*, is the node with the highest depth-first position. The *immediate prefix* of P is the sub-pattern of P obtained by deleting the *rml* from P. The *equivalence class* of P is the set of all the patterns of size k that have P as the immediate prefix. We denote the equivalence class of P as $[P]$.

Each element of $[P]$ can be represented by a pair (x, i), where x is the label of the *rml* of the k-pattern and i specifies the depth-first position of the parent of the *rml* in P. Any two members of $[P]$ differ only in their *rmls*. We use the notation P_x^i to denote the k-pattern formed by adding a child node labeled by x to the node with position i in P as the rightmost leaf node. We may use P_x^i and (x, i) interchangeably in the following paragraphs.

Given an equivalence class $[P]$, the equivalence class expansion is used to obtain equivalence classes containing the successors of the patterns in $[P]$. The main idea of this expansion is to join each pattern $P_x^i \in [P]$ with any other pattern in $[P]$ including itself (self expansion) to produce the patterns of the equivalence class $[P_x^i]$. There can be up to two outcomes for each pair of patterns to be joined. Formally, let P_x^i and P_y^j denote any two elements in $[P]$. The join operation $P_x^i \otimes P_y^j$ is defined as follows:

- if $j \leq i$, return the pattern Q_y^j where $Q = P_x^i$.
- if $j = i$, return the pattern Q_y^{k-1} where $Q = P_x^i$.

By joining P_x^i with all elements P_y^j of $[P]$, we generate all possible k-patterns in $[P_x^i]$. We call patterns P_x^i and P_y^j the *left-parent* and *right-parent* of a join outcome, respectively. An equivalence class expansion example is given in Fig. 2.

3.2   Support Computation

Recall that the support of a pattern P in the input data tree T is defined as the size of the embedded occurrence list L_R^e of the root R of P on T (Section 2). A straightforward method for computing L_X^e consists of first computing OC^e (the relation which stores the embedded occurrences of P under all possible embeddings of P to T), and then projecting OC^e on column R to get L_R^e. Unfortunately, the problem of finding an unordered embedding of P to T is NP-Complete [12]. On the other hand, it has been shown that finding all the homomorphic matches of P on T can be done in linear time in the sum of sizes of the input T and the output (set of matches) [3]. Inspired by this observation, we develop a funtion called *IsFrequent*, which first computes the relation OC of the occurrences of P under all possible homomorphisms of P to T, and then filters out non-embedded occurrences from OC to get OC^e. The outline of function *IsFrequent* is shown in Fig. 3. We discuss below this function in more detail.

A Holistic Twig-Join Approach for Computing Relation OC. In order to compute OC, we use a holistic twig-join algorithm (e.g., *TwigStack* [3]), the state of the art algorithm for computing all the occurrences of tree-pattern queries on tree data. Algorithm *TwigStack* joins multiple inverted lists at a time to avoid generating intermediate join results. It uses a stack for every tree pattern node, and works in two phases. In the first phase, it computes occurrences of the individual root-to-leaf paths of the pattern. In the second phase, it merge-joins the path occurrences to compute the results for the pattern (Function *Merge-AllPathOccurrences* in Fig. 3).

An important property of *TwigStack* is that whenever a data node x is pushed into the stack of a pattern node X having child nodes $Y_1, \ldots, Y_n$, the algorithm ensures that: (a) x has a descendant node on each of the inverted lists corresponding to the labels of nodes $Y_1, \ldots, Y_n$, and (b) each of these descendant nodes recursively satisfies this property. Thus, the algorithm can guarantee worst-case performance *linear* to the size of the data tree inverted lists (the input) and the size of relation OC (the output), i.e., it is optimal.

Nevertheless, the *TwigStack*-based method can still be expensive for computing a large number of candidates, since it needs to scan fully the inverted lists corresponding to every candidate pattern. We elaborate below on an incremental method, which computes relation OC of a pattern by leveraging the computation done at its parent patterns in the search space.

Computing Relation OC Incrementally. Let P be a pattern and X be a node in P labeled by a. Using *TwigStack*, P is computed by iterating over the inverted lists corresponding to every pattern node. If there is a sublist, say L_X, of L_a such that P can be computed on T using L_X instead of L_a, we say that node X can be *computed using L_X on T*. Since L_X is non-strictly smaller than L_a, the computation cost can be reduced. Based on this idea, we propose an incremental method that uses the occurrence lists of the two parent patterns of a given pattern to compute that pattern.

Let P be a pattern of size $k-1$, where $k > 1$, and P_x^i and P_y^j be two k-patterns in the class $[P]$. Recall that the equivalence class expansion operation $P_x^i \otimes P_y^j$ can have at most two outcomes. Let pattern Q denote an outcome of $P_x^i \otimes P_y^j$. Observing that: (i) the two parents (P_x^i and P_y^j) of Q share the same immediate prefix pattern P, (ii) the immediate prefix pattern of Q is its left-parent P_x^i, and (iii) the rightmost node of Q (i.e., the node labelled by y) is the rightmost node

---

Input: Pattern Q, Q's parents P_1 and P_2, OLs of P_1 and P_2, and support threshold *minsup*

Output: *true* if Q is a frequent pattern, *false* otherwise

1. $X_i :=$ the node of P_i corresponding to node X in Q, $i = 1, 2$;
2. **if** $(|L_{R_1} \cap L_{R_2}| < minsup)$ **then**
3.   **return** *false*;
4. $OL' := \{L_{X_1} \cap L_{X_2} \mid X \in Q\} \cup \{L_{rml_1}\} \cup \{L_{rml_2}\}$;
5. Compute path occurrences of Q using $TwigStack$ on OL';
6. $OC :=$ MergeAllPathOccurrences();
7. $OC^e := \{occ \in OC \mid \text{IsEmbOcc}(occ) \text{ is } true\}$;
8. Compute OL^e using OC^e, and discard OC^e afterwards;
9. **if** $(|L_r^e| < minsup)$ **then**
10.   **return** *false*;
11. **return** *true*;

Function MergeAllPathOccurrences()
1. **for** (each branching node X of Q in postorder) **do**
2.   Let $Y_1, \ldots, Y_m$ denote the list of children of node X in Q; Let $P(X)$ denote the subtree of P composed of the path from the root to X and the complete subtree rooted at X.
3.   Merge join occurrences of $P(Y_i)$ to produce occurrences for $P(X)$, $i = 1, \ldots, m$;
4. **return** the set of occurrences of Q;

Fig. 3. Function *IsFrequent*

of its right-parent P_y^j, we can easily identify a homomorphism from each parent of Q to Q. The following proposition can be shown.

Proposition 1. *Let X' be a node in a parent Q' of Q and X be the image of X' under a homomorphism from Q' to Q. The occurrence list L_X of X on T, is a sublist of the occurrence list $L_{X'}$ of X' on T.*

Recall that L_X is the inverted list of data tree nodes that participate in the occurrences of Q' to T. By Proposition 1, X can be computed using L_X instead of using the corresponding label inverted list. Further, if X is the image of nodes X_1 and X_2 defined by the homomorphisms from the left and right parent of Q, respectively, we can compute X using the *intersection*, $L_{X_1} \cap L_{X_2}$, of L_{X_1} and L_{X_2} which is the sublist of L_{X_1} and L_{X_2} comprising the nodes that appear in both L_{X_1} and L_{X_2} (line 4 in Algorithm *IsFrequent* of Fig. 3).

Using Proposition 1, we can compute Q using only the occurrence list sets of its parents. Thus, we only need to store with each frequent pattern its occurrence list set. Our method is space efficient since the occurrence lists can encode in linear space an exponential number of occurrences for the pattern [3]. In contrast, the state-of-the-art methods for mining embedded patterns [29, 31] have to store information about all the occurrences of each given pattern in T.

Another advantage offered by the incremental method is that it allows a quick identification of some non-frequent candidates before their occurrence relations are computed: let L_{R_1} and L_{R_2} denote the root occurrence lists of the left and right parents of a candidate Q, respectively. If $|L_{R_1} \cap L_{R_2}|$ is less than *minsup*, then Q is infrequent and should be excluded from further processing (lines 2-3 in Algorithm *IsFrequent* of Fig. 3). As verified by our experimental results, substantial CPU cost can be saved using this early-detection of infrequent candidate patterns.

Representing Occurrence Lists as Bitmaps. The occurrence list L_X of a pattern node X labeled by a on T can be represented by a bitmap on L_a that has a '1' bit at position i iff L_X comprises the tree node at position i of L_a. Then, the occurrence list set of a pattern is the set of bitmaps of the occurrence lists of its nodes. The last column of Table 1 shows an example of bitmaps for pattern occurrence lists. Clearly, storing the occurrence lists of multiple patterns as bitmaps results in important space savings. Moreover, as we explain below, the use of bitmaps also offers CPU and I/O cost savings.

The intersection of the occurrence lists of pattern nodes can be implemented by a bitwise operation on the corresponding bitmaps: first, the bitmaps of the operand pattern nodes are bitwise AND-ed. Then, the target occurrence list is constructed by fetching into memory the inverted list nodes indicated by the resulting bitmap. Exploiting bitmaps and bitwise operations results in time saving for two reasons. First, bitwise AND-ing bitmaps incurs less CPU cost than intersecting the corresponding occurrence lists. Second, fetching into memory the target occurrence list nodes indicated by the resulting bitmap incurs less I/O cost than fetching the entirety of the occurrence lists of the operand pattern nodes as this is required for the direct application of the intersection operation. Storing inverted lists as bitmaps is a technique initially introduced and exploited in [25–27] for materializing tree-pattern views and for efficiently answering queries using materialized views.

Identifying Embedded Occurrences. Let *occ* be an occurrence of P on T in OC, and X be a node in P. Let also *occ.X* denote the value (which is a node in T) associated with X in *occ*. Occurrence *occ* is an embedded occurrence iff for any pair of sibling nodes X and Y of P, *occ.X* and *occ.Y* are not on the same path in T. Recall that every node in T is associated with its positional representation which is a *(begin, end, level)* triple. The regional encoding allows for efficiently checking ancestor-descendant relationships between two nodes: node n_1 is an ancestor of node n_2 iff $n_1.begin < n_2.begin$, and $n_2.end < n_1.end$.

The checking procedure *IsEmbOcc* for a given occurrence *occ*, called in line 7 of Algorithm *IsFrequent*, works as follows: traverse the nodes of P in postorder; for each node X under consideration having children $Y_1, \ldots, Y_m$, check if no two *occ.Y$_i$*s are on the same path. If the condition is violated for some node X, we can conclude that *occ* is not an embedded occurrence.

The time complexity of procedure *IsEmbOcc* is $O(|P| \times P_f)$, where P_f denotes the maximum fan-out of the nodes of P. Thus, the time complexity of function *IsFrequent* is $O(|P| \times P_f \times |OC|)$, where $|OC|$ is $O(|T|^{|P|})$.

Input: inverted lists $\mathcal{L}$ of tree T and *minsup*.
Output: all the frequent embedded tree patterns in T.

1. $F_1 :=$ {frequent 1-patterns};
2. $F_2 :=$ {classes $[P]_1$ of frequent 2-patterns};
3. **for** (every $[P] \in F_2$) **do**
4. $MineEmbPatterns([P])$;

Procedure $MineEmbPatterns$(Equivalence class $[P]$)
1. **for** (each $(x, i) \in [P]$) **do**
2.   **if** (P_x^i is in canonical form) **then**
3. $[P_x^i] := \emptyset$;
4.   **for** (each $(y, j) \in [P]$) **do**
5.   **for** (each expansion outcome Q of $P_x^i \otimes P_y^j$) **do**
6.   **if** ($IsFrequent(Q, P_x^i, P_x^j, OL(P_x^i), OL(P_y^j))$) **then**
7.   add (y, j) to $[P_x^i]$;
8. $MineEmbPatterns([P_x^i])$

Fig. 4. Algorithm *EmbTPMBit* for Mining Embedded Tree Patterns

3.3   The Tree Pattern Mining Algorithm *EmbTPMBit*

We present now our embedded tree pattern mining algorithm called *EmbTPMBit* (Fig. 4). The first part of the algorithm computes the sets containing all frequent 1-patterns F_1 (i.e., nodes) and 2-patterns F_2 (lines 1-2). F_1 can be easily obtained by finding inverted lists of T whose size (in terms of number of nodes) is no less than *minsup*. The total time for this step is $O(|T|)$. F_2 is computed by the following procedure: let $X//Y$ denote a 2-pattern formed by two elements X and Y of F_1. The support of $X//Y$ is computed via algorithm *TwigStack* on the inverted lists $L_{lb(X)}$ and $L_{lb(Y)}$ that are associated with labels $lb(X)$ and $lb(Y)$, respectively. Notice that, since we do not need to check whether the occurrences of a 2-pattern are embedded (they all are), *TwigStack* can generate OL without explicitly generating OC for 2-patterns. The total time for each 2-pattern candidate is $O(|T|)$.

Then, the main loop starts by calling the procedure *MineEmbPatterns* for every frequent 2-pattern (lines 3-4). *MineEmbPatterns* is a recursive procedure that performs the equivalence class expansion to each element $(x, i) \in [P]$. It attempts to expand P_x^i with every element $(y, j) \in [P]$ and computes the support of each possible expansion outcome using Algorithm *IsFrequent*(lines 4-6). Any new pattern that turns out to be frequent is added to the new class $[P_x^i]$ (line 7). The frequent patterns at the current level form the elements of classes for the next level. The recursive process is repeated until no more frequent patterns can be generated.

Before expanding a class $[P]$, we make sure that P is in canonical form (line 2 in *MineEmbPatterns*). Our approach is independent of any particular canonical form; it can work with any systematic way of choosing a representative from isomorphic representations of the given pattern, such as those presented

Input: Pattern Q, Q's parents P_1 and P_2, OLs of P_1 and P_2, and support threshold *minsup*

Output: true if Q is a frequent pattern, *false* otherwise

1. Same as lines 1-5 of Function *IsFrequent* of Fig. 3;
2. $OL^e :=$ ComputeEmbOL();
3. **if** ($|L_r^e| < minsup$) **then**
4.   **return** *false*;
5. **return** *true*;

Function ComputeEmbOL()

1. **for** (each branching node X of Q in postorder) **do**
2.   Let $Y_1, \ldots, Y_m$ denote the list of children of node X in Q;
3.   **for** (each $x \in L_X$ in its preorder appearance in T) **do**
4. $L_{Y_i|x} := \{y \mid y \in L_{Y_i}$, and x is the parent of y in $T\}$, $i = 1, \ldots, m$;
5.   **if** ($\neg \exists (y_1, \ldots, y_m) \in L_{Y_1|x} \times \ldots \times L_{Y_m|x}$, s.t. any two y_is are not on the same path in T) **then**
6.   remove x from L_X;
7. **return** $\{L_X \mid X \in Q\}$;

Fig. 5. Function *IsFrequent2*

in [2,6,15,29]. Efficient methods for checking canonicity can also be drawn from [2,6,15,29].

3.4   An Improvement of *EmbTPMBit*: Algorithm *EmbTPMBit+*

Recall that in order to compute pattern support, Algorithm *EmbTPMBit* invokes Procedure *IsFrequent*, which first generates the occurrence relation OC. However, generating OC can incur high memory footprint when its size is large. Next, we introduce an improvement to Algorithm *EmbTPMBit*, called *EmbTPMBit+*. Algorithm *EmbTPMBit+* uses Procedure *IsFrequent2* shown in Fig. 5 to compute pattern support without explicitly generating relation OC.

In Procedure *IsFrequent2*, Function *MergeAllPathOccurrences* used in *IsFrequent* is replaced by Function *ComputeEmbOL*. At each branching node X of pattern P in postorder, *ComputeEmbOL* scans L_X and filters out nodes that do not participate in any embedded occurrences of a subpattern $P(X)$ of P, which is composed of the path from the root to X and the complete subtree rooted at X (Lines 3-6). In other words, any node x in L_X that participates in an embedded occurrence to $P(X)$ is retained.

Complexity. To compute the support for pattern P on T, *EmbTPMBit+* takes time $O(|P| \times P_f \times l^{P_f})$, where l is the maximum size of the inverted lists of T, and P_f is the maximum fan-out of the nodes of P. Also, since *EmbTPMBit+* avoids storing pattern occurrences in memory as intermediate results, it greatly improves over *EmbTPMBit* both in time and memory footprint.

In comparison, the-state-of-art embedded pattern mining algorithm *sleuth* needs $O(|P| \times |T|^{2|P|})$ to compute the support for pattern P on T [29].

4   Experimental Evaluation

We implemented our algorithms *EmbTPMBit* and *EmbTPMBit+* and we compare them with a state-of-the-art unordered embedded tree mining algorithm *sleuth* [29]. *sleuth* was designed to mine embedded patterns from a set of small trees. In order to allow the comparison, we adapted *sleuth* to a large single tree setting by making it to return as support of a pattern the number of root occurrences of this pattern in the data tree.

Our implementation was coded in Java. All the experiments reported here were performed on a workstation having an Intel Xeon CPU 3565 @3.20 GHz processor with 8GB memory running JVM 1.7.0 in Windows 7 Professional. The Java virtual machine memory size was set to 4GB.

Table 2. Dataset statistics

Dataset	Tot. #nodes	#labels	Max/Avg depth	#paths
Treebank	906337	191	36/8.4	521052
T10k	10000	10	21/20.7	8431
CSlogs	772188	13355	86/4.4	59691 (#trees)

Datasets. We ran experiments on three datasets with different structural properties[1]. Their main characteristics are summarized in Table 2.

$Treebank$[2] is a real XML dataset derived from computation linguistics. The dataset is deep and comprises highly recursive and irregular structures. The original XML tree has 2.4M nodes. To allow *sleuth* mine some patterns within a reasonable amount of time, we used a subtree which has 35% of the nodes of the original tree.

$T10K$[3] is a synthetic dataset generated by the tree generation program provided by Zaki [30]. The tree generation process by the program is based on different parameters.

$CSlogs$[3] is a real dataset provided by Zaki [30] and is composed of users access trees to the CS department website at RPI. The dataset contains 59,691 trees that cover a total of 13,355 unique web pages. The average size of each tree is 12.94.

4.1   Time Performance

Figures 6(a), 7(a), and 8(a) show the time spent by *sleuth*, *EmbTPMBit*, and *EmbTPMBit+*, respectively, under different support thresholds on the Treebank,

---

[1] We also ran experiments on another two datasets, the results of which are similar and are omitted in the interest of space.

[2] http://www.cis.upenn.edu/~treebank

[3] http://www.cs.rpi.edu/~zaki/software/

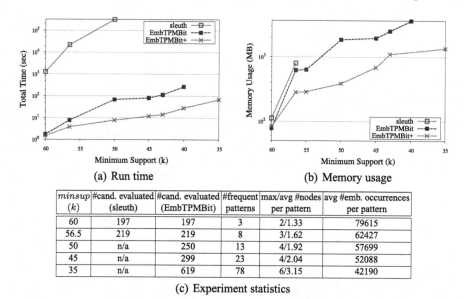

(a) Run time (b) Memory usage

minsup (k)	#cand. evaluated (sleuth)	#cand. evaluated (EmbTPMBit)	#frequent patterns	max/avg #nodes per pattern	avg #emb. occurrences per pattern
60	197	197	3	2/1.33	79615
56.5	219	219	8	3/1.62	62427
50	n/a	250	13	4/1.92	57699
45	n/a	299	23	4/2.04	52088
35	n/a	619	78	6/3.15	42190

(c) Experiment statistics

Fig. 6. Performance comparison on Treebank

CSlogs and T10K datasets. We observe that on all three datasets, *EmbTPMBit+* runs orders of magnitude faster than *sleuth* especially for low support levels. The rate of increase of the running time for *EmbTPMBit+* is slower than that for *sleuth* as the support level decreases. The reasons for the large performance gap are different for each dataset as we discuss below, but their common denominator is the efficiency by which our algorithms compute the support of the generated patterns.

On the Treebank dataset, the average number of pattern occurrences is very large. For instance, the number reaches more than 62k at support threshold 56.5k (see Fig. 6(c)). *sleuth* has to keep track of all possible occurrences of a candidate to a data tree, and to perform expensive join operations over these occurrences. On the other hand, *EmbTPMBit+* incrementally computes candidate pattern occurrence lists using an efficient stack-based algorithm. As shown in Fig. 6(a), it is not possible for *sleuth* to find all frequent patterns at support threshold 56.5k or lower within a reasonable amount of time, whereas *EmbTPMBit+* spends only 68 seconds at support threshold 35k.

On the real dataset CSlogs, the number of candidates that can be enumerated is substantially larger than the Treebank and T10K datasets. For example, the number of enumerated candidates is more than 16k at support threshold 0.13 percent on CSlogs (see Fig. 7(c)). This is because CSlogs contains a large number of distinct node labels (there are 13355 labels for 772188 nodes) allowing a very large number of candidate patterns to be constructed. As the number of candidates increases, *sleuth* suffers and has to compute a large number of occurrences.

For T10K, there is a large difference in the number of candidates evaluated by *sleuth* and *EmbTPMBit+* (see Fig. 8(c)). This is because, using bitmaps of already computed patterns, *EmbTPMBit+* is able to detect many infrequent

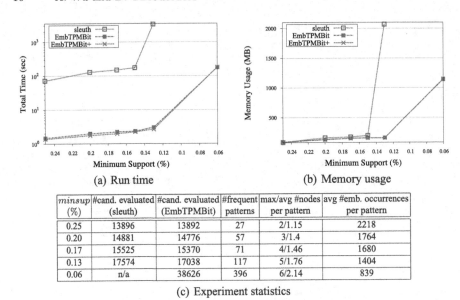

(a) Run time   (b) Memory usage

$minsup$ (%)	#cand. evaluated (sleuth)	#cand. evaluated (EmbTPMBit)	#frequent patterns	max/avg #nodes per pattern	avg #emb. occurrences per pattern
0.25	13896	13892	27	2/1.15	2218
0.20	14881	14776	57	3/1.4	1764
0.17	15525	15370	71	4/1.46	1680
0.13	17574	17038	117	5/1.76	1404
0.06	n/a	38626	396	6/2.14	839

(c) Experiment statistics

Fig. 7. Performance comparison on CSlogs

candidates before evaluating them on the data. This way, substantial CPU cost can be saved. In contrast, *sleuth* has to evaluate each candidate on the data.

Finally, we find that *EmbTPMBit+* and *EmbTPMBit* have similar time performance on CSlogs and T10K, while on Treebank, *EmbTPMBit+* outperforms *EmbTPMBit* by a factor of up to 8. The reason is that, as aforementioned, the average number of pattern occurrences on Treebank is very large (this number on the other two datasets is relatively small). Unlike *EmbTPMBit+*, *EmbTPMBit* has to explicitly generate all these occurrences to compute the pattern support.

4.2   Memory Usage

Figures 6(b), 7(b), and 8(b) show the memory consumption of *sleuth*, *EmbTPMBit*, and *EmbTPMBit+*, respectively, under different support thresholds for the Treebank, CSlogs and T10K datasets. Overall, *EmbTPMBit+* has the best memory performance, consuming substantially less memory than *sleuth* in almost all the test cases. This is mainly because *sleuth* needs to store in memory all the pattern occurrences for candidates under consideration, whereas both *EmbTPMBit*, and *EmbTPMBit+* avoid storing pattern occurrences and store only bitmaps of occurrence lists which are usually of insignificant size. *EmbTPMBit+* improves over *EmbTPMBit* by avoiding explicitly generating and storing occurrences of candidates as intermediate results (which are discarded anyway once bitmaps of occurrence lists are obtained).

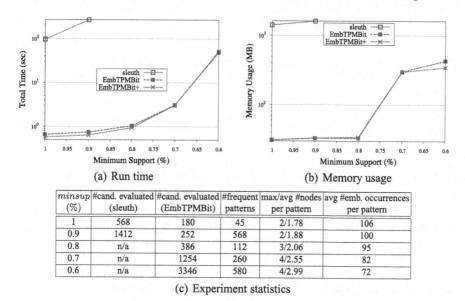

(a) Run time   (b) Memory usage

minsup (%)	#cand. evaluated (sleuth)	#cand. evaluated (EmbTPMBit)	#frequent patterns	max/avg #nodes per pattern	avg #emb. occurrences per pattern
1	568	180	45	2/1.78	106
0.9	1412	252	568	2/1.88	100
0.8	n/a	386	112	3/2.06	95
0.7	n/a	1254	260	4/2.55	82
0.6	n/a	3346	580	4/2.99	72

(c) Experiment statistics

Fig. 8. Performance comparison on T10k

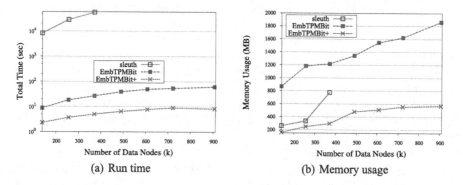

(a) Run time   (b) Memory usage

Fig. 9. Scalability comparison on Treebank with increasing size ($minsup = 5.5\%$)

4.3 Scalability Comparison on *Treebank*

In order to run scalability experiments, we created six fragments of our Treebank dataset of increasing size.

Fig. 9 shows how the three algorithms scale when the number of nodes increases from 140K to 906K, at support threshold 5.5%.

The results show that *EmbTPMBit+* always has the best time performance. As the input data size increases, we find a linear increase in the running time of *EmbTPMBit+*. The running time of *sleuth* grows much more sharply. *EmbTPMBit+* is up to 6 times faster than *EmbTPMBit*, which in turn outperforms *sleuth* by at least three orders of magnitude.

Not surprisingly, the memory consumption of all the three algorithms increases on trees with increasing number of nodes. *EmbTPMBit+* always has the smallest memory footprint. The growth of its memory consumption is slower than that of *EmbTPMBit* and *sleuth*. *EmbTPMBit* consumes more memory than *sleuth*. This is because on Treebank, the former needs to store temporarily a large set of occurrences for each candidate, whose size is a bit larger than the corresponding embedded occurrences stored by *sleuth*.

5 Related Work

The problem of mining tree patterns has been studied extensively in the past decade. Many frequent tree pattern mining algorithms [1, 2, 5, 7, 9, 11, 15, 16, 18–23, 28, 29, 31, 33] have been proposed in the literature. A majority of these works have focused almost exclusively on mining frequent isomorphic patterns from a set of small trees. We give a brief overview to algorithms that mine unordered embedded patterns [8, 20, 29].

TreeFinder [20] is the first algorithm for mining unordered embedded patterns. It uses Inductive Logic Programming and represents the trees using a relational encoding which captures all ancestor-descendant relationships. From these relational encodings, maximal frequent itemsets are computed. The frequent itemsets are used to cluster the input trees. Nevertheless, *TreeFinder* tends to miss many frequent patterns and is computationally expensive. Like *TreeFinder*, *WTIMiner* [8] transfers the frequent tree pattern mining to itemset mining. It first uses *FP-growth* [10] to find all the frequent itemsets, and then for each itemset found, it scans the database to count all the corresponding tree patterns. Although *WTIMiner* is complete, it is inefficient since the structural information is lost while mining for frequent itemsets. Further, the overhead for processing false positives may potentially reduce the performance.

sleuth [29] is developed on top of TreeMiner [31]. It associates with every pattern a *scope-list* to store the list of all its occurrences. These scope-lists are maintained to avoid repeated invocation of tree inclusion checking. The equivalence class pattern expansion method is used for generating candidates. To compute the support of a candidate a quadratic join operation is defined over the scope-lists of its two parent patterns. As illustrated in [4], the size of a pattern scope-list may be much larger than the size of a data tree. The redundant information stored in scope-lists greatly increases the memory usage of *sleuth*, especially when the pattern has a large number of occurrences. Further, the expensive join operation over large scope-lists significantly affects the runtime performance of *sleuth*. Our approach relies on an incremental stack-based approach that exploits bitmaps to efficiently compute the support.

6 Conclusion

In this paper, we have studied the important problem of discovering all embedded unordered frequent tree patterns from a single large tree. To address this pattern mining problem, we have designed a novel approach for efficiently computing the support of a candidate pattern which combines different techniques from tree databases: (a) answering and optimizing tree-pattern queries using materialized views, (b) materializing tree-pattern queries as bitmaps of inverted lists, and (c) employing holistic twig-join algorithms for efficiently finding all the homomorphisms of a tree pattern to a data tree. Our extensive experimental results show that compared to a state-of-the-art tree mining algorithm, our algorithms perform better by a wide margin in terms of time, space and scalability and indeed empower the mining of embedded tree patterns from large data trees.

We are currently working on extending our techniques in order to mine generalized graph patterns [24] though unconstrained homomorphisms from a large data tree.

References

1. Asai, T., Abe, K., Kawasoe, S., Arimura, H., Sakamoto, H., Arikawa, S.: Efficient substructure discovery from large semi-structured data. In: SDM (2002)
2. Asai, T., Arimura, H., Uno, T., Nakano, S.: Discovering frequent substructures in large unordered trees. In: Grieser, G., Tanaka, Y., Yamamoto, A. (eds.) DS 2003. LNCS (LNAI), vol. 2843, pp. 47–61. Springer, Heidelberg (2003)
3. Bruno, N., Koudas, N., Srivastava, D.: Holistic twig joins: optimal XML pattern matching. In: SIGMOD (2002)
4. Chi, Y., Muntz, R.R., Nijssen, S., Kok, J.N.: Frequent subtree mining - an overview. Fundam. Inform. **66**(1–2) (2005)
5. Chi, Y., Xia, Y., Yang, Y., Muntz, R.R.: Mining closed and maximal frequent subtrees from databases of labeled rooted trees. IEEE Trans. Knowl. Data Eng. **17**(2) (2005)
6. Chi, Y., Yang, Y., Muntz, R.R.: Canonical forms for labelled trees and their applications in frequent subtree mining. Knowl. Inf. Syst. **8**(2) (2005)
7. Dries, A., Nijssen, S.: Mining patterns in networks using homomorphism. In: SDM (2012)
8. Feng, Z., Hsu, W., Lee, M.-L.: Efficient pattern discovery for semistructured data. In: ICTAI (2005)
9. Goethals, B., Hoekx, E., den Bussche, J.V.: Mining tree queries in a graph. In: KDD (2005)
10. Han, J., Pei, J., Yin, Y.: Mining frequent patterns without candidate generation. In: SIGMOD Conference (2000)
11. Hido, S., Kawano, H.: Amiot: Induced ordered tree mining in tree-structured databases. In: ICDM (2005)
12. Kilpeläinen, P., Mannila, H.: Ordered and unordered tree inclusion. SIAM J. Comput. **24**(2), 340–356 (1995)
13. Miklau, G., Suciu, D.: Containment and equivalence for a fragment of xpath. J. ACM **51**(1), 2–45 (2004)
14. Mozafari, B., Zeng, K., D'Antoni, L., Zaniolo, C.: High-performance complex event processing over hierarchical data. ACM Trans. Database Syst. **38**(4), 21 (2013)
15. Nijssen, S., Kok, J.N.: Efficient discovery of frequent unordered trees (2003)
16. Nijssen, S., Kok, J.N.: A quickstart in frequent structure mining can make a difference. In: KDD (2004)
17. Ogden, P., Thomas, D.B., Pietzuch, P.: Scalable XML query processing using parallel pushdown transducers. PVLDB **6**(14), 1738–1749 (2013)

18. Tan, H., Hadzic, F., Dillon, T.S., Chang, E., Feng, L.: Tree model guided candidate generation for mining frequent subtrees from xml documents. TKDD **2**(2) (2008)
19. Tatikonda, S., Parthasarathy, S., Kurç, T.M.: Trips and tides: new algorithms for tree mining. In: CIKM (2006)
20. Termier, A., Rousset, M.-C., Sebag, M.: Treefinder: a first step towards xml data mining. In ICDM (2002)
21. Termier, A., Rousset, M.-C., Sebag, M., Ohara, K., Washio, T., Motoda, H.: Dryadeparent, an efficient and robust closed attribute tree mining algorithm. IEEE Trans. Knowl. Data Eng. **20**(3) (2008)
22. Wang, C., Hong, M., Pei, J., Zhou, H., Wang, W., Shi, B.-L.: Efficient pattern-growth methods for frequent tree pattern mining. In: Dai, H., Srikant, R., Zhang, C. (eds.) PAKDD 2004. LNCS (LNAI), vol. 3056, pp. 441–451. Springer, Heidelberg (2004)
23. Wang, K., Liu, H.: Discovering structural association of semistructured data. IEEE Trans. Knowl. Data Eng. **12**(3) (2000)
24. Wu, X., Souldatos, S., Theodoratos, D., Dalamagas, T., Vassiliou, Y., Sellis, T.K.: Processing and evaluating partial tree pattern queries on xml data. IEEE Trans. Knowl. Data Eng. **24**(12), 2244–2259 (2012)
25. Wu, X., Theodoratos, D., Kementsietsidis, A.: Configuring bitmap materialized views for optimizing xml queries. World Wide Web, pp. 1–26 (2014)
26. Wu, X., Theodoratos, D., Wang, W.H.: Answering XML queries using materialized views revisited. In: CIKM (2009)
27. Wu, X., Theodoratos, D., Wang, W.H., Sellis, T.: Optimizing XML queries: Bitmapped materialized views vs. indexes. Inf. Syst. **38**(6), 863–884 (2013)
28. Xiao, Y., Yao, J.-F., Li, Z., Dunham, M.H.: Efficient data mining for maximal frequent subtrees. In: ICDM (2003)
29. Zaki, M.J.: Efficiently mining frequent embedded unordered trees. Fundam. Inform. **66**(1–2) (2005)
30. Zaki, M.J.: Efficiently mining frequent trees in a forest: Algorithms and applications. IEEE Trans. Knowl. Data Eng. **17**(8) (2005)
31. Zaki, M.J., Hsiao. C.-J.: Efficient algorithms for mining closed itemsets and their lattice structure. IEEE Trans. Knowl. Data Eng. **17**(4) (2005)
32. Zhu, F., Qu, Q., Lo, D., Yan, X., Han, J., Yu, P.S.: Mining top-k large structural patterns in a massive network. PVLDB **4**(11) (2011)
33. Zou, L., Lu, Y.S., Zhang, H., Hu, R.: PrefixTreeESpan: a pattern growth algorithm for mining embedded subtrees. In: WISE (2006)

Cold-Start Expert Finding in Community Question Answering via Graph Regularization

Zhou Zhao[1]([⊠]), Furu Wei[2], Ming Zhou[2], and Wilfred Ng[1]

[1] The Hong Kong University of Science and Technology, Hong Kong, China
{zhaozhou,wilfred}@cse.ust.hk
[2] Microsoft Research, Beijing, China
{fuwei,mingzhou}@microsoft.com

Abstract. Expert finding for question answering is a challenging problem in Community-based Question Answering (CQA) systems such as Quora. The success of expert finding is important to many real applications such as question routing and identification of best answers. Currently, many approaches of expert findings rely heavily on the past question-answering activities of the users in order to build user models. However, the past question-answering activities of most users in real CQA systems are rather limited. We call the users who have only answered a small number of questions the cold-start users. Using the existing approaches, we find that it is difficult to address the cold-start issue in finding the experts.

In this paper, we formulate a new problem of cold-start expert finding in CQA systems. We first utilize the "following relations" between the users and topical interests to build the user-to-user graph in CQA systems. Next, we propose the *Graph Regularized Latent Model* (GRLM) to infer the expertise of users based on both past question-answering activities and an inferred user-to-user graph. We then devise an iterative variational method for inferring the GRLM model. We evaluate our method on a well-known question-answering system called Quora. Our empirical study shows encouraging results of the proposed algorithm in comparison to the state-of-the-art expert finding algorithms.

1 Introduction

Expert finding is an essential problem in CQA systems [4,25], which arises in many applications such as question routing [28] and identification of best answers [2]. The existing approaches [2,27,28,34,39] build a user model from their past question-answering activities, and then use the model to find the right experts for answering the questions. However, the past question-answering activities of most users in real CQA systems are rather limited. We call the users who have only answered a small number of questions the *cold-start users*. The existing approaches work well if the users have sufficient question-answering activities, while they may not provide satisfactory results for the cold-start users.

Z. Zhao—The work was done when the first author was visiting Microsoft Research, Beijing, China.

© Springer International Publishing Switzerland 2015
M. Renz et al. (Eds.): DASFAA 2015, Part I, LNCS 9049, pp. 21–38, 2015.
DOI: 10.1007/978-3-319-18120-2_2

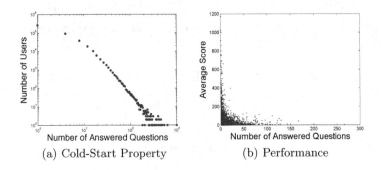

(a) Cold-Start Property   (b) Performance

Fig. 1. Cold-Start Users in Quora

In fact, a vast majority of existing users in real CQA systems, including many that have joined the system for a relatively long period of time, do not have sufficient activities. To illustrate this fact, we summarize the question-answering activities of the users in Quora in Figure 1(a). From the figure, we can see that the participation of most users in question-answering activities falls into the *long tail* part of the power-law curve. This indicates that the majority of the users only answered very few questions. Thus, it is difficult to build an effective user model for cold-start users by using existing methods. Let us call the problem of expert finding with the presence of many cold-start users in a CQA system *the cold-start expert finding problem.* Interestingly, we observe that CQA systems enjoy great benefits contributed by the cold-start users. To show this point, we summarize the performance of the users in Quora in Figure 1(b). Consider the thumbs-up/downs voted by the community as quality score for users on answering questions [28]. We can see that a significant number of cold-start users obtain high quality scores.

To address the cold-start expert finding problem, we incorporate the user-to-user graph in CQA systems to build a regularized user model. Currently, CQA systems such as Quora define thousands of topical interests, which are represented by keywords such as "startups" and "computer programming". The users may follow these keywords when they have the topical interests. If two users follow some common topical interests, we consider that there is a user-to-user relation (i.e. an edge) between them. The works [13,17] show that a user-to-user relation between two users provides a strong evidence for them to have common interests or preferences. Thus, we attempt to integrate both user-to-user graph and question-answering activities into a seamless framework that tackles the cold-start expert finding problem.

The main contributions of our work are summarized as follows:

- We illustrate that the question-answering activities of most users in real CQA systems are rather few and formally propose a new problem of cold-start expert finding in CQA systems.
- We explore the "following relations" between users and topical interests to build the user-to-user graph. We then propose the graph regularized latent

Fig. 2. An Illustration of User's Topical Interests

model by incorporating with the user-to-user graph and devise a variational method for inferring the model.

– We conduct extensive experiments on our proposed method. We demonstrate that, by incorporating with user-to-user graph, our method significantly outperforms other state-of-the-art expert finding techniques.

There exists some work addressing the cold-start problem in user-item recommendation systems [13,14,20,21,29,36]. However, most of them are not applicable in addressing the problem of cold-start expert finding in CQA systems. Even though finding an expert for a question seems to be analogous to recommending an item to a user, there are some subtle differences between them. First, the existing work incorporates with the social relations of users to improve the performance of recommending an item to a user. In the context of our work, there is no relation between the questions and thus the existing cold-start recommendation techniques cannot be applied to our problem. Second, the goal of expert finding is fundamental different from that of recommendation. The existing recommendation techniques focus on recommending existing items to the users while expert finding aims to select the right users to answer some new questions.

The rest of the paper is organized as follows. Section 2 introduces some notations and formulates the problem. We then propose a graph regularized latent model for cold-start expert finding in Section 3. We report the experimental results in Section 4. Section 5 surveys the related work. We conclude the paper in Section 6.

2   Background

In this section, we first introduce some notation of community-based question answering used in our subsequent discussion. The notation includes a data matrix of questions $\mathbf{Q}$, a data matrix of users $\mathbf{U}$, a question-answering activity set Ω and an observed quality score matrix $\mathbf{S}$. Then, we formulate the problem of cold-start expert finding. The summary of the notation is given in Table 1.

We represent the feature of questions by *bag of words*, which has been shown to be successful in many question answering applications [5,35,37]. Therefore, the feature of each question $\mathbf{q}_i$ is denoted by d-dimensional word vector.

Table 1. Summary of Notation

Group	Notation	Notation Description
Data	$\mathbf{Q}$	data matrix of questions
	$\mathbf{U}$	a data matrix of users
	$\mathbf{S}$	an observed quality score matrix
	$\mathbf{F}$	a set of topical interests
	$\mathbf{W}$	a similarity matrix of users
	Ω	a set of existing question-answering activities
	$\mathbf{I}_\Omega$	an indicator matrix for existing activities
	Θ	a topic matrix of questions
	$\mathbf{Z}$	a topic assignment matrix of words
Model	$Mult(\cdot)$	a multinomial distribution
	$Dir(\cdot)$	a dirichlet distribution
	$Norm_\delta(\cdot)$	a normal distribution with standard deviation δ
	λ	a graph regularization term
	K	a dimension of latent space

We then denote the collection of questions by $\mathbf{Q} = [\mathbf{q}_1, \ldots, \mathbf{q}_m] \in R^{d \times m}$ where m is the total number of the questions.

We denote by $\mathbf{U} = [\mathbf{u}_1, \ldots, \mathbf{u}_n] \in R^{d \times n}$ the collection of users in CQA systems, where n is the total number of the users. The parameter $\mathbf{u}_j$ represents a d-dimensional vector for modeling the j-th user. The terms in $\mathbf{u}_j$ indicate the strengths and weakness of the j-th user on the latent space of the questions.

We denote by score matrix $\mathbf{S} \in R^{m \times n}$ the quality of all users on answering the questions. The thumb-ups/downs value in $\mathbf{S}$ is voted by the users in a CQA community. The voting result indicates the community's long term view for the quality of users on answering the questions. Let Ω be the set of existing question-answering activities of users. The quality score S_{ij} exists in matrix $\mathbf{S}$ if activity $(i, j) \in \Omega$.

We observe that many users in CQA systems follow some topical interests. Figure 2 shows the set of topical interests followed by a Quora user. In this example, the user Adam (one of Quora co-founders) follows four topical interests, which are "startups", "google", "computer programming" and "major Internet companies". Let $\mathbf{F}_i$ be the set of topical interests followed by the i-th user. Consider the topical interests of the i-th user and the j-th user, $\mathbf{F}_i$ and $\mathbf{F}_j$. We use the *Jaccard Distance* to model the similarity between them, which is denoted by $W_{ij} = \frac{|\mathbf{F}_i \bigcap \mathbf{F}_j|}{|\mathbf{F}_i \bigcup \mathbf{F}_j|}$. The $\mathbf{F}_i \bigcap \mathbf{F}_j$ is the set of two users' common following topical interests and $\mathbf{F}_i \bigcup \mathbf{F}_j$ is the set of two users' total following topical interests. We note that the similarity value in matrix $\mathbf{W}$ is within the range $[0, 1]$. We therefore model the user-to-user graph based on the similarity between users by $\mathbf{W} \in R^{n \times n}$.

Using the notation given in Table 1, we now define the problem of cold-start expert finding with respect to a CQA system as follows.

Consider a data matrix of questions $\mathbf{Q}$, a quality score matrix $\mathbf{S}$ and a similarity matrix of users $\mathbf{W}$. Given a new question $\mathbf{q}$, we aim to choose the users with high predicted quality score for answering the question.

3 Cold-Start Expert Finding Algorithm

In this section, we present our algorithm for tackling the problem of cold-start expert finding in CQA systems. We first introduce the basic latent model, which has been widely used for addressing the problem of expert finding in [27,28,34, 39]. Next, we propose our graph regularized latent model (GRLM). The graphical representation of GRLM is illustrated in Figure 3. We then devise a variational method for solving the optimization problem in GRLM. Finally, we present the expert finding algorithm based on GRLM.

3.1 Basic Latent Model

The basic latent model tackles the problem of expert finding based on the past question-answering activities and quality score matrix. The latent topic model is first utilized to extract the feature of the questions. Then the user model is inferred from question features and a quality score matrix. The main procedure of basic latent model can be summarized as follows:

Question Feature Extraction. The topic modelling technique [23] has been widely used for question feature extraction in many recent work concerning the problem of expert finding [27,28,34,39]. Topic models provide an interpretable low-dimensional representation of the questions. In this work, we employ the famous latent dirichlet allocation model (LDA) [1] to extract the feature of the questions, which has been shown to be successful in [28,34,39]. The graphical representation of the LDA model is illustrated in the left box in Figure 3. Given K topics, the generative process of LDA is given as follows:
   For each question $\mathbf{q}_i$:

- Draw topic proportions $\theta_i \sim Dir(\alpha)$
  - For the j-th word in $\mathbf{q}_i$
    * Draw a topic assignment of j-th word $z_{ij} \sim Mult(\theta_i)$
    * Draw a word $q_{ij} \sim Mult(\beta_{z_{ij}})$

Therefore, the latent topic proportion θ_i is inferred for the feature of the i-th question. We denote the feature of the existing questions in CQA systems by Θ.

User Model Inference. Given latent feature of questions Θ and quality score matrix $\mathbf{S}$, we infer the latent feature of users $\mathbf{U}$. We assume that the quality score matrix $\mathbf{S}$ is generated by

$$\mathbf{S} \sim Norm_{\lambda_S^{-1}}(\Theta^T \mathbf{U}) \tag{1}$$

where $Norm(\cdot)$ is a normal distribution with mean $\Theta^T \mathbf{U}$ and standard deviation λ_S^{-1}. The graphical representation of this quality score generative model is illustrated in the upper box in Figure 3. For each question-answering activity (i,j), its quality score is generated by

$$S_{ij} \sim Norm_{\lambda_S^{-1}}(\mathbf{q}_i^T \mathbf{u}_j) = Norm_{\lambda_S^{-1}}(\sum_{k=1}^{K} q_{ik} u_{jk}). \tag{2}$$

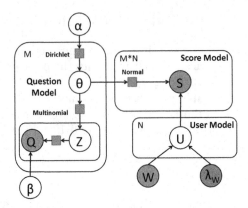

Fig. 3. Graphical Representation of the GRLM Model

The underlying idea of the quality score generative model is as follows. The quality score value is proportional to the dot-product of the question feature and user feature. We consider that the feature $\mathbf{U}$ represents the strongness and weakness of users on a specified topic.

Assume that the standard variance λ_S^{-1} is independent on and identical for different question-answering activities, we the problem of maximum likelihood inference for user feature $\mathbf{U}$ can be given by

$$\max_{\mathbf{U}} -||I_\Omega \otimes (\mathbf{S} - \mathbf{\Theta}^T \mathbf{U})||_F^2 \tag{3}$$

where $|| \cdot ||_F^2$ denotes the Frobenius norm, and $\otimes$ represents the Hadamard element-wise product. I_Ω is an indicator matrix with ones for the existing question-answering activities, and zeros for the missing ones.

Therefore, the user feature $\mathbf{U}$ can be inferred by solving the expression in Formula 3. Then, we can predict the quality score for new questions by Equation 1, and then choose those users who have high predicted scores for answering the questions.

Although a latent model is feasible for tackling the problem of expert finding, it may not be able to solve the cold-start problem well. In cold-start expert finding, there may be a number of users having only few question-answering activities. Under the framework of latent model, the inference for user feature may not be accurate, since there are many missing values in matrix $\mathbf{S}$. Thus, we propose to make use of a user-to-user graph to tackle the cold-start problem.

3.2 Graph Regularized Latent Model

In this section, we present our *Graph Regularized Latent Model* (referred to as GRLM) to tackle the problem of cold-start expert finding. First, we introduce the general idea of our model. Then, we present the detail of the generative process.

Algorithm 1. Generate Observed Question-Answering Activities

Input: a set of users, indicator matrix for existing activities I_Ω
Output: a data matrix of questions $\mathbf{Q}$, a quality score matrix $\mathbf{S}$

1: **for** each question $\mathbf{q}_i \in \mathbf{Q}$ **do**
2:     Draw topic proportions $\theta_i \sim Dir(\alpha)$.
3:     **for** each word $\mathbf{q}_{ij}$ **do**
4:         (a) Draw a topic assignment $z_{ij} \sim Mult(\theta_i)$
5:         (b) Draw a word $q_{ij} \sim Mult(\beta_{z_{ij}})$.
6:     **end for**
7: **end for**
8: Draw a data matrix of users $\mathbf{U}$ by Equation 4.
9: Draw a quality score matrix $\mathbf{S}$ by Equation 1.

Consider the similarity matrix of users $\mathbf{W}$ which is inferred from the following relation between users and topical interests. Based on the property of the user-to-user relation, it is natural to require the similar users in matrix $\mathbf{W}$ have similarity user feature, that is, $W_{ij}(\mathbf{u}_i - \mathbf{u}_j)^2$. Thus, the generation process for data matrix of users $\mathbf{U}$ with graph regularization can be achieved by

$$p(\mathbf{U}) = -\sum_i \mathbf{u}_i^T \mathbf{u}_i - \lambda_W \sum_{i,j} W_{ij}(\mathbf{u}_i - \mathbf{u}_j)^2 \qquad (4)$$

We denote by $\lambda_\mathbf{U}$ the collection of standard deviations for generating the data matrix of users $\mathbf{U}$. Thus, the prior distribution of the data matrix of users $\mathbf{U}$ is given by a product of normal distributions. Note that we set the standard deviation inversely proportional to the similarity of users with constant parameter λ_W. We illustrate the impact of parameter λ_W in the experimental study.

We denote a set of parameters α, β and λ_W as hyper parameters of our model. Referring to Figure 3, the whole generative procedure of our model is outlined in Algorithm 1. We then present the objective function for our graph regularized latent model below:

We observe that the joint distribution for generating quality score matrix $\mathbf{S}$, latent topics of the questions $\mathbf{\Theta}$, a data matrix of questions $\mathbf{Q}$ and a data matrix of users $\mathbf{U}$ can be factorized. Thus, we give the posterior distribution based on hyper parameters by

$$p(\mathbf{S}, \mathbf{\Theta}, \mathbf{Q}, \mathbf{U}, \mathbf{Z}, \mathbf{Q}|\lambda_S^{-1}, \lambda_Q^{-1}, \lambda_U^{-1}, \alpha, \beta, \Omega)$$
$$= p(\mathbf{\Theta}|\alpha)p(\mathbf{Z}|\mathbf{\Theta})p(\mathbf{Q}|\mathbf{Z}, \beta)$$
$$\times p(\mathbf{U}|\lambda_\mathbf{U}^{-1})p(\mathbf{S}|\mathbf{Q}^T\mathbf{U}, \Omega) \qquad (5)$$

where the generation for a data matrix of question is

$$p(\mathbf{\Theta}|\alpha) = \prod_{\theta_i \in \mathbf{\Theta}} Dir(\alpha)$$

$$p(\mathbf{Z}|\mathbf{\Theta}) = \prod_{z_{i,j} \in \mathbf{Z}} Mult(\theta_i)$$

$$p(\mathbf{Q}|\mathbf{Z}, \beta) = \prod_{\mathbf{q}_i \in \mathbf{Q}} \beta_{\mathbf{z}_i, \mathbf{q}_i}$$

and the generation for a data matrix of users and a quality score are

$$p(\mathbf{U}|\lambda_{\mathbf{U}}) = Norm_1(\mathbf{u}_i) \prod_{i,j} Norm_{\delta_{ij}}(\mathbf{u}_i - \mathbf{u}_j)$$

$$p(\mathbf{S}|\mathbf{Q}, \mathbf{U}, \Omega) = \prod_{(i,j) \in \Omega} Norm_{\lambda_S^{-1}}(S_{i,j}|\mathbf{q}_i^T \mathbf{u}_j).$$

We solve the probabilistic inference problem for GRLM by finding a maximum a posterior (MAP) configuration of the data matrix of questions $\mathbf{Q}$ and data matrix of users $\mathbf{U}$. The MAP is an objective function conditioning on the quality score matrix $\mathbf{S}$ and data matrix of questions $\mathbf{Q}$. That is, we aim to find

$$(\mathbf{Q}^*, \mathbf{U}^*) = \arg \max_{\mathbf{Q}, \mathbf{U}} p(\mathbf{Q}, \mathbf{U}|\mathbf{S}, \mathbf{Q}, \lambda_W, \alpha, \beta, \Omega).$$

We observe that maximization a posterior configuration is equivalent to maximizing the complete log likelihood of matrix $\mathbf{Q}$ and $\mathbf{U}$. Thus, we give the complete log likelihood by

$$\mathcal{L} = -\frac{\lambda_{\mathbf{U}}}{2} \sum_{\mathbf{u}_i \in \mathbf{U}} \mathbf{u}_i^T \mathbf{u}_i - \frac{\lambda_{\mathbf{Q}}}{2} \sum_{\mathbf{q}_j \in \mathbf{Q}} (\mathbf{q}_j - \theta_j)^T (\mathbf{q}_j - \theta_j)$$

$$+ \sum_{\mathbf{q}_i \in \mathbf{Q}} \sum_{q_{ij} \in \mathbf{q}_i} \log(\sum_{k=1}^{K} \theta_{i,k} \beta_{z_{i,j}=k, q_{i,j}})$$

$$- \frac{\lambda_W}{2} \sum_{\mathbf{u}_i, \mathbf{u}_j \in \mathbf{U}} \delta_{ij} (\mathbf{u}_i - \mathbf{u}_j)^T (\mathbf{u}_i - \mathbf{u}_j)$$

$$- \frac{\lambda_S}{2} \sum_{(i,j) \in \Omega} (S_{i,j} - \mathbf{q}_i^T \mathbf{u}_j) \tag{6}$$

where $\lambda_W \geq 0$ and $\lambda_S \geq 0$ are trade-off parameters. We assume that the prior latent topic distribution for questions α is a uniform distribution. We adopt this assumption from the topic-model based performance prediction work [24, 26].

3.3   The Optimization Method

In this section, we propose an optimization method for solving Problem (6). We take the partial derivative for parameters $\mathbf{Q}$, $\mathbf{U}$, $\mathbf{\Theta}$ and β in the complete log likelihood $\mathcal{L}$ in Problem (6) and set them to zero.

We first report the optimization result for data matrix of questions $\mathbf{Q}$ and data matrix of users $\mathbf{U}$ by

$$\mathbf{q}_i \leftarrow (\lambda_S \mathbf{U}^T \mathbf{U} + \lambda_Q I_K)^{-1}(\lambda_S \mathbf{U} \mathbf{S}_i^q + \lambda_Q \theta_{\mathbf{q}_i}) \tag{7}$$

$$\mathbf{u}_j \leftarrow (\lambda_S \mathbf{Q}^T \mathbf{Q} + \lambda_W \sum_{\mathbf{u}_k \in \mathbf{U}} \delta_{jk} + \lambda_U I_K)^{-1}$$

$$\times (\lambda_S \mathbf{Q} \mathbf{S}_i^u + \lambda_W \sum_{\mathbf{u}_k \in \mathbf{U}} \delta_{jk} \mathbf{u}_k) \tag{8}$$

where $\mathbf{S}_j^q$ and $\mathbf{S}_i^u$ are the diagonal quality score matrices for j-th question and i-th uesr, respectively.

We then present the optimization result for the latent topic of questions Θ and β, respectively. We first find that it is difficult to directly take the derivative for the complete log likelihood problem with respect to parameter Θ. This is due to the decoupling between β and topic assignment matrix of words $\mathbf{Z}$ [1]. Thus, we introduce a new variational parameter Φ for topic assignment matrix of words $\mathbf{Z}$ to derive a lower bound for the complete log likelihood, denoted by $\mathcal{L}'$. Consider the term $\sum_{q_{ij} \in \mathbf{q}_i} \log(\sum_{k=1}^K \theta_{i,k} \beta_{z_{ij},q_{ij}})$ in $\mathcal{L}$. We derive its lower bound such that the lower bound of $\mathcal{L}$ can also be obtained. The derivation is based on Jensen's Inequality.

By introducing the new variational parameter Φ, the lower bound of $\sum_{q_{ij} \in \mathbf{q}_i} \log(\sum_{k=1}^K \theta_{i,k} \beta_{z_{ij},q_{ij}})$ in $\mathcal{L}$ is given by

$$\sum_{q_{ij} \in \mathbf{q}_i} \log(\sum_{k=1}^K \theta_{i,k} \beta_{z_{ij},q_{ij}})$$

$$= \sum_{q_{ij} \in \mathbf{q}_i} \log(\sum_{k=1}^K \frac{\theta_{i,k} \beta_{z_{ij},q_{ij}} \phi_{(q_i,j),k}}{\phi_{(q_i,j),k}})$$

$$\geq - \sum_{q_{ij} \in \mathbf{q}_i} \sum_{k=1}^K \phi_{q_{ij},k} \log \phi_{q_{ij},k}$$

$$+ \sum_{q_{ij} \in \mathbf{q}_i} \sum_{k=1}^K \phi_{q_{ij},k} \log(\theta_{i,k}) \beta_{z_{ij}=k,q_{ij}}.$$

Thus, we can iteratively estimate the latent topic of questions Θ and β on $\mathcal{L}'$.

We then estimate the parameters Θ, Φ and β iteratively on $\mathcal{L}'$. We first report the optimization results for parameters Φ and β by

$$\phi_{q_{ij},k} \propto \theta_{i,k} \beta_{k,q_{ij}} \tag{9}$$

$$\beta_{k,q_{ij}} \propto \sum_{\mathbf{q}_i \in \mathbf{Q}} \sum_{q_{ij} \in \mathbf{q}_i} \phi_{q_{ij},k}. \tag{10}$$

We then estimate the latent topic of the questions Θ by using the root fining algorithm in numerical optimization tool in [7].

Algorithm 2. The Expert Finding Algorithm

Input: An i-th new question $\mathbf{q}$, data matrix of users $\mathbf{U}$ and β
Output: A ranked list of users $R(\mathbf{U})$

 1: Set latent topic $\theta_i \propto$ Uniform distribution
 2: **for** $t : 1 \rightarrow \tau_{max}$ **do**
 3:    **for** each word $q_{ij} \in \mathbf{q}_i$ **do**
 4:       **for** $k : 1 \rightarrow K$ **do**
 5:          Compute variational parameter $\phi_{q_{ij},k} \propto \theta_i \beta_{k,q_{ij}}$
 6:       **end for**
 7:    **end for**
 8:    Sample $\theta_i \propto \prod_{j=1}^{I} \sum_{k=1}^{K} \theta_{i,k} \phi_{q_{ij},k}$
 9: **end for**
10: Rank users by Equation 1
11: **return** A ranked list of users $R(\mathbf{U})$

3.4 The Expert Finding Algorithm

We now present a cold-start expert finding algorithm based on our proposed model GRLM in Algorithm 2.

Given an i-th new question $\mathbf{q}_i$, Algorithm 2 aims to choose the users with highly predicted score for answering this question. The main process of our algorithm can be divided into two parts. First, the algorithm estimates the data vector of the i-th question, denoted by $\mathbf{q}_i$. Second, the algorithm ranks the users for answering the i-th question based on both data matrix of users $\mathbf{U}$ and data matrix of questions $\mathbf{q}_i$ by Equation 1.

We now give the details of our expert finding algorithm as follows. First, Algorithm 2 iteratively estimates the latent topic of the i-th new question θ_i and variational parameter ϕ_i from Lines 1 to 9. Then, the algorithm ranks the users for question $\mathbf{q}_i$ in Line 11 and returns a ranked list.

4 Experimental Study

In this section, we conduct several experiments on the question-answering platform, Quora, and the social network, Twitter. The experiments are conducted by using Java, tested on machines with Linux OS Intel(R) Core(TM2) Quad CPU 2.66Hz, and 32GB RAM. The objectives of the study is to show the effectiveness of our proposed model GRLM for the problem of expert finding in CQA.

4.1 Datasets

We collect the data from Quora. Quora is a popular question-and-answer website, in which questions are posted and then answered by the community of its users. Quora was launched to the public in June, 2010 and it becomes very successful in terms of the number of users since then. We first crawl the questions posted between September 2012 and August 2013 and then crawl all the users who answered these questions. In total, we collect 444,138 questions, 95,915 users, 887,771 answers and 32,231 topical interests. In the following experiments, we evaluate our model GRLM on Quora instead of Yahoo Answer, since Quora provides the user specified topical interests.

We first sort the resolved questions by their posted timestamp and then split the resolved questions in Quora into the training dataset (i.e. first half of the questions) and the testing dataset (i.e. second half of the questions). Based on the number of the collected answers running from 1 to 6, we split the resolved questions into six groups denoted by $\mathbf{Q}_1$, $\mathbf{Q}_2$, ..., $\mathbf{Q}_6$. For example, group $\mathbf{Q}_1$ contains the questions with at least one answer. For each group $\mathbf{Q}_i$, we randomly sample 100 questions as the testing dataset, denoted by $\mathbf{Q}'_i$. In total, we have 600 testing questions. We then keep other questions in groups $\mathbf{Q}_1$, $\mathbf{Q}_2$, ..., $\mathbf{Q}_6$ as the training dataset. Therefore, we generate a pair of training and testing datasets. In this study, we generate ten pairs of training and testing datasets to evaluate the performance of the algorithms. We take the average of the experimental results of these algorithms on ten pairs of datasets. The summary of the datasets is given in Table 2. The dataset will be provided later.

Table 2. Summary of Datasets

Dataset	#Questions	Average #Answers
$\mathbf{Q}_1$	444k	2
$\mathbf{Q}_2$	178k	3.5
$\mathbf{Q}_3$	86k	5.0
$\mathbf{Q}_4$	48k	6.7
$\mathbf{Q}_5$	30k	8.3
$\mathbf{Q}_6$	20k	10.0

4.2 Evaluation Criteria

We now discuss how to evaluate our algorithm. The performance of the expert finding algorithm can be gauged by the following three metrics: **Precision**, **Recall** and **Cold-Start Rate**.

Precision. We evaluate the ranking quality of different algorithms for the users who answered the questions by the two measurements of *Accu* and *Precision@1*. Given a question, we consider the user whose answer receives the highest number of thumb-ups as the best answerer. Both *Accu* and *Precision@1* evaluate the ranking of the best answerer by different algorithms (i.e. whether the best answerer can be ranked on top). These measurements are widely used in existing work [27,34] to evaluate the performance of the expert finding algorithms in CQA systems.

Given a question $\mathbf{q}_i$, we denote by $R(\mathbf{U})^i$ the ranking of the users who answered this question. We denote by $|R(\mathbf{U})^i|$ the number of the users in the ranking $R(\mathbf{U})^i$. We denote by R^i_{best} the rank of the best answerer for question $\mathbf{q}_i$ by different algorithms. The formula of *Accu* is given by

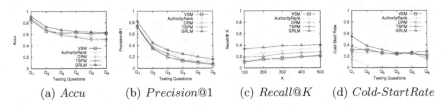

(a) *Accu*   (b) *Precision@1*   (c) *Recall@K*   (d) *Cold-StartRate*

Fig. 4. Performance Comparison of the Algorithms

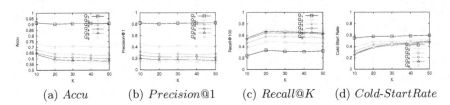

(a) *Accu*   (b) *Precision@1*   (c) *Recall@K*   (d) *Cold-StartRate*

Fig. 5. Effect on Dimension of Latent Space K

$$Accu = \sum_{\mathbf{q}_i \in \mathbf{Q}'} \frac{|R(\mathbf{U})^i| - R_{best}^i}{|R(\mathbf{U})^i||\mathbf{Q}'|},$$

where $\mathbf{Q}'$ is the set of the testing questions. The *Accu* illustrates the ranking percentage of the best answerer by different algorithms.

We now evaluate the precision of the experts ranked on top by different algorithms. We use *Precision@1* to validate whether the expert ranked on top is the best answerer by different algorithms. The formula of *Precision@1* is given by

$$Precision@1 = \frac{|\{\mathbf{q}_i \in \mathbf{Q}' | R_{best}^i \leq 1\}|}{|\mathbf{Q}'|}.$$

Recall. We employ the measurement *Recall@K* to evaluate the ranking quality for all users in CQA systems by different algorithms. Given the i-th new question $\mathbf{q}_i$, we denote by R_{TopK}^i the set of users ranked on $TopK$ by the algorithms. The formula of *Recall@K* is given by

$$Recall@K = \frac{|\{\mathbf{q}_i \in \mathbf{Q}' | j \in R_{TopK}^i \text{ and } (i,j) \in \Omega\}|}{|\mathbf{Q}'|}.$$

The *Recall@K* aims to choose the right experts from all the users in CQA systems.

Cold-Start Rate. We also investigate the types of the experts found by different algorithms (i.e. cold-start users or warm-start users). In this experimental study, we consider the users who answered less than τ questions as cold-start users, where τ is the threshold for cold-start users. We propose the measurement *Cold-Start Rate* to illustrate the type of the experts ranked on top (i.e. Top1), which is given by

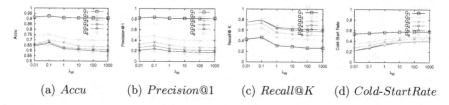

(a) *Accu*   (b) *Precision@1*   (c) *Recall@K*   (d) *Cold-StartRate*

Fig. 6. Effect on Regularization Term λ_W

$$\text{Cold-Start Rate} = \frac{|\{\mathbf{q}_i \in \mathbf{Q}' | j \in R^i_{Top1} \text{ and } 1_{(i,j)} \leq \tau\}|}{|\mathbf{Q}'|},$$

where R^i_{Top1} is the set containing the found expert ranked the top.

We compute the Precision, Recall and Cold-Start Rate of all the algorithms on $\mathbf{Q}'_1$, $\mathbf{Q}'_2$, ..., $\mathbf{Q}'_6$.

4.3 Performance Evaluation

We compare our model GRLM with the following state-of-the-art expert finding algorithms: Vector Space Model (VSM) [27], AuthorityRank [2], Dual Role Model (DRM) [27] and Topic Sensitive Probabilistic Model (TSPM) [34]. The underlying idea of using these algorithms for expert finding in CQA systems are highlighted below:

- **VSM.** The VSM constructs the feature of the users based on the past question-answering activities in a word level. Consider the word vector of the i-th question as $\mathbf{q}_i$. The word vector of the j-th user is constructed from the word vector of the answered questions, denoted by $\mathbf{u}_j$. Given a new question q, VSM ranks the relevance of the users based on the dot product of the word vectors of the i-th question and the j-th user u by $\hat{S}_{ij} = \mathbf{q}_i^T \mathbf{u}_j$.
- **AuthorityRank.** The AuthorityRank computes the expertise authority of the users based on the number of provided best answerers, which is an in-degree method. Given a new question, AuthorityRank ranks the users based on their expertise authority.
- **DRM.** The DRM discovers the latent expertise of the users from their past question-answering activities, which is based on the famous topic modeling technique called *probabilistic latent semantic analysis* (PLSA) [10]. Given a new question, DRM ranks the users based on their latent expertise.
- **TSPM.** The TSPM discovers the latent expertise of the users based on another famous topic modeling technique called *latent Dirichlet allocation* (LDA) [1] and ranks the users based on their latent expertise.

Figures 4(a) to 4(d) show the evaluation results based on *Accu*, *Precison@1*, *Recall@K* and *Cold-Start Rate*, respectively. The evaluation were conducted with different types of the questions. For each dataset, we report the performance of all methods.

The AuthorityRank method is based on the link analysis of the question-answering activities of users while DRM and TPSM models are based on topic-oriented probabilistic model. These experiments reveal a number of interesting points as follows:

- The topic-oriented probabilistic models DRM and TPSM outperform the authority-based model. This findings suggests that using the latent user model for tackling the problem of expert finding in CQA systems is effective.
- Our GRLM model achieves the best performance, which indicates that leveraging the user-to-user graph can further improve the performance of expert finding in CQA systems.
- The experimental study on $Recall@K$ indicates that our method can find the right experts where the candidate experts are all the users in CQA systems. We notice that our model GRLM is able to find the best answerer in the top 100 ranked users with the probability of 0.37 as shown in Figure 4(c).

There are two essential parameters in our model, which are the dimension of latent space K, and the graph regularization parameter λ_W. The parameter K represents the latent feature size of latent user model and latent topic space of questions. The parameter λ_W shows the obtained benefits of our method from the inferred user-to-user graph.

We first study the impact of parameter K by varying its value from 10 to 50, and present the experimental results in Figures 5(a) to 5(d). Figure 5(a) shows that the Cold-Start Rate of the experts found by GRLM increases and then becomes convergent with respect to the dimension of latent space. Figure 5(d) shows that the recall of the expert finding has 10% improvement by varying the parameter K. By transferring the knowledge to the cold-start users, both cold-start users and warm-start users can be selected such that the recall is improved. Figures 5(b) and 5(c) illustrate that the accuracy doesn't vary for the parameter K. From these results, we conclude that the setting $K = 10$ is good enough to represent the latent features of both users and questions in CQA systems.

We then study the impact of the regularization term λ_W on the performance of GRLM, which is illustrated in Figures 6(a) to 6(d). We vary the value of the regularization term λ_W from 0.01 to 1000. The success of graph regularized latent model for expert finding relies on the assumption that two neighboring users share the similar user model. When the value of λ_W becomes small, our model can be considered as the previous topic-oriented expert finding methods, which are only based on the past question-answering activities. We vary parameter λ_W to investigate the benefits of our methods from the idea of graph regularized latent model for the problem of expert finding. We notice that our method consistently performs on most of the varied values of parameter λ_W. To balance the inference of latent user model from both past question-answering activities and user-to-user graph, we set the value of parameter λ_W as a new regularization term. We report that the overall performance of GRLM with the new regularization term can also be improved by 3%, 3% and 10% on $Accu$, $Precision@1$ and $Recall@100$, respectively.

5   Related Work

In this section, we briefly review some related work on the problem of expert finding, cold-start recommendation in the literature.

Expert Finding. The problem of expert finding in CQA systems has attracted a lot of attention recently. Roughly speaking, the main approaches for expert finding can be categorized into two groups: the authority-oriented approach [2,11,15,30,39] and the topic-oriented approach [9,16,18,19,22,27,28,31–34].

The authority-oriented expert finding methods are based on link analysis of the past question-answering activities of the users in CQA systems. Bouguessa et al. [2] discover the experts based on the number of best answers provided by users, which is an in-degree-based method. Zhu et al. [38,39] select experts based on the authority of the users on the relevant categories of the questions. Jurczyk et al. [11] propose a HITS [12] based method to estimate the ranking score of the users based on question-answering activity graphs. Zhang et al. [30] propose an expertise ranking method and evaluated link algorithms for specific domains. Jing et al. [15] propose a competition model to estimate the user expertise score based on question-answering activity graphs.

The topic-oriented expert finding methods are based on latent topic modeling techniques. Deng et al. [3] and Hashemi et al. [9] tackle the problem of expert finding in bibliographic networks. Using the generative topic model, Xu et al. [27] propose a dual role model that jointly represents the roles of answerers and askers. Liu et al. [16] propose a language model to predict the best answerer. Guo et al. [8] and Zhou et al. [34] devise the topic sensitive model to build the latent user model for expert finding. Liu et al. [28] model both topics and expertise of the users in CQA for expert finding. Saptarshi et al. [6] utilize the crowdsourcing techniques to find the topic experts in microblogs. Fatemeh et al. [22] incorporate the topic modeling techniques to estimate the expertise of the users.

In contrast to the above-mentioned work, our emphasis is on cold-start expert finding in CQA systems. We suggests exploiting the "following relation" between users and topical interests to resolve the problem. The existing work mainly focus on the problem of expert finding based on the past question-answering activities of users.

Cold-Start Recommendation. Recently, the cold-start problem in user-item recommendation has attracted a lot of attention and several approaches [14,20, 21,29,36] are proposed to solve this problem. Park et al. [20] propose a latent regression model that leverages the available attributes of items and users to enrich the information. Zhou et al. [36] devise an interview process that iteratively enriches the profile of the new users. Yin et al. [29] propose a random walk based method to choose the right cold-start items for users. Purushotham et al. [21] utilize both the text information of items and social relations of users to user-item recommendation. Zhu et al. [14] extract the information of items from Twitter to overcome the difficulty of cold-start recommendation. However, the cold-start recommendation techniques cannot be applied to the context of this work.

6   Conclusion

We formulate the problem of cold-start expert finding and explore the user-to-user graph in CQA systems. We propose a novel method called graph regularized latent model. We consider the latent user model based on the topic of the questions which can be inferred from question feature and the quality score matrix. Our approach integrates the inferred user-to-user graph and past question-answering activities seamlessly into a common framework for tackling the problem of cold-start expert finding in CQA systems. In this way, our approach improves the performance of expert finding in the cold-start environment. We devise a simple but efficient variational method to solve the optimization problem for our model. We conduct several experiments on the data collected from the famous question-answering system, Quora. The experimental results demonstrate the advantage of our GRLM model over the state-of-the-art expert finding methods.

Acknowledgments. This work is partially supported by GRF under grant number HKUST FSGRF13EG22 and HKUST FSGRF14EG31.

References

1. Blei, D.M., Ng, A.Y., Jordan, M.I.: Latent dirichlet allocation. The Journal of machine Learning research **3**, 993–1022 (2003)
2. Bouguessa, M., Dumoulin, B., Wang, S.: Identifying authoritative actors in question-answering forums: the case of yahoo! answers. In: SIGKDD, pp. 866–874. ACM (2008)
3. Deng, H., King, I., Lyu, M.R.: Formal models for expert finding on dblp bibliography data. In: ICDM, pp. 163–172. IEEE (2008)
4. Dror, G., Koren, Y., Maarek, Y., Szpektor, I.: I want to answer; who has a question? yahoo! answers recommender system. In: Proceedings of SIGKDD, pp. 1109–1117. ACM (2011)
5. Figueroa, A., Neumann, G.: Learning to rank effective paraphrases from query logs for community question answering. In: Twenty-Seventh AAAI Conference on Artificial Intelligence (2013)
6. Ghosh, S., Sharma, N., Benevenuto, F., Ganguly, N., Gummadi, K.: Cognos: crowd-sourcing search for topic experts in microblogs. In: Proceedings of the 35th International ACM SIGIR Conference on Research and Development in Information Retrieval, pp. 575–590. ACM (2012)
7. GSL. https://www.gnu.org/software/gsl/
8. Guo, J., Xu, S., Bao, S., Yu, Y.: Tapping on the potential of q&a community by recommending answer providers. In: CIKM, pp. 921–930. ACM (2008)
9. Hashemi, S.H., Neshati, M., Beigy, H.: Expertise retrieval in bibliographic network: a topic dominance learning approach. In: CIKM, pp. 1117–1126. ACM (2013)
10. Hofmann, T.: Probabilistic latent semantic indexing. In: SIGIR, pp. 50–57. ACM (1999)
11. Jurczyk, P., Agichtein, E.: Discovering authorities in question answer communities by using link analysis. In: CIKM, pp. 919–922. ACM (2007)

12. Kleinberg, J.M.: Authoritative sources in a hyperlinked environment. Journal of the ACM (JACM) **46**(5), 604–632 (1999)
13. Li, W.-J., Yeung, D.-Y.: Relation regularized matrix factorization. In: IJCAI, pp. 1126–1131 (2009)
14. Lin, J., Sugiyama, K., Kan, M.-Y., Chua, T.-S.: Addressing cold-start in app recommendation: Latent user models constructed from twitter followers (2013)
15. Liu, J., Song, Y.-I., Lin, C.-Y.: Competition-based user expertise score estimation. In: SIGIR, pp. 425–434. ACM (2011)
16. Liu, X., Croft, W.B., Koll, M.: Finding experts in community-based question-answering services. In: CIKM, pp. 315–316. ACM (2005)
17. Ma, H., Zhou, D., Liu, C., Lyu, M.R., King, I.: Recommender systems with social regularization. In: WSDM, pp. 287–296. ACM (2011)
18. Miao, G., Moser, L.E., Yan, X., Tao, S., Chen, Y., Anerousis, N.: Generative models for ticket resolution in expert networks. In: Proceedings of the 16th ACM SIGKDD International Conference on Knowledge Discovery and Data Mining, pp. 733–742. ACM (2010)
19. Mimno, D., McCallum, A.: Expertise modeling for matching papers with reviewers. In: Proceedings of the 13th ACM SIGKDD International Conference on Knowledge Discovery and Data Mining, pp. 500–509. ACM (2007)
20. Park, S.-T., Chu, W.: Pairwise preference regression for cold-start recommendation. In: RecSys, pp. 21–28. ACM (2009)
21. Purushotham, S., Liu, Y., Kuo, C.-C.J.: Collaborative topic regression with social matrix factorization for recommendation systems (2012). arXiv preprint arXiv:1206.4684
22. Riahi, F., Zolaktaf, Z., Shafiei, M., Milios, E.: Finding expert users in community question answering. In: Proceedings of the 21st International Conference Companion on World Wide Web, pp. 791–798. ACM (2012)
23. Srivastava, A.N., Sahami, M.: Text mining: Classification, clustering, and applications. CRC Press (2009)
24. Wang, C., Blei, D.M.: Collaborative topic modeling for recommending scientific articles. In: Proceedings of the 17th ACM SIGKDD International Conference on Knowledge Discovery and Data Mining, pp. 448–456. ACM (2011)
25. Wang, G., Gill, K., Mohanlal, M., Zheng, H., Zhao, B.Y.: Wisdom in the social crowd: an analysis of quora. In: Proceedings of WWW, pp. 1341–1352. International World Wide Web Conferences Steering Committee (2013)
26. Wang, H., Chen, B., Li, W.-J.: Collaborative topic regression with social regularization for tag recommendation. In: Proceedings of the Twenty-Third International Joint Conference on Artificial Intelligence, pp. 2719–2725. AAAI Press (2013)
27. Xu, F., Ji, Z., Wang, B.: Dual role model for question recommendation in community question answering. In: Proceedings of SIGIR, pp. 771–780. ACM (2012)
28. Yang, L., Qiu, M., Gottipati, S., Zhu, F., Jiang, J., Sun, H., Chen, Z.: Cqarank: jointly model topics and expertise in community question answering. In: CIKM, pp. 99–108. ACM (2013)
29. Yin, H., Cui, B., Li, J., Yao, J., Chen, C.: Challenging the long tail recommendation. VLDB **5**(9), 896–907 (2012)
30. Zhang, J., Ackerman, M.S., Adamic, L.: Expertise networks in online communities: structure and algorithms. In: WWW, pp. 221–230. ACM (2007)
31. Zhao, Z., Cheng, J., Wei, F., Zhou, M., Ng, W., Wu, Y.: Socialtransfer: transferring social knowledge for cold-start cowdsourcing. In Proceedings of the 23rd ACM International Conference on Conference on Information and Knowledge Management, pp. 779–788. ACM (2014)

32. Zhao, Z., Ng, W., Zhang, Z.: Crowdseed: query processing on microblogs. In: Proceedings of the 16th International Conference on Extending Database Technology, pp. 729–732. ACM (2013)

33. Zhao, Z., Yan, D., Ng, W., Gao, S.: A transfer learning based framework of crowd-selection on twitter. In: Proceedings of the 19th ACM SIGKDD International Conference on Knowledge Discovery and Data Mining, pp. 1514–1517. ACM (2013)

34. Zhou, G., Lai, S., Liu, K., Zhao, J.: Topic-sensitive probabilistic model for expert finding in question answer communities. In: Proceedings of CIKM, pp. 1662–1666. ACM (2012)

35. Zhou, G., Liu, Y., Liu, F., Zeng, D., and J. Zhao. Improving question retrieval in community question answering using world knowledge. In: Proceedings of the Twenty-Third International Joint Conference on Artificial Intelligence, pp. 2239–2245. AAAI Press (2013)

36. Zhou, K., Yang, S.-H., Zha, H.: Functional matrix factorizations for cold-start recommendation. In: SIGIR, pp. 315–324. ACM (2011)

37. Zhou, T.C., Si, X., Chang, E.Y., King, I., Lyu, M.R.: A data-driven approach to question subjectivity identification in community question answering. In: AAAI (2012)

38. Zhu, H., Cao, H., Xiong, H., Chen, E., Tian, J.: Towards expert finding by leveraging relevant categories in authority ranking. In: CIKM, pp. 2221–2224. ACM (2011)

39. Zhu, H., Chen, E., Xiong, H., Cao, H., Tian, J.: Ranking user authority with relevant knowledge categories for expert finding. World Wide Web, pp. 1–27 (2013)

Mining Itemset-Based Distinguishing Sequential Patterns with Gap Constraint

Hao Yang[1], Lei Duan[1,2]([✉]), Guozhu Dong[3], Jyrki Nummenmaa[4],
Changjie Tang[1], and Xiaosong Li[2]

[1] School of Computer Science, Sichuan University, Chengdu, China
hyang.cn@outlook.com, {leiduan,cjtang}@scu.edu.cn
[2] West China School of Public Health, Sichuan University, Chengdu, China
lixiaosong1101@126.com
[3] Department of Computer Science and Engineering, Wright State University,
Dayton, USA
guozhu.dong@wright.edu
[4] School of Information Sciences, University of Tampere, Tampere, Finland
jyrki.nummenmaa@uta.fi

Abstract. Mining contrast sequential patterns, which are sequential patterns that characterize a given sequence class and distinguish that class from another given sequence class, has a wide range of applications including medical informatics, computational finance and consumer behavior analysis. In previous studies on contrast sequential pattern mining, each element in a sequence is a single item or symbol. This paper considers a more general case where each element in a sequence is a set of items. The associated contrast sequential patterns will be called itemset-based distinguishing sequential patterns (itemset-DSP). After discussing the challenges on mining itemset-DSP, we present iDSP-Miner, a mining method with various pruning techniques, for mining itemset-DSPs that satisfy given support and gap constraint. In this study, we also propose a concise border-like representation (with exclusive bounds) for sets of similar itemset-DSPs and use that representation to improve efficiency of our proposed algorithm. Our empirical study using both real data and synthetic data demonstrates that iDSP-Miner is effective and efficient.

Keywords: Itemset · Sequential pattern · Contrast mining

1 Introduction

Imagine you are a supermarket manager facing a collection of customers' shopping records, each of which is a sequence of all purchases by a customer over a fixed time period. (See Table 1 for illustration.) To provide specialized service

This work was supported in part by NSFC 61103042, SKLSE2012-09-32, and China Postdoctoral Science Foundation 2014M552371. All opinions, findings, conclusions and recommendations in this paper are those of the authors and do not necessarily reflect the views of the funding agencies.

M. Renz et al. (Eds.): DASFAA 2015, Part I, LNCS 9049, pp. 39–54, 2015.
DOI: 10.1007/978-3-319-18120-2_3

Table 1. A toy dataset of shopping records of married and unmarried customers

ID	Shopping records	Married
S1	<{bread, milk} {milk, towel} {coffee, beef, cola} {lipstick}>	
S2	<{bread, perfume} {book} {coffee, beef, cola} {lipstick} {milk}>	Yes
S3	<{towel, bread, perfume, beef, book} {coffee, beef, cola} {book} {milk}>	
S4	<{bread, perfume} {coffee, beef, cola} {lipstick, shaver} {milk}>	
S5	<{towel, bread} {bread} {cola, shaver} {coffee, beef, cola} {milk}>	
S6	<{bread} {book} {milk, shaver} {cola} {towel, book}>	No
S7	<{milk} {book, bread} {milk, shaver} {coffee, beef, cola}>	
S8	<{bread, cola} {coffee} {cola} {lipstick, cola} {milk, cola}>	

to married customers, you may want to find and utilize informative differences between the married and unmarried customers on their shopping preferences.

The above motivation scenario cannot be addressed well using existing sequential pattern mining [1] or contrast data mining [2] methods, and thus suggests a novel data mining problem. In a sequential dataset of two classes for this scenario, each sequence is an ordered list of itemsets; given a target class, we want to find the sequential patterns that are frequent in the target class but infrequent in the other class. We call such a pattern an *itemset-based distinguishing sequential pattern (itemset-DSP)* since each of its elements is an itemset instead of a single item. Itemset-DSP mining is an interesting problem with many useful applications. As another example in addition to the shopping application given above, when an analyst in a pharmaceutical company is investigating the effect of a new drug, she may record the symptoms of patients once every 12 hours after taking the drug over one week, then compare the observed data with similarly observed data of patients not taking the drug.

While there are many existing studies on distinguishing sequential pattern mining, they focus on distinguishing sequential patterns whose elements are single items. The itemset-DSP mining problem addressed here is different. It focuses on mining patterns from sequences whose elements are itemsets. Moreover, there is a serious need to represent the patterns concisely, to avoid combinative explosion. Due to these key differences, the potential application and the mining methods of this mining problem differ significantly from those for the case of single item based sequences. We will review the related work and explain the differences systematically in Section 3.

To tackle the problem of mining itemset-DSPs, we need to address several technical challenges. First, a brute-force method, which enumerates every nonempty itemset to generate candidate elements for sequence patterns is very costly on sequence sets with a large number of distinct items and a large maximum number of items in an element. We need an efficient method to avoid generating useless candidates.

Second, we need to have a concise yet complete way to represent sequential patterns satisfying the support thresholds, so that the number of discovered patterns can be reduced, the mining results are more comprehensible, and the algorithm can be efficient.

Third, we also need to find effective techniques to efficiently discover the itemset-DSPs with concise representations. This issue is also complicated because we also consider gap constraint in the discovery of itemset-DSPs.

This paper makes the following main contributions on mining minimal itemset-DSPs with gap constraint: (1) introducing a novel data mining problem of itemset-DSP mining; (2) presenting an efficient algorithm for discovering itemset-DSPs and presenting a concise representation for the mining results; (3) conducting extensive experiments on both real data and synthetic data, to evaluate our itemset-DSP mining algorithm, and to demonstrate that the proposed algorithm can find interesting patterns, and it is effective and efficient.

The rest of the paper is organized as follows. We formulate the problem of itemset-DSP mining in Section 2, and review related work in Section 3. In Section 4, we discuss the critical techniques of our method (called iDSP-Miner). We report a systematic empirical study in Section 5, and conclude the paper in Section 6.

2   Problem Formulation

We start with some preliminaries. Let Σ be the *alphabet*, which is a finite set of distinct items. An *element* (of sequences, defined below) is a subset e of Σ. The *size* of e is the number of items in e, denoted by $|e|$. Given two elements e and e' such that e' is not empty, if $e' \subseteq e$, then we say e' is a *sub-element* of e.

An *itemset-based sequence* S over Σ is an ordered list of elements with the form $S = < e_1 e_2 ... e_n >$, where e_i $(1 \leq i \leq n)$ is an element. The *length* of S is the number of elements in S, denoted by $||S||$. For brevity, below we will often call itemset-based sequences as sequences when it is clear what is meant.

We use $S_{[i]}$ to denote the i-th element in S $(1 \leq i \leq ||S||)$. For two elements $S_{[i]}$ and $S_{[j]}$ in S, satisfying $1 \leq i < j \leq ||S||$, the *gap* between $S_{[i]}$ and $S_{[j]}$, denoted by $Gap(S, i, j)$, is the number of elements between $S_{[i]}$ and $S_{[j]}$ in S. Thus, $Gap(S, i, j) = j - i - 1$.

Example 1. For $S = <\{bread, milk\} \{milk, towel\} \{coffee, beef, cola\} \{lipstick\}>$, we have $||S|| = 4$, $S_{[1]} = \{bread, milk\}$, and $|S_{[1]}| = 2$. The sub-elements of $S_{[1]}$ include $\{bread\}$, $\{milk\}$ and $\{bread, milk\}$. The gap between $S_{[1]}$ and $S_{[4]}$, i.e. $Gap(S, 1, 4)$, is 2.

For two sequences S and S' satisfying $||S|| \geq ||S'||$, if there exist integers $1 \leq k_1 < k_2 < ... < k_{||S'||} \leq ||S||$, such that $S'_{[i]} \subseteq S_{[k_i]}$ for all $1 \leq i \leq ||S'||$, then we say $< k_1, k_2, ..., k_{||S'||} >$ is an *occurrence* of S' in S (and we also say S contains S'). Observe that S' may have more than one occurrences in S.

The gap constraint γ is defined as an interval which consists of two nonnegative integers $[\gamma.min, \gamma.max]$. Given two sequences S and S', let $< k_1, k_2, ..., k_{||S'||} >$ be an occurrence of S' in S. We say S' is a *subsequence* of S (and S is a *super-sequence* of S'). Furthermore, if for all $1 \leq i < ||S'||$, $\gamma.min \leq Gap(S, k_i, k_{i+1}) \leq \gamma.max$, we say that S' is a *subsequence* of S with gap constraint γ, denoted by $S' \sqsubseteq_\gamma S$.

Example 2. Let S =<{*book, milk*} {*bread*} {*book, milk, coffee*} {*coffee*}> and S' =<{*book, milk*} {*coffee*}>. Since $S'_{[1]}$ ={*book, milk*} $\subseteq S_{[1]}$ = {*book, milk*} and $S'_{[2]}$ = {*coffee*} $\subseteq S_{[3]}$ = {*book, milk, coffee*}, $< 1,3 >$ is an occurrence of S' in S. Similarly, $S'_{[1]} \subseteq S_{[1]}$, $S'_{[2]} \subseteq S_{[4]}$, so $< 1,4 >$ is also an occurrence of S' in S. Let gap constraint $\gamma = [2,3]$. Because $2 \le Gap(S,1,4) = 2 \le 3$, $S' \sqsubseteq_{[2,3]} S$.

The *support* of a sequence P with gap constraint γ in sequence set D, denoted by $Sup(D,P,\gamma)$, is defined by Equation 1.

$$Sup(D,P,\gamma) = \frac{|\{S \in D \mid P \sqsubseteq_\gamma S\}|}{|D|} \qquad (1)$$

Please note that given an element e (i.e. a length-1 sequence), the support of e, which is the ratio of the number of sequences containing e to $|D|$, is independent of γ. Thus, we denote $Sup(D,e)$ the support of e in D for brevity.

Definition 1 (Minimal Itemset-Based Distinguishing Sequential Patterns with Gap Constraint). *Given two sets of itemset-based sequences, D_+ and D_-, two thresholds, α and β, and gap constraint γ, a sequence $P = < e_1 e_2 ... e_{\|P\|} >$ is a Minimal Itemset-based Distinguishing Sequential Pattern with Gap Constraint (itemset-DSP), if the following conditions are true:*

1. *(support contrast) $Sup(D_+,P,\gamma) \ge \alpha$ and $Sup(D_-,P,\gamma) \le \beta$;*
2. *(minimality) There does not exist a sequence $P' = < e_k e_{k+1} ... e_{k+m} >$ satisfying Condition 1 such that $k \ge 1$ and $k + m \le \|P\|$.*

Given α, β, and γ, the minimal itemset-based distinguishing sequential pattern mining problem is to find all the itemset-DSPs from D_+ and D_-.

Table 2. List of itemset-DSPs discovered in Table 1 ($\alpha = 0.6, \beta = 0.3, \gamma = [0,2]$)

ID	itemset-DSP P	$Sup(D_+,P,\gamma)$	$Sup(D_-,P,\gamma)$
P1	<{*coffee, beef, cola*} {*milk*}>	0.75	0.25
P2	<{*coffee, beef*} {*milk*}>	0.75	0.25
P3	<{*coffee, cola*} {*milk*}>	0.75	0.25
P4	<{*beef, cola*} {*milk*}>	0.75	0.25
P5	<{*beef*} {*milk*}>	0.75	0.25
P6	<{*lipstick*}>	0.75	0.25
P7	<{*beef, perfume*}>	0.75	0.0
P8	<{*perfume*}>	0.75	0.0

Example 3. Consider the sequences in Table 1. Let support thresholds $\alpha = 0.6$ and $\beta = 0.3$, and gap constraint $\gamma = [0,2]$. There are 8 itemset-DSPs (Table 2) discovered from the married customers (D_+) and the unmarried customers (D_-). Taking <{*beef, cola*} {*milk*}> for instance, we can see that 75% married customers (compared against 25% unmarried customers) buy beef and cola together, and will buy milk within the next three shopping purchases.

Table 3 lists the frequently used notations of this paper.

Table 3. Summary of notations

Notation	Description
Σ	alphabet (the set of items)
$\lvert e \rvert$	the size of e (the number of items in e)
$e' \subseteq e$	e' is a sub-element of e
$\lVert S \rVert$	length of S (the number of elements in S)
$S_{[i]}$	the i-th element in S ($1 \leq i \leq \lVert S \rVert$)
γ	gap constraint
$S' \sqsubseteq_\gamma S$	S' is a subsequence of S with gap constraint γ
$Sup(D, P, \gamma)$	support of sequence P with gap constraint γ in sequence set D
D_+, D_-	the positive, negative sequence sets resp.
α, β	the positive, negative support thresholds resp.

3   Related Work

Sequential pattern mining [3] is a significant task in data mining and has attracted extensive attention from both research and industry. Several types of sequential patterns, such as frequent sequential pattern [4], distinguishing sequential pattern [5], closed sequential pattern [6], (partial) periodic sequential pattern [7,8] and partial order pattern [9], have been proposed. There are quite a few successful applications of sequential pattern mining, such as protein and nucleotide sequence analysis [10,11], software bug feature discovery [12] and musical data analysis [13].

There are several studies on mining sequential patterns from sequences in which each element is an itemset. For example, Rabatel *et al.* [14] considered mining sequences with contextual information. Han *et al.* [7] considered mining sequences where each position contains an itemset. Feng *et al.* [15] proposed a language-independent keyword extraction algorithm based on mining sequential patterns without semantic dictionary. Chang *et al.* [16] used the length of time interval to measure the importance of patterns, and presented a framework to mine time-interval weighted sequential patterns. Recently, Low-Kam *et al.* [17] proposed a method to mine statistically significant, unexpected patterns, so that the number of discovered patterns is reduced. However, these studies are significantly different from our study, since we consider the support contrast measure instead of just the support measure.

There are several studies considering gap constraint in sequential pattern mining. For example, Antunes *et al.* [18] proposed an algorithm to handle the sequence pattern mining problem with gap constraint based on *PrefixSpan* [19]. Xie *et al.* [20] studied the discovery of frequent patterns satisfying one-off condition and gap constraint from a single sequence. Zhang *et al.* [21] solved the problem of mining frequent periodic patterns with gap constraint from a single sequence.

Distinguishing sequential patterns has many interesting applications, as it can describe contrast information between different classes of sequences. Ji *et al.* [5] proposed *ConsGapMiner* for mining minimal distinguishing subsequence patterns with gap constraints. Shah *et al.* [22] mined contrast patterns with gap constraint from peptide datasets, and applied patterns to build a supervised classifier for folding prediction. Deng *et al.* [23] built a classifier for sequence data based on contrast sequences. Wang *et al.* [24] introduced the concept of density into distinguishing sequential pattern mining problem, and designed a method to mine this kind of contrast patterns.

To the best our knowledge, there are no previous existing methods tackling exactly the same problem as itemset-DSP mining. The most related previous work is that of Ji *et al.* [5], which finds the minimal distinguishing subsequences. However, it is considerably different from our work since they focused on the sequences in which each element is a single item rather than an itemset. Moreover, Ji *et al.* [5] didn't consider the concise representation of patterns.

4   Design of iDSP-Miner

In this section, we present our method, iDSP-Miner, for mining itemset-DSPs from D_+ and D_-. In general, the framework of iDSP-Miner includes: candidate element generation, candidate pattern enumeration, support contrast checking and minimality test. Technically, the key issues of iDSP-Miner are the generation and effective representation of candidate elements.

4.1   Candidate Element Generation and Representation

Recall that an itemset-DSP is an ordered list of elements. For candidate itemset-DSP generation, the first step is enumerating the elements that can be used to compose a candidate itemset-DSP.

Naturally, we want to know "how to represent patterns in a concise way and how to generate candidate elements efficiently?" To answer this question, we first make some observations and then introduce some necessary definitions.

Observation 1. *Every element of an itemset-DSP must be a subset of a sequence element in the dataset.*

By Observation 1, a brute-force way is enumerating the subsets of all sequence elements that occur in the given data as candidate elements. Clearly this method is time-consuming, and the number of candidate elements can be massive.

Observation 2. *Some itemset-DSP may be a subsequence of some other itemset-DSP. For example, in Table 2, $P2, P3, P4$ and $P5$ are subsequences of $P1$.*

Definition 2 (Element Instance). *Given a set of sequences D, if there are sequences S, $S' \in D$ and integers i and j such that $1 \leq i \leq ||S||$ and $1 \leq j \leq ||S'||$, and $e = S_{[i]} \cap S'_{[j]}$ is non-empty, we call e an element instance in D.*

Example 4. Let $S = <\{bread, milk\} \{milk, towel\}>$ and $S' = <\{coffee, cola\} \{bread, coffee, book\}>$. Then, there are 7 element instances: $\{bread, milk\}$ ($S_{[1]} \cap S_{[1]}$), $\{milk, towel\}$ ($S_{[2]} \cap S_{[2]}$), $\{coffee, cola\}$ ($S'_{[1]} \cap S'_{[1]}$), $\{bread, coffee, book\}$ ($S'_{[2]} \cap S'_{[2]}$), $\{milk\}$ ($S_{[1]} \cap S_{[2]}$), $\{bread\}$ ($S_{[1]} \cap S'_{[2]}$), and $\{coffee\}$ ($S'_{[1]} \cap S'_{[2]}$).

Given an element instance e, the *position* of e in sequence S, denoted by $pos(e, S)$, is $\{i \mid e \subseteq S_{[i]}\}$. For a sequence set D, the *position list* of e, denoted by $posList(e, D)$, is the set of positions of e associated with all sequences of D. That is, $posList(e, D) = \{< S.id, pos(e, S) >\mid S \in D\}$, where $S.id$ is the index of S in D. We will ignore $pos(e, S)$ in $posList(e, D)$ if $pos(e, S) = \emptyset$.

Example 5. Consider Table 1 and element $\{book\}$. Then, $posList(\{book\}, D_+) = \{< S2, \{2\} >, < S3, \{1,3\} >\}$.

iDSP-Miner starts with computing all element instances in D_+ (denoted by E_+) and all element instances in D_- (denoted by E_-), respectively.

Theorem 1. *Given sequence set D, sequence P and gap constraint γ, we have $Sup(D, P', \gamma) \geq Sup(D, P, \gamma)$ for all $P' =< P_{[k]}P_{[k+1]}...P_{[k+m]} >$ such that $k \geq 1$ and $k + m \leq ||P||$.*

Proof (Outline). For all $S \in D$, if $P \sqsubseteq_\gamma S$, then we have $P' \sqsubseteq_\gamma S$. □

Corollary 1. *Let P be an itemset-DSP satisfying conditions in Definition 1. For each element e of P, we have $Sup(D_+, e) \geq \alpha$.*

Proof (Outline). By Theorem 1, we have $Sup(D_+, e) \geq Sup(D_+, P, \gamma) \geq \alpha$. □

It follows that every element of an itemset-DSP must be a subset of an element instance in E_+. Moreover, we have following pruning rule for E_+:

Pruning Rule 1. *Element instances $e \in E_+$ satisfying $Sup(D_+, e) < \alpha$ can be pruned from E_+.*

A practical observation is that otherwise similar sequences have sets and some of their subsets in equal positions in the result set. Thus, we need an efficient method to represent the set-subset structures and an efficient way to maintain those structures in our mining algorithm, at the same time avoiding repeated computations on the sequence elements. We will introduce the concept of equivalence element for the representation of the set-subset structures, and a split operation to maintain these structures in our algorithm.

Definition 3 (Element Closure). *Given an element e, the closure of e, denoted by $\mathcal{C}(e)$, is the set of all non-empty sub-elements of e.*

To concisely represent iDSPs and also to make iDSP-Miner efficient, we introduce a concept that is somehow similar to "borders" [25] and "equivalence class" [26] (which were used previously in data mining). Traditionally, both "border" and *equivalence class* were used to represent collections of itemsets that share some properties such as "always occur together in sequences of D". We are interested in a particular kind of borders each containing one longest itemset (similar to closed pattern) and several shorter itemsets; traditionally, such a border represents all itemsets that are subsets of the longest pattern and are supersets of some of the shorter patterns. In this paper, we use such borders in a new and different way, by having the shorter itemsets as excluders.

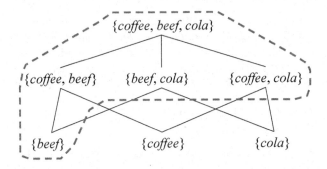

Fig. 1. Illustration of $[\{\{coffee\}, \{cola\}\}, \{coffee,\ beef,\ cola\}]$ (within the red dash line)

We define *equivalence element* to uniquely and concisely represent a set of elements, using (1) an element c (which we will call a *closed element*) and (2) a set $\mathcal{X}$ of elements (which we will call *excluders*), in the form of $[\mathcal{X}, c]$; here, $[\mathcal{X}, c]$ represents $\{e \mid e \subseteq c \wedge e \not\subseteq x$ for every $x \in \mathcal{X}\}$.

Observe that we have the following relationships between the closure and the equivalence element: $\mathcal{C}(c) = [\emptyset, c]$ and $[\mathcal{X}, c] = \mathcal{C}(c) \setminus \bigcup_{x \in \mathcal{X}} \mathcal{C}(x)$.

For example, as shown in Figure 1, $[\{\{coffee\}, \{cola\}\}, \{coffee,\ beef,\ cola\}] = \{\{coffee,\ beef,\ cola\}, \{coffee,\ beef\}, \{coffee,\ cola\}, \{beef,\ cola\}, \{beef\}\}$.

For element instances in E_+, we construct their closures, denoted by $EC_+^\tau = \{[\emptyset, e] \mid e \in E_+\}$. Please note that there may be some redundancy in EC_+^τ. For example, $[\emptyset, e'] \subset [\emptyset, e]$ if e and e' are two elements in E_+ satisfying $e' \subset e$.

To handle those subsets with different support, we define the *split operation* to divide an equivalence element into two disjoint parts. For equivalence element $\hat{e} = [\mathcal{X}, e] \in EC_+^\tau$, if there is an element $e' \in E_+$ such that $e' \in \hat{e}$ and $e' \neq e$, we split $\hat{e}$ by e', denoted by $\hat{e}|e'$, into two disjoint equivalence elements $[\mathcal{X}\widetilde{\cup}\{e'\}, e]$ and $[\{x \in \mathcal{X} \mid x \subset e'\}, e']$, where $\mathcal{X}\widetilde{\cup}\{e'\}$ denote $\mathcal{X} \cup \{e'\} \setminus \bigcup_{x \in \mathcal{X}} \{x \mid x \subset e'\}$.

Example 6. Let $\hat{e} = [\{\{coffee\}, \{beef\}\}, \{coffee,\ beef,\ cola\}]$, and $e' = \{coffee,\ cola\}$. The results of splitting $\hat{e}$ by e' are: $[\{\{coffee,\ cola\}, \{beef\}\}, \{coffee,\ beef,\ cola\}]$ and $[\{\{coffee\}\}, \{coffee,\ cola\}]$.

Obviously, $[\mathcal{X}, e] = [\mathcal{X}\widetilde{\cup}\{e'\}, e] \cup [\{x \in \mathcal{X} \mid x \subset e'\}, e']$. We use EC_+ to denote the set of all equivalence elements after removing the redundancy.

Corollary 2. *Given sequence set D, gap constraint γ, support threshold β, and element e in sequence P, if $Sup(D, e) \leq \beta$, then $Sup(D, P, \gamma) \leq \beta$.*

Proof (Outline). By Theorem 1, we have $Sup(D, P, \gamma) \leq Sup(D, e) \leq \beta$. $\square$

It follows that an element e that satisfies $Sup(D_-, e) > \beta$ may occur in an itemset-DSP. Thus, for equivalence element $\hat{e} = [\mathcal{X}, e] \in EC_+$, if there is an element $e' \in E_-$ such that $Sup(D_-, e') > \beta$, $e' \in \hat{e}$ and $e' \neq e$, we split $\hat{e}$ by e' into two equivalence elements $[\mathcal{X}\widetilde{\cup}\{e'\}, e]$ and $[\{x \in \mathcal{X} \mid x \subset e'\}, e']$. We denote EC the set of equivalence elements after this splitting process.

Given equivalence element $\hat{e} = [\mathcal{X}, e] \in EC$, for any element $e' \in \hat{e}$, we have $posList(e, D_+) = posList(e', D_+)$. iDSP-Miner takes each equivalence element in EC as a set of candidate elements to generate candidate patterns. This leads generally to a significant reduction in the number of candidate elements. The details will be discussed in Section 4.2.

Example 7. Consider Table 1. Given $\alpha = 0.6, \beta = 0.3, \gamma = [0, 2]$, we see that $EC = \{[\emptyset, \{milk\}], [\emptyset, \{lipstick\}], [\emptyset, \{beef\}], [\emptyset, \{cola\}], [\emptyset, \{bread\}], [\{\{bread\}\}, \{bread, perfume\}], [\{\{cola\}, \{beef\}, \{coffee\}\}, \{coffee, beef, cola\}], [\emptyset, \{coffee\}]\}$.

4.2 Pattern Mining

To ensure the completeness of candidate pattern enumeration, iDSP-Miner traverses the set enumeration tree [27] of equivalence elements in EC in a depth-first manner.

Given a node $\mathbb{N}$ in the set enumeration tree, let $\hat{P}$ be the list of equivalence elements that occur on the path from the root node to $\mathbb{N}$. Thus, $\hat{P}$ is a concise representation of $\{P \mid P_{[i]} \in \hat{P}_{[i]}$ for $1 \leq i \leq ||P||, ||P|| = ||\hat{P}||\}$. As the elements represented by an equivalence element occur together, given sequence set D and gap constraint γ, for $P, P' \in \hat{P}$ $(P \neq P')$, we have $Sup(D, P, \gamma) = Sup(D, P', \gamma)$.

We define $Sup(D, \hat{P}, \gamma) = Sup(D, P, \gamma)$, where $P \in \hat{P}$ and $P_{[i]}$ is the closed element of $\hat{P}_{[i]}$ for $1 \leq i \leq ||P||$.

Pruning Rule 2. *If $Sup(D_+, \hat{P}, \gamma) \geq \alpha$ and $Sup(D_-, \hat{P}, \gamma) \leq \beta$, according to the minimality condition in the problem definition (Definition 1), all descendants of $\mathbb{N}$ can be pruned.*

Pruning Rule 3. *If $Sup(D_+, \hat{P}, \gamma) < \alpha$, according to Theorem 1, all descendants of $\mathbb{N}$ can be pruned.*

If $Sup(D_+, \hat{P}, \gamma) \geq \alpha$ and $Sup(D_-, \hat{P}, \gamma) > \beta$, then, to search for superpatterns with lower support in D_-, we extend the set enumeration tree by appending another equivalence element (from EC) as a child of $\mathbb{N}$, to $\hat{P}$ to generate a new candidate.

For any pattern satisfying the support contrast condition, iDSP-Miner performs the minimality test. That is, it compares the pattern with the other discovered patterns to remove the non-minimal ones.

To further remove the redundant representation of itemset-DSPs, for the patterns satisfying the support contrast and minimality conditions, iDSP-Miner simplifies the representation of patterns as follows. Given two patterns $\hat{P}$ and $\hat{P}'$ with the same length, let $\hat{P}_{[k]} = [\mathcal{X}, e]$ and $\hat{P}'_{[k]} = [\mathcal{X}', e']$ $(k \in [1, ||\hat{P}||])$. If $e' \in \mathcal{X}$ and $\hat{P}_{[i]} = \hat{P}'_{[i]}$ for $1 \leq i \leq ||\hat{P}||, i \neq k$, then we say $\hat{P}$ and $\hat{P}'$ are *mergeable*. iDSP-Miner merges $\hat{P}$ with $\hat{P}'$ into a new pattern $\hat{P}''$, such that $\hat{P}''_{[i]} = \hat{P}_{[i]}$ and $\hat{P}''_{[k]} = [\mathcal{X} \cup \mathcal{X}' \setminus \{e'\}, e]$ $(1 \leq k, i \leq ||\hat{P}||, i \neq k)$.

Algorithm 1. iDSP-Miner($D_+, D_-, \alpha, \beta, \gamma$)

Input: D_+: a class of sequences, D_-: another class of sequences, α: minimal support for D_+, β: maximal support for D_-, γ: gap constraint

Output: Ans: the set of itemset-DSPs of D_+ against D_- with concise representation

1: initialize $Ans \leftarrow \emptyset$;
2: $E_+ \leftarrow$ element instances in D_+; $E_- \leftarrow$ element instances in D_-;
3: $EC \leftarrow \{[\emptyset, e] \mid e \in E_+\}$;
4: **while** $\exists \hat{e} = [\mathcal{X}, e] \in EC, e' \in E_+$ such that $e' \in \hat{e}$ and $e' \neq e$ **do**
5:    $EC \leftarrow EC \setminus \hat{e} \cup (\hat{e}|e')$;
6: **end while**
7: **while** $\exists \hat{e} = [\mathcal{X}, e] \in EC, e' \in E_-$ satisfying $Sup(e', D_-) > \beta$ such that $e' \in \hat{e}$ and $e' \neq e$ **do**
8:    $EC \leftarrow EC \setminus \hat{e} \cup (\hat{e}|e')$;
9: **end while**
10: **for** each candidate pattern $\hat{P}$ searched by traversing the set enumeration tree of EC in a depth-first manner **do**
11:    **if** $Sup(D_+, \hat{P}, \gamma) < \alpha$ **then**
12:       prune all super-sequences of $\hat{P}$;
13:    **end if**
14:    **if** $Sup(D_-, \hat{P}, \gamma) \leq \beta$ **then**
15:       prune all super-sequences of $\hat{P}$;
16:       **if** $\hat{P}$ is minimal **then**
17:          $Ans \leftarrow Ans \cup \{\hat{P}\}$;
18:       **end if**;
19:    **end if**
20: **end for**
21: **for** every mergeable pair of $\hat{P}, \hat{P}' \in Ans$ **do**
22:    merge $\hat{P}$ with $\hat{P}'$;
23: **end for**
24: **return** Ans;

Example 8. Let $\hat{P} = <[\{\{coffee, cola\}, \{beef\}\}, \{coffee, beef, cola\}] [\emptyset, \{milk\}]>$, $\hat{P}' = <[\{\{cola\}\}, \{coffee, cola\}] [\emptyset, \{milk\}]>$. Then, we can get $<[\{\{cola\}, \{beef\}\}, \{coffee, beef, cola\}] [\emptyset, \{milk\}]>$ by merging $\hat{P}$ with $\hat{P}'$.

Algorithm 1 gives the pseudo-code of iDSP-Miner. Again, taking Table 1 as an example, the results of iDSP-Miner include: $<[\{\{bread\}\}, \{bread, perfume\}]>$, $<[\emptyset, \{lipstick\}]>$, and $<[\{\{cola\}, \{coffee\}\}, \{coffee, beef, cola\}] [\emptyset, \{milk\}]>$. Compared with patterns listed in Table 2, we can see that our method can represent patterns more concisely.

5   Empirical Evaluation

In this section, we report a systematic empirical study using both real and synthetic sequence sets to test the effectiveness and efficiency of iDSP-Miner. All experiments were conducted on a PC computer with an Intel Core i7-3770

Table 4. Sequence set characteristics

Sequence set	DB	DM	IR
Num. of sequences	100	100	100
Num. of items	1921	1966	1477
Avg. element size	4.30	6.67	4.77
Min. element size	1	1	1
Max. element size	49	60	54
Avg. sequence length	30.54	20.12	20.91
Min. sequence length	5	10	7
Max. sequence length	44	37	37

3.40 GHz CPU, and 8 GB main memory, running Windows 7 operating system. All algorithms were implemented in Java and compiled using JDK 8.

5.1 Effectiveness

Arnetminer[1] groups computer science researchers by different research topics and computes the H-index score for each researcher. We apply iDSP-Miner to analyzing the differences of publication preferences among researchers in database (DB), data mining (DM) and information retrieval (IR). We fetch top 100 scholars in each topic sorted by the H-index score. For each researcher, we construct a sequence, in which an item is the title of a conference or a journal where the researcher published a paper, and an element is the set of items that are in the same year. We collect the publication information of each researcher until 2013 from DBLP[2]. Table 4 shows the characteristics of the sets of sequences. We use "D_+vsD_-" to denote the two sequence sets that we selected to analyze. For example, "DBvsIR" implies that we find itemset-DSPs from DB against IR.

Table 5 summarizes the characteristics of discovered patterns. We can see that with the increase of α, the number of itemset-DSPs, the average/maximum pattern length, and the average/maximum element size are typically decreased.

Table 6 lists the discovered itemset-DSPs with concise representation in IRvsDB when α (min support for IR) = 0.4, β (max support for DB) = 0.2, and γ (gap constraint) = [0, 5]. We can observe that, as shown by patterns <[{{CIKM}}, {SIGIR, CIKM}]> and <[∅, CIKM] [∅, {CIKM}] [∅, {CIKM}]>, researchers in IR prefer publishing in SIGIR and CIKM conferences.

Figure 2 shows the number of itemset-DSPs and the number of itemset-DSPs with concise representation. We can see that the number of discovered patterns is reduced by our equivalence element based representation. Especially, when the number of itemset-DSPs is large, using a small number of itemset-DSPs where equivalence items are used, one can represent many more detailed itemset-DSPs in a highly structured manner. Thus, the results of iDSP-Miner are easier to manage and easier to digest for user. Notably, if the average element size is close

---

[1] http://arnetminer.org/
[2] http://dblp.uni-trier.de/

Table 5. Characteristics of discovered patterns by iDSP-Miner ($\beta = 0.2$, $\gamma = [0, 5]$) (# iDSP(CR): the number of itemset-DSPs with concise representation, $||P||$: pattern length, $|e|$: element size)

Sequence sets	α	# iDSP(CR)	Avg. $||P||$	Max. $||P||$	Avg. $|e|$	Max. $|e|$
DBvsDM	0.25	208	2.31	5	1.05	3
	0.3	153	2.29	5	1.05	2
	0.35	109	2.26	5	1.05	2
	0.4	82	2.15	4	1.07	2
	0.45	52	2.31	4	1.03	2
DBvsIR	0.25	47	1.04	2	1.30	3
	0.3	30	1.03	2	1.25	2
	0.35	21	1	1	1.26	2
	0.4	17	1	1	1.26	2
	0.45	12	1	1	1.2	2
DMvsIR	0.25	96	1.16	3	1.40	3
	0.3	63	1.16	3	1.38	3
	0.35	46	1.20	3	1.33	3
	0.4	35	1.27	2	1.24	2
	0.45	27	1.28	2	1.17	2
DMvsDB	0.25	85	1.17	4	1.39	3
	0.3	54	1.17	4	1.37	3
	0.35	34	1.08	3	1.38	3
	0.4	25	1.09	3	1.29	2
	0.45	18	1.06	2	1.24	2
IRvsDB	0.25	38	1.06	3	1.56	3
	0.3	27	1.07	3	1.51	3
	0.35	22	1.10	3	1.32	3
	0.4	15	1.10	3	1.27	2
	0.45	8	1.11	3	1.29	2
IRvsDM	0.25	54	1.42	6	1.37	3
	0.3	37	1.41	5	1.35	3
	0.35	31	1.57	5	1.18	3
	0.4	23	1.66	4	1.13	2
	0.45	18	1.86	4	1.09	2

to 1.0, then the difference in the output size between the concise representation and listing all patterns explicitly is likely to be small, like in Figure 2 (a).

We note that iDSP-Miner is efficient. For example, the average runtime is 2.51 seconds when $\alpha = 0.4$, $\beta = 0.2$ and $\gamma = [0, 5]$. We will present more analysis on the efficiency of iDSP-Miner in the next section.

5.2 Efficiency

To the best of our knowledge, there were no previous methods tackling exactly the same mining problem as the one studied in this paper. Therefore, we evaluate the efficiency of only iDSP-Miner and the baseline method, which takes each subset of an element instance as a candidate element. Please note that the mining

Table 6. item-DSPs with concise representation for IR*vs*DB ($\alpha = 0.4$, $\beta = 0.2$, $\gamma = [0, 5]$)

<[∅, {TREC, SIGIR}]>	<[{{CIKM}}, {SIGIR, CIKM}]>
<[∅, {JASIST}]>	<[∅, {SIGIR, SIGIR Forum}]>
<[∅, {SIGIR, Inf. Retr.}]>	<[∅, {SIGIR, Inf. Process. Manage.}]>
<[∅, WWW]]>	<[∅, {ACM Trans. Inf. Syst.}] [∅, {CIKM}]>
<[{{CIKM}}, {TREC, CIKM}]>	<[∅, {TREC, Inf. Process. Manage.}]>
<[∅, {RIAO}]>	<[{{CIKM}}, {CIKM, SIGIR Forum}]>
<[∅, {JASIS}]>	<[∅, CIKM] [∅, {CIKM}] [∅, {CIKM}]>
<[∅, {SIGIR, ECIR}]>	

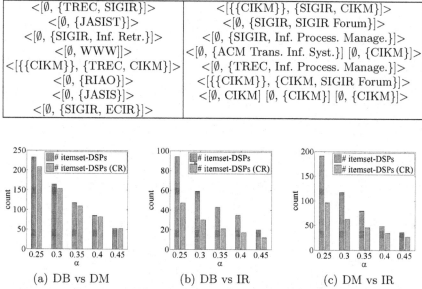

(a) DB vs DM   (b) DB vs IR   (c) DM vs IR

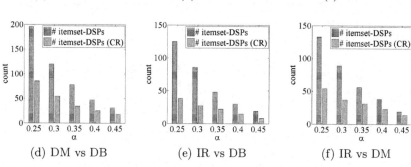

(d) DM vs DB   (e) IR vs DB   (f) IR vs DM

Fig. 2. Comparison of the number of itemset-DSPs and the number of itemset-DSPs with concise representation

result of iDSP-Miner is the set of itemset-DSPs with concise representation, while the mining result of the baseline method is the set of itemset-DSPs.

We generate synthetic sequence sets for efficiency test. There are four parameters for synthetic sequence set generation: the number of items (denoted by NI), the number of sequences (denoted by NS), the average sequence length (denoted by SL), and the average element size (denoted by ES).

Figure 3 shows the efficiency test of iDSP-Miner with respect to α, β and γ when $NI = 30$, $NS = 50$, $SL = 10$ and $ES = 4$. We can see that the runtime of both iDSP-Miner and the baseline method decrease with the increase of α and β, while the runtime of both iDSP-Miner and the baseline method increase with larger gap constraint. iDSP-Miner runs faster than the baseline method, since

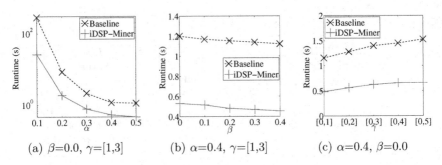

(a) β=0.0, γ=[1,3]   (b) α=0.4, γ=[1,3]   (c) α=0.4, β=0.0

Fig. 3. Efficiency evaluation

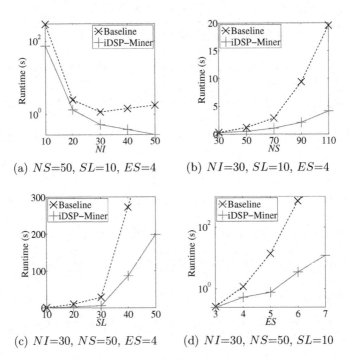

(a) NS=50, SL=10, ES=4   (b) NI=30, SL=10, ES=4

(c) NI=30, NS=50, ES=4   (d) NI=30, NS=50, SL=10

Fig. 4. Scalability evaluation

iDSP-Miner employs a concise representation for candidate patterns to avoid repeated computation.

Figure 4 illustrates the scalability of iDSP-Miner with respect to NI, NS, SL and ES when $\alpha = 0.2$, $\beta = 0.0$ and $\gamma = [1,3]$. When NI becomes larger, more elements can be generated. However, the number of elements/patterns satisfying the positive support threshold condition decreases. Thus, the runtime is reduced by Pruning Rules 1 and 3. On the other hand, more candidate patterns will be generated by increasing NS, SL and ES. Correspondingly, the runtime of iDSP-Miner will increase. Again, we can see that iDSP-Miner runs faster than the baseline method for all parameter settings.

Please note that in some cases in Figure 3 and Figure 4, logarithmic scale has been used for the runtime to better demonstrate the difference in the behavior between iDSP-Miner and the baseline. This should be clear from the figures.

6   Conclusions

In this paper, we propose and study a new problem of mining itemset-based distinguishing sequential patterns with a gap constraint. To mine these patterns and to present the result sets concisely, we propose an algorithm called iDSP-Miner. Our experiments verify the effectiveness and efficiency of iDSP-Miner.

In our work we apply the straightforward assumption that the user preference for output is a minimum distinguishing sequential pattern. It is interesting to explore strategies for incorporating domain constraints into itemset-DSP mining, and design a tuning mechanism for positive and negative support thresholds. Instead of the traditional gap constraint used in this paper, we are also considering the use of temporal gap constraints for sequences of timestamped events.

References

1. Dong, G., Pei, J.: Sequence Data Mining. Springer-Verlag, Berlin, Heidelberg (2007)
2. Dong, G., Bailey, J., eds.: Contrast Data Mining: Concepts, Algorithms, and Applications. CRC Press (2012)
3. Agrawal, R., Srikant, R.: Mining sequential patterns. In: Proceedings of the Eleventh International Conference on Data Engineering, pp. 3–14. IEEE Computer Society, Washington, DC (1995)
4. Zaki, M.J.: Spade: an efficient algorithm for mining frequent sequences. Mach. Learn. **42**(1–2), 31–60 (2001)
5. Ji, X., Bailey, J., Dong, G.: Mining minimal distinguishing subsequence patterns with gap constraints. Knowl. Inf. Syst. **11**(3), 259–286 (2007)
6. Yan, X., Han, J., Afshar, R.: Clospan: mining closed sequential patterns in large databases. In: SDM (2003)
7. Han, J., Dong, G., Yin, Y.: Efficient mining of partial periodic patterns in time series database. In: Proceedings of the 15th International Conference on Data Engineering, pp. 106–115. IEEE Computer Society, Washington, DC (1999)
8. Zhang, M., Kao, B., Cheung, D.W., Yip, K.Y.: Mining periodic patterns with gap requirement from sequences. ACM Trans. Knowl. Discov. Data **1**(2), August 2007
9. Pei, J., Wang, H., Liu, J., Wang, K., Wang, J., Yu, P.S.: Discovering frequent closed partial orders from strings. IEEE Trans. on Knowl. and Data Eng. **18**(11), 1467–1481 (2006)
10. Ferreira, P.G., Azevedo, P.J.: Protein sequence pattern mining with constraints. In: Jorge, A.M., Torgo, L., Brazdil, P.B., Camacho, R., Gama, J. (eds.) PKDD 2005. LNCS (LNAI), vol. 3721, pp. 96–107. Springer, Heidelberg (2005)
11. She, R., Chen, F., Wang, K., Ester, M., Gardy, J.L., Brinkman, F.S.L.: Frequent-subsequence-based prediction of outer membrane proteins. In: Proceedings of the Ninth ACM SIGKDD International Conference on Knowledge Discovery and Data Mining, pp. 436–445. ACM, New York, NY (2003)

12. Zeng, Q., Chen, Y., Han, G., Ren, J.: Sequential pattern mining with gap constraints for discovery of the software bug features. Journal of Computational Information Systems **10**(2), 673–680 (2014)
13. Conklin, D., Anagnostopoulou, C.: Comparative pattern analysis of cretan folk songs. Journal of New Music Research **40**(2), 119–125 (2011)
14. Rabatel, J., Bringay, S., Poncelet, P.: Contextual sequential pattern mining. In: Proceedings of the 2010 IEEE International Conference on Data Mining Workshops. ICDMW 2010, pp. 981–988. IEEE Computer Society, Washington, DC (2010)
15. Feng, J., Xie, F., Hu, X., Li, P., Cao, J., Wu, X.: Keyword extraction based on sequential pattern mining. In: Proceedings of the Third International Conference on Internet Multimedia Computing and Service. ICIMCS 2011, pp. 34–38. ACM, New York, NY (2011)
16. Chang, J.H.: Mining weighted sequential patterns in a sequence database with a time-interval weight. Know.-Based Syst. **24**(1), 1–9 (2011)
17. Cécile, L.K., Chedy, R., Mehdi, K., Jian, P.: Mining statistically significant sequential patterns. In: Proceedings of the 13th IEEE International Conference on Data Mining (ICDM2013). ICDM2013, pp. 488–497. IEEE Computer Society, Dallas, TX (2013)
18. Antunes, C., Oliveira, A.L.: Generalization of pattern-growth methods for sequential pattern mining with gap constraints. In: Perner, P., Rosenfeld, A. (eds.) MLDM 2003. LNAI 2734, vol. 2734, pp. 239–251. Springer, Heidelberg (2003)
19. Pei, J., Han, J., Mortazavi-asl, B., Pinto, H., Chen, Q., Dayal, U., Chun Hsu, M.: Prefixspan: mining sequential patterns efficiently by prefix-projected pattern growth. In: Proceedings of the 17th International Conference on Data Engineering, pp. 215–224. IEEE Computer Society, Washington, DC (2001)
20. Xie, F., Wu, X., Hu, X., Gao, J., Guo, D., Fei, Y., Hua, E.: MAIL: mining sequential patterns with wildcards. Int. J. Data Min. Bioinformatics **8**(1), 1–23 (2013)
21. Zhang, M., Kao, B., Cheung, D.W., Yip, K.Y.: Mining periodic patterns with gap requirement from sequences. ACM Transactions on Knowledge Discovery from Data (TKDD) **1**(2), 7 (2007)
22. Shah, C.C., Zhu, X., Khoshgoftaar, T.M., Beyer, J.: Contrast pattern mining with gap constraints for peptide folding prediction. In: FLAIRS Conference, pp. 95–100 (2008)
23. Deng, K., Zaïane, O.R.: Contrasting sequence groups by emerging sequences. In: Gama, J., Costa, V.S., Jorge, A.M., Brazdil, P.B. (eds.) DS 2009. LNCS, vol. 5808, pp. 377–384. Springer, Heidelberg (2009)
24. Wang, X., Duan, L., Dong, G., Yu, Z., Tang, C.: Efficient mining of density-aware distinguishing sequential patterns with gap constraints. In: Bhowmick, S.S., Dyreson, C.E., Jensen, C.S., Lee, M.L., Muliantara, A., Thalheim, B. (eds.) DASFAA 2014, Part I. LNCS 8421, vol. 8421, pp. 372–387. Springer, Switzerland (2014)
25. Dong, G., Li, J.: Efficient mining of emerging patterns: discovering trends and differences. In: Proceedings of the Fifth ACM SIGKDD International Conference on Knowledge Discovery and Data Mining, pp. 43–52 (1999)
26. Li, J., Liu, G., Wong, L.: Mining statistically important equivalence classes and delta-discriminative emerging patterns. In: Proceedings of the 13th ACM SIGKDD International Conference on Knowledge Discovery and Data Mining. KDD 2007, pp. 430–439 (2007)
27. Rymon, R.: Search through systematic set enumeration. In: Proc. of the 3rd Int'l Conf. on Principle of Knowledge Representation and Reasoning. KR 1992, pp. 539–550 (1992)

Mining Correlations on Massive Bursty Time Series Collections

Tomasz Kusmierczyk[(✉)] and Kjetil Nørvåg

Norwegian University of Science and Technology (NTNU), Trondheim, Norway
{tomaszku,noervaag}@idi.ntnu.no

Abstract. Existing methods for finding correlations between bursty time series are limited to collections consisting of a small number of time series. In this paper, we present a novel approach for mining correlation in collections consisting of a large number of time series. In our approach, we use bursts co-occurring in different streams as the measure of their relatedness. By exploiting the pruning properties of our measure we develop new indexing structures and algorithms that allow for efficient mining of related pairs from millions of streams. An experimental study performed on a large time series collection demonstrates the efficiency and scalability of the proposed approach.

1 Introduction

Finding correlations between time series has been an important research area for a long time [10]. Previously, the focus has mostly been on a single or few time series, however recently many new application areas have emerged where there is a need for analyzing a large number of long time series. Examples of domains where there is a need for detecting correlation between time series include financial data, data from road traffic monitoring sensors, smart grid (electricity consumption meters), and web page view counts.

Bursts are intervals of unexpectedly high values in time series and high frequencies of events (page views, edits, clicks etc.) in streams [6]. In contrast to the raw time series, bursts reduce the information to the most interesting by filtering out the low intensity background so that only information about regions with the highest values are kept (i.e., what would be the most visible on plots; see Fig. 1). In this paper, we introduce the problem of identifying streams of bursts that behave similarly, i.e., are correlated. We propose to use bursts as indicators of characteristics shared between streams. If there are many correlated bursts from two streams, it means that these streams are probably correlated too, i.e., respond to the same underlying events. However, different streams may vary in sensitivity, may be delayed or there might be problems in the bursts extraction process. As an example, consider Fig. 1 that shows two streams representing page views of two related Wikipedia articles; the first representing the TV show *The Big Bang Theory*, the second one represents one of the main characters from the show. Although the plots differ in intensity and bursts vary in heights, bursts

© Springer International Publishing Switzerland 2015
M. Renz et al. (Eds.): DASFAA 2015, Part I, LNCS 9049, pp. 55–71, 2015.
DOI: 10.1007/978-3-319-18120-2_4

locations match. Consequently, a standard way of correlating bursts is overlap relation, overlap operator or overlap measure [11,12]. In that sense, bursts are assumed to be related if they overlap and the measure of stream similarity proposed in this paper is based on this relation.

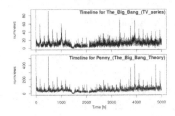

Fig. 1. Comparison of page views of two related Wikipedia articles

The main challenge of correlation analysis of many streams is the computational complexity of all-pairs comparison. Approaches proposed for general time series are not appropriate for streams of bursts, since binary intervals have different nature than real-value, continuous signals. Therefore, we propose novel indexing structures and algorithms devoted particularly to efficient mining of correlated bursty streams. The core of the approach is a Jaccard-like measure based on the overlap relation and using its pruning properties to limit the number of pairs of streams that must be compared. Additionally, the bursts are indexed in a hybrid index that provides for efficient matching of streams, further reducing the computation cost of the correlation analysis.

We provide an extensive evaluation of our approach where we study the effect of different similarity thresholds and collection sizes. Although our approach is generic and can be applied in any domain, in this paper we perform the experimental evaluation on burst streams extracted from time series representing the number of page views per hour for Wikipedia articles. This collection contains a large number of time series and is freely available, thus facilitating reproducibility of our experiments.

To summarize, the main contributions of the paper include:

- A measure and framework for correlating streams using bursts.
- Indexes and algorithms for efficient mining of correlated bursty streams.
- An experimental evaluation on a large collection of time series studying the efficiency and scalability of the approach.

The rest of this paper is organized as follows. Sect. 2 gives an overview of related work. In Sect. 3 we formulate the main task and introduce needed notions. In Sect. 4 we discuss issues related to measures of similarity. Sect. 5 describes the indexes and algorithms used in our approach. Our experimental results are presented in Sect. 6. Finally, in Sect. 7 we conclude the paper.

2 Related Work

There is a large amount of related work on time series and correlation. For example, Gehrke et al. [5] designed a single-pass algorithm that approximately calculates correlated aggregates of two or more streams over both landmark and sliding windows. In [4] correlations were used for measuring semantic relatedness in search engine queries. In [15] it is studied how to select leading series in context of lagged correlations in sliding windows. However, the main challenge of correlation analysis is computational complexity of all-pairs comparison. Zhu and Shasha [17] addressed the problem of monitoring thousands of data streams. Mueen et al. [9] considered computing correlations between all-pairs over tens of thousands of time series. In both papers, they used the largest coefficients of the *discrete Fourier transform* as the approximate signal representation. In other works, researchers were mining and correlating large numbers of time series (millions of streams) using symbolic representation [2,3]. However, binary intervals (bursts) cannot be effectively compressed (not loosing much information and without introducing artifacts) and indexed in that way. Similarly, we rejected wavelets as not matching bursts characteristics and efficient processing requirements.

Our task has similarities to clustering (a survey on clustering of related time series can be found in [8]). Those techniques are either not scalable, require embedding into Euclidean space, or provide only approximate results. Related works can be also found in the area of mining correlated sets (e.g. [1]) and also *sequential patterns*. However, bursts (with overlap relation) cannot be translated into set elements or items without additional assumptions or introducing transformation artifacts.

There are several papers exploiting bursts in a way similar to ours. Vlachos et al. [12] focused on indexing and matching single bursts. This task is different from our since bursts are considered independently within streams. Vlachos et al. adapted *containment encoded intervals* [16]. Although the proposed index is efficient in answering queries composed of sets of bursts, is not able to handle whole streams. An interesting approach to discover correlated bursty patterns containing bursts from different streams, can be found in [13,14]. The basic idea is to introduce a latent cause variable that models underlying events. A similar approach was applied in [7] where they used *hidden Markov models* with *Poisson* emission distributions instead. However, all these approaches are based on matching probabilistic models and are not scalable, the authors assume not more than some hundreds of streams.

3 Preliminaries

We assume a set of N raw streams of basic events (or time series). Time intervals of untypically high frequencies (or values) are called *bursty intervals* (bursts).

Definition 1. Burst b *is time interval* $[start(b), end(b)]$ *of high frequency of basic events, where* $start(b)$ *denotes starting time point of the burst and* $end(b)$ *stands for ending time point.*

As mentioned above, we do not consider bursts height or width but only the fact of occurrence. The bursts are extracted from the original streams either in on-line or offline manner, for example using the algorithm of Kleinberg [6]. Similar to [11,12], for the purpose of measuring similarity we use overlapping bursts. We define the overlap relation as follows:

Definition 2. Overlap relation between two bursts b and b': $b \circ b'$ $\iff$ $(start(b) \leq start(b') \wedge end(b) \geq start(b')) \vee (start(b') \leq start(b) \wedge end(b') \geq start(b))$. The overlap relation is reflexive and symmetric but not transitive.

The burst extraction process results in a set of bursty streams: $D = \{E^1, E^2, ..., E^N\}$ where $N = |D|$.

Definition 3. Streams of bursty intervals are defined as $E^i = (b_1^i, b_2^i, ...)$ where $b_j^i \circ b_k^i \iff j = k$ and $start(b_j^i) > start(b_k^i) \iff j > k$.

We define overlapping between streams as follows:

Definition 4. Set of bursts of E^i **overlapping** with E^j: $O_j^i = \{b : b \in E^i \wedge \exists_{b' \in E^j} b \circ b'\}$. **Set of non-overlapping bursts** of E^i when compared to E^j: $N_j^i = E^i \setminus O_j^i$. We denote $e^i = |E^i|$, $o_j^i = |O_j^i|$, $n_j^i = |N_j^i|$.

The main problem we address in this paper is how to efficiently mine interesting pairs of bursty streams.

Definition 5. Interesting correlated streams are defined as pairs of streams, which for some measure of similarity J have similarity no smaller than a threshold J_T.

Definition 6. We define a set S^n containing all **streams with exactly** n **bursts**: $S^n = \{E^i : E^i \in D \wedge e^i = n\}$.

4   Correlation of Bursty Streams

We aim at mining streams having correlated (overlapping) bursts. Neither set oriented measures such as the *Jaccard index*, contingency measures such as the *phi coefficient* nor general time series measures such as *dynamic time-warping* or *longest common subsequence* are directly applicable. Because of overlap relation properties, streams of bursts cannot be mapped to sets. Bursts could be grouped in equivalence classes according to underlying events, but such mapping is not available. Also, interpreting intervals (bursts) as continuous, real signals implies the need of adjustments which at the end introduce undesirable effects. For example, scaling time to maximize stream matching violates the assumption about simultaneous occurrence of related bursts. Consequently, we decided to focus on the overlap relation e.g., our measure should rely on o_i^j, o_j^i, n_i^j, and n_j^i. Below, we will discuss possible solution and desired properties.

We are interested in measures that can be efficiently tested against many pairs of streams. We would like to be able to prune pairs that have similarity below some threshold in advance. For that purpose, we introduce pruning property.

Definition 7. *Similarity measure s has* **pruning property** *if* $s \geq s_T \implies |e^i - e^j| \leq f(e^i, e^j, s_T)$ *where* s_T *is some constant value of similarity (threshold) and f is some function.*

For measures having this property, the number of pairs of streams to be considered can be reduced as only pairs having similar (difference is limited by f) number of bursts can be scored above the threshold s_T. Naturally, we would like f to be as small as possible. In practice, streams have limited burst counts. Then, only f obtained values below that limit are interesting and allow for pruning.

We adapted one of the measures used widely in areas such as information retrieval and data mining, the *Jaccard index*. Unfortunately, bursts are are objects with properties different from set elements. In the general case there is no simple mapping, so the measure needs to be defined for the new domain as shown below. If there is a one-to-one assignment between overlapping elements from both streams, our measure reduces to standard Jaccard index. In that case, overlapping pairs of bursts from both streams are treated as common elements of sets. This situation is obviously a special case, in general one interval can overlap with two or more intervals. Because of that, our measure also does not preserve triangle inequality.

Definition 8. *For the purpose of measuring similarity of bursty streams we define an adapted* **Jaccard index** *as:*

$$J(E^i, E^j) = \frac{\min(o_i^j, o_j^i)}{e^j + e^i - \min(o_i^j, o_j^i)} \in [0, 1]$$

Lemma 1. *J has* **pruning property** *with* $f(e^i, e^j, J_T) = \max(e^i, e^j) - \lceil \max(e^i, e^j) \cdot J_T \rceil$

The maximum value of J for a pair of two streams E^i, E^j is obtained when $\min(o_i^j, o_j^i) = \min(e^j, e^i)$. Then the measure reduces to: $J_{max} = \frac{\min(e^j, e^i)}{\max(e^j, e^i)} \rightarrow J \leq \frac{\min(e^j, e^i)}{\max(e^j, e^i)}$. Without loss of generality, we assume that $e^i \geq e^j$. This implies $J \leq \frac{e^j}{e^i}$ and for fixed J_T: $J_T \cdot e^i \leq e^j \leq e^i$. Consequently, to obtain related pairs, sets S^n, need to be compared only with streams in S^m where $m \in [\lceil n \cdot J_T \rceil, n]$.

Definition 9. *We define* **connected counts** *connected(n) as the set of such values m for some burst count n that* $m \in [n - f(n, m, s_T), n]$. *We denote n as the* **base count**.

5 Indexing and Mining

Mining related pairs is a computationally expensive task. The cost of similarity measure calculation for a single pair E^i, E^j is $O(\max(e^i, e^j))$. For all pairs, it sums up to $O(|D|^2 e)$, where e stands for average number of bursts per stream.

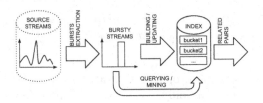

Fig. 2. Overview of data workflow

However, thanks to the pruning property of the measure, we do not need to consider all pairs: we can prune those differing much in number of bursts. Unfortunately, what is left can be also expensive to compute. Therefore further optimizations are needed.

A high level description of our approach is presented on Fig. 2. The workflow can be used to describe both an offline setting where source streams are stored in some local storage, and an on-line setting where streams of events are produced in a continuous manner. In the latter case we assume that the burst detection algorithm is applied immediately, and therefore the amount of data is significantly reduced. Although initially there can be millions of raw streams, in real-life contexts we expect no more than some of tens of bursty intervals per stream. Each interval can be stored using only two numbers (start and end time point), and then we are able to store in main memory all bursts and all streams from the interesting period of time.

Bursty streams are indexed in a high level index composed of buckets. Buckets are responsible for keeping track of subsets of streams (or pairs of streams) and can contain lower-level indexes. The partitioning into buckets is based on number of bursts per stream. Mining of correlated pairs is done by comparing buckets between themselves and/or against stored streams. However, details vary according to approaches described below.

5.1 Naive Approach

As baseline, which we call *Naive*, we compare all pairs that are not pruned. As described above, streams are partitioned to buckets according to number of bursts. I.e., set S^n is placed in bucket n, and there is no further lower-level indexes within the buckets. To obtain all related pairs, for each base count n we simply check all connected counts m and compute the correlation J between all possible pairs of streams from $S^n \times S^m$ (n-th bucket vs. each of m-th buckets).

5.2 Interval Boxes Index

The naive baseline can be improved by speeding up the matching within buckets. Each bucket n is assigned a lower-level index responsible for keeping track of S^n. During the mining, the bucket n index is queried with all streams having connected counts m, i.e., queried with streams from $\bigcup_{m \in connected(n)} S^m$. One approach that could be used is to apply some of the existing indexes designed for

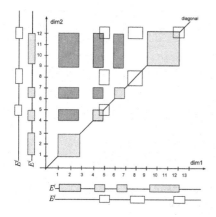

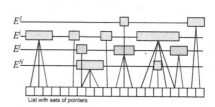

Fig. 3. Example of *IB* index: indexed stream E^i and querying stream E^j $(k = 2)$

Fig. 4. Structure of a single *LS* index

efficient retrieval of (single) overlapping bursts, e.g., *containment encoded intervals* [16] for selection of candidates. Then candidate streams returned by the index, already having at least one overlapping burst with the query stream, are validated using the similarity measure and those under the threshold are rejected. Unfortunately this does not scale well, because candidate sets increase proportionally to number of streams in data set. In order to overcome this, we propose to consider k-element, ordered subsets of bursts. Each k-subset is placed in a k-dimensional space as shown in Fig. 3. The first burst from the subset determines interval in the first dimension, second burst determines interval in the second dimension, and so on. A k-element subset determines a k-dimensional box. For example, in Fig. 3, we have all possible 2-dimensional boxes representing all possible ordered 2-subsets of streams E^i and E^j.

The idea of the *Interval Boxes (IB)* index is to have a k-dimensional box assigned to stream E^i in the index, that will match (overlap) some k-dimensional box assigned to E^j that the index will be queried with. Assuming that E^j, E^i have similarity above some similarity threshold J_T, there are o_i^j bursts of E^j overlapping with o_j^i bursts of E^i. Consequently, there are two boxes: μ-dimensional, where $\mu = \min(o_i^j, o_j^i)$, of E^j and μ-dimensional of E^j that overlap. What is more, all k-dimensional projections of these boxes also overlap. Now instead of single burst, we are matching k bursts at once. Because we are matching k bursts at the same time, the probability of spurious candidates (that will be later rejected as having less than the similarity threshold) is low. The higher k we choose, the lower that probability and the additional cost of validating.

In practice, boxes can be stored in any spatial index that supports overlap queries. In our approach R-trees were used. Because bursts are ordered, only the "top" half of the k-dimensional space is filled. Ordering of bursts is important

---

Algorithm 1. Generation of k-dimensional boxes

1: **function** BOXES(E, k, r)
2:     $C \leftarrow$ COMBINATIONS(E, k)
3:     $C' \leftarrow \emptyset$
4:     **for all** $r' = 1..\min(r, \rho - 1)$ **do**
5:         $k' \leftarrow k - r'$
6:         **for all** $c' \in$ COMBINATIONS(E, k') **do**
7:             **for all** $I \in$ COMBINATIONSREP($1..|c'|$, r') **do**
8:                 $c' \leftarrow c'$ with bursts of indices I repeated
9:                 $C' \leftarrow C' \cup c'$
10:     **return** $C \cup C'$

COMBINATIONS(S, k) - k-element ordered combinations of S
COMBINATIONSREP(S, k) - combinations with replacement

---

because of complexity issues. It significantly reduces number of subsets to be considered and inserted into the index. However, the guarantee that no potentially matching streams will be missed still holds. If some burst b from stream E^i overlaps with some b' from stream E^j it means that those later than b cannot overlap with these being earlier than b'.

One should also notice boxes on the diagonal. Pure subsets (without replacement) do not cover situations where one burst overlaps with many. In such cases, some burst must be repeated in several consecutive dimensions (bursts are ordered). For $k = 2$ each burst can be used once, twice or not used at all in the box. For $k = 3$ each burst can be repeated once, twice, three times or not at all as long as no more than 3 dimensions are used in total. Higher k implies many more combinations to be considered. This can significantly increase the number of boxes and decrease the efficiency. On the other hand multiple overlapping is not very probable. To prevent inserting and querying indexes with unnecessarily many boxes, we introduced an additional parameter ρ that limits how many times, i.e, in how many dimensions, each burst can be repeated. As a result, some pairs, i.e. relying on multiple overlaps between bursts, may be missing but the mining speed increases significantly.

Mining correlated pairs of streams is done by querying all the indexes with streams having connected counts. Index in bucket n, where n is the base count, is queried with all streams from S^m, for all possible connected counts m. The index itself is queried with all possible k-dimensional boxes generated from query stream. For each query box all overlapping boxes from the index are retrieved. For each of them candidate pair (query stream and matching stream from the index) is extracted. In the final step, candidate pairs are validated against similarity measure J and only these having no less than threshold value J_T are kept.

Algorithm 1 shows how k-dimensional boxes are generated. It is composed of two parts. In line 2, boxes without repetitions are generated. In lines, 3-9 boxes having up to r dimensions (given as the parameter) being repetitions of previous dimensions are computed (recall that bursts and dimensions are ordered). An important line is 4, where we additionally limit number of dimensions being

---

Algorithm 2. Querying the *IBHD* index

1: **function** QUERY(E^j, J_T)
2:     $n \leftarrow Index.baselevel$
3:     $k \leftarrow Index.dimensionality$
4:     $m \leftarrow |E^j|$ ▷ We assume $\lceil n \cdot J'_T \rceil \leq m \leq n$
5:     $r \leftarrow m - \lceil J_T \cdot (n + m)/(1 + J_T) \rceil$
6:     $CANDIDATES \leftarrow \emptyset$
7:     **for all** $B \in \text{BOXES}(E^j, k, r)$ **do**
8:         $MATCHING \leftarrow Index.\text{getOverlapping}(B)$
9:         $STREAMS \leftarrow B'.stream$ **for each** $B' \in MATCHING$
10:        $CANDIDATES \leftarrow CANDIDATES \cup STREAMS$
11:    $OUTPUT \leftarrow \emptyset$ ▷ Subset of interesting pairs with E^j on the first position
12:    **for all** $E^i \in CANDIDATES$ **do**
13:        **if** $J(E^j, E^i) \geq J_T$ **then**
14:            $OUTPUT \leftarrow OUTPUT \cup (E^j, E^i)$ ▷ Output pair found
15:    **return** $OUTPUT$

---

repetitions with ρ. For $\rho = 1$, the set of combinations with repetitions C' remains empty.

The number of k-subsets (and consequently k-boxes) without replacement ($\rho = 1$) of some stream E^i is equal to $\binom{e^i}{k}$. For higher values of ρ it is even more. To keep the number of generated boxes (both in the index and in queries) relatively small, k should be either very small or close to e^i. Consequently, we introduce two types of indexes: *IBLD* (low dimensional, for small k-s) and *IBHD* (high dimensional, for big k-s) that have very different properties.

IBLD Index. *IBLD* (low dimensional) indexes are ordinary k-dimensional (e.g., $k = 2$) R-trees. The dimensionality k is constant for indexes in all buckets. Consequently, insertion, deletion and querying require generation of all possible k-dimensional boxes.

What is important, *IBLD* cannot be used for streams having very small number of bursts (e.g. $\sim k$) and for very low values of threshold. The index does not work when the similarity for output pair (by output pair we mean the pair that has similarity above the threshold) of streams is obtained for number of overlapping bursts (measured with o^i_j, o^j_i) lower than k.

IBHD Index. *IBHD* (high dimensional) indexes require k to be as high as possible in order to reduce overall size. On the other hand, if $k > \mu$ we can miss matching between some pairs. From that we imply $k = \mu$. Unfortunately, μ depends on the measure threshold. Consequently, *IBHD* indexes are built for some threshold J'_T and cannot be used for finding pairs of similarity below this threshold. For index built for J'_T, only values $J_T \geq J'_T$ can be used. The border situation is when all bursts of stream having $m' = \lceil J'_T \cdot n \rceil$ (the lowest number of bursts that stream must have to be compared to streams having n bursts) overlap with bursts of stream having n bursts. It means $k = \lceil J_T' \cdot n \rceil$. To match streams having m bursts for $m > m'$, higher values of k would be better. Unfortunately, this would introduce additional costs both in computations and space needed. Therefore, we prefer to have a single index for whole range of connected counts $m \in [m', n]$ and for each n we choose $k = m'$. This value is the highest possible, guaranteeing not missing any pairs (holds when $\rho = \infty$; for $\rho < \infty$ some pairs may be missing but not because of selection of k and due to some boxes being skipped).

As mentioned earlier, one of the major issues influencing speed is the possibility of single burst overlapping with many. Fortunately, in *IBHD* indexes, only a limited number of dimensions needs to be covered with repeated (copied) bursts. For example, if $n = 10$ and $J_T = 0.7$, the lowest $m = 7$. It means that $min(o_i^j, o_j^i) = 7$ and at least 7 out of 10 bursts must be different. Only 3 dimensions can be repeated. In general, for base count n: $r = n - \lceil J_T \cdot n \rceil$. For connected counts m situation is slightly different (as $m \leq n$) and $r = m - \lceil J_T(n + m)/(1 + J_T) \rceil$.

Algorithm 2 shows how above ideas can be implemented in practice. The index keeps track of streams having n bursts and can handle queries of streams having $m \in [m', n]$ bursts. For input stream E^j all possible k-dimensional boxes are generated. Each of these boxes is used to query internal index (e.g., R-Tree). The internal index returns boxes overlapping with query boxes. For each returned box relevant stream identifier is extracted. Then, streams are appended to the candidates set. Finally, all generated candidate streams E^i are compared to E^j using similarity measure J. If the value is above the threshold J_T the pair (E^j, E^i) is included in the result set.

5.3   List-Based Index

For the *IB* approach, in bucket n we store the index responsible for keeping track of streams from S^n. An alternative is to use buckets indexed with two values: base count n and some connected count m. Each bucket contains a single *List-based* (*LS*) index covering both S^n and S^m. The number of connected counts to be considered depends on the predetermined threshold J_T'. Consequently, such an architecture is able to support mining for thresholds $J_T \geq J_T'$.

The structure of a single *LS* index is shown in Fig. 4. For *LS* we use the notion of discrete time where the timeline is divided with some granularity (e.g., hour, day, week; depending on data characteristics) into time windows. In such a setting bursts must be adjusted to window borders. For each window we store the information about all bursts overlapping with it. Consequently, the index is a list of sets of references pointing at bursts where a single set represents single time window.

For the *Naive* and *IB* approaches, mining was done by querying proper indexes with all of the input streams. Using *LS* index it is done directly. Algorithm 3 presents how values of o_j^i, o_i^j are computed. The main loop (lines 5-21) iterates over all time windows (sets of references) indexed with t. In each step (for each t), four sets are computed: *ACTIVE*, *NEW*, *OLD*, and *END*. *ACTIVE* is a set of bursts active in the current time window (taken directly from the index). *NEW* is a set of bursts that were not active in the previous $(t-1)$-th time window but are active in the current one, *OLD* contains those active both in the current and the previous window and *END* those active in the previous but not in the current one. A map (dictionary) *OVERLAPS* contains sets of streams overlapping with bursts active in the current time window. Keys are bursts and values are sets of streams. Maintenance is done in lines 11-12. When a burst ends, it is removed from the map. Pairs of overlapping bursts that were not seen previously (in the previous step of the main loop) are those from the set $NEW \times NEW \cup OLD \times NEW$. For each of these pairs the map *OVERLAPS* and the map o are updated in lines 16-21. Using the map o, candidate pairs of streams can be validated against the threshold J_T. The final step of the mining (lines 22-26) is then validation of all pairs included in the map o. Only pairs having at least one pair of bursts overlapping are considered (included in the o).

Algorithm 3. Candidates generation and validation in LS index

---

```
 1: function QUERY(J_T)
 2:     o ← ∅                          ▷ dictionary {(i, j) → current value of o_i^j }
 3:     PREV ← ∅                                           ▷ empty set of bursts
 4:     OVERLAPS ← ∅          ▷ dictionary {burst → set of overlapping streams}
 5:     for all t = 1...Index.length do           ▷ Iterate over consecutive windows
 6:         ACTIVE ← Index.interval[t]                          ▷ Bursts in t
 7:         NEW ← ACTIVE\PREV                                  ▷ New bursts
 8:         OLD ← ACTIVE\NEW                                   ▷ Old bursts
 9:         END ← PREV\ACTIVE                              ▷ Ending bursts
10:         PREV ← ACTIVE

11:         for all b ∈ END do
12:             delete OVERLAPS[b]

13:         for all b, b' ∈ NEW × NEW ∪ OLD × NEW do
14:             i ← b.streamindex
15:             j ← b'.streamindex
16:             if j ∉ OVERLAPS[b] then
17:                 OVERLAPS[b] ← OVERLAPS[b] ∪ {j}
18:                 o_j^i = o_j^i + 1                             ▷ Increase count
19:             if i ∉ OVERLAPS[b'] then
20:                 OVERLAPS[b'] ← OVERLAPS[b'] ∪ {i}
21:                 o_i^j = o_i^j + 1                             ▷ Increase count
22:     OUTPUT ← ∅                              ▷ Subset of interesting pairs
23:     for all (i, j) ∈ o do
24:         if  (min(o_j^i, o_i^j)) / (e^i + e^j − min(o_j^i, o_i^j))  ≥ J_T then
25:             OUTPUT ← OUTPUT ∪ (E^j, E^i)            ▷ Output pair found
26:     return OUTPUT
```

What is more, each pair is validated in constant time as o_j^i, o_i^j (and e^i, e^j) are already known.

The biggest disadvantage of the LS index is its memory use if there are many streams bursty in a particular index time window (bursts from many streams overlapping at once), i.e., when there are particularly many references in some list set. A solution to this problem is sharding, i.e., we use several indexes per each high-level bucket n, m. Each index covers only a subset of possible pairs of streams. Division of the pairs can be done in any way. Function QUERY guarantees that any pair of bursts, and any pair of streams having bursts overlapping, will be considered. We only need to make sure that all possible pairs of streams (that are not pruned) are at some point placed together in the same LS index.

5.4 Hybrid Index

The IB and LS indexes have different properties and are appropriate for data of different characteristics. IB efficiency depends mostly on the number of k-dimensional boxes in the index. This increases fast with number of bursts per stream and when threshold J_T' is decreasing (this applies only to $IBHD$). On the other hand, LS efficiency does not depend directly on either number of bursts per stream or J_T'. This two factors influence only number of buckets to be considered, but not LS processing time. What affects mostly the efficiency here is the size of sets of references. The bigger sets are, the more pairs need to be considered at each step.

Consequently, we propose the *Hybrid* index that exploits good properties of both approaches. It uses $IBHD$ for low base counts (and proper connected counts) and LS for high base counts. Switching count value depends mostly on data characteristics but some observations can be made. Number of boxes generated for $IBHD$ index for each base count e depends mostly on index dimensionality k. Assuming $\rho = 1$, the number of boxes generated per stream can be approximated as $\sim e^{e-k}$. This can be seen from the number of distinct k-element subsets $\binom{e}{k} = \frac{e!}{(e-k)!k!}$, when k is close to e, then $(e-k)!$ is small and $\binom{e}{k} \sim e \cdot (e-1) \cdot ... \cdot (k+1) \sim \prod_{l=1..(e-k)} e \sim e^{e-k}$. The exponent $e - k$ changes stepwise. For example for $J_T' = 0.95$ it changes for $e = 20, 40, 60, ...,$ and in ranges $e \in (0, 20), [20, 40), [40, 60), ...$ the efficiency of $IBHD$ is more or less constant. Consequently, it is enough to check only one switching count per range, starting from lower values toward higher ones.

5.5 On-Line Maintenance

The described approaches can be used both in the offline and on-line case. For offline mining, where we have information about number of bursts per stream available in advance, construction of the indexes is performed by assigning each stream to the proper bucket (or buckets for LS) and then by inserting it into the index (or indexes) within that bucket.

In the on-line case, bursts arrive and number of bursts per stream changes in a continuous way. When a new burst is extracted, the relevant stream is moved from the previous bucket (or buckets) to a new one matching the new number of bursts. In addition, indexes need to be updated accordingly. Although this introduces additional maintenance cost these operations are distributed over time. In contrast to offline mining, where the whole dataset must be processed at once, for the online case the cost is dispersed among burst arrival moments.

For the *IB* index, deletion and insertion require generation of all k-dimensional boxes. First, all boxes generated for the old version of the stream (without the new burst) are deleted from the proper spatial index. Then, all boxes generated for the stream with the new burst included are inserted into the index responsible for keeping track of streams with higher number of bursts.

For the *LS* index, insertion and removal are performed by adding and deleting relevant burst references. This is done over all indexes matching the stream burst count (recall that each stream can be indexed in several buckets). First, all sets of references assigned to time windows overlapping with old bursts are updated by deletion of proper references. Then, references to all bursts including the new one are inserted to sets into indexes matching the new number of bursts.

6 Experimental Evaluation

In this section, we present the results of the experimental evaluation. All our experiments were carried out on a machine with two Intel Xeon X5650 2.67GHz processors and 128GB RAM. However, experiments were limited to using one thread and 24GB of main memory.

6.1 Experimental Setup

Dataset. For the experimental evaluation we used a dataset based on English Wikipedia page view statistics[1] ($\sim 4.5M$ articles) from the years 2011-2013. These statistics have a granularity equal to 1 hour, so that the time series for each page covers 26304 time points. Bursts extraction was performed using Kleinberg's off-line algorithm [6]. In post-processing, we reduced the hierarchical bursts that are produced by the algorithm to a flat list of non-overlapping bursts, and we also filtered out bursts that had average height (measured in original page views) lower than twice the average background level. After applying this bursts extraction procedure the dataset had $\sim 47M$ bursts in total.

Key features that influence efficiency of our approach are number of streams, number of bursts per stream, and length and distribution of bursts. Fig. 5 shows the distribution of number of bursts per article (stream) in our dataset. One can see that pages with low number of bursts dominate and that number of streams decreases fast when number of bursts per stream increases (notice the use of logarithmic scale). Streams having a low number of bursts have a higher probability of spurious correlations, therefore we filtered out streams having less than 5 bursts. After that, we are left with $\sim 2.1M$ streams, having in total $\sim 43M$ bursts.

---

[1] https://dumps.wikimedia.org/other/pagecounts-raw/

Fig. 6 presents the distribution of length of bursts in the dataset after filtering. In the dataset short bursts dominate. The average length is equal to 28 hours but median is only 10 hours. Nevertheless, one should notice there is non-negligible number of long bursts (100-1000 hours) that significantly increases the sizes of candidate sets and computation time for indexes.

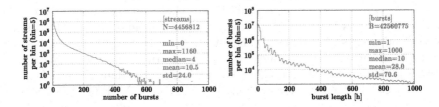

Fig. 5. Number of bursts per stream   **Fig. 6.** Length of bursts

Algorithms and Measurements. In the experiments, we studied the *Naive*, *LS*, 2-dimensional *IBLD* (denoted as *IBLD2*), and *Hybrid* approaches. For *LS* the number of streams in bucket processed at once was limited to $50k$, and for the *Hybrid* approach *IBHD* was used for *burst count* < 40 and *LS* was used for *burst count* ≥ 40). In the experiments we set $\rho = 1$ and $J_T = J_T'$. Mining time is the actual run time for mining, i.e., excluding index building time (which was measured separately).

6.2   Experimental Results

Fig. 7 presents the time for querying streams for each base count, i.e., for each stream having a particular base count (and stored in the index), the streams having related counts (wrt. to the particular base count) are used to query the stored streams. The efficiency of both the *Naive* and *LS* approaches increases with increasing burst count. This is caused by a decreasing number of streams having high number of bursts. *IBLD2* behaves worse than the *Naive* and *LS* approaches for almost any base count. The reason is that the cost of matching of all possible 2-dimensional boxes dominates the benefit of reduced number of pairs to be validated. Notice that the *IBHD* index approach has a stepped plot. Whenever the difference between index dimensionality and base count increases, the computation time also increases (about 10 times). The observations in this figure also gives a justification for the *Hybrid* approach, where *IBHD* is used for low burst counts and then *LS* for the higher ones. It also shows the threshold for switching strategy, i.e., with burst count of 40.

Fig. 8 shows the cost for mining correlations for random subsets of varying cardinalities (the cost of *IBHD* is not shown since even for small datasets the number of boxes per stream can be extremely high). As can be seen, *IBLD2* is not scalable and even behaves worse than *Naive* in all cases. In contrast, both the *LS* and *Hybrid* approaches scale well. However, for large volumes *Hybrid* outperform all the other approaches, as it combined the benefits of *IBHD* and *LS*. Fig. 9 shows the cost of building the indexes. As can be seen, this cost is insignificant (less than 10%) compared to the cost of the correlation mining.

Fig. 10 presents the behavior of the *Hybrid* approach for different thresholds J'_T. With higher threshold, the cost of mining reduces. There are two reasons for that. First, the number of counts connected to each base count is expressed by this value. Second, the *IBHD* dimensionality is also related to it. Consequently, for lower J'_T more pairs need to be considered and using lower dimension indexes.

As shown above, the *Hybrid* approach is scalable wrt. computation time. Regarding memory requirements, the indexes in this approach fit in main

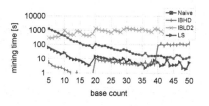

Fig. 7. Querying efficiency for different base counts ($J'_T = 0.95$, $N = 100k$)

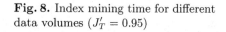
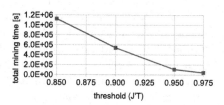

Fig. 8. Index mining time for different data volumes ($J'_T = 0.95$)

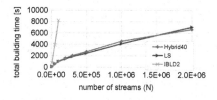

Fig. 9. Index building time for different data volumes ($J'_T = 0.95$)

Fig. 10. *Hybrid* mining time for different thresholds ($N \sim 2.1M$)

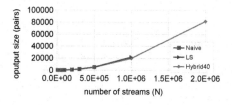
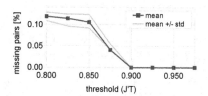

Fig. 11. Number of generated pairs for different data volumes ($J'_T = 0.95$)

Fig. 12. Number of missing pairs for different thresholds ($N=100k$, $\rho=1$, *Hybrid*)

memory. The reason is that the *LS* index size can be easily expressed by the number of pointers plus some additional space for the main list. The number of pointers is equal to *total number of bursts* times *average burst length*. *IBHD* uses spatial index and memory usage depends mostly on the size of that index, and is proportional to *number of boxes inserted* times *insertion cost*. For the dataset used in this evaluation, the size of the indexes is in the order of a few GB.

Fig. 11 illustrates how the number of generated pairs increases with the size of the dataset. The output size increases quadratically with input size, up to $\sim 82k$ for the whole dataset. The *Naive* and *LS* approaches guarantee generation of all pairs above the selected threshold. This does not hold for *Hybrid*. However, in our experiments the number of missing output pairs was small e.g., for $N = 500k$ streams it was always less than 15%. Fig. 12 shows the influence of $\rho = 1$ on $N = 100k$ streams and for different thresholds. We can observe that even for low thresholds, e.g., $J_T' = 0.8$, the number of missing pairs is smaller than 15%. Furthermore, for higher thresholds the matching streams must have smaller differences in number of bursts and therefore the influence of ρ decreases.

7   Conclusions

With emerging applications creating very large numbers of streams that needs to be analyzed, there is a need for scalable techniques for mining correlations. In this paper, we have presented a novel approach based on using bursts co-occurring in different streams for the measurement of their relatedness. An important contribution of our approach is the introduction and study of a new Jaccard-like measure for correlation between streams of bursts, and exploiting the pruning properties of this measure for reducing the cost of correlation mining. Combined with a new hybrid indexing approach, the result is an approach that allows for efficient mining of related pairs from millions of streams.

In the future we plan to work further on improvements of our algorithms. One interesting issue is border cases that not necessarily follow the assumptions of the algorithms, i.e., when there is many streams bursting often and almost always in overlapping time moments. We also would like to investigate how to reduce the cost of online index maintenance.

References

1. Alvanaki, F., Michel, S.: Tracking set correlations at large scale. In: Proceedings of the 2014 ACM SIGMOD International Conference on Management of Data (2014)
2. Camerra, A., Palpanas, T., Shieh, J., Keogh, E.J.: iSAX 2.0: indexing and mining one billion time series. In: Proceedings of the 2010 IEEE International Conference on Data Mining (2010)
3. Camerra, A., Shieh, J., Palpanas, T., Rakthanmanon, T., Keogh, E.J.: Beyond one billion time series: indexing and mining very large time series collections with iSAX2+. Knowl. Inf. Syst. **39**(1), 123–151 (2014)

4. Chien, S., Immorlica, N.: Semantic similarity between search engine queries using temporal correlation. In: Proceedings of the 14th International Conference on World Wide Web (2005)
5. Gehrke, J., Korn, F., Srivastava, D.: On computing correlated aggregates over continual data streams. SIGMOD Rec. **30**(2), 13–24 (2001)
6. Kleinberg, J.: Bursty and hierarchical structure in streams. In: Proceedings of the Eighth ACM SIGKDD International Conference on Knowledge Discovery and Data Mining (2002)
7. Kotov, A., Zhai, C., Sproat, R.: Mining named entities with temporally correlated bursts from multilingual web news streams. In: Proceedings of the Fourth ACM International Conference on Web Search and Data Mining (2011)
8. Liao, T.W.: Clustering of time series data - a survey. Pattern Recognition **38**, 1857–1874 (2005)
9. Mueen, A., Nath, S., Liu, J.: Fast approximate correlation for massive time-series data. In: Proceedings of the 2010 ACM SIGMOD International Conference on Management of Data (2010)
10. Ratanamahatana, C., Lin, J., Gunopulos, D., Keogh, E., Vlachos, M., Das, G.: Mining time series data. In: Data Mining and Knowledge Discovery Handbook. CRC Press (2010)
11. Vlachos, M., Meek, C., Vagena, Z., Gunopulos, D.: Identifying similarities, periodicities and bursts for online search queries. In: Proceedings of the 2004 ACM SIGMOD International Conference on Management of Data (2004)
12. Vlachos, M., Wu, K.-L., Chen, S.-K., Yu, P.S.: Correlating burst events on streaming stock market data. Data Mining and Knowledge Discovery **16**(1), 109–133 (2008)
13. Wang, X., Zhai, C., Hu, X., Sproat, R.: Mining correlated bursty topic patterns from coordinated text streams. In: Proceedings of the 13th ACM SIGKDD International Conference on Knowledge Discovery and Data Mining (2007)
14. Wang, X., Zhang, K., Jin, X., Shen, D.: Mining common topics from multiple asynchronous text streams. In: Proceedings of the Second ACM International Conference on Web Search and Data Mining (2009)
15. Wu, D., Ke, Y., Yu, J.X., Yu, P.S., Chen, L.: Detecting leaders from correlated time series. In: Proceedings of the 15th International Conference on Database Systems for Advanced Applications (2010)
16. Wu, K.-L., Chen, S.-K., Yu, P.S.: Query indexing with containment-encoded intervals for efficient stream processing. Knowl. Inf. Syst. **9**(1), 62–90 (2006)
17. Zhu, Y., Shasha, D.: StatStream: statistical monitoring of thousands of data streams in real time. In: Proceedings of the 28th International Conference on Very Large Data Bases (2002)

Data Streams and Time Series

Adaptive Grid-Based k-median Clustering of Streaming Data with Accuracy Guarantee

Jianneng Cao[1][✉], Yongluan Zhou[2], and Min Wu[1]

[1] Institute for Infocomm Research at Singapore, Singapore, Singapore
{caojn,wumin}@i2r.a-star.edu.sg
[2] University of Southern Denmark, Odense, Denmark
zhou@imada.sdu.dk

Abstract. Data stream clustering has wide applications, such as online financial transactions, telephone records, and network monitoring. Grid-based clustering partitions stream data into cells, derives statistical information of the cells, and then applies clustering on these much smaller statistical information without referring to the input data. Therefore, grid-based clustering is efficient and very suitable for high-throughput data streams, which are continuous, time-varying, and possibly unpredictable. Various grid-based clustering schemes have been proposed. However, to the best of our knowledge, none of them provides an accuracy guarantee for their clustering output. To fill this gap, in this paper we study grid-based k-median clustering. We first develop an accuracy guarantee on the cost difference between grid-based solution and the optimum. Based on the theoretical analysis, we then propose a general and adaptive solution, which partitions stream data into cells of dynamically determined granularity and runs k-median clustering on the statistical information of cells with an accuracy guarantee. An extensive experiment over three real datasets clearly shows that our solution provides high-quality clustering outputs in an efficient way.

1 Introduction

A data stream is a massive sequence of data objects, where each object can be described by a d-dimensional attribute vector. Data streams are common in many applications, which include online financial transactions, telephone records, web applications, and sensor network. Analyzing data streams brings unique opportunities to optimize the operation of companies and organizations in a more responsive fashion. Typically, data streams are continuous, time-varying, unpredictable and possibly with a high throughput. Such unique characteristics have attracted great attention of the research community, and numerous techniques have been proposed to process data streams. Examples include query processing [27], sampling [32], clustering [12,35], and classification [23].

In this work we study continuous k-median clustering on data streams. Informally, clustering is to divide data into groups, such that objects in a same group are more similar to each other than those from other groups. k-median clustering

© Springer International Publishing Switzerland 2015
M. Renz et al. (Eds.): DASFAA 2015, Part I, LNCS 9049, pp. 75–91, 2015.
DOI: 10.1007/978-3-319-18120-2_5

is known to be NP-hard, and it is thus time-consuming to compute, especially for a large dataset. As such, grid-based clustering [17] was proposed. It partitions data space into (usually) uniform cells, and computes statistical information for each cell. Cell size, max/min/mean value, and standard deviation are the typical statistical information for a cell. Clustering is then run on the statistical information without a need to access the raw input data. Since statistical information is much smaller than the raw input dataset, grid-based clustering is much more efficient than clustering on the raw input data. In addition, grid-based clustering processes data streams on the fly. That is, it can process streaming tuples one by one without a need of considering their dependencies. Thus, it has been regarded as a good candidate solution for clustering continuous streaming data. Various grid-based solutions have been proposed thus far [17,21,31]. However, as far as we know none of them has given an accuracy bound, which clearly specifies the difference between their clustering output and the optimum.

Coreset-based clustering has also been proposed to continuously cluster data streams [4,20,24]. Given a clustering algorithm and a dataset, a coreset is a summary of the dataset, such that the output of the clustering algorithm running on the coreset can approximate that of the same algorithm running on the whole dataset. Since coreset is smaller than the input data, clustering algorithms based on coresets also have high efficiency. However, coresets cannot be computed on the fly as grid cells. To compute the coreset of a set of tuples, the whole set of data needs to be available. Batch-based approaches like [5,34] split data streams into segments. Still, they require that a whole segment of streaming tuples is available before the coreset corresponding to the segment can be computed. Therefore, Coreset-based clustering has an output latency and a high storage consumption, especially when the size and time span of the segment is big.

In this paper, we propose a general and adaptive grid-based k-median clustering solution with an accuracy guarantee. We adopt the sliding-window model, where only streaming tuples in a recent window contribute to the clustering output. Such a model is suitable for many real-world applications that care for more recent data instead of the whole unbounded data stream. Our solution leverages approximation algorithms [10,16] to cluster data. An approximation k-median algorithm achieves a clustering cost at most α times of the optimum, where $\alpha > 1$ is a parameter given by the algorithm. Our solution is general, and can be built on any approximation algorithm. It adaptively partitions tuples in sliding windows into cells with a dynamically determined granularity, and applies approximation k-median algorithm on the statistical information of cells to output clustering results. In summary, the contributions of our work are as follows.

- We present a theoretical analysis of approximation clustering algorithms running on statistical information of cells, and provide accuracy guarantee for their clustering output.
- Based on the theoretical analysis, we provide a general and adaptive solution, which dynamically determines the granularity of cells, and runs k-median clustering on the cells with an accuracy guarantee.

– Our extensive experimental results over real datasets show that our solution
  has clustering cost close to that of approximation algorithms running on the
  raw input data, but is 2 orders of magnitude faster.

We organize the remaining of our work as follows. We present the background
knowledge in the next section and then perform a theoretical study of grid-based
clustering in Section 3, based on which we propose our solution in Section 4.
We carry out experimental evaluation in Section 5 and survey related work in
Section 6. We conclude our work in Section 7.

2 Preliminaries

2.1 Data Stream

We model data stream $\mathcal{DS}$ as an infinite append-only array of tuples, and denote
its tuple at timestamp t by $\mathcal{DS}\,[t]$, where t is from 1 to ∞. We assume a data
stream has d dimensions.

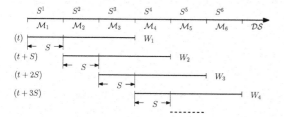
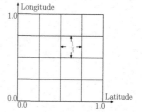

Fig. 1. An illustration of sliding windows   **Fig. 2.** A 4×4 uniform grid

Data stream has time-varying data distribution. To capture its evolving characteristics, recent data need to be assigned higher weight than the outdated when
carrying out clustering analysis. In this paper we adopt sliding window model,
which considers only the tuples in a recent window (i.e., from the current timestamp up to a certain timestamp of the past). A sliding window is characterized
by two parameters: WS window size and S step size. The window slides forward
at every time interval of S. As the sliding continues, a sequence of windows
is generated. Figure 1 gives an example. The first window W_1 at timestamp t
contains tuples in the time span $[t - (WS - 1),\ t]$. At the timestamp $t + S$,
window W_1 slides forwards to window W_2, which contains tuples in the time
span $[(t + S) - (WS - 1),\ t + S]$. At the timestamp $t + 2S$, window W_2 slides
forwards to W_3, and so on. For information of other window-based models (e.g.,
damped window and landmark window), please refer to [19].

Sliding window moves forwards every time interval of S. For clear presentation of our solution, we mark the steps. In particular, we partition the data
stream into segments, each having the length of S. Then, we take the first segment as the first step and label it by S^1, the second segment as the second step
and label it by S^2, and so on. Figure 1 illustrates this.

2.2   k-medians Clustering and Its Approximation

Given a set of tuples $X = \{x_1, x_2, \ldots, x_N\}$ in a d-dimensional space, k-medians clustering is to find k centers $\mathcal{O} = \{O_1, O_2, \ldots, O_k\}$ in the same data space, such that the cost (i.e., the average distance between a tuple and its nearest center)

$$\mathsf{Cost}\,(X, \mathcal{O}) = \frac{1}{N} \sum_{i=1}^{N} \mathsf{Dist}\,(x_i, \mathsf{NN}\,(x_i, \mathcal{O})) \tag{1}$$

is minimized, where Dist is a distance function defined in a metric space satisfying triangular inequality, and $\mathsf{NN}\,(x_i, \mathcal{O})$ returns tuple x_i's nearest neighbor in $\mathcal{O}$ based on Dist. In this work we consider Euclidean distance.

k-medians clustering is NP-hard. The research community has proposed various approximation algorithms [10,15,28,29]. An approximation algorithm is characterized by two parameters α and β, where $\alpha > 1$ and $\beta \geq 1$. Specifically, let $\mathcal{O}^\circ$ be a set of k cluster centers output by an optimal k-median clustering algorithm on dataset X, and $\mathcal{O}$ be a set of cluster centers output by an (α, β)-approximation clustering algorithm running on the same dataset. Then,

$$\frac{\mathsf{Cost}\,(X, \mathcal{O})}{\mathsf{Cost}\,(X, \mathcal{O}^\circ)} \leq \alpha \quad (2) \qquad \text{and} \qquad \frac{|\mathcal{O}|}{k} \leq \beta \quad (3)$$

The α value in Equality 2 shows the difference of clustering cost between the approximation algorithm and the optimal one. The smaller the α value is, the closer the approximation is to the optimum. The β value in Equality 3 bounds the maximum number of cluster centers needed in the approximation. Some approximation algorithms [28,29] require $\beta > 1$, while others [10,15] set $\beta = 1$. For an (α, β)-approximation algorithm with $\beta = 1$, we denote it by α-approximation.

We will study continuous k-medians clustering on sliding windows to learn the evolution of clusters under observation. We will dynamically partition streaming tuples into cells (i.e., d-dimensional rectangles), and run (α, β)-approximation clustering algorithms on the statistical information of cells. Let $\overline{\mathcal{O}}$ and $\mathcal{O}^\circ$ be the sets of cluster centers output by our solution and the optimal one, respectively. Our theoretical analysis in the next section proves that

$$\mathsf{Cost}\,(X, \overline{\mathcal{O}}) \leq \alpha \mathsf{Cost}\,(X, \mathcal{O}^\circ) + \delta. \tag{4}$$

where $\delta > 0$ is a function, showing the difference between (α, β)-approximation algorithm running on cells from its running on input data.

3   Theoretical Analysis

This section investigates the accuracy guarantee of (α, β)-approximation clustering algorithms running on statistical information of cells.

3.1   Accuracy Analysis of Grid-Based k-median Clustering

Let X be the input dataset in a d-dimensional data space. We uniformly partition each dimension into ω segments of equal length. Let $\overline{X}$ be the resultant cells from the partitioning. For each cell, we compute a representative, which is flexible – it can be a tuple in the cell, the cell mean, or a certain function. Given a cell c, we denote its representative by $\mathsf{R}\,(c)$. All the tuples $x \in c$ share the same representative, so we also use $\mathsf{R}\,(x)$ to denote the representative. For brevity, when the context is clear, we will also use $\overline{X}$ to denote the statistical information of the cells. Let $\overline{\mathcal{O}}$ be a set of cluster centers output by an (α, β)-approximation clustering algorithm running on $\overline{X}$. We define its clustering cost (i.e., the average distance between cell representatives and their nearest cluster centers in $\overline{\mathcal{O}}$) as

$$\mathsf{Cost}\left(\overline{X}, \overline{\mathcal{O}}\right) = \frac{1}{|X|} \sum_{c \in \overline{X}} |c| \cdot \mathsf{Dist}\left(\mathsf{R}\,(c), \mathsf{NN}\left(\mathsf{R}\,(c), \overline{\mathcal{O}}\right)\right), \qquad (5)$$

where function $\mathsf{NN}\left(\mathsf{R}\,(c), \overline{\mathcal{O}}\right)$ returns the nearest neighbor of $\mathsf{R}\,(c)$ in $\overline{\mathcal{O}}$. The cost is a weighted average with the cell size as the weight.

Lemma 1. *Let X and $\overline{X}$ be an input dataset and its cells, respectively. Let $x \in X$ be a tuple and $r = \mathsf{R}\,(x)$ be its representative in $\overline{X}$. Suppose that $\mathcal{O}$ is a set of cluster centers output by any k-median clustering scheme running on X. Then, $\left|\mathsf{Dist}\,(r, \mathsf{NN}\,(r, \mathcal{O})) - \mathsf{Dist}\,(x, \mathsf{NN}\,(x, \mathcal{O}))\right| \leq \mathsf{Dist}\,(r, x)$.*

Proof. Let $O' = \mathsf{NN}\,(r, \mathcal{O})$ and $O = \mathsf{NN}\,(x, \mathcal{O})$ be the cluster centers in $\mathcal{O}$ nearest to r and x, respectively. Then,

$$\mathsf{Dist}\,(x, O) \leq \mathsf{Dist}\,(x, O') \leq \mathsf{Dist}\,(x, r) + \mathsf{Dist}\,(r, O'). \qquad (6)$$

$$\mathsf{Dist}\,(r, O') \leq \mathsf{Dist}\,(r, O) \leq \mathsf{Dist}\,(r, x) + \mathsf{Dist}\,(x, O). \qquad (7)$$

Combining the two inequalities, we reach the conclusion of the lemma.

On the basis of Lemma 1, we develop the relationship between optimal clustering cost and that of approximation algorithm running on cells.

Theorem 1. *Let X and $\overline{X}$ be an input dataset and its cells, respectively. Suppose that $\overline{\mathcal{O}}$ is the cluster centers output by an (α, β)-approximation k-median algorithm on $\overline{X}$, and $\mathcal{O}^{\circ}$ is the cluster centers output by an optimal solution on X. Then*

$$\mathsf{Cost}\left(\overline{X}, \overline{\mathcal{O}}\right) \leq \alpha \cdot \left(\mathsf{Cost}\,(X, \mathcal{O}^{\circ}) + \frac{D}{|X|}\right), \qquad (8)$$

where $D = \sum_{x \in X} \mathsf{Dist}\,(x, \mathsf{R}\,(x))$.

Proof. Let $X = \{x_1, x_2, \ldots, x_N\}$, and $r_i = \mathsf{R}\,(x_i)$ be the representative of x_i. Let $\overline{\mathcal{O}}^{\circ}$ be the cluster centers output by an optimal k-median algorithm on $\overline{X}$. Then,

$$\mathsf{Cost}\left(\overline{X}, \overline{\mathcal{O}}^{\circ}\right) \leq \frac{1}{N} \sum_{i=1}^{N} \mathsf{Dist}\,(r_i, \mathsf{NN}\,(r_i, \mathcal{O}^{\circ})). \qquad (9)$$

Since $\overline{\mathcal{O}}$ is the output of an (α, β)-approximation k-medians algorithm on $\overline{X}$, it follows that

$$\text{Cost}\left(\overline{X}, \overline{\mathcal{O}^\circ}\right) \geq \frac{1}{\alpha} \cdot \text{Cost}\left(\overline{X}, \overline{\mathcal{O}}\right). \tag{10}$$

Combining Inequalities 9 and 10 gives

$$\text{Cost}\left(X, \mathcal{O}^\circ\right) = \frac{1}{N} \sum_{i=1}^{N} \text{Dist}\left(x_i, \text{NN}\left(x_i, \mathcal{O}^\circ\right)\right)$$

$$\geq \frac{1}{N} \sum_{i=1}^{N} \left(\text{Dist}\left(r_i, \text{NN}\left(r_i, \mathcal{O}^\circ\right)\right) - \text{Dist}\left(r_i, x_i\right)\right)$$

$$\geq \frac{1}{\alpha} \cdot \text{Cost}\left(\overline{X}, \overline{\mathcal{O}}\right) - \frac{D}{N},$$

where the first inequality holds by Lemma 1. This concludes the proof.

In Theorem 1, the clustering cost of approximation algorithm is $\text{Cost}\left(\overline{X}, \overline{\mathcal{O}}\right)$, which is relative to the cells. The next theorem will investigate the cost relative to the input dataset.

Theorem 2. *Let X and $\overline{X}$ be an input dataset and its cells, respectively. Suppose that $\overline{\mathcal{O}}$ is the cluster centers output by an (α, β)-approximation k-median algorithm on $\overline{X}$, and $\mathcal{O}^\circ$ is the cluster centers output by an optimal solution on X. Then*

$$\text{Cost}\left(X, \overline{\mathcal{O}}\right) \leq \alpha \cdot \text{Cost}\left(X, \mathcal{O}^\circ\right) + (1 + \alpha)\frac{D}{|X|}, \tag{11}$$

where $D = \sum_{x \in X} \text{Dist}\left(x, \text{R}\left(x\right)\right)$.

Proof. Let $X = \{x_1, x_2, \dots, x_N\}$, and $r_i = \text{R}\left(x_i\right)$ be the representative of x_i. Then,

$$\text{Cost}\left(X, \overline{\mathcal{O}}\right) = \frac{1}{N} \sum_{i=1}^{N} \text{Dist}\left(x_i, \text{NN}\left(x_i, \overline{\mathcal{O}}\right)\right)$$

$$\leq \frac{1}{N} \sum_{i=1}^{N} \text{Dist}\left(x_i, \text{NN}\left(r_i, \overline{\mathcal{O}}\right)\right)$$

$$\leq \frac{1}{N} \sum_{i=1}^{N} \left[\text{Dist}\left(x_i, r_i\right) + \text{Dist}\left(r_i, \text{NN}\left(r_i, \overline{\mathcal{O}}\right)\right)\right]$$

$$= \frac{D}{N} + \frac{1}{N} \sum_{i=1}^{N} \text{Dist}\left(r_i, \text{NN}\left(r_i, \overline{\mathcal{O}}\right)\right)$$

$$= \frac{D}{N} + \text{Cost}\left(\overline{X}, \overline{\mathcal{O}}\right). \tag{12}$$

Plugging Inequality 12 into Inequality 8, we reach Inequality 11.

Theorem 2 thus gives the accuracy guarantee. It has also computed the δ function in Inequality 4.

$$\delta = (1 + \alpha) \cdot \frac{D}{|X|}. \tag{13}$$

3.2   Determining Grid Granularity

Our work adopts uniform grid. We normalize each dimension to $[0.0,\ 1.0]$, so as to treat each dimension equally for its contribution to the clustering cost. Every cell in the grid is a d-dimensional rectangle with edge length equal to $\frac{1}{\omega}$. Thus,

$$\mathsf{Dist}\,(x, \mathsf{R}\,(x)) \leq \frac{\sqrt{d}}{\omega} \quad (14) \quad \text{and} \quad \frac{D}{|X|} \leq \frac{\sqrt{d}}{\omega} \quad (15)$$

Figure 2 illustrates an example, in which *Latitude* and *Longitude* are the 2 dimensions, $\omega = 4$, and the distance between a tuple and its representative is at most $\frac{\sqrt{2}}{4}$. Plugging Inequality 15 in Inequality 11 gives

$$\mathsf{Cost}\,(X, \overline{\mathcal{O}}) \leq \alpha \mathsf{Cost}\,(X, \mathcal{O}^\circ) + (1 + \alpha) \cdot \frac{\sqrt{d}}{\omega}. \tag{16}$$

The above Inequality implies a trade off between efficiency and accuracy. When the cells are of coarse granularity, the number of cells is small. Thus, the clustering is efficient, but $(1+\alpha) \cdot \frac{\sqrt{d}}{\omega}$ is big, leading to low clustering accuracy. As the cell granularity is finer, the number of cells increases. The efficiency decreases, but the accuracy improves. When each cell has a single tuple, our solution is the same as (α, β)-approximation algorithm running directly on input dataset.

We now study how to decide the ω value. Both $\mathsf{Cost}\,(X, \overline{\mathcal{O}})$ and $\mathsf{Cost}\,(X, \mathcal{O}^\circ)$ are data dependent – their values change with data distribution. We thus also dynamically compute ω according to data distribution. Our computation starts from an analysis of relative distance between $\mathsf{Cost}\,(X, \overline{\mathcal{O}})$ and $\mathsf{Cost}\,(X, \mathcal{O}^\circ)$.

Corollary 1. *Let X and $\overline{X}$ be an input dataset and its cells, respectively. Suppose that $\overline{\mathcal{O}}$ is the cluster centers output by an (α, β)-approximation k-median algorithm on $\overline{X}$, and $\mathcal{O}^\circ$ is the cluster centers output by an optimal solution on X. If the following inequality holds*

$$\frac{D}{|X|} \leq \frac{(\gamma - \alpha)\mathsf{Cost}\,(\overline{X}, \overline{\mathcal{O}})}{\alpha(\gamma + 1)}, \tag{17}$$

then $\mathsf{Cost}\,(X, \overline{\mathcal{O}}) \leq \gamma \mathsf{Cost}\,(X, \mathcal{O}^\circ)$, where $\gamma > \alpha$ and $D = \sum_{x \in X} \mathsf{Dist}\,(x, \mathsf{R}\,(x))$.

Proof. Combining Inequality 17 with Inequality 8, it follows that

$$\frac{\mathsf{Cost}\,(\overline{X}, \overline{\mathcal{O}})}{\mathsf{Cost}\,(X, \mathcal{O}^\circ)} \leq \frac{\alpha(\gamma + 1)}{\alpha + 1},$$

which (if combined with Inequalities 17 and 11) gives $\mathsf{Cost}\,(X, \overline{\mathcal{O}}) \leq \gamma \mathsf{Cost}\,(X, \mathcal{O}^\circ)$.

Now consider Inequalities 15 and 17 together. Clearly, Inequality 17 holds, if

$$\omega \geq \frac{\sqrt{d}}{\eta \cdot \mathsf{Cost}\,(\overline{X}, \overline{\mathcal{O}})},$$

where $\eta = \frac{(\gamma-\alpha)}{\alpha(\gamma+1)}$. In our experiments, we empirically set $\eta = 0.5$, which gives good results. Our solutions runs k-median clustering on sliding windows continuously. Let W_i (for $i = 1$ to ∞) be the sequence of sliding windows, $\overline{X_i}$ be the statistical information of cells in sliding window W_i and $\overline{O_i}$ be the cluster centers derived from $\overline{X_i}$. Then, we set

$$\omega_{i+1} = \frac{\sqrt{d}}{\eta \cdot \mathsf{Cost}\left(\overline{X_i}, \overline{O_i}\right)}. \tag{18}$$

Note that our grid-based clustering solution only has the accuracy guarantee in Inequality 16. It does not guarantee that the relative difference between $\mathsf{Cost}\left(X, \overline{\mathcal{O}}\right)$ and $\mathsf{Cost}\left(X, \mathcal{O}^\circ\right)$ is at most γ, since the required ω value in Corollary 1 might be bigger than the one used to partition the data space. In brief, Equation 18 is a heuristic, which determines the ω value adaptively based on data distribution. Furthermore, as the dimensionality of the dataset increases, the data easily becomes sparse. As the ω value increases, the number of cells with small number of tuples increases fast. To address this, we set a threshold, which is the maximum allowed ω value for a data stream.

4   The Adaptive Grid-Based Solution

We now develop our continuous k-median clustering solution for stream data based on the theoretical analysis in Section 3. For the clarity of presentation, we first present an algorithm, for which the sliding window size is equal to the step size (i.e., tumbling window). We will later extend it to the general case, where neighboring sliding windows may overlap.

Algorithm 1 is the solution for tumbling windows. Lines 1 and 2: an iterative search is applied on historical data to find an initial ω value, such that the clustering cost of the approximation algorithm running on the cells (of historical data) determined by the ω value is 'similar' to the clustering cost of the same approximation algorithm running on the input historical data. In the experiments, we use 10% tuples in a dataset as historical data, and the remaining ones as stream data. The streaming tuples are processed one by one on the fly (Lines 4 to 21). Given a tuple, the cell containing it is computed (Line 6). If the cell exists, then simply update the statistical information of the cell (Lines 7 to 8). Otherwise, a new cell is created (Lines 9 to 10). Our cell management strategy stores only non-empty cells, and thus ensures that the total number of cells is never larger than the total number of tuples. Once a new sliding window is generated, approximation algorithm runs on the statistical information of the cells (Lines 12 to 20), and ω is updated for the next sliding window (Line 19).

In the algorithm we maintain a map $\mathcal{M}$ to efficiently locate the cells containing incoming streaming tuples. We use the cell's boundary as the key of the map. Each cell has simple statistical information, which is the summation of tuple values in the cell, the cell size, and the cell mean. We use the mean to represent the cell. Other statistical information can be easily added if needed,

Algorithm 1. adapGridKM

Input: data stream $\mathcal{DS}$, window size WS, and step size S
Output: the continuous update of βk cluster centers

1: Let HD be a set of history data of $\mathcal{DS}$
2: $\omega \leftarrow$ findOmega (HD)
3: Initialize an empty map $\mathcal{M}$
4: **for** $i \leftarrow 1$ **to** $|\mathcal{DS}|$ **do**
5:     Let $\mathcal{DS}$ [i] be the i-th tuple in $\mathcal{DS}$ ▷ *Lines 5-10: Insert a tuple to a cell.*
6:     $c \leftarrow$ Compute$(\mathcal{DS}[i], \omega)$
7:     **if** $c \in \mathcal{M}$ **then**
8:         $c.sum = c.sum + \mathcal{DS}[i], \quad c.size = c.size + 1$
9:     **else**
10:        $c.sum = \mathcal{DS}[i], \quad c.size = 1, \quad$ Insert c into $\mathcal{M}$
11:    **end if**
12:    **if** $(i \geq WS) \wedge (i \mod S == 0)$ **then**
13:        **for** each cell $c \in \mathcal{M}$ **do**   ▷ *Lines 13-19: Cell clustering and ω update.*
14:            $c.mean = \frac{c.sum}{c.size}$
15:        **end for**
16:        Run (α, β)-approximation clustering on the means of the cells
17:        Update the cluster centers
18:        Let $\mathcal{C}$ be the cost of clustering on the cells
19:        $\omega \leftarrow \frac{\sqrt{d}}{\eta \cdot \mathcal{C}}, \quad$ Empty $\mathcal{M}$
20:    **end if**
21: **end for**

---

and the cell representative can also be changed to cell center or a streaming tuple in the cell or some function as required.

We now extend Algorithm 1 to the general case, where neighboring sliding windows may overlap. The extension is simple. In Algorithm 1, each sliding window has an ω value, and all the tuples in the window are partitioned into cells only once based on the ω value. In the general case, overlapping windows share common tuples, which need to be partitioned into cells for multiple times, each for one sliding window. As such, we maintain an ω value for each step, and partition all the tuples in a window based on the ω value of the first step in the window. We use the following example to better illustrate our ideas.

Example 1. *In Figure 1, the sliding window is 3 times as big as the step. Suppose that at timestamp t we have the first sliding window W_1, which includes steps S^1, S^2, and S^3. Furthermore, suppose that the ω values of these three steps are ω_1, ω_2, and ω_3, respectively. Then, the tuples in windows W_1, W_2, and W_3 are inserted into cells determined by ω_1, ω_2, and ω_3, respectively. We run clustering on the cells of W_1. An ω value is then computed by Equation 18. Denote the value by ω_4, and assign it to step S^4. All the tuples in window W_4 will then be inserted into cells determined by ω_4. At timestamp $t+S$, window W_1 slides forward to window W_2. Clustering is run on the cells of W_2, and another*

ω value (denoted by ω_5) is computed and assigned to step S^5. All the tuples in window W_5 (not shown in Figure 1) will be inserted into cells determined by ω_5. At timestamp $t + 2S$, window W_2 slides forwards to window W_3. The tuple insertion into cells, clustering, and window sliding then continue iteratively like the above.

Storage Complexity. Consider first a single cell. In the d-dimensional data space, the storage cost of its boundary is $2d$. Its statistical information (i.e., summation, size, and mean) is $2d + 1$. Thus, the storage cost of a cell is $4d + 1$. The number of active windows we need to maintain is WS/S, where WS and S are the window size and step size, respectively. Therefore, the storage cost is $\Theta(\frac{n \cdot d \cdot WS}{S})$, where n is the average number of cells in a window.

Time Complexity. The time cost consists of inserting tuples into cells and k-median clustering on cells. Given a tuple, locating the cell containing it takes $2d$. Updating the cell size and the summation takes $d + 1$. Therefore, the time complexity of inserting all the tuples of a sliding window into cells is $\Theta(WS \cdot d)$. The time cost of clustering varies from one approximation algorithm to another. For the simplicity we consider one of the core operations – computing the clustering cost (i.e., distances between tuples and their nearest centers as defined in Equation 1). For this operation, the cost of an approximation algorithm on input data is $\Theta(WS \cdot k \cdot C_d)$, where C_d is the cost of computing the distance between two points in the d-dimensional space, $k \cdot C_d$ is the cost of finding the nearest center out of the given k centers for a given tuple, and WS is the window size. For this operation, the cost of the same approximation algorithm running on the cells is $\Theta(n \cdot k \cdot C_d)$, where n is the number of cells in the sliding window. Since $n \leq WS$, the approximation algorithm running on cells is faster than that on input data. The experimental results in the next section conform this.

5   Experimental Evaluation

5.1   Experimental Setup

We adopt the $(5+\epsilon)$-approximation k-median algorithm [10], which is based on local search. It first randomly selects k cluster centers as the initial solution. It then iteratively improves the centers, until a better solution with an improvement factor of at least $\frac{\epsilon}{5k}$ cannot be found. We denote this algorithm applied on input stream tuples by locSearch. We build our grid-based k-median clustering algorithm on top of [10] – we use it to cluster cells generated from streaming tuples. We denote our algorithm by adapGridKM, representing adaptive grid-based k-median clustering. Furthermore, we include a benchmark fixedGridKM, whose only difference from adapGridKM is fixing the ω value.

We use three real datasets in the experiments. The first is the 2006 Topologically Integrated Geographic Encoding and Referencing (TIGER) dataset [1]. We extract the GPS coordinates of road intersections in the states of Washington and New Mexico. We randomly sample 1 million tuples, and set $k = 2$. The second is the Gowalla dataset [2], which records the check-in information of users of a social network. The information includes user ID, check-in time, check-in location (by latitude and longitude), and location ID. We keep check-in time, latitude, and longitude, and randomly sample 2 million tuples. For this dataset, we set $k = 5$. The last is Power dataset [3], which measures the electric power consumption in one household with a one-minute sampling rate over a period of almost 4 years. After removing tuples with missing values, 2,049,280 tuples are left. We keep all its 9 attributes, and set $k = 4$.

We treat each dimension equally for its contribution to the clustering cost (Equation 1) by normalizing its values to [0.0, 1.0]. Our reported experimental results are window-based. We compute various measures (e.g., clustering cost and elapsed time) for each sliding window, and report the average for all the sliding windows. All the experiments were conducted on an Intel dual core i7-3770 CPU machine with 8G RAM running windows 7.

5.2 Effectiveness and Efficiency

We first investigate the experimental results of the TIGER dataset. In the first row of Figure 3, we fix the step size to 30,000, and vary the window size from 30,000 to 210,000. Figure 3(a) compares the three approaches with respect to effectiveness (i.e., clustering cost by Equation 1). The tuning on historical data suggests $\omega = 128$. We thus set the initial ω value to 128 for adapGridKM, and fix the ω value of fixedGridKM to 128. Figure 3(a) shows that the two grid-based approaches, adapGridKM and fixedGridKM, are as effective as locSearch, i.e., their clustering cost is very close to that of locSearch.

Figure 3(b) compares the approaches in terms of efficiency (i.e., average clustering time for a window). The results show that adapGridKM is consistently 4 orders of magnitude faster than locSearch. The fixedGridKM approach is also more efficient than locSearch, but less efficient than adapGridKM. A careful study of the experimental output shows – the ω values computed by adapGridKM are mostly 32 and 64, and only a few 128. Thus, on average the number of cells in a sliding window of adapGridKM is smaller than that of fixedGridKM. This explains why adapGridKM is more efficient. As the window size increases, the number of tuples (cells) grows. Thus, the time cost for all the approaches grows.

Considering Figures 3(a) and 3(b) together, we can see that the two grid-based approaches have similar clustering cost with fixedGridKM having slightly lower cost, but adapGridKM is much faster than fixedGridKM. The reason behind is that fixedGridKM overestimates the ω value based solely on historical data. This result highlights the importance of adaptively determining the ω value by stream data distribution.

Next, for the TIGER dataset, we fix window size to 180,000, and vary the step size from 10,000 to 180,000 (the second row of Figure 3). Again, in Figure 3(c) the clustering cost of adapGridKM is very close to that of locSearch. For locSearch, the number of tuples for clustering is equal to the fixed window size. Hence, as the step size increases, the average time of clustering tuples does not change obviously. For adapGridKM (fixedGridKM), on average the number of cells generated from each sliding window is almost the same when varying the step size. Thus, the clustering time does not vary obviously either.

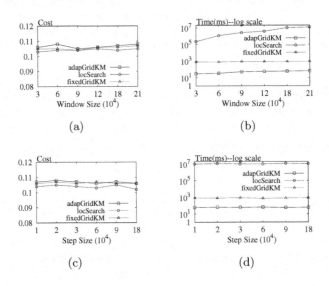

Fig. 3. The evaluation on TIGER dataset

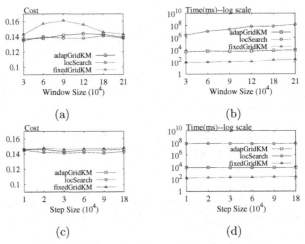

Fig. 4. The evaluation on Gowalla dataset

We now study the experimental results on the Gowalla dataset. The tuning on the historical data suggests $\omega = 8$. We thus set the initial ω value of adapGridKM to 8, and fix that of fixedGridKM to 8. Figure 4(a) reports the clustering cost.

Clearly, adapGridKM outperforms fixedGridKM. Through a careful study of experimental output, we find that most ω values generated by adapGridKM are 32. This shows that fixedGridKM underestimates the grid granularity. The experimental results thus once again prove the importance of adaptively adjusting grid granularity based on stream data distribution. Since fixedGridKM has a coarser granularity, it is faster than adapGridKM in Figure 4(b).

Figure 5 reports the results on the Power dataset. This dataset has 9 dimensions; its data is much sparser than the other two lower-dimensional datasets. Thus, an increase of ω may increase the number of non-empty cells dramatically. To address this issue, we set an upper bound (i.e., 8) for the ω value. Such a setting converges the two grid-based approaches – they both use $\omega = 8$ to partition data into cells. Again, grid-based approaches are comparable to locSearch in terms of clustering cost, but are 2 orders of magnitude more efficient.

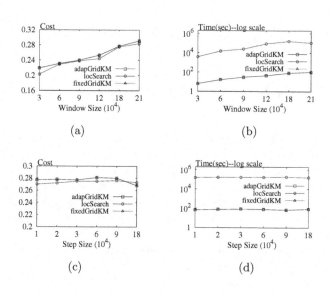

Fig. 5. The evaluation on Power Consumption dataset

5.3 Storage and Tuple-Processing Cost

We now compare the storage cost. For a d-dimensional tuple, we take its storage cost as d, i.e., 1 unit for 1 coordinate. The storage cost of a cell as analyzed in Section 4 is $4d + 1$. Figure 6 shows the results as we vary the window size. Clearly, grid-based approaches need less storage. Take the TIGER dataset as an example (Figure 6(a)). When step size is 180,000, the storage cost of locSearch is 360,000, while those of adapGridKM and fixedGridKM are less than 17,000 and 130,000, respectively. The storage cost of grid-based approaches is up to the number of cells. For the TIGER dataset, the grid granularity of adapGridKM is coarser than that of fixedGridKM. For the Gowalla dataset, it is the opposite.

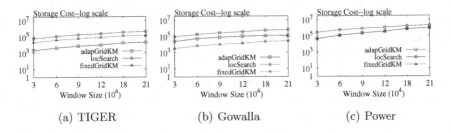

Fig. 6. The storage cost when varying window size

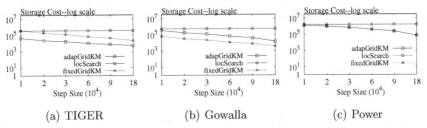

Fig. 7. The storage cost when varying step size

Therefore, in Figure 6(a) adapGridKM has lower storage cost, while having higher cost in Figure 6(b). When window size increases, the number of tuples (and also the cells generated from them) grows. Therefore, the storage cost of all the approaches increases. In Figure 7 we fix the window size to 180,000, and vary the step size. When the step size increases, the number of overlapping windows we need to maintain is smaller. Therefore, the storage cost of the two grid-based approaches decreases.

The tuple-processing time of our approach adapGridKM consists of: a) *Insertion* – the time cost of inserting tuples into cells, and b) *Clustering* – k-median clustering on the cells. Figure 8 and Figure 9 report the results, when varying window size and step size, respectively. In both figures, the elapsed time of *Insertion* is much smaller than that of *Clustering*. For example, in the Gowalla dataset, when the window size and the step size are 60,000 and 30,000, respectively, the elapsed time for *Insertion* and *Clustering* are approximately 26 ms and 6,400 ms, respectively.

6   Related Work

Grid-based clustering is closely related to our work. It was first proposed for static dataset. Representative solutions include STING [33] and CLIQUE [7]. Grid-based clustering has also been applied to the context of data stream to efficiently cluster continuous stream data. The proposed solutions include but not limited to D-Stream [17], cell-tree [30,31], and top-m most frequent cells [21].

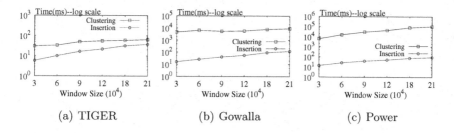

(a) TIGER   (b) Gowalla   (c) Power

Fig. 8. Clustering and tuple-insertion time when varying window size

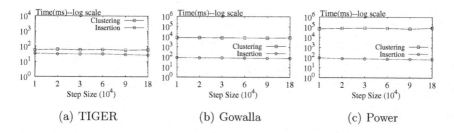

(a) TIGER   (b) Gowalla   (c) Power

Fig. 9. Clustering and tuple-insertion time when varying step size

However, all the above solutions do not give accuracy guarantee of their clustering output. Although our solution is also grid-based, it has clear accuracy bound, thus distinguishing it from existing ones.

Coreset-based clustering [4,5,20,24,25,34] is also related to our work. However, as stated in the introduction, such solutions cannot process streaming tuples on the fly. To compute the coreset of a set of tuples, the whole set needs to be available. Grid-based clustering solutions [17,31] and ours instead can process stream tuples on the fly, i.e., processing tuples independently one by one. Besides the grid-based and coreset-based approaches, data stream clustering has also been studied in [6,8,12,18,22,35]. Refer to [19] for more related work.

k-median clustering is NP-hard. Plenty of (α, β)-approximation algorithms have been proposed [9–11,13–15,26,28,29]; they ensure that the difference between the output and that of the optimal solution is within a given bound. Of the algorithms, [11,13,28,29] have approximation factor α, which is dependent on the dataset size and/or the k parameter. Constant-factor approximation algorithms [10,14,15,26] instead have a constant α. Charikar et al. [15,16] proposed a $6\frac{2}{3}$-approximation algorithm. Jain and Vazirani [26] applied the primal-dual schema to k-median problem, and developed a 6-approximation algorithm. Charikar and Guha [14] refined algorithm [26] and developed a 4-approximation solution. Arya et al. [10] proposed a $(5+\epsilon)$-approximation k-median solution using local search heuristics, where ϵ is a tunable parameter controlling the convergence rate of the solution. Note that our solution is general. It can leverage any constant-factor approximation k-median clustering algorithm discussed above to build a grid-based clustering solution for data stream.

7   Conclusion

In this paper we have proposed a general and adaptive sliding-window-based k-median clustering solution. Our solution dynamically determines the granularity of cells, and runs clustering efficiently on the statistical information of cells. It has a theoretical accuracy bound between its clustering cost and the optimum. The extensive experimental results show that the proposed solution is efficient and effective.

Acknowledgments. This work has been partially supported by Energy Market Authority in Singapore with grant NRF2012EWT-EIRP002-044. We thank Zhenjie Zhang, Yin Yang, and Hong Cao for their valuable comments.

References

1. https://www.census.gov/geo/maps-data/data/tiger.html
2. https://snap.stanford.edu/data/loc-gowalla.html
3. https://archive.ics.uci.edu/ml/datasets/Individual+household+electric+power+consumption
4. Ackermann, M.R., Blömer, J.: Coresets and approximate clustering for bregman divergences. In: SODA
5. Ackermann, M.R., Märtens, M., Raupach, C., Swierkot, K., Lammersen, C., Sohler, C.: Streamkm++: a clustering algorithm for data streams. ACM Journal of Experimental Algorithmics **17**(1) (2012)
6. Aggarwal, C.C., Han, J., Wang, J., Yu, P.S.: A framework for clustering evolving data streams. In: VLDB, pp. 81–92 (2003)
7. Agrawal, R., Gehrke, J., Gunopulos, D., Raghavan, P.: Automatic subspace clustering of high dimensional data for data mining applications. In: SIGMOD Conference, pp. 94–105 (1998)
8. Ailon, N., Jaiswal, R., Monteleoni, C.: Streaming k-means approximation. In: NIPS, pp. 10–18 (2009)
9. Arora, S., Raghavan, P., Rao, S.: Approximation schemes for euclidean k-medians and related problems. In: STOC, pp. 106–113 (1998)
10. Arya, V., Garg, N., Khandekar, R., Meyerson, A., Munagala, K., Pandit, V.: Local search heuristic for k-median and facility location problems. In: STOC, pp. 21–29 (2001)
11. Bartal, Y.: Probabilistic approximations of metric spaces and its algorithmic applications. In: FOCS, pp. 184–193 (1996)
12. Cao, F., Ester, M., Qian, W., Zhou, A.: Density-based clustering over an evolving data stream with noise. In: SDM, pp. 328–339 (2006)
13. Charikar, M., Chekuri, C., Goel, A., Guha, S.: Rounding via trees: deterministic approximation algorithms for group steiner trees and k-median. In: STOC, pp. 114–123 (1998)

14. Charikar, M., Guha, S.: Improved combinatorial algorithms for the facility location and k-median problems. In: FOCS, pp. 378–388 (1999)
15. Charikar, M., Guha, S., Tardos, É., Shmoys, D.B.: A constant-factor approximation algorithm for the k-median problem (extended abstract). In: STOC, pp. 1–10 (1999)
16. Charikar, M., Guha, S., Tardos, É., Shmoys, D.B.: A constant-factor approximation algorithm for the k-median problem. J. Comput. Syst. Sci. 65(1), 129–149 (2002)
17. Chen, Y., Tu, L.: Density-based clustering for real-time stream data. In: KDD, pp. 133–142 (2007)
18. Cormode, G., Muthukrishnan, S., Zhuang, W.: Conquering the divide: continuous clustering of distributed data streams. In: ICDE, pp. 1036–1045 (2007)
19. de Andrade Silva, J., Faria, E.R., Barros, R.C., Hruschka, E.R., de Carvalho, A.C.P.L.F., Gama, J.: Data stream clustering: a survey. ACM Comput. Surv. 46(1), 13 (2013)
20. Feldman, D., Schmidt, M., Sohler, C.: Turning big data into tiny data: constant-size coresets for k-means, pca and projective clustering. In: SODA
21. Gama, J., Rodrigues, P.P., Lopes, L.M.B.: Clustering distributed sensor data streams using local processing and reduced communication. Intell. Data Anal. 15(1), 3–28 (2011)
22. Guha, S., Meyerson, A., Mishra, N., Motwani, R., O'Callaghan, L.: Clustering data streams: theory and practice. IEEE Trans. Knowl. Data Eng. 15(3), 515–528 (2003)
23. Guo, T., Zhu, X., Pei, J., Zhang, C.: Snoc: streaming network node classification. In: ICDM (2014)
24. Har-Peled, S., Kushal, A.: Smaller coresets for k-median and k-means clustering. In: Proceedings of the Twenty-first Annual Symposium on Computational Geometry
25. Har-Peled, S., Mazumdar, S.: On coresets for k-means and k-median clustering. In: STOC, pp. 291–300 (2004)
26. Jain, K., Vazirani, V.V.: Primal-dual approximation algorithms for metric facility location and k-median problems. In: FOCS, pp. 2–13 (1999)
27. Koudas, N., Ooi, B.C., Tan, K.-L., Zhang, R.: Approximate nn queries on streams with guaranteed error/performance bounds. In: VLDB, pp. 804–815 (2004)
28. Lin, J., Vitter, J.S.: Approximation algorithms for geometric median problems. Inf. Process. Lett. 44(5), 245–249 (1992)
29. Lin, J., Vitter, J.S.: Epsilon-approximations with minimum packing constraint violation (extended abstract). In: STOC, pp. 771–782 (1992)
30. Park, N.H., Lee, W.S.: Statistical grid-based clustering over data streams. SIG-MOD Record 33(1), 32–37 (2004)
31. Park, N.H., Lee, W.S.: Cell trees: an adaptive synopsis structure for clustering multi-dimensional on-line data streams. Data Knowl. Eng. 63(2), 528–549 (2007)
32. Tao, Y., Lian, X., Papadias, D., Hadjieleftheriou, M.: Random sampling for continuous streams with arbitrary updates. IEEE Trans. Knowl. Data Eng. 19(1), 96–110 (2007)
33. Wang, W., Yang, J., Muntz, R.R.: Sting: a statistical information grid approach to spatial data mining. In: VLDB, pp. 186–195 (1997)
34. Zhang, Q., Liu, J., Wang, W.: Approximate clustering on distributed data streams. In: ICDE, pp. 1131–1139 (2008)
35. Zhang, Z., Shu, H., Chong, Z., Lu, H., Yang, Y.: C-cube: elastic continuous clustering in the cloud. In: ICDE, pp. 577–588 (2013)

Grouping Methods for Pattern Matching in Probabilistic Data Streams

Kento Sugiura[1]([✉]), Yoshiharu Ishikawa[1], and Yuya Sasaki[2]

[1] Graduate School of Information Science, Nagoya University, Nagoya, Japan
[2] Institute of Innovation for Future Society, Nagoya University, Nagoya, Japan
{sugiura,yuya}@db.ss.is.nagoya-u.ac.jp, ishikawa@is.nagoya-u.ac.jp

Abstract. In recent years, *complex event processing* has attracted considerable interest in research and industry. *Pattern matching* is used to find complex events in data streams. In probabilistic data streams, however, the system may find multiple matches in a given time interval. This may result in inappropriate matches, because multiple matches may correspond to a single event. We therefore propose *grouping methods* of matches for probabilistic data streams, and call such merged matches a *group*. We describe the definitions and generation methods of groups, propose an efficient approach for calculating an occurrence probability of a group, and compare the proposed approach with a naïve one by experiment. The results demonstrate the properties and effectiveness of the proposed method.

Keywords: Complex event processing · Pattern matching · Grouping · Probabilistic data streams

1 Introduction

In recent years, *complex event processing* (CEP) has been a topic of great interest in research and industry. *Pattern matching* is of particular interest because of its usefulness [2–4,10–14,16,19]. The majority of the existing research, however, does not consider data source uncertainty. Data sources such as sensor devices are uncertain because they may contain measurement error, communication error, or both. A data stream has a probabilistic nature when the data source is uncertain. Figure 1 shows an example of a stream that corresponds to an uncertain data source. We call such a data stream a *probabilistic data stream*. In our research, we investigate pattern matching in probabilistic data streams.

However, pattern matching in probabilistic data streams is difficult because the system may find multiple matches in a given interval. For example, Fig. 2 shows results of pattern matching over the stream in Fig. 1 when the pattern $\langle a\ b^+c \rangle$ is given. Methods presented in the existing research remove such matches with low probability because such matches are not important [9]. Such an approach, however, may not be appropriate because every match implies the possibility that the pattern occurred in the interval. We therefore propose

© Springer International Publishing Switzerland 2015
M. Renz et al. (Eds.): DASFAA 2015, Part I, LNCS 9049, pp. 92–107, 2015.
DOI: 10.1007/978-3-319-18120-2_6

time		1	2	3	4	5	6
	a	1.0	0.3	0.1	0.1	0	0
event	b	0	0.7	0.8	0.7	0.9	0
	c	0	0	0.1	0.2	0.1	1.0

Fig. 1. A probabilistic data stream

match	time						probability
	1	2	3	4	5	6	
m_1	a	b	c				0.07
m_2	a	b	b	b	b	c	0.3528
m_3		a	b	b	c		0.0168
m_4				a	b	c	0.09

Fig. 2. Pattern matching result for pattern $\langle a\, b^+ c \rangle$

grouping methods for matches in a given interval. We call such a set of matches a *group*. For example, we merge all matches in Fig. 1 into one group and calculate the probability that the pattern $\langle a\, b^+ c \rangle$ exists in the time interval $[1, 6]$.

The remainder of the paper is organized as follows. In Sect. 2, we describe the background of our research. Section 3 describes the definition of a group and Sect. 4 explains how to generate groups. In Sect. 5, we introduce an effective approach for calculating probabilities of groups. Section 6 describes the settings and results of experiments. Section 7 introduces related work and Sect. 8 concludes the paper.

2 Preliminaries

Assumptions. We make two assumptions here:

1. Each event occurs every unit time and arrives in a data stream engine in order.
2. A probability of an event at time t_i is independent of that of an event at time t_j $(i \neq j)$.

For example, in the probabilistic data stream in Fig. 1, probability $P(a_2 \wedge b_3) = P(a_2) \times P(b_3) = 0.24$ according to the second assumption.

Probability Space. We first define a *probabilistic event* as an entry of a probabilistic data stream.

Definition 1. *A probabilistic event e_t is an event with its probability. The probability that the value of e_t is α is denoted as $P(e_t = \alpha)$. For a discrete domain of events V, the properties*

$$\forall \alpha \in V, 0 \leq P(e_t = \alpha) \leq 1$$

and

$$\sum_{\alpha \in V} P(e_t = \alpha) = 1$$

hold.

For example, in Fig. 1 the occurrence probability of e_3 is $\sum_{\alpha \in \{a,b,c\}} P(e_3 = \alpha) = P(a_3) + P(b_3) + P(c_3) = 1$. We may use $P(\alpha_t)$ as a shorthand of $P(e_t = \alpha)$.

Next, we define a *probabilistic data stream* in our research.

Definition 2. *A* probabilistic data stream $PDS = \langle e_1, e_2, ..., e_t, ... \rangle$ *is a sequence of probabilistic events.*

For instance, the probabilistic data stream in Fig. 1 is represented by $PDS = \langle e_1, e_2, e_3, e_4, e_5, e_6 \rangle$.

Then, we define the notion of *sequence* $s_{[t_i, t_j]}$.

Definition 3. $s_{[t_i,t_j]} = \langle \alpha_{t_i}, \alpha_{t_i+1}, ..., \alpha_{t_j} \rangle$ *is a sequence of events from* t_i *to* t_j. *A probability of* $s_{[t_i,t_j]}$ *is defined as the product of the probabilities of the events in* $s_{[t_i,t_j]}$: $P(s_{[t_i,t_j]}) = \prod_{k=t_i}^{t_j} P(e_k = \alpha_k)$.

For example, one of the sequences in Fig. 1 is $s_{[1,3]} = \langle a_1, a_2, b_3 \rangle$ and the probability of $s_{[1,3]}$ is $P(s_{[1,3]}) = P(a_1) \times P(a_2) \times P(b_3) = 0.24$.

If a *window* is specified as $w = [t_i, t_j]$, we denote $s_{[t_i,t_j]}$ as s_w. In addition, we represent the universal set of s_w as S_w. For instance, Fig. 1 is a data stream for the window $w = [1, 6]$ and examples of the elements of S_w are $\langle a_1, a_2, a_3, a_4, a_5, a_6 \rangle$ and $\langle a_1, a_2, a_3, a_4, a_5, b_6 \rangle$.

Next, we define a probability space using S_w.

Definition 4. *Given a window* w, $(2^{S_w}, P)$ *is the probability space for a probabilistic data stream, where* 2^{S_w} *is the power set of* S_w. P *gives a probability* $P(x)$ *to each element* $x \in 2^{S_w}$ *by summing the probabilities of all sequences in* x *such as* $P(x) = \sum_{s_w \in x} P(s_w)$.

Query Pattern. We use a *regular expression* for representing a query pattern. For example, we use $\langle a\ b\ c \rangle$ if we want to find matches that include a, b, and c with this order: events a, b, and c must be contiguous in the stream. In a query pattern, we can use the Kleene plus ($^+$) as an option for the regular expression. For instance, for the pattern $\langle a\ b^+ c \rangle$, we accept matches such as $\langle a_t\ b_{t+1}\ c_{t+2} \rangle$ and $\langle a_t\ b_{t+1}\ b_{t+2}\ c_{t+3} \rangle$.

Matches. A *match* is an instance of a pattern found in the target probabilistic stream. For example, in Fig. 1 one of the matches for the pattern $\langle a\ b\ c \rangle$ is $\langle a_1\ b_2\ c_3 \rangle$. We define the notion of a match and its probability in a consistent manner with the probability space.

Definition 5. *A match* m *is a set of sequences that include the pattern occurrence as a subsequence. A probability of match* m *is given as* $P(m) = \sum_{s_w \in m} P(s_w)$.

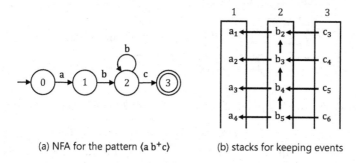

(a) NFA for the pattern ⟨a b⁺c⟩   (b) stacks for keeping events

Fig. 3. An NFA and stacks for generating matches in the stream in Fig. 1

For instance, we consider the probability of $m_1 = \langle a_1\, b_2\, c_3 \rangle$ in Fig. 2. Suppose the window $w = [1, 4]$ is specified for the stream in Fig. 1. In this case, S_w holds sequences such as $\langle a_1, a_2, a_3, a_4 \rangle$ and $\langle a_1, a_2, a_3, b_4 \rangle$. In S_w, there are three sequences that include m_1:

$$s_1 = \langle a_1, b_2, c_3, a_4 \rangle$$
$$s_2 = \langle a_1, b_2, c_3, b_4 \rangle$$
$$s_3 = \langle a_1, b_2, c_3, c_4 \rangle$$

Thus, the probability of m_1 is $P(m_1) = P(s_1) + P(s_2) + P(s_3) = 0.07$.

We follow the NFA-based approach to generate matches [1]. This approach represents a pattern as a non-deterministic finite automaton (NFA) and manages events and matches using stacks. For example, Fig. 3 shows an NFA and stacks for generating matches over the probabilistic data stream in Fig. 1 for the pattern $\langle a\, b^+c \rangle$. The stacks correspond to the respective states of the NFA and store each event that has an occurrence probability. In this example, stack 1 stores events $\{a_1, a_2, a_3, a_4\}$ and does not contain $\{a_5, a_6\}$ because their probabilities are 0. We connect the events using pointers according to the edges of the NFA. We can generate matches by tracing the pointers from the events in the stack of the final state. For example, $\langle a_1\, b_2\, b_3\, c_4 \rangle$ and $\langle a_2\, b_3\, c_4 \rangle$ are generated by tracing the pointers from c_4. In the following, we call a candidate of matches under construction a *run*.

3   Grouping Policies

In our framework, a group is defined by a *grouping policy*. In this section, we introduce two policies. Intuitively, it is natural to merge matches in a given time interval into a group. Thus, we consider the time intervals of matches to decide whether to merge them. The time interval of a match is given by its start and end times. For example, the time interval of m_1 in Fig. 2 is $[1, 3]$.

For considering grouping policies, we use the complete link method and the single link method in hierarchical clustering [6]. The complete link method

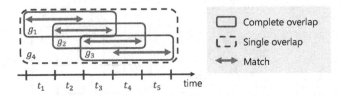

Fig. 4. Group generation based on complete overlap and single overlap

requires that every document is similar to (linked to) all other documents in the same cluster. In contrast, the single link method requires that every document is similar to at least one other document in the same cluster. For our context, we propose *complete overlap* and *single overlap*, inspired by the two methods, and give their definitions below. In the following definition, $ts\_overlap(m, m')$ is a predicate that is true when the time interval of m overlaps with that of m'.

Definition 6. *A set of matches M has a property of* complete overlap *when M satisfies the following condition:*

$$\forall m, m' \in M, ts\_overlap(m, m'). \tag{1}$$

Definition 7. *A set of matches M has a property of* single overlap *when M satisfies the following condition:*

$$\forall m \in M, \exists m' \in M, m \neq m' \land ts\_overlap(m, m'). \tag{2}$$

Now, we define a group using overlaps:

Definition 8. *A group g is a set of matches. g should have a property of complete overlap or single overlap and g should not be a subset of other groups. A group is represented as a tuple $g = (t_s, t_e, p)$ that contains the starting time, the end time, and the corresponding probability.*

Complete overlap ensures that all matches in a group overlap with each other. In contrast, single overlap ensures that each match overlaps with at least one other match in the same group. Figure 4 shows an example of group generation using complete overlap and single overlap. For the case of complete overlap, three groups $g_1 = (t_1, t_3, p_1), g_2 = (t_2, t_4, p_2)$, and $g_3 = (t_3, t_5, p_3)$ are generated. In contrast, one group $g_4 = (t_1, t_5, p_4)$ is generated for the case of single overlap.

The above example shows a tendency of group generation based on complete overlap and single overlap. In this case, complete overlap may generate more useful groups in general, because single overlap excessively merges matches. Suppose the first and last matches have high occurrence probability in Fig. 4. We should distinguish among them in such a case, but single overlap merges them into g_4. On the other hand, complete overlap can distinguish among them as g_1 and g_3. Single overlap is, however, more useful than complete overlap in some ways. For example, it may be appropriate to merge all the matches in Fig. 2 but complete overlap cannot merge them. In contrast, single overlap can merge all the matches into a group. Therefore, we should selectively use complete or single overlap according to the usage scenario.

Next we define the probability of a group. A group is a set of matches according to Definition 8. Moreover, a match is a set of sequences according to Definition 5. A group is therefore also a set of sequences, so we can define the probability of a group as follows:

Definition 9. *A probability of a group g is given as*

$$P(g) = P\left(\bigcup_{m_i \in g} m_i\right) = \sum_{s_w \in \bigcup_{m_i \in g} m_i} P(s_w). \tag{3}$$

In the following, we use the term *group probability* for simplicity.

4   Algorithms for Generating Groups

In this section, we explain how to generate groups for each type of overlap. Moreover, we introduce the use of a probability threshold for efficient group generation.

4.1   The Case of Complete Overlap

As described in Definition 6, complete overlap requires that all matches overlap with each other. Thus, small groups such as g_1, g_2, and g_3 in Fig. 4 are generated. We can identify such groups when a group finds a match for the first time. For example, we consider the groups $g_1 = \{m_1, m_2, m_3\}$, and $g_2 = \{m_2, m_3, m_4\}$ in Fig. 2. g_1 does not have m_4 because m_4 does not overlap with m_1. In other words, all matches in g_1 are fixed when we detect m_1 at time 3. In more detail, g_1 has $\langle a_1\, b_2\, c_3\rangle$, $\langle a_1\, b_2\, b_3\rangle$, and $\langle a_2\, b_3\rangle$ at time 3, and any runs and matches are not added to g_1 after time 3 due to the condition of complete overlap. Thus, we can distinguish g_1 and the other groups such as g_2 at time 3.

Figure 5 shows the algorithm for generating groups based on complete overlap. Note that we omit explanation of lines 10 and 15 in this section; they are covered in Sect. 5. Suppose the pattern $\langle a\ b\ c\rangle$ is given for the stream in Fig. 1. First, we initialize R and G (lines 2 and 3). R is a temporal set of runs and G holds the candidates of groups. We process the events in order (line 4) and add new runs to R to generate candidates. At time 1, a new run $r_1 = \langle a_1\rangle$ is generated and added to R (line 5). The conditions at lines 6, 11, and 14 are not satisfied in this iteration. Then R becomes $\{r_1 = \langle a_1\, b_2\rangle, r_2 = \langle a_2\rangle\}$ at time 2 and $\{m_1 = \langle a_1\, b_2\, c_3\rangle, r_2 = \langle a_2\, b_3\rangle, r_3 = \langle a_3\rangle\}$ at time 3 at line 5. As we find a match m_1, we generate a copy of R as g_1 and add g_1 to G (line 7). Hereafter, we update only $\{r_2, r_3\}$, the remaining runs of g_1 (line 5). Note that we remove $m_1 = \langle a_1\, b_2\, c_3\rangle$ from R (line 8) because R cannot get a new run like $r_4 = \langle a_4\rangle$ for the condition of complete overlap. We output groups and remove them from G when they have no runs (lines 16 and 17). In this example, g_1 is output at

```
 1: procedure GenerateGroupsForCompleteOverlap(PDS)
 2:    R = ∅ // set of runs
 3:    G = ∅ // candidates of groups
 4:    for each e_t ∈ PDS do
 5:       update runs and generate a new run ⟨e_t⟩ then add it to R
 6:       if R has a match then
 7:          generate a copy g_copy of R and add g_copy to G
 8:          remove matches from R
 9:       end if
10:       update the group probability of R using (5)
11:       if R has runs that are out of the window next time then
12:          remove such runs from R
13:       end if
14:       for each g ∈ G do
15:          update the group probability of g using (5)
16:          if g does not have a run then
17:             output g and remove it from G
18:          else if g has runs or matches that are out of the window next time then
19:             output g
20:             remove such runs and matches from g
21:          end if
22:       end for
23:    end for
24: end procedure
```

Fig. 5. Group generation based on complete overlap

time 5 because $g_1 = \{\langle a_1\, b_2\, c_3 \rangle, \langle a_2\, b_3\, c_4 \rangle, \langle a_3\, b_4\, c_5 \rangle\}$ does not have a run. This process continues until the data stream terminates. Note that lines 11 to 13 and lines 18 to 21 are for window processing. We remove runs and matches that are out of the window next time (lines 12 and 20). When a group has matches, we output it (line 19).

4.2 The Case of Single Overlap

In a group based on single overlap, each match should overlap with at least one of the other matches in the same group. The group formation process is not simple, as described below. Consider the situation where the match and the runs in Fig. 6 are generated for the pattern $\langle a\, b^+ c \rangle$.

In this case, we can formulate the groups $g_1 = \{m_1, r_1\}, g_2 = \{r_2\}$, and $g_3 = \{r_3\}$. However, we cannot yet merge g_1 and g_2 because r_2 overlaps with only r_1, and r_2 may not overlap with g_1 if r_1 is rejected in the future. Similarly we cannot merge g_1 and g_3, nor g_2 and g_3, because they overlap with the runs only. Then, we merge groups into one group when an overlap between them is confirmed. For example, when r_2 becomes $m_2 = \langle a_4\, b_5\, c_6 \rangle$ at time 6, we can merge g_2 and g_3 into $g = \{m_2, r_3\}$. Note that we can merge only the groups generated after g_2 because g_2 overlaps with only the run in g_1. If r_2 becomes $m_3 = \langle a_1\, b_2\, b_3\, b_4\, b_5\, c_6 \rangle$ at time 6, we can merge all the groups into one group.

match / run	time				
	1	2	3	4	5
m_1	a	b	c		
r_1	a	b	b	b	b
r_2				a	b
r_3					a

Fig. 6. A match and runs for the pattern $\langle a\, b^+ c\rangle$

Figure 7 shows the algorithm for generating groups based on single overlap. Suppose that the pattern $\langle a\, b\, c\rangle$ is given for the stream in Fig. 1. A new run $r_1 = \langle a_1\rangle$ is generated at time 1 (line 4). As G is an empty set, we do not execute lines 5 to 19 in this iteration. At line 21, we generate a new group $g_1 = \{r_1\}$ and add it to G because r_1 is not added to any group at time 1. At time 2, a new group $g_2 = \{r_2 = \langle a_2\rangle\}$ is generated at line 21 because $g_1 = \{r_1 = \langle a_1\, b_2\rangle\}$ does not have matches. g_2 is merged into g_1 at time 3 because g_1 gets the match $m_1 = \langle a_1\, b_2\, c_3\rangle$ (lines 6 to 11). We can merge g_1 and g_2 because $g_1 = \{\langle a_1\, b_2\, c_3\rangle\}$ and $g_2 =$

```
 1: procedure GenerateGroupsForSingleOverlap(PDS)
 2:    G = ∅ // candidates of groups
 3:    for each e_t ∈ PDS do
 4:       update runs and generate a new run r_new = ⟨e_t⟩
 5:       for each g_i ∈ G do // subscript means the generation order
 6:          if g_i found a match this time then
 7:             for each g_j ∈ G (j > i) do
 8:                g_i = g_i ∪ g_j and remove g_j from G
 9:             end for
10:             g_i = g_i ∪ {r_new}
11:          end if
12:          update the group probability of g_i using (5)
13:          if g_i does not have a run then
14:             output g_i and remove it from G
15:          else if g_i has runs or matches that are out of the window next time then
16:             output g_i
17:             remove such runs and matches from g_i
18:          end if
19:       end for
20:       if r_new is not added to any group then
21:          generate a new group g_new = {r_new} and add it to G
22:          update the group probability of g_new using (5)
23:       end if
24:    end for
25: end procedure
```

Fig. 7. Group generation based on single overlap

$\{\langle a_2\, b_3 \rangle\}$ certainly overlaps. We output groups that have no runs and remove them (line 14). In this example, $g_1 = \{\langle a_1\, b_2\, c_3 \rangle, \langle a_2\, b_3\, c_4 \rangle, \langle a_3\, b_4\, c_5 \rangle\ \langle a_4\, b_5\, c_6 \rangle\}$ is output at time 6. This process continues until the data stream terminates.

4.3 Use of Threshold of Match Probability

We consider the threshold of a match probability to generate groups efficiently. The runtime of group generation is large when we generate all matches. Thus, we remove matches whose probabilities are lower than the specified threshold. We can remove runs and matches at line 5 in Fig. 5 and line 4 in Fig. 7.

For instance, suppose the pattern $\langle a\, b^+ c \rangle$ is given and the match threshold is $\theta = 0.1$ for single overlap matching for the stream in Fig. 1. Ten matches are found from Fig. 1, but matches satisfying θ are $\langle a_1\, b_2\, b_3\, c_4 \rangle$, $\langle a_1\, b_2\, b_3\, b_4\, b_5\, c_6 \rangle$, and $\langle a_2\, b_3\, b_4\, b_5\, c_6 \rangle$. Therefore, we construct a group from these three matches.

Although we prune matches with low probabilities, we do not ignore those probabilities. That is, we remove matches such as $\langle a_1\, b_2\, c_3 \rangle$ and $\langle a_1\, b_2\, b_3\, b_4\, c_5 \rangle$ in our example, but we compute the group probability according to (3). We explain the details in the next section.

5   Efficient Calculation of Group Probability

Using (3), we can calculate a group probability by summing probabilities of all sequences in the group. However, such an approach is not efficient because the number of sequences increases with order $O(|V|^w)$. Therefore, we propose an efficient method using a transducer.

A *finite state transducer* is a finite state automaton which produces output as well as reading input. Figure 8 shows an example of a transducer for the pattern $\langle a\, b^+ c \rangle$. This transducer is generated to accept all sequences that contain the pattern as a subsequence. Thus, a probability of arriving at the final state is the sum of probabilities of the sequences. That is, the probability of arriving at the final state becomes the group probability.

We can generate the transducer by adding edges to the NFA in Fig. 3. The rules for adding edges are as follows:

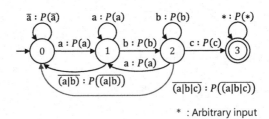

* : Arbitrary input

Fig. 8. Transducer for the pattern $\langle a\, b^+ c \rangle$

1. We add an edge from the final state to itself with arbitrary inputs.
2. If the state does not have edges with the first event of the pattern, we add an edge that shifts the state to state 1 with the first event.
3. We add edges that shift each state to the initial state with other inputs.

For example, we add an edge with the label "$*: P(*)$" to the final state according to rule 1. This edge enables the transducer to keep accepted sequences in the final state. As the first event of $\langle a\ b^+ c\rangle$ is a, we add edges with the label "$a: P(a)$" to state 1 and 2 according to rule 2. We add those edges to accept sequences that contain a part of the pattern such as $\langle a_1, b_2, a_3, b_4, c_5\rangle$. According to rule 3, we add edges with the labels "$\overline{a}: P(\overline{a})$", "$\left(\overline{a|b}\right): P\left(\left(\overline{a|b}\right)\right)$", and "$\left(\overline{a|b|c}\right): P\left(\left(\overline{a|b|c}\right)\right)$." Those edges enable rejected sequences to start again from the initial state.

In the following, we explain how to use a transducer for single and complete overlaps.

5.1   The Case of Single Overlap

A group $g = (t_s, t_e, p)$ based on single overlap has all matches in the interval $[t_s, t_e]$. Figure 4 shows an example where g_4 consists of all the matches in $[t_1, t_5]$. Therefore, in single overlap, a group probability is equal to the sum of the probabilities of all sequences that contain the pattern as a subsequence in the interval $[t_s, t_e]$. As described above, such a probability is the probability of arriving at the final state of the transducer. Thus, we can compute a group probability using a transducer instead of computing the probabilities of all sequences.

We use a *transition matrix* of a transducer to calculate the probability of arriving at the final state. Equation (4) is an example of the transition matrix of the transducer in Fig. 8:

$$T_{t_{i-1},t_i} = \begin{bmatrix} P(\overline{a_{t_i}}) & P(a_{t_i}) & 0 & 0 \\ P\left(\overline{a|b_{t_i}}\right) & P(a_{t_i})\ P(b_{t_i}) & 0 \\ P\left(\overline{a|b|c_{t_i}}\right) & P(a_{t_i}) & P(b_{t_i}) & P(c_{t_i}) \\ 0 & 0 & 0 & 1 \end{bmatrix}. \tag{4}$$

Each row corresponds to the previous states and each column corresponds to the present states. For example, $T_{t_{i-1},t_i}(0,1) = P(a_{t_i})$ means a probability that shifts state 0 to state 1 is $P(a_{t_i})$. Let St be a vector that contains state probabilities. Then we can update the probabilities as follows:

$$St_{t_i} = St_{t_{i-1}} \times T_{t_{i-1},t_i}. \tag{5}$$

In other words, we can calculate the probability of the final state by updating the vector at each time. This process corresponds to lines 12 and 22 in Fig. 7. For example, Fig. 9 shows the change of vector St for computing the group probability of $g = (1, 6, p)$ in Fig. 1. Note that g is the set of all matches found

time		init	1	2	3	4	5	6
state	$St[0]$	1.0	0	0	0.03	0.047	0.0563	0.0563
	$St[1]$	0	1.0	0.3	0.10	0.093	0	0
	$St[2]$	0	0	0.7	0.80	0.630	0.6507	0
	$St[3]$	0	0	0	0.07	0.230	0.2930	0.9437

Fig. 9. Updating state vector St

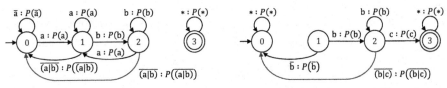

(a) Transducer for a group without matches   (b) Transducer for a group with matches

Fig. 10. Additional transducers to calculate a group probability in complete overlap

in Fig. 1. When a group is generated, we initialize the vector such that $St_{init}[0] = 1.0$ and the others are 0. We update the vector using (5) at each time until the group is output. In this case, the group probability is 0.9437 because we output the group at time 6.

This approach can compute a group probability even if we prune matches using a match threshold. As we use a transducer and its transition matrix, we can compute the sum of probabilities of all sequences that contain matches regardless of whether matches are generated or not. Let us continue the above example with $g = (1, 6, p)$. Suppose the matches such as m_1 in Fig. 2 are pruned by a match threshold. In such a case, however, we can compute the same transition matrix at each time using (4). For example, $T_{1,2}$ does not change regardless of whether the matches are generated or not. As we use the transition matrix for calculating the group probability, we can compute the same group probability even if the matches are pruned.

5.2   The Case of Complete Overlap

We cannot calculate a group probability using the former approach for complete overlap. Recall Fig. 2, where two groups $g_1 = \{m_1, m_2, m_3\}$ and $g_2 = \{m_2, m_3, m_4\}$ are generated. If we use the transducer in Fig. 8 to calculate the group probability of g_1, the calculated probability is not correct because it contains the probability of m_4. Similarly, the probability of g_2 is also not correct because of m_1.

Therefore, we use two additional transducers to solve the problem. Figure 10 shows the transducers for the pattern $\langle a\ b^+c \rangle$. The transducer (a) accepts no matches because it does not have edges that shift states to the final state. On the other hand, the transducer (b) generates no runs because it does not have edges that shift states to state 1. We use the transducer (a) while the group does not have matches (line 10 in Fig. 5). The transducer (b) is used after the group finds matches (line 15 in Fig. 5). Note that we use the transducer in Fig. 8 only once, when a group finds matches for the first time (line 15 in Fig. 5).

Let us continue the above example with $g_1 = \{m_1, m_2, m_3\}$ and $g_2 = \{m_2, m_3, m_4\}$. Let us denote the transition matrix of the transducer in Fig. 8 as T. Similarly, we represent the transition matrices of transducers (a) and (b) in Fig. 10 as T^A and T^B, respectively. We consider the case of calculating the probability of g_1. The system uses T^A before time 3 and T at time 3. T^B is used after time 3 to avoid including the probability of m_4. At $t = 6$, the vector St of g_1 is as follows:

$$St_6 = St_{init} \times T^A_{init,1} \times T^A_{1,2} \times T_{2,3} \times T^B_{3,4} \times T^B_{4,5} \times T^B_{5,6}.$$

Similarly, for g_2 we use T^A before time 5 to avoid including the probability of m_1. Then T is used at time 5 and T^B is used after time 5. Thus, St of g_2 is as follows:

$$St_6 = St_{init} \times T^A_{init,1} \times T^A_{1,2} \times T^A_{2,3} \times T^A_{3,4} \times T_{4,5} \times T^B_{5,6}.$$

If the system uses a match threshold, our approach can compute group probabilities as well as the case of single overlap. In the case of complete overlap, the system decides whether to use the transducers according to the time t_f that a group gets matches for the first time. Thus, the system can compute the group probability of $g = (t_s, t_e, p)$ only if it has a tuple (t_s, t_f, t_e). t_f is easily determined in the process of group generation, because the system can recognize it at line 6 in Fig. 5. Therefore, our approach can calculate correct group probabilities for complete overlap.

6   Experiments

In this section, we analyze the performance of our approach. We constructed a system that generates groups and computes group probabilities using the described approach. The system is an extension of SASE+ [1], a Java-based system for pattern matching queries in a non-uncertain data stream. We performed all measurements on a computer with an Intel Core i7-2600 3.40 GHz CPU, 4.0 GB main memory, and the Windows 7 Professional 64-bit operating system. The system runs under Java Hotspot VM 1.5 with the JVM allocation pool set to 1.5 GB.

The experiments are performed based on simulations. We generate a synthetic probabilistic data stream and use it as an input stream. The generation process is as follows. First, we generate a non-uncertain data stream $\langle \alpha_1, \alpha_2, ..., \alpha_{10000} \rangle$ with 10,000 events. Each event value α_t is taken from the domain $V = \{a, b, c, d\}$. The probability distribution for each event is set as follows. Consider the case of $\alpha_1 = a_1$. We randomly choose the occurrence probabilities of b_1, c_1, and d_1 from the range $[0, 0.1]$. Then the probability of a_1 is given as $P(a_1) = 1 - \sum_{\alpha_1 \in \{b_1, c_1, d_1\}} P(\alpha_1)$. For $t = 2, 3, ...$, we follow the same procedure.

We evaluate the performance of the proposed method using throughput (the number of events processed per second). In the experiments, we use the pattern $\langle a\ b^+ c \rangle$ and the parameters in Table 1. In the following, we represent the setting

Table 1. Parameters in the experiments

Parameters	Values
o: overlap type	*comp*: complete overlap
	sing: single overlap
θ: threshold of match probabilities	$\{0, 0.01, 0.02, 0.04, 0.06, 0.08, 0.1\}$
w: window size	$\{5, 10, 15, 20, 25, 50\}$

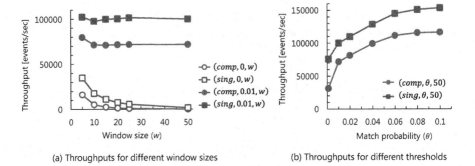

(a) Throughputs for different window sizes

(b) Throughputs for different thresholds

Fig. 11. Throughputs for different window sizes and thresholds

of the parameters as a tuple (o, θ, w). For example, $(comp, 0.01, 50)$ means that the system uses complete overlap, the match threshold is 0.01, and the window size is 50.

6.1 Effect of Parameters

Figure 11(a) shows the throughputs for different window sizes. We show the cases of only $\theta = 0$ and 0.01 because the tendencies for larger thresholds are similar to that of $\theta = 0.01$. We can observe three properties. First, the throughput decreases rapidly if we do not use a threshold ($\theta = 0$). This is due to the number of generated matches. When we use the pattern $\langle$a b$^+$c$\rangle$, the number of matches increases with order $O(w^2)$. Second, the throughput is independent of the window size if we use a threshold, because most matches are pruned early by the threshold. As described above, the number of matches increases rapidly, but most matches do not have high probabilities. High throughput is achieved because many matches are pruned before their generation. Furthermore, the throughput of single overlap is larger than that of complete overlap due to the number of generated groups. As shown in Fig. 4, complete overlap generates more groups than single overlap. Thus, we need more computation time for complete overlap.

We next study the effect of the threshold setting. Figure 11(b) shows the throughputs for different thresholds. We show the case of only $w = 50$ because

Table 2. Throughputs of the transducer-based approach and the naive approach for different window sizes

w	5	10	15
$(comp, 0.01, w)$+proposed	79,581	71,356	71,076
$(comp, 0.01, w)$+naïve	265	14	Out of memory
$(sing, 0.01, w)$+proposed	102,389	97,575	99,865
$(sing, 0.01, w)$+naïve	267	14	Out of memory

the tendencies for other window sizes are similar to those of $w = 50$. The throughput increases as the threshold becomes higher. This is also due to the number of generated matches. As described above, the higher the threshold, the more we can prune matches. Therefore, we can process each event rapidly if we use a high threshold.

6.2   Effect of Transducer-Based Approach

In this experiment, we study the efficiency of our method with transducers. We compare the proposed method with a naïve one. In the naïve method, we generate all sequences in a window and summarize their probabilities according to (3).

Table 2 shows the throughput measurements for different window sizes, where "proposed" means that we use the proposed method to compute group probabilities. Similarly, "naïve" means the use of the naïve method. Table 2 shows that the throughput of the naïve method decreases exponentially. The naïve method generates all sequences in a window, but their number increases with order $O(|V|^w)$. Therefore, the naïve method cannot compute group probabilities due to memory shortage for $w = 15$. On the other hand, the proposed method can compute group probabilities regardless of the window size. The proposed method uses (5) to compute group probabilities. The computational complexity of (5) is $O(|p|^2)$, where $|p|$ is the length of pattern p. Note that the length of a pattern means the number of events in the pattern (e.g., $|\langle a\ b^+c\rangle| = 3$). Thus, the computation time of the proposed method is much smaller than that of the naïve one.

7   Related Work

In the literature of non-uncertain data streams, many methods for pattern matching are proposed [1–5,10–14,16–19]. The SASE project [1,5,17–19], proposes an NFA-based approach for finding matches as described in Sect. 2. Moreover, they propose a method to efficiently process the Kleene closure [1,5]. We have implemented our system by extending their CEP system SASE+ [1] and their methods. Including the SASE project, however, all the methods do not consider and process uncertain data streams.

Some researchers have taken on pattern matching in uncertain data streams [7–9,15]. The Lahar project [7,8,15] considers correlated probabilistic data streams. In correlated streams, every event has a conditional probability as an occurrence probability because an underlying Markov process is assumed. To compute match probabilities, they use an NFA translated from a query. They keep probabilities of the states and regard the probability of the final state as the match probability. Their approach, however, merges only simultaneously accepted matches. On the other hand, our approach considers a time interval of matches and merges them according to complete overlap or single overlap. Therefore, we can group matches more flexibly. We do not consider correlations between events in this paper, but we will be able to extend our approach to correlated streams.

[9] proposes a method to find top k matches in probabilistic streams. In [9], probabilistic streams are generated by their system using an error model that translates a non-uncertain event to a probabilistic event. Moreover, the system computes probabilities for the top k matches using the error model, and can merge them. Their approach, however, merges only matches that are among the top k simultaneously accepted matches. On the other hand, we can merge all probabilities in a group using transducers.

8   Conclusion

We proposed a grouping method of matches for probabilistic data streams. We proposed complete overlap and single overlap and defined a group using them. Then, we explained the two algorithms for generating groups. To compute a group probability efficiently, we proposed an approach that uses transducers. We evaluated the efficiency of our approach in simulation-based experiments. Future work will include refinement of the grouping policy and method for group generation, extension to correlated probabilistic data streams, support of other options for the regular expression such as negation, and re-evaluation of our approach using real data sets.

Acknowledgments. This research is partially supported by KAKENHI (25280039, 26540043) and the Center of Innovation Program from the Japan Science and Technology Agency (JST).

References

1. Agrawal, J., Diao, Y., Gyllstrom, D., Immerman, N.: Efficient pattern matching over event streams. In: Proc. ACM SIGMOD, pp. 147–160 (2008)
2. Akdere, M., Çetintemel, U., Tatbul, N.: Plan-based complex event detection across distributed sources. Proc. VLDB Endow. 1(1), 66–77 (2008)
3. Chandramouli, B., Goldstein, J., Maier, D.: High-performance dynamic pattern matching over disordered streams. Proc. VLDB Endow. 3(1–2), 220–231 (2010)
4. Demers, A., Gehrke, J., Panda, B.: Cayuga: A general purpose event monitoring system. In: Proc. CIDR, pp. 412–422 (2007)

5. Gyllstrom, D., Agrawal, J., Diao, Y., Immerman, N.: On supporting Kleene closure over event streams. In: Proc. ICDE, pp. 1391–1393 (2008)
6. Jain, A.K., Murty, M.N., Flynn, P.J.: Data clustering: A review. ACM Comput. Surv. **31**(3), 264–323 (1999)
7. Letchner, J., Ré, C., Balazinska, M., Philipose, M.: Access methods for Markovian streams. In: Proc. ICDE, pp. 246–257 (2009)
8. Letchner, J., Ré, C., Balazinska, M., Philipose, M.: Approximation trade-offs in Markovian stream processing: An empirical study. In: Proc. ICDE, pp. 936–939 (2010)
9. Li, Z., Ge, T., Chen, C.X.: ε-matching: Event processing over noisy sequences in real time. In: Proc. ACM SIGMOD, pp. 601–612 (2013)
10. Liu, M., Golovnya, D., Rundensteiner, E.A., Claypool, K.T.: Sequence pattern query processing over out-of-order event streams. In: Proc. ICDE, pp. 784–795 (2009)
11. Majumder, A., Rastogi, R., Vanama, S.: Scalable regular expression matching on data streams. In: Proc. ACM SIGMOD, pp. 161–172 (2008)
12. Mei, Y., Madden, S.: ZStream: A cost-based query processor for adaptively detecting composite events. In: Proc. ACM SIGMOD, pp. 193–206 (2009)
13. Mozafari, B., Zeng, K., Zaniolo, C.: High-performance complex event processing over XML streams. In: Proc. ACM SIGMOD, pp. 253–264 (2012)
14. Qi, Y., Cao, L., Ray, M., Rundensteiner, E.A.: Complex event analytics: Online aggregation of stream sequence patterns. In: Proc. ACM SIGMOD, pp. 229–240 (2014)
15. Ré, C., Letchner, J., Balazinksa, M., Suciu, D.: Event queries on correlated probabilistic streams. In: Proc. ACM SIGMOD, pp. 715–728 (2008)
16. Woods, L., Teubner, J., Alonso, G.: Complex event detection at wire speed with FPGAs. Proc. VLDB Endow. **3**(1–2), 660–669 (2010)
17. Wu, E., Diao, Y., Rizvi, S.: High-performance complex event processing over streams. In: Proc. ACM SIGMOD, pp. 407–418 (2006)
18. Zhang, H., Diao, Y., Immerman, N.: Recognizing patterns in streams with imprecise timestamps. Proc. VLDB Endow. **3**(1–2), 244–255 (2010)
19. Zhang, H., Diao, Y., Immerman, N.: On complexity and optimization of expensive queries in complex event processing. In: Proc. ACM SIGMOD, pp. 217–228 (2014)

Fast Similarity Search of Multi-dimensional Time Series via Segment Rotation

Xudong Gong[1], Yan Xiong[1], Wenchao Huang[1(✉)], Lei Chen[2],
Qiwei Lu[1], and Yiqing Hu[1]

[1] University of Science and Technology of China, Hefei, China
lzgxd@mail.ustc.edu.cn, {yxiong,huangwc}@ustc.edu.cn
[2] Hong Kong University of Science and Technology, Hong Kong, China
leichen@cse.ust.hk

Abstract. Multi-dimensional time series is playing an increasingly important role in the "big data" era, one noticeable representative being the pervasive trajectory data. Numerous applications of multi-dimensional time series all require to find similar time series of a given one, and regarding this purpose, Dynamic Time Warping (DTW) is the most widely used distance measure. Due to the high computation overhead of DTW, many lower bounding methods have been proposed to speed up similarity search. However, almost all the existing lower bounds are for general time series, which means they do not take advantage of the unique characteristics of higher dimensional time series. In this paper, we introduce a new lower bound for constrained DTW on multi-dimensional time series to achieve fast similarity search. The key observation is that when the time series is multi-dimensional, it can be rotated around the time axis, which helps to minimize the bounding envelope, thus improve the tightness, and in consequence the pruning power, of the lower bound. The experiment result on real world datasets demonstrates that our proposed method achieves faster similarity search than state-of-the-art techniques based on DTW.

1 Introduction

Multi-dimensional time series are playing an increasingly important role in this "big data" era. For example, with the rapid development of wireless communication and location positioning technologies, we can easily acquire the location of a moving object, e.g. a person, a vehicle, or an animal, at different time. Such movements are generally recorded as a series of triples (x, y, t), where x and y are coordinates and t is the sample time. When talking about the dimensionality of time series, the sample time is often omitted, so such trajectory is regarded as "two dimensional" time series, which is a representative of multi-dimensional time series. Various time series data has enabled many interesting applications, such as finding potential friends according to similar trajectories [12], human activity recognition [20], and climate change prediction [15] etc.

A basic and important operation in various applications of multi-dimensional time series is to find similar time series of a given one, which is a *similarity search*

© Springer International Publishing Switzerland 2015
M. Renz et al. (Eds.): DASFAA 2015, Part I, LNCS 9049, pp. 108–124, 2015.
DOI: 10.1007/978-3-319-18120-2_7

problem. The similarity between two time series is often decided by the distance between them. According to the thorough experiments carried out in [19], among all the proposed distance measures for time series data, DTW may be potentially the best one, and it has achieved great success in highly diverse domains, such as DNA sequence clustering [13], query by humming [26], RFID tag location [18] etc. The straightforward computation of DTW takes quadratic time, which renders it unacceptably slow for applications involving large datasets. In the past years, many techniques have been proposed to prune unqualified candidates by first computing a lower bound, thus reduce the number of required DTW computations [9,11,21,26]. To the best of our knowledge, all the proposed lower bounding techniques are for general time series, which means they don't care the dimensionality of the data, although there are efforts to extend some lower bounding methods to multi-dimensional time series, such as [14,17].

We notice that these general lower bounding techniques can be further improved if we consider some unique characteristics of time series in higher dimensional space. For example, when a time series is in two or more dimensional space, it can be rotated around the time axis without changing its geometrical property, thus the distance between two time series will not be affected. This feature of multi-dimensional time series can be utilized to get a tighter lower bound for candidate pruning.

Inspired by this observation, we introduce a new lower bound called *LB_rotation* to speed up similarity search process. The basic idea is to, for each time series, rotate it by an appropriate angle to reduce the volume of its envelope, because it has been pointed out in [9] that "the envelope is wider when the underlying query sequence is changing rapidly, and narrower when the query sequence plateaus". In such a way, we can improve the tightness of the lower bound, and prune more unqualified candidates to reduce the required DTW computations.

In order to get a satisfactory lower bound, we need to solve several problems:

- **Directly rotate the whole time series may not be a good idea.** As we will show later, the more straight a time series is, the better improvement we can get. Thus we first perform segmentation on the target time series to divide it into several segments as straight as possible, then deal with each segment respectively.
- **Deciding the rotation angle.** This is a key factor affecting the effectiveness of LB_rotation. Rather than directly reducing the volume of the envelope, we aim at reducing the volume of its bounding hypercube, since the envelope is included in the hypercube. For every time series segment, we can find the direction of its major axis by least square fitting. The rotation angle is just the included angle between this direction and the x axis.
- **Computing the lower bound.** After segmentation, the warping range of a point in the candidate time series may intersect with several segments of the query time series, thus we have to compute the distance between the point and its matching point in each segment, and sum up the minimal distance to get final lower bound. We first construct extended envelope for each segment, then locate the corresponding matching point for distance computation.

To demonstrate the superiority of LB_rotation, we compare it with LB_Keogh [9], which is the most widely used lower bound for constrained DTW, and LB_Improved [11], which is recognized as the only lower bound that actually improves LB_Keogh [19], through experiments on real world datasets. The executable and datasets we used are freely available at [1]. We will show that increasing warping constraint has smaller impact on the tightness of LB_rotation, while it may hurt the tightness of LB_Keogh and LB_Improved considerably.

Our major contribution can be summarized as follows:

– We propose a new lower bound *LB_rotation* for constrained DTW based on time series rotation to achieve fast similarity search on multi-dimensional time series. It can shrink the envelope of time series, thus improve the tightness of lower bound, which helps to reduce similarity search time.
– We improve the effectiveness of LB_rotation by dividing the time series into several segments and rotating each segment respectively, rather than directly rotating the whole time series. The experiment result on real world datasets shows that LB_rotation is more effective than existing lower bounds.

The rest of this paper is organized as follows: Section 2 reviews related work, and introduces some necessary extensions. Then we demonstrate the details of LB_rotation in Section 3. Experiment results and discussions are presented in Section 4. Section 5 concludes this paper.

2   Preliminaries

2.1   Related Work

Since retrieval of similar time series plays an important role in many applications, such as time series data mining, a lot of effort has been devoted to solving this problem, and many distance measures have been proposed, such as Euclidean Distance [7], Dynamic Time Warping (DTW) [21], Longest Common Subsequences (LCSS) [5], Edit Distance on Real sequence (EDR) [4], Edit distance with Real Penalty (ERP) [3], etc. Among them, DTW is the most widely used distance measure on time series, because of its effectiveness and robustness.

Due to the high computation complexity of DTW, there are many techniques developed for it to speed up the distance computation. These techniques can be mainly divided into two categories: a) directly speed up DTW computation; b) reduce the number of DTW computations via lower bounding. Lower bounding DTW is a widely used technique, because it can filter a large part of candidate time series using relatively cheap computation. Generally, for any lower bounding algorithm, the nearest neighbor searching process is shown in Algorithm 1. It's clear that the tighter a lower bound is, the higher its pruning power will be, since more candidates will be discarded in Algorithm 1.

The early attempts to lower bound DTW are LB_Yi[21] and LB_Kim[10]. Since they only use the global information of the time series, such as the maximal and minimal values to compute the lower bound, their results are relatively

---

Algorithm 1. Nearest time series search using lower bounding method

---

Input: Q ▷ query time series
Input: $\mathcal{C}$ ▷ database of candidate time series
Output: the index of nearest time series regarding Q
  1: **function** NEARESTNEIGHBOR($Q,\mathcal{C}$)
  2:     $dist_{min} = \infty$
  3:     **for** $i \leftarrow 1$ to $|\mathcal{C}|$
  4:         $lb \leftarrow lower\_bound(Q, C_i)$
  5:         **if** $lb < dist_{min}$
  6:             $true\_dist \leftarrow DTW(Q, C_i)$ ▷ C_i is the ith candidate in the database
  7:             **if** $true\_dist < dist_{min}$
  8:                 $dist_{min} \leftarrow true\_dist$
  9:                 $index \leftarrow i$
 10:     **return** $index$

---

loose. Keogh et al. [9] took advantage of the warping constraint to construct an envelope for the query time series, and proposed the first non-trivial lower bound LB_Keogh for DTW, which greatly eliminates the number of required DTW computations.

There are several extensions of LB_Keogh, e.g. [11,16,25,26]. Among them, LB_Improved [11] is recognized as the only improvement that has reproducible result to reduce searching time [19], thus we compare LB_rotation with LB_Keogh and LB_Improved in Section 4. LB_Improved is built upon LB_Keogh, which improves the tightness through a second pass. It is computed as $LB\_Improved(Q, C) = LB\_Keogh(Q, C) + LB\_Keogh(Q, H(C, Q))$, where $H(C, Q)$ is the projection of C on Q [11].

Generally for almost all the non-trivial lower bounding techniques, there are two prerequisites: a) DTW must be compliant to a constraint enforced on the warping path; b) the trajectories should be of the same length. If not otherwise stated, we assume these conditions are already met hereafter. For more details of lower bounding DTW, please refer to [9,11,19].

We note that these lower bounding techniques are all for general time series, without considering the unique characteristics of high dimensional time series. Actually, when it comes to time series in two or more dimensional space, we can rotate the time series to "flatten" them, thus reduce the volume of their bounding envelopes, which will improve the tightness of the lower bound, as we will show in the following of this paper.

2.2 Extending LB_Keogh and LB_Improved to Multi-dimensional Time Series

Now we introduce the extended lower bounds for multi-dimensional time series, and use them for experimental comparison in Section 4.

Originally, LB_Keogh and LB_Improved are proposed to deal with one dimensional (univariate) time series. We start extending them to multi-dimensional time series by introducing multi-dimensional bounding envelopes.

Definition 1. *The bounding envelope of a time series Q of length n in l dimensional space, with respect to the global warping constraint c, is defined as*

$$Env(Q) = (U_1, U_2, \ldots, U_n, L_1, L_2, \ldots, L_n), \tag{1}$$

where $U_i = (u_{i,1}, u_{i,2}, \ldots, u_{i,l})$, $L_i = (l_{i,1}, l_{i,2}, \ldots, l_{i,l})$, and $u_{i,p} = \max\{q_{i-c,p} : q_{i+c,p}\}$, $l_{i,p} = \min\{q_{i-c,p} : q_{i+c,p}\}$, where q_i is the ith point in Q.

For LB_Keogh, we adopt the extension introduced in [14].

Definition 2. *The multi-dimensional extension of LB_Keogh is defined as*

$$LB\_MV(Q,C) = \sqrt{\sum_{i=1}^{n}\sum_{p=1}^{l} \begin{cases} (c_{i,p} - u_{i,p})^2, & if\ c_{i,p} > u_{i,p} \\ (c_{i,p} - l_{i,p})^2, & if\ c_{i,p} < l_{i,p} \\ 0, & otherwise \end{cases}} \tag{2}$$

where Q is the query time series, C is the candidate time series, c_i is the ith point in C, u_p and l_p are the maximum and minimum values of dimension p, with respect to Q. n is the length of the time series, and l is the dimensionality of each point in the time series.

The proposition below is proved in [14].

Proposition 1. *For any two sequences Q and C of the same length n, for any global constraint on the warping path of the form $j - c \le i \le j + c$, the following inequality holds: $LB\_MV(Q,C) \le DTW(Q,C)$.*

Following [14], we extend LB_Improved to multi-dimensional time series. We only need to extend the projection function (equation (1) in [11]) as follows.

Definition 3. *The projection of C on Q in multi-dimensional LB_Improved is defined as*

$$H(C,Q)_{i,p} = \begin{cases} u_{i,p} & if\ c_{i,p} \ge u_{i,p} \\ l_{i,p} & if\ c_{i,p} \le l_{i,p}\ ,\ 1 \le p \le l \\ q_{i,p} & otherwise \end{cases} \tag{3}$$

Similarly, we can prove the following proposition.

Proposition 2. *For any two sequences Q and C of the same length n, for any global constraint on the warping path of the form $j - c \le i \le j + c$, the following inequality holds: $LB\_Improved(Q,C) \le DTW(Q,C)$.*

The proof of Proposition 2 is a straightforward extension of Proposition 1, since LB_Improved simply uses LB_Keogh twice; we omit it for brevity.

For the succinctness of notations, hereafter we use LB_Keogh and LB_Improved to refer to the multi-dimensional extension of the original version respectively. By convention, we also use *time series* and *trajectory* interchangeably when referring to two or more dimensional time series.

3   LB_rotation

As discussed in Section 1, when it comes to two or more dimensional space, the query time series can be rotated by a certain angle to minimize the volume of its envelope, thus improves the tightness of the lower bound. We will present the details in this section.

3.1   Intuitive Explanation

First we use an example to show the idea. For simplicity, we only plot the projections of the two time series as well as the envelope of the query time series (time series T_1) in the $x - y$ plane. Note that in Figure 1a the four envelopes are partly overlapped.

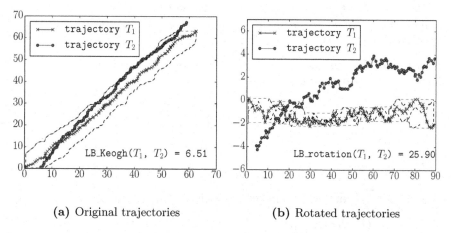

(a) Original trajectories   (b) Rotated trajectories

Fig. 1. Two sample trajectories of length $n = 128$, with warping width $c = 0.1n$. The true DTW distance $DTW(T_1, T_2) = 36.84$.

In Figure 1a are the original trajectories, and in Figure 1b are the rotated trajectories. The y axis is scaled with respect to the coordinate range. After rotation, the y axis has a very small span, because those points in time series T_1 almost lie in the same straight line. It's clear that before rotation, time series T_2 is almost wholly inside the envelope of time series T_1, while only a small part of it is contained in the envelope of time series T_1, after rotation. So, if we rotate the trajectories by an appropriate angle, we can reduce the volume of their envelopes, thus get a tighter lower bound.

However, we should note that, by rotating the time series, we can only reduce the area of the envelopes in either the $t - y$ plane or the $t - x$ plane; we cannot achieve area reduction in both planes. This is because the geometrical shape of the time series is rotation-invariant. If it becomes flat in one direction after rotation, it will surely become steep in the perpendicular direction. In the above example, the area of envelope in the $t - x$ plane actually increases, which will counteract the gain in the $t - y$ plane. Nonetheless, if we can reduce enough area of the envelopes in one direction, the result is still preferable, as we can see in the experiment result in Section 4. To achieve this, we need to divide the time series into several segments that are as straight as possible, because we can see from the example in Figure 1 that straight time series will greatly reduce the volume of envelope after rotation. The details will be introduced later.

3.2 Formal Definition of LB_rotation

Based on above observation, we propose a new lower bound for the constrained DTW, which we call *LB_rotation*. To formally define LB_rotation, we first define time series segmentation, and the distance from a point to the envelope of a segment.

Definition 4. *The segmentation of a time series Q is to divide Q into consecutive and non-overlapping segments $s_i = Q[s_i.start, s_i.end]$ where $\cup_{s_i} = Q \wedge \forall i \neq j : s_i \cap s_j = \emptyset$.*

Definition 5. *The distance from a point q to the envelope $Env(s)$ of a time series segment s is defined as*

$$d(q, Env(s)) = d(q, Env(s)_i) = \sum_{p=1}^{l} \begin{cases} (q_p - u_{i,p})^2 & \text{if } q_p > u_{i,p} \\ (q_p - l_{i,p})^2 & \text{if } q_p < l_{i,p} \\ 0 & \text{otherwise} \end{cases} \qquad (4)$$

where i is the index of the matching point in s with respect to p.

How to decide this matching point will be deferred to Algorithm 2.

Definition 6. (LB_rotation). *The lower bound LB_rotation of two time series Q and C of length n is defined as*

$$LB\_rotation(Q, C) = \sum_{i=1}^{n} \min_{s_j \in \mathcal{S}_i} \{d(c_i, Env(s_j))\} \qquad (5)$$

where $Env(s_j)$ is the bounding envelope of segment s_j, and $\mathcal{S}_i = \{s_k \mid s_k \in Q \wedge [s_k.start, s_k.end] \cap [i - c, i + c] \neq \emptyset\}$, c is the warping constraint.

For each point $c_i \in C$, we compute the distance from c_i to the segments in Q that overlaps with $Q[i - c, i + c]$ respectively, and sum up the minimal distance regarding each point as the final lower bound. This ensures that no matter which point $q_i \in Q$ is matched by c_i, the contribution of c_i to the lower bound will never exceed $d(c_i, q_i)$.

We can prove the following proposition.

Proposition 3. *For any two sequences Q and C of the same length n, for any global constraint on the warping path of the form $j - c \leq i \leq j + c$, the following inequality holds: $LB\_rotation(Q, C) \leq DTW(Q, C)$.*

Proof. Sketch: $\forall c_i \in C, 1 \leq i \leq n$, it may match the points in the range $Q[i - c, i + c]$, and its contribution to LB_rotation is $d_i = \min_{s_j \in S_i}\{d(c_i, Env(s_j))\}$. Suppose there are m segments of Q intersecting with this range, and the real matching point q_i belongs to segment s_k, then $d_i \leq d(c_i, Env(s_k))$. Based on Equation (4) we have $d(c_i, Env(s_k)) \leq d(c_i, p_j), \forall p_j \in s_k \wedge j \in [i - c, i + c]$. By transitivity, $d_i \leq d(c_i, q_i)$. Since $DTW(Q, C) \geq \sum_{i=1}^{n} d(c_i, q_i) \geq \sum_{i=1}^{n} \min_{s_j \in S_i}\{d(c_i, Env(s_j))\} = LB\_rotation(Q, C)$, we can conclude that LB_rotation lower bounds DTW.

3.3   Detailed Steps of LB_rotation

It takes 4 steps to compute LB_rotation:

1. **Time series segmentation.** As noted in Section 3.1, if we want to achieve satisfactory lower bound via time series rotation, we need to apply segmentation on the query time series, then deal with each segment respectively. We want each segment to be as straight as possible, so intuitively we should partition the time series at those "turning points". The classic Douglas-Peucker algorithm [6] is used for the segmentation, since each resulted splitting point is exactly such a turning point.

   We demonstrate the result of segmentation in Figure 2a, where a time series extracted from the *Character Trajectories* dataset is divided into 8 segments. We can see that each segment is almost straight, with different length.

2. **Segment rotation.** After segmentation, we need to find the rotation angle that best reduces the volume of envelopes. For each segment, we use least square fitting to compute the direction of the corresponding major axis, then we rotate each point $p \in s$ around the origin by $-s.\theta$ to get the rotated segment s', where $s.\theta$ is the included angle between the major axis and the x axis. Thus after rotation, the major axis of s' is aligned with the x axis, and the points in the time series segment will have a smaller span around the x axis, which leads to a narrower envelope.

3. **Extended envelope computation.** The next step is to compute the envelope for each rotated segment of query time series Q, which is almost the same as the envelope computation of LB_Keogh. The only difference is that, in the original envelope, each point will cover at least c points of the time series, however, after segmentation, the matching range may only intersect with the beginning or ending $k(1 \leq k \leq c)$ points of a certain segment. Covering extra points will hurt the tightness of LB_rotation.

   To solve this problem, we pad c points at the start and end of segment s respectively. When computing the upper bounding envelope, we fill the first and last c points of the padded s with a value that is smaller than all the values in s (e.g. $-\infty$), while fill with a value that is larger than all the values in s (e.g. $+\infty$) when computing the lower bounding envelope.

   We illustrate the extended envelope in Figure 2b. The original envelopes are between the two dashed vertical lines, while the extended parts lie outside.

4. **Lower bound computation.** Now we have a series of rotated segments of the query time series Q, we will describe for a candidate time series C in the database, how to compute $LB\_rotation(Q, C)$ using these segments.

First, we need to find the corresponding matching point for the points in C, in order to apply Equation (4). Given a point $c_i \in C$, it may match any point in $Q[i-c, i+c]$ with respect to a warping constraint c. Since Q is divided into a series of segments, the points in $Q[i-c, i+c]$ may belong to different segments, thus we should take care of different conditions. Specifically, if a segment s intersects with $Q[i-c, i+c]$, there are four possible situations.

(a) $(s.start \le i-c) \land (s.end \ge i+c)$, i.e. s contains $Q[i-c, i+c]$.
(b) $(s_j.start \ge i-c) \land (s_j.end \le i+c)$, i.e. $Q[i-c, i+c]$ contains s.
(c) $(s_j.start \le i+c) \land (s_j.end > i+c)$, i.e. $Q[i-c, i+c]$ contains the head of s.
(d) $(s_j.start < i-c) \land (s_j.end \ge i-c)$, i.e. $Q[i-c, i+c]$ contains the tail of s.

The index of corresponding matching point is computed as in Algorithm 2, which gives the final procedures of LB_rotation.

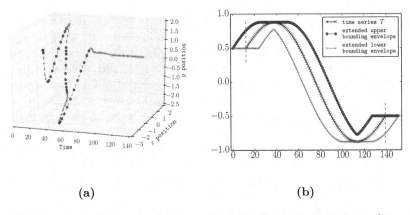

(a)   (b)

Fig. 2. (a) The segmentation result of a time series from *Character Trajectories* dataset, with $m = 8$. (b) The extended envelope of a time series T with length $n = 128$, and warping constraint $c = 0.1n$.

3.4   Performance Analysis

We briefly analyze the time complexity of each step in Section 3.3.

1. **Time series segmentation.** For a time series of length n, the Douglas-Peucker algorithm costs on average $O(n \log n)$, and $O(n^2)$ in the worst case.
2. **Segment rotation.** For a time series segment of length k, it costs $O(k)$ to compute the inclination angle of its major axis, and $O(k)$ to rotate each point. So it costs totally $\sum_{i=1}^{m} O(k_i) = O(n)$ in this step.

Algorithm 2. Lower bound computation for LB_rotation

Input: $\{\mathcal{S}_i\}$: each $\mathcal{S}_i$ contains rotated segments of the query time series Q that intersect with $Q[i-c, i+c]$
Input: C: candidate time series in the database
Output: d: the lower bound distance

1: **function** LB_ROTATION($\{\mathcal{S}_i\}$, C)
2:    $d \leftarrow 0$;
3:    **for** $c_i \in C$
4:       $dist_{min} \leftarrow \infty$
5:       **for** $s_j \in \mathcal{S}_i$
6:          **if** $(s_j.start \leq i-c) \wedge (s_j.end \geq i+c)$
7:             $index \leftarrow i+c-s_j.start$
8:          **else if** $(s_j.start \geq i-c) \wedge (s_j.end \leq i+c)$
9:             $index \leftarrow c+(s_j.end-s_j.start)/2$
10:         **else if** $(s_j.start \leq i+c) \wedge (s_j.end > i+c)$
11:            $index \leftarrow i+c-s_j.start$
12:         **else if** $(s_j.start < i-c) \wedge (s_j.end \geq i-c)$
13:            $index \leftarrow s_j.end-i$
14:         $c'_i \leftarrow c_i$ rotated by $-s_j.\theta$ ▷ $s_j.\theta$ is the inclination angle of the major axis of s_j
15:         $t \leftarrow d(c'_i, Env(s_j)_{index})$ ▷ Equation (4)
16:         **if** $dist_{min} > t$
17:            $dist_{min} \leftarrow t$
18:       $d \leftarrow d + dist_{min}$
19:    **return** d

---

3. **Extended envelope computation.** For a time series segment of length k, it costs $O(k+2ck/n)$ to compute the extended envelope using the streaming algorithm introduced in [11], so in total $\sum_{i=1}^{m} O(k_i + 2ck_i/n) = O(n+2c)$.
4. **Lower bound computation.** With warping constraint c, and the number of segment m, on average the matching range of each point will cover $\min\{m, \lceil 2cm/n \rceil\}$ segments, thus the time complexity of LB_rotation is asymptotically $O(\min\{m, \lceil 2cm/n \rceil\}n)$.

The first three steps can be precomputed before entering the **for** loop in Algorithm 1 of Algorithm 1, so the cost will be amortized. If there are enough candidate time series, this amortized overhead is negligible, just as what we observed in the experiment. While for the last step, as m is generally fixed, the time complexity increases with c. However, since generally LB_rotation will produce tighter lower bound, it requires fewer expensive DTW computations, thus the overall time needed to perform nearest neighbor search will be reduced.

Because the actual performance of all the lower bounding techniques is data-dependent, we only give a rough analysis here, and compare LB_Keogh, LB_Lemire and LB_rotation through experiments on different datasets.

4   Experiment

4.1   Setup and Datasets

We implemented the algorithms in C++, compiled by g++ 4.9.1. The platform is a ThinkPad X220 running Arch Linux, with 8GB of RAM and a 2.6GHz Intel Core i7 processor.

We use two real world datasets for experiments.

- The *Character Trajectories*[1] dataset from the UCI Machine Learning Repository [2], which contains 2858 trajectories of writing 20 different letters with a single pen-down segment. The length of each trajectory varies between 109 and 205, and we rescaled them to the same length of 128, using Piecewise Aggregate Approximation [8] (for longer trajectories) or linear interpolation (for shorter trajectories).
- The *GeoLife*[2] dataset [22–24] from MSRA, which contains 17,621 trajectories of 182 users in a period of over three years. We extracted those trajectories containing at least 1000 sample points for experiment, and rescaled them to length 256.

The time series in both datasets are all z-normalized [13]. We assume the datasets are already loaded into memory before running following experiments, and the true DTW distance is computed using the standard $O(mn)$ dynamic programming algorithm, subjected to the corresponding warping constraint c.

The compiled executable and preprocessed datasets are freely available at [1], including the python script to compute the accuracy of 1 Nearest Neighbor classification on *Character Trajectories* dataset.

4.2   Evaluation Metrics

The effectiveness of a lower bound is usually reflected in the tightness, pruning power and the overall wall clock time. The first two metrics are independent of implementation details, while the last one may vary. Nonetheless, the wall clock time is still an important metric, since although some lower bound may be tighter than others, it actually will cost much more time to compute, which largely nullifies its effectiveness [19].

Following [9], we define the tightness of a lower bound as

$$T = \frac{\text{Lower Bound of DTW Distance}}{\text{True DTW Distance}} \tag{6}$$

and define pruning ratio as

$$P = \frac{\text{Number of Omitted Full DTW Computation}}{\text{Number of Objects}}. \tag{7}$$

Both T and P are in the range $[0, 1]$, and the larger the better.

---

[1] http://archive.ics.uci.edu/ml/datasets/Character+Trajectories
[2] http://research.microsoft.com/en-us/projects/geolife/

To evaluate the tightness of each lower bounding method, we randomly sampled 100 time series from the dataset, then computed the three lower bounds as well as the true DTW distance for each pair of them (in total 9900 pairs), and recorded the corresponding tightness. The average over 9900 pairs is reported. Note that we have to compute the lower bound between each pair, since these lower bounds are not symmetric, i.e. $LB(T_1, T_2) = d \not\Rightarrow LB(T_2, T_1) = d$.

To evaluate the pruning power of each lower bounding method, we randomly sampled 100 time series from the dataset, then for each time series, we performed 1-Nearest Neighbor search on the rest 99 time series, using Algorithm 1, by plugging in each lower bounding method in Algorithm 1, and recorded the corresponding pruning ratio. The average over 100 time series is reported.

To evaluate the efficiency of each lower bounding method, for each dataset, we randomly sampled 1000 time series from it, then performed 1-Nearest Neighbor (1NN) search for 50 randomly sampled time series from the same dataset, using Algorithm 1, by plugging in each lower bounding method in Algorithm 1. In order to rule out the influence of random factors, the 1NN search time for each time series is reported as the average over 10 runs. We repeated above experiments with various parameter combinations.

4.3   The Effect of Segment Number m

First, we inspect how the number of segments will affect the tightness and pruning power of LB_rotation. Since this parameter is only used in LB_rotation, we do not compare with the other two lower bounding methods.

We randomly sample 1000 trajectories from the *GeoLife* dataset, then compute pair-wise lower bound using LB_rotation as well as the true DTW distance, and record the corresponding tightness. The pruning ratio and 1NN search time are gathered through 1NN search for 50 random sampled time series. The averages are reported in Figure 3. For the *Character Trajectories* dataset, we observe similar results, and we only report one of them for brevity.

We can see that the tightness and pruning power increases with m, however, for larger m the 1NN search time becomes longer, because the saved DTW computation cannot break even the time needed to compute LB_rotation. We empirically find that $m = 8$ achieves a good compromise between the pruning ratio and extra computation overhead. In the following, if not otherwise stated, we set $m = 8$ for all the experiments.

4.4   The Effect of Warping Constraint c on Tightness and Pruning Power

In the following, we present the tightness and pruning power of the three lower bounding techniques, with respect to varying warping constraint, on different datasets. The warping constraint c varies from 0 (corresponding to Euclidean distance) to n (corresponding to unconstrained DTW distance), with step size $0.05n$. The results are presented in Figure 4a through Figure 4d.

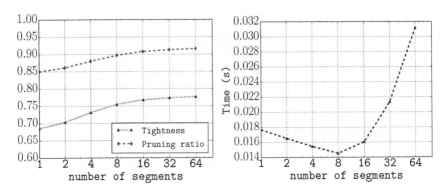

Fig. 3. Tightness, pruning power and 1NN search time vs. number of segments on *GeoLife* dataset. Warping constraint $c = 0.1n$.

First of all, we need to point out that for the *Character Trajectories* dataset, under the optimal constraint $c = 0.4n$, LB_rotation is 2× as tight as LB_Keogh, and 1.3× as tight as LB_Improved. It prunes 30% and 16% more unqualified candidates respectively when compared with LB_Keogh and LB_Improved. This optimal constraint is obtained by testing different warping constraint on the *Character Trajectories* dataset, since the trajectories are labeled, they can be used to test the classification accuracy. We used 1-Nearest Neighbor classification, and validated the result by leave-one-out validation. As for the *GeoLife* dataset, due to the lack of labels, we cannot decide the optimal warping constraint for 1NN classification, however we noticed there are trajectories that are almost identical, while largely shifted along the time axis (about half of the trajectory length), which indicates that large warping constraint should be used to correctly align these trajectories.

From Figure 4a through Figure 4d we can observe that LB_rotation consistently achieves higher tightness and pruning power than LB_Keogh, which demonstrates the effectiveness of time series rotation.

For $c \geq 0.4n$, LB_rotation outperforms LB_Improved in terms of both tightness and pruning ratio. Because as c increases, the volume of the envelope will also increase, since it will cover more data points, and intuitively, the probability of including points with large values is proportional to the covering range of the envelope, thus the envelope will be enlarged. On the other hand, if we rotate each segment respectively, recall Figure 1, even the covering range increases, the volume of the envelope won't increase much. As a result, although larger c will hurt the tightness and pruning ratio of all the three lower bounds, the influence on LB_rotation is obviously smaller.

When c is relatively small ($< 0.4n$), LB_rotation generally achieves comparable or even higher tightness than LB_Improved, although the pruning ratio of the latter is sometimes better. This is because with small warping constraints, LB_Keogh has considerably good tightness, thus it requires only a few second pass computations for LB_Improved, which will not improve the tightness much.

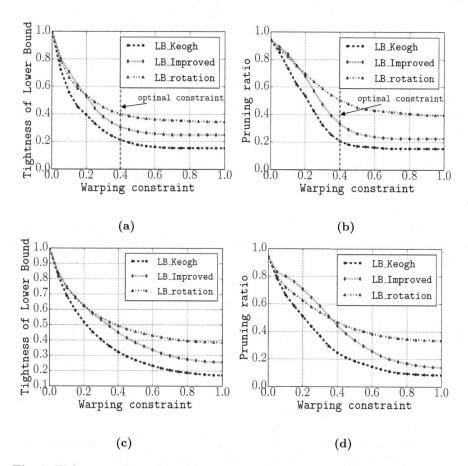

Fig. 4. Tightness and pruning ratio w.r.t. varying warping constraint c on different datasets. (a)-(b): *Character Trajectories* dataset; (c)-(d): *GeoLife* dataset.

However, these second pass computations do help to prune more candidates, so the pruning ratio of LB_Improved will increase.

We also note that even in the extreme situation where $c = n$ (unconstrained DTW), LB_rotation can still achieve pruning ratio around 40%, while both LB_Keogh and LB_Improved have pruning ratio hardly exceeds 20%.

4.5   The Effect of Warping Constraint c on Search Time

In this experiment, we compare the wall clock time for 1NN search on different datasets, with respect to varying warping constraint from 0 to n, increasing at step size $0.05n$.

From Figure 5a and Figure 5b we find that the 1NN search time agrees with tightness and pruning ratio we just depicted in Section 4.4 very well. The result indicates that on all the datasets, LB_rotation will achieve the fastest 1NN

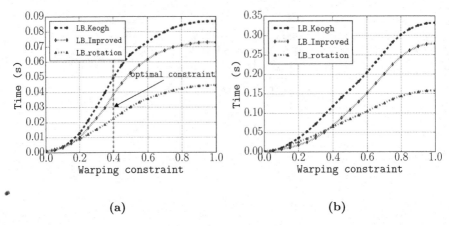

Fig. 5. 1NN search time w.r.t. varying warping constraint c on (a) *Character Trajectories* dataset; (b) *GeoLife* dataset

search once the warping constraint c exceeds $0.4n$. We have shown that warping constraints this large are realistic for some real world applications. Figure 5a shows that for the *Character Trajectories* dataset, under the optimal warping constraint $c = 0.4n$, LB_rotation costs about half the time of LB_Keogh, and about 60% of LB_Improved. Actually this trend starts once c exceeds $0.2n$.

5   Conclusion and Future Work

In this paper, we propose a new lower bounding technique *LB_rotation* for constrained DTW, which is based on the observation that if the time series is in multi-dimensional space, it can be rotated around the time axis to reduce the volume of its bounding envelope, and as a consequence, the tightness and pruning power of the lower bound will increase. Then we notice that if we divide the time series into several segments as straight as possible, then treat them separately, the effectiveness can be further improved, so we use a greedy algorithm to achieve this. We carried out experiments on real world datasets, which demonstrate the superiority of LB_rotation over state-of-the-art lower bounding techniques.

With more and more high dimensional time series being generated nowadays, it is of significant importance to effectively process them. In the future, we intend to further investigate how to utilize the characteristics of multi-dimensional time series to achieve even better result.

Acknowledgments. This work was supported by National Natural Science Foundation of China under Grant No.61202404, No.61170233, No.61232018, No.61272472, No.61272317 and the Fundamental Research Funds for the Cerntral Universities, No. WK0110000041.

References

1. Executable and datasets used in the experiment. https://www.dropbox.com/s/gkmcy9up73y5vmo/data.tar.gz?dl=0
2. Bache, K., Lichman, M.: UCI machine learning repository (2013). http://archive.ics.uci.edu/ml
3. Chen, L., Ng, R.: On the marriage of lp-norms and edit distance. In: Proceedings of the Thirtieth International Conference on Very Large Data Bases - Volume 30, VLDB 2004, pp. 792–803. VLDB Endowment (2004)
4. Chen, L., Özsu, M.T., Oria, V.: Robust and fast similarity search for moving object trajectories. In: Proceedings of the 2005 ACM SIGMOD International Conference on Management of Data, SIGMOD 2005, pp. 491–502. ACM, New York (2005)
5. Das, G., Gunopulos, D., Mannila, H.: Finding similar time series. In: Komorowski, J., Zytkow, J. (eds.) Principles of Data Mining and Knowledge Discovery. LNCS, vol. 1263, pp. 88–100. Springer, Heidelberg (1997)
6. Douglas, D.H., Peucker, T.K.: Algorithms for the reduction of the number of points required to represent a digitized line or its caricature. Cartographica: The International Journal for Geographic Information and Geovisualization 10(2), 112–122 (1973)
7. Faloutsos, C., Ranganathan, M., Manolopoulos, Y.: Fast subsequence matching in time-series databases. SIGMOD Rec. 23(2), 419–429 (1994)
8. Keogh, E., Chakrabarti, K., Pazzani, M., Mehrotra, S.: Dimensionality reduction for fast similarity search in large time series databases. Knowledge and Information Systems 3(3), 263–286 (2001)
9. Keogh, E., Ratanamahatana, C.A.: Exact indexing of dynamic time warping. Knowledge and Information Systems 7(3), 358–386 (2005)
10. Kim, S.-W., Park, S., Chu, W.: An index-based approach for similarity search supporting time warping in large sequence databases. In: Proceedings of the 17th International Conference on Data Engineering, 2001, pp. 607–614 (2001)
11. Lemire, D.: Faster retrieval with a two-pass dynamic-time-warping lower bound. Pattern Recognition 42(9), 2169–2180 (2009)
12. Li, Q., Zheng, Y., Xie, X., Chen, Y., Liu, W., Ma, W.-Y.: Mining user similarity based on location history. In: Proceedings of the 16th ACM SIGSPATIAL International Conference on Advances in Geographic Information Systems, GIS 2008, pp. 34:1–34:10. ACM, New York (2008)
13. Rakthanmanon, T., Campana, B., Mueen, A., Batista, G., Westover, B., Zhu, Q., Zakaria, J., Keogh, E.: Searching and mining trillions of time series subsequences under dynamic time warping. In: Proceedings of the 18th ACM SIGKDD International Conference on Knowledge Discovery and Data Mining, KDD 2012, pp. 262–270. ACM, New York (2012)
14. Rath, T.M., Manmatha, R.: Lower-bounding of dynamic time warping distances for multivariate time series. Technical Report MM-40, Center for Intelligent Information Retrieval, University of Massachusetts Amherst (2002)
15. Sefidmazgi, M.G., Sayemuzzaman, M., Homaifar, A.: Non-stationary time series clustering with application to climate systems. In: Jamshidi, M., Kreinovich, V., Kacprzyk, J. (eds.) Advance Trends in Soft Computing WCSC 2013. STUDFUZZ, vol. 312, pp. 55–63. Springer, Heidelberg (2014)
16. Vlachos, M., Gunopulos, D., Das, G.: Rotation invariant distance measures for trajectories. In: Proceedings of the Tenth ACM SIGKDD International Conference on Knowledge Discovery and Data Mining, KDD 2004, pp. 707–712. ACM, New York (2004)

17. Vlachos, M., Hadjieleftheriou, M., Gunopulos, D., Keogh, E.: Indexing multi-dimensional time-series with support for multiple distance measures. In: Proceedings of the Ninth ACM SIGKDD International Conference on Knowledge Discovery and Data Mining, KDD 2003, pp. 216–225. ACM, New York (2003)

18. Wang, J., Katabi, D.: Dude, where's my card?: RFID positioning that works with multipath and non-line of sight. In: Proceedings of the ACM SIGCOMM 2013 Conference on SIGCOMM, SIGCOMM 2013, pp. 51–62. ACM, New York (2013)

19. Wang, X., Mueen, A., Ding, H., Trajcevski, G., Scheuermann, P., Keogh, E.: Experimental comparison of representation methods and distance measures for time series data. Data Mining and Knowledge Discovery **26**(2), 275–309 (2013)

20. Yang, A.Y., Jafari, R., Sastry, S.S., Bajcsy, R.: Distributed recognition of human actions using wearable motion sensor networks. Journal of Ambient Intelligence and Smart Environments **1**(2), 103–115 (2009)

21. Yi, B.-K., Jagadish, H., Faloutsos, C.: Efficient retrieval of similar time sequences under time warping. In: Proceedings of the 14th International Conference on Data Engineering, 1998, pp. 201–208 (1998)

22. Zheng, Y., Li, Q., Chen, Y., Xie, X., Ma, W.-Y.: Understanding mobility based on GPS data. In: Proceedings of the 10th International Conference on Ubiquitous Computing, UbiComp 2008, pp. 312–321. ACM, New York (2008)

23. Zheng, Y., Xie, X., Ma, W.-Y.: GeoLife: A collaborative social networking service among user, location and trajectory. IEEE Data Eng. Bull. **33**(2), 32–39 (2010)

24. Zheng, Y., Zhang, L., Xie, X., Ma, W.-Y.: Mining interesting locations and travel sequences from GPS trajectories. In: Proceedings of the 18th International Conference on World Wide Web, WWW 2009, pp. 791–800. ACM, New York (2009)

25. Zhou, M., Wong, M.-H.: Boundary-based lower-bound functions for dynamic time warping and their indexing. In: IEEE 23rd International Conference on Data Engineering, ICDE 2007, pp. 1307–1311 (2007)

26. Zhu, Y., Shasha, D.: Warping indexes with envelope transforms for query by humming. In: Proceedings of the 2003 ACM SIGMOD International Conference on Management of Data, SIGMOD 2003, pp. 181–192. ACM, New York (2003)

Measuring the Influence from User-Generated Content to News via Cross-Dependence Topic Modeling

Lei Hou[1]([✉]), Juanzi Li[1], Xiao-Li Li[2], and Yu Su[3]

[1] Tsinghua National Laboratory for Information Science and Technology,
Department of Computer Science and Technology,
Tsinghua University, Beijing 100084, China
houl10@mails.tsinghua.edu.cn, lijuanzi@tsinghua.edu.cn
[2] Institute for Infocomm Research, A*STAR, Singapore 138632, Singapore
xlli@i2r.a-star.edu.sg
[3] Communication Technology Bureau, Xinhua News Agency, Beijing 100803, China
suyu@xinhua.org

Abstract. Online news has become increasingly prevalent as it helps the public access timely information conveniently. Meanwhile, the rapid proliferation of Web 2.0 applications has enabled the public to freely express opinions and comments over news (user-generated content, or UGC for short), making the current Web a highly interactive platform. Generally, a particular event often brings forth two correlated streams from news agencies and the public, and previous work mainly focuses on the topic evolution in single or multiple streams. Studying the inter-stream influence poses a new research challenge. In this paper, we study the mutual influence between news and UGC streams (especially the UGC-to-news direction) through a novel three-phase framework. In particular, we first propose a cross-dependence temporal topic model (CDTTM) for topic extraction, then employ a hybrid method to discover short and long term influence links across streams, and finally introduce four measures to quantify how the unique topics from one stream affect or influence the generation of the other stream (e.g. UGC to news). Extensive experiments are conducted on five actual news datasets from Sina, New York Times and Twitter, and the results demonstrate the effectiveness of the proposed methods. Furthermore, we observe that not only news triggers the generation of UGC, but also UGC conversely drives the news reports.

Keywords: News stream · User-generated content · Cross dependence · Influence · Response

1 Introduction

Nowadays, social media are ubiquitous, offering many opportunities for people to access and share information, to create and distribute content, and to

© Springer International Publishing Switzerland 2015
M. Renz et al. (Eds.): DASFAA 2015, Part I, LNCS 9049, pp. 125–141, 2015.
DOI: 10.1007/978-3-319-18120-2_8

interact with more traditional media [13]. According to the report [14] from *Pew Research Center*, over half of the social users in U.S. access news online, as well as actively express their opinions and comments on daily news, either directly from the online news (comments following the news) or through other services such as blogging, Twitter, which produces the rich user-generated content (UGC). Digital storytelling and consistently available live streaming is fuelling the news with different events from different perspectives [19], indicating the public voice, e.g. their opinions, concerns, requests, debates, reflections, can spur additional news coverage in the event. This comes as no surprise as the main function of the news is to provide updates on the public voice, and the latest measures taken by the involved organizations. Therefore, investigating and responding the public voice is of great benefit to valuable news clues acquisition for news agency, crisis monitoring and management for functional departments.

When a particular event happens, news articles typically form a news stream that records and traces the event's beginning, progression, and impact along a time axis. Meanwhile, the UGC stream is also naturally formed by the public to reflect their views over news reports. These two different streams are highly interactive and inter-dependent. On one hand, the news stream has big influence on the UGC stream as the public posts their comments based on the corresponding news articles and they are typically interested in certain aspects or topics in the news stream. On the other hand, the UGC stream, containing the public opinions, voice and reflection, could potentially influence and even drive the news reports, which is the focus of this paper.

Example. Fig. 1 presents the news and comment streams about *U.S. Federal Government Shutdown* from New York Times (NYT). At the beginning, the news reported the budget bill proposed by the *House of Representatives*, leading to the public debate as well as *Cruz's* speech. Then the public turned their attention to the following *vote* raised by *the Senate*. They also encouraged the president after the *shutdown*, which might affect the final decision. Meanwhile, the news agencies preferred to report what the public cared most (e.g. *vote, insurance*). As such, it is interesting to systematically study how the topics from two correlated streams interact with each other and co-evolve over time.

Recently, many research efforts have been put on topic evolution within news stream, e.g. [1,12]. Morinaga et al. proposed a framework for tracking topical dynamics using a finite mixture model [24]. A representative work from Mei et al. employed adapted PLSA for topic extraction in text streams, KL-divergence for discovering coherent topics over time, and HMM for analyzing the lifecycle in [22]. However, they only detect how the topics evolve but we further explore what factors could drive their evolution. Another line of research focuses on simultaneously modeling multiple news streams, such as mining common and private features [15,29], and identifying the characteristics of the social media and news media via a modified topic model [32]. However, they still did not investigate if the inter-stream influence could lead to their co-evolution.

We apparently expect to study the interactions between news and UGC streams, and address the problem of influence quantification. To the best of our knowledge,

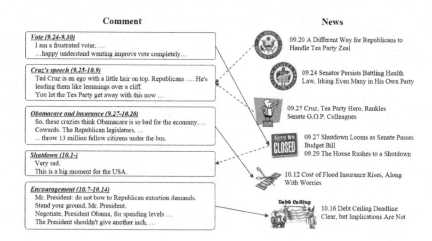

Fig. 1. News and UGC Interaction in *U.S. Federal Government Shutdown*

it is the first research that focuses on investigating the mutual influence between these two streams. However, the novel task brings new challenges to conventional mining approaches. Firstly, it proves utilizing both streams can significantly benefit the topic discovery process than each individual stream alone [30] and we are dealing with two highly interactive streams, which requires us to consider the inter-stream dependence and temporal dynamics during topic extraction. Secondly, if there appears a new topic in UGC stream and it is mentioned in the subsequent news, we assume news is potentially driven by UGC and vice versa. The influence could be short-term (people talked about *Cruz* when his speech just ended), long-term (the discussion about *the budget bill* lasted throughout the event) or none, and we need to detect and distinguish different types of influence, as well as quantify the mutual influence. Thirdly, news is responsible for dealing with the controversial (or influential) topics in UGC, and we need to figure out what is an appropriate response from the processed events, such as how many topics the news respond to, how fast the response is and whether the public accept it.

To tackle the issues above, we introduce a three-phase framework: we first propose a novel cross dependence temporal topic model (CDTTM) to organize news and UGC into two dependent topic streams. The core idea of CDTTM is employing the dependent and temporal correlation across streams for building up two correlated generative processes with mutual reinforcement [16,30]. Then we develop a hybrid method to build links among the topics across streams based on KL-divergence and dynamic time wraping, which can effectively distinguish short or long-term influence. Finally, we systematically propose four statistical measures to quantify the mutual influence between two streams. Specifically, we introduce *topic progressiveness* to determine whether UGC goes ahead of news in some topics, *response rate*, *response promptness*, and *response effect* to evaluate how the news responds to UGC. Our main contributions include:

– We propose and formalize a novel problem to measure the mutual influence across two text streams and address it through a three-phase framework.
– We introduce a novel CDTTM model for topic extraction from two correlated text streams, which utilizes both temporal dynamics and mutual dependence.
– We propose a hybrid topic linking method, which can effectively discover the short-term and long-term influence links across streams.
– We define four metrics to quantify the influence between news and UGC streams. Experiments on five real news datasets show that the influence is bidirectional, namely news can trigger the generation of UGC and vice versa.

The rest of the paper is organized as follows. In Section 2, we formally define the problem of news and UGC influence analysis, and then demonstrate our methods. Our experimental results are reported in Section 3. Section 4 reviews the related literatures, and finally Section 5 concludes this paper with future research directions.

2   Problem and the Proposed Method

In this section, we formally define the problem of analyzing the influence between news and UGC streams, and then present our methods on topic extraction, link discovery and influence quantification.

2.1   Preliminaries and Problem Definition

Whenever an important *event* happens, it often brings forth two correlated text streams, namely news from media forms a news stream NS and users' voice from different social applications converges into a UGC stream $UGCS$. Each *news* d_i or user *post* p_i is represented by a content-time pair $(\mathbf{w_i}, t_i)$, where $\mathbf{w_i}$ denotes the words from a vocabulary V and t_i is the time stamp from a time collection T. Meanwhile, they both talk about several *topics* z_d or z_p, and the topics themselves keep changing along the timeline.

Definition 1. *News and UGC Influence Analysis.* *Given news stream NS and UGC stream $UGCS$, our goal is to extract time-ordered topic sets for both streams, namely Z_n and Z_u, discover the influence link collection $L = \{l_i = (z_x, z_y, \zeta)\}$ and characterize the mutual influence as several well-designed measures over the influence links. Note that $z_x \in Z_n$ and $z_y \in Z_u$ are topics from different streams NS and $UGCS$ respectively and ζ is a real number stating the link strength.*

Fig. 1 gives the NYT news and comment streams about *U.S. Federal Government Shutdown*, both talking about common topics like *budget bill*, *vote* and *debt ceiling*. The influence analysis aims to detect the dynamic topics for each stream, link the topics across streams, and evaluate how they influence each other.

According to the definition above, we present the solution in three phases: 1) *topic extraction*: identify the topics from two correlated streams. 2) *influence link discovery*: it is required to consider the influence types (i.e. short-term or long-term) when linking the common topics across streams. 3) *influence quantification*: measure the intrinsic relations between UGC and news.

2.2 Topic Extraction from Two Text Streams

In this section, we extract the topics Z_n and Z_u from news and UGC streams. We observe that topics in these two streams are *cross-dependent*: comments in $UGCS$ are typically formed by the public to reflect their opinions on the topics published in news or provide new information about the progress of events, while the subsequent news reports in NS often provides additional information or clarifications to respond to the public comments in $UGCS$. To capture the temporal and cross-dependent information across streams, we expand the document comment topic model in [16] and design the cross-dependence temporal topic model (CDTTM) in Fig. 2.

We first introduce the notations. θ_d and θ_p are topic distributions for news and UGC and ϕ denotes the word distribution of each topic; x is a binary variable indicating whether the generation of the current word is influenced by the previous news (or UGC) ($x = 1$) or not ($x = 0$); α, β are the Dirichlet hyper parameters; λ is the *Bernoulli* parameter for sampling x and γ_d, γ_p are its hyper parameters.

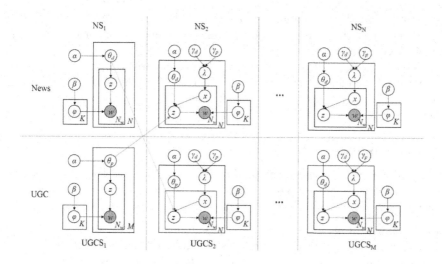

Fig. 2. Cross-Dependence Temporal Topic Model

For the generative process, we partition both steams into disjoint substreams with fixed time intervals, e.g. $NS = NS_1 \cup \ldots \cup NS_N$ where NS_i is with time

$[t_i, t_{i+1})$, and initialize two standard LDA models in the first substreams NS_1 and $UGCS_1$. As for the subsequent substreams, if there is a previous substream in the other stream (e.g. NS_{i+1} has a previous UGC substream $UGCS_i$), a coin x is tossed according to $p(x|d) \sim beta(\gamma_d, \gamma_p)$ to decide whether w_d inherits from the previous substream, otherwise (namely when there is no previous substream in the other stream) a standard LDA model is employed for word sampling. For example, we are currently sampling the word *majority* in news substream, and it appears frequently in the topic *budget bill* in previous UGC substream, which could serve as prior knowledge, namely, it has a higher probability to be assigned as topic *budget bill* in the current news substream as well.

For parameter estimation, we take Gibbs sampling technique for its ease of implementation. Suppose the previous UGC model is given, we sample the coin x and topic assignment of word w in the current news substream separately. For x, we derive the posterior probability:

$$p(x_i = 0|\mathbf{x}_{\neg i}, \mathbf{z}, \cdot) = \frac{n_{dx_0}^{\neg di} + \gamma_d}{n_{dx_0}^{\neg di} + n_{dx_1}^{\neg di} + \gamma_d + \gamma_p} \times \frac{n_{z_{di}}^{\neg di} + \alpha}{\sum_z (n_z^{\neg di} + \alpha)} \tag{1}$$

where n_{dx_0}, $n_{z_{di}}$ are the number of times that coin $x = 0$ and topic z has been sampled from d, and $\neg$ means exclusion. Then the posterior probability of topic z for word w_{di} when the coin $x = 1$ is derived as follows:

$$p(z_{w_{di}} = j|x_{w_{di}} = 1, \mathbf{z}_{\neg w_{di}}, \cdot) = \frac{n_{jw_{di}}^{\neg i} + m_{jw_{di}} + \beta}{\sum_w (n_{jw}^{\neg i} + m_{jw} + \beta)} \frac{n_{dj}^{\neg i} + m_{dj} + \alpha}{\sum_z (n_{dz}^{\neg i} + m_{dz} + \alpha)} \tag{2}$$

where $m_{jw_{di}}$ denotes the times that word w_{di} has been generated by topic j in current news substream and previous UGC substream respectively, $n_{dj}^{\neg di}$ and m_{dj} are the times that topic j has been sampled independently or influenced by previous UGC substream.

Readers who are interested in the solution can refer [16,26] for details. Through topic modeling, we turn news and UGC streams into two correlated and time-ordered topic streams Z_n and Z_u, where $Z_n^i \subset Z_n$ is the news topic set in time interval $[t_i, t_{i+1})$.

2.3   Topic Influence Link Discovery

In this section, we present our method to link topics across streams. Normally, we are dealing with three types of influence links, short-term, long-term and no influence.

Link Measurement. To measure if there is influence link between topics across streams, we calculate their distance using Kullback-Leibler(KL) divergence. Given two topics z_1 and z_2 associated with two distributions ϕ_1 and ϕ_2, the influence distance from z_1 to z_2 is defined as the additional new information in z_2 compared to z_1:

$$\zeta_{z_1 \to z_2} = KL(z_2||z_1) = \sum_{i=1}^{\nu} \phi_{2i} \times \log \frac{\phi_{2i}}{\phi_{1i}} \tag{3}$$

where a larger $\zeta_{z_1 \rightarrow z_2}$ value indicates a weaker influence from topic z_1 to z_2. Note that the KL-divergence is asymmetric, but it makes sense in our scenario because the influence link strengths between two topics are not equal, i.e. news topics usually have higher influence to UGC topics than the other way around.

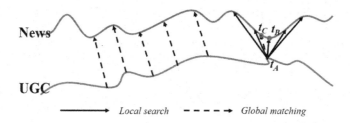

Fig. 3. Local search and global matching methods

Link Discovery. To accurately discover the influence between topics across streams, we perform both local and global search. Considering the short-term response between news and UGC, we set a time window $[-2, +2]$ for each topic in one stream to detect its local influence links, e.g. given a topic z in UGC stream, we search the topics locally that are most likely to influence z in the current and its previous two news substreams, and the topic that it may influence in next two news substreams. The linked topic is denoted as z_s representing the short-term influence. In Fig. 3, we link the topic in UGC stream at time t_A to its nearest point t_B in news stream within ± 2 time intervals. To find the topic z_l with long-term influence, we first calculate the topic hotness along time for each topic, then employ dynamic time wraping (DTW) due to its high efficiency on dealing with time series data [11]. DTW is a class of widely-used dynamic algorithms which take two time series data as input, stretch or compress them in order to make one resemble the other as much as possible. As shown in Fig. 3, although t_B is the most appropriate point locally, we still link t_A and t_C and store it into z_l after making global matching on these two topic series.

Link Filtering. For each topic z at a specified time, z_s presents us a microcosmic view of the influence between topics within a pre-defined time interval while z_l discovers the macroscopic view of the influence. We compare z_s and z_l, and take the one with smaller KL-divergence as the final result. Finally, we sort all the linked pairs by distances and keep pairs whose distances are lower than the median value to remove those noisy or no influence cases, which constitute the result influence link collection L.

2.4   Influence Quantification

In this section, we infer the influence between news and UGC streams by defining four metrics in terms of news communication [21]. On the UGC side, we evaluate whether UGC influences news by introducing a novel concept of topic progressiveness, while on the news side, we evaluate how news reacts to UGC topics through adapting three popular measures in public opinion analysis.

Definition 2. *Topic Progressiveness tells whether UGC topics could trigger news topics. Inspired by [23], we consider the time difference between topics across streams. Specifically, for each topic z_p in UGC stream with linked topics $L(z_p)$ in the news stream, we compute the cross-entropy values to find the minimum one z_d^{min}, and then the topic progressiveness is defined as:*

$$Prog(z_p) = T(\underset{z_d \in L(z_p)}{\arg\min} H(z_p, z_d)) - T(z_p) \qquad (4)$$

where $H(z_p, z_d)$ is the cross-entropy of z_p with its linked topic z_d, and $T(.)$ returns the time stamp of the input topic.

While it is clear and verified that news guides the generation of UGC [16, 30], the influence from UGC stream to news stream is our focus in this paper. Under this notation, if UGC topic z_p comes before news topic z_d^{min}, the topic progressiveness would be positive, meaning that z_p gives contribution to the z_d in news stream, and z_p is called as *progressive topic*. On the other hand a negative value of topic progressiveness indicates that news z_d^{min} has led to topic z_p in UGC stream.

Oftentimes, there emerges several hot or controversial topics in UGC stream, which might become public opinion crisis if they are not handled properly. Several models of cognitive determinants of social behaviors, e.g. the theory of planned behavior (TPB) [2,3], prove that intention is the most reliable predictor of behavior, but there is still a substantial gap between peoples intentions and their subsequent behavior due to many factors. Besides *perceived behavioral control*, the mediators play critical roles in the intention-behavior relation [27,28]. Since news is the most important mediator between the emergent topics and the public, we want to know whether and how much does news respond to the public topics, and if the public accept the response. Therefore, we quantify the following three metrics in public opinion management [9].

For convenience, we first classify the discovered links in previous steps into two groups. For a UGC topic z_p, if its linked news topic z_d appears earlier, we call z_d *previous linked topic* of z_p; otherwise it is *future linked topic* of z_p. The previous/future linked topics of z_d are denoted as $PL(z_p) = \{z_d | z_d \in L(z_p)$ and $T(z_d) < T(z_p)\}$ and $FL(z_p) = \{z_d | z_d \in L(z_p)$ and $T(z_d) > T(z_p)\}$.

Definition 3. *News Response Rate (NRR) defines how many UGC topics that draw news' attention. It is obtained through computing the percentage of topics that appear in UGC before news media:*

$$NRR(NS) = \frac{|\{z_p | FL(z_p) \neq \emptyset \&\& PL(z_p) = \emptyset\}|}{|\{z_p | PL(z_p) = \emptyset\}|} \qquad (5)$$

Note that we only consider those topics without previous links, namely, they appear in UGC stream first.

Definition 4. *News Response Promptness (NRP)*, *which evaluates how fast news responds to the UGC topics, is calculated by the average time difference between the first appearance time in UGC and the response time in news:*

$$NRP(NS) = \frac{\sum_{z_p \ and \ z_d \in FL(z_p)}[T(z_d) - T(z_p)]}{\sum_{z_p}|FL(z_p)|} \qquad (6)$$

Since z_d is *future linked topic* of z_p, $T(z_d) - T(z_p)$ should be positive consistently.

When a piece of news with topic z_p is published at a particular time t, it could address the voice and concerns in UGC stream via providing additional relevant information, and subsequently the public might answer whether they are satisified. In general, users often express their sentiments about certain topic using opinion words.

Definition 5. *News Response Effect* *is defined by checking if the number of opinion words has largely been reduced after news response. If so, it means that news has effectively address the concerns from the users.*

$$NRE(z_p, t) = \frac{C(z_p^{t-}) - C(z_p^{t+})}{C(z_p)} \qquad (7)$$

where $C(z) = |\{w|w \in KW(z) \cap OP\}|$ with a pre-defined opinion words set OP, $KW(\cdot)$ representing the keywords in given topics, and t the time stamp of the linked news topic.

Remarks. *Progressiveness* presents a microcosmic view of the influence between topics across streams, while the other three metrics demonstrate the news response macroscopically. Particularly, larger *rate* and smaller *promptness* express that news responds to public topics actively and promptly, and large *effect* value indicates effective news response.

3   Experiments

In this section, we evaluate the proposed methods for topic extraction, influence link discovery and analysis. We first briefly introduce our datasets, and then present the detailed experimental results.

3.1   Data Preparation

To the best of our knowledge, no public existing benchmark data is available for analyzing the influence between news and UGC. Therefore, we have prepared five data of different events from influential news portals and social media platforms (e.g. NYT, Twitter), including *the Federal Government Shutdown* (cFGS/eFGS) in two languages, *Jang Sung-taek's* (Jang), *The Boston Marathon Booming* (Boston) and *India Election* (India). Particularly, we crawled news and comments about the first three events from specific pages[1,2] or through keyword search[3]. While for the last event from Twitter, we collected 2,890,801 related tweets using keyword filtering, then recognized all the tweets accompanied with URLs (there are 5,949 URLs found in 784,237 tweets), and finally news URLs whose frequencies are greater than 5 and corresponding tweets were selected.

For each dataset, we kept the continuous news reports and comments (or tweets), and further sorted them by published time (*last update time* for comments), and performed some cleaning work, such as removing low-frequency(≤ 3) words and stop words. The basic statistics after preprocessing are summarized in Table 1.

Table 1. Datasets: sources of two streams (Twitter news comes from various news websites, users post the shortened URLs along with their comments in tweets), duration, numbers of comments and news articles, max and average number of comments per news

Source	Event	Days	Comments	News	Com./News	
					max	avg
Sina-Weibo	cFGS	35	12,995	97	7,818	134
	Jang	43	3,291	84	467	39.2
NYT-Comment	eFGS	53	17,295	136	1,112	127
	Boston	46	7,521	211	518	29.4
News-Twitter	India	66	4,723	88	113	53.7

3.2 Topic Extraction

For the topic extraction process, we first compare the proposed method with several baseline models in perplexity, and then demonstrate the model result through a case study.

Perplexity Evaluation. To evaluate the topic model, we split the news and UGC by date to apply our proposed CDTTM model, and compare the results with the following methods:

- **DTM**: dynamic topic model which is proposed in [6]. We use the released version[4] on news and comments separately since it does not consider the cross dependence between the two streams.
- **DCT**: document comment topic model [16] which models a single news article and associated comments by considering the news-comment dependence. We implement this model, and adapt it to model multiple documents.

---

[1] http://news.sina.com.cn/zt/

[2] http://www.nytimes.com/pages/topics/index.html

[3] http://query.nytimes.com/search/sitesearch/

[4] https://code.google.com/p/princeton-statistical-learning/

- **TCM**: temporal collection model introduced in [15] which models temporal dynamics through associating the Dirichlet hyper parameter α with a pre-defined time-dependent function, and we implement the algorithm described in their paper.

We employ perplexity [7] of the held-out test data as our goodness-of-fit measure. A model with lower perplexity indicates it has good generalization performance.

As for the parameters, we set the number of topics $K = 5$, fix the hyper parameters $\alpha = 50/K$, $\beta = 0.1$, $\gamma_d = 5$, $\gamma_p = 0.2$ for news, and swap the values of γ_d and γ_p for UGC as recommended in [16, 26]. Since we are modeling temporal data, we just perform the experiments by taking the prepositive several days for training and the rest for testing instead of random separation or cross validation. The results in Table 2 show that CDTTM performs better than the three state-of-the-arts. The reason for those with minor higher perplexity is that the test data is so sparse that CDTTM degenerates into standard LDA.

Table 2. Perplexity of four different topic models: the experiment settings include the duration, numbers of news articles and comments/tweets for both the *Train* and *Test* data, and the last four columns present the perplexity of different methods

Event	Train		Test		Perplexity			
	days	*docs*	*days*	*docs*	*DTM*	*DCT*	*TCM*	*CDTTM*
cFGS	28	74+10,175	7	23+2,820	19,717	19,204	18,071	17,923
Jang	10	71+1,979	33	13+1,312	17,203	16,211	17,146	17,307
eFGS	43	98+11,900	10	38+5,395	28,074	26,557	26,831	25,835
Boston	16	173+6,222	30	38+1,299	17,129	16,317	17,294	16,677
India	42	68+3,246	24	20+1,477	17,444	16,863	17,208	17,153

Case Study. Table 3 shows the results for eFGS. Note we have grouped the event into 5 stages according to their topic similarities and removed the common topic words (like *obama*) across multiple stages. We observe the topic trends from both streams (in the following description, **n** and **u** denote the information sources, i.e. news and UGC): [**n&u**]after the discussion (Aug. 29 to Sep. 19) on the Obamacare, [**n**]NYT published the news about the *Senate* and *House of representatives* discussing if they should support this program on Sep. 20. [**n**]Then *Ted Cruz* delivered an extremely long *speech* to argue that *Obamacare* was a disaster on Sep. 24. [**n&u**]They *voted, debated* and *voted* again (Sep. 30), [**n**]but the U.S. government still shut down on Oct. 1. During this period, interestingly, [**u**]the UGC from the public played crucial roles, e.g. they wanted a decision for the *budget* program (keywords *cost, debt*), and accelerated the vote (keyword *majority* on Sep. 25). [**u**]Public further appealed the *obamacare* should be *negotiated* and *passed* before the *shutdown*, which potentially led to [**n**]government re-opening on Oct. 17.

Table 3. An example for topic extraction on *eFGS* dataset: due to the space limitation, we list the top words with generative probability greater than 0.01 in each stage for both news and UGC, and those words that might link news and UGC are highlighted in bold

	8.29~9.19	9.20~9.25	9.26~10.2	10.3~10.17	10.18~10.21
News	*health* .021	*health* .038	*health* .037	*shutdown* .021	*medicaid* .023
	debt .019	*senate* .021	*care* .022	**debt** .020	*care* .019
	senate .017	**vote** .017	*insurance* .012	*health* .014	**budget** .014
	deficit .011	*congress* .012	**cruz** .011	**debate** .012	*national* .013
UGC	*care* .029	**cost** .024	*gop* .026	**debt** .032	*health* .025
	vote .016	*congress* .017	**obamacare** .018	**believe** .020	*care* .025
	job .015	**majority** .017	**pass** .017	*right* .015	**tax** .024
	debt .014	*coverage* .012	**negotiate** .016	**vote** .013	*money* .011

3.3   Influence Link Discovery

We evaluate the link discovery from two aspects, namely the number and the correctness of the discovered links.

Table 4 shows the number of discovered links using different topic models on the eFGS data, and we can observe that the number of local links is much more than that of global links for all topic models indicating that most UGC influence to news is more timely than slowly; the introduction of (mutual) dependence benefits the link discovery and that's why other three methods tend to find more links than DTM.

Table 4. Number of discovered influence links on *eFGS* data, including the numbers of local/global links and the average links per day

	DTM	DCT	TCM	CDTTM
Local	77	89	94	97
Global	13	13	13	14
Average	1.698	1.925	2.019	2.094

To obtain the ground truth for correctness evaluation, we invite three annotators to build links (*Total*) between two topics series and we only include those links that at least two of them agree (*Agree*) in the following evaluation. We use *Hybrid* to denote our proposed method, and compare it with the simplified version that only uses local search (denoted as *Local*) as well as random linking (*Random*). Table 5 shows the annotated statistics and the comparison results. We can see that: the annotated agreement ratio is around 50% which indicates that it is really a tough task; the random linking quality is very poor, while both local search and hybrid method are 10 times better; the hybrid method can achieve comparable (even better) results to the human annotation.

Table 5. Link correctness comparison: number of all distinct annotated links (*Total*), number of links that at least two annotator agrees (*Agree*), and the performance of different strategies

Event	Annotation		Comparison in F1		
	Total	*Agree*	*Random*	*Local*	*Hybrid*
cFGS	177	84	4.27%	43.4%	46.4%
Jang	201	89	4.66%	42.0%	45.5%
eFGS	239	123	3.93%	46.4%	51.3%
Boston	213	117	4.08%	43.3%	45.6%
India	291	124	4.11%	43.6%	44.9%

3.4 Influence Quantification

Fig. 4 shows how the four measures change over time in different events (to reflect the general trends, we normalize all time spans into [0,1]). For the topic progressiveness, we can see that: 1) UGC indeed has guidance to the news report throughout event life cycles, and the influence mainly falls into the beginning part and decreases along the timeline. 2) The highest values often come from the key points of events (e.g. the shutdown day in eFGS). The reason why there are two peaks in Jang is that Kim Jong-un gave a speech on Jan. 1 which captured many user-concerned topics.

Then we evaluate the news response to the topics in UGC stream, and have the following observations:

Macroscopically. 1) The response rate increases initially and then decreases with time whereas the promptness follows an opposite trends. This indicates that UGC topics attract attention from the news (response rate) and news responds to them rapidly (response promptness), especially when the events are still *hot*. 2) For the response effect, at the beginning, the public increasingly use sentimental words to express their opinions. When the event reaches around halfway stages, the sentimental words have largely reduced, indicating that news is effective in responding to public opinions.

Microscopically. Very interestingly, we also observe that English news can respond more and faster than Chinese news for the event of Federal Government Shutdown, and the corresponding effect is more significant. A possible explanation is that it's a U.S. internal event and American people have a better understanding the gists of the event. This observation also tells us that the metrics themselves are correlated with each other, e.g. a larger progressiveness often leads to higher response rate and lower response promptness, and the consequent effect are more likely to be significant.

4   Related Work

Our work in this paper is related to several lines of research in text mining and streaming data process.

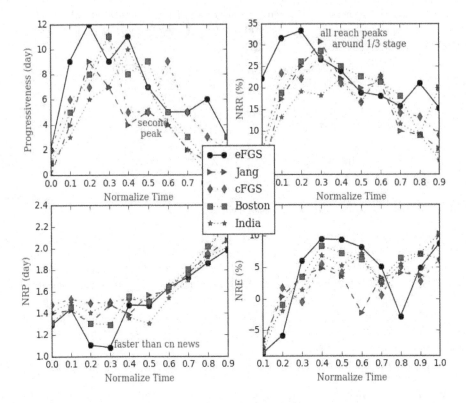

Fig. 4. Results Analysis for Influence Quantification

4.1   News and UGC Analysis

The rapid development of social media encourages many researchers to study its relationship between traditional news media. For example, Zhao et al. employed Twitter-LDA to compare the topic coverage of Twitter and NYT news and found Twitter actively helped spread news of important world events although it showed low interests in them [32]. Petrovic et al. examined the relation between Twitter and Newsfeeds and concluded that neither streams consistently lead the other to major events [25]. Liu et al. adapted the *Granger* causality to model the temporal dependence from large-scale time series data [5,8].

In this paper, we study the interplay of news and UGC in specific events where there should be more interactions than the general streams in the work mentioned above.

4.2   Streaming and Temporal Topic Model

For single stream model, Blei et al. proposed dynamic topic model to analyze the time evolution of topics in large document collections [6]. Wang et al. presented a non-Markov topic model named d Topic over Time (TOT), which jointly modeled both word co-occurrences and localization in continuous time [31]. Alsumait et al. proposed online-LDA to identify emerging topics and their changes over time in text streams [4]. Recently, Gao et al. derived topics in news stream incrementally using hierarchical dirichlet process, and then connected them using splitting and merging patterns [10].

For topic extraction in multiple streams, Wang et al. tried to extract common topics from multiple asynchronous text streams [29]. Hong et al. focused on analyzing multiple correlated text streams, allowing them to share common features and preserve their own private topics. Hou et al. proposed DCT model, which employed news as kind of prior to guide the generation of users' comments [16].

Compared with these models, the advantage of our model is that it captures the temporal dynamics and the mutual dependence between news and UGC streams.

4.3   Topic Evolution and Lifecycle

Mei et al. discovered evolutionary theme patterns in single text stream [22]. Wang et al. aimed at finding the burst topics from coordinated text streams based on their proposed coordinated mixture model [30]. Hu et al. modeled the topic variations through time and identifies the topic breakpoints [17] in news stream. Lin et al. formalized the evolution of an arbitrary topic and its latent diffusion paths in social community as an joint inference problem, and solved it through a mixture model (for text generation) and a Gaussian Markov Random Field (for user-level social influence) [20]. Jo et al. further captured the rich topology of topic evolution and built a global evolution map over the given corpus [18].

In this paper, we pay little attention on topic evolution and lifecycle, but more on analyzing the influence between two correlated text streams and try to figure out how news and UGC co-evolve along the time.

5   Conclusion and Future Work

In this paper, we study the mutual influence between news and UGC streams through a three-phase framework: extract topics from two correlated text streams, employ a hybrid method to discover short-term and long-term influence links across streams, introduce four metrics to measure the influence from UGC to news, as well as investigate how the news responds to the public opinion in UGC stream. Experiments on five news datasets confirm the existence of mutual influence, and present some interesting patterns.

There are several interesting directions to further extend this work. For example, our topic model returns a flat structure of topics and the topic number is pre-defined; it would be interesting to explore the hierarchical methods and non-parameter methods [10]. In addition, we only discover the influence links without distinguishing their different effects (e.g promote or suppress), so we will investigate the deeper semantics on the influence links which is another challenging but interesting problem.

Acknowledgments. Thank anonymous reviewers for their valuable suggestions that help us improve the quality of the paper. Thanks Prof. Chua Tat-Seng from National University of Singapore for discussion. The work is supported by 973 Program (No. 2014CB340504), NSFC-ANR (No. 61261130588), Tsinghua University Initiative Scientific Research Program (No. 20131089256) and THU-NUS NExT Co-Lab.

References

1. Ahmed, A., Xing, E.P.: Timeline: A dynamic hierarchical dirichlet process model for recovering birth/death and evolution of topics in text stream. In: Proceedings of the 26th Conference on Uncertainty in Artificial Intelligence. pp. 20–29 (2010)
2. Ajzen, I.: From intentions to actions: a theory of planned behavior. Springer, Heidelberg (1985)
3. Ajzen, I.: The theory of planned behavior. Organizational behavior and human decision processes **50**(2), 179–211 (1991)
4. AlSumait, L., Barbará, D., Domeniconi, C.: On-line lda: Adaptive topic models for mining text streams with applications to topic detection and tracking. In: Proceedings of the 8th IEEE International Conference on Data Mining, pp. 3–12 (2008)
5. Arnold, A., Liu, Y., Abe, N.: Temporal causal modeling with graphical granger methods. In: Proceedings of the 13th ACM SIGKDD International Conference on Knowledge Discovery and Data Mining, pp. 66–75 (2007)
6. Blei, D.M., Lafferty, J.D.: Dynamic topic models. In: Proceedings of the 23rd International Conference on Machine Learning, pp. 113–120 (2006)
7. Blei, D.M., Ng, A.Y., Jordan, M.I.: Latent dirichlet allocation. Journal of Machine Learning Research, 993–1022 (2003)
8. Cheng, D., Bahadori, M.T., Liu, Y.: Fblg: A simple and effective approach for temporal dependence discovery from time series data. In: Proceedings of the 20th ACM SIGKDD International Conference on Knowledge Discovery and Data Mining, pp. 382–391 (2014)
9. Daily, C.Y.: Chinese monthly public opinion index. Tech. rep, China Youth Daily, December 2013
10. Gao, Z., Song, Y., Liu, S., Wang, H., Wei, H., Chen, Y., Cui, W.: Tracking and connecting topics via incremental hierarchical dirichlet processes. In: Proceedings of the 11th IEEE International Conference on Data Mining, pp. 1056–1061 (2011)
11. Giorgino, T.: Computing and visualizing dynamic time warping alignments in r: The dtw package. Journal of Statistical Software **31**(7), 1–24 (2009)
12. Gohr, A., Hinneburg, A., Schult, R., Spiliopoulou, M.: Topic evolution in a stream of documents. In: the 9th SIAM International Conference on Data Mining, pp. 859–872 (2009)
13. Hänska-Ahy, M.: Social media & journalism: reporting the world through user generated content. Journal of Audience and Reception Studies **10**(1), 436–439 (2013)
14. Holcomb, J., Gottfried, J., Mitchell, A.: News use across social media platforms. Tech. rep., Pew Research Center, November 2013
15. Hong, L., Dom, B., Gurumurthy, S., Tsioutsiouliklis, K.: A time-dependent topic model for multiple text streams. In: Proceedings of the 17th ACM International Conference on Knowledge Discovery in Data Mining, pp. 832–840 (2011)
16. Hou, L., Li, J., Li, X., Qu, J., Guo, X., Hui, O., Tang, J.: What users care about: a framework for social content alignment. In: Proceedings of the 23rd International Joint Conference on Artificial Intelligence, pp. 1401–1407 (2013)

17. Hu, P., Huang, M., Xu, P., Li, W., Usadi, A.K., Zhu, X.: Generating breakpoint-based timeline overview for news topic retrospection. In: Proceedings of the 11th IEEE International Conference on Data Mining, pp. 260–269 (2011)
18. Jo, Y., Hopcroft, J.E., Lagoze, C.: The web of topics: discovering the topology of topic evolution in a corpus. In: Proceedings of the 20th International World Wide Web Conference, pp. 257–266 (2011)
19. Jönsson, A.M., Örnebring, H.: User-generated content and the news: Empowerment of citizens or interactive illusion? Journalism Practice 5(2), 127–144 (2011)
20. Lin, C.X., Mei, Q., Han, J., Jiang, Y., Danilevsky, M.: The joint inference of topic diffusion and evolution in social communities. In: Proceedings of the 11th IEEE International Conference on Data Mining, pp. 378–387 (2011)
21. McCombs, M., Holbert, L., Kiousis, S., Wanta, W.: The news and public opinion: Media effects on civic life. Polity (2011)
22. Mei, Q., Zhai, C.: Discovering evolutionary theme patterns from text: an exploration of temporal text mining. In: Proceedings of the 11th ACM International Conference on Knowledge Discovery in Data Mining, pp. 198–207 (2005)
23. Mizil, C.D.N., West, R., Jurafsky, D., Leskovec, J., Potts, C.: No country for old members: user lifecycle and linguistic change in online communities. In: Proceedings of the 22nd International World Wide Web Conference, pp. 307–318 (2013)
24. Morinaga, S., Yamanishi, K.: Tracking dynamics of topic trends using a finite mixture model. In: Proceedings of the 10th ACM International Conference on Knowledge Discovery and Data Mining, pp. 811–816 (2004)
25. Petrovic, S., Osborne, M., McCreadie, R., Macdonald, C., Ounis, I., Shrimpton., L.: Can twitter replace newswire for breaking news? In: Proceedings of the 7th international AAAI Conference on Weblogs and Social Media (Poster) (2013)
26. Rosen-Zvi, M., Griffiths, T.L., Steyvers, M., Smyth, P.: The author-topic model for authors and documents. In: Proceedings of the 20th Conference on Uncertainty in Artificial Intelligence, pp. 487–494 (2004)
27. Sheeran, P., Abraham, C.: Mediator of moderators: Temporal stability of intention and the intention-behavior relation. Personality and Social Psychology Bulletin 29(2), 205–215 (2003)
28. Sheeran, P., Orbell, S., Trafimow, D.: Does the temporal stability of behavioral intentions moderate intention-behavior and past behavior-future behavior relations? Personality and Social Psychology Bulletin 25(6), 724–734 (1999)
29. Wang, X., Zhang, K., Jin, X., Shen, D.: Mining common topics from multiple asynchronous text streams. In: Proceedings of the 2nd ACM International Conference on Web Search and Data Mining, pp. 192–201 (2009)
30. Wang, X., Zhai, C., Hu, X., Sproat, R.: Mining correlated bursty topic patterns from coordinated text streams. In: Proceedings of the 13th ACM International Conference on Knowledge Discovery in Data Mining, pp. 784–793 (2007)
31. Wang, X., McCallum, A.: Topics over time: a non-markov continuous-time model of topical trends. In: Proceedings of the 12th ACM International Conference on Knowledge Discovery and Data Mining, pp. 424–433 (2006)
32. Zhao, W.X., Jiang, J., Weng, J., He, J., Lim, E.P., Yan, H., Li, X.: Comparing twitter and traditional media using topic models. In: Proceedings of the 33rd European Conference on Information Retrieval, pp. 338–349 (2011)

Database Storage and Index I

SASS: A High-Performance Key-Value Store Design for Massive Hybrid Storage

Jiangtao Wang, Zhiliang Guo, and Xiaofeng Meng[✉]

School of Information, Renmin University of China, Beijing, China
{jiangtaow,zhiliangguo,xfmen}@ruc.edu.cn

Abstract. Key-value(KV) store is widely used in data-intensive applications due to its excellent scalability. It supports tremendous working data set and frequent data modifications. In this paper, we present SSD-assisted storage system (SASS), a novel high-throughput KV store design using massive hybrid storage. SASS meets three exclusive requirements of enterprise-class data management: supporting billions of key-value pairs, processing thousands of key-value pairs per second, and taking advantage of the distinct characteristics of flash memory as much as possible. To make full use of the high IOPS of sequential write on the SSD, all modification operations are packaged as operation logs and appended into SSD in the time order. To handle the tremendous number of key-value pairs on hard disk, a novel sparse index, which can be always kept in the SSD, is proposed. Moreover, we also propose an in-memory dense index for the operation logs on SSD. Our evaluation mainly characterizes the throughput of read and write, namely the ops/sec(**get-set** operations per second). Experiments show that our SASS design enjoys up to 96806 write ops/sec and 3072 read ops/sec over 2 billion key-value pairs.

Keywords: Key-value · Solid state disk · Cache · IOPS

1 Introduction

With the rapid development of Internet technologies, many web applications, such as internet services, microblogging network, and multi-player gaming, need to consistently meet the service requests of user within fast response time. The traditional disk-based relational database systems can hardly support the high-concurrent access gracefully. Recently, a lot of server-side applications have preferred to use noSQL databases implemented by key-value stores to provide high-throughput performance. Compared to the traditional relational database, key-value storage exhibits better scalability, efficiency and availability. Without

This research was partially supported by the grants from the Natural Science Foundation of China (No. 61379050,91224008); the National 863 High-tech Program (No. 2013AA013204); Specialized Research Fund for the Doctoral Program of Higher Education(No. 20130004130001), and the Fundamental Research Funds for the Central Universities, and the Research Funds of Renmin University(No. 11XNL010).

© Springer International Publishing Switzerland 2015
M. Renz et al. (Eds.): DASFAA 2015, Part I, LNCS 9049, pp. 145–159, 2015.
DOI: 10.1007/978-3-319-18120-2_9

complex command parse or execution plan optimization, key-value storage systems can enjoy excellent ops/sec performance. Hence, a key-value storage system is a better choice for the web applications which need to meet the data durability and high performance requirements. Furthermore, the technology of flash memory offers an alternative choice for storage system designers.

Over the past decades, flash-based solid state disk(SSD) is making deep inroads into enterprise applications as its increasing capacity and dropping price. Many web service providers have used flash memory to improve their system performance. However, the comparatively small capacity and high price hinder flash memory from a full replacement of hard disks. Although SSD RAID technology[1,2] makes the high-capacity flash memory device possible, the hybrid storage is still a prevalent mode. A key challenge in the hybrid storage is how to take full advantage of the flash memory to maximize the system performance.

In this paper, we present the design and evaluation of SSD-assisted storage system (SASS), a key-value store design supporting massive hybrid storage and high throughput. SASS has a three-part storage architecture that integrates main memory, SSD, and hard disk, in which we take flash memory as the write cache for the hard disk. In the main memory, we allocate a cluster of separate log buffers. All data modification operations (*insert*, *delete*, and *update*) are not immediately written back to the hard disk. Instead, they are stored in these log buffers as *operation logs*. When these buffers become full, these logs are appended to the log file on the SSD, and eventually merged with the original data on hard disk under certain conditions. As a result, we can take advantage of the high IOPS of sequential write on SSD and maximize the write throughput. In order to process random *get* query, we propose a sparse index, a hierarchical bloom filter residing in the SSD, to manage the tremendous number of key-value pairs on hard disks. Furthermore, we also design an in-memory index, *operation list*, for the operation logs on SSD. In general, a random *get* query can be answered by one flash read in the best case and one extra hard disk read in the worst case. The contributions of this paper are summarized as follows:

1. We present a novel key-value store design, called SSD-assisted storage system (SASS), which aims to support large scale data-intensive applications. In SASS, SSD serves as a write cache for the hard disk, and the key-value pairs are organized into blocks on hard disks, while the recent modifications of the key-value pairs are buffered in the SSD as operation logs. The query processing procedure is further accelerated by two novel index mechanisms.
2. We implement an industry-strength SASS system and conduct extensive experiments on it. The evaluation results demonstrate that SASS enjoys up to 96806 write IOPS with key-value pairs log buffers, which outperforms BerkeleyDB to 9.46x. Moreover, as the introduction of SSD and hierarchical bloom filter index, SASS provides 2.98x speedup over BerkeleyDB when measuring the IOPS of read operation.

The rest of the paper is organized as follows. In Section 2, we present an overview of SASS and some critical components of SASS, including data organization, hierarchical bloom filter and operation list. Some system maintenance operations and how to achieve concurrency control are explained in Section 3. Section 4 gives the results of our experiment evaluation and Section 5 surveys the related work. Finally, Section 6 concludes this paper.

2   SASS Design and Implementation

2.1   Overview of SASS

Fig. 1 gives an overview of SASS, which employs a three-part storage architecture integrating main memory, SSD, and hard disk. To support massive data volume, we take hard disk as the main storage medium considering its large data capacity. SSD is used to the write cache for the hard disk due to its high IOPS performance. Some recently modified key-value pairs are cached in the SSD, which will be merged to hard disk eventually. Thus, for a query request, SASS always check the SSD at first to see if the most fresh key-value pair exists. If it does, SASS just reads it from SSD. Otherwise, SASS checks the hard disk.

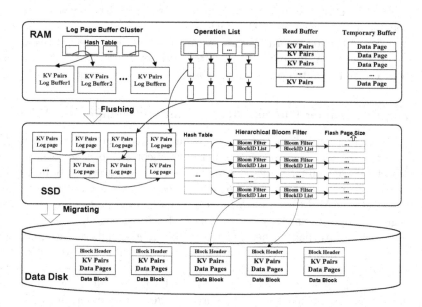

Fig. 1. An overview of SSD-assisted storage system (SASS)

2.2   Block, Page and Record

In SASS, the disk-resident key-value pairs is managed at a block granularity. All the incoming key-value pairs are distributed into different data blocks by a hash function. However, the big block size introduces a problem, that is, we have to read a block even if we just want to get a single key-value pair, which is inefficient and memory-consuming. So, we introduce data page into SASS. A block consists of multiple data pages, and a block header structure is designed for each block to summarize the key-value pair distribution in a data block. We can get the requested key-value pair through checking its block header. The key-value pair is stored as a variable-length record. Fig. 2(a) gives the layout of a record, which consists of three fields: *Record Size* stores the size of the record, *Key Size* stores the size of the key, and *Data Value* stores the key-value pair.

2.3   Operation Log, Operation Log Page

In SASS, all key-value pair modification operations are first stored in the SSD as operation logs. The operation logs are organized into log pages following the order of timestamp. As more and more operation logs are flushed to the SSD, the first few log pages of the list will be merged and moved to the hard disk over time. By calculating the hash value of a given *key*, the key-value pairs which share the same hash value are accumulated to form a data block. All the data blocks with the same hash value are clustered into a partition. In log page, a key-value pair is stored as a variable-length record. Fig. 2(b) gives the layout of an operation log record on SSD, which also consists of three fields: *Log Size* stores the size of the log record, *Partition_ID* identifies the target partition and *Record* stores the key-value pair. We store the new key-value pairs in the *Record* field for insert and update operations and *null* for delete operation.

2.4   Log Page Buffer Cluster, Read Buffer and Temporary Buffer

In the main memory, there are three types of buffers: log page buffer, read buffer and temporary buffer. The log page buffer cluster consists of a set of log page buffers, each log page buffer is a fixed-size data structure that is used to buffer the dedicated key-value pairs by a hash function. Specifically, when a new key-value pair is generated, SASS uses a hash function to locate an associated partition, and assigns a proper log page buffer to hold it. Each log page buffer shares the same size with a data page, when the log page buffer is full, the accumulated logs will be appended to the SSD as a log page. Read buffer is also a fixed-size data structure to cache the recently read data, including data pages, log pages and block headers. The least recently used (LRU) pages will be evicted when read buffer runs out of free space. Temporary buffer, as its name suggests, is a temporary data structure used for the merge operations.

2.5   Operation List

In SASS, we store all the recent updated data on SSD as operation logs, just depicted in Fig. 2(b). Whenever a query request arrives, SASS firstly checks the operation logs cached in the SSD. Upon a miss on the SSD, the query continues to look up the key-value pairs on hard disks. Consequently, an index for the operation logs on SSD is definitely necessary to accelerate the check. We design an index for the operation logs on SSD, namely operation list. Fig. 3 shows the structure of operation list. Operation list is an array of doubly linked lists of operation elements, which uses a mapping table to maintain all the operation logs on SSD. In the mapping table, the key is the *Partition_ID* while the mapped value is OpHeader. As soon as an operation log is flushed to the SSD, a new operation element pointed to that operation log will be created and linked to the corresponding doubly linked list. Actually, we just need one doubly linked list for a partition. However, we make an array of double linked lists for each partition to avoid a double linked lists with too many elements which can be a nightmare for query. For each operation element to be linked, we compute a HashCode using division method with the key at first and then link the element to the double linked list with the same HashCode. In this way, we can transform a long list into many short lists and reduce the query cost. Fig. 2(c) describes the layout of an operation element, which contains five fields: *Operation Type* marking the type of the operation, *Log Address* keeping the exact address of the operation log record on SSD, *PrevOpElement* and *NextOpElement* pointing to the previous and next OpElements respectively, *Key* representing the key of the operated record. In this way, we can arrange the operation log records targeting on a certain partition to a doubly linked list from the head to the tail in the order of their arrival time. For a query with specific key, we can find the corresponding list and traverse the list from tail to head to look up the first operation element with the same key.

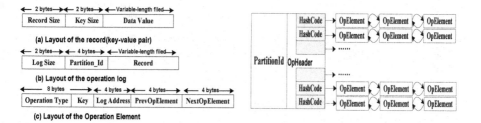

Fig. 2. Layout of record, operation log and operation element

Fig. 3. Structure of operation list

2.6 Hierarchical Bloom Filter

Hierarchical bloom filter is a sparse index, it is designed for indexing the key-value pairs migrated to hard disk. Because the in-memory index needs to take up a considerable amount of buffer space, so, we use SSD to store the hierarchical bloom filter. Bloom filter is a space-efficient probabilistic data structure which supports set membership queries. One single bloom filter designed for flash memory may suffer from the drawback that inserting a key almost always involves a lot of flash page writes, since the k bit positions may fall into different flash pages. Considering the poor random write of flash memory, we design a hierarchical bloom filter index structure. Fig. 4 shows the hierarchical bloom filter, which can be taken as a bloom filter tree. On the lowest level of the tree, namely the leaf level, each leaf node contains an independent bloom filter which occupies a single flash page. A leaf node summarizes the membership of the keys which are scattered in multiple disk-resident blocks. That is, each independent bloom filter is responsible for indexing one or more specific blocks. To insert or lookup an element, we employ a hash function to locate the sub-bloom filter that the requested key-value pair may reside in. Then, k bit positions are identified within the sub-bloom filter for setting or checking the bits. Thus, this design requires only one flash page read per element lookup. To further to accelerate the key lookup, we also add a block header list for each independent bloom filter. When a key is identified to fall into some sub-bloom filter, we can locate the block which the requested key-value pair resides in by searching the block header list.

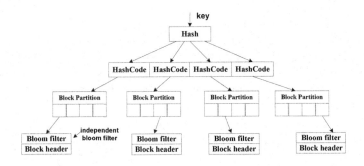

Fig. 4. Structure of hierarchical bloom filter

2.7   Key Set and Get Operations

SASS supports the following basic operations: *insert, update, delete, get,* as well as *merge.* In this section, we will explain how they work in SASS.

1)set: SASS transforms all the insert, update and delete operations into operation logs and appends them to the SSD. Subsequently, the corresponding operation elements are created and linked to the tails of the target list. In this way, SASS transforms the random write into sequential write and maximize the write throughput. Actually, all these operations will not return until the operation logs are flushed to SSD for guaranteeing the durability of data.

2) get: A *get* query uses a key to retrieve a key-value pair. Given the key in the *get* query, the id of the partition that contains the key is determined at first by checking the in-memory hash table. Then, we can get the corresponding double linked list to the partition id from the operation list. The first element with the same key can be found by traversing the list of operation elements from tail to head. Upon a hit on the list, we will check the operation type, *insert* or *update* indicates the record is resident on SSD, and *delete* indicates the record had been eliminated recently. Hence, the data address will be returned for inserting or updating and *null* for deleting. Otherwise, we will search the hierarchical bloom filter to locate the page in which the key-value pair resides.

3   Advanced Issues

This section discusses some advanced and important issues for SASS.

Algorithm 1. Evict_SSDPage()

Require: The operation list *list* that triggers the merge operation
Ensure: Merge the operation log with the key-value pair residing hard disk
1: *PartitionId*=Lookup_OP(*list*);/*locate the *partitionID* of the given list*/
2: *MergeBlock*=GetBlock(*PartitionId*);/*get the block with the *PartitionId**/
3: **for** (*ele* = *list.head;ele*! = *list.tail;ele* = *ele* − >*nextOpElement*) **do**
4:    **if** (*ele.type* == *insert*) **then**
5:        *data*=GetData(*ele.dataAddress*);
6:        **if** (MergeBlock is full) **then**
7:            allocate a new *MergeBlock* for the incoming key-pair page;
8:        InsertData(*MergeBlock,data*);
9:    **if** (*ele.type* == *update*) **then**
10:        *data*=GetData(*ele.dataAddress*);
11:        **if** (MergeBlock is full) **then**
12:            allocate a new *MergeBlock* for the incoming key-pair page;
13:        UpdateData(*MergeBlock,data*);
14:    **if** (*ele.type* == *delete*) **then**
15:        DeleteData(*MergeBlock,ele.key*);
16: FlushAllBlockData();/*migrate the SSD-resident KV pairs to disk*/;
17: *deltaindex*=BulidDiskDataIndex();/*build index for the evicted KV pairs*/;
18: update *deltaindex* to the hierarchical bloom filter;
19: RemoveOplistFromSSDIndex(*list*);/*remove the list from the operation list*/;
20: return;

3.1   Merge

As the data modifications consume the space of SSD steadily, the operation log records in the oldest log pages have to be merged with their original data on the hard disk periodically. This process is managed by a *merge* operation. In general, two conditions will trigger a merge operation: the number of operation elements in an operation list exceeds a threshold and the flash memory usage exceeds a certain threshold. During the merging process triggered by a large number of operation elements, the corresponding log pages will be read into the temporary buffer and the list will be traversed from head to tail to execute the merge operation. During the merging process triggered by the flash usage, the oldest log blocks are chosen to be recycled. The maximum number of operation elements in an operation list is very a critical configuration parameter. A relatively small operation element number setting reduces the traversal time on the list but makes the merge operation more frequent. A relatively large operation elements number accelerates the data access but reduces the space that can be used by SSD itself, which is considered necessary for some internal operations (e.g., garbage collection). Algorithm 1 gives the detailed description of the merge operation.

3.2 Concurrency Control

To achieve high throughput, SASS must support multi-thread operations, which require an effective concurrency control mechanism. Temporary buffer is a temporary structure allocated for each thread exclusively, so there is no need to protect temporary buffer. For other shared data structures, Table 1 lists the related operations and lock strategies.

Read Buffer: A *get* query may check the read buffer to see if the target data pages are already cached. Upon a miss, the least recently used (LRU) pages will be evicted and then the query thread reads the target pages from SSD or hard disk. In fact, we do not write the data pages back to hard disk, since they are never modified. The only thing should be ensured is that the data pages being accessed by some threads cannot be evicted by other threads. Consequently, each query thread must hold a read lock on the target pages and cannot evict any pages until it holds the write locks on them.

KV Pairs Log Buffers: KV pairs log buffers collect the operation log records created by *insert*, *update* and *delete* operations in the order of their timestamps. When the operation log records fill up the buffer, all the write threads will be blocked until all log records in the buffer are flushed to the SSD. As the write threads and the flush thread have a producer-consumer relationship on both buffers, we use a producer-consumer lock to protect them.

Operation List: For operation list, *get* traverses the list to find the target operation element and all the elements on the list will be checked upon a miss. *Insert*, *update* and *delete* always add an element to the tail of the target list. *Merge* traverses and frees a part of or the entire list. For this situation, we must prevent these operations from being disturbed by each other. So, we choose to use the reader-writer lock.

4 Experimental Evaluation

We implement a key-value store with about 20,000 lines of C code and perform a serial of experiments on this system. As SASS aims to be a high throughput storage system, our evaluation mainly characterizes the throughput of read and write, namely the ops/sec. Our experiments run on a server powered by Intel Xeon CPU E5645 at 2.40GHz with 16GB of DRAM. We use the Seagate hard

Table 1. Lock Strategy Of Shared Data Structure

Share Data Structure	Operation	Lock Strategy
Read buffer	*get*	Reader-Writer
KV pairs log buffer	*insert, update, delete*	Producer-Consumer
Operation list	*get, insert, update, delete, merge*	Reader-Writer

disk store all the key-value pairs. The storage capacity of hard disk is set to 10TB. A 256G GB Intel SSD serves as the write cache.

We pick two different data sets, i.e., post messages and pictures, as our evaluation datasets. The data items in the former dataset are mostly small(100bytes ~ 1000bytes) while the data items in the latter one are relatively large(10KB ~ 100KB). To make a thorough evaluation of SASS, we chose two extreme data traces, Random Set and Random Get, to squeeze SASS for its maximum write and read performance. In addition, we also chose a typical data trace, Canonical, which is a normalized read-world workload. Table 2 describes the properties of each test set, in which a suffix L stands for large data (i.e, thumbnail pictures).

Table 2. Experimental Data Trace

Trace	Number of Operations	get:set:update:delete	Value Size(KB)
Random Set	4billions	0:1:0:0	0.1 ~ 1
Random Set-L	4billions	0:1:0:0	10 ~ 100
Random Get	4billions	1:0:0:0	0.1 ~ 1
Random Get-L	4billions	1:0:0:0	10 ~ 100
Canonical	2.5billions	64:8:4:1	0.1 ~ 1
Canonical-L	2.5billions	64:8:4:1	10 ~ 100

4.1 Set and Get Performance

We compare SASS with a popularly used database system, BerkeleyDB or BDB. BDB is a software library that provides a high-performance embedded database for key-value data, we select hash table to build the index for BerkeleyDB. To make a fair performance comparison, we implement the BerkeleyDB with a non-transactional data store mode, both SASS and BDB run on the same machines that we described above. We compare the performance of BDB and SASS using the workloads listed in Table 2.

Fig. 5 and Fig. 6 show the random set ops/sec of BDB, SASS over two datasets. In both figures, the random set ops/sec of BDB and SASS decrease when the number of concurrent threads increases. An exception is that the random set ops/sec of SASS does not decrease until the number of test threads exceeds 128. Especially for the microblog messages dataset(shown in Fig. 5), SASS provides a speedup of 9.5 times relative to BDB when the number of test

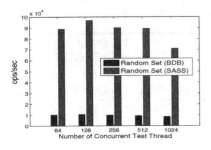

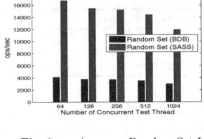

Fig. 5. ops/sec over Random Set   **Fig. 6.** ops/sec over Random Set-L

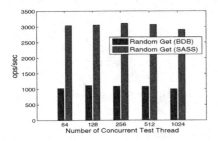

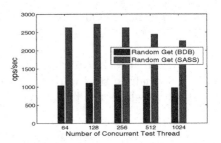

Fig. 7. ops/sec over Random Get   **Fig. 8.** ops/sec over Random Get-L

threads is set to 512. In addition, as SASS transforms all the random set into sequential write, SASS exhibits a higher ops/sec than that of BDB when dealing with the relatively large data items (shown in Fig. 6).

Fig. 7 and Fig. 8 show the random get ops/sec of BDB and SASS over two datasets. From both figures, we can see that both the random get ops/sec of BDB over microblog messages dataset and thumbnail pictures dataset display little difference. That is, the random get performance of BDB is non-sensitive to different datasets. In contrast, SASS shows better random get ops/sec over microblog messages dataset. Concerning the data items in microblog messages dataset are short, much more key value pairs can resident on SSD, so the performance improvement over microblog messages dataset makes sense. In general, a random get operation in SASS can be answered by one flash read for the best case and one extra hard disk read for the worst. However, a random get operation in BDB requires one hard disk read for the best case. Hence the random get performance of SASS always outperforms that of BDB.

Both random set and random get are extreme workloads and can hardly happen in practice, so we choose another typical workload, canonical. Fig. 9 and Fig. 10 exhibit the canonical ops/sec of BDB, SASS. For the canonical dataset, SASS gains the maximum of ops/sec when the number of test thread is set to 128, and provides a speedup of 8.49X compared to BDB.

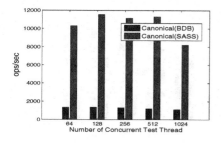
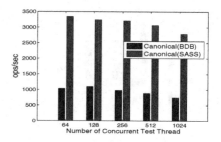

Fig. 9. ops/sec over Canonical   **Fig. 10.** ops/sec over Canonical-L

4.2   Impact of the Length of Operation List

Merge is the most expensive in all of the operations and it also affects the execution of other operations. As stated in Section 3, the two conditions trigger merge operation: the number of elements in an operation list exceeds a threshold and the amount of log pages usage on SSD as well exceeds another predefined threshold. Considering the relatively small capacity of SSD, the value of flash usage threshold is set to 90%. Consequently, we just tune the maximum number of elements in a OpHeader in the following evaluation.

With different maximum number of elements in a OpHeader, we conduct canonical workload again over two datasets. Besides, we count the ops/sec of set operations and get operations in canonical workload respectively, so that we can figure out how much the merge operations affect set operation and get operation. We vary the number of element in operation list from 512 to 2048. Fig. 11 and Fig. 12 display the ops/sec of set operation in canonical workload over two datasets. From the figures we can determine that the bigger number of elements the better ops/sec of set operation. Bigger number of elements in a OpHeader means less merge operations triggered by operation list. Merge operation holds an exclusive lock to prevent subsequent set operations, hence a bigger number of elements setting can improve the ops/sec of set operation.

Fig. 13 and Fig. 14 display the ops/sec of get operation in canonical workload over two datasets. From these figures, we can say that the bigger number of elements the lower ops/sec of set operation. Although we make an array of double linked list, the number of elements can be large if we choose a large number of elements setting. For every get operation, we have to choose a list from the operation list and traverse it. Accordingly, a bigger number of elements setting increase the overhead of traverse. Merge operations do not affect the get operation, because they hold share lock for each other.

4.3   Impact of Merge Operation

As shown above, the number of elements affects the set and get operation in different ways. We count the number of merge operation triggered as SASS processes more and more requests in canonical workload. In Fig. 15, we can

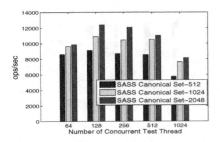

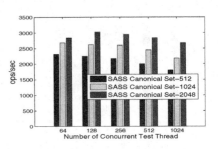

Fig. 11. Set ops/sec over Canonical

Fig. 12. Set ops/sec over Canonical-L

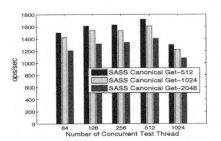

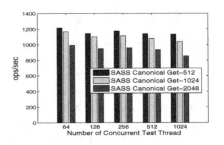

Fig. 13. Get ops/sec over Canonical

Fig. 14. Get ops/sec over Canonical-L

see that as the data size accumulates gradually, SASS with maximum number of elements setting 512 incurs most merge operations. SASS with maximum number of elements setting 1024 incurs the least merge operations.

For the best case, SASS with maximum number of elements setting 1024, we analyze the merge operations and count the number of merge operations triggered by operation list and the number of merge operations triggered by SSD respectively. From the Fig. 16, we can see that there is no merge operations triggered by SSD until the size of test data reaches up to 200GB. The reason is that the SSD has a 256GB capacity, so it can hold about 200GB operation logs and trigger few merge operations. However, as the size of test data exceeds 400GB, more and more merge operations are triggered by SSD.

We also vary the size of data block from 2MB to 16MB, and appreciate their impact on the throughput performance when the number of test thread is set to 128. We find that a block with the size of 8MB provides the maximum of throughput. We discuss the reason why different block sizes can exhibit different performance improvements. When we select a small block size, the size of SSD-resident index grows rapidly, which increases the cost of maintaining the index on SSD. If we use a larger data block, SFHS can reduce the seek latency of the hard disk, which can improve the I/O performance significantly. However, for a given key, SFHS has to spend more system resource to answer the requested key-value pair in a block.

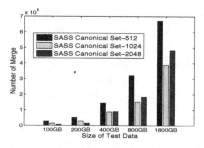
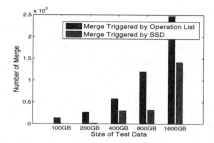

Fig. 15. Number of merges over Canonical **Fig. 16.** Number of merges triggered by operation list and SSD

5 Related Work

As a novel storage medium that is totally different from magnetic disk, flash memory is getting more and more attention in recent years. Lots of work emerged to solve different problems. Some work focus on an intrinsic component of the flash, namely flash translation layer *(FTL)*, which is an important firmware in flash-based storage[3, 4]. Some work focus on the measurements on flash memory [5, 6], and some other work focus on how to adjust the traditional methods in DBMS to take full advantage of the unique characteristics of flash memory [7–9]. We can't cover all those excellent work here, so we just give a brief review for the work related to key-value storage in this section.

FAWN[10] is a cluster architecture for low-power data intensive computing. It uses an array of embedded processors together with small amounts of flash memory to provide efficient power performance. FAWN uses an in-memory hash table to index key-value pairs on flash while SASS adopts hierarchical bloom filter for blocks on hard disk and doubly linked list for operation logs on SSD.

FlashStore[11] is a high throughput persistent key-value store that uses flash memory as a non-volatile cache between RAM and hard disk. It stores the working set of key-value pairs on flash and indexes the key-value pairs by a hash table stored in RAM. FlashStore organizes key-value pairs in a log-structure on flash to obtain faster sequential write performance. So it needs one flash read per key lookup. The hash table stores compact key signatures instead of full keys to reduce RAM usage. FlashStore provides high performance for random query. However, FlashStore employs an in-memory index to record the key-value pair residing in the hard disk, the memory overhead for implementing the index may exceed the available memory when handling the billion-scale keys.

ChunckStash[12] is a key-value store designed for speeding up inline storage deduplication using flash memory. It builds an in-memory hash table to index data chunks stored on flash. And the hash table can help to identify duplicate data. ChunkStash organizes the chunk data in a log-structure on flash to exploit fast sequential writes and needs one flash read per lookup.

SkimpyStash[13] is a RAM space skimpy key-value store on flash-based storage designed for server applications. It uses a hash table directory in RAM to index the key-value pairs stored on flash. To reduce the utility of RAM, SkimpyStash stores most of the pointers that locate each key-value pair on flash. It means that SkimpyStash uses linear chaining to resolve hash table collisions, and stores the link list on flash. SkimpyStash may need multiple flash read for one key lookup. In addition, because the flash memory is more expensive than the hard disk, the cost of using flash memory to handle the large-scale key-value pairs is very huge.

From the related work stated above, we can conclude that most existing key-value store designed for hybrid storage adopt in-memory data structure to index the key-value pairs. However, with 2 billion or more key-value pairs stored on each machine, these designs consumes memory excessively. Furthermore, the design of SASS is based on a slim SSD capacity(256GB) and a chubby hard disk capacity(10TB) while most other designs are based on a comparative SSD capacity with hard disk. By placing the hierarchical bloom filter index on SSD, SASS reduces the memory consumption and answers any query request with one flash memory read for the best case or an extra hard disk read for the worst case.

6   Conclusion

We propose SASS, a key-value store design supporting massive data set and high throughput. Experiment results show that SASS takes full advantage of the flash memory and enjoys excellent read/write throughput. Actually, there are still some interesting problems to be studied and researched. For example, what on earth the role that flash memory should play, using for logging or caching. Furthermore, how to setup a hybrid storage strategy in a distributed system, hybrid on each node or some are flash nodes(only flash memory adopted in the machines) and some others are hard disk nodes(only hard disk adopted in the machines). All these problems are realistic in industry and valuable for researchers.

References

1. Jeremic, N., Mühl, G., Busse, A., Richling, J.: The pitfalls of deploying solid-state drive RAIDs. In: 4th Annual Haifa Experimental Systems Conference, pp. 14:1–14:13. ACM Press, Haifa (2011)
2. Balakrishnan, M., Kadav, A., Prabhakaran, V., Malkhi, D.: Differential RAID: Rethinking RAID for SSD reliability. In: 5th European Conference on Computer Systems, pp. 15–26. ACM Press, Paris (2010)
3. Gupta, A., Kim, Y., Urgaonkar, B.: DFTL: A flash translation layer employing demand-based selective caching of page-level address mappings. In: 14th International Conference on Architectural Support for Programming Languages and Operating Systems, pp. 229–240. ACM Press, Washington (2009)
4. Lee, S., Shin, D., Kim, Y.J., Kim, J.: Last: locality-aware sector translation for nand flash memory-based storage systems. ACM SIGOPS Operating Systems Review. **42**(6), 36–42 (2008)

5. Bouganim, L., Jnsson, B., Bonnet, P.: uFLIP: Understanding flash IO patterns. In: Online Proceedings of the 4th Biennial Conference on Innovative Data Systems Research, pp. 1–12, Asilomar (2009)
6. Chen, F., Koufaty, D.A., Zhang, X.D.: Understanding intrinsic characteristics and system implications of flash memory based solid state drives. In: 11th International Joint Conference on Measurement and Modeling of Computer Systems, pp. 181–192. ACM Press, Seattle (2009)
7. Chen, S.M.: FlashLogging: exploiting flash devices for synchronous logging performance. In: ACM SIGMOD International Conference on Management of Data, pp. 73–86. ACM Press, Rhode Island (2009)
8. Nath, S., Kansal, A.: FlashDB: dynamic self-tuning database for NAND flash. In: 6th International Conference on Information Processing in Sensor Networks, pp. 410–419. ACM Press, Massachusetts (2007)
9. Trirogiannis, D., Harizopoulos, S., Shah, M.A., Wiener, J.L., Graefe, G.: Query processing techniques for solid state drives. In: ACM SIGMOD International Conference on Management of Data, pp. 59–72. ACM Press, Rhode Island (2009)
10. Andersen, D.G., Franklin, J., Kaminsky, M., Phanishayee, A., Tan, L., Vasudevan, V.: FAWN: a fast array of wimpy nodes. In: 22nd Symposium on Operating Systems Principles, pp. 1–14. ACM Press, Montana (2009)
11. Debnath, B., Sengupta, S., Li, J.: FlashStore: high throught persistent key-value store. Proceedings of the VLDB Endowmen. 3(2), 1414–1425 (2010)
12. Debnath, B., Sengupta, S., Li, J.: ChunkStash: speeding up inline storage deduplication using flash memory. In: 2010 USENIX Conference on USENIX Annual Technical Conference, pp. 1–12. USENIX Association, Boston (2010)
13. Debnath, B., Sengupta, S., Li, J.: SkimpyStash: RAM space skimpy key-value store on flash-based storage. In: ACM SIGMOD International Conference on Management of Data, pp. 25–36. ACM Press, Athens (2011)

An Efficient Design and Implementation of Multi-level Cache for Database Systems

Jiangtao Wang, Zhiliang Guo, and Xiaofeng Meng[✉]

School of Information, Renmin University of China, Beijing, China
{jiangtaow,zhiliangguo,xfmeng}@ruc.edu.cn

Abstract. Flash-based solid state device(SSD) is making deep inroads into enterprise database applications due to its faster data access. The capacity and performance characteristics of SSD make it well-suited for use as a second-level buffer cache. In this paper, we propose a SSD-based multilevel buffer scheme, called flash-aware second-level cache(FASC), where SSD serves as an extension of the DRAM buffer. Our goal is to reduce the number of disk accesses by caching the pages evicted from DRAM in the SSD, thereby enhancing the performance of database systems. For this purpose, a cost-aware main memory replacement policy is proposed, which can efficiently reduce the cost of page evictions. To take full advantage of the SSD, a block-based data management policy is designed to save the memory overheads, as well as reducing the write amplification of flash memory. To identify the hot pages for providing great performance benefits, a memory-efficient replacement policy is proposed for the SSD. Moreover, we also present a light-weight recovery policy, which is used to recover the data cached in the SSD in case of system crash. We implement a prototype based on PostgreSQL and evaluate the performance of FASC. The experimental results show that FASC achieves significant performance improvements.

Keywords: Solid state driver · Flash · Cost · Cache

1 Introduction

Large-scale data intensive applications have gained tremendous growth in recent years. Since these applications always contain massive random and high-concurrent accesses over large datasets, traditional disk-based database systems(DBMSs) cannot support the system gracefully. Compared with magnetic hard disks, flash memory has a myriad of advantages: higher random read performance, lighter weight, better shock resistance, lower power consumption, etc. Today, flash-based solid

This research was partially supported by the grants from the Natural Science Foundation of China (No. 61379050,91224008); the National 863 High-tech Program (No. 2013AA013204); Specialized Research Fund for the Doctoral Program of Higher Education(No. 20130004130001), and the Fundamental Research Funds for the Central Universities, and the Research Funds of Renmin University(No. 11XNL010).

© Springer International Publishing Switzerland 2015
M. Renz et al. (Eds.): DASFAA 2015, Part I, LNCS 9049, pp. 160–174, 2015.
DOI: 10.1007/978-3-319-18120-2_10

state device(SSD) is making deep inroads into enterprise applications with its increasing capacity and decreasing price. The unique features of flash memory make it the best storage media for DBMSs. However, the comparatively small capacity and high price hinder flash memory from full replacement of hard disks. Flash memory will be used along with hard disks in enterprise-scale data management system for a long time[1, 2]. Therefore, the hybrid storage may be an economical way to deal with the rapid data growth.

In this paper, we propose a flash-aware second-level cache scheme(FASC). The FASC takes flash-based SSD as a non-volatile caching layer for hard disk. When a pages is evicted from main memory, the FASC adopts a data admission policy to selectively cache the evicted page in the SSD for subsequent requests. In this way, the cost of disk access can be reduced due to the faster read/write performance of flash memory. Unlike the traditional single-level buffer management system, the multilevel cache design results in a complicated page flow across the different storage devices. Meanwhile, flash memory has some inherent physical characteristics[3, 4], such as erase-before-write limitation, poor random write and limited lifetime. The data scheduling policies designed and optimized for the single-level buffer cache must be carefully reconsidered when switching to a SSD-based multilevel buffer cache. Otherwise, the performance improvement may be suboptimal. How to take full advantages of the flash memory is the key issue and challenge in the design of SSD-based multilevel buffer system.

The main contributions of the paper are summarized as follows:

1. We present a SSD-based multilevel cache strategy, called FASC, which aims to improve I/O efficiency of DBMSs by reducing the penalty of disk accesses. FASC is a hybrid storage system which uses flash-based SSD as an extension of main memory. To reduce the I/O cost for each page eviction, we discuss the different states of the memory pages, and propose a cost-aware buffer replacement policy for the DRAM buffer.
2. We design a memory-efficient data management policy for the SSD. The data cached in the SSD is organized into blocks. Our block-based policy can not only reduce the memory overheads for maintaining the metadata, but also the number of scattered random write operation on flash memory. We also propose a novel hot page identification method to find the pages that are accessed more frequently. In addition, we propose a lightweight metadata recovery scheme to ensure transaction durability.
3. We implement a prototype system based on PostgreSQL, and evaluate its performance using TPC-C and TPC-H benchmarks. The experimental results show that FASC not only provides significant performance improvements, but also outperforms other cache policies presented in recent studies.

The rest of the paper is organized as follows: Section 2 describes the related works about SSD-based hybrid storage system. We present the system overview and some critical components related to FASC in Section 3, while Section 4 details our cost-aware memory replacement policy. In Section 5, we elaborate the data management policy for the SSD buffer. The data recovery scheme designed for FASC is described in Section 6. We give the results of our experiment evaluation in Section 7. Finally, Section 8 concludes this paper.

2 Related Work

Flash memory based SSD has drawn more and more attention in recent years. Many research works have been done to solve different problems. Some works focus on how to adjust the traditional methods in DBMSs to take full advantage of the unique characteristics of SSDs[5–7]. Recently, some works have attempted to use SSD store frequently accessed data to enhance transactional DBMS performance[8–10]. We cannot cover all the excellent works here, so we just give a brief review for the works related to FASC in this section.

TAC[8] is a SSD-based bufferpool extension scheme to accelerate reads and writes from hard disks by caching frequently accessed pages in SSDs. This work adopts a temperature-based data admission policy to cache the pages evicted by the bufferpool layer. Whenever a page request arrives, TAC computes the regional temperatures of the extent(32 contiguous pages) where the requested page resides. If the extent is identified as a hot region, the page will be written to both SSD and main memory synchronously. TAC adopts a *write-through* caching policy, the dirty page evicted by the DRAM will be directly written to disk. Therefore, it does not reduce the total number of write operation to the disk.

The benefits of using SSD in an extended cache manner have been discussed in work[9]. In that work, the authors propose a lazy cleaning(LC) method to handle the pages evicted by the DRAM buffer. Unlike the TAC, LC adopts a *write-back* caching policy to handle the dirty pages evicted from the main memory. All the dirty pages are only written to the SSD first, and these dirty pages are not flushed to the hard disk until the SSD is full. The SSD data replacement policy adopted by the LC is based on LRU-2 algorithm, which triggers many expensive random writes when replacing a page in SSD. In addition, considering that each mapping record occupies 88 bytes memory space, LC suffers huge memory overheads, which may potentially result in a low hit ratio.

Kang et al. [10] propose a flash-based second-level cache policy, called FaCE, to improve the throughput of DBMS. FaCE adopts a multi-version policy to manage the pages cached in SSD. Whenever a page is evicted from the DRAM buffer, it is written to SSD sequentially in an append-only fashion. Stale copies of the page, if exist, will be invalidated. FaCE manages the SSD buffer in a first-in-first-out(FIFO) fashion. In this way, the expensive random writes on SSD are avoided. However, the FIFO policy is hard to capture the pages that contribute more accesses, which degrades the performance of extended buffer cache system.

Previous works mainly focus on how to flush pages from the DRAM to SSD. However, for a SSD-based multilevel cache, the asymmetric speed of read/write operations on SSD is an important factor to speed up I/O access, and we must reduce the cost of page migration among SSD and DRAM. On the other hand, considering the large capacity of SSD, it is necessary to use memory-efficient data structures and algorithms to manage the page cached in the SSD. In this paper, we analyze the different page states from the data flow perspective, and discuss the critical issues related to the performance of bufferpool extension system. Based on our observations, we propose a cost-aware data management policy to reduce the overhead of I/O accesses. We also design a lightweight data recovery policy to minimize the negative impact of maintaining the mapping table.

3   System Architecture

The FASC design aims to exploit the high-speed random read of flash memory to enhance the performance of DBMSs. To support large-scale data processing, we take hard disk as the main storage medium. Figure 1 gives an overview of FASC, which employs a three-part storage architecture integrating main memory, SSD, and hard disk. We take SSD as the extension of DRAM due to its fast access speed. When the DRAM buffer is full, a page evicted from the DRAM buffer is admitted to be cached in the SSD buffer only when it is not present in the SSD. If the evicted page is dirty, the evicted page is cached in the SSD until it is written back to the disk. The *write-back* policy can reduce write traffic to disk if the dirty pages cached the SSD are frequently updated. Upon a buffer fault, FASC first checks the SSD to see if the requested page exists. If it does, FASC just reads it from the SSD. Otherwise, the request is forwarded to the disk.

3.1   Architectural Components

The main components of FASC are as the follows:

1)SSD buffer manager: The SSD buffer manager is in charge of data transfer between the SSD and main memory. Specifically, when the DRAM buffer receives a page request, if the requested page does not reside in main memory, FASC will invoke the SSD buffer manager to search the requested page in the SSD buffer. If the DRAM buffer has no available space, the SSD buffer manager is invoked to accommodate the victim pages evicted from main memory. With the increase of incoming evicted pages, the buffer manager needs to reclaim rarely accessed pages periodically.

2) Victim pages buffer: This is an in-memory data structure. It is responsible for accumulating the pages evicted by the DRAM buffer. If the evicted page is clean, the page is admitted into the victim pages buffer only when it is not resident in the SSD. The dirty pages are admitted to the SSD buffer. Note that if an old version of the evicted dirty page exists in the SSD, the old copy of data is marked as invalid. The invalid pages will be recycled by the cleaner thread. When the victim pages buffer is full, FASC flushes the victim pages to the SSD.

3) Temporary buffer: It is a temporary data structure used for migrating the dirty pages to the hard disk. The temporary buffer is sized to 8 times the page size. In the actual implementation, we use double buffers. Once one buffer is full, all the write threads will not be blocked but directed to the second buffer.

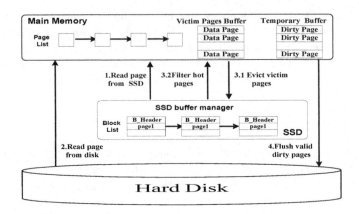

Fig. 1. FASC System Overview

4   Cost-Aware Buffer Management Policy

Buffer is one of the most fundamental component in modern storage systems. The goal of existing buffer replacement policies is to save the retrieval of data from disk by minimizing the buffer miss ratio. Unfortunately, this performance metric is ineffective on a SSD-based multilevel buffer scheme. This is because flash memory provides relatively slow performance for write, and the cost of evicting a dirty page is higher than that of evicting a clean page. This case indicates that maintaining a high hit ratio does not necessarily bring a higher I/O performance. We must revisit the replacement policies to make them fit well in SSD-based multilevel cache system. In fact, the memory pages present different states due to the complicated page flow, and the cost difference on evicting the page with various states is significant. Considering the inconsistency between cache hit ratio and I/O performance, we design a cost-aware replacement policy which is aware of the read/write characteristics of flash memory. Our cost model includes two parts: accessing a SSD page and writing a dirty page to the disk. Except for the latency for reading a page from the SSD, the cost of accessing a SSD page also includes the latency for writing a page to the SSD(if the page is not currently cached in the SSD). Considering that some dirty pages need to be migrated to disk, the latency for data migration is introduced in our cost model.

Based on the analysis of the state of the page, we classifies the memory pages into two types: *SSD-present* page and *SSD-absent* page. Here, a page which has the same copy in the SSD is called a *SSD-present* page, while a *SSD-absent* page represents the page that is not currently cached in the SSD. Considering the asymmetric cost of page migration, we further divide the *SSD-present* pages into two groups: the R_{rc} which keeps the clean pages and the R_{rd} which keeps the dirty pages. Because the R_{rd} page needs to be migrated to disk, its replacement cost is always expensive than that of the R_{rc} page. We also classify the *SSD-absent* pages into two types, say M_{mc} and M_{md}. M_{mc} keeps the clean pages while M_{md} keeps the dirty pages. Note that evicting a M_{md} page which has an invalid version in the SSD may incur an expensive data transfer overhead, since it may be updated or dirtied multiple times within the buffering period. To reduce write traffic to flash memory, it should have the priority to reside in main memory when executing data eviction. In the following discussion, the cost of reading a page from the SSD is C_r, while the cost of writing a page to the SSD is C_w. D_w represents the cost of fetching a page from the hard disk.

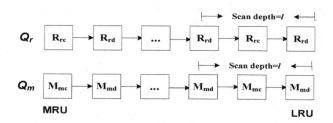

Fig. 2. Cost-based Memory Management

Based on the above analysis, we split the DRAM buffer into two LRU queues: Q_r and Q_m. As shown in Figure 2, Q_r is responsible to maintain the *SSD-present* pages, while Q_m is used to manage the *SSD-absent* pages. The size ratio of the two queues(i.e.,Q_r and Q_m) is dynamically adjusted using the Formula 1:

$$\beta = \frac{C(Q_r) + \theta * M(Q)}{C(Q_m) + \theta * M(Q)} = \frac{C_r + \theta * (C_r + D_w)}{C_r + C_w + \theta * (C_r + D_w)} \tag{1}$$

In Formula 1, the cost of fetching a *SSD-present* page, denoted by $C(Q_r)$, only needs to spend a SSD read; while for a *SSD-absent* page, we must write the page to the SSD first. So, the cost of fetching a *SSD-absent* page from the SSD is $C_r + C_w$(denoted by $C(Q_m)$). $M(Q)$ represents the cost of migrating a page to the hard disk(i.e., $C_r + D_w$). To ensure that the size ratio of the two queues can dynamically adapt to different workloads, we add a tuning parameter θ in Formula 1($0 < \theta < 1$). If the workload is update-intensive, there are a lot of dirty pages in main memory, which need to be merged to the disk. In this case, the FASC will reduce the value of θ to keep more memory for the Q_m list.

When the main memory becomes full, if the size ratio of Q_r and Q_m is greater than β, the FASC will select a page from Q_r as the victim page. The cost of searching the victim page in Q_r or Q_m is bounded by a parameter l called the scan depth. Specifically, if Q_r serves as the victim queue, FASC selects the R_{rc} page closest to the end of Q_r region as the victim page. If there is no R_{rc} page within a scan depth, the R_{rd} page closest to the end of Q_r will serve as the victim. If the victim queue is Q_m, the procedure of replacement is similar to the generation of victim page in Q_r. If we do not find a M_{mc} page within a scan depth, the M_{md} page residing in the end of list is selected as victim page.

5   SSD Data Management

Using flash memory as an extended cache is a cost-effective way to improve the performance of system. However, flash memory is not perfect due to its inherent characteristics, such as poor random write. To obtain substantial performance benefits, we need to carefully design the SSD data management policy to reduce the small random write traffic to flash memory.

5.1   Block-Based Data Admission Policy

The small random write pattern is not an ideal one for flash memory due to its erase-before-write limitation. Hence it must be reduced or avoided. The previous study has shown that flash memory can handle random writes with larger I/O request more efficiently[4]. So, we select a larger page size for the SSD buffer. Specifically, when the in-memory victim page buffer is full, all the pages residing in this buffer are gathered into a large data block and then written to the SSD buffer. To reduce the overhead of read operation, we provide an asymmetric I/O pattern. In FASC, each block consists of multiple data pages, and we employ a block header to summarize the pages distribution in the block. The read unit size is a data page, and we can get the requested page through checking its block header. The block-based data admission policy has advantages over a page level policy. First, the capacity of SSD is always much larger than that of the DRAM buffer. In order to keep track of the pages cached on SSD, the memory consumption for the in-memory data structures is very large. Using a large block size can save a lot of main memory by reducing the size of DRAM directory, which in turn increases the hit ratio. Second, using a large block can crease sequentiality in write traffic, which reduces write amplification and improves the performance of SSDs[11]. Third, our block-based policy can reduce the number of erase/write cycles on SSD, which indirectly extends the lifetime of the SSD.

5.2   Memory-Efficient Replacement Policy

As the incoming data blocks consume the space of SSD, the cold pages need to be migrated from the SSD to the disk. In order to find the rarely-accessed pages, the data replacement policy always has to maintain some in-memory data structures to record the access information for the SSD pages(e.g., recency or frequency). Considering the large capacity of SSD, if we directly apply the existing caching policies to the SSD buffer, the overhead of in-memory index is expensive. Therefore, we must provide a memory-efficient cache replacement policy for the SSD buffer.

Hot data identification: The FASC aims to use flash memory as buffer caches to speed-up I/O accesses. If we can identify frequently accessed pages and buffer them in the SSD buffer, the number of disk accesses can be greatly reduced. So, hot data identification plays a crucial role in improving the performance of SSD-based multilevel cache system. Most of existing hot data identification policies typically employ a locality-based algorithm. However, because all the accesses to the extended buffer are misses from the DRAM buffer, the second-level caches exhibit poor temporal locality. The previous work has shown that pages referenced more than twice contribute more accesses for the second-level buffer cache[12]. Replacement algorithm needs to give higher priorities for the pages that are accessed more frequently.

According to above analysis, we propose a memory-efficient hot data identification algorithm for the SSD buffer. Our hot identification method not only offers a tradeoff between temporal locality and access frequency, but also realizes a low memory consumption used for maintaining the access information. The thorough consideration of our policy makes it suitable for implementing a large SSD-based extended cache. Our hot page identification policy can be divided into two phases: victim block identification and hot page filtration.

(a) **Finding victim block:** In order to identify and retain the hot pages, the FASC dynamically maintains a hotness-based replacement priority for each block. When the SSD buffer is low in available space, the block with the lowest priority is selected as a victim block. For the block b, its replacement weight *replace_pri(b)* is computed according to Formula 2:

$$replace\_pri(b) = \frac{block\_fre(b) * \theta(t)}{invalidpage\_c(b)} * dirtypage\_c(b) \qquad (2)$$

In this formula, the function $block\_fre(b)$ denotes the access frequency of a block b. The aging function, denoted by $\theta(t)$, is used to decrease the replacement priority of a block that has not been accessed for a long time. The FASC dynamically adjusts the value of $\theta(t)$ by scanning the access information maintained in two time windows. We will further elaborate the two time windows in Section(b). The function $invalidpage\_c(b)$ represents the number of invalid pages in block b. If there exists many invalid pages in a block due to frequent updates, the FASC will assign a low priority for it according to Formula 2. $dirtypage\_c(b)$ denotes the number of dirty page in block b. The benefits of keeping dirty page in the SSD tend to be higher than that of keeping clean page.

(b) **Filtering hot page:** The disadvantage of block-based replacement policy is that some recently accessed pages of the victim block may be evicted from the SSD. We must pick recently accessed pages from the victim block. For this purpose, we need to keep track of the recency of data. In order to reduce the overhead of maintaining access information, we propose a window-based hot page identification policy for the SSD, where a window is defined as a predefined number of consecutive page requests. In this paper, the size of a time window is set to 4096. Unlike previous caching strategies, we only maintain the pages that are accessed within two time windows. For ease of presentation, both time windows are denoted as *Wpre* and *Wcur*, respectively. We maintain two bitmaps to mark the recency of the accessed pages during the two time windows. Each bit in this bitmap indicates whether the page has been recently accessed. A bitmap, called *Bcur*, is used to mark the recently accessed pages within the time window *Wcur*, and all the pages that are accessed during the time window *Wpre* are maintained in another bitmap(called *Bpre*). Periodically, we shift two bitmaps: discard *Bpre* and replace it with the current bitmap (i.e., *Bcur*), reset *Bcur* to capture the recency within the next window. All the recently accessed pages within two time windows will be found by scanning the two bitmaps. To facilitate fast lookups, we also employ a bloom filter to keep track of the blocks which are accessed within a time window. In this way, the search for a recently accessed page can be avoided when lookups are done on a non-existing block.

Algorithm 1. Evict_SSDPage()

Require: The SSD buffer *SSDBuffer*
Ensure: Evict some pages from *SSDBuffer*
 1: **if** the current time window is finished **then**
 2:     **for** each data block $B \in SSDBuffer$ **do**
 3:         **if** B is not found in *BFcur* or *BFpre* **then**
 4:             calculate the value of tuning parameter f;
 5:             decay the frequency of B;
 6: update the weight heap according to formula 2;
 7: find the victim block with the lowest priority *victimB*;
 8: **for** each page $bp \in victimB$ **do**
 9:     **if** bp can be found in two recency bitmaps **then**
10:         write bp to the victim pages buffer ;
11: write the cold dirty pages of *victimB* back to disk;
12: return;

Replacement Algorithm: We present the details of replacement policy for the SSD buffer in Algorithm 1. When a read request arrives, if the requested page hits the SSD buffer. FASC updates the recency of the page. But if the SSD buffer receives a write request, our replacement policy checks if the available space on SSD exceeds a predefined threshold. If so, FASC will evict a block out of the SSD to accommodate the incoming pages. To find a proper victim block, FASC performs an aging mechanism for the elder data blocks first, and updates the replacement priority according to Formula 2. The block with the lowest replacement priority serves as the victim block. Next, the FASC checks every page in the victim block and picks out recently accessed pages. The hot pages are retained in the victim page buffer to serve the subsequent requests.

6 Data Recovery for System Failure

In FASC, the dirty pages evicted by the DRAM buffer will be temporarily cached in the SSD. As a result, data consistency and recovery become a concern. Therefore, we also introduce a recovery policy that ensures the system can reuse the pages cached in the SSD during crash recovery. In fact, because the process of transaction recovery always incurs an intensive random read/write operations, the crash recovery can operate with a warm cache if the pages cached in the SSD are recovered. For this purpose, we always need to maintain an accurate SSD mapping table to keep track of the pages cached in the SSD. Considering the non-volatility of flash memory, the SSD buffer mapping table can be stored in the SSD. To guarantee the validity of the SSD mapping table, a simple approach is to flush, at all times, the updates that belong to the SSD mapping table to the SSD. Such scheme ensures that the SSD buffer manager can quickly recover the data cached in the SSD. However, the drawback of this approach is that it may generate a large amount of additional I/O traffic. In fact, the SSD mapping table may be updated frequently with the increase of incoming pages, and the SSD buffer has to spend an effort to flushing these updates. This may hinder the overall performance of transaction processing.

Based on the above analysis, we must design an efficient method to manage the SSD mapping table. In FASC, each mapping table entry includes block_ID, validation flag that indicates the page whether is valid or not, dirty flag, frequency information, some pointers such as next block pointer, and etc. The FASC organizes the mapping table into a series of fixed-size chunks, and each chunk consists of multiple flash pages. To reduce the overhead of hardening the mapping table, we assign an exclusive flash page for each chunk as the log page. Whenever the most recent update to some mapping table chunk is finished, the SSD buffer manager writes a log record describing the change to the mapping table to its log page. When a log page runs out of free storage space, it is merged with the associated mapping table chunk. Overwriting a mapping table record will most likely incur a random write. By combining the modifications of the mapping table chunk into a single batch update, we avoid a number of small random writes on flash memory, and reduce the negative effect on system throughput.

7   Performance Evaluation

In this section, we implement the FASC policy in PostgreSQL open source DBMS, and evaluate its performance using TPC-C and TPC-H benchmarks. Our experiments are run on a 2.00 GHz Intel Xeon(R) E5-26200 processor with 16GB of DRAM. Operating system (Ubuntu Linux with kernel version 3.2.0) is installed on a Seagate 15K RPM 146 GB hard disk. We use a 128GB Sumsung 840 Pro Series SSD as the second-level buffer cache.

7.1   Prototype Design and Implementation

We implement a prototype based on PostgreSQL, and perform a serial of experiments. A *SSD-buffer-manager* is added to the original buffer manager for PostgreSQL. It is an independent module which provides a rich interface layer to match the buffer manager. We modify the processing of buffer allocation and page eviction. The functions added to the prototype can ensure that all the evicted pages are written to SSD instead of hard disk. We also modify the *BufferAlloc* function, and add a *page flag logic* to indicate the state of a page. An *Adaptive_VictimBuffer* module is introduced in our prototype, which is responsible for searching a victim page in Q_r or Q_m. We also introduce a recovery component which ensures the durability of transactions in case of system crash. When the data cached in SSD exceeds a certain threshold, a cleaner module is in charge of reclaiming invalid pages. The block size of our policy is set to 1MB.

7.2   TPC-C Evaluation

In the experiments, we implement a throughput test according to the TPC-C specifications, and measure the number of new orders that can be fully processed per minute (tpmC). We run the benchmark with 500 warehouses. The size of database files is approximately 60GB. To achieve a steady-state throughput, each design runs for 1 hour. The PostgreSQL buffer pool size, including 512MB share memory, is set to 1GB. We vary the size of the SSD from 8GB and 16GB.

1) Transaction Throughput: To demonstrate the effectiveness of our cache policy, we compare FASC with two state-of-the-art SSD caching strategies—FaCE[10] and lazy cleaning(LC)[9]. In order to compare the performance, our experiments include the cases where the PostgreSQL is stored on an entire hard disk (denoted by disk-only) or an entire SSD (denoted by SSD-only). Figure 3 shows the transaction throughput achieved by different buffer extension designs. As the figure shows, except for the SSD-only case, our method outperforms all other designs. For example, with a 16GB SSD cache, we observed up to 3.78X improvement in throughput compared to the noSSD design(i.e.,Disk-only). In the following, we analyze the behavior of each caching design. The LC method uses LRU-2 replacement algorithm to manage the SSD pages, and the eviction operation always incurs many costly random write on flash memory due to overwriting a page. This results in a degradation in throughput. Under the FaCE design, although the FIFO-based data management policy ensures that all pages are written to SSD sequentially. However, the replacement policy is hard to capture the frequently accessed pages that contribute more accesses on the SSD. In addition, the multi-version method adopted by FaCE incurs a lot of invalid pages that are scattered over the entire SSD buffer. The cleaner module needs to reclaim the invalid data constantly, which in turn generates a lot of expensive write operations, and seriously aggravates the performance of SSD-based I/O processing. For a 16GB SSD buffer cache, the SSD-only case outperforms our caching design, by roughly 20%, although the database engine is running entirely on the SSD. Therefore, using SSD as the extended cache is a cost-effective and attractive solution to improve the performance of transaction throughput.

To gain a thorough comparative analysis of different SSD designs, we describe the average throughput with a 6 minutes interval, and present the cumulative distribution in terms of tpmC. As shown in Figure 4, our policy gains its peak faster than other policies. After 30 minutes, the performance of FASC slightly decreases, and an explanation for this behavior is that the accumulative pages evicted from main memory soon use up the available SSD space. Thus, the SSD has to consume a significant I/O throughput to handle the page eviction. We can see that the throughput provided by the LC design is very close to that of the FaCE design in the initial 20 minutes, while the performance gap between

LC and FaCE becomes more obvious after 36 minutes. This is because the LC design uses a LRU-2 replacement policy manage the pages cached in the SSD. Hence, evicting a page may incur an expensive random write.

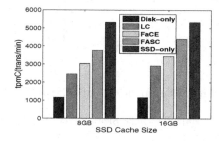

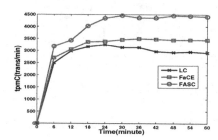

Fig. 3. Transaction Throughput   **Fig. 4.** Throughput Curves over Time

Table 1. SSD Write Ratio and Disk Access Reduction.

(Measured in %)	FASC	FaCE	LC
Ratio of SSD Writes	33.63	41.32	38.41
Disk Access Reduction	62.12	51.39	55.47

2) Memory-efficient SSD Data Management: To demonstrate that our SSD replacement policy is a memory-efficient method, we reduce the size of main memory assigned for PostgreSQL, and repeat the transaction throughput experiment using the same TPC-C instance described above. In this evaluation, 512MB of DRAM is dedicated to the DBMS buffer, and the share memory pool size is limited to 256MB. As we can see from Figure 5, the FASC shows the best performance with respect to transaction throughput, and a speedup of 3.8X is achieved in terms of the tpmC figures when the SSD buffer size is set to 16GB. In contrast, the performance efficiency of the LC design is the worst. We suspect that the main reason lies in the fact that implementing the LC design consumes too much memory, which in turn results in a low DRAM hit ratio. From the results of experimentation, we can see that the metadata overhead is closely related to the throughput of database. Our replacement policy designed for the SSD buffer offers an efficient hot pages identification with low memory overhead, and improves transaction processing performance dramatically.

3) I/O Traffic Reduction: Our cost-aware replacement policy aims to reduce I/O accesses to and from the hard disk, as well as reducing the write traffic to flash memory. In this section, we compare FASC with existing SSD-based caching methods with respect to I/O traffic reduction. As shown in Table 1, the ratio of the SSD write requests to the total SSD I/O requests is lower than those of other policies. An important reason is that our policy considers the asymmetric read/write cost of flash memory when evicting the victim pages, and tends to

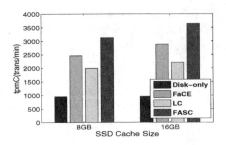
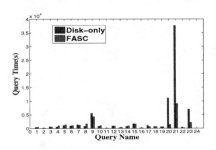

Fig. 5. Throughput with 512MB memory   **Fig. 6.** Concurrency stream(SF=30)

Table 2. Time Taken to Recovery the System(SSD Cache Size=16GB)

Restart Design	recovery time(second)	sustained throughput(tpmC)
FASC	189	4409
FASC_T	203	4122
FaCE	192	3445
Disk-only	847	1166

minimize the data transmission cost between main memory and SSD. In addition, our block-based SSD management policy reduces the memory consumption for maintaining mapping table, which improves the cache hit rate. The low write traffic to SSD contributes to the performance improvement of the FASC. We can see that FASC is approximately 6 to 10 percent higher than FaCE and LC in disk I/O reduction. The LC slightly outperforms the FaCE with respect to I/O traffic reduction. However, the FaCE achieves a higher transaction throughput than the LC method. We suspect that an important reason is that the FaCE adopts an append-only fashion to manage the SSD, which reduces the number of random writes on SSD. The I/O behavior dominated by random write degrades the performance of the LC method. These experimental results explain well why the FASC obtains higher throughput compared to other policies, and the I/O cost reduction is the major factor for performance improvement.

4) Performance of Data Recovery: To evaluate the recovery performance of FASC, we implement FASC with another recovery method(denoted by FASC_T). In FASC_T, an accurate SSD mapping table is stored in the SSD at all times. That is, any update on the mapping table is immediately written in the SSD. Because the LC design employs an in-memory mapping table, it cannot provide a warm SSD cache in case of system crash, and we omit the recovery times evaluation for this design. We compare the three caching designs concerning time taken to restart from a crash. As shown in Table 2, the FASC provides 4.5x speedup over Disk-only(the noSSD cache restarting design). This is because the SSD extension buffer provides a warm cache for DBMS, which accelerates the process of transaction recovery. The recovery times of FASC is similar to other SSD caching methods. However, we can see in Table 2 that the FASC obtains a higher sustained throughput than FASC_T. An important reason for this is that the FASC implements a light-weight method to harden the mapping table, and our policy reduces the SSD write traffic triggered by continuous updates for the SSD mapping table.

7.3 TPC-H Evaluation

In this section, we evaluate our SSD caching design by conducting experiments with TPC-H benchmark. In our evaluation, the scale factor is set to 30, which occupies a total of 70 GB of disk space. The query stream consists of 22 TPC-H queries and two update queries. As for the platform, we select 8GB SSD and 1GB main memory. We first run a power test by the TPC-H specifications. Then, we run a throughput test by running four query streams concurrently.

Table 3. Power and Throughput test

Scale Factor=30	Disk-only	FASC	FaCE	LC	SSD-only
Power test	336	442	409	374	589
Throughput test	94	316	232	209	402
QphH@30SF	177	374	308	280	487

1) Query performance evaluation: We run the TPC-H evaluation entirely on different designs. Table 3 shows the detailed results. In terms of the throughput test, FASC provides a speedup of 3.36X relative to the Disk-only. The throughput test always needs to handle multiple concurrent queries simultaneously, which creates a lot of random disk accesses. These random disk accesses degrade the system performance. Under the SSD caching scheme, part of the page requests that misses in DRAM can be serviced from the SSD, and thus considerable random disk accesses can be saved. The performance improvement is similar on three caching designs. This is because TPC-H is a read-intensive workload, while each design makes the best use of the excellent read performance of flash memory. The FASC is slightly better than FaCE and LC. An important reason is that our block level data management policy can effectively reduce the metadata overhead. In addition, our policy tends to keep the dirty page in the SSD, which reduces the transmission cost between SSD and disk.

2) Concurrency stream query : In this section, we look into the execution time of each query during the throughput test. As we can see from Figure 6, some queries achieve significant performance improvements on query latency. For example, for Q20 and Q21, the speedup is 8.1 and 4.2 respectively, We analyze the two queries closely, and find that both queries generate intensive random data accesses. Using the SSD buffer can significantly amortize the high cost of disk seeks, and shorten the executing time. It is worth noting that some queries are better than those of Disk-only(but no clear winner is found). This is because that the dominant I/O pattern of these queries is sequential read. In this case, considering the additional transmission costs between SSD and disk, there is no significant performance gain over the SSD caching scheme.

8 Conclusions

In this paper, we propose FASC, a SSD-based multilevel cache scheme, to improve the performance of DBMSs. Following this design principles, we propose a cost-aware replacement algorithm for main memory. To reduce the I/O traffic to the SSD, we implement a block-based data management policy for SSD. We develop a prototype system based on PostgreSQL. The experiment results show that FASC enjoys substantial performance improvements.

References

1. Do, J., Zhang, D.H., Patel, J.M., DeWitt, D.J.: Fast peak-to-peak behavior with SSD buffer pool. In: 30th International Conference on Data Engineering, pp. 1129–1140. IEEE Press, Brisbane (2013)
2. Koltsidas, I., Viglas, S.D.: Designing a Flash-Aware Two-Level cache. In: Eder, J., Bielikova, M., Tjoa, A.M. (eds.) ADBIS 2011. LNCS, vol. 6909, pp. 153–169. Springer, Heidelberg (2011)
3. Agrawal, N., Prabhakaran, V., Wobber, T., Davis, J.D., Manasse, M.S., Panigrahy, R.: Design tradeoffs for SSD performance. In: 2008 USENIX Annual Technical Conference, pp. 57–70. USENIX Association, Boston (2008)
4. Bouganim, L., Jnsson, B.T., Bonnet, P.: uFLIP: Understanding flash IO patterns. In: Online Proceedings of the 4th Biennial Conference on Innovative Data Systems Research, pp. 1–12, Asilomar (2009)
5. Chen, S.M.: FlashLogging: exploiting flash devices for synchronous logging performance. In: ACM SIGMOD International Conference on Management of Data, pp. 73–86. ACM Press, Rhode Island (2009)
6. Tsirogiannis, D., Harizopoulos, S., Shah, M.A., Wiener, J.L., Graefe, G.: Query processing techniques for solid state drives. In: ACM SIGMOD International Conference on Management of Data, pp. 59–72. ACM Press, Rhode Island (2009)
7. Lee, S.W., Moon, B., Park, C., Kim, J.M., Kim, S.W.: A case for flash memory ssd in enterprise database applications. In: ACM SIGMOD International Conference on Management of Data, pp. 1075–1086. ACM Press, Vancouver (2008)
8. Canim, M., Mihaila, G.A., Bhattacharjee, B., Ross, K.A., Lang, C.A.: SSD Bufferpool Extensions for Database Systems. Proceedings of the VLDB Endowment 3(2), 1435–1446 (2010)
9. Do, J., Zhang, D.H., Patel, J.M., DeWitt, D.J., Naughton, J.F., Halverson, A.: Turbocharging DBMS buffer pool using SSDs. In: ACM SIGMOD International Conference on Management of Data, pp. 1113–1124. ACM Press, Athens (2011)
10. Kan, W.H., Lee, S.W., Moon, B.: Flash-based Extended Cache for Higher Throughput and Faster Recovery. Proceedings of the VLDB Endowment 5(1), 1615–1626 (2012)
11. Hu, X.Y., Eleftheriou, E., Haas, R., Iliadis, I., Pletka, R.: Write amplification analysis in flash-based solid state drives. In: Israeli Experimental Systems Conference 2009, pp. 82–90. ACM Press, Haifa (2009)
12. Zhou, Y.Y., Chen, Z.F., Li, K.: Second-Level Buffer Cache Management. IEEE Transactions on Parallel and Distributed Systems 15(6), 505–519 (2004)

A Cost-Aware Buffer Management Policy for Flash-Based Storage Devices

Zhiwen Jiang[(✉)], Yong Zhang, Jin Wang, and Chunxiao Xing

RIIT, TNList, Department of Computer Science and Technology, Tsinghua
University, Beijing, China
{jiangzw14,wangjin12}@mails.tsinghua.edu.cn,
{zhangyong05,xingcx}@tsinghua.edu.cn

Abstract. Flash devices has become an important storage medium in
enterprise hybrid storage systems. Buffer manager is a central component
of database systems. However, traditional disk-oriented buffer replacement strategies are suboptimal on flash memory due to the read-write
asymmetry. In this paper, we propose a cost-aware buffer management
policy CARF for flash memory. We devise a novel cost model with low
computational overhead to make more accurate decisions about page
eviction. Moreover, this cost model can distinguish read and write operations as well as have better scan resistance. Experiments on synthetic
and benchmark traces show that CARF achieves up to 27.9% improvement than state-of-art flash-aware buffer management strategies.

Keywords: Buffer management · SSD · Cost-aware

1 Introduction

With dropping cost and increasing capacity of flash device, flash-based Solid
State Disks (SSD) have become an important storage medium in enterprise
hybrid storage systems. In the hybrid storage systems, the performance-critical
applications and data can be placed in the SSD to improve the performance.
Compared with hard disk drives, flash memory has two unique characteristics,
read-write asymmetry and erase-before-write mechanism: once a page is written, the only way of overwrite it is to erase the entire block where the page
resides. The average cost of erase operation is one to two orders of magnitude
higher than read operation. Besides, flash memory blocks have limited erase
cycles (typically around 1,000,000). To minimize the cost of erase operation, a
software layer called Flash Translation Layer (FTL) is implemented. Its main
purpose is to perform logic block mapping, garbage collection and wear leveling
[6]. The distinguished I/O characteristics of flash memory make it important
to reconsider the design of I/O intensive software, such as DBMS, to achieve
maximized performance.

© Springer International Publishing Switzerland 2015
M. Renz et al. (Eds.): DASFAA 2015, Part I, LNCS 9049, pp. 175–190, 2015.
DOI: 10.1007/978-3-319-18120-2_11

Buffer Manager is the central component of database systems as it narrows down the access gap between main memory and disk. Traditional disk-oriented buffer replacement strategies utilize the temporal locality of page requests to reduce the number of disk accesses. So the primary goal of traditional buffer management policy is to minimize miss rate. However, minimizing miss rate doesn't necessarily lead to better I/O performance on flash memory due to its inherent read-write asymmetry. An early previous work [5] has shown that whether a page is read only or modified should be taken into consideration when designing buffer management policy. This principle is more important for flash memory. Since the cost of write operation is much higher, the cost of evicting a dirty page is correspondingly more expensive. Thus, we should use total I/O cost rather than hit rate as the primary criterion to measure the effectiveness of a buffer management policy on flash-based storage devices.

A number of previous studies have focused on designing buffer management policy to address the I/O asymmetry on flash devices [16,19–21]. What makes them flash-aware is that they try to keep dirty pages in the buffer and reduce the number of expensive write operations. The basic idea of them is to use page state to determine victim pages for eviction. The state of a page includes whether a page is modified (clean/dirty) and whether a page is frequently or recently referenced (hot/cold). Intuitively, the primary goal is to keep hot dirty page in the buffer and evict cold clean pages as soon as possible. However, all above policies depend highly on LRU algorithm, which fail to take the frequency of page access into consideration. When the page access pattern is with weak locality, these LRU-based algorithms would be suboptimal. Besides, previous work [5] has shown that LRU algorithm has a problem of weak scan resistance on disk. This problem will still exist on flash-based storage devices. In addition, most of the previous flash-aware algorithms propose heuristics-based approaches to determine the victim page. It is difficult for them to decide the priority of cold dirty and hot clean pages. Therefore, we use accurate weight instead of page state to select victim page. To out best knowledge, there is no previous work using cost-aware replacement policy except FOR [16]. FOR and its approximate version FOR+ has devoted to provide a page weight metric considering both page state and future operations to select victim page. But its page weight metric brings too much computational overhead as is shown in our experiment.

To address these problems, we propose a cost-aware buffer replacement policy, named **C**ost-aware **A**lgorithm combining **R**ecency and **F**requency (**CARF**). CARF uses a novel cost model to select victim page according to the accurate cost instead of ad hoc way. This cost model is with low computational overhead so that it is practical to be applied. We implement the light-weight data structure according to the cost model.

In this paper we present the following contributions:

- We propose a cost-aware buffer management policy CARF for flash memory that selects victim page for eviction according to accurate page weight.
- We design a novel cost model as the weigh metric for CARF considering both frequency and recency of previous page access.

– We implement CARF with low computational overhead and good ability of scan resistance.

We evaluate our design with trace-driven simulation experiments on a flash chip and a representative SSD. We compare CARF with the traditional disk-oriented buffer management policies LRU and state-of-the-art flash-based algorithms. Experimental results show that CARF has much better scan resistance than LRU-based algorithms. Besides, CARF achieves up to 27.9% improvement of I/O cost than all the other flash-aware algorithms.

The rest of the paper is organized as follows: Section 2 describes the problem definition. Section 3 presents the cost model of CARF. Section 4 presents the implementation of CARF. Section 5 presents the results of evaluation. Related work is introduced in Section 6. Finally, conclusions are drawn in Section 7.

2   Problem Definition

In this section, we briefly introduce the preliminary about our work and some problems to be solved. The notations used throughout this paper are stated in Table 1.

Table 1. Notations used in this paper

Notations	Description
C_r/C_w	The average cost of one read/write operation
N_r/N_w	The total number of read/write operations
R	The IO asymmetry factor of flash memory
m	The buffer size
N	The working set of pages in the DBMS
t_i	The ith request time of a given page
x_i	The ith time span of a given page
t_c	The current global time

2.1   I/O Model for Flash Devices

Since average I/O cost is the weight metric for buffer management policies on flash memory, we will firstly look at the I/O model for flash memory. The traditional two-level model in [1] is designed to simulate the data transfer between main memory and disk. However, due to the characteristics of I/O asymmetry, the traditional model no longer works on flash memory. Here we modify the model in [1] and offer an I/O model for flash memory. Suppose in one read operation, B_r continuous blocks transfer from disk to main memory. In one write operation, B_w continuous blocks transfer from main memory to disk. Since a write operation costs more time than a read operation, we have $B_r = R * B_w$, where R is the asymmetry factor. Therefore, we have the following formula to calculate total I/O cost according to this model:

$$Cost_{io} = C_r * N_r + C_w * N_w = C_r * (N_r + R * N_w) \,. \tag{1}$$

2.2   Combine Frequency and Recency

According to previous studies [5,16], a good buffer management policy should take both recency and frequency of page access into consideration to take advantage of the spatial and temporal locality. This is more important on flash memory since the access pattern significantly influences the I/O performance of flash-based devices [2].

Now the key issue is how to make full use of the known information to determine the victim page for eviction. To achieve this goal, we build a cost model to measure the "hotness" of each page in the buffer, and then design a buffer replacement algorithm combining the flash-aware I/O model and this cost model to make page replacement decision. Here are the notations to be used in the following discussion:

- *Request Sequence:* A sequence of requested pages, denoted as $\sigma = \sigma(1), \sigma(2),$ $\cdots, \sigma(n)$.
- *Global Time:* An integer value that shows the number of requested pages since the system starts.
- *Time Set:* Suppose there are k requests for page p in the request sequence, the ith request is at global time t_i^{p}. Then we define the time set of page p as $T = \{t_1^{\text{p}}, t_2^{\text{p}}, \cdots, t_k^{\text{p}}\}$. The time set reflects the frequency of request to a page.
- *Time Span:* Suppose the time set of page p is $T = \{t_1, t_2, \cdots, t_k\}$. Then the time span of t_i is the length of time interval between the request time t_i and current time, denoted as $x_i = t_{\text{c}} - t_i (1 \le i \le k)$. The time span reflects the recency of a page request.

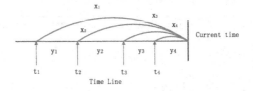

Fig. 1. An example of notations for page weight calculation

A specific example of these notations is shown in Fig. 1. The notation y_i denotes the time interval between t_i and t_{i+1}. Using these notations, we define the concept of page weight as follows to build our cost model:

Definition 1. *(Page Weight) Page Weight, denoted as $PW_t(p)$, is a value associated with each page in the buffer at the current time. It reflects the likelihood that the given page is requested again in near future.*

We use the definition of **weight function W(x)** to calculates the contribution of the time span x to the page weight at the current time. And then we can calculate the weight of page p using the information of *time set* and *time span*. To decide the frequency of page access, we should consider each request in the time set to calculate the weight. Suppose the time set of page p is $T = \{t_1, t_2, \cdots, t_k\}, (t_1 < t_2 < \cdots < t_k \le t_c)$, its *page weight* can be calculated as formula (2).

$$PW_t(p) = \sum_{i=1}^{k} W(x_i) = \sum_{i=1}^{k} W(t_c - t_i) . \tag{2}$$

3   Cost Model

In this section, we present a cost model named *RF Model* to calculate the page weight. RF Model defines the basic weight metric for CARF. For RF Model, we don't take read-write asymmetry into account.

3.1   Requirements for Cost Model

With the definitions in Section 2, we can easily build our cost model. When a buffer fault occurs at current time t_c, we just need to calculate the page weight for each page in the buffer and evict the one with least weight for replacement. We need to carefully choose the weight function, as it determines the ratio of contribution that recency and frequency makes to page weight. A good weight function should have the following properties:

Firstly, $\forall x \in N, W(x) > 0$. The value of weight function should always be positive to have practical meanings.

Secondly, $\forall x \in N, W(x)' \le 0$. The weight function should be monotonically non-increasing since the longer time span of page means that the page is less likely to be requested.

Finally, $W(0) = c$, where c is a constant. When time span is 0, the value of weight should be a constant as a baseline for comparison.

3.2   Page Weight Calculation

The formula (2) seems to perfectly fulfill the task of calculating the page weight. However, it is unpractical to implement such an algorithm because it brings too much overhead to maintain the required information. On one hand, for a given page with a time set of k requests, the space overhead to record all the requests is $\mathcal{O}(k)$. On the other hand, it needs to calculate the weight for each time span, which incurs $\mathcal{O}(k)$ computational overhead. In order to minimize both space and computational overhead to calculate the page weight, we need to make more limitation of the weight function $W(x)$. One feasible solution is to use the idea of recurrence, in which we can calculate the page weight at time t_i using the result at time t_{i-1}. In this way, we can reduce both the computational and space overhead to $\mathcal{O}(1)$ for one calculation. Thus, we define a property of recurrence of $W(x)$ as is shown in Lemma 1:

Lemma 1. *If the weight function $W(x)$ satisfies the property: $\forall x, y \in N, W(x + y) = W(x) * W(y)$, the weight of page p at the ith time p is requested, denoted as $PW_t(p)$, can be derived from the weight $PW_{t_{i-1}}(p)$ at the $(i-1)^{\text{th}}$ time p is requested.*

From the special case in Lemma 1, we can easily conclude a more general case:

Lemma 2. *We can derive $PW_{t_i}(p)$ from $PW_{t_{i-1}}(p)$ using the following formula with $y_i = t_i - t_{i-1}$:*

$$PW_{t_i}(p) = W(0) + W(y_i) * PW_{t_{i-1}}(p)$$

Combine the requirements and Lemma 1, we can find the following function that satisfies the requirements:

$$W(x) = a^x, (0 < a \leq 1) . \tag{3}$$

Notice that when $a = 1$, $W(x) = 1$ for any time span x. In this case, the recency of page request makes no contribution to weight calculation and RF Model becomes LFU; when $a \in (0, \frac{1}{2}]$, the weight function $W(x)$ satisfies the following property: $\forall i, W(i) > \sum_{j=i+1}^{\infty} W(j)$. In this case, the frequency of page access makes no contribution to weight calculation and RF Model is equal to LRU. Thus when $a \in (\frac{1}{2}, 1)$, RF Model takes both recency and frequency into consideration to calculate the page weight. Therefore, we can conclude that RF Model combines recency and frequency, which could make better decision for CARF on page eviction than other flash-aware algorithms [19–21].

4   Details of Implementation

In this section, we will introduce our buffer management policy, named CARF, which stands for Cost-aware Algorithm combining Recency and Frequency. We make CARF flash-aware by combing the RF Model with the I/O model for flash memory.

4.1   Combing the two models

We have developed two models in above sections: An I/O model for flash memory to consider the read-write asymmetry and a cost model to estimate the page weight according to the past request sequence. We implement the buffer management policy by combining the two models.

To consider the read-write asymmetry of flash memory when calculating the page weight, we need to take the page state into account, like previous studies did. If a page is clean at its last request time but at current time it is modified, it will make greater contribution to the page weight. Therefore, we need a tag named *Lastdirty* to keep the page status of the given page at its last request. Finally, we define a function named *asymmetry function* to represent the influence of read-write asymmetry on the page weight calculation. The value of the asymmetry function at time t for page p is denoted as:

$$R_{t_c}(p) = \begin{cases} R, \; lastdirty(p) = false \wedge \text{ modified at } t_c \\ 1, \; \text{otherwise} \end{cases}$$

Based on above consideration, we need a histogram to record some necessary information to help calculate the page weight of a page in $\mathcal{O}(1)$ time. The histogram of page p is updated when p is requested. Parameters contained in the histogram are as follows:

- *Lasttime(p)*: the time of last request for page p.
- *Lastweight(p)*: the weight of page p at the last time it is requested.
- *Lastdirty(p)*: whether page p is modified at the last time it is requested.

Therefore, with the consideration of read-write asymmetry the *flash-aware weight* of page p in the buffer at current time t_c can be calculated as:

$$FPW_{t_c}(p) = R_{t_c}(p) * [W(0) + W(t_c - Lasttime(p)) * Lastweight(p)] . \quad (4)$$

When buffer fault occurs, we just need to select the page with minimum *flash-aware page weight* in the buffer for eviction according to formula (4). This weight metric can also solve the problem of determining the priority of cold dirty and hot clean pages left by previous studies. The information needed for calculation only includes the last request instead of all the former requests for one page, so the space overhead of CARF is $\mathcal{O}(1)$ for each page.

However, it is not proper to directly use formula (4) when buffer fault occurs because the weights of all pages will be updated after each time step. Fortunately, we have the following lemma due to the property of $W(x)$:

Lemma 3. *For page p and q, if $FPW_{t_c}(p) > FPW_{t_c}(q)$ and neither of p nor q are requested after t_c, then for $\forall t > t_c, FPW_t(p) > FPW_t(q)$ holds.*

Since the relative order between pages will not change if neither of them is requested, we just need to update weight of the page that is requested and select the page with minimum *Lastweight* as the victim page.

4.2   Data Structure and Algorithms of CARF

Intuitively, since CARF makes the buffer replacement decision according to the order of pages by the value their weights, it can be implemented with a priority queue to maintain the order of pages in the buffer. As CARF selects the page with minimum weight when a buffer fault occurs, we can just use a min-heap to implement CARF. Each time when buffer fault occurs, the root page of this heap is selected as victim and replaced with the new page. Then the heap is adjusted top-down to maintain the order of *Lastweight*. Each time when a page in the heap is hit, the weight of the hit page is updated and the sub-heap with this page is adjusted top-down. In this way, it costs $\mathcal{O}(1)$ time to select victim page and $\mathcal{O}(\log m)$ time to maintain the structure of min-heap. The logarithm time complexity of CARF is the same as LFU, but is suboptimal than LRU-based algorithms. Therefore, we need to reduce the time complexity while taking the page weight into consideration.

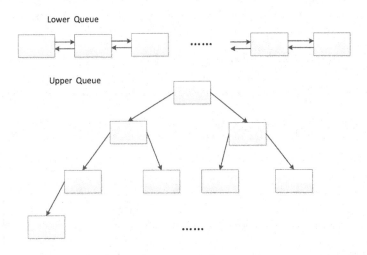

Fig. 2. Data structure of CARF

We can solve this problem by doing competitive analysis of our RF model, which is stated in Theorem 1:

Theorem 1. *When $a = 1$, RF Model is not competitive; when $a \in [\frac{1}{2}, 1)$, RF Model is $m - 1 + \lceil \frac{\log(1-a)}{\log a} \rceil$-competitive.*

Details of the proof is omitted here due to space limitation, they are shown in the full version [8]. Accordingly, we can get Theorem 2.

Theorem 2. *Suppose the most recent request for page p is at time t. There exists a minimum value $T_{min} = \lceil \frac{\lg(1-a)}{\lg a} \rceil$. Once $t_c - t > T_{min}$, $PW_{t_c}(p) < 1$ holds no matter how large the value of $PW_t(p)$ is.*

Then we can take advantage of Theorem 2 to solve this problem. According to the definition of page weight, when a page is fetched after buffer fault occurs, its weight equals to $W(0) = 1$. So we can conclude from Theorem 2 that if a page has a weight larger than 1, the most recent request for it must be within T_{min} time units. So we only need to maintain T_{min} pages in the priority queue. As is shown in Fig. 2, we can just divide the buffer into two parts to reduce computational overhead: the *Upper Queue* is a min-heap with maximum capacity of T_{min}; the

---

Algorithm 1. Read Buffer

     Input: p: The requested page
     Output: The content of page p
1 **begin**
2 | glbtime += 1;
3 | pos=lookup(p);
4 | **if** *pos!=INVALID* **then**
5 | HitTrigger(p);
6 | return pool[pos];
7 | **if** *buffer pool is not full* **then**
8 | Allocate a free slot for p at the position of pos;
9 | **else**
10 | Select victim page from the tail of L and put page p in this slot;
11 | **if** *victim page is dirty* **then**
12 | write victim page out to disk;
13 | update the information of bufferpool;
14 | MissTrigger(p);
15 | return pool[pos];
16 **end**

Lower Queue is a list of remaining $m - T_{\min}$ pages managed in a LRU way. Thus, when the Upper Queue needs to be adjusted top-down, the time complexity of maintaining the whole structure will be reduced to $\mathcal{O}(\log T_{\min})$. And the value of a can be chosen empirically.

However, Theorem 2 no longer holds on flash memory because of the *asymmetry function*. According to formula (4), the weight of a dirty page in the Lower Queue may be larger than 1. For simplicity, we just make an approximation according to Theorem 2: a pointer is maintained in the Lower Queue which points to the most recent used dirty page, denoted as p_{d}. This pointer will be updated each time a page request is served. If there is not dirty page in the Lower Queue, the pointer will be set as the head of the Lower Queue. If the weight of p_{d} is larger than one at a buffer fault or larger than the page being hit in the Lower Queue, the root page of Upper Queue will be replaced by p_{d}. In this way, although we don't guarantee all the pages in Lower Queue are strictly sorted according to their weight, we can move as many pages with larger weight as possible to the Upper Queue so as to improve the accuracy of our algorithm.Thus this approximation will not have obvious influence on CARF algorithm. The time complexity of CARF depends on the value of a, varying from $O(1)$ to $O(\log \lceil \frac{\log(1-a)}{\log a} \rceil)$, and the value of a can be chosen empirically.

The algorithm to answer a read request is shown in Algorithm 1. Upper Queue and Lower Queue are denoted as U and L in following algorithms. The function *lookup* [1] is used to check the position of a given page in the buffer. If the page doesn't exist, it will return INVALID. The process to respond to a write request is similar to Algorithm 1. The only difference is to mark the written

---

[1] Its time complexity is $\mathcal{O}(1)$ using hash table.

page as dirty and update its *Lastdirty* information in the histogram. So we omit it here due to space limitation.

Once a page is hit, the *HitTrigger* function is needed. At first, the histogram of the requested page is updated. We calculate and update the values of Lastweight(p) and Lasttime(p). Then the data structure of CARF should be adjusted. If the page is in Upper Queue, it just to adjust the sub-heap rooted by the page. Otherwise, the page with the largest weight is first selected from Lower Queue. Then it is compared with the root page of Upper Queue. If the selected page has larger weight, the root page will be replaced by it and the whole Upper Queue needs to be adjusted top-down. Otherwise, we just need to put the selected page at the head of Lower Queue.

Once a buffer fault occurs, the *MissTrigger* function will be called. Its process is similar to that of the *HitTrigger* function. One thing we should notice is that once a page is evicted from the buffer, its histogram will be deleted. If the same page is fetched into buffer again, it will have a new histogram. In this way, the space overhead of histogram can be limited to $\mathcal{O}(m)$. The details of *MissTrigger* and *HitTrigger* are shown in the full version [8].

5   Evaluation

In this section, we evaluate CARF with a trace-driven simulation using both synthetic and benchmark workloads. All the following experiments are set up on a server machine powered by a 2.40GHz Intel(R) Xeon E5620 CPU with 2 cores on Windows Server 2008 with 32GB RAM. We use two kinds of flash devices: a 16GB SamSung SD card (DCJH251GE337) and a 240GB SamSung 840 Series SSD. We tested the latencies of both devices and showed them in Table 2.

Table 2. The average latencies of flash-based devices

Block size: 8KB	SSD	SD Card
Sequential Read	$33\mu s$	$202\mu s$
Sequential Write	$46\mu s$	$284\mu s$
Random Read	$265\mu s$	$1.5ms$
Random Write	$90\mu s$	$5ms$

5.1   Experiment Setup

To evaluate the performance of CARF, we compared it against several state-of-art buffer management policies (LRU, FOR+ [16] and FD-Buffer [17] under synthetic workloads and three realistic benchmark workloads: TATP[2], TPC-B[3], and TPC-C[4]. There are also other algorithms like CFLRU [21], CASA [19],and

---

[2] http://tatpbenchmark.sourceforge.net/

[3] http://www.tpc.org/tpcb/

[4] http://www.tpc.org/tpcc/

CFDC [20]) But as previous studies [16, 17] has shown that FOR+ and FD-Buffer outperforms them, we don't compare our policy with them. All the baselines are well tuned according to the referenced papers. For TPC-C benchmark, the number of warehouses is set to 20. For TPC-B benchmark, we use 150 branches for test. And the number of subscribers in TATP is set to 1 million. Details of workloads are shown in Table 5.1.

Table 3. Details of traces in the experiment

	Database size (GB)	# of referenced pages (million)	Write ratio
TATP	0.4	2.5	4.8%
TPC-B	2.2	12.7	3.6%
TPC-C	2.4	16.8	16.3%
Synthetic	4	22	20%-40%

To get the page access traces for simulation, we ran each benchmark on PostgreSQL 9.3.1 with default setting (the page size is 8KB). We ran the test for around 3 hours for each benchmark as previous study did [16]. To make simulation, we use the collected traces above the postgreSQL buffer pool as input and record the response time under each buffer management policy. We implemented all the buffer management policies on top of OS file system. We disabled the buffer functionality of operating system using Windows API in order to avoid the disturbance from system buffer. We also made more experiments to test the impact of parameter a, please see [8] for more details.

5.2 Results on Synthetic Workload

We first show the performance of CARF under two synthetic workloads. To demonstrate that CARF outperforms LRU-based algorithms under workloads with weak locality, we tested two synthetic workloads SCAN and Mixed Skew. We ran the experiments several times and reported the best result. The description of the two workloads is as follows:

– SCAN: We generate a synthetic page access trace conforming to Zipf distribution with the write ratio of 40%. After a particular amount of time, table scans are periodically injected into the above trace. This workload is used to test the scan resistance of each policy. Since file scanning is a common operation in database system, scan resistance is important for buffer management policy.
– Mixed Skew: The Mixed Skew workload is a sequence of page requests conforming to Zipf distribution, which is a common distribution in realistic occasions. The ratio of write operation is set to 20% as previous work did [16].

We only report the results with the buffer size 256 for SCAN workload because as buffer size becomes larger, the results have the same trend. In case of smaller buffer size, all six algorithms have similar behavior. The reason is when buffer size is smaller than T_{min}, the size of Lower Queue in CARF is 0;

Table 4. Results of SCAN workload

	CARF	LRU	FD-Buffer	FOR+
Time Cost	2224	2807	3346	2395
Benefit		24%	50.4%	7.7%

and the scan resistance of CARF is the same as LRU. As we can see from Tab. 4, CARF has better scan resistance than other algorithms. It is about 10.5% to 50.4% better than other competitors under SCAN workload. This is because other LRU-based algorithms only take recency of page request into account. And CARF considers frequency of page request at the same time. Besides, the Lower Queue of CARF can also help filter out pages that are only requested one time.

The result of the skew workload is shown in Fig. 3(a). With the help of its cost model, CARF is able to utilize "deeper" information in the histogram. Thus, CARF has a better performance under the skew workload in most cases as well.

5.3   Results on Benchmarking

In this section, we evaluate CARF with its competitors under benchmark workloads. The experiments are run on both the SD card and SSD. Because results of the two experiments have similar trends, we only report the results on the SD card here due to page limitation.

Figure 3(b) shows the I/O time of different algorithms under TATP workload. When the buffer size is larger than 64, the performance of CARF is comparable with and finally outperforms other algorithms. Figure 3(c) shows the I/O time of each algorithm under the TPC-B workload. The experimental result is similar with the result under the TATP benchmark.

Figure 3(d) is about the results under TPC-C workloads. From these figures, we observe that CARF has the best performance. It achieves up to 27.9% gains over other flash-aware algorithms. In most cases, CARF has on average around 10% better performance than other baselines. Under the TPC-C benchmark, the performance of flash-aware buffer management policies is much better than that of TATP and TPC-B benchmark. For all the three workloads, the write locality is very well. So the write clustering technique in FD-Buffer fails to play a significant role in improving write performance.

The primary reason for CARF to outperform other flash-aware algorithms is that CARF can accurately define the priorities of pages in the buffer using its cost model. Thus, CARF could avoid the problem that cold dirty pages remain in buffer for a long period time. FOR+ solves this problem by moving some hot clean pages into its priority region. But dirty pages in its priority regions may also face similar problems. FD-Buffer divides the buffer pool into two sub-pools according to page state. Pages in the clean pool will be evicted first. However, some cold pages in the dirty pool will stay in buffer for a long period of time and lead to unnecessarily high miss rate. This phenomenon will also influence the adaptive mechanism for adjusting pool sizes.

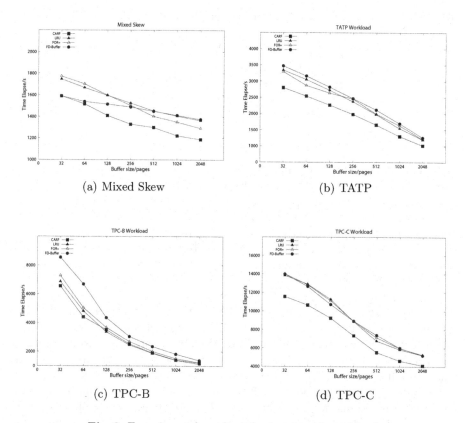

Fig. 3. Experimental result under benchmark workload

5.4   Evaluate CPU Time

In this section, we examine the computational time for each algorithm. The result of average computational time per operation under three benchmark workloads is shown in Table 5. We can see that LRU is surely to be with least computational overhead. Compared with other algorithms, CARF has slightly more computational overhead. The reason is that CARF needs to adjust its Upper Queue, which is a procedure with $\mathcal{O}(\log T_{min})$ time complexity. But this overhead is negligible compared with the saving in I/O time. So its computational overhead is acceptable. Moreover, the computational overhead of CARF relies only on the size of its Upper Queue. FD-Buffer needs to dynamically adjust the sizes of the two pools, so it also involves heavy computational overhead. Unlike other flash-aware buffer management policies, it will not increase drastically along with the buffer size. So the computational time of CARF has good scalability as well.

Table 5. The average computation time per operation (μs, Buffer size:1024 pages)

	LRU	FD-Buffer	FOR+	CARF
TATP	0.84	2.13	3.66	1.58
TPC-B	0.79	1.35	2.87	1.53
TPC-C	0.90	1.86	3.84	1.42

6   Related Work

Data management over flash memory has attracted much attention in recent years. A large number of studies have been proposed to deal with the unique characteristics of flash memory. Lee et al. [12] proposed the In-Page Logging to optimize write performance. Tsirogiannis et al. [22] make optimization of query processing in flash-aware DBMS. Li et al. [13] proposed a tree index to address the read-write symmetry of flash memory. Debnath et al. [3] designed a key-value store on flash-based storage devices. Li [14] optimized the efficiency of join processing on flash memory.

Buffer management is also a fruitful research area. Traditional buffer management policies devote to maximize the hit rate so as to get optimal IO performance. The most widely used policy is LRU, which chooses the least recently used page for eviction. Some variants of LRU [7,11,18] are proposed to combine the advantage of frequency and recency. Achieving better scan resistance [9] and lower computational overhead are also considered in previous studies.

For buffer management policy on flash memory, the primary goal is to address the read-write asymmetry. They can be divided into three kinds: embedded [10], buffer extensions [4,15] and page state [16,19–21]. BPLRU [10] is a buffer management policy for embedded system with limited resources. Policies in [4] and [15] are designed for hybrid storage systems which use flash memory as extensions of buffer pool, while the storage system uses only flash memory as external memory. Page state is crucial for addressing read-write asymmetry. CFLRU [21],CFDC [20],CASA [19] are heuristic-based methods using page state to select the victim page on flash memory. FOR [16] is a cost-based buffer replacement policy. It considers the spatial and temporal locality of read and write operation and selects victim page based on cost calculation. FD-Buffer [17] is a latest buffer management policy on flash memory. FD-Buffer separates the buffer into two pools and use independent management policy for each pool. It has an adaptive scheme to adjust the size of the two pools. The management policy of the two pools is independent from each other.

7   Conclusion

In this paper, we proposed CARF, a cost-aware buffer management policy on flash-based storage devices. We build a cost model to select victim page on the basis of page weight calculation. We also make competitive analysis of this cost model and prove that CARF combines frequency and recency from a theoretical view. We utilize the conclusion from the above analysis. Experimental study on both synthetic and benchmark workloads shows that CARF outperforms state-of-art flash-aware algorithms with 9.8-27.9% improvements.

Acknowledgements. Our work is supported by National Basic Research Program of China (973 Program) No.2011CB302302, the National High-tech R&D Program of China under Grant No. SS2015AA020102, Tsinghua University Initiative Scientific Research Program.

References

1. Aggarwal, A., Vitter, J.S.: The input/output complexity of sorting and related problems. Comm. of ACM **31**(9), 1116–1127 (1988)
2. Bouganim, L., Jonsson, B., Bonnet, P.: uflip: Understanding flash io patterns. In: CIDR (2009)
3. Debnath, B.K., Sengupta, S., Li, J.: Skimpy stash: ram space skimpy key-value store on flash-based storage. In: SIGMOD, pp. 25–36 (2011)
4. Do, J., Zhang, D., et al. J.M.P.: Turbocharging dbms buffer pool using ssds. In: SIGMOD, pp. 1113–1124 (2011)
5. Effelsberg, W., Harder, T.: Principles of database buffer management. TODS **9**(4), 560–595 (1984)
6. Gal, E., Toledo, S.: Algorithms and data structures for flash memories. ACM Computing Survey (2005)
7. Jiang, S., Zhang, X.: Lirs: an efficient low inter-reference recency set replacement policy to improve buffer cache performance. In: SIGMETRICS, pp. 31–42 (2002)
8. Jiang, Z., Zhang, Y., Wang, J., Xing, C.: A cost-aware buffer management policy for flash-based storage devices. Technical Report (2015)
9. Johnson, T., Shasha, D.: 2q: A low overhead high performance buffer management replacement algorithm. In: PVLDB, pp. 439–450 (1994)
10. Kim, H., Ahn, S.: Bplru: A buffer management scheme for improving random writes in flash storage. In: FAST (2008)
11. Lee, D., Choi, J., Kim, J.H., et al.: Lrfu: A spectrum of policies that subsumes the least recently used and least frequently used policies. IEEE Trans. on Computers **50**(12), 1352–1361 (2001)
12. Lee, S.W., Moon, B.: Design of flash-based dbms: an in-page logging approach. In: SIGMOD, pp. 55–66 (2007)
13. Li, Y., He, B., Yang, R.J., Luo, Q., Yi, K.: Tree indexing on solid state drives. PVLDB **3**(1), 1195–1206 (2010)
14. Li, Y., On, S.T., Xu, J., Choi, B., Hu, H.: Optimizing non-indexed join processing in flash storage-based systems. IEEE Trans. on Computers **62**(7), 1417–1431 (2012)
15. Liu, X., Salem, K.: Hybrid storage management for database systems. Proceedings of the VLDB Endowment **6**(8), 541–552 (2013)
16. Lv, Y., Cui, B., He, B., Chen, X.: Operation-aware buffer management in flash-based systems. In: SIGMOD, pp. 13–24 (2011)
17. On, S.T., Gao, S., He, B., Wu, M., Luo, Q., Xu, J.: Fd-buffer: A cost-based adaptive buffer replacement algorithm for flash memory devices. IEEE Trans. on Computers (2013)

18. O'Neil, E.J., O'Neil, P.E., Weikum, G.: The lru-k page replacement algorithm for database disk buffering. In: SIGMOD, pp. 297–306 (1993)
19. Ou, Y., Harder, T.: Clean first or dirty first? a cost-aware self-adaptive buffer replacement policy. In: IDEAS (2010)
20. Ou, Y., Harder, T., Jin, P.: Cfdc: A flash-aware replacement policy for database buffer management. In: DaMoN (2009)
21. Park, S.Y., Jung, D., Kang, J.U., Kim, J.S., Lee, J.: Cflru: a replacement algorithm for flash memory. In: CASES, pp. 234–241 (2006)
22. Tsirogiannis, D., Harizopoulos, S., Shah, M.A., Wiener, J.L., Graefe, G.: Query processing techniques for solid state drives. In: SIGMOD, pp. 59–72 (2009)

The Gaussian Bloom Filter

Martin Werner[✉] and Mirco Schönfeld

Ludwig-Maximilians-Universität in Munich, Munich, Germany
{martin.werner,mirco.schoenfeld}@ifi.lmu.de

Abstract. Modern databases tailored to highly distributed, fault tolerant management of information for big data applications exploit a classical data structure for reducing disk and network I/O as well as for managing data distribution: The Bloom filter. This data structure allows to encode small sets of elements, typically the keys in a key-value store, into a small, constant-size data structure. In order to reduce memory consumption, this data structure suffers from false positives which lead to additional I/O operations and are therefore only harmful with respect to performance. With this paper, we propose an extension to the classical Bloom filter construction which facilitates the use of floating point coprocessors and GPUs or additional main memory in order to reduce false positives. The proposed data structure is compatible with the classical construction in the sense that the classical Bloom filter can be extracted in time linear to the size of the data structure and that the Bloom filter is a special case of our construction. We show that the approach provides a relevant gain with respect to the false positive rate. Implementations for Apache Cassandra, C++, and NVIDIA CUDA are given and support the feasibility and results of the approach.

Keywords: Bloom filter · Database design · Data structures

1 Introduction

Nowadays, the Internet has become one of the most important information hubs of our societies. While in the past, the Internet was used mainly as a source of information, it is becoming more and more a hub for user-generated and sensor data. The Internet-of-Things paradigm envisions that more and more devices of daily life transmit information to the Internet and consume information retrieved over the Internet for flexible service delivery.

Due to the wide adoption of large-scale cloud computing approaches in data management and due to the rise of "NoSQL" databases for solving the problems of management of large collections of data in a distributed way with linear scaling, a lot of approaches have been discussed in order to index and manage the high amount of data in a more flexible way. The data flow can be subsumed as a process and the most important aspects for big data applications stem from this data flow which can be enumerated as follows:

© Springer International Publishing Switzerland 2015
M. Renz et al. (Eds.): DASFAA 2015, Part I, LNCS 9049, pp. 191–206, 2015.
DOI: 10.1007/978-3-319-18120-2_12

1. *Data Collection:* The process of moving the data into a big data infrastructure.
2. *Data Distribution:* The process of distributing the data reasonably to several instances for performance and error-tolerance.
3. *Storage and Retrieval:* The operations used to store and retrieve data from persistent storage.
4. *Searching and Indexing:* The operations making data access flexible and reasonably fast for applications.
5. *Big Data Analytics:* The area of analyzing the meaning of large datasets.
6. *Visualization:* The challenges of making these outcomes visible and understandable for humans.

It is estimated that the amount of data on the Internet doubles roughly every two years and this data is the commodity of information society.

For adressing the first three challenges, a well-known data structure has been used a lot: The Bloom filter. This data structure is able to model the containment relation of variable-sized small sets with high accuracy, constant memory, small computation cost, and some false positives. With respect to big data infrastructures, it has been widely used in order to be able to skip disk reads or network transmission and manage access to information in key-value stores. With respect to data distribution, it can be applied for set reconciliation between two large datasets [4,5]. With respect to data collection it has been widely applied to network routing and related problems [8,13].

With this paper, we provide an extension to the classical Bloom filter which can be used in big data applications. The extension is able to use additional memory in order to reduce the false-positive rate of a Bloom filter while the underlying classical Bloom filter can be extracted easily. Additionally, our approach is based on using floating point calculations, which can be shifted to a graphics card or other co-processor exploiting parallel computation capabilities.

We provide an implementation[1] of our approach for the well-known NoSQL database Apache Cassandra and highlight the performance from two perspectives: A moderate increase in running time for key insertion is traded against a higher rate of skipping irrelevant data blocks especially for data blocks for which the number of elements stored is smaller than the expected number of elements.

The remainder of this paper is structured as follows: We first review the needed background on classical Bloom filters in Section 2. Then, we detail our extension to this construction, the "Gaussian Bloom Filter" in Section 3. In Section 4, we evaluate our approach with respect to its false positive rate, the number of bits used for floating point representation, the performance of our implementation for the well-established key-value store Apache Cassandra, and the performance of different variants implemented for the CPU and the GPU in C++. Section 5 concludes the paper.

---

[1] supplementary material is available at http://trajectorycomputing.com/gaussian-bloom-filter/

2   Bloom Filter

The Bloom filter, introduced by B. H. Bloom in 1970, is a probabilistic data structure for representing small sets of strings in a space-efficient way [3]. A Bloom filter can be used to rapidly check if the underlying set contains a certain element. A Bloom filter's main asset is the absence of false negative filter results: A Bloom filter will never claim that an element is not in the filter while it has been inserted to the filter. But, a Bloom filter does report false positive results stating that the underlying set contains the inquired element although it has not been added. The amount of expectable false positive filter responses depend on the filter's configuration, mainly.

Bloom filters are a central element of several modern distributed systems such as Google BigTable [6], Apache Cassandra [1], and even BitCoin [9]. Their manifold application led to a lot of research investigating the specialization of the basic variant towards specific application scenarios. Up to now, there are counting Bloom filters, compressed Bloom filters, time-decaying Bloom filters, stable Bloom filters, coded Bloom filters and several more [4,10].

The basic filter itself consists of a binary hash field of fixed size supporting the insertion of and querying for single elements. The following definition gives the details.

Definition 1 (Bloom Filter). *A Bloom filter is given by a binary hash field F of fixed length m. Fix a set h_i of k pairwise independent hash functions mapping the universe U to the set $\underline{m} = \{1, 2, \ldots, m\}$. In this situation, the following set of operations defines a data structure describing sets. The empty set is represented by an all-zero hash field $F = 0$.*

1. *Insert(F, e): Set all bits of F to the value 1, which are indexed by the results of all k hash functions h_i applied to the element e.*
2. *Test(F, e): Return true if and only if all bits in F, which would have been set by the corresponding Insert operation, equal 1.*

From this definition, one can see why a Bloom filter does not allow for false negative filter responses but may report false positives: the test operation uses k hash functions to check for the addressed bits being set to 1 – these are the same k hash values the insert operation would have used to set those bits. So, if there is only one position of the addressed bits $F[x] = 0$ the element could not have been inserted into the filter. On the other hand, all tested bits could have been set to one by the insertion of various other elements due to hash collisions. In this case, the test operation would respond falsely positive.

Since false positive responses often lead to costly operations their occurences should be minimized. But, the probability of false positives can be calculated. Since all k hash functions are assumed to be pairwise independent, the probability of a false positive can be expressed as the probability that all k hash functions hit a one for an inquired element. With some simplifications, this can be approximated via the fraction of zeros in the filter.

The probability for a slot still being unset after evaluating one single hash function is

$$1 - \frac{1}{m}.$$

For the insertion of n different elements hashing is performed kn times since every insertions uses k different hash functions. Therefore, the probability p for a slot being zero after n insertions can be expressed as

$$p \approx \left(1 - \frac{1}{m}\right)^{kn} \approx \exp\left(-\frac{kn}{m}\right).$$

The probability of a false positive $P(fp)$ can now be given as the complementary event to p since a false positive occurs if all k hash functions hit a bucket "not being zero":

$$P(fp) \approx (1 - p)^k$$

To obtain the minimal false positive probability the first derivative of $P(fp)$ with respect to k is taken and set to zero resulting in an optimal parameter k_{Opt}:

$$k_{\mathrm{Opt}} = \frac{m}{n} \log 2.$$

This leads to a fraction of zero of

$$p_{\mathrm{Opt}} = \frac{1}{2}$$

and a false-positive probability of

$$P_{\mathrm{Opt}}(fp) = \left(\frac{1}{2}\right)^{\frac{m}{n}\log 2} \approx 0.6185^{\frac{m}{n}}.$$

An optimal Bloom filter configuration leads to roughly half of the bits being zero and half of the bits will be one. This motivates from an information-theoretic perspective that it is impossible to spill in more information into the hash-field: The entropy in this case is maximal. The optimal configuration depends on the number of elements being inserted, the size of the bit field and the number of hash functions being used.

When thinking about the Bloom filter, one quickly realizes that only zero-valued slots contribute information for reducing false-positives: A false-positive is rejected, if at least one of the addressed cells is zero. From this observation, we motivate our extension: We want to introduce more information into each "zero-valued" slot, namely some information about the ordering of hash functions. In this way, the surrounding fields store some information on which hash function has actually set the enclosed bit. This helps to further reduce the occurence of false positives since the test operation is now able to decide if a bit was set by storing the inquired element or a different one and additionally reject some of the false-positives in which the ones are addressed from different hash functions. The following section gives details on this construction.

3   Gaussian Bloom Filter

For the construction of the Gaussian Bloom filter, we first replace all binary slots of the hash field F with small floating point numbers. We further extend the insert and test operation as follows:

Definition 2 (Gaussian Bloom Filter). *A Gaussian Bloom filter is given by a hash field F of fixed length m composed of small cells storing floating point numbers. Fix a set h_i of k pairwise independent hash functions mapping the universe U to the set $\underline{m} = \{1, 2, \ldots, m\}$. In this situation, the following set of operations defines a data structure describing sets. The empty set is represented by an all-zero hash field $F = 0$.*

1. *Insert(F, e): For each of the k hash functions h_i create a Gaussian probability density function*

$$\mathcal{N}_i = \mathcal{N}(h_i(e), i)$$

   with mean $\mu = h_i(e)$ given by the hash function value and standard deviation $\sigma = i$ given by the hash function index. With respect to the hash field F, we set all entries to the maximum of the current value in the slot and the value in the normalized signature function

$$\tilde{\mathcal{N}}_i = \frac{\mathcal{N}_i}{\max(\mathcal{N}_i)}$$

   which attains its maximal value 1 at $h_i(e)$ and values between 0 and 1 elsewhere:

$$F[t] := \max\left(F[t], \tilde{\mathcal{N}}_1(t), \ldots, \tilde{\mathcal{N}}_k(t)\right)$$

2. *Test(F, e): Create the same normalized signature functions $\tilde{\mathcal{N}}_i$ and combine them into the* element signature *S by selecting the maximal value for each slot.*

$$S[t] = \max\left(\tilde{\mathcal{N}}_1(t), \ldots, \tilde{\mathcal{N}}_k(t)\right)$$

   Return true, if and only if $F[i] \geq S[i]$ for all $i \in \underline{m}$.

In this construction, the non-maximal slots contain information about the hash function index of the one causing the value of the slot to have changed. For filters with few elements, this can be a lot of information as a lot of non-maximal slots are available to encode such information. Towards an optimally filled filter, the number of available slots reduces.

Choosing the Gaussian kernel for this approach has several reasons: First of all, any bounded function could have been used which can be parametrized with an additional parameter encoding the index of the hash function. Choosing a function which has a single maximum at the location indexed by the hash functions, however, allows for easily extracting the underlying Bloom filter:

Definition 3 (Underlying Bloom Filter). *Given a Gaussian Bloom filter* F_G, *the underlying Bloom filter* F_B *can be retrieved by comparing all slots with the maximal value* M *of the signature functions:*

$$F_B[i] = (F_G[i] == M)$$

This recovers the classical Bloom filter with identical parameters (size, number of hash functions) which would have been generated by directly using the classical construction. This creates an important compatibility when it comes to network applications: We can use the Gaussian Bloom filter without prescribing its use to other components of a distributed system. Moreover, we can use the Gaussian Bloom filter only as long as it actually helps and the additional memory is available.

One subtlety with this definition of the underlying Bloom filter is given by the fact that the signature functions and the resolution and coding of the floating point numbers must be chosen in a way such that no non-maximal cells are rounded to the maximal value. Therefore, we propose to encode the numbers in the cells of a Gaussian Bloom filter in a special way. However, when using floating point processing units such as GPUs in order to perform the filtering operations, keep in mind that rounding can become a problem introducing a novel type of false-positives for the underlying Bloom filter.

Furthermore, the choice of Gaussian kernels can be motivated as making clear the extension nature of our approach: In the space of distributions, the family of Gaussian kernels with smaller and smaller standard deviation converges to the Dirac distribution which is zero everywhere except at 0.

Finally, there are very good speedups for calculating an approximation to the Gaussian function and the calculation can be localized in the array by calculating the function only in a local neighborhood of several multiples of the standard deviation around the indexed hash cells limiting the number of floating point operations per hash function activation.

But before we start discussing implementation issues and scalability of the given approach, we first fix some basic results on the filter, its performance and its configuration.

3.1 Properties

In order to discuss the properties of the Gaussian Bloom filter, we start with the following observation that the Gaussian Bloom filter does not allow false negatives:

Lemma 1. *The Gaussian Bloom filter has no false negatives.*

Proof. At any point in time the filter structure is larger than the largest element signature inserted. As the Test operation is based on comparing the signature using greater than or equal to the filter, the query pattern can not be larger than the filter in any slot unless the element has not been inserted to the filter.

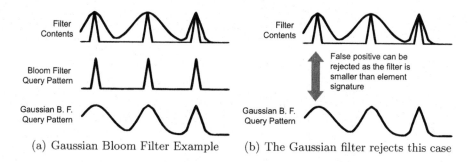

(a) Gaussian Bloom Filter Example   (b) The Gaussian filter rejects this case

Fig. 1. Gaussian Bloom Filter rejecting a Bloom filter false positive

The Gaussian Bloom filter allows for false positives just as a Bloom filter does. However, a Gaussian Bloom filter false-positive implies a false positive for the associated underlying Bloom filter.

Lemma 2. *The false positive rate of a Gaussian Bloom filter is smaller than or equal to the false positive rate of a Bloom filter.*

Proof. In the situation of a Gaussian Bloom filter false positive, all slots of the filter are larger than the element signature pattern. This is especially true for the slots, where this pattern attains its maximal value one. However, these slots have been addressed by hash functions of other elements directly.

The following gives a concrete example of a situation in which a Bloom filter reports a false-positive while the Gaussian Bloom filter is able to reject this case.

Example 1. Consider the following situation: Assume the element e_1 with associated hash values $h_1(e_1) = 50$, $h_2(e_1) = 10$, and $h_3(e_1) = 30$ has been added to an empty filter. Assume further that the query element e_2 has hash values $h_1(e_2) = 50$, $h_2(e_2) = 30$, and $h_3(e_2) = 10$

This situation is depicted in Figure 1 on the left: Since all three indexed hash addresses have been set, a Bloom filter returns true for element e_2. A Gaussian Bloom filter on the other hand returns false for the query element. It is able to distinguish the hash functions that addressed the underlying filter index. The right hand side of the figure shows how the query pattern differs from the filter content. Namely, the encoded Gaussian functions for hash addresses 10 and 30 differ in their standard deviation. Therefore, the relevant bits must have been set by different hash functions and, consequently, come from different elements.

Finally, we have to discuss how a Gaussian Bloom filter can be configured. A central element of the analysis of the Bloom filter was the assumption of independence of each bit of information used from other bits. However, in the case of a Gaussian Bloom filter this independence does not exist anymore as different hash values are used when rejecting false positives. Therefore, we are left with the fact that the Gaussian Bloom filter is not worse than the Bloom filter and have to use the Bloom filter analysis in order to choose the optimal configuration in terms of size, number of hash functions, and number of elements.

Performance

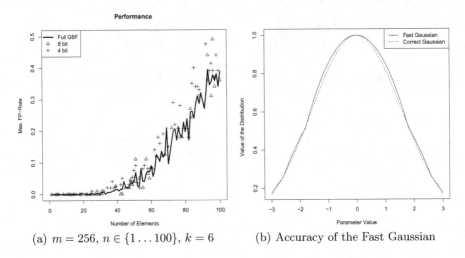

(a) $m = 256$, $n \in \{1 \ldots 100\}$, $k = 6$ (b) Accuracy of the Fast Gaussian

Fig. 2. False-positive rate for small floating points and the accuracy of the fast Gaussian

3.2   Efficient Representation with Small Counters

Depending on the application domain and the way in which the floating point calculations are performed in practice, it can become quite questionable, whether the amount of memory for example by utilizing single or double precision floating point numbers is reasonable. Furthermore, it is unclear how to prevent rounding up to the maximal value effectively. However, we provide a bit coding tailored to the situation in which even small sizes of 4 bit per slot reach a similar false-positive rate as compared to using an implementation based on an array of `double` values.

In order to encode the floating point values efficiently into a small bit field with t entries, we observe that only values between 0 and 1 need to be modelled. Furthermore, we observe that the maximal value 1 definitely needs a unique bit representation. We chose to model this maximal value by an array of t ones. If the value is not maximal, then we are left with $2^t - 1$ bit combinations which we can use. If we multiply the value with the following scaling factor, then usual rounding will be as expected:

$$\alpha = \frac{2^t - 1}{2^t}$$

In order to obtain the final bit pattern representing the value, let each bit model a fraction of two, the first bit models $\frac{1}{2}$, the second bit $\frac{1}{4}$, and so on. When decoding from this representation, we first check for the distinguished pattern for the maximal value, decode and rescale with α^{-1}.

In order to evaluate the efficiency of the representation, three different filters with equal random sets of elements of varying size are created and then queried with 100 random elements, which have not been inserted into the filters. Figure 2(a) depicts the result of comparing the false-positive rate of a filter based on full-sized double values as provided by the CPU compared to four bit and eight bit representation. You can clearly see that there is only a small quality loss in comparison between CPU double (bold line) and our binary representations. This is due to the fact that our binary representation uses the bits more efficiently by exploiting the domain limitation to modeling only the interval $[0,1]$. As to be expected, increasing the number of bits from 4 to 8 bit makes a difference: in general, the larger bit slots tend to reject more false positives.

3.3   Efficient and Local Evaluation of Gaussian Function

For inserting an element to the Gaussian Bloom filter, we have defined to use a Gaussian distribution with two variable parameters. In order to sucessfully apply this in a big data environment with possibly large Bloom filters, we have to optimize the situation into two directions: First of all, it is infeasible to evaluate this function for each cell of the filter and, secondly, it is infeasible to evaluate the correct formula for the Gaussian function containing the exponential function.

In order to be able to evaluate the Gaussian function for a smaller number of slots, we observe that the tails of the Gaussian function attain zero after rounding quickly. We employ the "3σ-rule" which states that evaluating operations involving the Gaussian functions can be skipped after three sigma due to neglectible size of the function. Outside the interval $[\mu - 3\sigma, \mu + 3\sigma]$ the Gaussian function is smaller than 0.0045, which is small enough to be neglected.

This step, in summary, makes the operations Insert and Test independent from the filter size. They, then, only depend on the number of hash functions used: For each hash function, the number of slots is determined by its index as 3σ determines the amount of slots being accessed. Therefore, the overall number of hash field operations scales with the number of hash functions. Note that even further reduction is possible at the cost of loosing some additional information by imposing a maximal allowed standard deviation. This is especially useful when very high standard deviations dominate too many cells of the filter removing useful ordering information.

In order to be able to create the Gaussian Bloom filter quickly, we adopt a well-known approximation to the exponential function consisting of a integer multiplication, an addition and a binary shift. This method is due to Schraudolph [11]. The speedup of this approximation is astonishing: In our experiments, this implementation was about 30 times faster than the library routine of Java. Figure 2(b) depicts the Gaussian function as computed using the library exponential function compared to the Gaussian implemented using the fast exponential function for the values from the interval $[-3\sigma, 3\sigma]$ in which we actually use the function. Note that outside this interval, the approximation quickly degrades.

4   Evaluation

This section evaluates our new data structure for filtering small sets using an extension to classical Bloom filters with respect to the false-positive rate and in

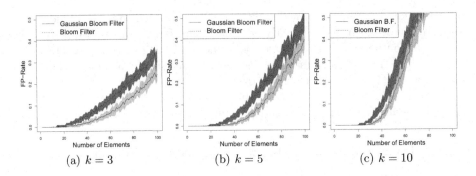

(a) $k = 3$ (b) $k = 5$ (c) $k = 10$

Fig. 3. Comparison of Bloom Filter and Gaussian Bloom Filter Performance ($m = 256$, $k \in \{3, 5, 10\}$ and $n \in \{1, \ldots 100\}$)

comparison to the underlying Bloom filter. We performed a lot of experiments with rather high false-positive rates by using small filters and medium numbers of hash functions. These situations serve as example for larger filters with much more elements to insert. In these experiments, a set of random strings has been generated and either SHA-1 [7] or the Murmur Hash [2] have been used to generate hash data. The different hash functions were generated using a different prefix for the hash argument.

Figure 3 compares the Gaussian Bloom filter with an equivalent Bloom filter. In all these figures, the experiments were performed with random string sets and varying parameters many times in order to generate the median performance as the median of the instances measured false-positive rate. Additionally, the first and third quartile area is shaded for both cases. In Figure 3 we see that the Gaussian Bloom filter clearly outperforms the Bloom filter for filters, which have enough zeroes. In the left figure, 256 slots were used together with 3 hash functions. The two figures to the right consider the same situation with an increasing number of hash functions ($k = 5, k = 10$) which leads to more ones in the filter and thereby less possibility to encode ordering information into non-maximal slots. This results in the overall increasing false-positive rate for both filters and the effect that the gain of the Gaussian Bloom filter relative to the Bloom filter gets smaller.

As the gain for the Gaussian Bloom filter construction is larger for underfull filter, we evaluate the false positive rate of a filter with respect to the number of hash functions. Figure 4 depicts the false-positive rate for several situations with different numbers of hash functions. We clearly see that the false-positive rate of the Gaussian Bloom filter is smaller for the number of hash functions and scales similar with the number of hash functions as compared to the Bloom filter.

In summary, this supports the recommendation of using the standard Bloom filter configuration for a Gaussian Bloom filter. If needed, one can also use a slightly smaller number of hash functions for the Gaussian Bloom filter in order to reduce the amount of consumed hash bits. This is especially important, if

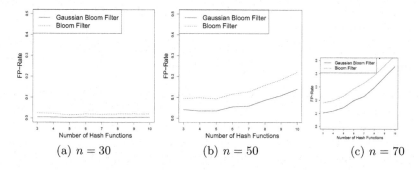

(a) $n = 30$ (b) $n = 50$ (c) $n = 70$

Fig. 4. The effect of the number of hash functions ($m = 256$, $n \in \{30, 50, 70\}$ and $k \in \{3, \ldots 10\}$)

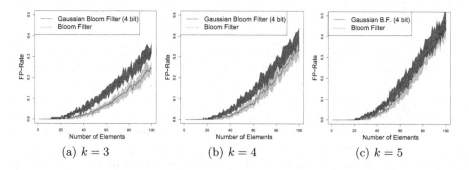

(a) $k = 3$ (b) $k = 4$ (c) $k = 5$

Fig. 5. Performance of a 4-bit Gaussian Bloom Filter ($m = 256$, $n \in \{1 \ldots 100\}$ and $k \in \{3, 4, 5\}$)

hashing of the elements is time consuming, for example in file synchronization applications.

As a third aspect, we have to discuss the performance of the Bloom filter with 4 bit floating point representations. We have already seen in Figure 2(a) that the false-positive rate keeps comparable to a double-based filter for a fixed number of hash functions. Figure 5 depicts the scaling behavior with respect to the number of hash functions. This figure makes clear that the gain of the construction is more sensitive to the number of hash functions as compared to the double approach as the choice of using the hash index as the standard deviation leads to the inclusion of equal non-maximal values in many non-maximal slots (e.g., the largest non-maximal value of the representation) which reduces the overall discriminative gain of the Gaussian signature function. It can be seen that for $k = 3$ the positive effect is clearly visible, but it degrades more quickly for increasing $k \in \{4, 5\}$ as compared to the implementation using double values depicted in Figure 3. In general, the smaller the number of bits in the floating point representation, the fewer false positives can be rejected. However, the system is keeping between the bounds given by a binary, classical Bloom filter and a filter based on an error-free representation.

4.1 Gaussian Bloom Filter in Apache Cassandra

In order to test our approach in a real-world environment, we decided to extend the open source Apache Cassandra database [1]. Apache Cassandra is based on the Hadoop distributed file system (HDFS) [12] and has been created at Facebook in order to tackle the inbox search problem. Up to the middle of 2011, Facebook was using it for this task. Meanwhile, other large and widely recognized Internet services started using Cassandra inside their backend systems for several tasks including Twitter, Digg, Reddit and others.

From a technological perspective, Cassandra starts with the block-based Hadoop distributed file system and manages a key-value store with which values can be put into the database and retrieved by giving a key. This data is then organized into so-called SSTables and Bloom filters are used to keep track, whether specific keys could be inside specific SSTables in order to reduce I/O. Therefore, each time a key is inserted into an SSTable, the associated Bloom filter is updated. We replace the classical Bloom filter of Cassandra with our modified version and still keep track of the behavior of the original system by managing a classical filter as an instance variable of our new filter. Therefore, performance information from the stress test can not be used to compare the filters, as they contain the overhead of calculating each operation for both filters. Cassandra uses a sophisticated Bloom filter subsystem in which filters are dynamically reconfigured to match the amount of information that needs to be indexed.

We performed experiments on a single-node cluster running our modified version of Cassandra based on Version 2.0.4. We added a new mixed filter containing the Gaussian Bloom filter as well as a copy of the original Bloom filter implementation. Furthermore, logging code has been added to track the performance of each individual Bloom filter operation. For Cassandra, the Bloom filter false positive rate can be configured and we set it to 10% for the experiments in order to have many cases in which both filters will fail in order to reliably count the number of cases in which the Gaussian filter outperformed the classical filter. We logged for a complete run of the default Cassandra stress test the number of situations in which either of both filters was able to reject a false positive with the expected outcome that the proposed Gaussian Bloom filter have no additional false positives and rejects several false positives which kept undetected by the classical Bloom filter. The stress test consists of first writing one million keys into the database, then recalculating all filters using `nodetool scrub` and finally reading one million random keys back from the database.

During this evaluation on a single-node cluster, 23 Bloom filter were generated out of which nine were used in order to manage the stress data. The remaining filters were managing organizational key spaces and registered only few operations. Table 1 collects the number of situations in which both filters

Table 1. Numbers of SSTable reads for each filter during the stress test

Filter	Bloom SSTable Reads	Gaussian SSTable Reads	Fraction
1	263,937	141,598	0.54
2	625,879	524,120	0.84
3	639,631	539,161	0.84
4	1,230,417	1,163,531	0.94
5	440,584	336,882	0.76
6	271,404	146,835	0.54
7	255,446	138,246	0.54
8	305,535	172,626	0.56
9	260,190	138,611	0.53
Total	4,293,023	3,301,610	0.67

were unable to skip an SSTable read. This includes positives as well as false positives.

In summary, our novel data structure of a Gaussian Bloom filter was able to skip 991,413 additional SSTable reads as compared to the equally configured Bloom filter, an overall gain of 33% on average. Furthermore, one observes that the reconstruction of Bloom filters in the middle of the experiment before starting reading leads to similar gains of nearly 50%. That is, when the database is restructured and no new data arrives, our structure – with the default configuration of Cassandra – outperforms the given data structure by rejecting 50% of the SSTable reads not rejected by the classical Bloom filter.

The Bloom filter operation performance was also monitored by aggregating the running time of the Insert and Test operation for both, the original Bloom filter and the modified Gaussian Bloom filter. As is to be expected, the Gaussian Bloom filter needs more computations, but still moderately. For the stress test, we observed an increase in running time for the Insert operation by 42%. This is due to the fact that the classical Bloom filter does only access the indexed bits while the Gaussian Bloom filter has to access many more slots, especially when many hash functions are in use such that large standard deviations occur. For the Test operation, however, the measured overhead was only 8.82%. This is due to the fact that testing can be stopped as soon as a single contradiction occurs. This makes our filtering structure more sensible for database applications in which the read-write ratio tends clearly to more reads than writes.

4.2   Performance of a GPU implementation

In order to further motivate the use of Gaussian Bloom filters, we completed another implementation of Gaussian Bloom filters in C++, both for the CPU as well as supported by the GPU via the NVIDIA CUDA programming model.

In this situation, the additional memory of the Gaussian Bloom filter allows for full parallel access without any synchronization and therefore allows for boosting the performance of the Gaussian Bloom filter even more. For the GPU case, we create a thread for each hash function index and calculate the Murmur

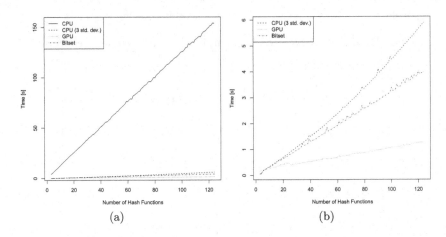

Fig. 6. Performance of various C++ implementations inserting 1,000 random strings

hash inside the GPU. When this finishes, we create one thread for each slot of the Bloom filter and fill in the original exponential function signature into each slot. Note that this opens up the Gaussian Bloom filter construction to non-local kernels, for which all slots have to be calculated and no optimization comparable to the three sigma rule is available.

Figure 6 depicts the performance of this approach. It shows the time in seconds which is used to insert thousand random strings into a filter. The implementations are pure CPU with and without three sigma rule, the GPU variant, and a bit-packing filter. Note that all implementations except the CPU version without three sigma rule are reasonably fast and scale linearly with the amount of hashing. The small peaks show situations in which the Murmur hash was more complex based on the data. It is interesting that this peak can be observed in all implementations while it is much smaller for the GPU. This is due to the fact the even the calculation of the d hash functions is performed in parallel on the GPU.

The GPU implementation results in several welcome facts: Firstly, the additional memory overhead is not taken from system memory, instead, system memory consumption is reduced as the filter memory is on the GPU. Secondly, the CPU is free for other operations while the hashing and calculation is deferred to the GPU and, finally, exploiting the strong parallelism of typical GPUs allows for calculating kernels globally over the complete hash field and opens up the construction for more complicated kernels.

In order to show the reduction of CPU demand, we performed an experiment on a desktop computer (Intel Xeon E5620 @ 2.4 GHz, 2 CPUs, NVIDIA Quadro 4000, 256 CUDA cores, Windows 7) as follows: Eight threads are started performing a typical workload of randomly generating 10^{12} integer numbers, sorting them and reversing their order in a best effort manner while another thread is performing 400 insertions into a Bloom filter per second using 50 hash functions and 10,000 slots. This resulted in $1.18 \cdot 10^{18}$ integer workloads completed after 10 seconds for the CUDA implementation compared to $1.15 \cdot 10^{18}$ for the Gaussian Bloom filter on the CPU and $1.14 \cdot 10^{18}$ operations for the binary Bloom filter.

In summary, this approach allowed to free up CPU performance and system memory for applications while maintaining Gaussian Bloom filter in the graphics card.

5 Conclusion

With this paper, we have proposed a novel approach to tuning the classical Bloom filter construction. While the classical Bloom filter performance is optimal when configured with the optimal values, a lot of real-world applications either have to recalculate Bloom filters often or make use of Bloom filters that are configured for many more elements than they actually contain most of the time. In this situation, we propose to model additional information into the Bloom filter data structure into the zero slots. This approach is an extension to the Bloom filter in the sense that maximal values still recover the Bloom filter of the same number of slots, the same hash functions and the same elements. However, we make use of small floating point slots, which can encode from which index of a hash function their value was most influenced. We show in theory and practice that this Bloom filter variant can be used sucessfully. Even for the modern database system Cassandra, which contains sophisticated techniques to optimally configure Bloom filters during operation, a clear gain was visible.

One drawback of this construction is the additional space used to maintain the small floating point numbers. Still, this memory can be on a coprocessing unit such as a GPU freeing up primary memory and CPU capacity for applications. Still, for storage-only backend systems, the additional CPU time and the amount of RAM did not negatively impact the performance of the standard Cassandra stress test of inserting and reading a million keys from a table. We suspect that in many storage backends based on Apache Cassandra the additional amount of memory and CPU is well-invested when the number of complete recalculations of the Bloom filters can be reduced. This is an important direction for future work which, however, can only be performed sensibly with real workloads for the cluster and with a distributed cluster deployment.

The most important advantage consists, however, of the fact that unlike other approaches to provide more efficient filter structures for small sets, our construction is compatible with the original structure and therefore with communication protocols as the underlying Bloom filter can be extracted easily.

References

1. The Apache Cassandra database (2014). http://cassandra.apache.org/
2. Appleby, A.: Murmurhash (2009). http://code.google.com/p/smhasher/
3. Bloom, B.H.: Space/time trade-offs in hash coding with allowable errors. Communications of the ACM **13**(7), 422–426 (1970)

4. Broder, A., Mitzenmacher, M.: Network applications of bloom filters: A survey. Internet Mathematics **1**(4), 485–509 (2004)
5. Byers, J., Considine, J., Mitzenmacher, M.: Fast approximate reconciliation of set differences. Tech. rep., Boston University Computer Science Department (2002)
6. Chang, F., Dean, J., Ghemawat, S., Hsieh, W.C., Wallach, D.A., Burrows, M., Chandra, T., Fikes, A., Gruber, R.E.: Bigtable: A distributed storage system for structured data. ACM Transactions on Computer Systems (TOCS) **26**(2), 4 (2008)
7. Eastlake, D., Jones, P.: US Secure Hash Algorithm 1 (SHA1) (2001)
8. Feng, W.C., Kandlur, D.D., Saha, D., Shin, K.G.: Stochastic fair blue: a queue management algorithm for enforcing fairness. In: Proceedings of the Twentieth Annual Joint Conference of the IEEE Computer and Communications Societies, INFOCOM 2001, vol. 3, pp. 1520–1529. IEEE (2001)
9. Nakamoto, S.: Bitcoin: A peer-to-peer electronic cash system (2008)
10. Schönfeld, M., Werner, M.: Node wake-up via ovsf-coded bloom filters in wireless sensor networks. In: Ad Hoc Networks, pp. 119–134. Springer (2014)
11. Schraudolph, N.N.: A fast, compact approximation of the exponential function. Neural Computation **11**(4), 853–862 (1999)
12. Shvachko, K., Kuang, H., Radia, S., Chansler, R.: The hadoop distributed file system. In: 26th IEEE Symposium on Mass Storage Systems and Technologies (MSST), pp. 1–10 (2010)
13. Whitaker, A., Wetherall, D.: Forwarding without loops in icarus. In: 2002 IEEE Open Architectures and Network Programming Proceedings, pp. 63–75. IEEE (2002)

Spatio-Temporal Data I

Detecting Hotspots From Trajectory Data in Indoor Spaces

Peiquan Jin[1,2(✉)], Jiang Du[1], Chuanglin Huang[1],
Shouhong Wan[1,2], and Lihua Yue[1,2]

[1] School of Computer Science and Technology,
University of Science and Technology of China, Hefei 230027, China
[2] Key Laboratory of Electromagnetic Space Information,
Chinese Academy of Sciences, Hefei 230027, China
jpq@ustc.edu.cn

Abstract. The increasing deployment of indoor positioning technologies like RFID, Wi-fi, and Bluetooth offers the possibility to obtain users' trajectories in indoor spaces. In this paper, based on indoor moving-object trajectories, we aim to detect hotspots from indoor trajectory data. Such information is helpful for users to understand the surrounding locations as well as to enable indoor trajectory mining and location recommendation. We first define a new kind of query called *indoor hotspot query*. Then, we introduce a pre-processing step to remove meaningless locations and obtain indoor stay trajectories. Further, we propose a new approach to answering indoor hotspot queries w.r.t. two factors: (1) users' interests in indoor locations, and (2) the mutual reinforcement relationship between users and indoor locations. Particularly, we construct a *user-location* matrix and use an iteration-based technique to compute the hotness of indoor locations. We evaluate our proposal on 223,564 indoor tracking records simulating 100 users' movements over a period of one month in a six-floor building. The results in terms of *MAP*, *P@n*, and *nDCG* show that our proposal outperforms baseline methods like *rank-by-visit*, *rank-by-density*, and *rank-by-duration*.

Keywords: Indoor space · Hotspot · Trajectory

1 Introduction

With the development of indoor positioning technologies and various portable devices, it becomes necessary to provide location based services in indoor spaces [1, 2]. For example, shopping malls intend to find the hotspots in the buildings and thereby can adjust the deployment of shops and further provide better services for customers. Generally, hotspots in indoor spaces can mostly reflect people's interests in those locations. Therefore, we can find out the hottest shops in a shopping mall and even mine the hot routes so as to provide better shopping guide and recommendation for customers.

© Springer International Publishing Switzerland 2015
M. Renz et al. (Eds.): DASFAA 2015, Part I, LNCS 9049, pp. 209–225, 2015.
DOI: 10.1007/978-3-319-18120-2_13

In this paper, we focus on detecting hotspots from trajectory data in indoor spaces. We define a new type of query for indoor spaces, which is called *indoor hotspot query*, and propose an effective approach. An indoor hotspot query aims to find out the hottest locations from a set of moving trajectories in indoor spaces. A typical application is to find the hottest shops and products in shopping malls based on users' trajectories, which can be further utilized for product recommendation and sales promotions. Note that the traditional way for this purpose is only based on the statistics of sales. However, people's moving trajectories in a shopping mall can also reflect their buying interests, which can be another aspect that could be more useful for product recommendation. This is simply because many people will not buy the same product (e.g., a wedding ring, a watch, etc.) again in a long time after they have bought it. Thus, it can be helpful to utilize people's moving trajectories to find the potential buying interests, because most trajectories can be obtained before actual buying behaviors. As a result, if we can extract hotspots from people's indoor trajectories, we are able to detect the locations (shops in a shopping mall) that most people are interested in. This can be then used for many other applications, e.g., improving the spatial deployment of shops, providing product recommendation in shopping malls, offering better tour guide in a museum, etc.

Previous studies on hotspots queries mainly focus on outdoor space, such as the Euclidean space [3, 4] and the road-network space [5-7]. There is also one existing research towards constrained movements in a conveyor-based constrained space in an airport [8], but this work cannot be applied to other types of indoor spaces, because the conveyor-based indoor space only considers conveyors, which is much different from common indoor environments such as office buildings, museums, metro stations, etc. Following [8], [20] uses a density-based approach to find hotspots in semi-constrained indoor space, focusing on employing a density-based index to improve time performance. However, this paper proposes a new approach differing from density or visiting-count based ones. Further, our study focuses on general indoor spaces that are composed of common indoor elements such as rooms, doors, hallways, stairs, elevators, etc.

The challenges of processing indoor hotspot queries are two-fold. First, indoor spaces are usually three-dimensional, and different floors and locations have different possibilities to be visited. For example, in a shopping mall, the first floor usually has much more visits than other ones, because all people have to visit the first floor when entering into the shopping mall. Also some special locations such as elevator and stair rooms are often visited. Therefore, it is not suitable to simply count the visits to each location to identify hotspots. Second, users' interests in an indoor location do not simply rely on the number of users visiting the location, but also depend on users' stay time in the location as well as users' activeness and travel experiences in the related indoor space. In traditional road-network space, we usually use visiting count or visiting duration to identify hot road segments [9, 10]. However, in indoor spaces we have to consider more factors. For instance, the stay time of users in each indoor location should be considered. Besides, active users and inactive ones have different influences on the hotness of indoor locations. For example, in a shopping mall, active users are more likely to be potential buyers and therefore we need to consider their impacts on hotspots more than other users. Therefore, how to model the mutual reinforcement relationship between users and indoor locations is a critical issue in indoor hotspot detection.

In this paper, we propose a new approach to indoor hotspot queries and make the following contributions:

(1) We define a new kind of query on indoor moving objects, which is called *indoor hotspot query*.

(2) We propose a new approach to processing indoor hotspot queries. In particular, we consider two factors to identify indoor hotspots, namely users' interests in indoor locations and the mutual reinforcement relationships between users and indoor locations.

(3) We conduct experiments on a RFID-based indoor trajectory data set consisting of 223,564 trajectories, and compare our proposal with several baseline methods including *rank-by-visit*, *rank-by-density*, and *rank-by-duration*. The performance results w.r.t. *MAP*, *P@n*, and *nDCG* demonstrate the superiority of our proposal.

The rest of the paper is organized as follows. Section 2 defines the problem. Section 3 explains our proposal of processing indoor hotspot queries. Section 4 presents the experimental results. We discuss the related work in Section 5 and conclude the paper in Section 6.

2   Problem Statement

2.1   Indoor Space

Modeling indoor spaces has been studied for years. Many models have been proposed, such as object-feature-based model [11], geometric model [12], and symbolic model [13-15]. Without the loss of generality, in this paper, we define indoor spaces as follows.

Definition 1 (Indoor Space). An indoor space is represented as a triple:

$$IndoorSpace = (Cell, Sensor, Deployment)$$

Here, *Cell* is a set of cells in the indoor space. According to previous researches, an indoor space can be partitioned into cells [13-15]. *Sensor* is a set of positioning sensors deployed in the indoor space. Typical sensors are RFID readers, Bluetooth detectors, and Wi-Fi signal receivers. *Deployment* records the placement information of the sensors in the indoor space

In real applications, rooms can be regarded as cells and sensors are usually used to identify cells in indoor space, e.g., to identify shops in a shopping mall. Thus, the *Deployment* information of sensors can be pre-determined and maintained in a database or a file. For example, in the indoor space shown in Fig. 1, the rooms denoted by r_1 to r_7 are regarded as cells. Then, we can use the following mappings to represent the *Deployment* information of sensors $(s_1 - s_7)$:

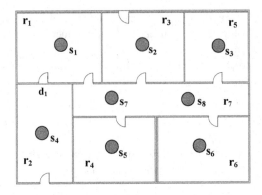

Fig. 1. An example of indoor spaces

$Deployment = \{<s_1, r_1>, <s_2, r_3>, <s_3, r_5>, <s_4, r_2>, <s_5, r_4>, <s_6, r_6>, <s_7, r_7>, <s_8, r_7>\}$

In order to introduce semantics into the model, we assign a semantic label to each sensor. As a result, the set *Sensor* can be represented as follows.

$$Sensor = \{s|s = (sensorID, location, label)\}$$

The *label* of a sensor provides descriptions on thematic attributes of the location identified by *sensorID*. For instance, we can use labels like *"elevator"* and *"stair"* to indicate the functions of the cell identified by a sensor. We can also use other labels like *"Starbucks"* and *"Burge King"* to annotate semantic features of the cell. The location of a sensor is a self-defined three-dimensional coordinate (x, y, z), which reflects the relative position of the sensor inside the indoor space where the sensor is deployed. In real applications, indoor maps are usually designed by AutoCAD [21]. Thus, if we import an indoor map into a database system, we can simply use the coordinates and floors in the map to represent the locations of sensors.

2.2   Indoor Hotspot Query

In outdoor space, a trajectory is a series of GPS locations, while a trajectory in indoor spaces is a series of sensor signals. Thus, we first define the indoor location of a moving object as follows:

Definition 2 (Indoor Location). An indoor location *LOC* of a moving object *mo* is defined as LOC_{mo}:

$$LOC_{mo} = \{p|p = (s, [t_s, t_e]) \wedge (t_s < t_e) \wedge s \in Sensor\},$$

where s refers to a sensor, *mo* is the identifier of a moving object, t_e is the instant that the object enters the sensor's range, and t_s is the instant that the object leaves the sensor range.

Then, we define the indoor trajectory for a moving object.

Definition 3 (Indoor Trajectory). An indoor trajectory TR_{mo} of a moving object mo is defined as a sequence of indoor locations of mo:

$$TR_{mo} = \{(mo, \langle p_1, p_2, \ldots, p_n, \rangle) | (\forall p_i) \left(p_i \in LOC_{mo} \wedge (p_i.t_e < p_{i+1}.t_s)\right)\}$$

Then, we define indoor trajectory similarity search in Definition 4.

Definition 4 (Indoor Hotspot Query). Given a set of indoor trajectories T and an integer k, an *indoor hotspot query* retrieves a list $H \subseteq Sensor$ that consists of the top-k hottest indoor locations, such that:

$$H = \left\{ \begin{array}{c} \langle s_1, s_2, \ldots, s_k \rangle | \\ s_i \in Sensor \wedge \left(hottness(T, s_i) > hottness(T, s_j) \; if \; 1 \leq i < j \leq k\right) \end{array} \right\}$$

Here, the function $hottness(T, s_i)$ returns the measure on the hotness of the indoor location s_i according to the indoor trajectory set T.

3 Indoor Hotspot Query Evaluation

In this section, we explain the details of evaluating indoor hotspot queries. We first present the general idea of the proposed algorithm in Section 3.1. Then, we discuss the pre-processing of indoor trajectories in Section 3.2. Finally, we give the algorithm for indoor hotspot query processing in Section 3.3.

3.1 General Idea

Generally, indoor hotspots refer to those locations that users are mostly interested, e.g., the hottest shops in a shopping mall. Indoor hotspots are very helpful to many indoor location-based services, such as spatial deployment optimization for indoor spaces, sales promotion, personalized route recommendation, etc. However, the problem is how to measure the hotness of indoor locations. It is not effective to simply count the number of moving objects visiting a location, or to aggregate the stay time of moving objects in a location. As a consequence, due to the complex structure of indoor spaces and the mutual reinforcement relationship between moving objects and indoor locations, we have to devise a new approach to evaluate indoor hotspot queries.

Our general idea is to consider users' interests in indoor locations and the mutual reinforcement relationship between users and indoor locations. First, the activeness of users is classified. This is simply because active users are more familiar with indoor locations than others, and thus we should consider more on their votes in indoor hotspots. This is very similar to the situation in travel recommendation, in which we always want to know the opinions from those who have many experiences in traveling to our interested places before. Thus, if more active users are focused on an indoor location, this location is more likely an indoor hotspot. On the other side, if one user visits more indoor hotspots, he or she is more active. As a result, there is a mutual reinforcement relationship between users and indoor locations.

Besides, indoor spaces as three-dimensional spaces have some special features. One feature is that users have to pass by low floors as well as some special locations (elevator room, stair room, hallway, etc.) in order to reach high floors. Thus, locations in low floors or common areas will be more frequently visited by users. However, this is simply because of the special structure of indoor spaces and we cannot regard those locations as indoor hotspots, even they are frequently visited. Rather, we have to consider the floor influence and remove meaningless locations in indoor hotspot detection. In this paper, we propose to use stay time to pre-process indoor trajectory data to detect indoor stay locations as well as indoor stay trajectories before identifying indoor hotspots.

In summary, the general idea of our proposal in indoor hotspot query processing can be summarized as follows:

(1) We consider users' interests in indoor locations and the mutual reinforcement relationship between users and indoor locations when identifying indoor hotspots. In particular, we introduce a *user-location* matrix to compute users' interests in indoor locations and conduct iteration-based computation on the matrix to reflect the mutual reinforcement relationship between users and indoor locations.

(2) We consider stay time to pre-process original indoor moving trajectories, so that we can remove meaningless locations when identifying indoor hotspots.

Figure 2 shows the basic procedure to detect indoor hotspots from indoor trajectory data. First, we conduct a pre-processing step to detect indoor stay locations and further indoor stay trajectories. Next, we build a *user-location* matrix to represent users' interests in indoor locations. In order to reflect the mutual reinforcement relationship between users and indoor locations, we iterate the *user-location* matrix to aggregate the users' contributions to the hotness of indoor locations. Finally, we rank and output top-k indoor locations according to the aggregated hotness.

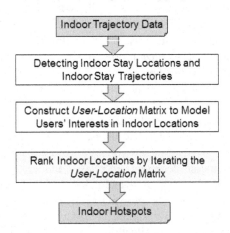

Fig. 2. The basic procedure to detect indoor hotspots

In the following, we first discuss the pre-processing technique for removing meaningless locations in Section 3.2, and then present the indoor hotspot query processing algorithm based on users' interests in locations and mutual reinforcement relationship in Section 3.3.

3.2   Pre-processing Trajectories: Detecting Indoor Stay Trajectories

In the pre-processing step, we use stay time to detect indoor stay trajectories of indoor moving objects. An indoor stay trajectory is a sequence of indoor stay locations with stay time, which is defined as follows.

Definition 5 (Indoor Stay Location). An indoor stay location SL_{mo} of an indoor moving object *mo* is defined as SL_{mo}:

$$SL_{mo} = \{p.s|p \in LOC_{mo} \wedge (p.t_e - p.t_s > \varphi)\},$$

where φ is a pre-defined time threshold.

Definition 6 (Indoor Stay Trajectory). An indoor stay trajectory ST_{mo} of an indoor moving object *mo* is defined as a trajectory that only travels the indoor stay locations of *mo*:

$$ST_{mo} = \left\{ \begin{array}{c} \langle p_1, p_2, \dots, p_n \rangle| \\ (\forall p_i)(p_i \in LOC_{mo} \wedge (p_i.t_e - p_i.t_s > \varphi) \wedge (p_i.t_e < p_{i+1}.t_s)) \end{array} \right\}$$

Algorithm 1: *StayTrajectory*

Input: TR: the original trajectory set; MO: the set of moving objects; φ.

Output: ST: the set of stay trajectories, SL: the set of stay locations.

Preliminary: $ST = (ST_{obj1}, ST_{obj2}, \dots, ST_{objn}), SL = (SL_{obj1}, SL_{obj2}, \dots, SL_{objn})$

1:   $SL \leftarrow \emptyset, ST \leftarrow \emptyset$;

2:   **for** each $obj \in MO$ **do**

3:        **for** each $tr \in TR$ **do**

4:              **if** $tr.mo = obj$ **then**

5:                    $temp \leftarrow staylocation(tr, \varphi)$; //*save the stay locations in tr as a list*

6:                    $SL_{obj} \leftarrow SL_{obj} \cup temp$;

            //*merge the stay locations into the current stay trajectory of the moving object*

7:                    $ST_{obj} \leftarrow merge(ST_{obj}, temp)$;

8:              **end if**

9:        **end for**

10:      $ST \leftarrow ST \cup ST_{obj}$;

11:      $SL \leftarrow SL \cup SL_{obj}$;

12: **end for**

13: **return** ST, SL;

End *StayTrajectory*

Fig. 3. Pre-processing trajectories into stay trajectories

An indoor stay trajectory of a moving object in indoor spaces remains the indoor stay locations that the object stays for a given time duration. The purpose of extracting indoor stay trajectories is to remove the meaningless locations and further improve the performance of indoor hotspot detection. In real scenarios, locations like gates and stair rooms are not appropriate to be recognized as hotspots even though they are frequently visited. Hence, we introduce stay time as a basic criteria to filter those meaningless locations.

Figure 3 shows the detailed procedure of pre-processing trajectories. After the pre-processing, we get a set of indoor stay trajectories and a set of indoor stay locations. The latter then become the candidates of indoor hotspots.

The sub-routine $staylocation(tr, \varphi)$ in Algorithm 1 returns the locations that the object stays over the time threshold φ in the trajectory tr, as well as the stay time. Then, we extract the locations in the list $temp$ as the indoor stay locations for the object obj (Line 6). The function $merge(ST_{obj}, temp)$ is used to merge the indoor stay locations and stay time in $temp$ with the existing elements in ST_{obj}.

3.3   The Algorithm for Indoor Hotspot Query Processing

Based on the indoor stay trajectories of each moving object detected by Algorithm 1, we are able to present the algorithm for indoor hotspot query processing.

As we discussed in Section 3.1, our algorithm is based on users' interests and the mutual reinforcement relationship between users and indoor locations. After the pre-processing step, we get a set of indoor stay locations. For the purpose of simplicity, in the following discussions, we only use the term "*indoor location*" to denote "*indoor stay location*". Also we do not distinguish the term of "*user*" and "*indoor moving object*". They have the same meaning in the context. Then, we first define the users' interests in indoor locations as follows.

Definition 7 (Users' Interests in Indoor Locations). Given a set of users $MO = \{MO_1, MO_2, \ldots, MO_m\}$ and a set of indoor locations $SL = \{SL_1, SL_2, \ldots, SL_n\}$, the interests of MO_i in an indoor location SL_j are defined as follows:

$$ID_j^i = w_s \cdot \frac{AD(SL_j)}{\sum_{k=1}^{|SL|} AD(SL_k)} + (1 - w_s) \frac{FL(SL_j)}{\sum_{k=1}^{n} k \cdot N_{FL(SL_j)}},$$

where $AD(SL_j)$ represents the average time duration of an indoor moving object MO_i staying on the indoor location SL_j, $\sum_{k=1}^{|SL|} AD(SL_k)$ is the total stay time of MO_i on all the indoor locations that it visits. $FL(SL_j)$ refers to the floor that SL_j is located, and we suppose that FL is 1 for the first floor, 2 for the second floor, etc. n is the number of floors, and the count of indoor locations in the floor numbered $FL(SL_j)$ is denoted as $N_{FL(SL_j)}$. w_s is a parameter for balancing the weights of stay time and floor influence on the computation.

Then, we construct a *user-location* matrix to maintain users' interests in indoor locations, as shown in Fig. 4.

The *user-location* matrix initially presents each user's interests in indoor locations. As we explained in Section 3.1, in this paper we consider the mutual reinforcement relationship between users and indoor locations. Basically, if a user shows more interests in more hot locations, the user should be more active. On the other side, if a location is visited by more active users, it should be hotter than others. According to this idea, we define the activeness of a user and the hotness of an indoor location as follows.

	SL_1	SL_2	$\cdots$	SL_j	$\cdots$	$\cdots$	SL_n
MO_1	ID_1^1	ID_2^1	$\cdots$	ID_j^1	$\cdots$	$\cdots$	ID_n^1
MO_2	ID_1^2	ID_2^2	$\cdots$	ID_j^2	$\cdots$	$\cdots$	ID_n^2
	$\vdots$	$\vdots$	$\cdots$	$\vdots$	$\cdots$	$\cdots$	$\vdots$
$\vdots$	$\vdots$	$\vdots$	$\cdots$	$\vdots$	$\cdots$	$\cdots$	$\vdots$
MO_m	ID_1^m	ID_2^m	$\cdots$	ID_j^m	$\cdots$	$\cdots$	ID_n^m

Fig. 4. The *user-location* matrix maintaining users' interests in indoor stay locations

Definition 8 (User Activeness). The activeness of a user in indoor spaces is defined as the sum of the hotness of all the indoor locations visited by the user.

Definition 9 (Location Hotness). The hotness of an indoor location is defined as the sum of the activeness of all the users who visited the location.

Figure 5 shows the illustration of the mutual reinforcement relationship between users and indoor locations.

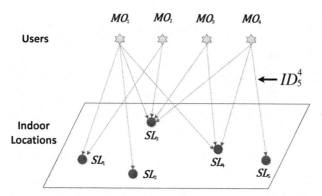

Fig. 5. The mutual reinforcement relationship between users and indoor locations

Suppose that we have a set of users $MO = \{MO_1, MO_2, \ldots, MO_m\}$, a set of indoor locations $SL = \{SL_1, SL_2, \ldots, SL_n\}$, and the *user-location* matrix M, the activeness of a user and the hotness of an indoor location according to a given set of indoor stay trajectories (ST) can be computed by the following equations.

$$activeness(ST, MO_i) = \sum_{j=1}^{n} ID_j^i \cdot hotness(ST, SL_j) \tag{3.1}$$

$$hotness(ST, SL_j) = \sum_{i=1}^{m} ID_j^i \cdot activeness(ST, MO_i) \tag{3.2}$$

Then, given an initial state of the activeness of each user as well as the hotness of each indoor location, we can iterately compute the final activeness and hotness. Let H represents the vector of hotness and A is the vector of activenss, we can transform Equation 3.1 and 3.2 into the matrix form.

$$H = M^T \cdot A \tag{3.3}$$

$$A = M \cdot H \tag{3.4}$$

Algorithm 2: *Hotspots*

Input: $ST = \{ST_1, ST_2, \cdots, ST_m\}$, indoor stay trajectories

Output: $H = \{H_1, H_2, \ldots, H_n\}$, vector of hotness of all indoor locations. $A = \{A_1, A_2, \ldots, A_m\}$, vector of activeness of all indoor moving objects.

Preliminary: w_s is the pre-defined weights of stay time in computing users' interests in locations. ε is the pre-defined parameter to control the ending of iterations on the user-location matrix

1.	**for** $i = 0$ to n **do** $H_i \leftarrow 1/n$; //initializing H
2.	**for** $i = 0$ to m **do** $A_i \leftarrow 1/m$; //initializing A
3.	$M \leftarrow buildingMatrix(ST, w_s)$; //constructing the initial user-location matrix
4.	$M \leftarrow normalizing(M)$; // normalizing M into $[0,1]$
5.	$k \leftarrow 1$;
6.	**repeat**
7.	$\quad k \leftarrow k + 1$;
8.	$\quad H_k = M \cdot M^T \cdot H_{k-1}$;
9.	$\quad A_k = M^T \cdot M \cdot A_{k-1}$;
10.	$\quad H_k \leftarrow normalizing(H_k)$; // normalizing H_k
11.	$\quad A_k \leftarrow normalizing(A_k)$; // normalizing A_k
12.	**until** $|H_k - H_{k-1}|_1 < \varepsilon$ or $|A_k - A_{k-1}|_1 < \varepsilon$
13.	**return** H, A

End *Hotspots*

Fig. 6. The algorithm of detecting indoor hotspots

Next, we can get the vector of hotness after kth iteration, i.e., H_k, as well as the vector of activeness by the following equations.

$$H_k = M^T \cdot M \cdot H_{k-1} \tag{3.5}$$

$$A_k = M \cdot M^T \cdot A_{k-1} \tag{3.6}$$

Based on the above explanation, now we can present the final algorithm of computing indoor hotspots, which is shown in Fig. 6. The algorithm *Hotspots* returns a vector recording the hotness of each indoor location, based on which we can answer indoor hotspot queries defined in Definition 4 in Section 2.2.

4   Performance Evaluation

4.1   Experimental Settings

Data Set. We simulate the building of the department of computer science in our university, and generate indoor tracking data using an indoor data generator called *IndoorSTG* [16]. *IndoorSTG* can simulate different indoor spaces consisting of various elements including rooms, doors, corridors, stairs, elevators, and virtual positioning devices such as RFID or Bluetooth readers. The simulated building has six floors, and there are totally 94 RFID readers deployed to represent different types of indoor elements. We simulate 100 moving objects in such an indoor space during a time period of 30 days, and finally generate 223,564 moving tracking records. These records are then used as the experimental data. Table 1 shows an example of indoor moving tracking data generated by *IndoorSTG*.

Table 1. An example of indoor moving tracking data

Reader_ID	Object_ID	Enter_Time	Leave_Time
1	1	2014-03-06 07:38:00	2014-03-06 07:48:24
5	1	2014-03-06 07:59:24	2014-03-06 08:09:36
12	1	2014-03-06-08:11:32	2014-03-06-08:18:35

Metrics. We mainly focus on the measurement of the effectiveness of indoor hotspot detection. Note that the time performance is not a crucial metric for indoor hotspot detection, because indoor hotspot detection is not an online job and can be executed as a background program in the server storing trajectories. For instance, it is possible for a shopping mall to run the program every weekend to find the hotspots during the past week. Therefore, in our experiments we will not present the results on time performance. In order to measure the effectiveness of indoor hotspot detection, we first ask 10 students at the department of computer science in our university, who are very familiar with the building of computer science, to manually annotate the hotness of each location, using the following scores shown in Table 2. The locations with an average score over 1.5 are annotated as hot locations.

Table 2. Users' interests in an indoor location

Ratings	Description
2	*Very interested*
1	*Interested*
0	*Neutral*
−1	*Uninterested*
−2	*Unknown*

Then, we use three different metrics to measure the effectiveness of indoor hotspot detection, namely *MAP* (Mean Average Precision), *P@n* (Precision for the top-*n* results), and *nDCG* (normalized Discounted Cumulative Gain) [17]. MAP stands for the average precision of each hot location. Suppose that a location with hotness over 1.5 is recognized as a hotspot, and the average hotness of the top-10 locations is <2.0, 0.5, 1.8, 0, 0.8, 1.9, 0.7, 1.2, 0.2, 0.3>, then we can compute MAP as follows:

$$ MAP = \frac{1 + 2/3 + 3/6}{3} = 0.722 $$

nDCG is used to measure the relative-to-ideal performance of information retrieval. The computation of DCG is defined as follows, where $G[i]$ is the hotness of the *i*-th location in the results:

$$ DCG[i] = \begin{cases} G[1] & if\ i = 1 \\ DCG[i-1] + G[i] & if\ i < 3 \\ DCG[i-1] + \frac{G[i]}{log_3 i} & if\ i \geq 3 \end{cases} \tag{4.1} $$

For example, given the hotness of the top-10 results <2.0, 0.5, 1.8, 0, 0.8, 1.9, 0.7, 1.2, 0.2, 0.3>, we can first compute the corresponding *DCG* for the results based on Equation 4.1. Next, we compute the ideal *DCG* (*IDCG*) of the ideal list for the results, which is <2.0, 1.9, 1.8, 1.2, 0.8, 0.7, 0.5, 0.3, 0.2, 0>. Then, the *nDCG* score at the *i*-th position can be computed as $nDCG[i] = DCG[i]/IDCG[i]$.

Configuration. All the involved algorithms are implemented in Java on Windows 7, and run on a PC with an Intel(R) Core(TM) i3-3220 CPU @3.30GHz and 4GB DDR2 memories. The time threshold φ for detecting stay trajectories is set to 10 minutes by default. The influence of φ on the performance will be measured in the following experiments.

Baseline Algorithms. We implement three existing algorithms as the baseline methods, including *rank-by-visit*, *rank-by-density* [8], and *rank-by-duration* [9, 10]. Regarding the *rank-by-visit* algorithm, the hotness of a location is based on the count of visits to the location. For the *rank-by-density* algorithm, the hotness of a location is based on the count of distinct users visiting the location. The *rank-by-duration* method uses the aggregated visiting time duration of a location to rank the hotness of the location.

4.2   Results

We first measure the impact of the weight-parameter W_s on the performance of our algorithm. This parameter is used to balance the influence of stay time and floor level on computing users' interests on locations. Basically, a larger W_s implies that stay time is more important in users' interests on locations. As Fig. 7 shows, when the value of W_s exceeds 0.7, the performance decreases with the increasing of the parameter. We can infer from this result that it is helpful to introduce the influence of floor level into indoor hotspot detection. Especially when $W_s = 1$, our proposal gets the worst performance. Here, our algorithm actually becomes the *rank-by-duration* method that does not consider floor influence. As a result, we find that our algorithm has the best overall performance at $W_s = 0.7$. Thus, we use this setting in the next experiment.

Table 3 shows the effectiveness of our method as well as the three baseline methods, w.r.t. five metrics. We can see that our method outperforms all the baseline methods. Especially when considering *nDCG*, our method get pretty good performance. This is owing to that our method has taken into account the mutual reinforcement relationship between users and indoor locations. Suppose that three users have visited one location for 100 times, our method does not simply count the visits or distinct users (density). In contrast, we consider the activeness of users as well as their interests in each location. For example, if two of the three users are active ones, they will contribute more to the hotness of the location, and if all the three users are inactive users, the hotness of the location will lower than that in the previous case. Our experimental results demonstrate that it is effective to integrate users' activeness, users' interests in indoor locations, and the mutual reinforcement relationships between users and indoor locations, into the indoor hotspot query evaluation.

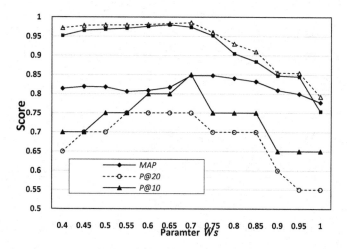

Fig. 7. Impact of the parameter W_s

Table 3. Effectiveness of all the methods

Method Metric	rank-by-visit	rank-by-density	rank-by-duration	**Our Method**
MAP	0.43	0.51	0.57	**0.84**
P@10	0.35	0.47	0.65	**0.85**
P@20	0.25	0.44	0.6	**0.75**
nDCG@10	0.675	0.712	0.773	**0.985**
nDCG@20	0.681	0.728	0.782	**0.973**

Figure 8 shows the influence of the time threshold φ on the performance of our algorithm. The results indicate that the time threshold is influential to the final performance. This is mainly due to the feature of indoor moving trajectories. As the trajectories used in the experiment are generated by *IndoorSTG* [16], which can configure the stay time of users in indoor locations during the data generation. In our experiment, most stay time is configured around 10 minutes, simulating a user's typical stay time when visiting an interested location, e.g., a shop inside a shopping mall. Thus, we can see that when a large time threshold is used in the experiment, i.e., a value over 15 minutes, the precision of indoor hotspot detection becomes relatively low, as many indoor locations are likely to be removed in the pre-processing step. However, in practical applications, e.g., a location-based service system in a shopping mall, we can first conduct a survey or historical data analysis to find out the typical stay time of a user staying in an indoor location, and then use it as the time threshold in the algorithm. Fig. 8 also shows that the pre-processing step is helpful to improve the performance of indoor hotspot detection, owing to its eliminating many frequently-visited but meaningless locations such as gates and stair rooms.

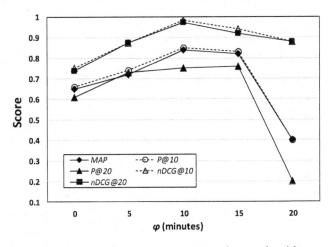

Fig. 8. Influence of the time threshold φ in our algorithm

5   Related Work

Hotspots detection has been a research focus in recent years. However, previous studies mainly focus on outdoor space, e.g., Euclidian space and road-network space. To the best of our knowledge, there are very few researches towards indoor hotspot query.

In 2007, Giannotti et al. first introduced the concept of hot regions among moving trajectories [6]. Alvares et al. proposed the concepts of Stops and Moves, and further presented a new approach to moving trajectory analysis called SMOT [7]. However, Stops were restricted to pre-defined locations such as hotels, sightseeing places, and road connection points, thus other hot regions cannot be found by SMOT. As an improvement to SMOT, CB-SMOT was proposed which can find hot regions that are not pre-defined [9].

There were some other works focusing on detecting moving patterns from trajectories [5, 17, 18]. The main idea of those works was identifying hot regions to find the similarity among users' traveling interests, in which the count of visits to different locations was considered. However, they did not take into account the time duration of each stop in the trajectories. Zheng et al. considered the correlations between users and locations and proposed a tree-based hierarchical graph for modeling locations as well as the computation of location hotness [18]. Further, in [19], Cao et al. proposed a two-layer graph to compute location hotness, which considered many aspects of users' visits, including the count of visits, the distance between locations, the semantic descriptions of locations, and the stop time of users in locations. Uddin et al. proposed a density-based approach to detecting hot regions on moving object trajectories [10]. The density of a location was measured by the number of distinct objects visiting the location. However, these previous works are all towards outdoor space. As indoor spaces introduce many special features, e.g., low-floor locations usually have more visits than high-floor locations, thus we have to devise new approaches for indoor hotspot detection.

So far, the only work close to indoor hotspot detection is conducted by Ahmed et al. [8, 20]. This work aims to detect hotspots from the bags' movement on the conveyers in an airport. They suggested using a density-based algorithm to detect hotspots. However, the conveyor-based indoor space only considers conveyors, which is much different from typical indoor environments such as office buildings, airports, etc. In addition, the proposed density-based approach in [8, 20] only considered the number of objects visiting a location, neglecting other special features of indoor spaces such as location type as well as the relations between users and locations. In this paper we finally compare our proposal with this work and the results have shown that our proposal outperforms it w.r.t. various metrics.

6   Conclusions

In this paper, we studied a new kind of query on indoor moving objects trajectories, which is called indoor hotspot query. We devised a new approach to evaluating such queries, based on users' interests in indoor locations and the mutual reinforcement relationship between users and indoor locations. We also introduced a pre-processing step to remove those frequently-visited but meaningless locations from the original trajectories. To the best of our knowledge, this work is the first one aiming at detecting hotspots in common indoor spaces consisting rooms, doors, hallways. Our experiments on a data set including 223,564 indoor moving trajectories showed that our proposal outperformed several baseline algorithms w.r.t. various metrics.

Our future work will be focused on testing our algorithm on real scenarios. We are currently collaborating with a shopping mall and planning to deploy Wi-fi infrastructure to collect real data from customers. Another future work is to develop effective and efficient personalized recommendation solutions for indoor spaces.

Acknowledgements. This work is supported by the National Science Foundation of China (61379037, 61472376, & 61272317) and the OATF project funded by University of Science and Technology of China.

References

1. Jensen, C.S., Lu, H., Yang, B.: Indoor - A New Data Management Frontier. IEEE Data Engineering Bulletin **33**(2), 12–17 (2010)
2. Li, Q., Jin, P., Zhao, L., Wan, S., Yue, L.: IndoorDB: extending oracle to support indoor moving objects management. In: Meng, W., Feng, L., Bressan, S., Winiwarter, W., Song, W. (eds.) DASFAA 2013, Part II. LNCS, vol. 7826, pp. 476–480. Springer, Heidelberg (2013)
3. Hadjieleftheriou, M., Kollios, G., Gunopulos, D., Tsotras, V.J.: On-line discovery of dense areas in spatio-temporal databases. In: Hadzilacos, T., Manolopoulos, Y., Roddick, J., Theodoridis, Y. (eds.) SSTD 2003. LNCS, vol. 2750, pp. 306–324. Springer, Heidelberg (2003)
4. Jensen, C.S., Lin, D., Ooi, B.C., Zhang, R.: Effective density queries on continuously moving objects. In: Proc. of ICDE, p. 71 (2006)
5. Li, X., Han, J., Lee, J.-G., Gonzalez, H.: Traffic density-based discovery of hot routes in road networks. In: Papadias, D., Zhang, D., Kollios, G. (eds.) SSTD 2007. LNCS, vol. 4605, pp. 441–459. Springer, Heidelberg (2007)
6. Giannotti, F., Nanni, M., Pinelli, F., Pedreschi, D.: Trajectory pattern mining. In: Proc. of KDD, pp. 330–339 (2007)
7. Alvares, L.O., Bogorny, V., Kuijpers, B., et al.: A model for enriching trajectories with semantic geographical information. In: Proc. of GIS, p. 22 (2007)
8. Ahmed, T., Pedersen, T.B., Lu, H.: Capturing hotspots for constrained indoor movement. In: Proc. of GIS, pp. 462–465 (2013)
9. Palma, A.T., Bogorny, V., Kuijpers, B., Alvares, L.O.: A clustering-based approach for discovering interesting places in trajectories. In: Proc. of SAC, pp. 863–868 (2008)
10. Uddin, M.R., Ravishankar, C.V., Tsotras, V.J.: Finding regions of interest from trajectory data. In: Proc. of MDM, pp. 39–48 (2011)
11. Dudas, P., Ghafourian, M., Karimi, H.: ONALIN: Ontology and algorithm for indoor routing. In: Proc. of MDM, pp. 720–725 (2009)
12. Kim, J., Kang, H., Lee, T., et al.: Topology of the prism model for 3D indoor spatial objects. In: Proc. of MDM, pp. 698–703 (2009)
13. Wang, N., Jin, P., Xiong, Y., Yue, L.: A multi-granularity grid-based graph model for indoor space. International Journal of Multimedia and Ubiquitous Engineering **9**(4), 157–170 (2014)
14. Jensen, C.S., Lu, H., Yang, B.: Graph model based indoor tracking. mobile data management. In: Proc. of MDM, pp. 17–24 (2008)
15. Jin, P., Zhang, L., Zhao, J., Zhao, L., Yue, L.: Semantics and modeling of indoor moving objects. International Journal of Multimedia and Ubiquitous Engineering **7**(2), 153–158 (2012)
16. Huang, C., Jin, P., Wang, H., Wang, N., Wan, S., Yue, L.: IndoorSTG: a flexible tool to generate trajectory data for indoor moving objects. In: Proc. of MDM, pp. 341–343 (2013)
17. Manning, C.D., Raghavanm, P., Schütze, H.: An Introduction to Information Retrieval. Cambridge University Press (2008)

18. Zheng, Y., Zhang, L., Xie, X., Ma, W.-Y.: Mining interesting locations and travel sequences from GPS trajectories. In: Proc. of WWW, pp. 791–800 (2009)
19. Cao, X., Cong, G., Jensen, C.S.: Mining significant semantic locations from GPS data. PVLDB 3(1), 1009–1020 (2010)
20. Ahmed, T., Pedersen, T.B., Lu, H.: Finding dense locations in indoor tracking data. In: Proc. of MDM, pp. 189–194 (2014)
21. Schafer, M., Knapp, C., Chakraborty, S.: Automatic generation of topological indoor maps for real-time map-based localization and tracking. In: Proc. of IPIN, pp. 1–8. IEEE CS (2011)

On Efficient Passenger Assignment for Group Transportation

Jiajie Xu[1,5]([✉]), Guanfeng Liu[1,5], Kai Zheng[2], Chengfei Liu[3],
Haoming Guo[4], and Zhiming Ding[4]

[1] Department of Computer Science and Technology,
Soochow University, Suzhou, China
{xujj,gfliu}@suda.edu.cn
[2] School of Information Technology and Electrical Engineering,
The University of Queensland, Brisbane, Australia
kevinz@itee.uq.edu.au
[3] School of Software and Electrical Engineering,
Swinburne University of Technology, Melbourne, Australia
cliu@swin.edu.au
[4] Institute of Software, Chinese Academy of Sciences, Beijing, China
{haoming,zhiming}@iscas.ac.cn
[5] Collaborative Innovation Center of Novel Software Technology and
Industrialization, Nanjing, China

Abstract. With the increasing popularity of LBS services, spatial assignment has become an important problem nowadays. Nevertheless most existing works use Euclidean distance as the measurement of spatial proximity. In this paper, we investigate a variant of spatial assignment problem with road networks as the underlying space. Given a set of passengers and a set of vehicles, where each vehicle waits for the arrival of all passengers assigned to it, and then carries them to the same destination, our goal is to find an assignment from passengers to vehicles such that all passengers can arrive at earliest together. Such a passenger assignment problem has various applications in real life. However, finding the optimal assignment efficiently is challenging due to high computational cost in the fastest path search and combinatorial nature of capacity-constrained assignment. In this paper, we first propose two exact solutions to find the optimal results, and then an approximate solution to achieve higher efficiency by trading off a little accuracy. Finally, performances of all proposed algorithms are evaluated on a real dataset.

1 Introduction

Consider a set of passengers P and a set of vehicles V, all distributed on road networks. Each vehicle waits for a group of passengers (assigned to it) and carries them to a common destination d. Our objective is to find the passenger assignment $A \subseteq P \times V$ that can enable all passengers as a whole to arrive destination d at the earliest possible time, and vehicles are not allowed to carry

© Springer International Publishing Switzerland 2015
M. Renz et al. (Eds.): DASFAA 2015, Part I, LNCS 9049, pp. 226–243, 2015.
DOI: 10.1007/978-3-319-18120-2_14

more passengers than their capacity limits. Such a problem is called passenger assignment for group transportation, which can find various applications in real life.

An illustrative example is shown in Figure 1 to describe the passenger assignment problem. Assume that a set of passengers (i.e. $p_1 - p_6$), a set of vehicles (i.e. v_1, v_2) and the final destination d are distributed on road network as Figure 1(a), and each vehicle can carry no more than three people. All passengers are required to go to d (by taking vehicles) as soon as possible for some purposes. To save time, it adopts an assemble-based-group-transportation fashion, i.e., passengers go to assemble on vehicle (by private car or taxi) like Figure 1(b), and then vehicles carry them to the final destination. The key issue is how to find the optimal assignment from passengers to vehicles efficiently such as Figure 1(b), such that all passengers can arrive the destination and start the activities together as early as possible. Likewise, the passenger assignment problem can be widely used in applications like logistic control, resource supply and other group transportation recommendation systems, etc.

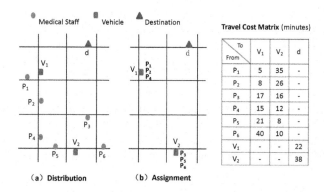

Fig. 1. An example of passenger assignment

Motivated by the above example, this paper studies the group transportation oriented passenger assignment problem, which tends to have high computational overhead for two main reasons: first, the large number of combinations of assignment from P to V makes the search space to be extremely large, particularly when the size of passenger or vehicle sets goes up; secondly, numerous fastest path queries (FPQ) need to be processed for deriving the travel cost from passengers to vehicles (i.e. information in travel cost matrix in Figure 1).

Capacity constrained assignment is a classical research problem that has been well studied in literature [4–6]. Recently lots of efforts have been made to address the capacity constrained spatial assignment (CCSA) problems [17,19], by using Euclidean distance as the measurement of spatial proximity between two objects. However, existing CCSA solutions cannot be simply applied to our problem for two reasons: (1) The goal of optimization is different. Most CCSA algorithms are mainly designed to minimize the sum of Euclidean distance of all assigned object pairs, while the passenger assignment problem seeks to minimize the travel cost for those most time-consuming passengers, which provides us opportunities to further speed up the assignment processing; (2) The spatial proximity measurement is different. The network-based spatial proximity used in the passenger assignment problem causes a large number of fastest path searches, which is generally regarded as much more computationally expensive than the evaluation of Euclidean distance.

In this paper, we present a novel strategy to find the optimal assignment based on maximum bipartite matching. It utilizes a set of pruning mechanisms sensitive to the most-time-consuming-passenger, so that the search can be constrained in small bipartite sub-graphs and in less number of loops. However, though the bounds used in the strategy ensure it has a fairly good performance on assignment step, the overall efficiency is still not satisfactory because of the vast processing cost of the scalable FPQs. Therefore, an approximate solution is further proposed to trade off a little accuracy for higher efficiency. The main contribution of this paper can be summarized as follows:

- We define a problem called passenger assignment for group transportation, which could potentially benefit many applications such as emergency response, supply chain management and traffic planning.
- We propose a novel assignment strategy based on maximum bipartite matching, by which the optimal passenger assignment can be efficiently found.
- We propose an approximate FPQ querying based assignment strategy, which utilizes bounded travel cost computation to reduce search space. It significantly improve the efficiency with a little sacrifice on accuracy.
- We implement the proposed algorithms and conduct experiments on real dataset to evaluate the performances of our proposed solutions.

The rest of the paper is organized as follows. Section 2 presents the related work and Section 3 formally defines the passenger assignment problem. Afterwards, we introduce two exact passenger assignment algorithms in Section 4, and an approximate solution called TN-FPS in Section 5. After discussions on experimental results in Section 6, the paper is concluded in Section 7.

2   Related Works

Assignment is a classical problem that has been studied intensively, with many classical matching problems of bipartite graph, such as the maximum perfect matching (MPM) and minimum weight perfect matching (MWPM) problem. So far, there exist a lot of literatures towards bipartite graph matching in the operational research area. A well-known solution is the classical Hungarian algorithm and its variations [1,10,13], both having $O(mn + n^2logn)$ time complexity. Later, increasing attention are paid to assignment with capacity constraints, with many algorithms proposed to address the capacity constrained assignment (CCA) problem, such as successive shortest path algorithm [3][17].

More recently, with the fast development of mobile computing, the problem of assignment and matching on spatial objects becomes popular [14,18,20]. Specifically, [18,20] studied the spatial matching problem, which is reduced to the stable marriage problem for spatial objects, and [17,19] studied the CCA problems for objects in Euclidean space. However many CCA applications are road network constrained, and they cannot be well supported by [17,19] because of their criterions and Euclidean distance measure adopted, like the passenger assignment problem of this paper. Therefore, new solutions are needed to support efficient assignment processing for applications in road networks like resource dispatch, evacuation in disaster, intelligent services and supply chain management.

In addition, another key technique our passenger assignment problem relies on is spatial and distance query processing. The basic type of distance query is the shortest path query (SPQ). Typical solutions for SPQ include Dijkstra and A* algorithm, which traverses the road network nodes in ascending order of their distances from the query position, and runs in $O(nlogn+m)$ time by using Fibonacci heap. [9] discussed SPQ problem on complex terrain space. Recently, many literatures tried to exploit the hierarchical structure of road map in a pre-processing step, and then properly use it to accelerate SPQ processing, such indexes mainly include the highway hierarchies (HH) [15], contraction hierarchies (CH) [7], and transit-node routing (TNR) [8] algorithms. More recently, the work [16] proposed to pre-compute certain shortest path distances called *path oracles* to answer approximate SPQ in $O(log|V|)$ time and $O(\frac{|V|}{\epsilon^2})$ space with an error bound of ϵ.

Passenger Assignment requires scalable FPQ to find the time cost between objects in P and V. Compared to SPQ, the FPQ is much more challenging because real-time traffic condition needs to be considered, resulting road network with dynamic travel cost on the edges. All the above algorithms rely on heavy pre-processing, and are thus not suitable for dynamic scenarios where the road network topology or edge weights may change frequently. As a result, we have to use some approximate FPQ algorithms, so as to improve the efficiency of query processing. Moreover it is necessary to design robust matching algorithms that can find good solution based on approximate FPQ results.

3   Problem Definition

3.1   Spatial Networks

A spatial network is modeled by a connected and undirected graph $G = (N, E)$, where N is a set of vertices and E is a set of edges. A vertex $n_i \in N$ indicates a road intersection or an end of a road. An edge $e \in E$ is defined as a pair of vertices and represents a segment connecting the two adjacent vertices. For example, edge $e = \{n_i, n_j\}$ represents a road segment that enables travel between vertices n_i and n_j. We use $time(e)$ to denote the time required to pass a road segment e based on real-time traffic condition. Given two locations a and b in spatial network, its fastest path $FP_{a,b}$ is a sequence of edges linking a and b with the minimal total travel cost. We use $TC_{a,b} = \sum_{e \in FP_{a,b}} time(e)$ to represent the travel cost between a and b. Note that the travel cost between any two vertexes may update with the road condition changes.

All passengers in P and vehicles in V are embedded in networks and they may be located on edges. If the network distances to the two end vertices of an edge are known, it is straightforward to derive network distance to any point in this edge. Thus, we assume that all data points are on vertices for simplification purpose.

3.2   Passenger Assignment for Group Transportation

Let P be a set of passengers and V a set of vehicles, all distributed on a spatial network G. Each $p \in P$ denotes a passenger with his or her geographical location $p.l$, and each $v \in V$ denotes a vehicle that can carry up to $v.c$ passengers. Each passenger in P moves to an assigned vehicle in V first. Each vehicle starts when its passengers arrive and transport them to a common destination d.

For example in disaster management scenario, medical staff members go to a special vehicle first, and special vehicles then carry them to rescue mission place (where might be dangerous) after all staff members assigned to this vehicle have assembled. To plan the routes for all medical staffs, we need to assign them to special vehicles, and the result is called an assignment. We first define the notion of a valid assignment to judge if an assignment is a qualified result.

Definition 1. *(Valid Assignment) An assignment $A \subseteq P \times V$ is said to be a valid assignment if it satisfies:*

(1) capacity constraint, i.e. each vehicle $v \in V$ appears at most $v.c$ times in assignment A due to its capacity constraint;

(2) assignment A must be full to P, i.e. each passenger $p \in P$ must appear and only appear once in A, i.e. $|A| = |P|$, and $p \neq p'$ for any two pairs $(p, v) \in A$ and $(p', v') \in A$.

For each pair $(p, v) \in A$ in assignment A, passenger p moves from $p.l$ toward the location of the assigned vehicle $v.l$ first. Later, each vehicle v carries $\{p' | (p', v) \in A\}$ to destination d. Obviously, the arrival time of all passengers in P is determined by the one having maximum total travel cost, so we need to define two important notions 'pair cost' and 'critical pair'.

Definition 2. *(Pair Cost) For each pair (p, v) where $p \in P$ and $v \in V$, the cost of this pair $PC_{p,v}$ is measured as the total travel cost from $p.l$ to destination d via $v.l$ on spatial network G such that*

$$PC_{p,v} = TC(p.l, v.l) + TC(v.l, d)$$

The pair cost of (p_2, v_1) in Figure 1 is thus computed as $PC_{p_2,v_1} = 8 + 22 = 30$.

Definition 3. *(Critical Pair of an Assignment) The critical pair $CP(A)$ of an assignment A is the pair $(p, v) \in A$ that has the maximum pair cost $PC_{p,v}$, and we call the passenger p as critical passenger.*

To ensure that all passengers in P can arrive destination in earliest, e.g. for the rescue missions or goods supply in emergency management scenarios, we further define the cost of a valid assignment based on its critical pair. For example in Figure 1, if assignment is made as Figure 1(b), then the pair (p_3, v_2) is the critical pair because of its greatest pair cost $PC_{p_3,v_2} = 54$.

Definition 4. *(Cost of an Assignment) The cost* $\psi(A) = max(\{PC_{p,v} \mid (p, v) \in A\})$ *of a valid assignment* A *is quantified to be the pair cost of its critical pair, meaning the maximal travel cost of all passengers in* P *on the spatial network.*

Problem Formalization. Given a spatial network G, a passenger set P and a vehicle set V as input. Among all valid assignments from set P to set V, we aim to find the one A with the minimal assignment cost $\psi(A)$, which is determined by the pair cost (i.e. total travel cost) of A's critical pair based on G.

4   Exact Algorithms

This section presents two exact methods called IPA and MBMA for computing the optimal passenger assignment.

4.1   Integer Programming Based Assignment (IPA)

Integer programming is known to be a widely used processing model for optimization problems. In this section, we introduce an effective integer programming based assignment solution.

As passenger assignment here is a road network constrained problem, we process the FPQs (between passengers/vehicles and vehicles/destination) first to derive all necessary travel cost information by the Dijkstra algorithm. After that, we find the valid assignment A with minimum value of $\psi(A)$ in the assignment step based on the pair cost information derived from the previous step. Valid assignment means to satisfy the capacity constraint of vehicles, and ensures each passenger to appear in A once and only once. Based on above constraints and goals, we notice that the passenger assignment problem can be expressed as the following integer program, where $x = (p, v)$ represents the 0-1 integer vector of the assignment between $p \in P$ and $v \in V$:

$$\text{minimize} \quad \tau(A) = max(\{PC_{p,v} \times x(p, v) \mid (p, v) \in A\})$$

$$\text{subject to} \quad x(p, v) = 0 \ or \ 1 \qquad \forall p \in P, v \in V$$

$$\sum_{p \in P} x(p, v) \leq v.c \quad \forall v \in V$$

$$\sum_{v \in V} x(p, v) \leq 1 \qquad \forall p \in P$$

In general, the purpose of above 0-1 program is to find the passenger assignment A such that: (1) to maximize value of $\tau(x)$ for the returned assignment; (2)

all of the subject conditions can be satisfied. Particularly, only part those pairs (p, v) having that $x(p, v) = 1$ are included in assignment A, meaning that the assignment from p to v is made. It is obvious that the final result A derived by above program is an assignment being valid and optimal. But the IPA algorithm is inefficient in most cases because of the 'max' operator used in optimization goal (i.e. to get the maximum pair cost), which leads to no bound to be easily found and used to stop the iterations. In contrast, the pruning effect would be greatly improved for cases that to minimize linear expression or using 'sum' operator, because an upper bound (i.e. current minimal value) is utilized to help us break in the middle. Therefore, we further propose another method to find optimal assignment in a more efficient manner.

4.2   Maximum Bipartite Matching Based Assignment (MBMA)

In this section, we introduce an MBMA strategy that can return optimal results efficiently. By constructing a bipartite graph $BG = (P, V, E)$, where the weight of an edge $e(p, v) \in E$ in bipartite graph $w(e) = PC_{p,v}$ is the pair cost between passenger p and vehicle v. We aim to find an valid assignment $A \subseteq P \times V$ on bipartite graph BG with minimum $\psi(A)$. Existing algorithms (e.g. Hungarian algorithm for maximum matching, and Kuhn-Munkres algorithm for minimum weight matching) cannot be applied because of the different optimization criterion and capacity constraints we face in our work. In this paper, we design a bipartite matching based algorithm that is more suitable for passenger assignment. The MBMA algorithm works with the following steps:

Step 1. Assignment Initialization Step. It first initializes a valid assignment A based on input bipartite graph using a given maximum bipartite matching strategy tailored for our problem;

Step 2. Assignment Improvement Step. We prune some high weighted bipartite edges, and re-assigns some matches in A for better assignment on a bipartite sub-graph;

Step 3. Iterative Step. Assignment improvement is processed in loop until it is not possible to perform the assignment improvement step.

4.2.1 Assignment Initialization

Given an input bipartite graph BG, we conduct the assignment operation to derive a valid assignment. Among the existing maximum bipartite matching (MBM) solutions, the Hungarian algorithm is the most famous one, and it finds the maximum matching by finding an augmenting path from each $v \in V$ to V' and adding it to the matching if it exists. We know that each augmenting path can be found in $O(|E| + |V \cup V'|)$ time, and we need to find $|V \cup V'|$ times of the augmenting path. Therefore, we can find maximum matching on $O(mn + n^2 log n)$ time, where $m = |E|$ and $n = |V \cup V'|$. It is a useful technique to our problem in finding a valid assignment from P to V.

Compared to the classical MBM, the assignment for the passenger assignment in this paper must be capacity aware. As stated in Section 3.4, capacity constraint can be addressed by vehicle-to-capacity-unit transformation, but such transformation incurs a much greater bipartite edge set, which in turn leads to more time cost. To handle this problem, we propose an assignment initialization algorithm tailored for our problem as shown in Algorithm 1.

The basic idea of assignment here is similar to Hungarian algorithm, i.e. to explore maximum matching on BG by finding augmenting path from each $p \in P$ to V and adding it to the matching if it exists. Specifically, let A be the matching of BG, a vertex $p \in P$ or $v \in V$ is matched if it is endpoint of edge in A, and it is free otherwise. A path is said to be alternating if its edge alternate between A and $E - A$. An alternating path is augmenting if both endpoints are free, and it has one less edge in A than in $E - A$. The assignment algorithm continuously replace the edges in augmenting path from A by the ones in $E - A$ to increase the size of the matching until it cannot be enlarged.

Algorithm 1. Assignment Initialization Algorithm

  input : input bipartite graph BG
  output: assignment result A
1 $A = \phi$;
2 **do**
3     $p = \text{FIND-AUGMENTING-PATH}(BG, A)$;
4   **if** $p \neq NIL$ **then**
5        $A = A \oplus p$;
6   **end**
7 **while** $p = NIL$;
8 **return** A;

The function FIND-AUGMENTING-PATH(BG, A) of Algorithm 1 (Line 3) means to find an augmenting path based on A. This is the key issue of this algorithm as it determines both accuracy and efficiency: (1) from accuracy perspective, it must be capacity aware. Therefore, a counter $v.cn$ is created for each $v \in V$ to record the times it has been assigned to, and augmenting path via it would be denied if $v.cn > v.c$; (2) from efficiency perspective, we hope the selected augmenting path can be covered by a valid assignment (i.e. feasibility), and also to reduce its assignment cost.

Particularly, we use some heuristics to speed up the assignment processing. To reduce the cost of assignment, edges with lower weight are encouraged to be chosen. To ensure the augmenting path to be part of valid assignment cover, we tend to avoid using $v \in V$ with low flexibility value $F(v)$ to reserve it for possible future use, and its flexibility is defined as:

$$F(v) = \frac{|v.c - v.cn|}{\sum_{(p,v) \in E - A_{temp}} PR(p, v)}$$

where A_{temp} is a temporal assignment result in processing, and $PR(p, v)$ is the probability of $(p, v) \in A$. In computation, we calculate it as $PR(p, v) = \frac{1}{degree(p)}$, because p has $degree(p)$ candidates for assignment in total. Obviously, $\sum_{(p,v) \in E - A_{temp}} PR(p,v)$ is the total possibility of v to be assigned accordingly. After assignment initialization, a valid assignment can be found if there is any, but the assignment may not be optimal. Therefore, we further seek to improve the assignment result in the next section.

4.2.2 Assignment Improvement

The general idea of assignment improvement is to find a better assignment on a subgraph with edges $E_{sub} \subseteq E$ of bipartite graph BG. The pruning of bipartite edge is thus vital to determine the accuracy and efficiency. Basically, we hope to filter out bipartite graph edges to form such a bipartite sub-graph: firstly, the size of bipartite edge set E_{sub} is supposed to be less for efficient matching purpose; secondly, all edges (i.e. passenger-vehicle pairs) in optimal assignment must be preserved in the sub-graph.

To achieve above two goals, we conduct *(1) relevance driven edge pruning* and *(2) improvement driven edge pruning* in sequential order, to filter out hopeless bipartite edges and edges unlikely to improve assignment result respectively. Then *(3) improved assignment search* is made on the sub-graph after pruning.

(1) Relevance Driven Edge Pruning. In relevance driven pruning, we try to prune out hopeless bipartite edges, e.g. edges with weights greater than the upper bound of assignment cost, as they are not relevant to query processing anymore. Let function $minW(E)$ take input as an edge set E of bipartite graph BG, and return the minimal weight of the edges in E. To facilitate the derivation of upper bound of assignment cost, we require the edge set E_{sub} of sub-graph to be *weight − bounded* as defined below.

Definition 5. *(Weight-bounded) An edge set $E_{sub} \subseteq E$ is said to be* weight-bounded *if it satisfies:*

$$minW(E - E_{sub}) \leq w(e) \qquad \forall e = (p,v) \in E_{sub}$$

Therefore, a weight-bounded edge set E_{sub} contains those and only those edges in E that have weight less than or equal to a threshold $minW(E - E_{sub})$. Conversely, all remaining edges in $E - E_{sub}$ have weight (i.e. pair cost) greater or equal to that threshold. Suppose that we are given a weight-bounded edge set E_{sub}, and a valid assignment can be found based on E_{sub}, the following theorem determines the upper bound of optimal assignment, and can be used to filter out non-relevant bipartite edges (to optimal assignment).

Lemma 1. *If a valid assignment A' is found from weight-bounded edge set $E_{sub} \subseteq E$, then the upper bound of the cost of optimal assignment A is $minW(E - E_{sub})$, and we have $A \subseteq E - E_{sub}$.*

Proof. Consider the edges in E_{sub}. First, their edge weights are less or equal to $minW(E - E_{sub})$. Second, for any bipartite edge $e = (p,v) \in E$ its edge weight is defined as the pair cost $w(e) = PC_{p,v}$. Given $A' \subseteq E_{sub}$, and we have $\psi(A') \leq minW(E - E_{sub})$ accordingly. As A is the optimal assignment, i.e. $\psi(A) \leq \psi(A')$, it must also hold that $\psi(A) \leq minW(E - E_{sub})$. Therefore, optimal assignment A has an upper bound of assignment cost at $minW(E - E_{sub})$. For $A \subseteq E - E_{sub}$, We can prove it by contradiction, i.e. UB would not be the upper bound if there exists an edge $(p, v) \in (E - E_{sub}) \cap A$ as we have $UB < minW(E_{sub}) \leq PC_{p,v}$ in such case.

Lemma 1 informs us how to conduct bipartite edge pruning based on the upper bound of cost assignment: assume that we can find a valid assignment A at cost $\psi(A)$ from edge set E_i at loop i, then the upper bound of assignment cost becomes $UB = \psi(A)$, and all of the hopeless bipartite edges $\{e \mid e \in BG.E \wedge w(e) > UB\}$ are pruned from the valid edge set E_V (Line 1). Such an upper bound based pruning is definitely meaningful, but not enough yet, because it is unlikely to find an assignment much better than A in the next loop that carries out on E_V. In contrast, we do hope the optimal assignment can be detected in just a few loops for efficiency purpose.

(2) **Improvement Driven Edge Pruning.** After the relevance driven edge pruning, additional edges (especially those with higher weight) must be pruned as well, so that the assignment result can be improved. Based on E_V after relevance pruning, the problems we face are: (1) the priority of edges for pruning; (2) the ratio of bipartite edges to be preserved. The first problem is relatively easy, as edges with higher weight (i.e. greater pair cost) tend to be removed. We focus on discussing the second problem here.

As each of the bipartite edges in E_V has the potential to be part of optimal assignment, in reality, the ratio of filtering is a trade-off between accuracy and efficiency: the higher the ratio is, the better assignment result we tend to have (as higher weight edges are pruned), but the less possible to be able to successfully find one (as the less edge candidates we have); on the other hand, the lower the ratio is, the more likely to find a valid assignment, even though the improvement tends to be not significant. How to find a good balance is an important but challenging problem here.

We notice that the trade-off balance of improvement driven pruning is subject to the lower bound of assignment cost in reality. The lower bound can be derived by two lemmas. Let function $\xi(p)$ to denote the minimal pair cost of all possible pairs $\{(p, v) \mid v \in V\}$ associated to a passenger $p \in P$, we have the following lemmas to find a static lower bound of assignment cost.

Lemma 2. *Given a bipartite graph BG, $min(\xi(p \in P))$ is a lower bound of assignment cost.*

Proof. From the view of P, each $p \in P$ must be assigned. If there is an assignment A that $\psi(A) < min(\xi(p \in P))$, then the $p \in P$ leading to $min(\xi(p \in P))$ is not assigned as no associated edge can be used. Therefore A must not be a valid assignment, and we thus have $\psi(A) \geq min(\xi(p \in P))$.

As we can see, Lemma 2 can give us a static lower bound of assignment cost, which can be computed based on the input data. Furthermore, Lemma 3 can help us to update the lower bound along with the assignment processing.

Lemma 3. *If a valid assignment cannot be found on a weight-bounded edge set* $LB = E_{sub} \subseteq E$, *then* $minW(E - E_{sub})$ *is a lower bound of assignment cost.*

Proof. For any valid assignment A, we know $A \subsetneq E_{sub}$, so there must be an edge $e = (p, v)$ such that $e \in E - E_{sub}$ and $e \in A$. Given that $\psi(A) \geq w(e) \geq minW(E - E_{sub})$, then we know $LB = minW(E - E_{sub})$ is the lower bound of assignment cost.

The lower bound of assignment cost is an important parameter, which is used to divide edges in E_V into two sets, i.e. one set $E_C = \{e|e \in E_V \wedge w(e) \in [LB, UB]\}$ and another set $E_V - E_C$ (weight bounded to E_C), towards which different criterions are used. For $E_V - E_C$, we preserve all of them because their edge weights are even less than LB (definitely not critical pair). In contrast, edges in E_C are potential critical pair, so improvement driven pruning on E_C is necessary. A straightforward pruning approach is the ϵ ($0 < \epsilon < 1$)cut pruning method (ϵCP-method), through which we only keep a ratio of ϵ edges in E_C with minimum edge weights. Though practical, its performance relies on parameter ϵ that cannot be set in a rational and automatic way.

To reduce the loops in assignment processing, a more intelligent method is thus highly sought after to set ϵ in rational. Basically, the value of ϵ is subject to two factors: (1) the abundance of choices, measured as the square of ratio between the number of requested edges to that of valid edges in E_V; (2) the ratio of E_C in E_V, where we tend to be more aggressive (smaller ϵ) if their ratio is greater, and to be conservative otherwise (e.g. binary cut $\epsilon = 0.5$). Putting the two factors together, we can normalize $\epsilon = \sqrt{\frac{|P|}{|E_V|}} \times \frac{|E_V|-|E_C|}{2 \cdot |E_V|}$, where $\frac{|E_V|-|E_C|}{2 \cdot |E_V|}$ means to be aggressive when majority of valid edges falls in E_C, and the value of ϵ is in the range of $[0, 1]$.

(3) Improved Assignment Search. Based on the sub-graph BG' after pruning, we try to find an improved assignment result by bipartite matching. But the improvement driven pruning may fail to find a valid assignment. We say assignment improvement is successful if a valid assignment can be found, and unsuccessful otherwise. In cases it is unsuccessful, we can have a more precise lower bound of assignment cost from Lemma 3, to adjust the bipartite sub-graph to find a valid assignment in the next round. We thus further discuss how to execute iteratively to find the optimal assignment.

4.2.3 Iterative Processing

In this part, we discuss how to iteratively improve the assignment until an optimal assignment can be derived, particularly about the iterative processing procedure, the reuse of intermediate results, and stop condition. Algorithm 2 shows the mechanism of MBMA algorithm that put together assignment initialization and improvement steps. The processing starts from the assignment initialization (Line 2), and moves to improvement step if a valid assignment is found. In improvement step, we use the upper and lower bounds to guide assignment evolution. The upper bound and lower bound are initialized and updated based on the Lemmas in the previous section. If an improvement is successful, the upper bound is updated (Line 12); Otherwise, we adjust the lower bound such that it gets increased (Line 16). Improvement is made in loop until the lower bound equals to the upper bound, indicating the optimal assignment is found.

Algorithm 2. MBMA Algorithm

---

    input   : Passenger set P, vehicle set V, and road network G
    output: optimal assignment A
 1 bipartite graph $BG = construct(O, V, G)$;
 2 $A = assignment(BG)$;
 3 **if** A *is NIL* **then**
 4 | return NIL;
 5 **end**
 6 **else**
 7 | $LB = min(\xi(p \in P))$; $UB = \psi(A)$;
 8 | $E_{sub} = \{e \mid w(e) \leq UB\}$;
 9 | **do**
10 | | $A = assgnImpr(BG, UB, LB)$;
11 | | $E_{bcp} \leftarrow$ set of edges pruned by binary cut pruning;
12 | | **if** A *is not NIL* **then**
13 | | | $UB = \psi(A)$; $E_{sub} = \{e|w(e) \leq UB\}$;
14 | | **end**
15 | | **else**
16 | | | $E_{sub} = E_{sub} - E_{bcp}$; $LB = minW(BG.E - E_{sub})$;
17 | | **end**
18 | **while** $LB \geq UB$;
19 | **return** A;
20 **end**

In addition, we further optimize the assignment processing by two points. In assignment improvement, we only adjust the assignment result in the previous loop based on the new bipartite sub-graph (after pruning). It is thus not necessary to do the complete matching on bipartite graph in each loop, as an improved assignment tends to be found in just a few operations.

Complexity Analysis. The computational overhead of MBMA algorithm mainly comes from spatial query processing and assignment computation. Identical to IPA algorithm, we need $|P| + |V|$ times of fastest path search, hence the time cost of spatial query processing is $O((|P| + |V|) \times (|G.E| \times |G.V| + |G.V|^2 log|G.V|))$; As for assignment processing, assume m and n are the sizes of edge and vertex sets of the used bipartite graph or subgraph respectively, we go through up to $O(m)$ loops for assignment improvement according to Algorithm 4. In each loop, we find the maximum weight matching, with a time complexity $O(mn + n^2 log\, n)$. The MBMA algorithm thus costs $O(m^2n + mn^2 log\, n)$ time in the worst case.

5 TN-FPS Based Algorithm

Above solutions tend to be time consuming because of the high cost on scalable FPQs processing by A* or Dijkstra, so we further use approximate FPQ querying techniques to speed up the execution. Motivated by [8], this paper adopts a Transit Node based Fastest Path Search (TN-FPS) algorithm to find approximate FPQs results based on the *transit nodes* (i.e. important traffic intersections) selected by historical trajectory data. Particularly, the approximations of TN-FPS are only allowed if they have no or few effect on assignment accuracy. Though TN-FPS may also affect assignment precision, the effect is usually trivial because of the *observation* that passengers tend to be assigned to close vehicles, rather than those far-away.

Above observation provides us important guidelines for algorithm design: for close passenger and vehicle pairs, to return their exact fastest path for assignment precision purpose; for long distance pairs, to derive approximate fastest path for efficiency purpose. Given two far-away locations a/b (that close to n/n'), the travel cost can thus be approximated by the following equation:

$$apprTC(a, b) = TC_{a,n} + TC_{n,n'} + TC_{n',b}$$

$apprTC(a, b)$ is the travel cost if a passenger goes to b from a via n and n'. Obviously, it is an upper bound of the accurate travel cost, and the assignment cost is thus not under-estimated when approximation occurs.

The number of transit nodes (tn) is in reality an efficiency and accuracy trade-off: the more transit nodes are, the better accuracy can be achieved despite of more computational cost. We set $tn = 20$ by experience, and users can revise it for extra precision or efficiency requirements. Two major problems of TN-FPS are: (1) the selection of transit nodes; (2) how the queries are processed.

(1) Transit nodes selection. In the selection of transit nodes, we apply the Trajectory Analysis (TA) based method, which seeks to identify important intersections from trajectory data to ensure that: firstly, the transit nodes are evenly distributed over the road network; secondly, transit nodes are meaningful intersections passengers likely to pass. Basically, trajectory data is the motion history of moving objects, and it is modeled as a sequence of time stamped geo-locations $Tr = (p_1, p_2, ...p_n)$, and each point p_i has its location $p_i.loc$ and $p_i.time$, and it can be aligned to the vertexes on the road network by some map-matching algorithms [2,11]. That means, each trajectory can be converted to a network constrained model $Tr_N = (n_i, n_j, ...n_k)$. Through a large trajectory dataset D, the importance of each vertex can be seen as the frequency to be passed in trajectories, e.g. an intersection passed by a large number of moving objects is always an important junction, and it can be formalized as

$$Freq(n_i) = \frac{\sum_{Tr \in D} NUM(n_i, Tr_N)}{\sum_{Tr \in D} |Tr|}$$

Motivated by [12], the network space is partitioned into $|tn|$ of grids in our approach. For each grid g, we formally measure the weight of each intersection n inside grid region g as $weight(n) = \frac{Freq(n)}{D_{EU}(n,g.c)}$, where $D_{EU}(n, g.c)$ is the

Euclidean distance between n and $g.c$ (i.e. the center of g). Then we select out the intersection that has the greatest weight value as a transit node. In this way, the selected transit nodes are thus rational in both spatial and importance domains.

(2) Approximate Assignment Processing. The TN-FPS algorithm computes the (possibly approximate) all-pair fastest pathes based on the transit nodes as input, and finds the assignment result by the MBMA strategy.

For each passenger p, the Dijkstra based traverse on road network is carried out but only limited to a small spatial space, and its accurate travel cost to all vehicles and transit nodes that it reaches are recorded in a $TC = P \times V$ matrix. To avoid impact from approximations to accuracy of assignment, the network traversal terminates if it meets two conditions: (1) the number of reached vehicles is larger than a threshold λ. As passengers tend to be assigned to a close vehicle, the accuracy of assignment can be guaranteed if a number of closest vehicles are found; (2) the traverse must reach no less than two or more transit node, so that approximate distances to non-reached vehicles can be calculated. Similarly, the distance cost from vehicles to transit nodes and the destination are also computed. By integrating all of traversal results, we can derive all the needed (accurate and approximate) travel cost information, by which a good assignment can be computed by the MBMA strategy.

6   Experimental Study

In this section, we conduct extensive experiments on a real spatial data set to demonstrate the performance of the proposed algorithms. The data set used in our experiments are Beijing Road network, which contains $226,238$ undirected edges (road segments) and $171,187$ vertices (intersections), and Figure 2(a) shows their distribution on the spatial space. The used trajectory data compose over 300,000 moving objects trajectories in Beijing. All algorithms were implemented in JAVA and tested on a HP Compaq 8180 Elite (i5 650) computer with 2-core CPUs at 3.2GHz and 1.12 GHz, 4GB RAM and running Windows XP operating system.

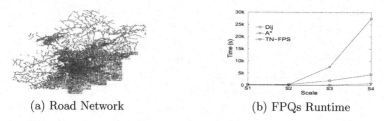

(a) Road Network   (b) FPQs Runtime

Fig. 2. Road Network, FPQs Runtime

Experiments are based on 100 test cases in four scale settings shown as Table 1. In each test case, we generate the given number of passengers and vehicles based on random distribution over road network. The efficiency and accuracy of different FPQs solutions are compared first, then we overlook the performances of the assignment algorithms, and finally evaluate the final performances by integrating them together.

Table 1. Setting of Test Cases

Scale	No. of Passengers	No. of Vehicles
S1	50	10
S2	100	20
S3	500	100
S4	5000	1000

Performances on processing FPQs. Figure 2(b) shows the performances of FPQs processing using the A*, Dijkstra and TN-FPS algorithms. In comparison, the efficiency of TN-FPS significantly outperforms both of the Dijkstra and A* based algorithm according to Figure 2(b). In contrast to A* and Dijkstra algorithms that finds the accurate FPQs results, the result returned by TN-FPS may not be accurate because of the approximation. We thus further evaluate TN-FPS under different setting of tn and k.

Figure 3(a) and Figure 3(b) are the efficiency and accuracy comparison in different tn (i.e. number of transit nodes). From Figure 3(a), we can easily observe that the runtime of FPQs processing is in reverse proportion to the number of tn, which can be explained by the fact a closer transit node tends to exist, so the network traversal of FPQ processing can stop early. Also, Figure 3(b) confirms the assumption that the greater number of transit nodes tends to improve accuracy. In terms of accuracy, we measure the error of the computed results as the ratio such that $Error = \frac{\sum_{p \in P} \sum_{v \in V} apprTC_{p,v}}{\sum_{p \in P} \sum_{v \in V} TC_{p,v}}$.

Figure 3(c) and Figure 3(d) show the comparison of algorithm efficiency and accuracy in different k (i.e. number of reached vehicles in search). The efficiency varies only for scale settings S1 and S2 because finding k vehicles for each passenger may traverse a large space of the road network in such cases. In contrast for S3 and S4, the network traverse space becomes no longer sensitive to k given lots of vehicles. Figure 3(d) shows larger k tends to improve the accuracy of FPQs processing because more traverse on road network, but the improvement is only significant for long distance FPQs.

Performances on Assignment. Based on the FPQs results, we further evaluate the performances of the proposed assignment algorithms. Figure 4(a) indicates that the IPA algorithm has poor efficiency and scalability performances, so only suitable for cases with small number of passenger and vehicle. This phenomenon is cause by the unsatisfactory pruning effects towards the cost measure of passenger assignment problem. In contrast, the MBMA based algorithm is much more efficient, particularly when the scale of passengers and vehicles are large. Figure 4(b) shows that the evolution of the iterative assignment loops based on different settings of ε (in scale S3). Towards MBMA, we further evaluate its performance in different values of parameter ε.

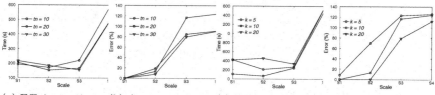

(a) Efficiency via tn (b) Accuracy via tn (c) Efficiency via k (d) Accuracy via k

Fig. 3. FPQs Processing Performances via tn and k

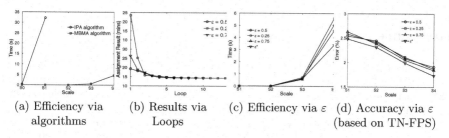

(a) Efficiency via   (b) Results via   (c) Efficiency via ε (d) Accuracy via ε
    algorithms   Loops   (based on TN-FPS)

Fig. 4. Assignment Processing Performances via selected algorithms, loops, ε

In the MBMA algorithm, ε is an important parameter to determine how we filter out bipartite edges for result improvement. Figure 4(c) and (d) show how ε impacts the efficiency of accuracy, especially when the FPQs results returned by TN-FPQS are used as the input weights in assignment step. Regarding to the efficiency comparisons in Figure 4(c), we can easily observe that the different thresholds tend to have similar efficiency performances, and the adaptive setting of ε leads to the minimum processing cost. Also, Figure 4(d) shows that passenger assignment by means of the MBMA algorithm integrating FPQ results from TN-FPQS can achieve high accuracy result (near optimal with less than 3% average error rate) in different ε settings.

To sum up, the experimental results implies that the MBMA algorithm can support us to find the optimal assignment efficiently, but the time cost for computing accurate FPQs is usually much greater when classical algorithms like A* or Dijkstra are used. If TN-FPS is applied for approximate FPQs processing instead, the efficiency will be improved greatly and the final passenger assignment result is near-optimal.

7   Conclusion and Future Work

In this paper, we have defined the problem of passenger assignment with road network as the underlying space, and devised two exact assignment algorithms based on integer programming and maximum bipartite matching techniques respectively. To reduce the high computational cost for scalable FPQs, an approximate solution has been further introduced to find the near-optimal results in a much more efficient way. Comprehensive experiments have been carried out to evaluate the performance of different algorithms.

In the future, we would like to improve the current approach by incorporating speed patterns, so that pre-computed traffic knowledge and the pattern aware spatial indexes can be used to further speed up the assignment.

Acknowledgments. This work was partially supported by Chinese NSFC project under grant numbers 61402312, 91124001, 61303019, 61202064, Australian ARC project under grant number DP140103499, and Collaborative Innovation Center of Novel Software Technology and Industrialization.

References

1. Balinski, M.L., Gomory, R.E.: A primal method for the assignment and transportation problems. Management Sci. **10**(3), 578–593 (1964)
2. Brakatsoulas, S., Pfoser, D., Salas, R., Wenk, C.: On map-matching vehicle tracking data. In: Proceedings of VLDB, pp. 853–864 (2005)
3. Brunsch, T., Cornelissen, K., Manthey, B., Röglin, H.: Smoothed analysis of the successive shortest path algorithm. In: Proceedings of SODA, pp. 1180–1189 (2013)
4. Dantzig, G., Ramser, J.: The truck dispatching problem. Management Science **6**, 80–91 (1959)
5. Desrochers, M., Jones, C., Lenstra, J.K., Savelsbergh, M., Stougie, L.: Towards a model and algorithm management system for vehicle routing and scheduling problems. Decision Support Systems **2**(25), 109–133 (1999)
6. Duan, R., Su, H.-H.: A scaling algorithm for maximum weight matching in bipartite graphs. In: Proceedings of SODA, pp. 1413–1424 (2012)
7. Geisberger, R., Sanders, P., Schultes, D., Delling, D.: Contraction hierarchies: faster and simpler hierarchical routing in road networks. In: McGeoch, C.C. (ed.) WEA 2008. LNCS, vol. 5038, pp. 319–333. Springer, Heidelberg (2008)
8. Matijevic, D., Sanders, P., Bast, H., Funke, S., Schultes, D.: In transit to constant time shortest-path queries in road networks. In: Proceedings of ALENEX (2007)
9. Kaul, M., Wong, R.C.-W., Yang, B., Jensen, C.S.: Finding shortest paths on terrains by killing two birds with one stone. PVLDB **7**(1), 73–84 (2013)
10. Kuhn, H.W.: The hungarian method for the assignment problem. Naval Research Logistics Quarterly **2**, 83–97 (1955)
11. Liu, K., Li, Y., He, F., Xu, J., Ding, Z.: Effective map-matching on the most simplified road network. In: Proceedings of ACM SIGSPATIAL GIS, pp. 609–612 (2012)
12. Ma, S., Zheng, Y., Wolfson, O.: T-share: A large-scale dynamic taxi ridesharing service. In: Proceedings of ICDE, pp. 410–421 (2013)
13. Munkres, J.: Algorithms for the assignment and transportation problems. J. Soc. Indust. Appl. Math. **5**, 32–38 (1957)
14. Pournajaf, L., Xiong, L., Sunderam, V.S., Goryczka, S.: Spatial task assignment for crowd sensing with cloaked locations. In: Proceedings of MDM, pp. 73–82 (2014)

15. Sanders, P., Schultes, D.: Engineering highway hierarchies. In: Azar, Y., Erlebach, T. (eds.) ESA 2006. LNCS, vol. 4168, pp. 804–816. Springer, Heidelberg (2006)
16. Sankaranarayanan, J., Samet, H.: Query processing using distance oracles for spatial networks. IEEE TKDE **22**(8), 1158–1175 (2010)
17. U, L.H., Mouratidis, K., Mamoulis, N.: Continuous spatial assignment of moving users. VLDBJ **19**(2), 141–160 (2010)
18. U, L.H., Yiu, M.L., Mouratidis, K., Mamoulis, N.: Capacity constrained assignment in spatial databases. In: Proceedings of SIGMOD, pp. 15–28 (2008)
19. U, L.H., Yiu, M.L., Mouratidis, K., Mamoulis, N.: Optimal matching between spatial datasets under capacity constraints. ACM TODS **35**(2), 1–43 (2010)
20. Wong, R.C.-W., Tao, Y., Fu, A.W.-C., Xiao, X.: On efficient spatial matching. In: Proceedings of VLDB, pp. 579–590 (2007)

Effective and Efficient Predictive Density Queries for Indoor Moving Objects

Miao Li[✉], Yu Gu, and Ge Yu

Institute of Computer Software, Northeastern University, Shenyang 110819, China
limiao@research.neu.edu.cn, {guyu,yuge}@ise.neu.edu.cn

Abstract. Density queries are defined as querying the dense regions that include more than a certain number of moving objects. Previous research studies mainly focus on how to answer the snap-shot density queries over historical trajectories. However, the real applications usually tend to predict whether a region is a dense region. Especially in indoor environments, such predictive density queries are valuable for high-level analysis but face tremendous challenges. In this paper, by leveraging the Markov correlations, we effectively predict the future locations of moving objects and conduct the density queries accordingly. In particular, we present an optimized framework which contains three phases to tackle this problem. First, we design an index structure based on the transition matrix to facilitate the search process. Second, we propose the space and probability pruning techniques to improve the query efficiency significantly. Finally, we apply an accurate method and an approximate sampling method to verify whether each unpruned region is a dense region. Extensive experiments on real datasets demonstrate that the proposed solutions can outperform the baseline algorithm by up to 2 orders of magnitudes in running time.

Keywords: Density queries · Markov correlations · Predictive queries · Indoor environment

1 Introduction

In recent years, the exploding developments of sensing and communication devices combined with location-based service techniques have been significantly facilitating some novel applications. For example, in indoor environments the effective users' motion prediction and localization technology play an important role in all aspects of people's daily lives, including tourist management, emergency detection, flow control, etc [6]. Due to the high density of possible locations and short transition distances between these locations, predictive density queries in indoor environments face tremendous challenges.

Although a series of efforts have been devoted in literatures [4,5,9] toward the density query model and optimization problems over historical data, these proposed methods cannot be adapted to conduct effective and efficient indoor predictive queries. Given a set of query regions, predictive density queries aim

© Springer International Publishing Switzerland 2015
M. Renz et al. (Eds.): DASFAA 2015, Part I, LNCS 9049, pp. 244–259, 2015.
DOI: 10.1007/978-3-319-18120-2_15

to return those query regions which include more objects than a threshold ρ at a future time point (or in a future time period). In this paper, we propose new methods to tackle relative problems of predictive density queries. We use the following example to illustrate our motivation.

Example: Tourist flow management. In an indoor museum monitoring scenario, the locations of visitors can be automatically detected, tracked and analyzed. At the same time, excessive visitors may gather into some popular exhibition areas leading to overburdened space and intolerable queuing time. Therefore, if we can predict those *dense regions* in advance, effective warning and scheduling measures can be taken.

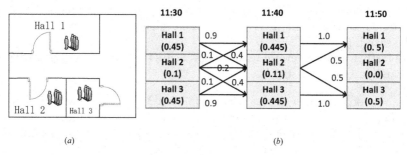

(a)   (b)

Fig. 1. (a) A schematic view of a deployment. (b) A Markov correlation representing a distribution over paths through the environment in (a).

To implement predictive queries, a key issue is to effectively predict the locations of moving objects. Most early-stage models assume that some deterministic movement functions can be defined [1,15,16] and thus the locations will be evaluated based on such functions. In fact, real-world moving objects such as visitors usually display complex and stochastic movement patterns. Specifically, as a statistical technique to predict the probability of event occurrence, the Markov model has been proved to be an effective tool to predict the locations of moving objects in a lot of scenarios [7,11,12] and thus gains more and more adoptions in the latest solutions. Let us review the example scenario by using Figure 1. Figure 1(a) shows a small portion of one indoor museum environment. In Figure 1(b), it illustrates that a visitor was in either Hall 1 or Hall 3 with the equal probability at time 11 : 50. The Markov correlation is shown to be temporally correlated: the distribution over the visitor's location at time $t + 1$ depends on its uncertain locations at time t. Figure 1(b) shows that, given that a visitor was in Hall 1 at 11 : 40, he/she stayed in Hall 1 at time 11 : 50 with probability 1.0. In our motivating scenario, because there exist obvious correlations between locations at two consecutive time points, a Markov transition matrix can be learned and leveraged to infer future locations. In this paper, we propose to build Markov correlations as our prediction model and accordingly design efficient density query techniques.

Predictive density queries over Markov correlated objects return the dense regions which have high probabilities. In terms of efficiency, two-fold challenges should be solved. First, the future locations of moving objects have to be inferred along the Markov chains considering massive possible paths from the current locations which may potentially render a long response time. Second, at the query time, all the possible locations of involved objects need be enumerated with corresponding probabilities evaluated, which may incur huge running costs. To achieve considerable efficiency enhancement, we present some index and pruning techniques to reduce the search space. Specifically, our major contributions are summarized as follows.

- We design a novel index structure and develop an efficient method for retrieving the correlations between two continuous timestamps.
- We propose several pruning techniques based on space and probability properties which can dramatically reduce the evaluation overhead and facilitate the search process.
- We further propose several refined approaches to speed up verifying whether a candidate region is the final result.

The rest of this paper is organized as follows. We first review related works in Section 2. Section 3 formally defines density queries on Markov correlated objects, introduces an index structure and develops an efficient method for retrieving the correlations between two continuous timestamps. In Section 4, we propose some effective pruning techniques which can reduce the search space for probability computation. Section 5 explores the refinements step to verify each region that has not been pruned. We discuss the experiment results in Section 6 and conclude the paper in Section 7.

2   Related Work

Density Queries. Hadjieleftherious et al. [4] first introduced the problem of density queries, and presented several techniques to evaluate such queries. Specifically, they proposed a method to partition the data space into disjoint cells, and the cell regions were returned as the approximate results instead of arbitrary regions. Further, Jensen et al. [5] presented an exact density query evaluation method and several pruning techniques based on the temporal histograms of counters. Ni et al. [9] proposed a new scheme of density queries based on the concept of pointwise-dense regions(PDRs) to answer density queries consistently and completely, regardless of the shape and size. However, all these methods for density queries are based on historical moving objects trajectories with accurate positions rather than predicted information with probabilistic locations.

Indoor Spatial Queries. In recent years, some typical solutions were proposed to handle indoor spatial queries efficiently. For example, Yang et al. [13] proposed a complete set of techniques for computing probabilistic threshold kNN queries in indoor environments. They proposed the minimum indoor walking distance (MIWD) as the distance metric for indoor spaces and designed a hash-based indexing scheme for indoor moving objects. Yu et al. [14] introduced the particle filter-based location inference method for evaluating indoor range and

kNN queries. Particularly, indoor walking graph model and anchor point indexing model were presented to improve effectiveness and efficiency. However, to the best of our knowledge, none of existing schemes were designed for predictive indoor density queries.

Query on Correlated Data. Recently, the significant amount of efforts have been devoted to support various queries over temporally and spatially correlated objects. For example, Emrich et al. [3] presented a framework for efficiently modeling and querying probabilistic spatio-temporal data. They built the Markov model for the spatio-temporal data and integrate pruning approaches into the Markov chain matrices, which contributes to reducing the search space and the computational costs during query evaluations. Niedermayer et al. [10] dealt with historical snapshot and continuous NN queries for objects with uncertain locations and proposed three different semantics of NN queries. They also addressed probabilistic nearest neighbor queries in databases with uncertain trajectories modeled by the Markov chain model. Emrich et al. [2] addressed the problem of probabilistic reverse NN queries on uncertain spatio-temporal data following the possible world's semantics. Ré et al. [11] proposed a scheme to support complex event queries on Markov correlated data streams. Accordingly, optimized algorithms were proposed to enhance the efficiency. Considering spatial correlations, Lian et al. [8] focused on answering queries on locally correlated uncertain data and proposed a cost-model-based offline pre-computation technique to enable online filtering. However, all these solutions do not consider handling the query processing in the dense regions over Markov correlated objects.

3   Problem Definition and Index Structure

3.1   Problem Definitions

In this work, we follow the related literature to denote the discrete space and the future time domain using S and T respectively, where a future time domain is expressed as $T = \{0, \cdots, n\}$ and a discrete state space of possible locations is $S = \{s_1, s_2, \cdots, s_{|S|}\}$. Let D be a spatio-temporal database containing n moving objects $D = \{o_1, o_2, \cdots, o_n\}$. Following [2], database D stores triples (o_i, t, s), where o_i is a unique object identifier, $t \in T$ is a time point (or time slot) and $s \in S$ is a position in space. Semantically, each such triple corresponds to an observation that object o_i has been seen at some location s at some time t. And each object o_i can be described by a function $o_i(t) : T \rightarrow S$ that maps each point in time to a location in space. We call this function as *trajectory*. Similarly, given the predictive density queries, the objects in a query region are associated with a probabilistic trajectory as well.

Therefore, we employ the first-order Markov chain model as a specific instance of a stochastic process. The state space of the model is the spatial domain S. State transitions are defined over the time domain T. In addition, the Markov chain model is based on the assumption that the position $o(t + 1)$ of an object o at time $t + 1$ only depends on the position $o(t)$ of o at time t. Specifically, the locations of an object are Markov correlated over adjacent time points, which are formally defined as follows.

Definition 1 (Markov Correlations). In the database $\mathcal{D}$, an object o is associated with a probabilistic trajectory in which the locations of this object are Markov correlated if and only if $\Pr(o(t + 1) = s_j | o(t) = s_i, o(t - 1) = s_k, \cdots, o(0) = s_m,) = \Pr(o(t + 1) = s_j | o(t) = s_i)$.

The conditional probability $\Pr(o(t + 1) = s_j | o(t) = s_i)$ is the transition probability of the moving object o from the location s_i to the location s_j at the time t, which describes the likelihood that an object is located at each possible location.

In order to obtain the transition probabilities from a location to another. We compute the transition matrix based on some statistical information of historical training data. By referring to such a transition matrix, we can predict the possible locations of moving objects with corresponding probabilities and furthermore conduct density queries.

Definition 2 (Density Queries over Markov Correlated Objects). For a spatial database $\mathcal{D}$ of Markov correlated moving objects, assume that we have a transition matrix M, a future query time t, a density threshold ρ, a probability threshold α and a query region set $\mathcal{A} = \{A_1, A_2, \cdots A_n\}$. If region A_j meets the following condition, A_j is a dense region.

$$\Pr(\frac{Num(A_j^t)}{Area(A_j)} \geq \rho) \geq \alpha \tag{1}$$

where, $Num(A_j^t)$ denotes the number of objects in a query region A_j at the query time t and $Area(A_j)$ is the area of query region A_j. Since the density threshold ρ is given by the user, and $Area(A_j)$ is a constant for a specified region, $\Pr(\frac{Num(A_j^t)}{Area(A_j)} \geq \rho)$ can be expressed as $\Pr(Num(A_j^t) \geq \lceil \rho \cdot Area(A_j) \rceil)$.

Next, we illustrate the problem by an example. Assume there are 5 exhibition halls and 7 visitors. The areas of these exhibition halls are $\{3, 4, 4, 3, 6\}$ (m^2) and all the visitors are located in the region A_1, A_5, A_2, A_4, A_5, A_1, A_3, respectively at the initial time. The transition matrix M is given as follows in Figure 2(a). We aim to evaluate the predictive density queries at $t = 3$ with the thresholds $\rho = 1$ and $\alpha = 0.35$. Considering all the exhibition halls, we could compute the probabilities, $\Pr(Num(A_1^3) \geq \lceil \rho \cdot Area(A_1) \rceil) = 0.0041$, $\Pr(Num(A_2^3) \geq \lceil \rho \cdot Area(A_2) \rceil) = 0.041$, $\Pr(Num(A_3^3) \geq \lceil \rho \cdot Area(A_3) \rceil) = 0.0476$, and $\Pr(Num(A_4^3) \geq \lceil \rho \cdot Area(A_4) \rceil) = 0.437$. According to α, we conclude that region A_4 is a dense region.

Note that a Markov chain state can be represented by different spatial granularities and we model an indoor room as a state in this paper. Similarly, the query regions are also defined as targeted indoor rooms in this paper. Our model and pruning methods can be extended to handle other types of states and query regions.

3.2   Markov Chain Index

Because the number of state transitions in indoor environments is limited in the two successive timestamps, the transition matrix M is a sparse matrix. To reduce the query and traverse time accordingly. We propose a Markov chain index that facilitates retrieving the correlations between two successive timestamps.

The proposed Markov chain index is a loop-structured index that provides efficient lookup and/or computation of the conditional probability relating any two Markovian continuous timestamps. The index is organized in the loop structure shown in Figure 2(b). The lower level of the tree (the timestamps t) is the set of all the states of Markov chain and each state stores the non-zero conditional probabilities of M. The upper level only stores all the states of Markov chain at the timestamps $t-1$. And the connecting lines between the lower level and the upper level denote the state transition between the two Markovian continuous timestamps.

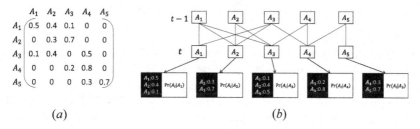

$$(a) \qquad\qquad\qquad\qquad (b)$$

Fig. 2. Transition Matrix and Markov China Index structure

The illustrated index in Figure 2(b) is constructed for the transition matrix M in Figure 2(a). Taking the state (region) A_1 as an example, we know that the states (regions) A_1 and A_3 at $t-1$ can transit to the state (region) A_1 at t. Similarly the states (regions) A_1 , A_2 and A_3 at t can be transferred from the state (region) A_1 at $t-1$.

4   Filtering Step

4.1   Spatial Pruning

First of all, we judge whether a query region is a dense region according to the density threshold ρ and the area of A_j i.e., $Area(A_j)$. We search M to find those states A_m^0 at the initial time which can reach A_j by backtracking and obtain $Num(A_m^0)$. Then, $N_{max}(A_j^t)$ can be inferred as Equations 2.

$$N_{max}(A_j^t) = \sum_m Num(A_m^0) \qquad (2)$$

When $N_{max}(A_j^t) < \lceil \rho \cdot Area(A_j) \rceil$ at the query time t, we can get $Pr(Num(A_j^t) \geq \lceil \rho \cdot Area(A_j) \rceil) = 0 < \alpha$. So, region A_j is pruned safely.

4.2   Efficient Pruning of Probability Bounds

In order to return dense regions, we must finish two high-cost computation procedures. The first procedure is to calculate the probability from some initial state (region) A_m^0 to a query region A_j. In the second procedure, all these obtained probabilities and corresponding object numbers for all the involved initial states are considered based on possible world semantics for final results. In this section, we first analyze the bounds of the first procedure, which are denoted as $\Pr^\perp(A_m^0 \rightsquigarrow A_j)$ and $\Pr^\top(A_m^0 \rightsquigarrow A_j)$ separately, where $\Pr(A_m^0 \rightsquigarrow A_j)$ denotes the probability that a moving object in A_m^0 at the initial time ($t = 0$) reaches A_j at the query time t.

For each query region A_j, based on the Markov Chain Index and M, if we give the query time t, we can infer that n states could reach state A_j at the initial time. For A_j, we find a preceding state $A_{j'}^{t-1}$ with a maximal inbound transition probability to A_j^t. We can prove an upper bound of $\Pr(A_m^0 \rightsquigarrow A_j)$ as follows.

$$\Pr^\top(A_m^0 \rightsquigarrow A_j) = max\{\Pr(A_j^t|A_{j'}^{t-1})\}$$

where $\Pr(A_m^0 \rightsquigarrow A_j) \neq 0$.

Similarly, the following formula describes the lower bound of $\Pr(A_m^0 \rightsquigarrow A_j)$. At the backtracking path, we always choose a preceding state $A_{j'}^{t-1}$ with a maximal inbound transition probability to A_j^t as the next state to explore. This is a recursive process. Finally, we acquire a transition path with the maximal probability. Because only one path is considered, the obtained probability can be proved to be a lower bound of $\Pr(A_m^0 \rightsquigarrow A_j)$ as follows. Note that in order to search more A_m^0 which can reach A_j, we record all initial inbound states for the final iteration.

$$\Pr^\perp(A_m^0 \rightsquigarrow A_j) = max\{\Pr(A_j^t|A_{j'}^{t-1})\}\times$$

$$max\{\Pr(A_{j'}^{t-1}|A_{j''}^{t-2})\} \times \cdots \times \Pr(A_{j(t-1)'}^1|A_m^0)$$

According to $\Pr^\top(A_m^0 \rightsquigarrow A_j)$, we can infer $\Pr^\top(Num(A_j^t) \geq \lceil \rho \cdot Area(A_j)\rceil)$ by considering all the involved A_m^0. Similarly, we can infer $\Pr^\perp(Num(A_j^t) \geq \lceil \rho \cdot Area(A_j)\rceil)$ by only considering all A_j^t and those corresponding A_m^0 found by backtracking. But on the grounds that huge costs will be incurred when evaluating $\Pr(Num(A_j^t) \geq \lceil \rho \cdot Area(A_j)\rceil)$ directly in the second procedure by enumerating all the possible world instances, we cannot bear the extra overhead to use $\Pr^\top(A_m^0 \rightsquigarrow A_j)$ and $\Pr^\perp(A_m^0 \rightsquigarrow A_j)$ to estimate the result probability bounds straightforward. Therefore, we further propose a grouping method for filtering to simplify the final enumeration process.

Notice that we study the predictive density queries in indoor environments, so the query time t would not be far from the initial time. We do not need to concern about the problem of a huge number of possible paths.

4.3   Probability Pruning by Grouping

At the query time t, we group all the states A_m^0 that can reach the query region A_j by the probability $\Pr(A_m^0 \rightsquigarrow A_j)$. We divide $[0,1]$ into $1/\theta$ equal intervals, where θ is the step size and $\theta \in [0,1]$. For each region A_j, $N_{max}(A_j^t)$ is the maximum number of moving objects which can reach A_j at the query time t. Besides, we define a variable g_k to represent the k-th group with k from 0 to $1/\theta - 1$. Given ρ and $Area(A_j)$, we aim to estimate $\Pr(Num(A_j^t) \geqslant \lceil \rho \cdot Area(A_j) \rceil)$ by grouping. We define $N_{max}(g_k^t)$ as the maximum number of moving object which can reach A_j in g_k at the query time t and $Num(g_k^t)$ as the number of object in g_k at the query time t. And thus, we have $\sum_{k=0}^{\frac{1}{\theta}-1} N_{max}(g_k^t) = N_{max}(A_j^t)$, and $\sum_{k=0}^{\frac{1}{\theta}-1} Num(g_k^t) = Num(A_j^t)$

By separately calculating $N_{max}(g_k^t)$, we have

$$\Pr(Num(A_j^t) \geqslant \lceil \rho \cdot Area(A_j) \rceil) =$$

$$\sum_{Num(g_0^t)}^{Num(A_j^t)} \sum_{Num(g_1^t)}^{Num(A_j^t)-Num(g_0^t)} \cdots \sum_{Num(g_{\frac{1}{\theta}-1}^t)}^{Num(A_j^t)-(\sum_{k=0}^{\frac{1}{\theta}-2} Num(g_k^t))} \quad (3)$$

$$\prod_{k=0}^{\frac{1}{\theta}-1} \Pr(C_{N_{max}(g_k^t)}^{Num(g_k^t)})$$

where $\Pr(C_{N_{max}(g_k^t)}^{Num(g_k^t)})$ is the probability that $Num(g_k^t)$ objects are in the g_k group in this region at the query time t.

Since the minimum and maximum probability of $\Pr(A_m^0 \rightsquigarrow A_j)$ in each group g_k are $(k-1)\theta$ and $k\theta$, respectively. We analyze the bounds of $\Pr(Num(A_j^t) \geqslant \lceil \rho \cdot Area(A_j) \rceil)$ as follows.

Lemma 1. *Let $\Pr(Num(A_j^t) \geqslant \lceil \rho \cdot Area(A_j) \rceil)$ be the probability of a region containing the $Num(A_j^t)$ objects at the query time t. Then, the lower bound of this probability is*

$$\Pr^{\perp}(Num(A_j^t) \geqslant \lceil \rho \cdot Area(A_j) \rceil) =$$

$$\sum_{Num(g_0^t)}^{Num(A_j^t)} \sum_{Num(g_1^t)}^{Num(A_j^t)-Num(g_0^t)} \cdots \sum_{Num(g_{\frac{1}{\theta}-1}^t)}^{Num(A_j^t)-(\sum_{k=0}^{\frac{1}{\theta}-2} Num(g_k^t))}$$

$$\prod_{k=0}^{\frac{1}{\theta}-1} C_{N_{max}(g_k^t)}^{Num(g_k^t)} ((k-1)\theta)^{Num(g_k^t)} (1-k\theta)^{N_{max}(g_k^t)-Num(g_k^t)}$$

and the upper bound of this probability is

$$\Pr^{\top}(Num(A_j^t) \geqslant \lceil \rho \cdot Area(A_j) \rceil) =$$

$$\sum_{Num(g_0^t)}^{Num(A_j^t)} \sum_{Num(g_1^t)}^{Num(A_j^t)-Num(g_0^t)} \cdots \sum_{Num(g_{\frac{1}{\theta}-1}^t)}^{Num(A_j^t)-(\sum_{k=0}^{\frac{1}{\theta}-2} Num(g_k^t))}$$

$$\prod_{k=0}^{\frac{1}{\theta}-1} C_{N_{max}(g_k^t)}^{Num(g_k^t)} (k\theta)^{Num(g_k^t)} (1-(k-1)\theta)^{N_{max}(g_k^t)-Num(g_k^t)}$$

Consequently, based on the upper bound and the lower bound given in lemma 1, when $\Pr^\top(Num(A_j^t) \geqslant \lceil \rho \cdot Area(A_j) \rceil) < \alpha$ at time t, region A_j can be pruned safely. Similarly, when $\Pr^\perp(Num(A_j^t) \geqslant \lceil \rho \cdot Area(A_j) \rceil) > \alpha$ at time t, the region A_j is the dense region and thus can be returned in the result set.

However, exactly calculating $\Pr(o_i^0 \rightsquigarrow A_j)$, i.e., $\Pr(A_m^0 \rightsquigarrow A_j)$, where $o_i^0 \in A_m^0$ in the first procedure, is also time consuming due to massive paths of Markov chains involved. And therefore, by dividing $\Pr^\perp(A_m^0 \rightsquigarrow A_j)$ and $\Pr^\top(A_m^0 \rightsquigarrow A_j)$ proposed in subsection 4.2 into different groups and adopting the techniques used in this subsection, we can get relatively loose bounds for $\Pr^\top(Num(A_j^t) \geqslant \lceil \rho \cdot Area(A_j) \rceil)$ and $\Pr^\perp(Num(A_j^t) \geqslant \lceil \rho \cdot Area(A_j) \rceil)$ at a smaller cost, and employ these bounds to conduct pruning first. For those unpruned regions, the proposed pruning based on Lemma 1 will be further performed.

5   Refinement Step

In order to verify whether a candidate region is a dense region, we present two exact methods and an approximate method during the refinement phase.

5.1   Exact Methods

We use the complete binary tree (CBT) to design an algorithm for identifying whether the unpruned regions are dense regions. The algorithm scans all unpruned regions in the space. There are at most $N_{max}(A_j^t)$ moving objects in the unpruned region A_j at the query time t. For each moving object o_i, its probability satisfies $\Pr(o_i^t \rightsquigarrow A_j) > 0$. Therefore, we can divide $\Pr(Num(A_j^t) \geqslant \lceil \rho \cdot Area(A_j) \rceil)$ into different parts in equation (4), and process each part with a branch B_i, where

$$\Pr(Num(A_j^t) \geqslant \lceil \rho \cdot Area(A_j) \rceil) =$$

$$\Pr(B_1) + \Pr(B_2) + \cdots + \Pr(B_n) + \cdots \qquad (4)$$

where, $\Pr(B_1) = \underbrace{[\Pr(o_i^t) \times \cdots \times \Pr(o_k^t)]}_{\lceil \rho \cdot Area(A_j) \rceil} \cdot \underbrace{[(1 - \Pr(o_m^t)) \times \cdots \times (1 - \Pr(o_h^t))]}_{\leq N_{max}(A_j^t) - \lceil \rho \cdot Area(A_j) \rceil}$

$i \neq k \neq m \neq h$, $i, k, m, h \in [0, N_{max}(A_j^t)]$ and each $\Pr(o_i^t) = \Pr(o_i^t \rightsquigarrow A_j)$ for simplicity here.

In equation (4), $\Pr(B_1)$ is the probability of a branch in the binary tree. As an example in Figure 3(a), we can calculate the probabilities of the moving objects in region A_1 at the query time t. Assuming the parameter $\lceil \rho \cdot Area(A_1) \rceil = 3$ and the $Num(A_1^t) = 4 > 3$, we have $\Pr(Num(A_1^t) \geqslant \lceil \rho \cdot Area(A_1) \rceil) = \Pr(B_1) + \Pr(B_2) + \cdots = \Pr(o_1^t) \cdot \Pr(o_2^t) \cdot \Pr(o_3^t) \cdot (1 - \Pr(o_4^t)) = \Pr(o_1^t) \cdot \Pr(o_2^t) \cdot (1 - \Pr(o_3^t)) \cdot \Pr(o_4^t) = \Pr(o_1^t) \cdot (1 - \Pr(o_2^t)) \cdot \Pr(o_3^t) \cdot \Pr(o_4^t) = (1 - \Pr(o_1^t)) \cdot \Pr(o_2^t) \cdot \Pr(o_3^t) \cdot \Pr(o_4^t)$. At last, we compare $\Pr(Num(A_1^t) \geqslant \lceil \rho \cdot Area(A_1) \rceil)$ with α. If $\Pr(Num(A_1^t) \geqslant \lceil \rho \cdot Area(A_1) \rceil) \geq \alpha$, region A_1 is a dense region.

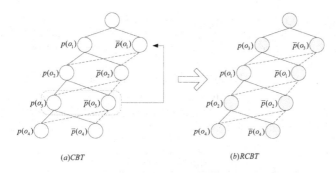

(a)CBT   (b)RCBT

Fig. 3. Example of accurate methods

Furthermore, we propose an improved refinement approach $RCBT$. In $RCBT$, we rank the probabilities of the moving objects in descending order. In the above example, we can see o_3 is the object with the maximum probability $\Pr(o_3^t)$, followed by objects o_1, o_2, o_4 and o_5 as shown in Figure 3 (b). When we calculate $\Pr(Num(A_j^t) \geqslant \lceil \rho \cdot Area(A_j^t) \rceil)$, the efficiency can be further improved.

Note that during the refinement, we adopt an alternative transformed model for the probability evaluation. We can infer that when $\lceil \rho \cdot Area(A_j) \rceil \leqslant \frac{N_{max}(A_j^t)}{2}$, the function $\Pr(Num(A_j^t) \geqslant \lceil \rho \cdot Area(A_j) \rceil)$ could be expressed as $\Pr(Num(A_j^t) \geqslant \lceil \rho \cdot Area(A_j) \rceil) = 1 - \Pr(Num(A_j^t) < \lceil \rho \cdot Area(A_j) \rceil) = 1 - \Pr(Num(A_j^t) = 1) - \Pr(Num(A_j^t) = 2) - \cdots - \Pr(Num(A_j^t) = (\lceil \rho \cdot Area(A_j) \rceil) - 1)$. In this case, the computation costs can be reduced compared to straightforward enumeration on possible worlds.

5.2   Approximate Sampling Methods

By trading accuracy with efficiency, we provide an approximate method based on Chernoff polynomials (CAM). Chernoff polynomials are widely used with a rich theory. We can leverage Chernoff bound to estimate our problem as follows.

There are $N_{max}(A_j^t)$ moving objects which may reach region A_j at the query time t. Consider accuracy parameters ϵ and δ, and a number of sample r. Let $\{I_i\}$, $1 \leq i \leq r$, be a set of r samples. We will sample the moving objects in a given region A_j. By selecting $r \geq \frac{(N_{max}(A_j^t)-1)^2}{2\epsilon^2} ln(\frac{2}{\delta})$ according to Chernoff Bound, we can prove

$$\Pr(\frac{1}{r}\sum_{i=1}^{r} I_i - \lceil \rho \cdot Area(A_j) \rceil \geq \epsilon) \leq \delta \ , \quad I_i = \begin{cases} 1 & p(o_i) > \lambda \\ 0 & otherwise, \end{cases}$$

where λ is a random number in $[0, 1]$.

The computational cost is then reduced from $2^{N_{max}(A_j^t)}$ to r. It indicates that the bound provides a good estimate for the expected reliable dense regions.

5.3   Predictive Density Queries Algorithms

The process of predictive density queries is described in algorithm 1. The steps $2 - 6$ demonstrate the spatial pruning method. Then, we process the unpruned regions by grouping the probabilities of the states (Steps $7 - 8$). Next, we use the upper bound and lower bound of $\Pr(Num(A_j^t) \geqslant \lceil \rho \cdot Area(A_j) \rceil)$ to verify whether the regions are the dense regions (Steps $9 - 13$). Finally, we use CBT, $RCBT$ or CAM to refine the result set (Steps $14 - 17$).

Algorithm 1.   Algorithm of Predictive Density Queries

Input   : Query Region Set $R = \{A_1, A_2, \cdots, A_j, \cdots\}$;
　　　　　　Query Time t; ρ; α; M; $Area(A_j)$; θ

Output: Result Set R';

1 $R' = \varnothing$;
2 **for** $\forall A_j \in R$ **do**
3 　│　Compute $N_{max}(A_j^t)$ by the formula (2)
4 　│　**if** $\Pr(N_{max}(A_j^t) \geqslant \lceil \rho \cdot Area(A_j) \rceil) = 0$ **then**
5 　│　└　Prune A_j; /* Remove A_j from R */
6 　│　**else**
7 　│　│　Compute $\Pr^\top(A_m^0 \rightsquigarrow A_j)$ and $\Pr^\perp(A_m^0 \rightsquigarrow A_j)$ by M
8 　│　│　Group $\Pr^\top(A_m^0 \rightsquigarrow A_j)$ and $\Pr^\perp(A_m^0 \rightsquigarrow A_j)$
9 　│　│　Compute $\Pr^\perp(Num(A_j^t) \geqslant \lceil \rho \cdot Area(A_j) \rceil)$ and
   　│　│　$\Pr^\top(Num(A_j^t) \geqslant \lceil \rho \cdot Area(A_j) \rceil)$ by the lemma 1
10 　│　│　**if** $\Pr^\perp(Num(A_j^t) \geqslant \lceil \rho \cdot Area(A_j) \rceil) > \alpha$ **then**
11 　│　│　└　$R' \leftarrow R' \cup A_j$
12 　│　│　**if** $\Pr^\top(Num(A_j^t) \geqslant \lceil \rho \cdot Area(A_j) \rceil) < \alpha$ **then**
13 　│　│　└　Prune A_j /* Remove A_j from R */

14 **for** $\forall A_j \in R$ **do**
15 　│　**if** $\Pr(Num(A_j^t) \geqslant \lceil \rho \cdot Area(A_j) \rceil) > \alpha$ **then**
16 　│　└　$R' \leftarrow R' \cup A_j$

17 Return R'

---

6   Experiment

In this section, we conduct a set of experiments to verify the efficiency of our proposed methods, using a desktop computer with an Intel $E5 - 1620$ CPU at 3.60 GHz and 8GB of RAM. All algorithms were implemented in C++.

6.1 Experiment Setting

All experiments are conducted on two sets of Markovian streams inferred from real-world RFID data[1]. Two real datasets were collected using building-wide RFID deployment. The data were collected by volunteers and these volunteers carried some sensors as they went through one-hour versions of typical daily routines (working in their offices, having meetings, taking coffee breaks, etc.). These datasets are given in the form of conditional probability. In particular, we infer the transition matrix M based on the conditional probabilities. According to the actual situation, the visitors stay at an exhibition hall for about 10 minutes and then go to another exhibition hall. So, we generate a transition matrix with a time interval of 10 minutes between two consecutive ticks. Finally, we obtain a common transition matrix M by taking the average of these matrices.

In real-world scenarios, we assume the area of each hall is randomly between $50m^2$ and $500m^2$. The state space of the two datasets are 160 rooms and 352 rooms respectively. Objects depart from random positions in indoor halls. In this paper, we predict a hall whether is a dense region after one hour, $i.e.$, the prediction time is 6 ticks from the initial time.

The default parameter values used in our tests are summarized in Table 1.

Table 1. Parameters and their values

Parameter	Setting
Density threshold: ρ	0.1, 0.3, **0.5**, 1,
Number of moving objects	500, **1000**, 1500, 3000
Probability threshold: α	0.3, 0.4, **0.5**, 0.6, 0.7
Step size: θ	0.1, **0.2**, 0.33, 0.5

6.2 Experimental Results

At first, we aim to evaluate the effects of our proposed pruning methods over the above two datasets separately. Particularly, three combinations of pruning methods, $i.e.$, the spatial pruning method (denoted as S pruning), spatial pruning followed by the probability pruning only based on the grouping of the bounds of $\Pr(A_m^0 \rightsquigarrow A_j)$ proposed in section 4.2 (denoted as $SP1$-pruning) and the spatial pruning combined with all the proposed methods in section 4.3. ($i.e.$, first filtering by the grouping of the probability bounds of $\Pr(A_m^0 \rightsquigarrow A_j)$ and then the grouping of the exact value of $\Pr(A_m^0 \rightsquigarrow A_j)$) (denoted as SP-pruning).

From the results measured by pruning ratio illustrated in Figure 4, we can observe that the proposed pruning techniques can achieve progressive filtering powers. Especially for SP which incorporates all the pruning techniques, a pruning ratio beyond 80% is satisfactory. Also, because a larger number of regions in dataset 2 can lead to more regions pruned, better filtering effectiveness is obtained accordingly.

---

[1] http://rfid.cs.washington.edu/

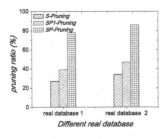

Fig. 4. Pruning ratios

Fig. 5. recall ratio for sampling

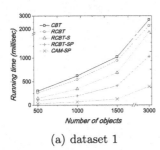

(a) dataset 1

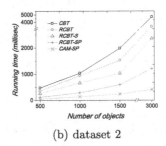

(b) dataset 2

Fig. 6. The effect of the object number on the running time

Next we exhibit in Figure 5 the recall ratio for the approximate sampling method on the above two datasets separately. From the experiment results, we can see that the proposed approximate sampling method can provide a good estimate for the expected reliable dense regions.

Furthermore, we investigate the impact of the number of moving objects on the running time of our algorithms, *i.e.*, CBT algorithm, $RCBT$ algorithm, $RCBT - S$ pruning algorithm, $RCBT - SP$ pruning algorithm and $CAM - SP$ algorithm. As shown in Figure 6, the running time of all the methods increases along with the increasing number of objects because of a higher workload. $RCBT$ can improve CBT in query time by constructing the optimization method, and $RCBT - S$ and $RCBT - SP$ substantially outperform CBT and $RCBT$ in efficiency on the grounds that a large portion of regions have been filtered beforehand. In addition, $CAM - SP$ can respond in realtime with approximate feedbacks by employing sampling techniques. Given a fixed sampling rate, more processing time will also be incurred due to the increase of involved objects.

Moreover, we test the effect of the density threshold ρ on the running time of these methods on different datasets. The experimental results are illustrated in Figure 7. We can observe that the running time of the $RCBT-SP$ algorithm also increases as the density threshold ρ increases. Because $Area(A_j)$ is fixed, when the density threshold ρ increases, the value of $\lceil \rho \cdot Area(A_j) \rceil$ becomes increasingly larger. Particularly, for $CAM - SP$, the running time remains steady, because the same number of objects will be sampled and evaluated regardless of ρ.

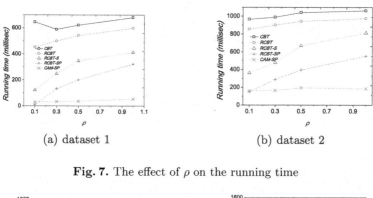

(a) dataset 1   (b) dataset 2

Fig. 7. The effect of ρ on the running time

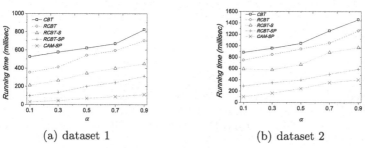

(a) dataset 1   (b) dataset 2

Fig. 8. The effect of α on the running time

Further, the effect of α on the running time is illustrated in Figure 8. The running time increases as the probability threshold α increases. This is because when we verify whether the regions are dense regions in the refinement phase, the higher the probability threshold α is, the more branches $\Pr(B_i)$ are calculated in the CBT and $RCBT$.

Finally, we investigate the effect of the step size θ on the running time and the result is showed in Figure 9. The running time of all the methods increases along with the increasing number of grouping. Because more number of grouping results in better pruning effect.

7   Conclusions

In this paper, we investigate the predictive density queries based on Markov correlations in indoor environments. Firstly, we presented a novel index structure, retrieving the correlations between two continuous timestamps, which can be adopted for efficient computation. Given a set of indoor moving objects that exhibits Markov correlations, we develop some pruning-refinement methods to answer predictive density queries on these moving objects. The pruning methods including one spatial pruning rule and two probability pruning rules are presented. These pruning methods can reduce the number of candidate regions to improve the query efficiency. Also, an exact method and an approximate method

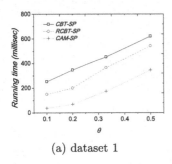

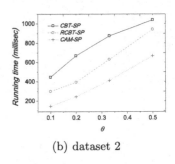

(a) dataset 1   (b) dataset 2

Fig. 9. The effect of θ on the running time

are proposed during the refinement phase. Finally, we identify the efficiency of the proposed methods through the extensive experiments by comparing these methods on real datasets.

Acknowledgments. This research is supported by the National Basic Research Program of China (973 Program) under Grant No. 2012CB316201, the National Natural Science Foundation of China (61472071, 61202086) and the Fundamental Research Funds for the Central Universities(N130404010).

References

1. Aggarwal, C.C., Agrawal, D.: On nearest neighbor indexing of nonlinear trajectories. In: Proceedings of the Twenty-Second ACM SIGMOD-SIGACT-SIGART Symposium on Principles of Database Systems, PODS 2003, pp. 252–259. ACM, New York (2003)
2. Emrich, T., Kriegel, H.-P., Mamoulis, N., Niedermayer, J., Renz, M., Züfle, A.: Reverse-nearest neighbor queries on uncertain moving object trajectories. In: Bhowmick, S.S., Dyreson, C.E., Jensen, C.S., Lee, M.L., Muliantara, A., Thalheim, B. (eds.) DASFAA 2014, Part II. LNCS, vol. 8422, pp. 92–107. Springer, Heidelberg (2014)
3. Emrich, T., Kriegel, H.-P., Mamoulis, N., Renz, M., Züfle, A.: Querying uncertain spatio-temporal data. In: ICDE, pp. 354–365 (2012)
4. Hadjieleftheriou, M., Kollios, G., Gunopulos, D., Tsotras, V.J.: On-line discovery of dense areas in spatio-temporal databases. In: Hadzilacos, T., Manolopoulos, Y., Roddick, J., Theodoridis, Y. (eds.) SSTD 2003. LNCS, vol. 2750, pp. 306–324. Springer, Heidelberg (2003)
5. Jensen, C.S., Lin, D., Ooi, B.C., Zhang, R.: Effective density queries on continuouslymoving objects. In: ICDE, p. 71 (2006)
6. Kolodziej, J., Khan, S.U., Wang, L., Min-Allah, N., Madani, S.A., Ghani, N., Li, H.: An application of markov jump process model for activity-based indoor mobility prediction in wireless networks. In: FIT 2011, pp. 51–56 (2011)
7. Letchner, J., Ré, C., Balazinska, M., Philipose, M.: Access methods for markovian streams. In: ICDE, pp. 246–257 (2009)
8. Lian, X., Chen, L.: A generic framework for handling uncertain data with local correlations. PVLDB 4(1), 12–21 (2010)

9. Ni, J., Ravishankar, C. V.: Pointwise-dense region queries in spatio-temporal databases. In: ICDE, pp. 1066–1075 (2007)
10. Niedermayer, J., Züfle, A., Emrich, T., Renz, M., Mamoulis, N., Chen, L., Kriegel, H.-P.: Probabilistic nearest neighbor queries on uncertain moving object trajectories (2013). arXiv preprint arXiv:1305.3407
11. Ré, C., Letchner, J., Balazinska, M., Suciu, D.: Event queries on correlated probabilistic streams. In: SIGMOD Conference, pp. 715–728 (2008)
12. Soliman, M.A., Ilyas, I.F., Ben-David, S.: Supporting ranking queries on uncertain and incomplete data. VLDB J. **19**(4), 477–501 (2010)
13. Yang, B., Lu, H., Jensen, C.S.: Probabilistic threshold k nearest neighbor queries over moving objects in symbolic indoor space. In: Proceedings of the 13th International Conference on Extending Database Technology, pp. 335–346. ACM (2010)
14. Yu, J., Ku, W.-S., Sun, M.-T., Lu, H.: An rfid and particle filter-based indoor spatial query evaluation system. In: Proceedings of the 16th International Conference on Extending Database Technology, pp. 263–274. ACM (2013)
15. Zhang, J., Mamoulis, N., Papadias, D., Tao, Y.: All-nearest-neighbors queries in spatial databases. In: SSDBM, pp. 297–306 (2004)
16. Zhang, M., Chen, S., Jensen, C.S., Ooi, B.C., Zhang, Z.: Effectively indexing uncertain moving objects for predictive queries. PVLDB **2**(1), 1198–1209 (2009)

Efficient Trip Planning for Maximizing User Satisfaction

Chenghao Zhu[1], Jiajie Xu[1,3](✉), Chengfei Liu[2], Pengpeng Zhao[1,3],
An Liu[1,3], and Lei Zhao[1,3]

[1] School of Computer Science and Technology, Soochow University, Suzhou, China
{xujj,ppzhao,anliu,zhaol}@suda.edu.cn
[2] Faculty of ICT, Swinburne University of Technology, Melbourne, Australia
cliu@swin.edu.au
[3] Collaborative Innovation Center of Novel Software Technology
and Industrialization, Nanjing, China

Abstract. Trip planning is a useful technique that can find various applications in Location-Based Service systems. Though a lot of trip planning methods have been proposed, few of them have considered the possible constraints of POI sites in required types to be covered for user intended activities. In this paper, we study the problem of multiple-criterion-based trip search on categorical POI sites, to return users the trip that can maximize user satisfaction score within a given distance or travel time threshold. To address this problem, we propose a spatial sketch-based approximate algorithm, which extracts useful global information based on spatial clusters to guide effective trip search. The efficiency of query processing can be fully guaranteed because of the superior pruning effect on larger granularity. Experimental results on real dataset demonstrate the effectiveness of the proposed methods.

1   Introduction

Location-Based Services (LBS) are one of the most frequently used tools in our life nowadays, and most of them rely on the trip planning techniques [1–4] to provide traffic related services. People can easily find the routes they need by accessing Google Map or Microsoft MapPoint through their personal computers or smart phones. Some systems are capable of suggesting users the trips that can cover their intended activities. Obviously, the trip planning algorithms play a key role to improve the quality and efficiency of our travel, and a lot of work [3,5–11] has been done to return users the rational trips and routes.

Given that people usually seek to find trips according to their intended activities, this paper investigates the problem of multiple-criterion-based trip search on categorical POI sites. Consider the example in Figure 1 where all POI sites are distributed on a road network, and each of them belongs to a type (e.g., a restaurant or a cafe) and has a satisfaction score marked by previous guests. A tourist issues a query to find a route from location B to location E via a

© Springer International Publishing Switzerland 2015
M. Renz et al. (Eds.): DASFAA 2015, Part I, LNCS 9049, pp. 260–276, 2015.
DOI: 10.1007/978-3-319-18120-2_16

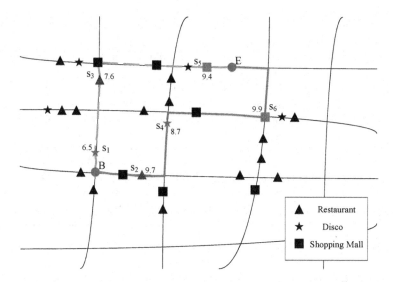

Fig. 1. Trip Search on Categorical POI Sites

cinema, a shopping mall and a restaurant within a distance threshold. Here, most of the route search methods [5,6] cannot be applied because they require a POI site in each type to be selected before the trip search; The categorical trip search methods proposed in [7,9] are able to return a shortest trip (i.e., Tr_1 in the dashed line), but it only supports single-criterion-based trip search, while in reality users would prefer to choose such a trip (i.e., Tr_2 in the full line) that can maximize the satisfaction of their intended activities within the given distance threshold.

Motivated by the above example, this paper studies the problem of multiple-criterion-based categorical trip search, which has high computational overhead for two main reasons: firstly, it requires to select some POI sites from database to meet the activity requirements from users, and the combinations between categorical POI sites are obviously high; secondly, the trip search requires to compare the scalable candidate trips that pass all selected POI sites, and the validation of each candidate trip often relies on a large number of shortest path queries on road network. As far as we know, no existing work can well address all above challenges to find the optimal trip in Figure 1 efficiently.

In this paper, we present a spatial sketch-based approximate algorithm to cope with the challenges mentioned above. It utilizes the spatial clustering to manage the scalable POI sites in larger granularity, then explores the sketched information in a global point of view to understand the spatial relations between clusters, as well as the overall execution plan to avoid unnecessary shortest path queries based on the cost analysis. Finally the extracted information is used to guide effective trip search. Based on a set of theoretical bounds and search space reduction techniques, trip search can be processed efficiently because of the much superior pruning effects over the required dimensions. In summary, the main contributions of this paper can be summarized as follows:

- We define a problem called efficient trip planning for maximizing user satisfaction, which could find applications in various LBS systems.
- We propose the satisfaction-score-first trip planning algorithm as the baseline
  algorithm to return the exact result.
- We propose the spatial sketch-based approximate trip planning algorithm to
  achieve a high performance with a little loss of accuracy.
- We implement the proposed algorithms and conduct experiments on real
  datasets to evaluate the performances of our proposed solutions.

The remainder of this paper is organized as follows: We firstly introduce
the related work in Section 2 and then define the problem in Section 3. Afterwards, an exact solution and an approximate solution are presented in Section 4
and Section 5 respectively. We carry out experimental analysis to evaluate and
compare the proposed methods in Section 6. Finally, we conclude this paper in
Section 7.

2   Related Work

Trip planning is a classical research problem that has been investigated for many
decades, and existing efforts can be generally classified into two main categories:
the point-based trip search and the categorical trip search.

The point-based trip search aims to find a good path that can pass some
user specified locations over the spatial network, and the evaluation of the trips
can be measured by a single factor [5,6] or multiple factors [12,13], e.g., distance, expense or time. Some indexing structures such as arterial hierarchy [14],
contraction hierarchies [15], path oracle [16], transit-node routing [17], detour
based [18] as well as some trajectory based methods [19] are developed to speed
up the searching process by the shortcuts embedded in the indexing structure.
However users tend to have some intended activities in the trip (e.g., to have
dinner) in some cases, the trip returned by those methods cannot be ensured to
cover the POI sites in required types (e.g., restaurant).

To address the above problem, some effort [3,7–9] has been made in recent
years toward the categorical trip search, i.e., to find the good paths with all
required categorical POI sites to be covered. Specifically, Li et al. proposed a
trip planning query (TPQ) in [3,7] to find the shortest path through at least
one POI site of each required type. Later, Sharifzadeh et al. [9] defined and
solved a variant problem of TPQ that additionally ensures the POI sites appear
in a correct order consistent with user specification. However, those methods
mainly take single criterion into account, while the users tend to have multiple
requirements (e.g., have distance threshold and prefer total satisfaction score to
be high). More recently, some work [20–23] was done to recommend trips to users
under different criteria based on the historical trajectory data. An assumption
was made that there exist one or more trajectories in database that can meet the
requirement of each user query. Unfortunately, this assumption is not realistic in
many cases especially we need to control the size of trajectory data for querying
efficiency concerns.

In this paper, we investigate the problem of multiple-criterion-based trip search over categorical POI sites. Compared to TPQ and its variants, our problem requires to balance different factors and thus incurs much greater complexity in query processing. Therefore, our methods need to adopt the spatial clustering techniques and explore useful knowledge in a global view to prune the search space in greater granularity.

3 Problem Definition

In this section, we formalize the problem of this paper based on some notions defined as follows.

Definition 1. *(Road Network) A spatial network is modeled as a connected and undirected graph $G = (V, E)$, where V is a set of vertices and E is a set of edges. A vertex $v_i \in V$ indicates a road intersection or an end of a road. An edge $e \in E$ is defined as a pair of vertices and represents a segment connecting the two constituent vertices.*

For example, (v_i, v_j) represents a road segment that enables travel between vertices v_i and v_j. We use $|v_i, v_j|$ to denote the shortest path between vertices v_i and v_j, and their network distance $||v_i, v_j||$ is the accumulated length of all edges on $|v_i, v_j|$, i.e., $||v_i, v_j|| = \Sigma_{e_k \in |v_i, v_j|} length(e_k)$.

Definition 2. *(Site) A site is a POI where users can take some activities, and it is stored in the form of $(loc, sc, type)$. Given a site s, $s.loc$ denotes the geographical location of the site; $s.sc$ is the satisfaction score of site s, which is computed as the average of all scores marked by historical guests, and thus implies the overall satisfaction level of guest experience; $s.type$ specifies the type of site s, e.g., 'shopping mall' and 'restaurant', based on a pre-defined type vocabulary.*

The location of each site can be mapped onto a given road network by map-matching algorithms. In this paper, we assume all sites have already been assigned to a vertex on the road network (i.e., $s.loc \in V$) for the sake of clear description. Also, we assume that each site belongs to only one type.

Obviously, users desire to obtain a trip (ordered) as the result returned by the system rather than a collection of sites (unordered), so a concept of our trip is necessary to be defined.

Definition 3. *(Trip) We use $Tr = \{B, S, E\}$ to denote our trip, which is from beginning point B to ending point E via a sequence of sites S (Duplicate site type is not allowed in S). Between any two adjacent sites $s_i, s_j \in S$, the trip always follows the shortest path connecting s_i and s_j. Therefore, the distance of a trip Tr can be calculated as*

$$||Tr|| = ||B, f_s|| + \sum_{s_i, s_j \in S} ||s_i, s_j|| + ||l_s, E|| \tag{1}$$

where f_s and l_s denote the first and the last sites in sequence S respectively. Regarding to user experience, we evaluate satisfaction score of a trip Tr as the sum of satisfaction scores of all sites in its site sequence S such that

$$score(Tr) = \sum_{s_i \in Tr.S} s_i.sc. \tag{2}$$

Definition 4. *(Query) Given a spatial network G, a collection of sites and a type vocabulary, we can define our query as $Q = \{B, E, Max_d, T\}$, where B and E specify the beginning point and the ending point respectively; Max_d is the threshold of the trip distance; and T is the subset of type vocabulary, which contains the activity types that the user desires to access.*

Definition 5. *(Candidate Trip) Given a query Q, a trip Tr is said to be a candidate trip if it satisfies the following conditions:*

(1) It starts at beginning point $Q.B$ and ends up at ending point $Q.E$;

(2) Every type in $Q.T$ corresponds to only one site in $Tr.S$, i.e., $Q.T = \{s_i.type | s_i \in Tr.S\}$;

(3) The distance of the trip is within the given distance threshold $Q.Max_d$, i.e., $||Tr|| \leq Q.Max_d$.

Problem Definition. Given a query Q, the road network G and a set of sites, from the set of all candidate trips $\mathbb{S}_{Tr}$, this paper intends to find and return the candidate trip $Tr \in \mathbb{S}_{Tr}$ that has the maximum value of satisfaction score, such that $\forall Tr' \in \mathbb{S}_{Tr}$, $score(Tr') \leq score(Tr)$.

4   Satisfaction-Score-First Trip Planning Algorithm

In this section, we propose a baseline algorithm, the satisfaction-score-first trip planning algorithm (S^2F) to process the user query. It adopts the best first method and returns the exact result. The main idea of S^2F is to verify the candidate sites (covering all required types) by the descending order of total satisfaction score. For each *candidate sites* as a group, we compute the candidate trips that pass all those sites, and a trip is returned if its distance is less than or equal to the given threshold. For the convenience of description, we define the notion candidate sites ($\mathbb{C}_s$) first.

Definition 6. *(Candidate Sites) Given a query Q, a set of sites $\mathbb{C}_s$ is called candidate sites if they can satisfy the following two conditions: (1) It covers all the required types in $Q.T$, i.e., $Q.T \subseteq \{s_i.type | s_i \in \mathbb{C}_s\}$; (2) If any site is removed from $\mathbb{C}_s$, the above condition would fail. The satisfaction score of $\mathbb{C}_s$ is measured as the total score of all sites in $\mathbb{C}_s$ such that $score(\mathbb{C}_s) = \sum_{s_i \in \mathbb{C}_s} s_i.sc$.*

We maintain some necessary data structures to support query processing. For each type in $Q.T$ (e.g., $'restaurant'$), we utilize a list in the form of $< siteId, score >$ to store all sites in this type, sorted by their value of score. Then a bi-dimensional list STL is used to cover all types, e.g., $STL[i][0]$ indicates the

---

Algorithm 1. Satisfaction-Score-First Trip Planning Algorithm

---

Input: a query Q, the inverted list STL
Output: the optimal planned trip Tr
 1: $Tr \leftarrow \emptyset$;
 2: $PQ_{cs} \leftarrow \{STL[i][0] \mid t_i \in Q.T\}$;
 3: **while** *true* **do**
 4:     candidate sites $\mathbb{C}_s \leftarrow PQ_{cs}.poll()$;
 5:     candidate trips $\mathbb{S}_{Tr} \leftarrow deriveCandidateTrips(\mathbb{C}_s)$;
 6:     **for** each candidate trip $Tr' \in \mathbb{S}_{Tr}$ **do**
 7:         **if** $\|Tr'\| \leq Q.Max_d$ **then**
 8:             $Tr \leftarrow Tr'$;
 9:             *break*;
10:         **else**
11:             $PQ_{cs} = UpdatePriorityQueue(PQ_{cs}, STL)$;
12:         **end if**
13:     **end for**
14: **end while**
15: **return** Tr;

---

first record in the list belonged to type t_i (i.e., the site that has the highest satisfaction score among all sites in type t_i). Based on STL, a priority queue PQ_{cs} is used in implementation to dynamically maintain a list of $\mathbb{C}_s$ for access in the descending order of their satisfaction scores.

Algorithm 1 shows how the S^2F algorithm works. At first, the $\mathbb{C}_s$ that has maximum satisfaction score in each required type is inserted to the priority queue PQ_{cs} (Line 2). Then we keep verifying all the $\mathbb{C}_s$ (Lines 3-14): for each round, we fetch the top record from PQ_{cs} to get the candidate sites $\mathbb{C}_s$ with highest satisfaction score (Line 4), and then verify it regarding to the candidate trips (Lines 5-13). If a valid trip is found, then we simply return this trip to the user (Lines 7-9); otherwise, we update the priority queue PQ_{cs} to guarantee the next best candidate sites appear in PQ_{cs} (Line 11). This procedure repeats until the optimal result can be found.

Given a candidate sites $\mathbb{C}_s$, it corresponds to a number of $|\mathbb{C}_s|!$ candidate trips (via sites in different sequences), so it may require to verify all of the candidate trips unless a valid trip can be found. By $Q.E$ is the intermediate destination in the way back to $Q.B$, it can be easily reduced to the famous Traveling Salesman Problem in NP Complete complexity. Fortunately, in the real applications, the number of required types tends to be relatively small, such that we can normally process it in an acceptable time. However, for those extreme cases with a large value of $|\mathbb{C}_s|$, we adopt a greedy method to iteratively choose the closest unvisited site in $\mathbb{C}_s$ to check the promising candidate trips with more priority, so as to stop early and thus accelerate the query processing.

In overall, the best first methods of the satisfaction-score-first trip planning algorithm contribute to reach the optimal trip earlier than the search by random. But obviously, it is potentially engaged in traversing all candidate trips as well. In such cases, the tremendous combinations would cause the algorithm to be extremely slow especially when the number of sites goes up.

5   Spatial Sketch-Based Approximate Trip Planning Algorithm

In this section, we further introduce a spatial sketch-based approximate trip planning algorithm (S^2A in short) to support efficient trip search. We first overview the structure of S^2A and then describe the work mechanism in detail.

5.1   Overview

In the S^2A algorithm, we process the trip search by three main steps as shown in Figure 2. The first step is the preprocessing step, in which the sites are clustered to spatial clusters, and the useful inter and intra cluster information are extracted as well. Figure 2 (a) illustrates the preprocessing step.

Afterwards, the second step is to search on the granularity of clusters, without touching at the site level. As shown in Figure 2 (b), we first discuss the priority of the candidate clusters to be checked. Then we consider the issue of sequence of the candidate clusters to go through (sketched trips in Figure 2 (b)), so that the hopeless and unpromising ones can be filtered out. Furthermore, we analyze the contents of the sketched trips to derive the execution plan based on the cost analysis of the required shortest path search, which contributes to avoid vast unnecessary computational cost in trip search.

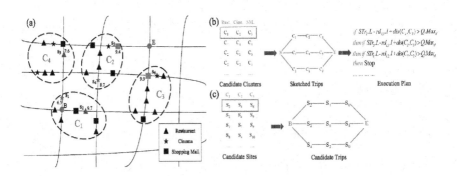

Fig. 2. An example of S^2A

At last, the trip search goes to the site level as Figure 2 (c) shows. Given a candidate clusters, we scan the corresponding candidate sites based on the execution plan derived from the previous step, and some bounds are used to guarantee the effectiveness and efficiency of trip search in S^2A.

5.2 Preprocessing

As the amount of sites is large, it is not realistic to process the trip search on all sites directly. In the S^2A algorithm, we carry out the spatial clustering on sites first, so that it is possible to maintain a global picture to prune on larger granularity and to guide smarter trip search.

Towards the spatial clustering, classical algorithms can be generally classified into two categories, namely density-based method and partitioning-based method. As illustrated in Figure 3, for algorithms in density-based method like DBSCAN, they tend to find clusters with high spatial coherence (circles in the full line), but nevertheless, they do not guarantee all sites to be included into the clusters because of its density requirements, possibly leading good sites unable to be represented in clusters. In contrast, the partition-based approaches like k-means or k-medoids are capable of ensuring every site to appear in a cluster, while the clusters (circles in the dashed line) tend to be not compact enough.

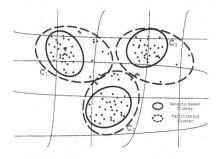
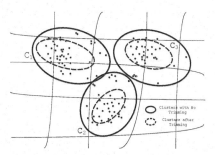

Fig. 3. Clustering By Classic Algorithms

Fig. 4. Clustering By the PTC

To address above problem, we present the partition-trimming clustering algorithm (PTC) to return the spatially coherent clusters like those in Figure 4. Specifically, the sites are clustered by the partitioning-based approach first, so all sites are covered by the clusters (circles in the full line). But the initial clusters are possibly too sparse, so we further trim the clusters in the second step: for each cluster C, we continuously remove the sites with satisfaction scores lower than that of an interior site (that has less distance to center of C) in the same type; other sites (circles in the dashed line) are reserved because they are likely to appear in the trip required by users. In this way, it achieves a good balance between the spatial coherence and good site coverage.

Based on the spatial clusters, we extract and store some necessary information to assist the pruning in trip search. Firstly, we compute and store the spatial relations of clusters, e.g., the minimum and the maximum possible distances between any two clusters, and the maximum distance between sites within a cluster, etc. Two variables $Cmin_{i,j}$ and $Cmax_{i,j}$ are used to represent the minimum and the maximum distance from cluster C_i to C_j respectively. Secondly, we extract the categorical information of each cluster, e.g., the types in the cluster, so we can have an overview of the cluster to help cluster selection.

5.3   Sketched Trip Search on Clusters

Based on the spatial clusters, we analyze the information extracted not only to prune the search space in a greater granularity, but also to find out guidelines of how to carry out trip search on scalable sites. Given a query Q, we commence with the *candidate clusters*, e.g., in Figure 2 (b) (C_1, C_2, C_3) if the user desires to get a trip via a restaurant, a cinema and a shopping mall. We formally give the definition of candidate clusters ($\mathbb{C}_c$) as follows:

Definition 7. *(Candidate Clusters) Given a query Q, a collection of clusters $\mathbb{C}_c$ is called candidate clusters if it can satisfy the following two conditions: (1) Each type in $Q.T$ corresponds to a cluster in $\mathbb{C}_c$, i.e., $\forall t_i \in Q.T$, $\exists C_j \in \mathbb{C}_c$, and $\exists s_k \in C_j$, such that $s_k.type = t_i$; (2) It's a one-to-one correspondence, i.e., each type in $Q.T$ is only bound with one cluster in $\mathbb{C}_c$. The satisfaction score of $\mathbb{C}_c$ is computed as:*

$$Score(\mathbb{C}_c) = \sum_{C_i \in \mathbb{C}_c} Max(s_j.sc | s_j \in C_i, s_j.type = C_i.type); \qquad (3)$$

where $C_i.type$ implies that C_i has to provide a site in $C_i.type$.

As our goal is to maximize the satisfaction score of the trip, those $\mathbb{C}_c$ with higher satisfaction scores have greater priority because they cover better trip candidates. Similar to S^2F, we use a priority queue to retrieve $\mathbb{C}_c$ in the descending order of satisfaction scores. For example, in Figure 2 (b), (C_1, C_2, C_3) has the greatest score value, so we fetch it to verify its feasibility first.

Given a $\mathbb{C}_c$ to verify, we devote to return a group of sketched trips from beginning point $Q.B$ to ending point $Q.E$ via the clusters in some sequences, e.g., in Figure 2 (b) $< B, C_1, C_2, C_3, E >, < B, C_1, C_3, C_2, E >$ and $< B, C_2, C_1, C_3, E >$. Defined as follows, a sketched trip in cluster level corresponds to a number of candidate trips in site level, through which we can implement the pruning on a larger granularity perspective and markedly improve the search efficiency.

Definition 8. *(Sketched Trip) A sketched trip STr is modeled as $STr=\{B, S_c, E\}$, where B and E are the beginning point and the ending point respectively; the S_c denotes a set of clusters with a sequence that appears in the trip (i.e., that has temporal order to visit), for example $S_c = < C_1, C_2, C_3 >$ in Figure 2 (b).*

In any two concessive clusters C_i and C_j in a sketched trip STr, the actual travel distance between the two clusters is in the range $[Cmin_{i,j}, Cmax_{i,j}]$ (can be derived from the matrix extracted in pre-processing step), and the distance of a sketched trip STr is thus in the range $[STr.L, STr.U]$ such that

$$STr.L = Cmin_{B,f_c} + \sum_{C_i, C_j \in S_c} Cmin_{i,j} + Cmin_{l_c,E} \qquad (4)$$

$$STr.U = Cmax_{B,f_c} + \sum_{C_i,C_j \in S_c} Cmax_{i,j} + Cmax_{l_c,E} \tag{5}$$

where f_c and l_c denote the first and the last clusters in the cluster sequence S_c respectively.

As a $\mathbb{C}_c$ can generate a large number ($|\mathbb{C}_c|!$) of sketched trips (via clusters in different sequences), it would be time-consuming if we scan all of them when the value $|\mathbb{C}_c|$ is high. Therefore it is not realistic to consider all sketched trips, and we seek to select out the top-k ones most likely to fulfill the distance requirement of the query for consideration. The problem is how to measure the possibility $P(STr)$ of an arbitrary sketched trip STr for satisfying the distance constraint. For a given distance threshold $Q.Max_d$, there are three **situations** of a sketched trip STr: for **situation 1** where $Q.Max_d < STr.L$, obviously STr is hopeless to meet the distance threshold ($P(STr) = 0$), so it is filtered; while in **situation 2** where $Q.Max_d \geq STr.U$, it can be guaranteed to satisfy the distance constraint ($P(STr) = 1$), and all trips covered by STr are valid trips; for the **situation 3** where $STr.L \leq Q.Max_d < STr.U$ holds, STr is said to be potentially valid, so we measure the valid possibility to be:

$$P(STr) = \frac{Q.Max_d - STr.L}{STr.U - STr.L} \tag{6}$$

where $STr.U - STr.L$ confines the range of possible distances of a trip covered by STr, and among the whole distance range, $Q.Max_d - STr.L$ confines the part of the range that is valid in terms of the distance constraint of query Q. Their ratio can thus represent the possibility of a trip (that can be sketched to STr) to be valid. Therefore given a $\mathbb{C}_c$, we only consider a sketched trip in situation 2 if it exists, or consider the top-k sketched trips regarding to their valid possibilities.

Based on the sketched trips to be considered, we further find out effective execution plan that requires less shortest path querying overhead based on our cost analysis model. For each sketched trip, we need to use a number of $|STr.S_c| + 1$ times of shortest path search (to calculate the distance between two sites in adjacent clusters, including the beginning/ending point), and we need to process all sketched trips of the candidate clusters in the worst case. Different cluster pairs should have varying priorities because of the difference of their capability on pruning for two main factors:

- *Frequency*. If a cluster pair appears in more sketched trips, e.g., (C_1, C_3) in Figure 2 (b), thus a single shortest path search may help us to prune multiple trips (e.g., on $< B, C_1, C_3, C_2, E >$ and $< B, C_2, C_1, C_3, E >$);
- *Impact*. Different cluster pairs have varying impacts on the pruning effect of each sketched trip it belongs to.

Therefore in this paper, we measure the priority $\omega_{i,j}$ of a given cluster pair $< C_i, C_j >$ as Equation 7:

$$\omega_{i,j} = \sum_{STr \in \tau} \frac{Cmax_{i,j} - Cmin_{i,j}}{STr.U - STr.L} \tag{7}$$

where τ is the set of relevant sketched trips (top-k sketched trips in valid possibility), and $\frac{Cmax_{i,j} - Cmin_{i,j}}{STr.U - STr.L}$ denotes the pruning effect of the cluster pair on a sketched trip STr. As a result, $\omega_{i,j}$ denotes the total pruning effect on all sketched trips the cluster pair $< C_i, C_j >$ belongs to. Once we have these computing priorities, we can use them to organize the verification of sketched trips and to guide the trip search on sites.

The execution plan implements the verification of sketched trips by some rules. The rules are made by the computing priorities and some global variables kept to maintain the current minimum distances of sketched trips. For example, in Figure 2 (b), for $STr_2 = < B, C_1, C_3, C_2, E >$, at the beginning its current minimum distance is initialized to its minimum distance $STr_2.L$, if the priority of $< C_1, C_3 >$ is the highest, we update the current minimum distance by subtracting $Cmin_{1,3}$ and adding the actual distance between a site in C_1 to a site in C_3. If it exceeds the distance threshold $Q.Max_d$ already, we prune it directly.

To sum up, sketched trip search on clusters implements a larger granularity search and markedly improve the search efficiency. And the execution plan avoids some unnecessary shortest path queries in trip search on sites.

5.4   Trip Search On Sites

If the sketched trip search on clusters cannot prune all the sketched trips, we must go further to the trip search on sites. In this subsection, we present the trip search on sites in detail. Algorithm 2 shows how the trip search is executed.

Given a query Q, we generate a list of $\mathbb{C}_c$ based on all requested types in $Q.T$ first. A priority queue PQ_{cc} is used to organize all the $\mathbb{C}_c$ in the descending order of their satisfaction scores (Line 4). Then we scan PQ_{cc}, each time we fetch a $\mathbb{C}_c$ to do the sketched trip search on clusters (Lines 9-11). If no valid sketched trip is found, we go on processing the next $\mathbb{C}_c$. Otherwise, we carry out the trip search on sites on basis of the sketched trips (Lines 12-24).

Given a group of sketched trips, the first step towards the trip search on sites is that we generate all the $\mathbb{C}_s$ derived from that group of sketched trips. Similar to before, we also adopt a priority queue PQ_{cs} to organize all the $\mathbb{C}_s$ obtained in the descending order by their satisfaction scores (Line 13). For example, In Figure 2 (c), $< s_2, s_4, s_6 >$ is the champion in satisfaction score.

The advantage that the execution plan brings to this stage is that we can check multiple trips with a single shortest path query. For example, in Figure 2 (c), we would launch a shortest path query from s_2 to s_6 if the priority of $< C_1, C_3 >$ is the highest. If the result distance exceeds the threshold that the execution plan gave, we can prune the two trips ($< B, s_2, s_6, s_4, E >$ and $< B, s_4, s_2, s_6, E >$) by only one shortest path query.

Once we find a feasible trip, we cannot directly return it as the result, because there may exist other solutions from the unprocessed $\mathbb{C}_c$ that have higher satisfaction scores. As such, a global bound is defined, S_{LB}, to guarantee the correctness of

---

Algorithm 2. Spatial Sketch-Based Approximate Trip Planning Algorithm

---

Input: a query Q
Output: a solution trip Tr

1: $Tr \leftarrow \emptyset$;
2: $S_{LB} \leftarrow 0$;
3: $Tr' \leftarrow \emptyset$;
4: $PQ_{cc} \leftarrow generateAllCandidateCluster()$;
5: **while** $(\mathbb{C}_c \leftarrow PQ_{cc}.poll())$!= $null$ **do**
6:    **if** $Score(\mathbb{C}_c) \leq S_{LB}$ **then**
7:      $break$;
8:    **end if**
9:    $STrList \leftarrow generateAllSTr(\mathbb{C}_c)$;
10:   $STrList.reduce()$;
11:   $STrList.executionPlan()$;
12:   **if** $STrList.isEmpty() == false$ **then**
13:     $PQ_{cs} \leftarrow generateAllCandidateSite(STrList)$;
14:     **while** $(\mathbb{C}_s \leftarrow PQ_{cs}.poll())$!= $null$ **do**
15:       **if** $Score(\mathbb{C}_s) < S_{LB}$ **then**
16:         $break$;
17:       **end if**
18:       $Tr' \leftarrow tripSearch(\mathbb{C}_s, STrList)$;
19:       **if** Tr' != $null$ **then**
20:         $S_{LB} \leftarrow Score(Tr')$;
21:         $Tr \leftarrow Tr'$;
22:         $break$;
23:       **end if**
24:     **end while**
25:   **end if**
26: **end while**
27: **return** Tr;

---

the trip search on sites. S_{LB} is designed to keep track of the best feasible solution's satisfaction score, which is computed as

$$S_{LB} = max_{Tr_i \in T_s}\{score(Tr_i)\} \tag{8}$$

where T_s is the set of feasible trips we have found. Obviously, S_{LB} is dynamically updated every time we find a new feasible trip, and it is the threshold in trip search processing: if the score of a trip falls below S_{LB} already, we need not continue to search as the remaining trips are hopeless to provide a higher score than S_{LB} (Lines 15-17). Since all the $\mathbb{C}_c$ are ranked in a descending order according to their satisfaction scores, **the entire search can be stopped** if a $\mathbb{C}_c$ can be found such that $Score(\mathbb{C}_c) \leq S_{LB}$ (Lines 6-8), because all of the trips derived from that $\mathbb{C}_c$ are hopeless to provide a higher score than S_{LB}.

6   Experimental Evaluation

In this section, extensive experiments are carried out on real spatial datasets to demonstrate the performance of the proposed algorithms. We use the real road network of Beijing, which contains 165,991 vertices and 225,999 edges respectively. Also, the POI dataset includes 100,000 Beijing POI sites that are classified into 11 categories, and each POI site is affiliated with a satisfaction score and mapped onto the road network. The number of POI sites in a categorical type varies between 5,835 to 13,639. Both algorithms are implemented in JAVA and tested on a DELL 7010MT (i5-3740) computer with 4-core CPUs at 3.2GHz and a 8GB RAM running Windows 7 (64 bits) operating system.

In the evaluations, we generate 100 representative sample queries based on the parameters shown in Table 1, and all results are derived by the average of the sample queries. Note that α is the parameter used to vary the distance threshold in a query, which is α times of the road network distance from beginning point to ending point via shortest path. We first compare the two proposed algorithms in the varying size of the POI sites, and then evaluate the effects of the varying parameters to the performance of the S^2A algorithm. Given that the algorithms are time-consuming for large dataset, we use a parameter ξ to evaluate the intermediate result at the ξ's iteration, so as to know the performance in the early iterations for scenarios requesting immediate response.

Table 1. Default Parameter Values

Parameter	Value	Description
k	4	Sketched trip reduction factor
t	4	Number of query site types
α	2	Shortest path distance multiplier
ξ	unlimited	Iteration threshold

Comparison of Proposed Algorithms. Figure 5 (a) and (b) show the efficiency and accuracy of the S^2F and S^2A algorithms in the varying size of the POI sites. From Figure 5 (a), it can be easily observed that the efficiency of the two algorithms are similar when the POI dataset is small. But when the POI dataset becomes large, the S^2A algorithm is much more efficient than S^2F, and this phenomenon can be explained by the much more superior pruning effect of S^2A (in the cluster granularity). In addition, the experimental results in Figure 5 (b) show that S^2A has fairly good accuracy, and particularly, the accuracy tends to increase when the size of POI sites goes up. The accuracy in small dataset cases is a lot worse than others, and this is is partly because of the top-k trips tend to be more skewed in the cases running on small POI dataset. Therefore the S^2A is a near optimal algorithm and much more efficient than S^2F.

Then we evaluate the impacts from parameters k, t and α to the performance of the S^2A algorithm. In each experiment, we compare the intermediate results (ξ = 10k, 30k, 50k candidate clusters to be checked) and the final result.

Effect of k. Figure 6 (a) and (b) show the efficiency and accuracy of the S^2A algorithm in the settings of k between 1 and 24 (as the default 4 types has

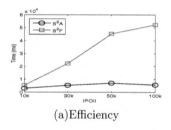
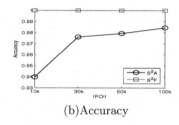

(a)Efficiency   (b)Accuracy

Fig. 5. Performances of S^2F and S^2A

24 trip sequences at most). When the value of k is greater than 4, the S^2A algorithm tends to be efficient (also stable) according to Figure 6 (a) and to be very accurate according to Figure 6 (b). In contrast when the value of k is less than 4, the accuracy tends to drop and it becomes a little more time-consuming (Especially, $\xi = 30k, 50k$). The reason is that, it would be likely to miss the valid trip if only very few top-k sequences are considered as candidate trip, leading us to scan much more candidate sites. Therefore we should avoid the value of k to be very small. The figures also indicate the intermediate result tend to have acceptable accuracy rate if users request the result in an immediate fashion.

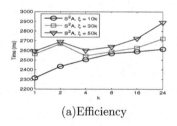
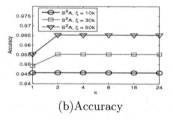

(a)Efficiency   (b)Accuracy

Fig. 6. Performances via k

Effect of t. Figure 7 (a) and (b) show the efficiency and accuracy comparisons in different number of required types in query Q. Given that the time complexity is in exponential time to t, we limit the size of t to be in a reasonable range of the applications (otherwise the results cannot be obtained in an acceptable time). As shown in Figure 7 (a), the efficiency tends to be stable when the number of t is relatively small, but it climbs up sharply when the number of t becomes larger because of the huge search space accordingly. Besides, Figure 7 (b) shows that the accuracy drops dramatically when the number of t increases due to insufficient iterations.

Effect of α. Figure 8 (a) and (b) show the efficiency and accuracy of the S^2A algorithm in varying distance thresholds based on the parameter α (the distance threshold is α times of the distance from beginning point to ending point of the query). We can easily observe from Figure 8 (a) that the efficiency is very sensitive to α, and the processing time significantly reduces when the distance

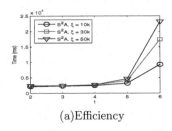 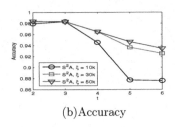

(a)Efficiency   (b)Accuracy

Fig. 7. Performances via t

threshold becomes large. Similarly, the accuracy of the algorithm is also in the direct proportion to the parameter α according to Figure 8 (b). Above phenomena can be well explained by the fact that it is easier to find a good result if the distance threshold is relatively loose.

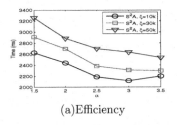 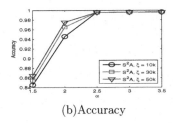

(a)Efficiency   (b)Accuracy

Fig. 8. Performances via α

To sum up, the experimental results imply that the S^2A algorithm can support us to find the trip with high frequency efficiently in different parameter settings, which are related to the algorithm logic or to query itself.

7   Conclusion

In this paper, we have investigated the problem of trip search on categorical POI sites, to return users the trip that can maximize user satisfaction score within a given distance threshold. A spatial clustering-based approximate algorithm S^2A has been proposed to support efficient trip search, in which the sketched information in a global point of view is extracted to prune the search space in greater granularity. Extensive experimental results based on real datasets have demonstrated the effectiveness of the proposed method.

In the future, we would like to improve the algorithm by applying hierarchial clustering techniques to further improve the accuracy of trip search.

Acknowledgments. This work was partially supported by Chinese NSFC project under grant numbers 61402312, 61232006, 61303019, 61440053, Australian ARC project under grant number DP140103499, and Collaborative Innovation Center of Novel Software Technology and Industrialization.

References

1. Roy, S.B., Das, G., Amer-Yahia, S., Yu, C.: Interactive itinerary planning. In: Proceedings of the 27th International Conference on Data Engineering, ICDE 2011, Hannover, Germany, 11–16 April, pp. 15–26 (2011)
2. Kanoulas, E., Du, Y., Xia, T., Zhang, D.: Finding fastest paths on A road network with speed patterns. In: Proceedings of the 22nd International Conference on Data Engineering, ICDE 2006, Atlanta, GA, USA, 3–8 April, p. 10 (2006)
3. Li, F., Hadjieleftheriou, M., Kollios, G., Cheng, D., Teng, S.: Trip planning queries in road network databases. In: Encyclopedia of GIS, pp. 1176–1181 (2008)
4. Yang, B., Guo, C., Jensen, C.S., Kaul, M., Shang, S.: Stochastic skyline route planning under time-varying uncertainty. In: IEEE 30th International Conference on Data Engineering, Chicago, ICDE 2014, IL, USA, 31 March–4 April, pp. 136–147 (2014)
5. Delling, D., Sanders, P., Schultes, D., Wagner, D.: Engineering route planning algorithms. In: Lerner, J., Wagner, D., Zweig, K.A. (eds.) Algorithmics of Large and Complex Networks. LNCS, vol. 5515, pp. 117–139. Springer, Heidelberg (2009)
6. Fredman, M.L., Tarjan, R.E.: Fibonacci heaps and their uses in improved network optimization algorithms. J. ACM **34**(3), 596–615 (1987)
7. Li, F., Cheng, D., Hadjieleftheriou, M., Kollios, G., Teng, S.-H.: On trip planning queries in spatial databases. In: Medeiros, C.B., Egenhofer, M., Bertino, E. (eds.) SSTD 2005. LNCS, vol. 3633, pp. 273–290. Springer, Heidelberg (2005)
8. Shang, S., Ding, R., Yuan, B., Xie, K., Zheng, K., Kalnis, P.: User oriented trajectory search for trip recommendation. In: Proceedings of the 15th International Conference on Extending Database Technology, EDBT 2012, Berlin, Germany, 27–30 March, pp. 156–167 (2012)
9. Sharifzadeh, M., Kolahdouzan, M.R., Shahabi, C.: The optimal sequenced route query. VLDB J. **17**(4), 765–787 (2008)
10. Yuan, J., Zheng, Y., Xie, X., Sun, G.: Driving with knowledge from the physical world. In: Proceedings of the 17th ACM SIGKDD International Conference on Knowledge Discovery and Data Mining, San Diego, CA, USA, 21–24 August, pp. 316–324 (2011)
11. Horvitz, E., Krumm, J.: Some help on the way: opportunistic routing under uncertainty. In: The 2012 ACM Conference on Ubiquitous Computing, Ubicomp 2012, Pittsburgh, PA, USA, 5–8 September, pp. 371–380 (2012)
12. Tian, Y., Lee, K.C.K., Lee, W.: Finding skyline paths in road networks. In: Proceedings of the 17th ACM SIGSPATIAL International Symposium on Advances in Geographic Information Systems, ACM-GIS 2009, Seattle, Washington, USA, 4–6 November, pp. 444–447 (2009)
13. Safar, M., El-Amin, D., Taniar, D.: Optimized skyline queries on road networks using nearest neighbors. Personal and Ubiquitous Computing **15**(8), 845–856 (2011)
14. Zhu, A.D., Ma, H., Xiao, X., Luo, S., Tang, Y., Zhou, S.: Shortest path and distance queries on road networks: towards bridging theory and practice. In: Proceedings of the ACM SIGMOD International Conference on Management of Data, SIGMOD 2013, New York, NY, USA, 22–27 June, pp. 857–868 (2013)
15. Geisberger, R., Sanders, P., Schultes, D., Delling, D.: Contraction hierarchies: faster and simpler hierarchical routing in road networks. In: McGeoch, C.C. (ed.) WEA 2008. LNCS, vol. 5038, pp. 319–333. Springer, Heidelberg (2008)

16. Sankaranarayanan, J., Samet, H., Alborzi, H.: Path oracles for spatial networks. PVLDB **2**(1), 1210–1221 (2009)
17. Bast, H., Funke, S., Matijevic, D., Sanders, P., Schultes, D.: In transit to constant time shortest-path queries in road networks. In: Proceedings of the Nine Workshop on Algorithm Engineering and Experiments, ALENEX 2007, New Orleans, Louisiana, USA, 6 January (2007)
18. Shang, S., Deng, K., Xie, K.: Best point detour query in road networks. In: Proceedings of the 18th ACM SIGSPATIAL International Symposium on Advances in Geographic Information Systems, ACM-GIS 2010, San Jose, CA, USA, 3–5 November, pp. 71–80 (2010)
19. Shang, S., Ding, R., Zheng, K., Jensen, C.S., Kalnis, P., Zhou, X.: Personalized trajectory matching in spatial networks. VLDB J. **23**(3), 449–468 (2014)
20. Zheng, K., Shang, S., Yuan, N.J., Yang, Y.: Towards efficient search for activity trajectories. In: 29th IEEE International Conference on Data Engineering, ICDE 2013, Brisbane, Australia, 8–12 April, pp. 230–241 (2013)
21. Chen, Z., Shen, H.T., Zhou, X., Zheng, Y., Xie, X.: Searching trajectories by locations: an efficiency study. In: Proceedings of the ACM SIGMOD International Conference on Management of Data, SIGMOD 2010, Indianapolis, Indiana, USA, 6–10 June, pp. 255–266 (2010)
22. Chen, Z., Shen, H.T., Zhou, X.: Discovering popular routes from trajectories. In: Proceedings of the 27th International Conference on Data Engineering, ICDE 2011, Hannover, Germany, 11–16 April, pp. 900–911 (2011)
23. Zheng, K., Trajcevski, G., Zhou, X., Scheuermann, P.: Probabilistic range queries for uncertain trajectories on road networks. In: Proceedings of the EDBT 2011, 14th International Conference on Extending Database Technology, Uppsala, Sweden, 21–24 March, pp. 283–294 (2011)

Modern Computing Platform

Accelerating Search of Protein Sequence Databases Using CUDA-Enabled GPU

Lin Cheng and Greg Butler$^{(\boxtimes)}$

Department of Computer Science and Software Engineering,
Concordia University, Montreal, Canada
`gregb@cs.concordia.ca`

Abstract. Searching databases of protein sequences for those proteins that match patterns represented as profile HMMs is a widely performed bioinformatics task. The standard tool for the task is HMMER version 3 from Sean Eddy. HMMER3 achieved significant improvements in performance over version 2 through the introduction of a heuristic filter called the Multiple Segment Viterbi algorithm (MSV) and the use of native SIMD instruction set on modern CPUs. Our objective was to further improve performance by using a general-purpose graphical processing unit (GPU) and the CUDA software environment from Nvidia.

An execution profile of HMMER3 identifies the MSV filter as a code hotspot that consumes over 75% of the total execution time. We applied a number of well-known optimization strategies for coding GPUs in order to implement a CUDA version of the MSV filter.

The results show that our implementation achieved 1.8x speedup over the single-threaded HMMER3 CPU SSE2 implementation on average. The experiments used a modern Kepler architecture GPU from Nvidia that has 768 cores running at 811 Mhz and an Intel Core i7-3960X 3.3GHz CPU overclocked at 4.6GHz.

For HMMER2 there was a significant speed-up of an order of magnitude obtained by implementations using GPUs. Such gains seem out of reach for HMMER3.

1 Introduction

A protein can be viewed as a sequence of amino acid residues. In Bioinformatics, the purpose of protein sequence search against databases is to identify regions of similarity that may be a consequence of functional, structural, or evolutionary relationships between the protein sequences. Such a similarity search produces an alignment of similar substrings of the sequences.

Classical sequence alignment algorithms such as Needleman-Wunsch [1], Smith-Waterman [2], and the BLAST family of programs [3] have long been used for searching by performing pairwise alignment of each query against every sequence in the database, thus identifying those sequences in the database that are most closely related to various regions of the query.

Besides the above pairwise comparison algorithms, another paradigm compares a sequence to a probabilistic representation of several proteins of the same family.

© Springer International Publishing Switzerland 2015
M. Renz et al. (Eds.): DASFAA 2015, Part I, LNCS 9049, pp. 279–298, 2015.
DOI: 10.1007/978-3-319-18120-2_17

Since all the sequences in a family are mostly similar to each other, it is possible to construct a common profile representing the *consensus sequence*, which simply reflects the most commonly occurring residue at each position. One such probabilistic representation is called the *profile HMM* (Hidden Markov Model) introduced by Anders Krogh and David Haussler [4] to improve the sensitivity of database-searching.

The profile HMM is a probabilistic finite state machine that assesses the probability of match, insert and delete at a given position of an alignment.

In this paper we address the application of a CUDA-enabled GPU (graphical processing unit) to accelerate the search routine in HMMER3, the most widely used software for profile HMM search of protein databases. Our approach is to target a performance hotspot in the search and to apply a range of optimization techniques. The hotspot is a dynamic programming algorithm, called the MSV filter, for Multiple (local, ungapped) Segment Viterbi. The optimization techniques were selected from related work that used a GPU to optimize the Smith-Waterman algorithm, which is also a variant of dynamic programming.

1.1   A Brief History of the HMMER Software

HMMER [5] is a free and commonly used software package for sequence analysis based on the profile HMM. HMMER aims to be significantly more accurate and to detect more remote homologs, compared to BLAST, and other sequence alignment tools and database search tools [6].

From 1992 to 1998, the HMMER1 series was developed by Sean Eddy. It includes a feature that is missing in HMMER2 and HMMER3: the *hmmt* program for training HMMs from initially unaligned sequences and hence creating multiple alignments. The final stable version of HMMER1 was released as 1.8.5 in 2006.

From 1998 to 2003, the HMMER2 series introduced the "Plan 7" profile HMM architecture (see Figure 1), which is still used in HMMER3, and is the foundation for Pfam and other protein domain databases. It includes local and global alignment modes that HMMER3 lacks, because HMMER3 currently implements only fully local alignment. HMMER2 lacks DNA comparison that was present in HMMER1. The final stable version of HMMER2 was released as 2.3.2 in 2003.

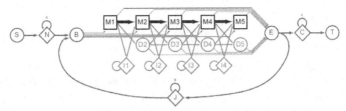

Fig. 1. Plan 7 Profile HMM Architecture [6]

In HMMER, the application *hmmbuild* is used to build a profile HMM using a multiple sequence alignment, or single sequence as input. The application *hmmsearch* is used to search a profile HMM against a sequence database, finding whether a sequence is member of the family described by the profile HMM. The *hmmsearch* application outputs a ranked list of the sequences with the most significant matches to the profile. Another similar application in HMMER, *hmmscan*, is the query of a single protein sequence of interest against a database of profile HMMs.

To compare a profile HMM with a protein sequence, HMMER uses the Viterbi algorithm, which evaluates the path that has the maximum probability of the profile HMM generating the sequence. The Viterbi algorithm is a dynamic programming algorithm. The fundamental task of the Viterbi algorithm for biological sequence alignment is to calculate three DP (Dynamic Programming) matrices: $M[\ ]$ for Match state, $I[\ ]$ for Insert state and $D[\ ]$ for Delete state. Each element value in the DP matrix is dependent on the value of previous element.

However, the widely used implementation of the Viterbi algorithm in HMMER2, has been slow and compute-intensive, on the order of more than 100 times slower than BLAST for a comparable search. In an era of enormous sequence databases, this speed disadvantage outweighs any advantage of the profile HMM method. With the exponential growth of protein databases, there is an increasing demand for acceleration of such techniques. HMMER has been a target of many acceleration and optimization efforts.

Specialized hardware architectures have been used to exploit coarse-grained parallelism in accelerating HMMER2. JackHMMer [7] uses the Intel IXP 2850 network processor. FPGAs (Field-Programmable Gate Arrays) were used by [8–10], and the Cell Broadband Engine developed by IBM was used by [11].

On traditional CPU architecture, MPI-HMMER [12] is a commonly used MPI implementation. A single master node is used to assign multiple database blocks to worker nodes for computing in parallel and is responsible for collecting the results. Landman et al. exploit MPI and Intel SSE2 intrinsics to accelerate HMMER2 [13].

ClawHMMer [14] is the first GPU-enabled *hmmsearch* implementation. Their implementation is based on the BrookGPU stream programming language, not the CUDA programming model. Since ClawHMMer, there has been several researches on accelerating HMMER for CUDA-enabled GPU, parallelizing the Viterbi algorithm [15–18]. However, these efforts have had limited impact on accelerating HMMER2 by only an order of magnitude.

In 2010, HMMER3.0 was released. It is the most significant acceleration of hmmsearch. The most significant difference between HMMER3 and HMMER2 is that HMMER3 uses a heuristic algorithm called the MSV filter, for Multiple (local, ungapped) Segment Viterbi, to accelerate profile HMM searches (see Figure 2). By using the Intel SSE2 intrinsics to implement programs, HMMER3 is substantially more sensitive, and 100 to 1000 times faster than HMMER2 [6].

HMMER3.1 beta was released in 2013. It has several new features that did not make them into 3.0, including *nhmmer* program for DNA homology searches with profile HMMs, the parallel search daemon *hmmpgmd* program underlying HMMER Web Services, and a new HMM file format called 3/f format.

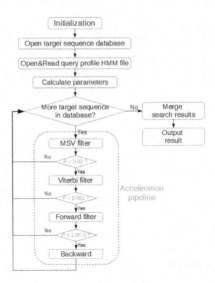

Fig. 2. The CPU serial version of hmmsearch

Although HMMER3 is much faster than HMMER2 and about as fast as BLAST for protein searches, it is still time-consuming. On an Intel Core i7-3930K high-power CPU, HMMER3 takes about 5 minutes to search a profile HMM with length 255 against the NCBI NR database.

Hotspot analysis using Intel VTune Analyzer [19] identified three individual algorithms: Forward, Backward and Viterbi [20] for which they carried out CUDA acceleration. According to the implementation of HMMER3 (see Figure 2), the MSV, Viterbi, Forward and Backward algorithms are implemented in the so-called "acceleration pipeline" at the core of the HMMER3 software package [6]. The MSV algorithm is the first filter of the acceleration pipeline and is the key hotspot of the whole process.

1.2   Organization of the Paper

The paper is organized with Section 2 presenting the Background on CUDA-enabled GPUs, techniques used for optimization for Smith-Waterman, and for HMMER. This is followed in Section 3 with a description of HMMER3 and its optimization for a CUDA-enabled GPU. Section 4 presents the evaluation benchmark and the results. Section 5 concludes the paper.

2   Background

2.1   CUDA-Enabled GPU

A GPU consists of one or more SMs (Streaming Multiprocessors). The Quadro K4000 used in our research has 4 SMs. Each SM contains the following specific features [21]: execution units to perform integer and single- or double-precision floating-point arithmetic, special function units to compute single-precision floating-point transcendental functions; thousands of registers to be partitioned among threads; shared memory for fast data interchange between threads; several caches, including constant cache, texture cache and L1 cache; and a warp scheduler to coordinate instruction dispatch to the execution units.

The SM has been evolving rapidly since the introduction of the first CUDA-enabled GPU device in 2006, with three major Compute Capability 1.x, 2.x, and 3.x, corresponding to Tesla-class, Fermi-class, and Kepler-class hardware respectively. Table 1 summarizes the features introduced in each generation of the SM hardware [21].

Table 1. Features per Compute Capability

Compute Capability	Features introduced
SM 1.x	Global memory atomics; mapped pinned memory; debuggable; atomic operations on shared memory; Double precision
SM 2.x	64-bit addressing; L1 and L2 cache; concurrent kernel execution; global atomic add for single-precision floating-point values; Function calls and indirect calls in kernels
SM 3.x	SIMD Video Instructions; Increase maximum grid size; warp shuffle; Bindless textures ("texture objects"); read global memory via texture; faster global atomics; 64-bit atomic min, max, AND, OR, and XOR; dynamic parallelism

CUDA memory spaces have different characteristics that reflect their distinct usages in CUDA applications as summarized in Table 2 [22]. The texture, constant and global memory can be allocated by the CPU host. Shared memory can only be shared and accessed by threads in a block. Registers and local memory are only available for one thread. Register access is the fastest and global memory access is the slowest. Since these memories have different features, one important aspect of CUDA programming is how to combine these memories to best suit the application.

Table 2. Salient Features of GPU Device Memory

Memory	Location	Cached	Access	Scope	Speed	Lifetime
Register	On chip	n/a	R/W	1 Thread	1	Thread
Local	Off chip	†	R/W	1 Thread	$\sim 2-16$	Thread
Shared	On chip	n/a	R/W	All threads in block	$\sim 2-16$	Block
Global	Off chip	†	R/W	All threads + host	200+	Host allocation
Constant	Off chip	Yes	R	All threads + host	$2-200$	Host allocation
Texture	Off chip	Yes	R	All threads + host	$2-200$	Host allocation

The Speed column is the relative speed in number of instructions. † means it is cached only on devices above Compute Capability 2.x.

2.2 CUDA Accelerated Smith-Waterman

The Smith-Waterman algorithm exploits dynamic programming for sequence alignment, which is also the characteristic of the HMM-based algorithms. In this section, we review the techniques used in parallelizing Smith-Waterman on a CUDA-enabled GPU. The strategies used include

1. task-based parallelism [23–29];
2. data-based parallelism [23,30];
3. use of GPU memory types [23–29]; and
4. use of SIMD vector instructions [24–27,29].

A pre-sorted sequence database is often used [23–27,29].

CUDASW++ in its various versions [23–25] is the most successful Smith-Waterman implementation for CUDA-enabled GPU. They use task-based parallelism to process each target sequence independently with a single GPU thread. CUDASW++3 [25] not only distributes tasks to many threads in the GPU kernel, but also balances the workload between the CPU and the GPU using a rate R of the number of residues assigned to GPUs:

$$R = \frac{N_G f_G}{N_G f_G + N_C f_C / C} \tag{1}$$

where f_C and f_G are the core frequencies of the CPU and the GPU respectively, N_C and N_G are the number of CPU cores and the number of GPU SMs respectively, and C is a constant derived from empirical evaluations. They find the sequence length deviation generally causes execution imbalance between threads, which in return can not fully utilize the GPU compute power. So they design two CUDA kernels based on two parallelization approaches: static scheduling and dynamic scheduling. These two kernels are launched based on the sequence length deviation of the database.

In data-based parallelism, each task is assigned to one or many thread block(s) and all threads in the thread block(s) cooperate to perform the task in parallel. The main target of [30] is to solve a single but very large Smith-Waterman problem for sequences with very long lengths. Their calculation works along anti-diagonals of the alignment matrix so that the calculations can be performed in parallel one row (or column) of the similarity matrix at a time. Row (or column) calculations allow the GPU global memory accesses to be consecutive and therefore high memory throughput is achieved.

CUDASW++ [23] sorts target sequences and arranges them in an array like a multi-layer bookcase to store into global memory, so that the reading of the database across multiple threads could be coalesced. They utilize the texture memory on the sorted array of target sequences in order to achieve maximum performance on coalesced access patterns. They also exploit constant memory to store the gap penalties, scoring matrix and the query sequence. CUDASW++2 [24] utilizes texture memory to store query profiles.

2.3   CUDA Accelerated HMMER

In this section, we review the techniques used in parallelizing HMMER on a CUDA-enabled GPU. The strategies used include

1. task-based parallelism [15, 16, 18, 20];
2. data-based parallelism [16, 17, 31];
3. use of GPU memory types [15, 17, 18]; and
4. use of SIMD vector instructions.

HMMER itself includes a MPI (Message Passing Interface) implementation of the search algorithms, which uses conventional CPU clusters for parallel computing.

GPU-HMM [15] ports the Viterbi function to a CUDA-enabled GPU with a variety of optimization approaches. Their implementation operates the GPU kernel on multiple sequences simultaneously, with each thread operating on an independent sequence. They found the number of threads that can be executed in parallel will be limited by two factors: one is the GPU memory which limits the number of sequences that can be stored, and another is the number of registers used by each thread which limits the number of threads that can execute in parallel. Registers are the most important resource in their implementation.

They split the inner loop for computing the dynamic programming matrix into three independent small loops. This approach requires fewer registers, resulting in higher GPU utilization. Further, splitting the loop provides an easy mechanism to exploit loop unrolling, which is a classic loop optimization strategy designed to reduce the overhead of inefficient looping. The idea is to replicate the loops inner contents such that the percentage of useful instructions in each statement of the loop increases. In their experiment, the performance improvement reaches 80%.

In order to achieve high efficiency for task-based parallelism, the run time of all threads in a thread block should be roughly identical. Therefore many studies often sort sequence databases by the length of the sequences.

Data-based parallelism applies the *wave-front* method [17, 31] which computes the cells along the anti-diagonal of the dynamic matrix in parallel, in order to accelerate the Viterbi algorithm. They apply a streaming method to process very long sequences.

[16] parallelizes the Viterbi algorithm by combining task-based and data-based parallelism. In order to accelerate the computation of the dynamic programming matrix rows, they partition each row into equal sized intervals of contiguous cells and calculate the dependencies between the partitions identically and independently in a data parallel setting.

GPU-HMM [15] found the most effective optimization for the Viterbi algorithm is to optimize CUDA memory layout and usage patterns within the implementation. They note that coalesced access of global memory can significantly improve hmmsearch speedup by up to 9x. They utilize texture memory to store the target sequences, and use texture memory and constant memory to store the query profile HMM depending on its size.

3   HMMER3 and Its Implementation on a CUDA-Enabled GPU

The CPU serial version of hmmsearch in HMMER3 is shown in Figure 2. The MSV and Viterbi algorithms are implemented in the so-called acceleration pipeline at the core of the HMMER3 software package [6]. One call to the acceleration pipeline is executed for the comparison of each query model and target sequence.

HMMER3 is a nearly total rewrite of the earlier HMMER2 package, with the aim of improving the speed of profile HMM searches. The main performance gain is due to a heuristic algorithm called the MSV filter, for Multiple (local, ungapped) Segment Viterbi. MSV is implemented in SIMD (Single-Instruction Multiple-Data) vector parallelization instructions and is about 100-fold faster than HMMER2.

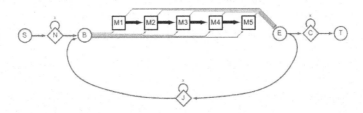

Fig. 3. MSV profile: multiple ungapped local alignment segments [6]

Figure 3 illustrates the MSV profile architecture. Compared with Figure 1, the MSV corresponds to the virtual removal of the delete and insert states. All match-match transition probabilities are treated as 1.0. The other parameters remains unchanged. So this model generates sequences containing one or more ungapped local alignment segments. The pseudocode of the MSV score algorithm is simplified and shown in Algorithm 1.

A Single-Instruction Multiple-Data (SIMD) instruction is able to perform the same operation on multiple pieces of data in parallel. The first widely-deployed desktop SIMD implementation was with Intel's MMX extensions to the x86 architecture in 1996. In 1999, Intel introduced Streaming SIMD Extensions (SSE) in Pentium III series processors. The modern SIMD vector instruction sets use 128-bit vector registers to compute up to 16 simultaneous operations. Due to the huge number of iterations in the Smith-Waterman algorithm calculation, using SIMD instructions to reduce the number of instructions needed to perform one cell calculation has a significant impact on the execution time. Several SIMD vector parallelization methods have been described for accelerating SW dynamic programming.

In 2000, Rognes and Seeberg presented an implementation of the SW algorithm running on the Intel Pentium processor using the MMX SIMD instructions [32]. They used a query profile parallel to the query sequence for each possible residue. A query profile was pre-calculated in a sequential layout just once before searching the database. A six-fold speedup was reported over an optimized non-SIMD implementation.

Algorithm 1. Pseudo code of the MSV algorithm

1: **procedure** MSV( )
2:     $N[0] \leftarrow 0; \quad B[0] \leftarrow tr(N, B)$
3:     $E[0] \leftarrow C[0] \leftarrow J[0] \leftarrow -\infty$
4:     **for** $i \leftarrow 1, L_t$ **do**   ▷ For every sequence residue i
5:         $N[i] \leftarrow N[i-1] + tr(N, N)$
6:         $B[i] \leftarrow max \begin{cases} N[i-1] + tr(N, B) \\ J[i-1] + tr(J, B) \end{cases}$
7:         $M[i, 0] \leftarrow -\infty$
8:         **for** $j \leftarrow 1, L_q$ **do**   ▷ For every model position j from 1 to L_q
9:             $M[0, j] \leftarrow -\infty$
10:            $M[i, j] \leftarrow e(M_j, S[i]) + max \begin{cases} M[i-1, j-1] \\ B[i-1] + tr(B, M_j) \end{cases}$
11:        **end for**
12:        $E[i] \leftarrow max\{M[i, j] + tr(M_j, E)\} \quad (j \leftarrow 0, L_q)$
13:        $J[i] \leftarrow max \begin{cases} J[i-1] + tr(J, J) \\ E[i-1] + tr(E, J) \end{cases}$
14:        $C[i] \leftarrow max \begin{cases} C[i-1] + tr(C, C) \\ E[i-1] + tr(E, C) \end{cases}$
15:    **end for**
16:    $Score \leftarrow C[L_t] + tr(C, T)$
17:    **return** $Score$
18: **end procedure**

In 2007, Farrar presented an efficient vector-parallel approach called stripped layout for vectorizing SW algorithm [33]. He designed a stripped query profile for SIMD vector computation. He used Intel SSE2 to implement his design. A speedup of 2-8 times was reported over the Rognes and Seeberg SIMD non-stripped implementations.

Inspired by Farrar, in HMMER3 [6], Sean R. Eddy used a remarkably efficient stripped vector-parallel approach to calculate the MSV alignment scores. To maximize parallelism, he implemented the MSV algorithm as a 16-fold parallel calculation with score values stored as 8-bit byte integers. He used SSE2 instructions on Intel-compatible systems and Altivec/VMX instructions on PowerPC systems. The pseudocode for the implementation is shown in Algorithm 2 Five pseudocode vector instructions for operations on 8-bit integers are used in the pseudocode. The instructions are vec_splat, vec_adds, vec_rightshift, vec_max and vec_hmax. Either scalars x or vectors v containing 16 8-bit integer elements numbered $v[0]...v[15]$. Each of these operations are either available or easily constructed in Intel SSE2 intrinsics as shown in Table 3.

3.1   GPU Implementation of the MSV Filter

A basic flow of the GPU implementation for the MSV filter is shown in Figure 4. The code is split up into two parts, with the left *host* part running on the CPU

Algorithm 2. Pseudo code of the SIMD vectorized MSV algorithm

---

1: **procedure** MSV-SIMD( )
2:    $xJ \leftarrow 0;\quad dp[q] \leftarrow vec\_splat(0)\ (q \leftarrow 0, L_Q - 1)$
3:    $xB \leftarrow base + tr(N, B)$
4:    $xBv \leftarrow vec\_adds(xB, tr(B, M))$
5:    **for** $i \leftarrow 1, L_t$ **do**   ▷ For every sequence residue i
6:       $xEv \leftarrow vec\_splat(0)$
7:       $mpv \leftarrow vec\_rightshift(dp[L_Q - 1])$
8:       **for** $q \leftarrow 0, L_Q - 1$ **do**
9:          $tmpv \leftarrow vec\_max(mpv, xBv)$ ▷ temporary storage of 1 current row value
10:          $tmpv \leftarrow vec\_adds(tmpv, e(M_j, S[i]))$
11:          $xEv \leftarrow vec\_max(xEv, tmpv)$
12:          $mpv \leftarrow dp[q]$
13:          $dp[q] \leftarrow tmpv$
14:       **end for**
15:       $xE \leftarrow vec\_hmax(xEv)$
16:       $xJ \leftarrow max \begin{cases} xJ \\ xE + tr(E, J) \end{cases}$
17:       $xB \leftarrow max \begin{cases} base \\ xJ + tr(J, B) \end{cases}$
18:    **end for**
19:    $Score \leftarrow xJ + tr(C, T)$
20:    **return** $Score$
21: **end procedure**

Table 3. SSE2 intrinsics for pseudocode in Algorithm 2

Pseudocode SSE2 intrinsic in C	Operation	Definition
v = vec_splat(x)		
v = _mm_set1_epi8(x)	assignment	$v[z] = x$
v = vec_adds(v1, v2)		
v = _mm_adds_epu8(v1, v2)	saturated addition	$v[z] = min(2^8 - 1, v1[z] + v2[z])$
v1 = vec_rightshift(v2)		$v1[z] = v2[z - 1](z = 15...1);$
v1 = _mm_slli_si128(v2, 1)	right shift	$v1[0] = 0;$
v = vec_max(v1, v2)		
v = _mm_max_epu8(v1, v2)	max	$v[z] = max(v1[z], v2[z])$
x = vec_hmax(v)		
-	horizontal max	$x = max(v[z]), z = 0...15$

The first column is pseudocode and its corresponding SSE2 intrinsic in C language. Because x86 and x86-64 use little endian, vec_rightshift() means using a left bit shift intrinsic _mm_slli_si128() to do right shift. No SSE2 intrinsic is corresponding to tbfvec_hmax(). Shuffle intrinsic _mm_shuffle_epi32 and _mm_max_epu8 can be combined to implement vec_hmax()

and the right *device* part running on the GPU. There is some redundancy as data needed by the GPU will be copied between the memories in the host and the device.

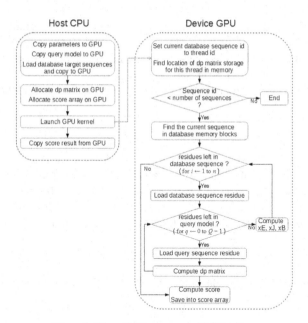

Fig. 4. The GPU porting of MSV filter

The CPU code mainly concerns allocating data structures on the GPU, loading data, copying data to the GPU, launching the GPU kernel and copying back the results for further steps.

The GPU kernel code corresponds to the MSV filter Algorithm 2. First, the thread's current database sequence is set to the thread id. Hence each thread begins processing a different neighbouring sequence. This thread id is a unique numeric identifier for each thread and the id numbers of threads in a warp are consecutive. Next, the location where each thread can store and compute its dp matrix is determined in the global memory. This is calculated also using the thread id for each thread. When processing the sequence, successive threads access the successive addresses in the global memory for the sequence data and dp matrix, i.e. using a coalesced access pattern. Execution on the GPU kernel is halted when every thread finishes its sequence.

We adopted a series of optimization techniques that led to a 135 times speedup over the straightforward GPU implementation of the MSV filter:

1. Using SIMD video instructions of the GPU;
2. Minimizing global memory access;
3. Using asynchronous memory copy and streams;
4. Coalescing global memory accesses;
5. Using texture memory for the matrix of the query profile HMM;
6. Sorting the database by decreasing length of sequence; and
7. Distributing workload between the CPU and the GPU.

GPU SIMD Instructions. CUDA provides scalar SIMD (Single Instruction, Multiple Data) video instructions. These are available on devices of compute capability 3.0. The SIMD video instructions enable efficient operations on pairs of 16-bit values and quads of 8-bit values needed for video processing. The SIMD video instructions can be included in CUDA programs by way of the assembler, $asm()$, statement. The basic syntax of an $asm()$ statement is:

asm("template-string" : "constraint"(output) : "constraint"(input));

The following three instructions are used in the implementation. Every instruction operates on quads of 8-bit signed values. The source operands ("op1" and "op2") and destination operand ("rv") are all unsigned 32-bit registers ("u32"), which is different from 128-bit CPU registers in SSE2. For additions and subtractions, saturation instructions ("sat") have been used to clamp the values to their appropriate unsigned ranges.

```
/* rv[z] = op1[z] + op2[z] (z = 0,1,2,3) */
asm("vadd4.u32.u32.u32.sat %0, %1, %2, %3;" : "=r"(rv) : "r"(op1),
"r"(op2), "r"(0));
/* rv = op1 + op2 */
asm("vsub4.u32.u32.u32.sat %0, %1, %2, %3;" : "=r"(rv) : "r"(op1),
"r"(op2), "r"(0));
/* rv = max(op1,op2) */
asm("vmax4.u32.u32.u32 %0, %1, %2, %3;" : "=r"(rv) : "r"(op1), "r"(op2),
"r"(0));
```

Switching to the SIMD video instructions achieved a speedup of nearly 2 times.

Minimizing Global Memory Access. The global memory is used to store most of data on the GPU. A primary concern in the optimization is to improve the efficiency of accessing global memory. One way is to reduce the frequency of access. Another way is coalescing access.

The elements of the dp matrix and the query profile matrix are 8-bit values. The $uint4$ and $ulong2$ (see the code below) are 128-bit CUDA built-in vector types. So the access frequency would be decreased 16 times by using $uint4$ or $ulong2$ to fetch the 8-bit values residing in global memory, compared with using 8-bit $char$ type.

```
struct __device_builtin__ uint4
{
    unsigned int x, y, z, w;
}
struct __device_builtin__ ulong2
{
    unsigned long int x, y;
};
```

Using Asynchronous Memory Copy and Streams. By default, any memory copy involving host memory is synchronous: the function does not return until after the operation has been completed. This is because the hardware cannot directly access host memory unless it has been page-locked or pinned and mapped for the GPU. An asynchronous memory copy for pageable memory could be implemented by spawning another CPU thread, but so far, CUDA has chosen to avoid that additional complexity.

Even when operating on pinned memory, such as memory allocated with *cudaMallocHost*(), synchronous memory copy must wait until the operation is finished because the application may rely on that behavior. When pinned memory is specified to a synchronous memory copy routine, the driver does take advantage by having the hardware use DMA, which is generally faster [21].

When possible, synchronous memory copy should be avoided for performance reasons. Keeping all operations asynchronous improves performance by enabling the CPU and GPU to run concurrently. Asynchronous memory copy functions have the suffix *Async*(). For example, the CUDA runtime function for asynchronous host to device memory copy is *cudaMemcpyAsync*().

Asynchronous memory copy works well only where either the input or output of the GPU workload is small and the total transfer time is less than the kernel execution time. By this means we have the opportunity to hide the input transfer time and only suffer the output transfer time.

A CUDA stream represents a queue of GPU operations that get executed in a specific order. We can add operations such as kernel launches, memory copies, and event starts and stops into a stream. The order in which operations are added to the stream specifies the order in which they will be executed. CUDA streams enable CPU/GPU and memory copy/kernel processing concurrency. For GPUs that have one or more copy engines, host to/from device memory copy can be performed while the SMs are processing kernels. Within a given stream, operations are performed in sequential order, but operations in different streams may be performed in parallel [34].

To take advantage of CPU/GPU concurrency, when performing memory copies as well as kernel launches, asynchronous memory copy must be used. Mapped pinned memory can be used to overlap PCI Express transfers and kernel processing.

CUDA compute capabilities above 2.0 are capable of concurrently running multiple kernels, provided they are launched in different streams and have block sizes that are small enough so a single kernel will not fill the whole GPU. By using multiple streams, we broke the kernel computation into chunks and overlap the memory copies with kernel execution.

Coalescing Global Memory Accesses. Coalescing access is the single most important performance consideration in programming for CUDA-enabled GPU architectures. Coalescing is a technique applied to combine several small and non-contiguous access of global memory, into a single large and more efficient contiguous memory access. A prerequisite for coalescing is that the words accessed by all threads in a warp must lie in the same segment. The memory spaces referred to by the same variable names (not referring to the same addresses) for all threads in a warp have to be allocated in the form of an array to keep them contiguous in address space.

For coalescing access, the target sequences are arranged in a matrix like an upside-down bookcase, where all residues of a sequence are restricted to be stored in the same column from top to bottom. When the sequence database is sorted, and all sequences are arranged in decreasing length order from left to right in the array, then this strategy is even more beneficial.

An alignment requirement is needed to fully utilize coalescing, which means any access to data residing in global memory is compiled to a single global memory instruction. The alignment requirement is automatically fulfilled for the built-in types like *uint4* [35].

The move to vertical alignment of *dp* matrix resulted in an improvement of about 44%.

Note. At the beginning, since *uint4* is a 16-byte data block, the traditional C/C++ memory block copy function *memcpy*() was used to copy data between global memory and register memory, as shown in the following code. The *dp* is the pointer to the address of global memory. The *mpv* and *sv* are *uint4* data type residing in register memory.

```
memcpy(&mpv, dp, sizeof(uint4));
memcpy(dp, &sv, sizeof(uint4));
```

However, in practice during CUDA kernel execution, the above *memcpy* involves $16 = sizeof(uint4)$ reads/writes from/to global memory respectively, not one read/write. Switching to the following direct assignment instruction will be one read/write and fully coalesce access to global memory, with 81% improvement over the above *memcpy*():

```
mpv = *(dp);
*(dp) = sv;
```

Using Texture Memory for the Matrix of the Query Profile HMM. The read-only texture memory space is a cached window into global memory that offers much lower latency and does not require coalescing for best performance. Therefore, a texture fetch costs one device memory read only on a cache miss; otherwise, it just costs one read from the texture cache. The texture cache is optimized for 2D spatial locality, so threads of the same warp that read texture addresses that are close together will achieve best performance [35].

Texture memory is well suited to random access. CUDA has optimized the operation fetching 4 values (RGB colors and alpha component, a typical graphics usage) at a time in texture memory. This mechanism is applied to fetch 4 read-only values from the query profile matrix *texOMrbv* with the *uint4* built-in type. Since the data of target sequences is read-only, it can also use texture memory for better performance.

Sorting the Database by Decreasing Length of Sequence. The MSV filter function is sensitive to the length of a target sequence, which determines the execution times of the main *for* loop in Algorithm 2. The target sequence database could contain many sequences with different lengths. This brings a problem for parallel processing of threads on the GPU: one thread could be processing a sequence of several thousands of residues while another might be working on a sequence of just a few. As a result, the thread that finishes first might be idle while the long sequence is being handled. Furthermore, unless care is taken when assigning sequences to threads, this effect might be compounded by the heavily unbalanced workload among threads.

In order to achieve high efficiency for task-based parallelism, the run time of all threads in a thread block should be roughly identical. Therefore the database is converted with sequences being sorted by length. Thus, for two adjacent threads in a thread warp, the difference value between the lengths of the associated sequences is minimized, thereby balancing a similar workload over threads in a warp.

Distributing Workload between the CPU and the GPU. After launching the GPU kernel, the CPU must wait for the GPU to finish before copying back the result. This is accomplished by calling *cudaStreamSynchronize(stream)*. We can get further improvement by distributing some work from the GPU to the CPU while the CPU is waiting. In a protein database, the sequences with the longest or the shortest length are very few. According to Swiss-Prot database statistics [36], the percentage of sequences with length > 2500 is only 0.2%. Considering the length distribution of database sequences and based on the descending sorted database, we assigned the first part of data with longer lengths to the CPU. By this way, we can save both the GPU global memory allocated for sequences and the overheads of memory transfer.

The compute power of the CPU and the GPU should be taken into consideration in order to balance the workload distribution between them. The distribution policy calculates a ratio R of the number of database sequences assigned to the CPU, which is calculated as

$$R = \frac{f_C}{N_G f_G + f_C}$$

where f_G and f_C are the core frequencies of the GPU and the CPU respectively, and N_G is the number of GPU multiprocessors.

4   Evaluation

The experiments used a Kepler architecture NVIDIA Quadro K4000 graphics card with 3 GB global memory, 768 Parallel-Processing Cores, 811 MHz GPU Clock rate, with CUDA Compute Capability 3.0; and a six-core Intel Core i7-3960X 3.3GHz CPU overclocked at 4.6GHz with 64GB RAM. The operating system used was Debian Linux v7.6; the CUDA toolkit used was version 6.0.

The dataset consisted of (1) the Pfam database of protein domains [37] which contains 14,831 profile HMMs of length from 7 to 2207 for a total of 2,610,332 states; and (2) the Swiss-Prot database of fasta protein sequences [38], which contains 540,958 sequences of length from 2 to 35,213 amino acids, and comprising 192,206,270 amino acids in total; a file more than 258MB in size. The databases are as of September 2013.

The measurement of performance used was GCUPS (GigaCell Units Per Second):

$$GCUPS = \frac{L_q * L_t}{T * 1.0e09} \tag{2}$$

where L_q is the length of query profile HMM, L_t is the total residues of target sequences in the database, and T is the execution time in seconds. The execution time of the application was timed using the C clock() instruction. All programs were compiled using GNU g++ with the -O3 option and executed independently in a 100% idle system.

Table 4. Speedup of using GPU relative to one CPU core

Benchmark		hmmsearch		hmmCUDAsearch		
profile HMMs	Sequences	Time(s)	GCUPS	Time(s)	GCUPS	Speedup
14831	540958	62332	8.05	35215	14.25	1.77

The table shows the performance of *hmmsearch* using one CPU core versus *hmmCUDAsearch* using one CPU core and the GPU when run on the benchmark. The columns show the total time in seconds, the performance in GCUPS, and speedup of *hmmCUDAsearch* relative to *hmmsearch*.

The results are presented in Figure 4 summarizing overall performance, and as a scatter plot Figure 5 of the length of the profile HMM versus the speedup of *cudaHmmsearch* relative to *hmmsearch* running on one core of the CPU. The speedup for the GPU over óne-core of the CPU ranged from 0.9x to 2.3x, with an average speedup of 1.8x.

4.1   Impact of Number of Cores

We investigated using one to six of the available cores of the CPU. Table 5 shows the performance of *hmmsearch* using either one, three, or six of the six cores of the CPU. So using multiple CPU cores is more effective than using a GPU with one CPU core.

4.2   Comparison with HMMER2

HMMER3 achieved significant improvements in performance over HMMER version 2 (HMMER2) through the introduction of a heuristic filter called the Multiple Segment Viterbi algorithm (MSV) and the use of native SIMD instruction set on modern CPUs.

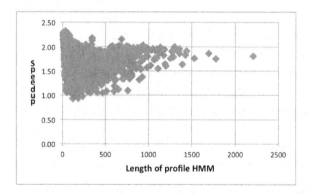

Fig. 5. Speedup vs Length of the profile HMM

The scatter shows the performance of *hmmsearch* using one CPU core versus *hmmCUDAsearch* using one CPU core and the GPU when run on the benchmark. The x-axis represents the length of the profile HMM, and the y-axis represents the speedup of *hmmCUDAsearch* relative to *hmmsearch*. The speedup for the GPU over one-core of the CPU ranged from 0.9x to 2.3x.

Table 5. Speedup using multiple CPU cores

CPUs	Time(s)	GCUPS	Speedup
1	62332	8.05	1.00
3	26185	19.16	2.38
6	18502	27.12	3.37

The table shows the performance of *hmmsearch* against the benchmark when using one, three, and six of the CPU cores. The columns show the total time in seconds, the performance in GCUPS, and speedup relative to the baseline of using one CPU.

In order to see the effect of those changes to HMMER, and the contributions of a GPU, we ran an experiment that searched the globins4 profile HMM of size 149 used in the HMMER Tutorial against the SwissProt database using four dif-

Table 6. Comparison with HMMER2 implementations

Application (Device)	HMMER2.3.2 (CPU)	GPU-HMMER2.3.2 (GPU)	HMMER3 (CPU)	cudaHmmsearch (GPU)
Performance (GCUPS)	0.14	0.95	8.47	17.36
Speedup (times)	1.00	6.79	60.50	124.00

The table shows the result of searching for the globins4 profile HMM against SwissProt database. The CPU implementations of hmmsearch were executed on one CPU core. The hmmsearch in GPU-HMMER2.3.2 and cudaHmmsearch were executed on one GPU using one CPU core. Speedup uses the performance of HMMER2.3.2 as baseline.

ferent versions of *hmmsearch*: HMMER 2.3.2, GPU-HMMER 2.3.2, HMMER3, and *cudaHmmsearch*. The results are shown in Table 6.

4.3 Contribution of Optimization Steps

We investigated the contribution made by each of the optimization techniques by again using the globins4 profile HMM of size 149 used in the HMMER Tutorial against the SwissProt database. The results are shown in Table 7.

Table 7. Performance of optimization approaches

Technique	Time (s)	Performance (GCUPS)	Improvement (%)
Initial implementation	227.178	0.126	-
SIMD Video Instruction	125.482	0.228	81
Minimizing global memory access	16.449	1.741	664
Async memcpy & Multi streams	9.463	3.026	74
Coalescing global memory	6.565	4.362	44
Texture memory	5.370	5.333	22
Sorting Database	2.346	12.207	129
Distributing workload	1.650	17.357	42

The table shows the result of using cudaHmmsearch to search the globins4 profile HMM against SwissProt. The fourth column Improvement is measured as a percentage compared with the previous approach. The row "Coalescing of global memory" is benchmarked only for the *dp* matrix. The row "Texture memory" is benchmarked only for the query profile texOMrbv 2D texture.

5 Conclusion

Searching databases of protein sequences for those proteins that match patterns represented as profile HMMs is a widely performed bioinformatics task. Our work used a general-purpose graphical processing unit (GPU) and the CUDA software environment to improve HMMER version 3, which is the standard tool for the task.

An execution profile of HMMER3 identified the heuristic Multiple Segment Viterbi algorithm (MSV) filter as a code hotspot that consumes over 75% of the total execution time. We followed a six-step process for tuning performance with CUDA programming: 1) assessing the application; 2) profiling the application; 3) optimizing memory usage; 4) optimizing instruction usage; 5) maximizing parallel execution; and 6) considering the existing libraries, that led us to apply well-known optimization strategies to implement a CUDA version *cudaHmmsearch* of *hmmsearch* by targetting the MSV filter.

Our experimental benchmark searched each Pfam domain against the SwissProt database. The results show that *cudaHmmsearch* achieved 1.8x speedup over the single-threaded HMMER3 CPU SSE2 implementation on average.

Acknowledgments. Funding in part provided by NSERC.

References

1. Needleman, S.B., Wunsch, C.D.: A general method applicable to the search for similarities in the amino acid sequence of two proteins. Journal of Molecular Biology **48**(3), 443–453 (1970)
2. Smith, T.F., Waterman, M.S.: Identification of common molecular subsequences. Journal of Molecular Biology **147**(1), 195–197 (1981)
3. Altschul, S.F., Gish, W., Miller, W., Myers, E.W., Lipman, D.J.: Basic local alignment search tool. Journal of Molecular Biology **215**(3), 403–410 (1990)
4. Krogh, A., Brown, M., Mian, I.S., Sjölander, K., Haussler, D.: Hidden markov models in computational biology: Applications to protein modeling. Journal of Molecular Biology **235**(5), 1501–1531 (1994)
5. Howard Hughes Medical Institute: HMMER (2014). http://hmmer.janelia.org/
6. Eddy, S.R.: Accelerated profile HMM searches. PLoS Computational Biology **7**(10) (2011)
7. Wun, B., Buhler, J., Crowley, P.: Exploiting coarse-grained parallelism to accelerate protein motif finding with a network processor. In: IEEE PACT, pp. 173–184. IEEE Computer Society (2005)
8. Maddimsetty, R.P.: Acceleration of profile-HMM search for protein sequences in reconfigurable hardware. Master thesis, Washington University in St. Louis (2006)
9. Derrien, S., Quinton, P.: Parallelizing HMMER for hardware acceleration on FPGAs. In: ASAP, pp. 10–17 (2007)
10. Oliver, T.F., Yeow, L.Y., Schmidt, B.: High performance database searching with HMMer on FPGAs. In: IPDPS, pp. 1–7. IEEE (2007)
11. Sachdeva, V., Kistler, M., Speight, E., Tzeng, T.H.K.: Exploring the viability of the Cell Broadband Engine for bioinformatics applications. Parallel Computing **34**(11), 616–626 (2008)
12. Walters, J.P., Qudah, B., Chaudhary, V.: Accelerating the HMMER sequence analysis suite using conventional processors. In: [39], pp. 289–294
13. Landman, J.I., Ray, J., Walters, J.P.: Accelerating HMMer searches on Opteron processors with minimally invasive recoding. In: [39], pp. 628–636
14. Horn, D.R., Houston, M., Hanrahan, P.: ClawHMMER: A streaming HMMer-search implementation. In: SC, p. 11. IEEE Computer Society (2005)
15. Walters, J.P., Balu, V., Kompalli, S., Chaudhary, V.: Evaluating the use of GPUs in liver image segmentation and HMMER database searches. In: [40], pp. 1–12
16. Ganesan, N., Chamberlain, R.D., Buhler, J., Taufer, M.: Accelerating HMMER on GPUs by implementing hybrid data and task parallelism. In: Zhang, A., Borodovsky, M., Özsoyoglu, G., Mikler, A.R. (eds.) BCB, pp. 418–421. ACM (2010)
17. Du, Z., Yin, Z., Bader, D.A.: A tile-based parallel viterbi algorithm for biological sequence alignment on GPU with CUDA. In: 2010 IEEE International Symposium on Parallel & Distributed Processing, Workshops and Phd Forum (IPDPSW), pp. 1–8. IEEE (2010)
18. Quirem, S., Ahmed, F., Lee, B.K.: CUDA acceleration of P7Viterbi algorithm in HMMER 3.0. In: Zhong, S., Dou, D., Wang, Y. (eds.) IPCCC, pp. 1–2. IEEE (2011)
19. Intel: Intel VTune amplifier XE 2013 (2014). https://software.intel.com/en-us/intel-vtune-amplifier-xe/
20. Ahmed, F., Quirem, S., Min, G., Lee, B.K.: Hotspot analysis based partial CUDA acceleration of HMMER 3.0 on GPGPUs. International Journal of Soft Computing and Engineering **2**(4), 91–95 (2012)

21. Wilt, N.: The CUDA Handbook: A Comprehensive Guide to GPU Programming. Addison-Wesley Professional (2013)
22. NVIDIA: CUDA C best practices guide (2013). http://docs.nvidia.com/cuda/cuda-c-best-practices-guide/index.html
23. Liu, Y., Maskell, D., Schmidt, B.: CUDASW++: optimizing Smith-Waterman sequence database searches for CUDA-enabled graphics processing units. BMC Research Notes 2(1) (2009)
24. Liu, Y., Schmidt, B., Maskell, D.L.: CUDASW++2.0: enhanced Smith-Waterman protein database search on CUDA-enabled GPUs based on SIMT and virtualized SIMD abstractions. BMC Research Notes 3(1) (2010)
25. Liu, Y., Wirawan, A., Schmidt, B.: CUDASW++ 3.0: accelerating Smith-Waterman protein database search by coupling CPU and GPU SIMD instructions. BMC Bioinformatics 14, 117 (2013)
26. Manavski, S.A., Valle, G.: CUDA compatible GPU cards as efficient hardware accelerators for Smith-Waterman sequence alignment. BMC Bioinformatics 9(Suppl 2), S10 (2008)
27. Akoglu, A., Striemer, G.M.: Scalable and highly parallel implementation of Smith-Waterman on graphics processing unit using CUDA. Cluster Computing 12(3), 341–352 (2009)
28. Ligowski, L., Rudnicki, W.R.: An efficient implementation of Smith Waterman algorithm on GPU using CUDA, for massively parallel scanning of sequence databases. In: [40], pp. 1–8
29. Kentie, M.: Biological Sequence Alignment Using Graphics Processing Units. Master thesis, Delft University of Technology (2010)
30. Saeed, A.K., Poole, S., Perot, J.B.: Acceleration of the Smith-Waterman algorithm using single and multiple graphics processors. Journal of Computational Physics 229(11), 4247–4258 (2010)
31. Aji, A.M., Feng, W., Blagojevic, F., Nikolopoulos, D.S.: Cell-SWat: modeling and scheduling wavefront computations on the Cell Broadband Engine. In: Ramírez, A., Bilardi, G., Gschwind, M. (eds.) Conf. Computing Frontiers, pp. 13–22. ACM (2008)
32. Rognes, T., Seeberg, E.: Six-fold speed-up of Smith-Waterman sequence database searches using parallel processing on common microprocessors. Bioinformatics 16(8), 699–706 (2000)
33. Farrar, M.: Striped Smith-Waterman speeds database searches six times over other SIMD implementations. Bioinformatics 23(2), 156–161 (2007)
34. Sanders, J., Kandrot, E.: CUDA by Example: An Introduction to General-Purpose GPU Programming, 1st edn. Addison-Wesley Professional (2010)
35. NVIDIA: CUDA C programming guide (2013). http://docs.nvidia.com/cuda/cuda-c-programming-guide/index.html
36. SIB Bioinformatics Resource Portal: UniProtKB/Swiss-Prot protein knowledgebase release 2014–05 statistics (2014). http://web.expasy.org/docs/relnotes/relstat.html
37. Wellcome Trust Sanger Institute and Howard Hughes Janelia Farm Research Campus: Pfam database (2013). ftp://ftp.sanger.ac.uk/pub/databases/Pfam/releases/Pfam27.0/Pfam-A.hmm.gz
38. Universal Protein Resource: UniProt release. Website (2014)
39. 20th International Conference on Advanced Information Networking and Applications (AINA 2006), Vienna, Austria, 18–20 April. IEEE Computer Society (2006)
40. 23rd IEEE International Symposium on Parallel and Distributed Processing, IPDPS 2009, Rome, Italy, 23–29 May. IEEE (2009)

Fast Subgraph Matching on Large Graphs
using Graphics Processors

Ha-Nguyen Tran[✉], Jung-jae Kim, and Bingsheng He

School of Computer Engineering, Nanyang Technological University,
Singapore, Singapore
{s110035,jungjae.kim,bshe}@ntu.edu.sg

Abstract. Subgraph matching is the task of finding all matches of a
query graph in a large data graph, which is known as an NP-complete
problem. Many algorithms are proposed to solve this problem using
CPUs. In recent years, Graphics Processing Units (GPUs) have been
adopted to accelerate fundamental graph operations such as breadth-
first search and shortest path, owing to their parallelism and high data
throughput. The existing subgraph matching algorithms, however, face
challenges in mapping backtracking problems to the GPU architectures.
Moreover, the previous GPU-based graph algorithms are not designed to
handle intermediate and final outputs. In this paper, we present a simple
and GPU-friendly method for subgraph matching, called *GpSM*, which is
designed for massively parallel architectures. We show that GpSM outperforms the state-of-the-art algorithms and efficiently answers subgraph
queries on large graphs.

1   Introduction

Big networks from social media, bioinformatics and the World Wide Web can be
essentially represented as graphs. As a consequence, common graph operations
such as breadth-first search and subgraph matching face the challenging issues
of scalability and efficiency, which have attracted increasing attention in recent
years. In this paper, we focus on *subgraph matching*, the task of finding all
matches or *embeddings* of a query graph in a large data graph. This problem
has enjoyed widespread popularity in a variety of real-world applications, e.g.,
semantic querying [1,2], program analysis [3], and chemical compound search [4].
In such applications, subgraph matching is usually a bottleneck for the overall
performance because it involves subgraph isomorphism which is known as an
NP-complete problem [5].

Existing algorithms for subgraph matching are generally based on the *filtering-
and-verification* framework [3,6–12]. First, they filter out all candidate vertices
which cannot contribute to the final solutions. Then the verification phase follows,
in which *backtracking*-based algorithms are applied to find results in an incremental fashion. Those algorithms, however, are designed to work only in small-graph
settings. The number of candidates grows significantly high in medium-to-large-
scale graphs, resulting in an exorbitant number of costly verification operations.

© Springer International Publishing Switzerland 2015
M. Renz et al. (Eds.): DASFAA 2015, Part I, LNCS 9049, pp. 299–315, 2015.
DOI: 10.1007/978-3-319-18120-2_18

Several indexing techniques have also been proposed for faster computation [3,9]; however, the enormous index size makes them impractical for large data graphs [14]. Distributed computing methods [13,14] have been introduced to deal with large graphs by utilizing parallelism, yet there remains the open problem of high communication costs between the participating machines.

Recently, GPUs with massively parallel processing architectures have been successfully leveraged for fundamental graph operations on large graphs, including breadth-first search [15,16], shortest path [15,17] and minimum spanning tree [18]. Traditional backtracking approaches for subgraph matching, however, cannot efficiently be adapted to GPUs due to two problems. First, GPU operations are based on *warps* (which are groups of threads to be executed in single-instruction-multiple-data fashion), and different execution paths generated by backtracking algorithms may cause a so-called *warp divergence* problem. Second, GPU implementations for coalesced memory accesses are no longer straightforward due to irregular access patterns [19].

To address these issues, we propose an efficient and scalable method called *GpSM*. GpSM runs on GPUs and takes on edges as the basic unit. Unlike previous *backtracking*-based algorithms, GpSM joins candidate edges in parallel to form partial solutions during the verification phase, and this procedure is conducted repeatedly until the final solution is obtained. An issue raised by such parallel algorithms is the considerable amount of intermediate results for joining operations, while backtracking algorithms only need to store less of such data during execution. We resolve this issue by adopting the pruning technique of [20], further enhancing it by ignoring *low-connectivity vertices* which have little or no effect of decreasing intermediate results during filtering.

To highlight the efficiency of our solution, we perform an extensive evaluation of GpSM against state-of-the-art subgraph matching algorithms. Experiment results on both real and synthetic data show that our solution outperforms the existing methods on large graphs.

The rest of the paper is structured as follows. Section 2 gives formal definitions and related works. In section 3, we introduce the filtering-and-joining approach to solve the problem. The filtering and joining phases are discussed in Section 4 and 5. Section 6 extends our method to deal with large graphs. Experiment results are shown in Section 7. Finally, Section 8 concludes our paper.

2   Preliminaries

2.1   Subgraph Matching Problem

We give a formal problem statement using *undirected labeled graphs*, though our method can be applied to *directed labeled graphs* as shown in the Experiment Results section.

Definition 1. *A* **labeled graph** *is a 4-tuple* $G = (V, E, L, l)$, *where* V *is the set of vertices,* $E \subseteq V \times V$ *is the set of edges,* L *is the set of labels and* l *is a labeling function that maps each vertex to a label in* L.

Definition 2. *A graph $G = (V, E, L, l)$ is **subgraph isomorphic** to another graph $G' = (V', E', L', l')$, denoted as $G \subseteq G'$, if there is an injective function (or a **match**) $f : V \to V'$, such that $\forall (u, v) \in E$, $(f(u), f(v)) \in E'$, $l(u) = l'(f(u))$, and $l(v) = l'(f(v))$.*

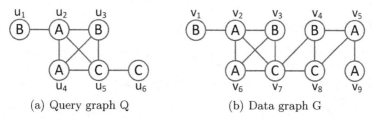

(a) Query graph Q   (b) Data graph G

Fig. 1. Sample query and data graph

Subgraph Matching Problem is defined as follows: Given a large data graph G and a query graph Q, we find all matches of Q in G. For example, the subgraph matching solution of the query graph Q in the data graph G in Figure 1 is $\{(u_1, v_1), (u_2, v_2), (u_3, v_3), (u_4, v_6), (u_5, v_7), (u_6, v_8)\}$.

Definition 3. *Given a query graph $Q = (V, E, L, l)$ and a data graph $G = (V', E', L', l')$, a vertex $v \in V'$ is called a **candidate** of a vertex $u \in V$ if $l(u) = l'(v)$, $degree(u) \leq degree(v)$ where $degree(u)$, $degree(v)$ are the number of vertices connected to edges starting vertex u and v respectively. The set of candidates of u is called **candidate set** of u, denoted as $C(u)$.*

The query vertex u_3 in Figure 1a has a label of B and a degree of 3. For the data graph vertex v_3 in Figure 1b, the label is also B and the degree is 3 which is equal to the degree of u_3. Therefore, v_3 is a candidate of u_3. The candidate set of u_3 is $C(u_3) = \{v_3, v_4\}$.

An *adjacency list* of a vertex u in a graph G is a set of vertices which are the destinations of edges starting from u, denoted as *adj(u)*. For example, the adjacency list of u_3 is $adj(u_3) = \{u_2, u_4, u_5\}$.

2.2 Subgraph Matching Algorithms

Most of state-of-the-art subgraph matching algorithms are based on backtracking strategies which find matches by either forming partial solutions incrementally or pruning them if they cannot produce the final results, as discussed in the works of Ullman [6,10], VF2 [7], QuickSI [8], GADDI [9], GraphQL [1] and SPath [3]. One of open issues in those methods is the selection of matching order (or visit order). To address this issue, TurboISO [11] introduces the strategies of candidate region exploration and combine-and-permute to compute a 'good' visit order, which makes the matching process efficient and robust.

To deal with large graphs, Sun et al. [14] introduce a parallel and distributed algorithm (which we call STW in this paper), in which they decompose the query graphs into 2-level trees, and apply graph exploration and join strategy to obtain solutions in a parallel manner over a distributed memory cloud. Unlike STW, our method uses GPUs in order to keep the advantages of parallelism during computation, while simultaneously avoiding high communication costs between participating machines.

2.3   General-Purpose Computing on GPUs

GPUs are widely used as commodity components in modern-day machines. A GPU consists of many individual multiprocessors (or *SMs*), each of which executes in parallel with the others. During runtime, threads on each multiprocessor are organized into thread blocks, and each block consists of multiple 32-thread groups, called *warps*. If threads within a warp are set to execute different instructions, they are called *diverged*; computations in diverged threads are only partially parallel, thus reducing the overall performance significantly. The GPU includes a large amount of device memory with high bandwidth and high access latency, called *global memory*. In addition, there is a small amount of *shared memory* on each multiprocessor, which is essentially a low latency, high bandwidth memory running at register speeds.

Due to such massive amounts of parallelism, GPUs have been adopted to accelerate data and graph processing [15,16,23,27]. Harish et al. [15] implement several common graph algorithms on GPUs including BFS, single source shortest path and all-pair shortest path. Hong et al. [23] enhance BFS by proposing a virtual warp-centric method to address the irregularity of BFS workload. Merrill et al. [16] propose a BFS algorithm which is based on fine-grained task management and built upon an efficient prefix sum; this work has generally been considered as one of the most complete and advanced works regarding BFS traversal on GPUs. Finally, Medusa is a general programming framework for graph processing in GPU settings [25], providing a rich set of APIs based on which developers can further build their applications.

3   GpSM Overview

We introduce a GPU-based algorithm called GpSM to solve the problem of subgraph matching. Unlike previous CPU methods with complicated pruning and processing techniques, our algorithm is *simple* and *designed for massively parallel architectures*.

3.1   Filtering-and-Joining Approach

We find that STW method [14] simultaneously filters out candidate vertices and matches of basic units (i.e. 2-level trees), and thus generates a large amount of irrelevant candidate vertices and edges. Our method performs the two tasks separately in order to reduce intermediate results. The main routine of the GPU-based algorithm is illustrated in Algorithm 1.

The inputs of GpSM are a connected query graph q and a data graph g. The vertex sets and edge sets of q and g are V_q, E_q and V_g, E_g respectively. The output is a set of subgraph isomorphisms (or matches) of q in g. In our method, we present a match as a list of pairs of a query vertex and its mapped data vertex. Our solution is the collection M of such lists.

Algorithm 1. GpSM(q,g)

Input: query graph q, data graph g
Output: all matches of q in g

1   C := InitializeCandidateVertices(q,g);
2   C := RefineCandidateVertices(q,g,C);
3   E := FindCandidateEdges(q,g,C);
4   M := JoinCandidateEdges(q,g,E);
5   **return** M

Our method uses a *filtering-and-joining* strategy. The filtering phase consists of two tasks. The first task filters out candidate vertices which cannot be matched to query vertices (Line 1). After this task there can still be a large set of irrelevant candidate vertices which cannot contribute to subgraph matching solutions. The second task continues pruning this collection by calling a refining function, *RefineCandidateVertices*. In the function, candidate sets of query vertices are recursively refined either until no more candidates can be pruned, or up to a predefined number of times (Line 2). The details of the filtering phase will be discussed in Section 4. In the joining phase, GpSM collects candidate edges based on the candidate vertices obtained in the previous phase (Line 3) and combines them to produce the final subgraph matching solutions (Line 4) which are finally returned to users. Section 5 gives the detailed implementation of the joining phase.

3.2   Graph Representation

In order to support graph query answering on GPUs, we use three arrays to represent a graph $G = (V, E)$: *vertices array*, *edges array*, and *labels array*. The edges array stores the adjacency lists of all vertices in V, from the first vertex to the last one. The vertices array stores the start indexes of the adjacency lists, where the i-th element of the vertices array has the start index of the adjacency list of the i-th vertex in V. The labels array maintains labels of vertices in order to support our method on labelled graphs. The first two arrays have been used in previous GPU-based algorithms [15,16,23]. Figure 2 shows the representation of the graph illustrated in Figure 1a in the GPU memory.

The advantage of the data structure is that vertices in the adjacency list of a vertex are stored next to each other in the GPU memory. During GPU execution, consecutive threads can access consecutive elements in the memory. Therefore, we can avoid the random access problem and decrease the accessing time for GPU-based methods consequently.

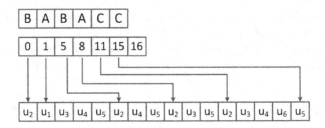

Fig. 2. Graph Representation in GPU Memory

4   Filtering Phase

This section describes the implementation of the filtering phase on GPUs. The purpose of this phase is to reduce the number of candidate vertices and thus decrease the amount of candidate edges as well as the running time of the joining phase. The filtering phase consists of two tasks: initializing candidate vertices and refining candidate vertices.

4.1   Initializing Candidate Vertices

The first step of the filtering phase is to initialize candidate sets of all query vertices. In the task, we take a spanning tree generated from the query graph as the input. This section presents a heuristic approach to selecting a good spanning tree among many spanning trees of the query graph. The approach is based on the observation that if the filtering starts from the query vertices with the smallest number of candidates, its intermediate results can be kept to the minimum. Since we do not know the number of candidates in the beginning, we estimate it by using a vertex ranking function $f(u) = \frac{deg(u)}{freq(u.label)}$ [11,14], where $deg(u)$ is the degree of u and $freq(u.label)$ is the number of data vertices having the same label as u.

We find a spanning tree T and a visit order O for a query graph as follows: Initially, we pick a query edge (u, v) such that $f(u) \geq f(v)$ and $f(u) + f(v)$ is the maximum among all query edges. We add u to the visit order O, and add the edges connected to u to the spanning tree T, except those whose endpoints are already in the vertices set of T, i.e. $V(T)$. The process continues to pick up another query edge connected to T and add to O and T until no edge remains. Figure 5a depicts the spanning tree of the Figure 1a graph. Also, the visit order is u_5, u_2.

Algorithm 2 outlines the task of finding candidate vertices of each query vertex from the data graph, following the visit order obtained earlier. For each query vertex u, GpSM first checks if each of data vertex is a candidate of u and keeps the candidacy information in the Boolean array $c\_set[u]$ in parallel ($kernel\_check$[1]; Line 7) in the case that its candidate set is not initialized (Line

---

[1] Note that all functions whose names start with $kernel$ are device functions that run on GPUs.

6). It then creates an integer array (c_array) that collects the indexes of candidates of u from $c\_set[u]$ ($kernel\_collect$; Line 9). GpSM calls another device function ($kernel\_explore$; Line 10) that prunes out all candidate vertices u' of u such that there is a vertex $v \in adj(u)$ which has no candidate vertex in $adj(u')$ (Lines 16-18), and explores the adjacency list of u in the spanning tree in order to filter the candidates of the vertices in $adj(u)$ (Lines 19-22). Thus, the final outputs are *Boolean* arrays c_set, which represent the filtered candidate sets of query vertices.

Algorithm 2. Initializing candidate vertices

Input: spanning tree T, data graph g
Output: candidate sets of vertices c_set

1 **Algorithm** InitializeCandidateVertices(T, g)
2 **foreach** *vertex* $u \in T$ **do**
3   $c\_set[u][v] := false; \forall v \in V_g$
4   $initialized[u] := false;$

5 **foreach** $u \in T$ *in the visit order* **do**
6   **if** *initialized[u] = false* **then**
7   kernel_check($c\_set[u]$, g);
8    $initialized[u] := true;$

9   $c\_array :=$ kernel_collect(u, $c\_set[u]$);
10   kernel_explore(u, c_array, c_set, T, g);
11   **foreach** $v \in adj(u)$ **do**
12    $initialized[v] := true;$

13 **return** c_set;

14 **Procedure** kernel_explore(u, c_array, c_set, T, g)
15 $u' :=$ GetCandidate(c_array, $warp\_id$);
16 **if** *exist* $v \in adj(u)$ *such that no* $v' \in adj(u')$ *is a candidate of* v **then**
17   $c\_set[u][u'] := false;$
18   **return**;

19 **foreach** $v \in adj(u)$ **do**
20   $v' :=$ GetAdjacentVertex(u', $thread\_id$);
21   **if** v' *is a candidate of* v **then**
22    $c\_set[v][v'] := true;$

GPU Implementation: We implement the two GPU device functions *kernel_collect* and *kernel_explore* in the first step of the filtering phase, based on two optimization techniques: *occupancy maximization* to hide memory access latency and *warp-based execution* to take advantage of the coalesced access and to deal with workload imbalance between threads within a warp. We skip details of the device function *kernel_check* since its implementation is straightforward.

1) *kernel_collect*. This function is to maximize the occupancy of the *kernel_explore* execution. At runtime, warps currently running in an SM are called

active warps. Due to the resource constraints, each SM allows a maximum number of active warps running concurrently at a time. Occupancy is the number of concurrently running warps divided by the maximum number of active warps. At runtime, when a warp stalls on a memory access operation, the SM switches to another active warp for arithmetic operations. Therefore, high-occupancy SM is able to adequately hide access latency.

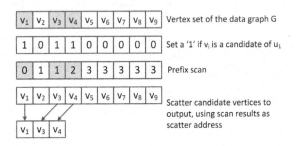

Fig. 3. Collect candidate vertices of u_1

A naive approach to executing *kernel_explore* is that only the warps corresponding to the *true* elements of $c\_set[u]$ continue filtering vertices in $adj(u)$. However, the approach suffers from the low-occupancy problem since warps with the *false* elements are idle. For example, we assume that the maximum number of active warps on the multiprocessor is 3. In the first 3 active warps, the occupancy is 66.66% because only the warps corresponding to v_1 and v_3 execute *kernel_explore* while the warp with v_2 is idle. For the next 3 concurrently running warps, the occupancy is only 33.33%. GpSM resolves the issue by adopting a stream compaction algorithm [26] to gather candidate vertices into an array c_array for those $c\_set[u]$ with true values. The algorithm employs prefix scan to calculate the output addresses and to support writing the results in parallel. The example of collecting candidate vertices of u_1 is depicted in Figure 3. By taking advantage of c_array, all 3 active warps are used to explore the adjacency lists of v_1, v_3 and v_4. As a result, our method achieves a high occupancy.

2) *kernel_explore.* Inspired by the warp-based methods used in BFS algorithms for GPUs [16,23], we assign to each warp a candidate vertex $u' \in C(u)$ (or c_array from *kernel_collect*). Within the warp, consecutive threads find the candidates of $v \in adj(u)$ in $adj(u')$. This method takes advantage of coalesced access since the vertices of $adj(u')$ are stored next to each other in memory. It also addresses the warp divergence problem since threads within the warp execute similar operations. Thus, our method efficiently deals with the workload imbalance problem between threads in a warp. Figure 4 shows an example of filtering candidate vertices of u_2 based on the candidate set of u_1, $C(u_1) = \{v_1, v_3, v_4\}$.

If a data vertex has an exceptionally large degree compared to the others, GpSM deals with it by using an entire block instead of a warp. This solution reduces the workload imbalance between warps within the block.

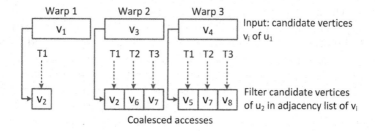

Fig. 4. Filter candidate vertices of u_2 based on adjacency lists of $C(u_1) = \{v_1, v_3, v_4\}$

4.2 Refining Candidate Vertices

After filtering out candidate vertices for the first time, there can be still a large number of candidate vertices which cannot be parts of final solutions. To address this issue, we propose a recursive filtering strategy to further prune irrelevant candidate vertices. The size of candidate edges and intermediate results are then reduced consequently.

We observe the followings: 1) Exploring non-tree edges (i.e. those that form cycles) can reduce the number of irrelevant candidates significantly; and 2) the more edges a vertex has, the more irrelevant candidates of the vertex the filtering techniques aforementioned can filter out. Based on the first observation, from the second round of the filtering process, our method uses the original query graph for exploration rather than a spanning tree of the query graph. Based on the second observation, our method ignores query vertices connected to small number of edges, called *low connectivity vertices*. For small-size query graphs, a low connectivity vertex has the degree of 1. As for big query graphs, we can increase the value of degree threshold to ignore more low connectivity vertices. The query graph obtained after removing low connectivity vertices from Q is shown in Figure 5b.

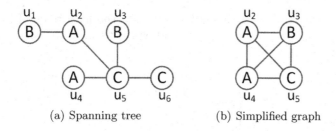

(a) Spanning tree   (b) Simplified graph

Fig. 5. Spanning tree and simplified graph of Q

GPU implementation: The main routine of the refining task is similar to the filtering in the previous section. The differences are as follows: 1) *kernel_check* is not necessary for the refining process and 2) we only use the pruning task (Lines 16-18) in the *kernel_explore* function. By taking advantage of the *c_set* array generated in the initialization step, the refinement can verify the candidate conditions easily and reduce the random accesses during the candidate verification.

Ideally, the optimal candidate sets of query vertices are obtained when the refinement is recursively invoked until no candidate is removed from the candidate sets. However, our experiments show that most of irrelevant candidates are pruned in the first few rounds. The later rounds do not prune out many candidates, but lead to inefficiency and reduce the overall performance. Therefore, the refining task terminates after a limited number of rounds.

In the tasks of initializing and refining candidate sets of query vertices, GpSM requires $O(|V_q| \times |V_g|)$ space to maintain *Boolean* arrays which are used to collect candidate vertices and $O(|V_g|)$ space to keep the collected set. Let S be the number of SMs. Each SM has P active threads. For each visited vertex, the prefix scan in *kernel_collect* executes in $O(|V_g| \times log(|V_g|)/(S \times P))$ time while *kernel_explore* runs in $O(|V_g| \times |d_g|/(S \times P))$, where d_g is the average degree of the data graph. Assume that the candidate refinement stops after k rounds, the total time complexity of the filtering phase is $O(|V_q| \times k \times (|V_g| \times log(|V_g|) + |V_g| \times |d_g|)/(S \times P))$.

5 Joining Phase

In the joining phrase, GpSM first gathers candidate edges in the data graph and then combines them into subgraph matching solutions.

The output of each query edge (u, v) in the task of gathering candidate edges is represented as a hash table, as depicted in Figure 6. The keys of this table are candidate vertices u' of u, and the value of a key u' is the address of the first element of the collection of candidate vertices v' of v such that $(u', v') \in E_g$. An issue of the step is that the number of the candidate edges is unknown, and thus that we cannot directly generate such a hash table. To address this issue, we employ the *two-step output scheme* [22] as follows: 1) Given a query edge (u, v), each warp is assigned to process a candidate vertex u' of u and counts the number of candidate edges starting with u' (designated as (u', v')). The system then computes the address of the first v' for u' in the hash table of (u, v). 2) It then re-examines the candidate edges and writes them to the corresponding addresses of the hash table.

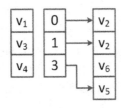

Fig. 6. Candidate edges of (u_1, u_2)

After finding the candidate edges, GpSM combines them to produce subgraph matching solutions as follows: Initially, we pick a query edge (u, v) with the smallest number of candidate edges, and mark as *visited* the vertices u, v and the edge (u, v). Here the candidates of (u, v) are partial subgraph matching solutions. We select the next edge among the unvisited edges of the query graph, denoted by (u', v'), such that 1) both u' and v' are visited vertices, or 2) if there is no such edge, either u' or v' is a visited vertex. If there are multiple such edges, we select the one with the smallest number of candidates. Candidate edges of (u', v') are then combined with the partial solutions. The procedure is conducted repeatedly until all query edges are visited.

GPU Implementation: The GPU implementation for the task of gathering candidate edges is similar to that of the filtering phase, except introducing the two-step output scheme. For the task of combining partial subgraph matching solutions, we apply the warp-based approach as follows: Each warp i is responsible for combining a partial solution $M_i(q)$ with candidate edges of (u, v), where u is already visited. First, the warp retrieves the candidate vertex of u from $M_i(q)$ (e.g., u'). It looks up the hash table storing candidate edges of (u, v) to find the key u' and retrieve the candidate vertices v' of v from the hash table. By using our data structure of candidate edges, this task can be done in logarithmic time. Threads within the warp then verify whether (u', v') can be merged to $M_i(q)$, in which GpSM again follows the two-step output scheme to write the merged results.

Shared Memory Utilization. The threads within the warp i should share the partial solution $M_i(q)$ and access them frequently. We thus store and maintain $M_i(q)$ in the shared memory instead of the device memory, which efficiently hides the memory stalls.

Let $C(e_i)$ be the candidate edges of the edge e_i. The joining phase is done in $O(\prod_{i=1}^{|E_q|} |C(e_i)| \times log(|V_g|)/(S \times P))$ time. Note that the running time of the joining phase highly depends on the number of candidates of query edges. Therefore, reducing the number of candidate vertices in the filtering phase plays an important role in decreasing both the running time and the memory used to maintain partial solutions.

6   Extended GpSM for Very Large Graphs

In real-world applications, the sizes of data graphs might be too large to be stored in the memory of a single GPU device. In general, such large graphs have many labels, and thus the vertices corresponding to the labels of a given query graph, together with their adjacency lists, are relatively small. Based on the observation, we make an assumption that the relevant data of query graph labels are small enough to fit into the GPU memory. Therefore, we can make GpSM work efficiently on large graphs by storing them with the *inverted-vertex-label* index data structure in CPU memory or hard disk and, given a query graph, by retrieving only relevant vertices and their adjacency lists to the GPU memory.

For each label l in the data graph G, we use three array structures. The first array contains all vertices that have the label of l (designated as V_l.) The other

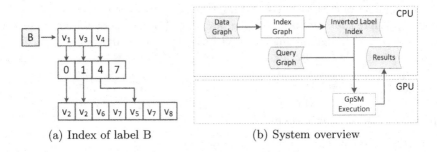

(a) Index of label B   (b) System overview

Fig. 7. GpSM Solution on large graphs

two arrays are the vertices and edges arrays corresponding to V_l, as defined in Section 3.2. An entry of the inverted-vertex-label index for label B of the data graph in Figure 1b is depicted in Figure 7a.

Figure 7b provides a system overview of our solution for large-graph subgraph matching using both GPUs and CPUs. Here rectangles indicate tasks while the others represent data structures used in our solution. The first task is to create an inverted-vertex-label index which is then stored in the hard disk. In order to decrease the time to transfer data from the hard disk, we keep the most frequent vertex labels in the main memory. Given a query graph, our solution retrieves all the data associated with the query labels from the main memory or hard disk and transfers them to the GPU memory to further search for subgraph matching solutions, which are finally returned to the main memory.

7   Experiment Results

We evaluate the performance of GpSM in comparison with state-of-the-art subgraph matching algorithms, including VF2 [7], QuickSI (in short, QSI) [8], GraphQL (in short, GQL) [1] and TuboISO [11]. The implementations of VF2, QuickSI and GraphQL used in our experiments are published by Lee and colleages [21]. As for TurboISO, we use an executable version provided by the authors.

Datasets. The experiments are conducted on both real and synthetic datasets. The real-world data include the Enron email communication network and the Gowalla location-based social network[2]. On the other hand, the synthetic datasets are generated by RMAT generator [24], and vertices are labeled randomly. As for query graphs, given the number of vertices N and the average number of edges per vertex D (called *degree*), we generate connected labeled query graphs of size N, randomly connecting the vertices and making their average degree D. Except for experiments with varying degrees, the query graphs always have the degree of 2.

Environment. The runtime of the CPU-based algorithms is measured using an Intel Core i7-870 2.93 GHz CPU with 8GB of memory. Our GPU algorithms

---

[2] These datasets can be downloaded from Stanford Dataset Collection website. See https://snap.stanford.edu/data for more details.

are tested using CUDA Toolkit 6.0 running on the NVIDIA Tesla C2050 GPU with 3 GB global memory and 48 KB shared memory per Stream Multiprocessor. For each of those tests, we execute 100 different queries and record the average elapsed time. In all experiments, algorithms terminate only when all subgraph matching solutions are found.

7.1   Comparison with State-of-the-art CPU Algorithms

The first set of experiments is to evaluate the performance of GpSM compared to the state-of-that-art algorithms. These comparisons are performed on both synthetic and real datasets. The input graphs are *undirected graphs* because the released version of TurboISO only works with undirected graphs.

Synthetic Datasets. The size of the synthetic data graphs of the first experiment set varies from 10,000 vertices to 100,000 vertices. All the data graphs have 10 distinct labels and the average degree of 16, and can fit into GPU memory. The query graphs contain 6 vertices and 12 edges. Figure 8 shows that GpSM clearly outperforms VF2, QuickSI and GraphQL. Compared to TurboISO, our GPU-based algorithm runs slightly slower when the size of the data graphs is relatively small (i.e. 10,000 vertices). However, when the size of data graphs increases, GpSM is more efficient than TurboISO. We thus make further comparisons with TurboISO in more experiment settings.

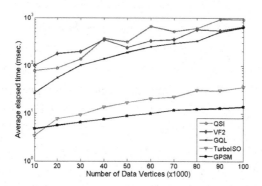

Fig. 8. Varying data sizes

Real Datasets. As for real-world datasets, Gowalla network consists of 196,591 vertices and 950,327 edges while Enron network has 36,692 vertices and 183,831 edges. In these experiments, we use 20 labels for Gowalla nerwork and 10 labels for Enron network. The number of query vertices varies from 6 to 13.

Figure 9a shows that TurboISO anwsers the subgraph matching queries against the Gowalla network efficiently when the size of query graphs is small. As the number of vertices increases, however, the processing time of TurboISO

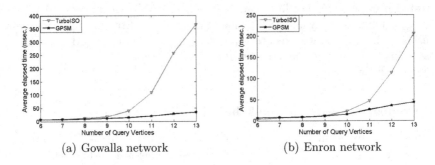

(a) Gowalla network (b) Enron network

Fig. 9. Experiment on real datasets

grows exponentially. In contrast, GpSM shows almost linear growth. The two methods show similar performance difference when evaluated against the Enron network, as plotted in Figure 9b.

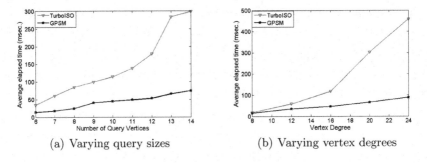

(a) Varying query sizes (b) Varying vertex degrees

Fig. 10. Comparison with TurboISO

Comparison with TurboISO. We also compare the performance of GpSM with TurboISO, varying the size of query graphs and the degree of the data graphs, as shown in Figure 10. The data graphs are synthetic undirected graphs with 100,000 vertices and 10 labels.

Figure 10a shows the performance results of GpSM and TurboISO on the query graphs whose numbers of vertices vary from 6 to 14. In the experiment, the degree of the data graph is 16. Figure 10b shows their performance results when the vertex degree increases from 8 to 24, where the query graph size is fixed to 10. As shown in the two figures, the performance of TurboISO drops significantly while that of GpSM does not. This may be due to the fact that the number of recursive calls of TurboISO grows exponentially with respect to the size of query graphs and the degree of the data graph. In contrast, GpSM takes advantage of the large number of threads in GPUs to handle candidate edges in parallel and thus keep the processing time rising slowly.

7.2 Scalability Tests

We test the extended GpSM against very large graphs. The data graphs are *directed graphs* which are generated using the RMAT generator. The number of data vertices varies from 1 million to 2 billion vertices while the number of labels varies from 100 to 2000 according to the vertex number. The average degree of the data vertices is 8. The data graph is stored as follows: When the data graph is small, i.e from 1 million to 25 million vertices, we store it in the GPU global memory. If the vertex number of the data graph is between 25 million and 100 million, CPU memory is used to maintain the data graph. For data graphs with 200 million vertices and above, we store them in both CPU memory and hard disk. The largest number of vertices per label of the 25-million-vertex graph is around 350,000 while that of the 2-billion-vertex graph is nearly 1,400,000. The query graphs used in the experiments consist of 10 vertices and 20 edges.

When the data graph size is 25 million vertices, we perform two experiments. The first one maintains the whole data graph in GPU memory and the second uses CPU memory. As shown in Figure 11a, the second experiment answers subgraph matching queries slower than the first one, due to the time for data transfer from CPU memory to GPU memory.

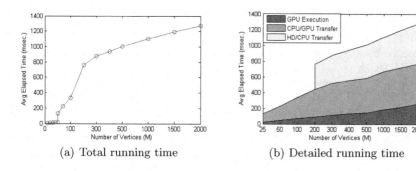

(a) Total running time   (b) Detailed running time

Fig. 11. Scalability Tests

The details of the running time are shown in Figure 11b, from which we observe the followings: 1) The time taken for GPU execution (i.e. subgraph matching) grows linearly as the data graph size increases, as expected from our time complexity analysis in Sections 4 and 5. 2) The GPU execution time takes around 11~20% of the total running time, while the rest is taken by data transfers between GPU and CPU or hard disk. 3) The data transfer time also grows almost linearly as the data graph size increases, though the transfer from hard disk adds additional running time.

8 Conclusions

In this paper, we introduce an efficient method which takes advantage of GPU parallelism to deal with large-scale subgraph matching problem. Our method called GpSM is simple and designed for massively parallel processing. GpSM is based on *filtering-and-joining* approaches, efficient GPU techniques of coalescence, warp-based and shared memory utilization, and a recursive refinement function for pruning irrelevant candidate vertices. Experiment results show that our method outperforms previous backtracking-based algorithms on CPUs and can efficiently answer subgraph matching queries on large graphs. In future, we will further improve the efficiency of GpSM for large graphs, for example, by dealing with large amount of intermediate results that do not fit into the GPU memory and also by adopting buffering techniques.

Acknowledgments. We thank to Prof. Wook-Shin Han and Dr. Jeong-Hoon Lee for sharing *iGraph* source code and executable files of $Turbo_{ISO}$ algorithm and providing clear explanations about $Turbo_{ISO}$. Bingsheng He is partly supported by a MoE. AcRF Tier 2 grant (MOE2012-T2-2-067) in Singapore.

References

1. He, H., Singh, A.K.: Graphs-at-a-time: query language and access methods for graph databases. In: SIGMOD, pp. 405–418 (2008)
2. Kasneci, G., Suchanek, F.M., Ifrim, G., Ramanath, M., Weikum, G.: Naga: Searching and ranking knowledge. In: ICDE, pp. 953–962 (2008)
3. Zhao, P., Han, J.: On graph query optimization in large networks. PVLDB **3**(1–2), 340–351 (2010)
4. Yan, X., Yu, P.S., Han, J.: Graph indexing: a frequent structure-based approach. In: SIGMOD, pp. 335–346 (2004)
5. Cook, S.A.: The complexity of theorem-proving procedures. In: STOC, pp. 151–158 (1971)
6. Ullmann, J.R.: An algorithm for subgraph isomorphism. JACM **23**(1), 31–42 (1976)
7. Cordella, L.P., Foggia, P., Sansone, C., Vento, M.: A (sub) graph isomorphism algorithm for matching large graphs. PAMI **26**(10), 1367–1372 (2004)
8. Shang, H., Zhang, Y., Lin, X., Yu, J.X.: Taming verification hardness: an efficient algorithm for testing subgraph isomorphism. PVLDB **1**(1), 364–375 (2008)
9. Zhang, S., Li, S., Yang, J.: GADDI: distance index based subgraph matching in biological networks. In: EDBT, pp. 192–203 (2009)
10. Ullmann, J.R.: Bit-vector algorithms for binary constraint satisfaction and subgraph isomorphism. JEA **15**, 1–6 (2010)
11. Han, W.S., Lee, J., Lee, J.H.: Turbo iso: towards ultrafast and robust subgraph isomorphism search in large graph databases. In: SIGMOD, pp. 337–348 (2013)
12. Kim, S., Song, I., Lee, Y.J.: An edge-based framework for fast subgraph matching in a large graph. In: Yu, J.X., Kim, M.H., Unland, R. (eds.) DASFAA 2011, Part I. LNCS, vol. 6587, pp. 404–417. Springer, Heidelberg (2011)
13. Brocheler, M., Pugliese, A., Subrahmanian, V.S.: COSI: Cloud oriented subgraph identification in massive social networks. In: ASONAM, pp. 248–255 (2010)
14. Sun, Z., Wang, H., Wang, H., Shao, B., Li, J.: Efficient subgraph matching on billion node graphs. PVLDB **5**(9), 788–799 (2012)
15. Harish, P., Narayanan, P.J.: Accelerating large graph algorithms on the GPU using CUDA. In: Aluru, S., Parashar, M., Badrinath, R., Prasanna, V.K. (eds.) HiPC 2007. LNCS, vol. 4873, pp. 197–208. Springer, Heidelberg (2007)

16. Merrill, D., Garland, M., Grimshaw, A.: Scalable GPU graph traversal. In: PPoPP, pp. 117–128 (2012)
17. Katz, G.J., Kider Jr., J.T.: All-pairs shortest-paths for large graphs on the GPU. In: GH, pp. 47–55 (2008)
18. Vineet, V., Harish, P., Patidar, S., Narayanan, P.J.: Fast minimum spanning tree for large graphs on the gpu. In: HPG, pp. 167–171 (2009)
19. Jenkins, J., Arkatkar, I., Owens, J.D., Choudhary, A., Samatova, N.F.: Lessons learned from exploring the backtracking paradigm on the GPU. In: Jeannot, E., Namyst, R., Roman, J. (eds.) Euro-Par 2011, Part II. LNCS, vol. 6853, pp. 425–437. Springer, Heidelberg (2011)
20. McGregor, J.J.: Relational consistency algorithms and their application in finding subgraph and graph isomorphisms. Information Sciences 19(3), 229–250 (1979)
21. Lee, J., Han, W.S., Kasperovics, R., Lee, J.H.: An in-depth comparison of subgraph isomorphism algorithms in graph databases. PVLDB 6(2), 133–144 (2012)
22. He, B., Fang, W., Luo, Q., Govindaraju, N.K., Wang, T.: Mars: a MapReduce framework on graphics processors. In: PACT, pp. 260–269 (2008)
23. Hong, S., Kim, S.K., Oguntebi, T., Olukotun, K.: Accelerating CUDA graph algorithms at maximum warp. In: PPoPP, pp. 267–276 (2011)
24. Chakrabarti, D., Zhan, Y., Faloutsos, C.: R-MAT: A Recursive Model for Graph Mining. In: SDM, pp. 442–446 (2004)
25. Zhong, J., He, B.: Medusa: Simplified graph processing on GPUs. TPDS 25(6), 1543–1552 (2013)
26. Harris, M., Sengupta, S., Owens, J.D.: Gpu gems 3. Parallel Prefix Sum (Scan) with CUDA, pp. 851–876 (2007)
27. Lu, M., He, B., Luo, Q.: Supporting extended precision on graphics processors. In: DaMoN, pp. 19–26 (2010)

On Longest Repeat Queries Using GPU

Yun Tian and Bojian Xu$^{(\boxtimes)}$

Department of Computer Science, Eastern Washington University,
Cheney, WA 99004, USA
{ytian,bojianxu}@ewu.edu

Abstract. Repeat finding in strings has important applications in sub-
fields such as computational biology. The challenge of finding the longest
repeats covering particular string positions was recently proposed and
solved by İleri et al., using a total of the optimal $O(n)$ time and space,
where n is the string size. However, their solution can only find the *leftmost* longest repeat for each of the n string position. It is also not known
how to parallelize their solution. In this paper, we propose a new solution
for longest repeat finding, which although is theoretically suboptimal in
time but is conceptually simpler and works faster and uses less memory
space in practice than the optimal solution. Further, our solution can
find *all* longest repeats of every string position, while still maintaining a
faster processing speed and less memory space usage. Moreover, our solution is *parallelizable* in the shared memory architecture (SMA), enabling
it to take advantage of the modern multi-processor computing platforms
such as the general-purpose graphics processing units (GPU). We have
implemented both the sequential and parallel versions of our solution.
Experiments with both biological and non-biological data show that our
sequential and parallel solutions are faster than the optimal solution by
a factor of 2–3.5 and 6–14, respectively, and use less memory space.

Keywords: String · Repeats · Longest repeats · Parallel computing ·
Stream compaction · GPU · CUDA

1 Introduction

Repetitive structures and regularities finding in genomes and proteins is important as these structures play important roles in the biological functions of genomes
and proteins [1,4,8,13,14,20–22]. It is well known that overall about one-third of
the whole human genome consists of repeated subsequences [17]; about 10–25%
of all known proteins have some form of repetitive structures [14]. In addition,
a number of significant problems in molecular sequence analysis can be reduced
to repeat finding [16]. Another motivation for finding repeats is to compress the
DNA sequences, which is known as one of the most challenging tasks in the data

---

Authors names are in alphabetical order.
B. Xu–Supported in part by EWU Faculty Grants for Research and Creative Works.

© Springer International Publishing Switzerland 2015
M. Renz et al. (Eds.): DASFAA 2015, Part I, LNCS 9049, pp. 316–333, 2015.
DOI: 10.1007/978-3-319-18120-2_19

compression field. DNA sequences consist only of symbols from {ACGT} and there-fore can be represented by two bits per character. Standard compressors such as gzip and bzip usually use more than two bits per character and therefore can-not achieve good compression. Many modern genomic sequence data compression techniques highly rely on the repeat finding in sequences [2, 15].

The notion of maximal repeat and super maximal repeat [1, 3, 8, 12] captures all the repeats of the whole string in a space-efficient manner, but it does not track the locality of each repeat and thus can not support the finding of repeats that cover a particular string position. For this reason, İleri et al. [9] proposed the challenge of longest repeat query, which is to find the longest repetitive substring(s) that covers a particular string position. Because any substring of a repetitive substring is also repetitive, the solution to longest repeat query effectively provides an effective "stabbing" tool for finding the majority of the repeats covering a string position. İleri et al. proposed an $O(n)$ time and space algorithm that can find the *leftmost* longest repeat of every string position. Since one has to spend $\Omega(n)$ time and space to read and store the input string, the solution of İleri et al. is optimal.

Our contribution. In this paper, we propose a new solution for longest repeat query. Although our solution is theoretically suboptimal in the time cost, it is conceptually simpler and runs faster and uses less memory space than the optimal solution in practice. Our solution can also find *all* longest repeats for every string position while still maintaining a faster processing speed and less space usage, whereas the optimal solution can only find the leftmost candi-date. Further, our solution can be parallelized in the shared-memory architec-ture, enabling it to take advantage of the modern multi-processor computing platforms such as the general-purpose graphics processing units (GPU) [6, 18]. We have implemented both the sequential and parallel versions of our solution. Experiments with both biological and non-biological data show that our solution run faster than the $O(n)$ optimal solution by a factor of 2–3.5 using CPU and 6–14 using GPU, and use less space in both settings.

Road map. After formulating the problem of longest repeat query in Section 2, we prepare some technical background and observations in Section 3 for our solutions. Section 4 presents the sequential version of our solutions. Following the interpretation in Section 4, it is natural and easy to get the parallel version of our solution, which is presented in Section 5. Section 6 shows the experimental results on the comparison between our solutions and the $O(n)$ solution using real-world data.

2 Problem Formulation

We consider a **string** $S[1 \ldots n]$, where each character $S[i]$ is drawn from an alphabet $\Sigma = \{1, 2, \ldots, \sigma\}$. A **substring** $S[i \ldots j]$ of S represents $S[i]S[i + 1] \ldots S[j]$ if $1 \leq i \leq j \leq n$, and is an empty string if $i > j$. String $S[i' \ldots j']$ is a

proper substring of another string $S[i \ldots j]$ if $i \leq i' \leq j' \leq j$ and $j' - i' < j - i$. The **length** of a non-empty substring $S[i \ldots j]$, denoted as $|S[i \ldots j]|$, is $j - i + 1$. We define the length of an empty string as zero. A **prefix** of S is a substring $S[1 \ldots i]$ for some i, $1 \leq i \leq n$. A **proper prefix** $S[1 \ldots i]$ is a prefix of S where $i < n$. A **suffix** of S is a substring $S[i \ldots n]$ for some i, $1 \leq i \leq n$. A **proper suffix** $S[i \ldots n]$ is a suffix of S where $i > 1$. We say the character $S[i]$ occupies the string **position** i. We say the substring $S[i \ldots j]$ **covers** the kth position of S, if $i \leq k \leq j$. For two strings A and B, we write $\mathbf{A} = \mathbf{B}$ (and say A is **equal** to B), if $|A| = |B|$ and $A[i] = B[i]$ for $i = 1, 2, \ldots, |A|$. We say A is lexicographically smaller than B, denoted as $\mathbf{A} < \mathbf{B}$, if (1) A is a proper prefix of B, or (2) $A[1] < B[1]$, or (3) there exists an integer $k > 1$ such that $A[i] = B[i]$ for all $1 \leq i \leq k - 1$ but $A[k] < B[k]$. A substring $S[i \ldots j]$ of S is **unique**, if there does not exist another substring $S[i' \ldots j']$ of S, such that $S[i \ldots j] = S[i' \ldots j']$ but $i \neq i'$. A substring is a **repeat** if it is not unique. A character $S[i]$ is a **singleton**, if it appears only once in S.

Definition 1. *For a particular string position $k \in \{1, 2, \ldots, n\}$, the **longest repeat (LR) covering position k**, denoted as LR_k, is a repeat substring $S[i \ldots j]$, such that: (1) $i \leq k \leq j$, and (2) there does not exist another repeat substring $S[i' \ldots j']$, such that $i' \leq k \leq j'$ and $j' - i' > j - i$.*

Obviously, for any string position k, if $S[k]$ is not a singleton, LR_k must exist, because at least $S[k]$ itself is a repeat. Further, there might be multiple choices for LR_k. For example, if $S = \texttt{abcabcddbca}$, then LR_2 can be either $S[1 \ldots 3] = \texttt{abc}$ or $S[2 \ldots 4] = \texttt{bca}$. In this paper, we study the problem of finding the longest repeats of every string position of S.

Problem (longest repeat query): For every string position $k \in \{1, 2, \ldots, n\}$, we want to find LR_k or the fact that it does not exist. If multiple choices for LR_k exist, we want to find all of them.

3   Preliminary

The **suffix array** $SA[1 \ldots n]$ of the string S is a permutation of $\{1, 2, \ldots, n\}$, such that for any i and j, $1 \leq i < j \leq n$, we have $S[SA[i] \ldots n] < S[SA[j] \ldots n]$. That is, $SA[i]$ is the starting position of the ith suffix in the sorted order of all the suffixes of S. The **rank array** $Rank[1 \ldots n]$ is the inverse of the suffix array. That is, $Rank[i] = j$ iff $SA[j] = i$. The **longest common prefix (lcp) array** $LCP[1 \ldots n + 1]$ is an array of $n + 1$ integers, such that for $i = 2, 3, \ldots, n$, $LCP[i]$ is the length of the lcp of the two suffixes $S[SA[i-1] \ldots n]$ and $S[SA[i] \ldots n]$. We set $LCP[1] = LCP[n + 1] = 0.$[1] Table 1 shows the suffix array and the lcp array of the example string $\texttt{mississippi}$.

---

[1] In the literature, the lcp array is often defined as an array of n integers. We include an extra zero at $LCP[n + 1]$ as a sentinel in order to simplify the description of our upcoming algorithms.

Table 1. The suffix array and the lcp array of an example string $S =$ mississippi

i	$LCP[i]$	$SA[i]$	suffixes
1	0	11	i
2	1	8	ippi
3	1	5	issippi
4	4	2	ississippi
5	0	1	mississippi
6	0	10	pi
7	1	9	ppi
8	0	7	sippi
9	2	4	sissippi
10	1	6	ssippi
11	3	3	ssissippi
12	0	–	–

Table 2. The number of walk steps in the LR's calculation using our solutions for several 50MB files from Pizza & Chili [2]

LLR array type	#walk steps	DNA	English	Protein
raw	Minimum	1	1	1
	Maximum	14,836	109,394	25,822
	Average (α)	48	4,429	215
compact	Minimum	1	1	1
	Maximum	14	10	35
	Average (β)	6	2	3

Definition 2. *For a particular string position $k \in \{1, 2, \ldots, n\}$, the **left-bounded longest repeat (LLR)** starting at position k, denoted as LLR_k, is a repeat $S[k \ldots j]$, such that either $j = n$ or $S[k \ldots j+1]$ is unique.*

Clearly, for any string position k, if $S[k]$ is not a singleton, LLR_k must exist, because at least $S[k]$ itself is a repeat. Further, if LLR_k does exist, there must be only one choice, because k is a fixed string position and the length of LLR_k must be as long as possible. Lemma 1 shows that, given the rank and lcp arrays of the string S, we can directly calculate any LLR_k or find the fact of its nonexistence.

Lemma 1 ([9]). *For $i = 1, 2, \ldots, n$:*

$$LLR_i = \begin{cases} S[i \ldots i + L_i - 1] , & \text{if } L_i > 0 \\ \text{does not exist} , & \text{if } L_i = 0 \end{cases}$$

where $L_i = \max\{LCP[Rank[i]], LCP[Rank[i] + 1]\}$.

Clearly, the left-ends of $LLR_1, LLR_2, \ldots, LLR_n$ strictly increase as $1, 2, \ldots, n$. The next lemma shows the right-ends of LLR's also monotonically increase.

Lemma 2. $|LLR_i| \leq |LLR_{i+1}| + 1$, *for every $i = 1, 2, \ldots, n - 1$.*

Proof. The claim is obviously correct for the cases when LLR_i does not exist ($|LLR_i| = 0$) or $|LLR_i| = 1$, so we only consider the case when $|LLR_i| \geq 2$. Suppose $LLR_i = S[i \ldots j]$, $i < j$. It follows that $i + 1 \leq j$. Since $S[i \ldots j]$ is a repeat, its substring $S[i+1 \ldots j]$ is also a repeat. Note that LLR_{i+1} is the longest repeat substring starting from position $i + 1$, so $|LLR_{i+1}| \geq |S[i + 1 \ldots j]| = |LLR_i| - 1$. □

Lemma 3 ([9]). *Every LR is an LLR.*

---

[2] http://pizzachili.dcc.uchile.cl/texts.html

4   Two Simple and Parallelizable Sequential Algorithms

We know every LR is an LLR (Lemma 3), so the calculation of a particular LR_k is actually a search for the longest one among all LLR's that cover position k. Our discussion starts with the finding of the leftmost LR for every position. In the end, an trivial extension will be made to find all LR's for every string position.

Algorithm 1. Sequential finding of every leftmost LR_k, using the raw LLR array.

Input: The rank array and the lcp array of the string S

```
    /* Calculate LLR₁, LLR₂, . . . , LLRₙ.                                      */
1   for i = 1, 2, . . . , n do
2   |   LLRr[i] ← max{LCP[Rank[i]], LCP[Rank[i] + 1]};              // |LLRᵢ|

    /* Calculate LR₁, LR₂, . . . , LRₙ.                                        */
3   for k = 1, 2, . . . , n do
4   |   LR ← ⟨−1, 0⟩ ;                    // ⟨start, end⟩: start and ending position of LRₖ.
5   |   for i = k down to 1 do
6   |   |   if i + LLRr[i] − 1 < k then    // LLRᵢ does not exist or does not cover k.
7   |   |   └ break;                                                   // Early stop
8   |   └ else if LLRr[i] ≥ LR.length then LR ← ⟨i, LLRr[i]⟩;
9   └ print LR;
```

4.1   Use the Raw LLR Array

We first calculate LLR_i, for $i = 1, 2, \ldots, n$, using Lemma 1, and save the result in an array $LLRr[1 \ldots n]$, where each $LLRr[i] = |LLR_i|$. We call $LLRr[1 \ldots n]$ the *raw* LLR array. Because the rightmost LLR that covers position k is LLR_k and the right boundaries of all LLR's monotonically increase (Lemma 2), the search for LR_k becomes simply a walk from LLR_k toward the left. The walk will stop when it sees an LLR that does not cover position k or it has reached the left end of the LLRr array. During this walk, we will record the longest LLR that covers position k. Ties can be broken by storing the leftmost such LLR. This yields the simple Algorithm 1, which outputs every LR as a $\langle start, length \rangle$ tuple, representing the starting position and the length of the LR.

Lemma 4. *Given the rank and lcp arrays, Algorithm 1 can find the leftmost LR_k for every $k = 1, 2, \ldots, n$, using a total of $O(n)$ space and $O(\alpha n)$ time, where α is the average number of LLR's that cover a string position.*

Proof. (1) The time cost for the $LLRr$ array calculation is obviously $O(n)$. The algorithm finds the LR of each of the n string positions. The average time cost for each LR calculation is bounded by the average number of walk steps, which is equal to the average number of LLR's that cover a string position. Altogether, the time cost is $O(\alpha n)$. (2) The main memory space is used by the rank, lcp, and $LLRr$ arrays, each of which has n integers. So altogether the space cost is $O(n)$ words.   □

Theorem 1. *We can find the leftmost LR_k for every $k = 1, 2, \ldots, n$, using a total of $O(n)$ space and $O(\alpha n)$ time, where α is the average number of LLR's that cover a string position.*

Proof. The suffix array of S can be constructed using existing $O(n)$-time and space algorithms (For example, [11]). After the suffix array is constructed, the rank array can be trivially created using another $O(n)$ time and space. We can then use the suffix array and the rank array to construct the lcp array using another $O(n)$ time and space [10]. Combining with Lemma 4, the claim in the theorem is proved. □

Extension: find all LR's for every string position. As we have demonstrated in the example after Definition 1, a particular string position may be covered by multiple LR's, but Algorithm 1 can only find the leftmost one. However, extending it to find all LR's for every string position is trivial: During each walk, we simply report all the longest LLR's that cover the string position, of which we are computing the LR. In order to do so, we will need to do the same walk twice. The first walk is to find the length of the LR and the second walk will actually report all the LR's. The pseudocode of this procedure can be found in the appendix of [23].

This algorithm certainly has another extra $O(\alpha)$ time cost on average for each string position's LR calculation due to the extra walk, but it still gives a total of $O(\alpha n)$ time cost and $O(n)$ space cost.

Corollary 1. *We can find all LR's covering every position $k = 1, 2 \ldots, n$, using a total of $O(n)$ space and $O(\alpha n)$ time, where α is the average number of LLR's that cover a position.*

Comment: (1) Algorithm 1 and its extension are cache friendly. Observe that the finding of every LR essentially is a linear walk over a continuous chunk of the LLRr array. For real-world data, the number of steps in every such walk is quite limited, as shown in the upper rows of Table 2. Note that the English dataset gives a much higher average number of walk steps, because the data was synthesized by appending several real-world English texts together, making many paragraphs appear several times. Because of the few walk steps needed for real-world data, the walking procedure can thus be well cached in the L2 cache, whose size is around several MBs in most nowadays desktops' CPU architecture, making our algorithm much faster in practice. Note that the optimal $O(n)$ algorithm [9] uses a 2-table system to achieve its optimality, which however has quite a pattern of random accessing the different array locations during its run and thus is not cache friendly. We will demonstrate the comparison with more details in Section 6. (2) Algorithm 1 and its extension are parallelizable in shared-memory architecture. First, each LLR can be calculated independently by a separate thread. After all LLR's are calculated, each LR can also be calculated independently by a separate thread going through an independent walk. This enables us to implement this algorithm on GPU, which supports massively parallel threads using data parallelism.

4.2 Use the Compact LLR Array

Observe that an LLR can be a substring (suffix, more precisely) of another LLR. For example, suppose $S = $ ababab, then $LLR_4 = S[4\ldots 6] = $ bab, which is a substring of $LLR_3 = S[3\ldots 6] = $ abab. We know every LR must be an LLR (Lemma 3). So, if an LLR_i is a substring of another LLR_j, LLR_i can never be the LR of any string position, because every position covered by LLR_i is also covered by at least another longer LLR, LLR_j.

Definition 3. *We say an LLR is* useless *if it is a substring of another LLR; otherwise, it is* useful.

Recall that in Algorithm 1 and its extension, the calculation of a particular LR_i is a search for the longest one among all LLR's that cover position i. This search procedure is simply a walk from LLR_i toward the left until it sees an LLR that does not cover position i or reaches the left end of the LLRr array. This search can be potentially sped up, if we have had all useless LLR's eliminated before any search is performed. We will use a new array LLRc, called the *compact* LLR array, to store all the useful LLR's in the ascending order of their left ends (as well as of their right ends, automatically).

By Lemma 2, we know if LLR_{i-1} is not empty, the right boundary of LLR_i is on or after the right boundary of LLR_{i-1}. So, we can construct the $LLRc$ array in one pass as follows. We will calculate every LLR_i using Lemma 1, for $i = 1, 2, \ldots, n$, and will eliminate every LLR_i if $|LLR_i| = 0$ or $|LLR_i| = |LLR_{i-1}| - 1$. Because of the elimination of the useless LLR's, we will have to save each LLR as a $\langle$start, length$\rangle$ tuple, representing the starting position and the length of the LLR, in the LLRc array. Figure 1 shows the geometric perspective of the elements in an example $LLRr$ array and its corresponding $LLRc$ array, where every LLR is represented by a line segment whose start and ending position represent the start and ending position of the LLR.

Note that, in the LLRc array, any two LLR's share neither the same left-end point (obviously) nor the same right-end point. In other words, the left-end points of all useful LLR's strictly increase, and so do their right-end points, i.e., all the elements in the LLRc array have been sorted in the strict increasing order of their left-end (as well as right-end) points. See Figure 1b for an example.

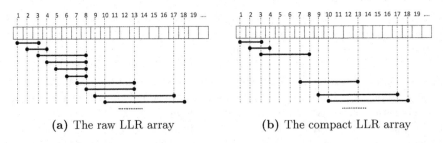

(a) The raw LLR array   (b) The compact LLR array

Fig. 1. The visualization of an example raw LLR array and its compact LLR array

Therefore, given a string position, we can find the leftmost useful LLR that covers that position using a binary search over the LLRc array and the time cost for such a binary search is bounded by $O(\log n)$. After that, we will simply walk along the LLRc array, starting from the LLR returned by the binary search and toward the right. The walk will stop when it sees an LLR that does not cover the string position or it has reached the right end of the LLRc array. During the walk, we will just report the longest LLR that covers the given string position. Ties are broken by picking the leftmost such longest LLR. This leads to the Algorithm 2.

Lemma 5. *Given the rank and lcp arrays, Algorithm 2 can find the leftmost LR_k for every $k = 1, 2, \ldots, n$, using a total of $O(n)$ space and $O(n(\log n + \beta))$ time, where β is the average number of useful LLR's that cover a string position.*

Proof. (1) The time cost for the $LLRc$ array calculation is obviously $O(n)$ time. The algorithm finds the LR of each of the n string positions. The average time cost for the calculation of the LR of one position includes the $O(\log n)$ time for the binary search and the time cost for the subsequent walk, which is bounded by the average number of useful LLR's that cover a string position. Altogether, the time cost is $O(n(\log n + \beta))$. (2) The main memory space is used by the rank, lcp, and $LLRc$ arrays. Each of the rank and lcp arrays has n integers. The $LLRc$ array has no more than n pairs of integers. Altogether, the space cost $O(n)$ words.   □

Theorem 2. *We can find the leftmost LR_k for every $k = 1, 2, \ldots, n$, using a total of $O(n)$ space and $O(n(\log n + \beta))$ time, where β is the average number of useful LLR's that cover a string position.*

Proof. The suffix array of S can be constructed by existing algorithms using $O(n)$ time and space (For example, [11]). After the suffix array is constructed, the rank array can be trivially created using another $O(n)$ time and space. We can then use the suffix array and the rank array to construct the lcp array using another $O(n)$ time and space [10]. Combining the results in Lemma 5, the theorem is proved.   □

Extension: find all LR's for every string position. Algorithm 2 can also be trivially extended to find all LR's for every string position by simply reporting all the longest LLR that covers the position during every walk. In order to do so, we will need to walk twice for each string position. The first walk is to get the length of the LR and the second walk will report all the actual LR's. The pseudocode of this procedure can be found in the appendix of [23]. This algorithm certainly has another extra $O(\beta)$ time cost on average for each LR's calculation due to the extra walk, but still gives a total of $O(n(\log n + \beta))$ time cost and $O(n)$ space cost.

Corollary 2. *We can find all LR's of every position $k = 1, 2, \ldots, n$, using a total of $O(n)$ space and $O(n(\log n + \beta))$ time, where β is the average number of useful LLR's that cover a position.*

Algorithm 2. Sequential finding of every leftmost LR_k, using the LLRc array.

Input: The rank array and the lcp array of the string S

```
/* Calculate the compact LLR array.                                         */
```
1   $j \leftarrow 1;\ prev \leftarrow 0;$
2   **for** $i = 1, 2, \ldots, n$ **do**
3      $L \leftarrow \max\{LCP[Rank[i]], LCP[Rank[i] + 1]\};$ `// |LLRᵢ|`
4   **if** $L > 0$ *and* $L \geq prev$ **then** $LLRc[j] \leftarrow \langle i, L \rangle;\ j \leftarrow j + 1$;
5      $prev \leftarrow L;$
6   $size \leftarrow j - 1$;   `// Size of the LLRc array.`

```
/* Calculate LR₁, LR₂, ..., LRₙ.                                            */
```
7   **for** $k = 1, 2, \ldots, n$ **do**
8      $LR \leftarrow \langle -1, 0 \rangle$;   `// ⟨start, end⟩: start and ending position of LRₖ.`
9      $start \leftarrow$ BinarySearch$(LLRc, k)$ `/* The index of the leftmost LLRc array element covering position k, if such element exists; otherwise, −1.           */`
10   **if** $start \neq -1$ **then**
11   **for** $i = start \ldots size$ **do**
12   **if** $LLRc[i].start + LLRc[i].length - 1 < k$ **then** `// LLRc[i] does not cover k.`
13   **break;**   `// Early stop`
14   **else if** $LLRc[i].length > LR.length$ **then** $LR \leftarrow LLRc[i];$
15   print $LR;$

Comment: (1) The binary searches that are involved in Algorithms 2 and its extension are not cache friendly. However, compared with Algorithms 1 and its extension, Algorithms 2 and its extension on average have much fewer steps (the β value) in each walk due to the elimination of the useless LLR's (see the bottom rows of Table 2). This makes Algorithms 2 and its extension much better choices rather than Algorithms 1 and its extension for run environments that have small cache size. Such run environments include the GPU architecture, where the cache size for each thread block is only several KBs. We will demonstrate this claim with more details in Section 6. (2) With more care in the design, Algorithms 2 and its extension are also parallelizable in shared-memory architecture (SMA), which is described in the next Section.

5   Parallel Implementation on GPU

In this section, we describe the GPU version of Algorithms 1 and 2 and their extensions. After we construct the SA, Rank, and LCP arrays on the host CPU [3], we transfer the Rank array and the LCP array to the GPU device memory. We start with the calculation of the raw LLR array in parallel.

Compute the Raw LLR Array. After the LCP and Rank arrays are loaded into GPU memory, we launch a CUDA kernel to compute the raw LLR array on GPU

---

[3] The SA, Rank, and LCP arrays can also be constructed in parallel on GPU [7,19], but due to the unavailability of the source code or executables from the authors of [7,19], we choose to construct these arrays on the host CPU, without affecting the demonstration of the performance gains by our algorithms.

device using massively parallel threads (Figure 2). Each thread t_i on the device computes a separate element $LLRr[i] = |LLR_i|$ using the following equation from Lemma 1.

$$LLR[i] = \max\{LCP[Rank[i]], LCP[Rank[i] + 1]\}$$

Since each LLR_i must start on position i, we only need to save the length of each LLR_i in $LLRr[i]$. After creating the raw LLR array, we have two options, which in turn lead to two different parallel solutions: using the raw LLR array or the compact LLR array.

5.1   Compute LR's Using the Raw LLR Array

The parallel implementation of Algorithm 1 using the raw LLR array is straightforward, as presented by the left branch of Figure 2. With the raw LLR array returned from the previous kernel launch on the GPU device, we launch a second kernel for LR calculation. Each CUDA thread t_i on the device is to find LR_i by performing a linear walk in the LLRr array, starting at $LLRr[i]$ toward the left. The walk continues until it finds an LLRr array element that does not cover position i or has reached the left end of the LLRr array. The leftmost or all LR_i can be reported during the walk, as discussed in Algorithm 1. Note that in this search, each CUDA thread checks a chunk of contiguous elements in the LLRr array and this can be cache-efficient.

Taking the calculation of LR_{10} using the raw LLR array shown in Figure 1a as an example. The corresponding CUDA thread t_{10} searches a contiguous chunk of the LLRr array starting from index 10 down to left in the $LLRr$ array. We do not search the LLRr elements that are to the right of index 10, because these elements definitely do not cover position 10 according to the definition of LLR. In particularly, thread t_{10} goes through $LLRr[10]$, $LLRr[9]$, $LLRr[8]$ and $LLRr[7]$ to find the longest one among the four of them as LR_{10}. Thread t_{10} stops the search at LLRr position 6, because $LLRr[6]$ and all LLR's to its left do not cover position 10 (Lemma 2).

5.2   Compute LR's Using the Compact LLR Array

LLR Compaction. The right branch of Figure 2 shows the second option in computing LR's on GPU. That is to use the compact LLR array. We first create the compact LLR array, named as LLRc, from the raw LLR array, which has been created and preserved on the device memory. To avoid the expensive data transfer between the host and the device and to achieve more parallelism, we perform the LLR array compaction on the GPU device in parallel. We launch three CUDA kernels to perform the compaction, denoted as $\mathcal{K}_1$, $\mathcal{K}_2$, and $\mathcal{K}_3$. As shown in Figure 3, after the LLRr array is constructed on the device, we first launch kernel $\mathcal{K}_1$ to compute a flag array $Flag[1 \ldots n]$ in parallel, where the value of each $Flag[i]$ is assigned by a separate thread t_i as follows: (1) $Flag[1] = 1$ iff

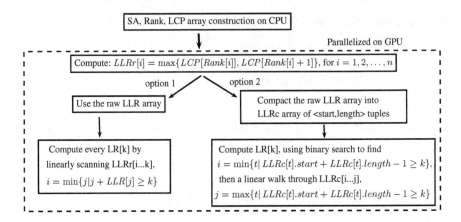

Fig. 2. Overview of the GPU Implementation

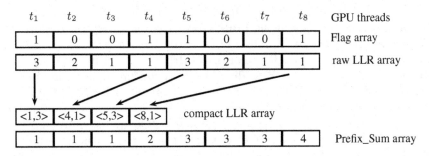

Fig. 3. LLR compaction on GPU

$LLRr[1] > 0$. (2) $Flag[i] = 1$, iff $LLRr[i] > 0$ and $LLRr[i] \geq LLRr[i - 1]$, for $i \geq 2$. $Flag[i] = 0$ means LLR_i is useless and thus can be eliminated.

After the Flag array is constructed from kernel $\mathcal{K}_1$, we launch kernel $\mathcal{K}_2$ to calculate the prefix sum of the Flag array on the device: $Prefix\_Sum[i] = \sum_{j=1}^{i} Flag[j]$. We modify and use the parallel prefix sum function provided by the CUDA toolkit, which is based on the parallel algorithm described in [5].

With the prefix sum array and the Flag array, we launch kernel $\mathcal{K}_3$ to copy the useful LLRr array elements into the LLRc array, as illustrated in Figure 3. Each thread t_i on the device moves in parallel the LLR_i to an unique destination $LLRc[Prefix\_Sum[i]]$, if $Flag[i] = 1$. That is, $LLRc[Prefix\_Sum[i]] = \langle i, LLRr[i] \rangle$, if $Flag[i] = 1$. Each element in the $LLRc$ array is a useful LLR and is represented by a tuple of $\langle start, length \rangle$, the start and ending position of the LLR.

Compute LR's. After the LLRc array is prepared, we calculate the LR for every string position in parallel. Recall that the calculation of each LR_k, for each $k = 1, 2, \ldots, n$, is a search for the longest useful LLR that covers position k. We also know all these relevant LLR's that we need to search comprise a continuous chunk of the LLRc array. The start position of the chunk can be found using a binary search as we have explained in the discussion of Algorithm 2. After that, a simple linear walk toward the right is performed. The walk continues until it finds an LLRc array element that does not cover position k or has reached the right end of the LLRc array.

To compute the LR's using the LLRc array, we launch another CUDA kernel, in which each CUDA thread t_k first performs a binary search to find the start position of the linear walk and then walk through the relevant LLRc array elements to find either all LR's or a single LR covering position k.

Referring to Figure 1b, we take the LR calculation covering the string position 9 as an example. Recall that we have discarded all useless LLR's in the LLRc array, so the LLRc array element at index 9 is not necessarily the rightmost LLR that cover string position 9. Therefore, we have to perform a binary search to locate that leftmost LLRc array element by taking advantage of the nice property of the LLRc array that both the start and ending positions of all LLR's in it are strictly increasing. After thread t_9 locates the LLRc element $LLRc[4]$, the leftmost useful LLR that covers the string position 9, it performs a linear walk toward the right. The walk will continue until it meets $LLRc[6]$, which does not cover position 9. Thread t_9 will return the longest ones among $LLRc[4 \ldots 6]$ as LR_9.

5.3 Advantages and Disadvantages: *LLRr* vs. *LLRc*

When the raw LLR array is used, the algorithm is straightforward and easy to implement, because there is no needs to perform the LLR compaction on the device. However, with a raw LLR array, we could have a large number of useless LLR's in the raw LLR array, especially when the average length of the longest repeats is quite large. For that reason, the subsequent linear walk for each CUDA thread can take many steps, making the overall search performance worse.

In contrast, under a compact LLR array, we have to perform the LLR compaction, which involves data coping and requires extra memory usage for the *Flag* and the prefix sum array on the device. In addition, a binary search, which is not present with a raw LLR array, is required to locate the first LLR for the linear walk. The advantage of a compact LLR array is that we remove the useless LLR's and dramatically shorten the linear walk distance.We provide more analysis and comparison between these two solutions in the experiment section.

6 Experimental Study

Experiment Environment Setup. We conducted our experiments on a computer running GNU/Linux with a kernel version 3.2.51-1. The computer is equipped with an Intel Xeon 2.40GHz E5-2609 CPU with 10MB Smart Cache and has 16GB RAM. We used a GeForce GTX 660 Ti GPU for our parallel tests. The GPU consists of 1344 CUDA cores and 2GB of RAM memory. The GPU is connected with the host computer with a PCI Express 3.0 interface. We install CUDA toolkit 5.5 on the host computer. We use C to implement our sequential algorithms and use CUDA C to implement our parallel solutions on the GPU, using gcc 4.7.2 with -O3 option and nvcc V5.5.0 as the compilers. We test our algorithms on real-world datasets including biological and non-biological data downloaded from the Pizza&Chili Corpus. The datasets we used are the three 50MB DNA, English, and Protein pure ASCII text files, each of which thus represents a string of $50 \times 1024 \times 1024$ characters.

Measurements. We measured the average time cost of three runs of our program. In order to better highlight the comparison of the algorithmics between the old and our new solutions, we did not include the time cost for the I/O operations that save the results. For the same purpose, we also did not include the time cost for the SA, Rank, and LCP array constructions, because in both the old and our new solutions, these auxiliary data structures are constructed based on the same best suffix array construction code available on the Internet[4]. Our source code for this work is also available on website.[5]

6.1   Time

In the left three charts of Figure 4, using three datasets, we compare different algorithms that return only the leftmost LR for every string position of the input data. In the right three charts, we present the performance of our algorithms that are able to find *all* LR's for every string position. We compare our new algorithms with the existing optimal sequential algorithm [9], which can only find the leftmost LR for every string position. Table 3 summarizes the speedup of our algorithms against the old optimal algorithm. From experiments, we are able to make the following observations.

Sequential algorithms on CPU. Our new sequential algorithm using the raw LLR is consistently faster than the old optimal algorithm by a factor of 1.97–3.44, while our new sequential algorithm that uses the compact LLR array is consistently slower. This observation is true in both finding the leftmost LR and all LR's. (Please note that the old optimal algorithm always finds the leftmost LR only.)

On the host CPU, three dominating factors contribute to the better performance of algorithms using a raw LLR array rather than using a compact LLR array. First, although the compact LLR array can still be constructed in one pass, but the construction involves a lot more computational steps than those needed in the construction of the raw LLR array. Second, sequential algorithms that use a compact LLR array require a binary search in order to locate the starting position of the subsequent linear walk in the calculation of every LR. However, binary searches are not required if we work with a raw LLR array. As

---

[4] http://code.google.com/p/libdivsufsort/
[5] http://penguin.ewu.edu/~bojianxu/publications

it is known, binary search over a large array is not cache friendly. Through profiling, we observe that the binary search operations consume from 63% to 73% of the total execution time. Third, even though for some datasets the search range size (or the number of walk steps) with a raw LLR array could be 10, 000 times larger than that using a compact LLR array, as shown in table 2, the L2 cache (10MB) of the host CPU is large enough to cache the range of *contiguous* LLR's that each linear walk needs to go through. Such efficient data caching helps all walks take less than a total of 100 milliseconds on the host CPU, accounting for less than 5% of the total execution time, even with the raw LLR array. In other words, given a large cache memory, the number of walk steps is no longer a dominating factor in the overall performance.

Parallel algorithms on GPU. Our new parallel algorithm on GPU using the compact LLR array is consistently faster than its counterpart that uses the raw LLR array, which is consistently faster than the old optimal algorithm by a factor of 8.32–14.62 in finding the leftmost LR and 6.36–10.35 in finding all LR's.

Unlike the sequential algorithm on the host CPU, the performance of the parallel algorithm on the GPU device is dominated by the number of LLR's (the number of walk steps) that each walk will go through. As we profile our GPU implementation, we observe that with the raw LLR array, all linear walks on the GPU take roughly a total of eight seconds for the *English* dataset. But, the walks take roughly 70 milliseconds only if using a compact LLR array on the GPU. This is because: (1) the small GPU L2 cache (384KB shared by all streaming multiprocessors) cannot host as many LLR's as what the CPU L2 cache (10MB) can host, resulting in more cache-read misses and more expensive global memory accesses. (2) The number of walk steps with a compact LLR array is less than that with a raw LLR array by a factor of up to four orders of magnitude (see Table 2). (3) The extra time cost for the LLR compaction that is needed when using the compact LLR array become much less significant in the total execution time on GPU. On the host CPU, our sequential solution takes roughly 1.3 seconds to perform the LLR compaction for datasets of 50MB and accounts for 20% of the total time cost on average. However, it takes less than 30 milliseconds on the GPU, accounting for only 9.5% of the entire time cost. We achieve more than 40 times speedup in the LLR compaction by utilizing GPU device.

The first two reasons above are reassured by the experimental results regarding the English dataset, which we purposely chose to use. The English file is synthesized by simply concatenating several English texts, and thus the text has many repeated paragraphs, which in turn creates many *useless* LLR's in the data. In this case, with the raw LLR array, each walk will have a large number of steps due to such useless LLR's. However, after we compact the raw LLR array, the number of walk steps can be significantly reduced (Table 2) and consequently the GPU code's performance is significantly improved (Figure 4).

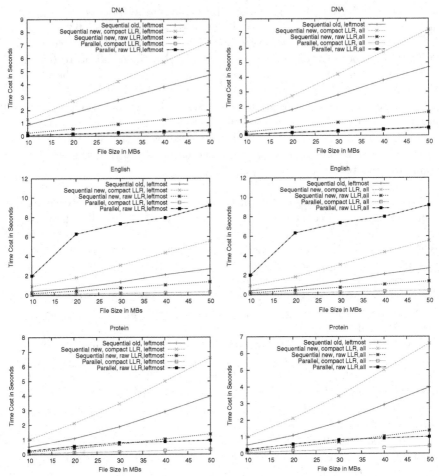

Fig. 4. Time cost vs. dataset size. The left and right charts show the experimental results on finding the leftmost and all repeats of every string position, respectively.

6.2   Space

Table 4 shows the peak memory usage of both the old and our new algorithms for datasets of size 50MBs. The memory usage of all of our algorithms is the same. This is because the space usage by the SA, Rank, and LCP array dominate the peak memory usage of all of our algorithms. On the other hand, due to its 2-table system that helps achieve the theoretical $O(n)$ time complexity, the old optimal algorithm's space usage is relevant to the dataset type and is higher than ours.

6.3   Scalability

Although our algorithms have a superlinear time complexity in theory, but they all scale well in practice as shown by Figure 4. As we increase the size of the test

Table 3. Speedup with 50MB Files

	Sequential No Compact Leftmost	Sequential No Compact All	Parallel Compact Leftmost	Parallel Compact All
DNA	2.91x	2.91x	13.48x	9.43x
English	1.97x	1.97x	8.32x	6.36x
Protein	3.44x	3.44x	14.62x	10.35x

Table 4. RAM Usage Comparison for 50MB Files

	Old (MBs)	Ours (MBs)	Space Saving
DNA	792.77	650.39	17.96%
English	654.02	650.39	0.56%
Protein	773.53	650.39	15.92%

data, we observe a consistent speedup. In addition, we *did* conduct experiments on datasets of 100MB on the GPU device by using a 2D grid of CUDA threads in order to create more than 100 million threads on the device. When finding the leftmost LR for each string position, we observed the same speedups as shown in Figure 4.

On the host CPU, the large cache size dramatically reduces the total number of memory reads during the linear walk in a raw LLR array and thus enables us to eliminate the expensive binary search operations by using a raw LLR array. On the GPU device, although all data is stored in the global memory, a compact LLR array helps greatly reduce the total number of global memory access; each thread linearly searches a smaller number of LLR's. As shown in Table 2, the average number of walk steps in a compact LLR array is no more than six, which enables the linear walk to be considered as a *constant*-time operation.

7 Conclusion and Future Work

We proposed conceptually simple and easy-to-implement solutions for longest repeat finding over a string. Our algorithm although is not optimal in time theoretically, but runs faster than the old optimal optimal algorithm and uses less space. Further, our algorithm can find all longest repeats of every string position, whereas the old solution can only find the leftmost one. Our algorithm can be parallelized in shared-memory architecture and has been implemented on GPU using data parallelism for further speedup.

Our GPU solution is roughly 4.5 times quicker than our *best* sequential solution on the CPU, and is up to 14.6 times quicker than the old optimal solution on the CPU. Also, we improve the LLR compaction performance by a factor of 40 on GPU. The multiprocessors in our current GPU have a built-in L1 and L2 cache, which help coalesce some global memory accesses. In the future, we will further optimize our parallel solution by utilizing the GPU shared memory or texture memory to reduce global memory access.

References

1. Becher, V., Deymonnaz, A., Heiber, P.A.: Efficient computation of all perfect repeats in genomic sequences of up to half a gigabyte, with a case study on the human genome. Bioinformatics 25(14), 1746–1753 (2009)
2. Behzadi, B., Le Fessant, F.: DNA compression challenge revisited: a dynamic programming approach. In: Apostolico, A., Crochemore, M., Park, K. (eds.) CPM 2005. LNCS, vol. 3537, pp. 190–200. Springer, Heidelberg (2005)
3. Beller, T., Berger, K., Ohlebusch, E.: Space-efficient computation of maximal and supermaximal repeats in genome sequences. In: Calderón-Benavides, L., González-Caro, C., Chávez, E., Ziviani, N. (eds.) SPIRE 2012. LNCS, vol. 7608, pp. 99–110. Springer, Heidelberg (2012)
4. Benson, G.: Tandem repeats finder: a program to analyze dna sequences. Nucleic Acids Research 27(2), 573–580 (1999)
5. Blelloch, G.E.: Prefix sums and their applications. Technical Report CMU-CS-90-190, Carnegie Mellon University (1990)
6. Che, S., Boyer, M., Meng, J., Tarjan, D., Sheaffer, J., Skadron, K.: A performance study of general-purpose applications on graphics processors using cuda. Journal of Parallel and Distributed Computing 68(10), 1370–1380 (2008)
7. Deo, M., Keely, S.: Parallel suffix array and least common prefix for the gpu. In: Proceedings of the 18th ACM SIGPLAN Symposium on Principles and Practice of Parallel Programming (PPoPP), pp. 197–206 (2013)
8. Gusfield, D.: Algorithms on strings, trees and sequences: computer science and computational biology. Cambridge University Press (1997)
9. İleri, A.M., Külekci, M.O., Xu, B.: On longest repeat queries. http://arxiv.org/abs/1501.06259
10. Kasai, T., Lee, G.H., Arimura, H., Arikawa, S., Park, K.: Linear-time longest-common-prefix computation in suffix arrays and its applications. In: Amir, A., Landau, G.M. (eds.) CPM 2001. LNCS, vol. 2089, pp. 181–192. Springer, Heidelberg (2001)
11. Ko, P., Aluru, S.: Space efficient linear time construction of suffix arrays. Journal of Discrete Algorithms 3(2–4), 143–156 (2005)
12. Kulekci, M.O., Vitter, J.S., Xu, B.: Efficient maximal repeat finding using the burrows-wheeler transform and wavelet tree. IEEE Transactions on Computational Biology and Bioinformatics (TCBB) 9(2), 421–429 (2012)
13. Kurtz, S., Schleiermacher, C.: Reputer: fast computation of maximal repeats in complete genomes. Bioinformatics 15(5), 426–427 (1999)
14. Liu, X., Wang, L.: Finding the region of pseudo-periodic tandem repeats in biological sequences. Algorithms for Molecular Biology 1(1), 2 (2006)
15. Manzini, G., Rastero, M.: A simple and fast dna compressor. Software-Practice and Experience 34, 1397–1411 (2004)
16. Martinez, H.M.: An efficient method for finding repeats in molecular sequences. Nucleic Acids Research 11(13), 4629–4634 (1983)
17. McConkey, E.H.: Human Genetics: The Molecular Revolution. Jones and Bartlett, Boston (1993)
18. Nickolls, J., Dally, W.J.: The gpu computing era. IEEE Micro 30(2), 56–69 (2010)
19. Osipov, V.: Parallel suffix array construction for shared memory architectures. In: Calderón-Benavides, L., González-Caro, C., Chávez, E., Ziviani, N. (eds.) SPIRE 2012. LNCS, vol. 7608, pp. 379–384. Springer, Heidelberg (2012)

20. Saha, S., Bridges, S., Magbanua, Z.V., Peterson, D.G.: Computational approaches and tools used in identification of dispersed repetitive dna sequences. Tropical Plant Biology **1**(1), 85–96 (2008)
21. Saha, S., Bridges, S., Magbanua, Z.V., Peterson, D.G.: Empirical comparison of ab initio repeat finding programs. Nucleic Acids Research **36**(7), 2284–2294 (2008)
22. Smyth, W.F.: Computing regularities in strings: A survey. European Journal of Combinatorics **34**(1), 3–14 (2013)
23. Tian, Y., Xu, B.: On longest repeat queries using gpu. http://arxiv.org/abs/1501.06663

Process-driven Configuration of Federated Cloud Resources

Denis Weerasiri$^{(\boxtimes)}$, Boualem Benatallah, and Moshe Chai Barukh

School of Computer Science and Engineering, The University of New South Wales,
Sydney, Australia
{denisw,boualem,mosheb}@cse.unsw.edu.au

Abstract. Existing cloud resource providers offer heterogeneous
resource deployment services to describe and deploy resource configurations. Describing and deploying federated cloud resource configurations
over such deployment services is challenging due to dynamic application
requirements and complexity of cloud environments. While solutions
exist to solve this problem, they offer limited facilities to cater for
resource provisioning over federated cloud services. This paper presents
a novel cloud resource deployment framework that leverages a unified
configuration knowledge-base where process-based notation is used to
describe complex configurations over federated cloud services. Based
on these notations, a deployment engine generates deployment scripts
that can be executed by external cloud resource deployment services
such as Puppet and Chef. The paper describes the concepts, techniques
and current implementation of the proposed system. Experiments on a
real-life federated cloud resource show significant improvements achieved
by our approach compared to traditional techniques.

1 Introduction

Cloud service deployment is evolving in the form of both public (deployed by IT
organizations) and private (deployed behind a company firewall) clouds. Both these
deployment environments provide virtualized and dynamically scalable resources.
Each resource is available as one of three layers of service offerings: Software-as-a-
Service (SaaS), Platform-as-a-Service (PaaS) and Infrastructure-as-a-Service
(IaaS). A third deployment form; federated clouds [1–4] are now emerging, where
resources are drawn from a subset of private and public clouds, configured at the
behest of its users. It is imperative that the federation of clouds leads to a unified
model, which represents a single cloud of multiple cloud platforms that can be used
as needed. Thus cloud federation requires the creation of an agile cloud-computing
environment, in which cloud capabilities can be procured, configured, deployed,
and managed on demand by consumers, regardless of whether cloud capabilities
are private or public.

Cloud resource-providers facilitate resource-consumers to describe, deploy and
manage resource configurations that satisfy users' application and resource

© Springer International Publishing Switzerland 2015
M. Renz et al. (Eds.): DASFAA 2015, Part I, LNCS 9049, pp. 334–350, 2015.
DOI: 10.1007/978-3-319-18120-2_20

requirements. Cloud resource providers offer proprietary configuration deployment languages (Command-Line interface (CLI) based, REST/SOAP based, Chef Cookbooks[1], Ubuntu Juju[2] etc.,) for users to support configuration and deployment tasks. Users automate their tasks using these languages, such that the language runtimes (i.e., cloud resource deployment services) procure, allocate and provision cloud resource configurations in a provider's environment based on users' demand. However, current configuration deployment solutions have two major limitations, which are significantly impeding the effective management of federated cloud resources, as follows.

(1) Heterogeneity of cloud resource deployment services: When a single provider cannot satisfy all application and resource requirements (e.g., when an Apache server, deployed in the private cloud reaches its maximum capacity, excess load can be outsourced to a replica of the Apache server that is deployed in a public cloud provider), users are inevitably responsible to describe, deploy and manage the component resources of a federated resource configuration in a segregated fashion because users have to deal with multiple configuration management languages (e.g., Chef cookbook[1] and Ubuntu Juju[2]). These languages possess different notations (e.g., json, xml, yaml); resource description models (e.g., Juju charms[3], Chef recipes); and capabilities (e.g., deployment, scaling, migration, monitoring) [5,6]. To describe a federated cloud resource configuration process, users need to understand the configuration description languages of all participating cloud resource providers. Furthermore to deploy the described configuration, users should understand configuration deployment service interfaces of all participating resource providers and implement ad-hoc scripts [7] to coordinate the deployment of component resources. This methodology is not scalable in dynamic environments where cloud resource providers appear and disappear over time.

(2) Adaptability to dynamic environments: A key distinguishing feature of cloud services is the elasticity, i.e., the power to dynamically scale resources up and down to adapt varying requirements. Elasticity is usually achieved through invocation of re-configuration tasks (e.g., add storage capacity, restart VM instances) that run as a result of events (e.g., service usage increases beyond a certain threshold) allowing the deployment engine to dynamically re-configure cloud resources. The automated and unified monitoring and control of federated services is still in the early stages [8].

In this paper we address above limitations by providing high-level abstractions for federated cloud resource configuration tasks, which replace existing, low-level and heterogeneous deployment services. The main contributions of this paper are:

(1) Unified representation of cloud resource configurations: We offer a unified resource configuration language for describe, deploy, reconfigure and undeploy federated cloud resources. Cloud resource providers and devOps compose,

---

[1] community.opscode.com/cookbooks

[2] juju.ubuntu.com

[3] jujucharms.com/

curate and publish unified resource descriptions in a knowledge-base by specifying the configuration attributes and available orchestration operations. Users discover those unified resource descriptions and consume them to satisfy users' federated resource requirements. Our language runtime handles the heterogeneity in resource description models, notations and capabilities of different provider-specific configuration languages and services. Furthermore, we offer a high-level policy language for specifying events and associated re-configuration behavior of cloud resources.

(2) Graphical process modeling notation for resource orchestration tasks: Deployment and orchestration activities of federated resource configurations can be modeled using a process-based language. But modeling orchestration processes by directly interacting with heterogeneous deployment services leads to ad-hoc scripts or manual tasks, which hinder the automation of orchestration activities in a dynamic cloud environment.

Hence it is desirable to provide productive and user-friendly modeling techniques for users to compose federated resource deployment and orchestration processes. We provide two high-level and process-based abstractions that facilitate users to describe, deploy and specify reconfiguration policies of their federated cloud resource configurations. Firstly, we propose the concept of *Cloud Resource Deployment Task* to simplify cloud resource configuration and foster independence between applications and cloud services. Secondly, we propose the concept of *Cloud Resource Re-configuration Policies* to endow resources with dynamic resource reconfigurations. We implemented these notations by extending BPMN [9], an open and graphical process modeling standard, which is already adapted by industry and academia for modeling cloud resource deployment and management tasks [10]. We provide mechanisms that automatically translate high-level deployment tasks and re-configuration policies into the corresponding BPMN resource configuration models. We decided to extend BPMN rather than use native BPMN to avoid orchestration tasks getting (1) complex; (2) error-prone; and (3) difficult to verify and manage later.

(3) A prototype implementation: We implemented **CloudBase**; a process-driven federated cloud resource deployment framework that leverages a unified configuration knowledge-base to model deployment tasks and dynamic re-configuration policies of federated cloud resources. *CloudBase* thus replaces time-consuming, frequently costly, naturally expert driven and manual cloud resource configuration and deployment tasks with a model-driven and unified cloud resource orchestration method and techniques. The paper presents an experiment using a real-life federated cloud resource that demonstrates the improvements achieved by *CloudBase*.

Together these contributions enable cloud resource consumers to focus on high-level application requirements, instead of low-level details related to dealing with heterogeneous deployment services. The following sections describe each of these high-level abstractions in more detail.

2   Unified Representation of Cloud Resource Configurations

Different cloud resource providers offer different and heterogeneous deployment languages; and services to configure and deploy cloud resources. Heterogeneity of configuration services can be stemmed from interface level, communication protocol level, service operation level or input/output message schema level. For an example, AWS and Rackspace use different formats of access credentials and service interfaces to deploy Virtual Machines (VM) (see Table 1).

Table 1. Types of different deployment service interfaces

Provider	Type	Names of resources	Types of deployment service interfaces
AWS	IaaS/PaaS	EC2, S3, RDS	CLI, SDKs, Web 2.0, REST and SOAP APIs
Rackspace	IaaS/PaaS	Server, Database	CLI, SDKs, Web 2.0, REST API
Puppet	IaaS/PaaS	Resource	CLI, Web 2.0, REST API
GitHub	PaaS	Repository	Web 2.0, REST API

Heterogeneity of configuration languages can be stemmed due to different notations (e.g., json, xml, yaml); resource description models (e.g., Juju charms, Chef recipes); and capabilities (e.g., deployment, scaling, migration, monitoring). Due to this heterogeneity, current practices of configuring and deploying federated cloud resources require ad-hoc integration techniques like low-level procedural scripting which are not scalable in dynamic environments and not intuitive for consumers who don't have configuration management knowledge of multiple providers.

We argue that a federated cloud resource deployment framework should abstract out interactions with heterogeneous deployment services. We propose an abstraction, called "Resource Configuration Service" (RCS), which allows cloud resource providers to expose their deployment interfaces through a unified interface. We also propose "Cloud Resource Configuration Description" (CRCD), which provides a unified and provider-independent data model for resource requirement descriptions. Together, these abstractions enable cloud resource consumers to model and execute cloud resource deployment tasks by specifying application-centric and provider-independent resource requirements.

2.1   Resource Configuration Service (RCS)

To abstract out the heterogeneity of different deployment services from users, we propose **Resource Configuration Services** that accept resource configuration descriptions and deploy concrete resources in a public, private or federated cloud infrastructure. In other words, *RCS*s provide a unified layer to represent provider-specific deployment knowledge over various configuration deployment languages like Chef Cookbooks[1] or low-level techniques like shell scripts. Cloud resource providers or devOps can implement and deploy *RCS*s.

We designed the *RCS* interface (see Table 2) by analysing common interface level characteristics of cloud configuration services of AWS, Rackspace, Puppet and GitHub. Every cloud deployment service exposes two basic operations: "deploy" and "undeploy" a cloud resource. The "deploy" operation requires a resource configuration description as the input. The output of the "deploy" operation returns an identification(id) value that uniquely represents the deployed cloud resource (e.g. "ImageId" value of an Amazon Machine Image(AMI)[4]). This id value is referred in managing and undeploying the particular resource. The "undeploy" operation takes the id value of an already deployed resource and returns optional information regarding the success/failure of the operation.

Table 2. Operations of *RCS* interface

deployResource(resource_meta_data): When a user requests to deploy a cloud resource, the runtime of *RCS* processes the incoming cloud resource description and selects a particular *RCS* that can satisfy the incoming resource description. Then the runtime invokes **deployResource** operation of the particular *RCS* along with the incoming resource description as **resource_meta_data**. This operation returns an object that consists of a resource_id and an optional resultant message.
undeployResource(resource_id): When a user requests to undeploy a deployed cloud resource, the runtime of *RCS* extracts the resource id of the deployed cloud resource and figures out the *RCS*, which was invoked to deploy this *CRCD*. Then runtime invokes **undeployResource** operation of the particular *RCS* along with the **resource_id**.

Re-configuration Policy. Cloud resources can be dynamically re-configured (e.g. restart VM instance) to satisfy varying resource requirements. Similarly, a subset of component resources of a federated cloud resource can be subjected to dynamic re-configurations when certain events (e.g. VM instance connection failure) occur. Cloud resource consumers should be able to specify a set of events and associated actions (i.e., re-configuration policies) for a cloud resource configuration.

We argue that a cloud resource deployment framework must support dynamic re-configuration policies that cater for flexible characterization and planning of varying resource needs overtime. We propose an abstraction, called **Re-configuration Policy** to specify how a cloud resource should behave when certain events occurs. For an example, a "restart" policy of a virtual machine (VM) is invoked whenever the deployment runtime detects a connection failure to the relevant VM. A *Re-configuration Policy* is attached to a *RCS* and performs some provider specific re-configuration operations such as requesting to restart the VM through the provider's deployment service interface. *RCS*s implement relatively complex, generic and customizable cloud resource configuration mechanisms. Our initial working assumption is that *RCS*s are created, verified, tested, reviewed and curated by experienced cloud resource configuration programmers or administrators based on available knowledge or experience, and may be reused in several cloud

---

[4] docs.aws.amazon.com/cli/latest/reference/ec2/create-image.html

applications. Cloud resource consumers can annotate the cloud resource description with those events and *Re-configuration Policies* to model reconfiguration policies of cloud resources without worrying about low-level scripting mechanisms or provider specific policy engines. It should be noted that, the issues of event specification and detection while important, they are complementary to research issues addressed in our work and outside the scope of this paper. See Table 3 for the interface design of a *Re-configuration Policy*.

Table 3. Operations of *Re-configuration Policy* interface

reconfigureResource(event_description): This operation accepts the event description that triggered this operation and specifies the re-configuration behavior in any configuration language. This operation returns an object that consists of the resource_id and an optional resultant message.

2.2   Cloud Resource Configuration Description(CRCD) Model

There are several reasons that led various cloud-resource providers expose proprietary configuration description language/data models. These reasons include lack of mature and open standards; gain of competitive advantage over other cloud resource providers. Mechanisms for describing a resource are model-driven (e.g., Ubuntu Juju Charms[3], TOSCA application topology [10]), template based (e.g., Chef[1] recipes), and hybrid (e.g., AWS OpsWorks [5]). Also some languages are only capable of describing specific types (e.g., VM, source-code) of resources. Hence it is challenging to come up with a description model that captures different language characteristics. In this setting we propose **Cloud Resource Configuration Description**(CRCD) model that lets users to describe cloud resource configurations in a unified and provider-independent manner.

The *CRCD* model follows an entity-relationship model. Each entity represents the configuration knowledge of a particular cloud resource and a relationship represents a one-way deployment dependency with another entity (e.g. deployment of a web application engine is depend upon the deployment of its component resources: a data storage and a language runtime). There are two types of entities, based on available relationships with other entities.

(1) Basic CRCD: An entity that represents configuration knowledge of a cloud resource, which does not rely on any other entity, is called a *Basic CRCD*. In other words, a *Basic CRCD* is an indivisible resource into component resources from the perspective of CRCD curators (those that primarily add/maintain configuration knowledge in CRCDs). For an example, a VM with 4GB RAM and 4GHz processing power, can be modeled as a *Basic CRCD* with two attributes to represent the memory and processing power. *Basic CRCD*s act as primary building blocks of *Federated CRCD*s.

(2) Federated CRCD: A *Federated CRCD* is an umbrella structure that brings together *Basic* and other *Federated CRCD*s to represent configuration knowledge of a federated cloud resource.

---

[5] aws.amazon.com/opsworks/

Basic and federated *CRCD*s are described using a set of "attributes" (i.e., name-value pairs) (see Listing 1.1). Every *CRCD* must include attributes named "name" and "tags" for indexing and discovery purposes of the configuration knowledge-base [11]. Other attributes are optional and should essentially represent configuration attributes of the resource. For example, the *CRCD* (as shown in Listing 1.1) contains two attributes "team-count" and "user-count-per-team" which represent the anticipated number of teams and users to be handled by the application.

Listing 1.1. Basic CRCD sample

```
1  {
2      "attributes": {
3          "name": "project-mgt-app",
4          "tags": "SaaS"
5          "team-count": "5",
6          "user-count-per-team": "10"
7      }
8  }
```

3 Modeling Cloud Resource Configuration Tasks

We introduced *RCS* and *CRCD* to abstract out the heterogeneity of cloud resource deployment services in Section 2. Users can discover those published resource descriptions (i.e., *CRCD*) and invoke available operations of associated configuration services (i.e., *RCS*) of resource descriptions to deploy, configure and undeploy a resource instance. To automatically deploy and orchestrate a federated resource configuration users require modeling an orchestration process that coordinates the deployment, configuration and undeployment tasks of several resources. We propose two high-level process based abstractions over RCS and CRCD to model federated cloud resource configuration tasks: Deployment Tasks and Reconfiguration Policies. In Section 4, we explain how these abstractions are implemented by extending BPMN.

Motivating Scenario: Consider a scenario, where a web-application developer needs to deploy a web-application in an Apache-Tomcat[4] based application server cluster. To distribute requests to a set of Tomcat application servers, the resource infrastructure includes an http load balancer (LB) like nginx[6]. The web application is deployed in each Tomcat server. When adding a new Tomcat server to the cluster, the web application also should be deployed within the new server. To add the newly deployed Tomcat server to the cluster, the routing table of the LB should be updated with details (e.g., IP and port) of the new server. Then more Tomcat servers can be deployed in the aforementioned manner until the cluster reaches the expected number of Tomcat servers.

The deployment tasks of the Tomcat cluster are depicted in Fig. 1. We excluded the deployment tasks of the LB from Fig. 1 for a simplified graphical representation. Once the Tomcat cluster is deployed, the application developer may release

---

[6] nginx.org/

new versions of the application. Hence each Tomcat server in the cluster must be updated with the new version of the application. The orchestration process in Fig. 1 should be updated by including additional orchestration tasks for continuously integrate web application updates. See the updated version in Fig. 2.

The complexity incurred by modeling even a single reconfiguration policy within the initial orchestration process points out the drawbacks of including reconfiguration policies as part of the initial orchestration process. Modeling further reconfiguration policies (e.g., a web application upgrade is rollbacked if it cannot happen on every Tomcat server in the cluster) within the initial orchestration process makes the resultant orchestration process (1) complex; (2) error-prone; and (3) difficult to verify and manage later. It is advantageous in such situations to separate the modeling of corresponding reconfiguration policies from the initial orchestration process and refer the relevant policies within initial tasks as in Fig. 3. Using our high-level process based abstractions, user is now capable of simply and clearly define high-level reconfiguration policies that apply to one or more resources in a federated resource configurations.

Fig. 1. Deployment plan for a web application in an Apache Tomcat cluster

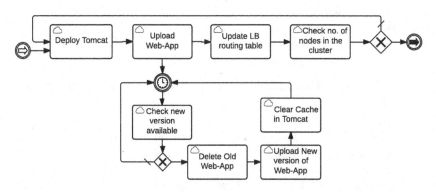

Fig. 2. Modeling application updates within the deployment plan in Fig. 1

Cloud Resource Deployment Tasks: We introduce "Cloud Resource Deployment Task" (CRD-Task) that allows users to model the deployment of cloud resources configurations. Every *CRD-Task* is associated with a *CRCD* and a potential *RCS* that can deploy the associated CRCD (see Fig. 4). We implemented a recommender service [11] that facilitates users to search for available *CRCDs*. During

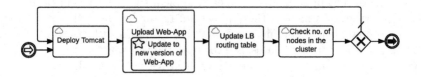

Fig. 3. Modeling application updates in Fig. 2 using CRR-Policy

the execution of this task, the "deployResource" operation of the associated *RCS* is triggered along with the "attributes" component of the associated *CRCD* as the input. For an example, the deployment of a HP-Cloud-Compute[7] VM is modeled using *CRD-Task* named "HP-Compute-VM", which triggers the "deployResource" operation of the *RCS* named "HP-Deployer" with the *CRCD* named "desc1".

Cloud Resource Re-configuration Policies: "Cloud Resource Re-configuration Policy" (CRR-Policy) allows users to specify a high-level and dynamic re-configuration policy for a cloud resource. A *CRR-Policy* is a pair of an event description and a *Re-configuration Policy* that was introduced in Section 2.1. Users can add any number of *CRR-Policies* to a *CRD-Task*, given that events and *Re-configuration Policies* are registered in the associated *RCS* of the *CRD-Task*. A *CRR-Policy* is triggered whenever its associated event occurs. The *RCS* is responsible to propagate the event to the runtime of the *CRR-Policy*. For an example, the *CRD-Task*, named "HP-Compute-VM" contains two *CRR-Policies*, called "CRR1" and "CRR2" (see Fig. 5). BPMN runtime triggers "CRR1" (i.e., IF {incoming Message=="restart-desc1"} THEN RUN {"restart-policy"}) whenever a user sends a request to restart the cloud resource. "CRR2" (i.e., IF {getDay()=="sunday"} THEN RUN {"backup-policy"}) is triggered weekly to backup the cloud resource.

4   Translating Cloud Resource Configuration Tasks into BPMN

The choice of BPMN, as the language for capturing configuration and deployment knowledge of federated cloud resources is motivated by several reasons. First, BPMN is an open, standardized and task-based service composition language that is heavily used in application layer. Next, BPMN is suitable to express executional dependencies among different deployment tasks. Furthermore, BPMN supports extension points which are crucial to model deployment workflows as BPMN doesn't support modeling deployment tasks out of the box.

Translating CRD-Tasks: Fig. 4 depicts a federated cloud resource deployment workflow. The "HP-Compute-VM" is a *CRD-Task*, which is annotated with a

---

[7] www.hpcloud.com/products-services/compute

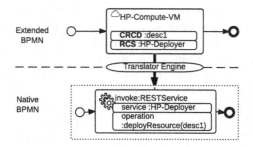

Fig. 4. A CRD-Task and its BPMN generation (within the dotted rectangle)

CRCD and *RCS*. During the runtime of the workflow, the *CRD-Task* is transformed into a BPMN sequence flow that includes a Service Task [9] that triggers the "deployResource" of *RCS* along with the CRCD as the input parameter.

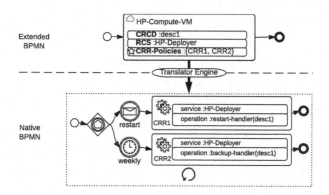

Fig. 5. A CRD-Task with two CRR-Policies and its BPMN generation (within the dotted rectangle)

Translating CRR-Policies: BPMN allows to model events and associated tasks as the business logic. We decided not to reuse the native BPMN events to implement *CRR-Policies*, because modeling several events and associated tasks within the deployment workflow makes the resultant deployment workflow complex. Also it enforces the workflow designer to know exactly where to inject those events and associated tasks within the workflow. Hence it is advantageous to provide a high-level abstraction to implement re-configuration policies in BPMN. We implemented *CRR-Policy* as an extension to BPMN to define re-configuration policies, which are linked with the *CRD-Tasks* while separating the original deployment workflow from re-configuration policies.

Fig. 5 depicts a federated cloud resource deployment workflow. The "HP-Compute-VM" is a *CRD-Task*, which is annotated with two *CRR-Policies* (i.e., event-policy pairs) that define re-configuration policies of the cloud resource configuration. During the runtime of the workflow, each *CRR-Policy* is transformed into a BPMN sequence flow that includes an Event and a Service Task that triggers the *Re-configuration Policy*. All the sequence flows are initiated from an Event-based Gateway within a Loop Task.

Our approach supports the automated generation of BPMN processes that deploy and re-configure the appropriate SaaS, PaaS or IaaS resources with respect to the introduced modeling abstractions, namely *CRD-Tasks* and *CRR-Policies*. A detailed description of generation techniques is outside of this paper.

5   Implementation and Evaluation

We built a proof-of-concept (POC) prototype of *CloudBase*[8]. We also implemented several POC prototypes of *RCS*s, which act as extension points of *CloudBase* architecture (see Fig. 6) that leverages heterogeneous deployment services (e.g., AWS, Rackspace and GitHub) to deploy federated cloud resources in a unified manner. In the current implementation, cloud resource providers or devOps implement and register *RCS*s as RESTful services in *CloudBase* via ServiceBus API [12].

We implemented a deployment workflow editor and engine by extending Activiti[9], a graphical BPMN editor and engine. Our workflow editor was extended to model deployment workflows with tasks named *CRD-Task* and *CRR-Policy* (see Section 3 and 3). Our deployment engine was extended to parse and generate code for *CRD-Tasks* and *CRR-Policies*; and execute them.

Evaluation: To evaluate our approach, we measured the overall productivity gained by three professional software engineers from business process and application development backgrounds. For the experiment we provided each testee a deployment specification and asked them to model an arbitrary deployment workflow to deploy a software development and distribution platform. This platform was intended for software engineers who want to manage the entire lifecycle of a project. Multiple projects can leverage this platform by just cloning the deployment multiple times. We enforced testees to limit their resource selection choices to a fixed set of cloud resources, which were currently supported by *CloudBase*[10]. We expect to increase selection choices as the design of *CloudBase* inherently supports to incrementally collaborate resource configuration knowledge in terms of *RCS* and *CRCD*. In the deployment specification, testees were instructed to deploy an AWS EC2 VM where Redmine [13], a project management service and a Git client [14] is installed. Testees were advised to deploy a new source code repository in GitHub and integrate it with the Redmine service such that the Redmine service

---

[8] github.com/ddweerasiri/Federated-Cloud-Resources-Deployment-Engine
[9] activiti.org/
[10] github.com/ddweerasiri/Federated-Cloud-Resources-Deployment-Engine

automatically extracts the latest commits from the repository via the Git client. Additionally testees were instructed to deploy an AWS S3 bucket which act as a software distribution repository.

For the evaluation purposes we implemented the same deployment specification in three languages; (i) Shell scripts, (ii) Docker and (iii) Juju. The main reason to choose Shell scripts was to estimate an upper bound of the result set. Docker and Juju were selected as their popularity among devOps and they are specifically designed for cloud resource configuration and deployment. We measured (i) the total number of lines-of-code ("actual" lines of code written and how many generated by *CloudBase*), excluding white spaces and comments; (ii) number of external dependencies/libraries required to describe and deploy each federated cloud resource; and (iii) time taken to complete the modeling task. We measured the correctness of the modeling tasks by executing each deployment workflow and checking whether the resultant deployment complied with the initial deployment specification. The benefits of *CloudBase* is further demonstrated in embracing the knowledge-sharing paradigm (inspired from industry[3]): Given that users in this scenario would not require the efforts of development, registration and maintenance of the *RCS* - since this could be pre-done once and re-used multiple times for the benefit of many.

Analysis and Discussion: Results of the experiment (see Table 4) show that lines-of-code; number of external dependencies; and time-to-modeling are improved

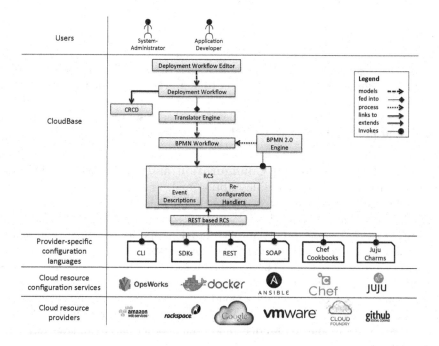

Fig. 6. System Overview

Table 4. Results of the experiment

Parameters	Shell Scripts	Docker	Juju	*CloudBase*
average time-to-modeling (min)	103	95	72	61
#lines-of-code	107	116	127 (generated)	541 (generated)
#dependencies	3	1	1	1
knowledge shareable without changing deployment workflows	no	yes	yes	yes

when using *CloudBase* over prevalent resource deployment techniques. More specifically, the time-to-modeling is reduced by 15.2% assuming the required *RCS*s for all component resources are registered in *CloudBase*. We argue that unified and provider-independent *CRCD* model; and tool-support like graphical deployment workflow editors improve the time-to-modeling. Comparing with proprietary languages like Juju, we argue that extending standards-based workflow languages like BPMN further improves the time-to-modeling for users like business process and application developers. Shell scripts, Docker and Juju required 116 lines-of-code on average to model the deployment plan. *CloudBase* generated 541 lines-of-code because *RCS*s were implemented using Java and BPMN based resource deployment tasks generated XML and JSON files which are more verbose compared to shell scripting based approaches like Docker and Juju. Therefore we determine the improved productivity of the process-driven federated resource deployment over a unified configuration knowledge rich layer.

6   Related Work

In this section, we briefly describe three areas of related works: (1) federated cloud resource configuration and orchestration frameworks; (2) unified representation and invocation of heterogeneous cloud resource configuration services; and (3) modeling dynamic orchestration of federated cloud resources. We compare and contrast our proposed approach with these related works.

Federated Cloud Resource Configuration and Orchestration Frameworks: Various cloud resource description and orchestration frameworks are proposed in industry and research. Market-leading cloud resource providers like AWS OpsWorks[6] and CA AppLogic[11] allow describing and deploying complete application stacks. These providers offer provider-specific resource representations while our approach allows multiple providers to describe and publish unified resource representations. These unified resource representations allow users to compose federated resource configurations provider-independently. In research, Konstantinous et al. [15] presents a cloud resource description and deployment model that first models a resource as a provider-independent resource configuration, called "Virtual Solution Model", and then another party can transform the

---

[11] www.ca.com/au/cloud-platform.aspx

provider-independent model to a provider-specific model called, "Virtual Deployment Model". This approach only allows users to compose federated resource configurations from a single provider for a single deployment, in contrast to our approach, which considers the resource federation from multiple providers as a first class citizen.

Configuration management tools, resource orchestration tools and virtualized container engines like (Puppet, Chef, Juju, Docker, Smartfrog etc.,) and other research initiatives provide different domain specific languages and tools to represent and manage resource configurations in a cloud environment [15–19]. These languages provide heterogeneous resource representation models, management operations and event descriptions. Hence deployment and orchestration of resources described using different languages can be only achieved as ad-hoc scripts or manual tasks. Our unified resource representation model allows resource providers and devOps to implement RCSs that encapsulate heterogeneity among different languages. Users then automate the deployment, orchestration and attach dynamic reconfiguration policies using our graphical modeling language without directly interacting with heterogeneous configuration services.

Unified Representation and Invocation of Heterogeneous Cloud Resource Configuration Services: In the domain of multi-cloud development, wrapping heterogeneous cloud resources has been researched [20] and implemented as language libraries (e.g., Apache jclouds[12]). However, the fact that providers furnish different offerings and change them frequently often complicates these approaches.

TOSCA [10] is an open standard for representation and orchestration of cloud resources. TOSCA facilitates to describe a federated cloud resource configuration using a "Service Template" that captures (1) topology of component resources; and (2) plans that orchestrate component resources. But TOSCA does not define (1) the implementation language of orchestration plans; and (2) how to specify dynamic reconfiguration policies. When developing TOSCA orchestration plans for multi-cloud environments, developers often need to deal with different invocation mechanisms (e.g., SOAP, REST and SSH) even though the resource representation is standardized. Hence our graphical process modeling language and Resource Configuration Service (RCS) can be leveraged for implementing high-level orchestration plans of a TOSCA "Service Template".

Different cloud resource providers offer resource configuration services with different interfaces (e.g., CLI, SDKs, REST/SOAP based and Scripts) for users who have different levels of comfort and experience with those interfaces. Methodologies for unified representation and invocation of heterogeneous web services are proposed in several research [12,21,22]. Those approaches only focus on application-level services instead of arbitrary services like CLI, SDKs and scripts. In the domain of cloud resource management [23] proposes a unified cloud resource deployment API using Web Service standards (e.g., WSDL and WS-BPEL [24]). However the authors leverage native WS-BPEL to describe federated resource

---

[12] jclouds.apache.org

deployment processes. Comparatively, we propose an extended BPMN based language rather than using native BPMN to avoid orchestration processes getting (1) complex; (2) error-prone; and (3) difficult to verify and manage later. Wettinger et al. [6] proposes a REST-based unified invocation API that abstracts out different invocation mechanisms, interfaces and tools available for cloud resource configuration and orchestration. However the authors have not focused on modeling orchestration processes and reconfiguration policies on top of that unified invocation API. Whereas our research focuses on (1) unified representation and invocation; and (2) modeling high-level orchestration processes and reconfiguration policies.

Modeling Dynamic Orchestration of Federated Cloud Resources: Cloud resource orchestration processes are composed of deployment and reconfiguration tasks, which require advanced abstractions over general business process modeling languages like BPMN and BPEL. BPMN and BPEL focus primarily on the application layer [9, 25]. However, orchestrating cloud resources requires rich abstractions to reason about application resource requirements and constraints, support exception handling, flexible and efficient scheduling of resources. Nonetheless a user can leverage native BPMN or BPEL to model deployment processes [23] with reduced design flexibility, increased modeling size and complexity. Furthermore, modeling dynamic reconfiguration policies (e.g., backup the MySQL database on every Sunday) within the same deployment process makes it difficult to verify and manage the workflow later. We propose two BPMN extensions to model deployment tasks and high-level reconfiguration policies while keeping the deployment workflow simple and modular.

Extending modeling languages to facilitate domain specific needs is a common practice. For an example, Sungur et al. [26] propose BPMN extensions to program wireless sensor networks. BPMN4TOSCA [27] extends BPMN to implement orchestration plans of resource representations described as TOSCA "Topology Templates" [10]. BPMN4TOSCA includes 4 BPMN extensions, which facilitate to model configuration and orchestration tasks associated with a Topology Template. Comparing to our graphical modeling language, BPMN4TOSCA does not model dynamic reconfiguration policies that are essential to model the elasticity of cloud resources. We propose a high-level process based notation to model dynamic reconfiguration policies, which trigger orchestration tasks when specific events happen, without complicating the initial deployment process model.

7   Conclusions and Future Work

In this paper, we have presented a federated cloud resource configuration orchestration framework, which leverages unified, customizable and reusable configuration knowledge representations. The framework consists of (1) a unified configuration knowledge representation model for federated cloud resources; and (2) a graphical and process-based notation to describe and automate the deployment tasks of complex configurations over federated clouds. To evaluate the feasibility and productivity of the proposed framework, we implemented our system as a proof-of-concept prototype. As future works, we plan to (1) extend the orchestration framework to feature as a cloud resource management framework; and (2) enable the orchestration framework to select the potential RCSs on arrival of deployment requests based on a user defined QoS (e.g., availability >95%) based policy.

References

1. Veeravalli, B., Parashar, M.: Guest editors' introduction: Special issue on cloud of clouds. IEEE Transactions on Computers **63**(1), 1–2 (2014)
2. Buyya, R., Ranjan, R., Calheiros, R.N.: InterCloud: utility-oriented federation of cloud computing environments for scaling of application services. In: Hsu, C.-H., Yang, L.T., Park, J.H., Yeo, S.-S. (eds.) ICA3PP 2010, Part I. LNCS, vol. 6081, pp. 13–31. Springer, Heidelberg (2010)
3. Elmroth, E., Larsson, L.: Interfaces for placement, migration, and monitoring of virtual machines in federated clouds. In: GCC 2009, pp. 253–260. IEEE (2009)
4. Villegas, D., et al.: Cloud federation in a layered service model. J. Comput. Syst. Sci. **78**(5), 1330–1344 (2012)
5. Papazoglou, M.P., van den Heuvel, W.J.: Blueprinting the cloud. IEEE Internet Computing **15**(6), 74–79 (2011)
6. Wettinger, J., et al.: Unified invocation of scripts and services for provisioning, deployment, and management of cloud applications based on TOSCA. In: CLOSER 2014, April 3–5, 2014, pp. 559–568. SciTePress, April 2014
7. Liu, C., Loo, B.T., Mao, Y.: Declarative automated cloud resource orchestration. In: Proceedings of the SOCC 2011, pp. 1–8. ACM (2011)
8. Ranjan, R., Benatallah, B.: Programming cloud resource orchestration framework: Operations and research challenges. CoRR abs/1204.2204 (2012)
9. OMG: Business Process Model and Notation (BPMN), Version 2.0 (2011)
10. OASIS: Topology and Orchestration Specifation for Cloud Applications (TOSCA), Version 1.0 (2013)
11. Weerasiri, D., Benatallah, B., Yang, J.: Unified representation and reuse of federated cloud resources configuration knowledge. Technical Report UNSW-CSE-TR-201411, Department of CSE, University of New South Wales (2014)
12. Barukh, M.C., Benatallah, B.: ServiceBase: a programming knowledge-base for service oriented development. In: Meng, W., Feng, L., Bressan, S., Winiwarter, W., Song, W. (eds.) DASFAA 2013, Part II. LNCS, vol. 7826, pp. 123–138. Springer, Heidelberg (2013)
13. Redmine. http://www.redmine.org/ accessed: October 28, 2014
14. Git -distributed-is-the-new-centralized. http://git-scm.com/ accessed: October 28, 2014
15. Konstantinou, A.V., et al.: An architecture for virtual solution composition and deployment in infrastructure clouds. In: Proceedings of the 3rd International Workshop on VTDC, pp. 9–18. ACM (2009)
16. Chieu, T.C., et al.: Solution-based deployment of complex application services on a cloud. In: 2010 IEEE International Conference on SOLI, pp. 282–287. IEEE (2010)
17. Goldsack, P., et al.: The smartfrog configuration management framework. ACM SIGOPS Operating Systems Review **43**(1), 16–25 (2009)

18. Delaet, T., Joosen, W., Vanbrabant, B.: A survey of system configuration tools. In: Proceedings of the 24th International Conference on LISA, pp. 1–8. USENIX Association (2010)

19. Wilson, M.S.: Constructing and managing appliances for cloud deployments from repositories of reusable components. In: Proceedings of the 2009 Conference on Hot-Cloud 2009. USENIX Association (2009)

20. Moscato, F., et al.: An analysis of mosaic ontology for cloud resources annotation. In: FedCSIS 2011, pp. 973–980. IEEE (2011)

21. Barukh, M.C., Benatallah, B.: A toolkit for simplified web-services programming. In: Lin, X., Manolopoulos, Y., Srivastava, D., Huang, G. (eds.) WISE 2013, Part II. LNCS, vol. 8181, pp. 515–518. Springer, Heidelberg (2013)

22. Barukh, M.C., Benatallah, B.: *ProcessBase*: a hybrid process management platform. In: Bhiri, S., Franch, X., Ghose, A.K., Lewis, G.A. (eds.) ICSOC 2014. LNCS, vol. 8831, pp. 16–31. Springer, Heidelberg (2014)

23. Mietzner, R., Leymann, F.: Towards provisioning the cloud: on the usage of multi-granularity flows and services to realize a unified provisioning infrastructure for saas applications. In: IEEE Congress on Services - Part I, pp. 3–10 (2008)

24. Juric, M.B., Weerasiri, D.: WS-BPEL 2.0 beginner's guide. Packt Publishing Ltd (2014)

25. De Alwis, B., Malinga, S., Pradeeban, K., Weerasiri, D., Perera, S., Nanayakkara, V.: Mooshabaya: mashup generator for xbaya. In: Proceedings of the 8th International Workshop on Middleware for Grids, Clouds and e-Science, p. 8. ACM (2010)

26. Sungur, C., et al.: Extending bpmn for wireless sensor networks. In: 2013 IEEE 15th Conference on Business Informatics (CBI), pp. 109–116, July 2013

27. Kopp, O., Binz, T., Breitenbücher, U., Leymann, F.: BPMN4TOSCA: a domain-specific language to model management plans for composite applications. In: Mendling, J., Weidlich, M. (eds.) BPMN 2012. LNBIP, vol. 125, pp. 38–52. Springer, Heidelberg (2012)

Social Networks I

An Integrated Tag Recommendation Algorithm Towards Weibo User Profiling

Deqing Yang[1], Yanghua Xiao[1]([⊠]), Hanghang Tong[2],
Junjun Zhang[1], and Wei Wang[1]

[1] School of Computer Science, Shanghai Key Laboratory of Data Science Fudan
University, Shanghai, China
{yangdeqing,shawyh,zhangjunjun,weiwang1}@fudan.edu.cn
[2] Arizona State University, Tempe, AZ, USA
hanghang.tong@gmail.com

Abstract. In this paper, we propose a tag recommendation algorithm for profiling the users in Sina Weibo. Sina Weibo has become the largest and most popular Chinese microblogging system upon which many real applications are deployed such as personalized recommendation, precise marketing, customer relationship management and etc. Although closely related, tagging users bears subtle difference from traditional tagging Web objects due to the complexity and diversity of human characteristics. To this end, we design an integrated recommendation algorithm whose unique feature lies in its *comprehensiveness* by collectively exploring the social relationships among users, the co-occurrence relationships and semantic relationships between tags. Thanks to deep comprehensiveness, our algorithm works particularly well against the two challenging problems of traditional recommender systems, i.e., *data sparsity* and *semantic redundancy*. The extensive evaluation experiments validate our algorithm's superiority over the state-of-the-art methods in terms of matching performance of the recommended tags. Moreover, our algorithm brings a broader perspective for accurately inferring missing characteristics of user profiles in social networks.

Keywords: Tag recommendation · User profiling · Tag propagation · Chinese knowledge graph

This paper was partially supported by the National NSFC(No.61472085, 61171132, 61033010), by National Key Basic Research Program of China under No.2015CB358800, by Shanghai STCF under No.13511505302, by NSF of Jiangsu Prov. under No. BK2010280, by the National Science Foundation under Grant No. IIS1017415, by the Army Research Laboratory under Cooperative Agreement Number W911NF-09-2-0053, by National Institutes of Health under the grant number R01LM011986, Region II University Transportation Center under the project number 49997-33 25. Correspondence author: Yanghua Xiao.

M. Renz et al. (Eds.): DASFAA 2015, Part I, LNCS 9049, pp. 353–373, 2015.
DOI: 10.1007/978-3-319-18120-2_21

1   Introduction

Sina Weibo[1] (Weibo in short), the largest counterpart of Twitter in China, is experiencing fast growth and becoming a world-widely used microblogging system. So far, Weibo has attracted more than 0.6 billion users in total and 5 million active users per day. The applications or services related to Weibo are creating a plenty of business opportunities since Weibo is attracting more and more users.

One of the most important services provided by Weibo is user tagging which allows a user to publish several tags to label themselves. These tags usually describe user profiles including hobby, career, education, religion and etc. Hence, Weibo tags are important for user understanding which is critical for many real industry applications, e.g., personalized recommendation, precise marketing and customer relationship management.

An effective tag recommendation algorithm is critical for Weibo. Weibo users can be divided into two groups: the groups are willing/or not to label themselves. For the group willing to, an effective tag recommendation mechanism can make it easy for them to 'label' themselves. The other group is not willing to label themselves with informative tags mostly out of the privacy concerns. An effective recommendation algorithm thus is critical for the accurate characterization of these users. Despite of its importance, current tagging service only attracts 55% of Weibo users to tag themselves. The remaining users do not label themselves with any tags either due to privacy concern or inconvenient tagging service.

In general, tag recommendation for Weibo user has been rarely studied. Although many tag-based recommender systems have been proposed, they generally can not be used for tagging Weibo users due to the following reasons.

- *First, the object to be tagged is different.* In this paper, we focus on tagging Weibo users, whereas most existing tag-based recommendation systems focused on tagging Web objects, such as photos in Flickr [25] or URLs [30,11]. In general, these systems make successful recommendations by utilizing abundant tagging activities on objects and users. However, much of these information in general is absent in Weibo setting (known as *data sparsity* problem), which poses a great challenge to accurately tag a Weibo user. Worse comes to worse, many users do not have any tag at all.
- *Second, the objective of tag recommendation is different.* Our recommendation aims to characterizing a user's *individual preference* of tags while many social tagging mechanisms were designed for *collective preference* of tags on the targeted object. Clearly, mining individual preference is different to mining collective preference since each user has his/er own unique taste. We should recommend not only diverse tags for a user but also satisfy a user's unique taste.

In this paper, we develop an effective and efficient algorithm to recommend tags for Weibo users. Although our algorithm is proposed for Weibo setting, the proposed recommendation schemes can also be imported into other social network platforms, such as Twitter and Facebook.

---

[1] http://weibo.com

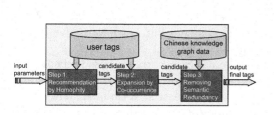
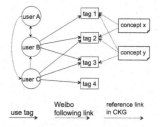

Fig. 1. Framework of tag recommendation algorithm

Fig. 2. The meta-graph in our recommendation framework

1.1 Requirements

First, *the recommendation should effectively handle data sparsity.* In our scenario, nearly 45% of Weibo users have no tag. This will disable many collaborative filtering (CF in short) based recommender systems [24,12] and co-occurrence based recommendations [25].

Second, *the recommended tags should be diversified enough to capture the* multi-facet *characteristics of a real person.* A user may publish several tags to characterize all of these aspects, e.g., education, career, hobbies, favorite idols and etc. How to recommend a set of diversified tags to a user is challenging.

Third, *the recommendation should be aware of the* semantic redundancy *in the recommended tags.* It is not suitable for real applications if too many tags are recommended. E.g., a Weibo user is restricted to use 10 tags at most. Hence a user generally expects that the recommended tags are expressive and contain no (near-)synonyms. In contrast, it is acceptable that different users use (near-)synonyms to tag the same object [8]. Thus, in those recommender systems towards tagging objects, semantic redundancy is not an issue.

To satisfy the above requirements, in this paper we first conduct empirical studies to understand the tagging behaviors of Weibo users. Our findings reveal two effective tag recommendation mechanisms:

1. *Homophily based recommendation.* Homophily is the tendency that birds with a feather flock together [19]. It also holds on Weibo. A Weibo user tends to use the same or similar tags as his/er friends, especially when the friend is simultaneously one of his followees and followers, i.e., mutual fan.
2. *Co-occurrence based recommendation.* If a tag is deserved to be recommended, the other tags that co-occurs with it are also deserved to be recommended.

Armed with these findings, we propose a tag recommendation algorithm to generate informative and personalized tags for profiling Weibo users. Our algorithm is an integrated algorithm consisting of three major steps. Each step aims to address one of the above requirements. Fig. 1 illustrates our algorithm's framework.

1. *Step 1: Recommendation by Homophily.* We recommend to a user with the most *frequent* and *informative* tags from the tags used by his/er friends. We import TF-IDF scheme to remove those frequent but less informative tags. We use this step to solve the data sparsity problem.
2. *Step 2: Expansion by Co-occurrence.* We use co-occurrence based scheme to enrich the recommended tag list so that the final tag list is diverse enough.
3. *Step 3: Removing Semantic Redundancy.* We construct a Chinese knowledge graph (CKG in short) from online Chinese encyclopedias. Then, we map Weibo user tags into CKG entities so that we can measure the semantic similarity of tags. Next, we use an ESA-based (explicit semantic analysis) [7] metric to remove the synonyms or near-synonyms from the recommended tag list. This step satisfies the third requirement.

Fig. 2 shows the entities and their relationships in our tag recommendation algorithm. We use this figure to illustrate our recommendation mechanism. In Step 1, we recommend to user A with tag 2 and tag 3 that are mostly used by user B and C because A follows B and C (which suggests that they have similar tag preferences). In Step 2, tag 1 and tag 4 are also recommended because they are co-used with tag 2 and tag 3. In Step 3, we remove either tag 1 or tag 2 because both concept x and y in CKG refer to them implying their redundant semantics. The detailed mechanisms will be introduced in the subsequent sections.

1.2   Contributions and Organization

In summary, the main contributions of this paper include:

1. *Empirical Findings.* We conducted extensive empirical studies to show statistical user tagging behaviors and unveil effective recommendation schemes for tagging Weibo users.
2. *Effective Algorithm.* We proposed an integrated algorithm to recommend a set of tags to Weibo users towards personalized and informative user profiling.
3. *Evaluations.* We conducted extensive evaluations to justify the effectiveness of our recommendation algorithm. The results show that our algorithm is useful in enriching user profiles as well as inferring the missing characteristics of Weibo users.

The rest of this paper is organized as follows. We first display our empirical results in Section 2 which are the basis of our recommendation algorithm. In Section 3, we elaborate the detailed procedure of tag recommendation algorithm. In Section 4, we present our experiments for evaluating algorithm performance. We survey the related works in Section 5 and conclude our paper in Section 6.

2   Empirical Study

In this section, we conduct empirical studies on the collective tagging behaviors of Weibo users. The empirical findings construct the basis of our tag recommendation algorithm. We first introduce our dataset.

Dataset: We first randomly selected 3,000 Weibo users as seeds, then crawled their followers and followees. Thus there are more than 2.1 million users and 875,186 unique user tags in total. Besides tags, the following relationships between the users were fetched. All data were crawled before Oct. 2013. The statistics show that only 55.01% of these users, i.e., about 1.15 million users have at least one tag.

2.1 Homophily in Tagging Behavior

Homophily is a tendency that an interaction between similar people occurs with a higher probability than among dissimilar people [19]. Homoplily was shown to be a universal phenomenon across a variety of social media platforms such as Twitter [28]. More specifically, the Twitter users following reciprocally (mutual fans) tend to share topical interests, have similar geographic and popularity [15]. Thus, an interesting question arises: *do close social relationships in Weibo also imply similar profiles or tags?* To answer this question, we first distinguish three important types of social relationships among Weibo users: following (follower), followed (followee) and following reciprocally (mutual fan)². Next, we will empirically study the effects of these three relationships on tag similarity. At first, we define two types of tags for a Weibo user u.

Definition 1 (Real Tags). If u originally labels him/erself with some tags, these tags are referred to as u's *real tags* and denoted by RT_u.

Definition 2 (Collective Tags). The tags that are most frequently used by u's friends are referred to as u's *collective tags* and denoted by CT_u.

To find the tags in CT_u, we define a score function $tf(t)$ to quantify the likelihood that tag t belongs to CT_u. The $tf(t)$ function is defined as

$$tf(t) = \frac{r(t)}{\sum\limits_{t' \in T(Neg(u))} r(t')} \qquad (1)$$

where $Neg(u)$ is u's friend group and $r(t)$ is the number of users in $Neg(u)$ who have used tag t. $T(Neg(u))$ represents the tag set used by the users in $Neg(u)$. We denote the score function as tf because it is equivalent to the term frequency in document retrieval. The larger the $tf(t)$ is, the more likely the tag t belongs to CT_u. If $|CT_u|$ is limited to k, we select the top-k tags from $T(Neg(u))$ according to $tf(t)$ value. In the following text, we refer to $tf(t)$ as the *frequency* based tag ranking score.

Metrics of Evaluation: To justify the homophily in Weibo Tagging behavior, we compare CT_u with RT_u for those users having real tags. If the matching of CT_u and RT_u is more evident than the matching of RT_u and a random tag set, the homophily in tagging behavior is evident. In this paper, we use the following

---

² In this paper, we often refer to these three social relationships in Weibo as friend.

three metrics to evaluate matching performance of generated/recommended tags to a user's real tags (ground truth).

Precision (P@k): It is defined as the proportion of top-k recommended tags that are matched to the ground truth (i.e., they are in real tag set), averaged over all samples.

Mean Average Precision (MAP@k): It is the mean of the *average precision score* (AP) of top-k recommended tags for all samples. AP is defined as

$$AP@k = \frac{\sum_{i=1}^{k}(P(i) \times rel(i))}{H} \tag{2}$$

where $rel(i)$ is an indicator function equaling 1 if the i-th tag is matched, 0 otherwise. $P(i)$ is the matched proportion of top-i tags and H is total number of matched tags in all top-k tags.

Normailzed Discounted Cumulative Gain (nDCG@k): It is a famous metric to measure relevance level of search results to the query in IR systems [14]. For top-k recommended tags, the nDCG score can be calculated as

$$nDCG = \frac{1}{Z} \sum_{i=1}^{k} \frac{2^{rel(i)} - 1}{log_2(i + 1)} \tag{3}$$

where $rel(i)$ is the same as Eq. 2 and Z is the normalized factor. Compared with MAP, nDCG is more sensitive to rank position of recommended tags. In general, a user pays less attention to the tags listed behind, hence nDCG is better to evaluate recommendation performance.

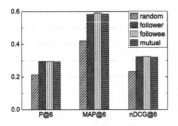

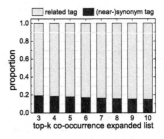

Fig. 3. Matching performance of CT_u to RT_u show that tag similarity is more evident for social friends than general users

Fig. 4. The proportion of (near-) synonyms in top-k expanded list. Some expanded tags are (near-) synonyms of the parent tags, but most of them are complementary in semantic.

Results: Next, we show our empirical results which in general justify that *the users in Weibo who have close social relationships with each other tend to share similar tags*. Since the mean tag number of a Weibo user is 5.69 by our statistics, we only list the results of $|CT_u|=6$ ($k=6$) in Fig. 3 due to space limitation.

We got consistent results under other sizes of CT_u. For comparison, we also compare RT_u with a random tag set. We randomly selected some users from the universal user set and used the most frequent tags of these users as the random tag set. The figure displays that under all metrics, random tag set have the worst matching performance and the collective tags from followees have the best performance. These results imply that *homophily is effective in tagging behaviors of Weibo users*. That is, Weibo friends tend to share similar tags. These results also justify the rationality of homophily-based tag recommendation, which is used as the basic scheme in our tag recommendation algorithm.

2.2   Co-Occurrence in Tagging Behavior

From our dataset, we found that many Weibo users have more than one real tag. It inspires us to use tag cooccurrence for tag recommendation. That means, if two tags t_1 and t_2 co-occur with each other in many persons' real tag lists and t_1 has been recommended to a user, then t_2 also deserves to be recommended to this user. Tag co-occurrence was shown to be an effective mechanism for tag recommendation for photos in Flickr [25]. Next, we first give the ranking scheme of tag t' that co-occurs with t, then we justify the co-occurrence based tag recommendation for Weibo users by empirical studies.

Ranking: For a tag t recommended to a user, we first need to measure *the extent to which we recommend another tag t' that co-occurs with t to the user*. We may directly measure it by t''s co-occurrence frequency with t, denoted as $tf_t(t')$. Thus, the direct implementation of co-occurrence based tag recommendation is recommending tag t' with largest $tf_t(t')$ if t is recommended. The direct solution clearly favors those general tags with high occurrence frequency, such as 'music' and 'movie'. We need to suppress them to select informative tags. We import an *idf* factor to reflect this requirement. As in [10,27], *idf* factor generally is defined as

$$idf(t') = log \frac{M - n(t') + 0.5}{n(t') + 0.5} \qquad (4)$$

Table 1. Co-occurrence tags ranked by tf-idf score

machine learning	tour	advertisement
data mining	food	media
NLP	movie	marketing
recommender sys.	fashion	communication
information retrieval	music	design
computer vision	listen to music	photography
pattern recognition	80s	Internet
A.I.	freedom	innovation
big data	travel	movie
search engine	photography	art
Internet	indoorsy	fashion

where $n(t')$ is the frequency of tag t''s co-occurrence with t. M is the user number of universal user set. Then, similar to TF-IDF in IR systems, we define a tf-idf score to measure the extent to which tag t' co-occurs with t as

$$s_t(t') = tf_t(t') \times idf(t') \tag{5}$$

Given this score function, we can enrich a tag list by homophily based recommendation.

Results: Next, we justify the co-occurrence expansion by case studies on three typical tags 'machine learning', 'tour' and 'advertisement', which are called as *parent tags* of their co-occurring tags. In Table 1, we list the top-10 tags ranked by $s_t(t')$ that co-occurs with the three parent tags. These tags are the candidates to enrich a recommended tag list and called as *expanded tags*. From the table, we can see that most expanded tags are semantically related but different from their parent tags. All these related tags often tend to be co-used by users, e.g., 'machine learning' is very related to 'data mining' and 'A.I.', 'design'ing an 'advertisement' needs 'innovation'. It is desirable to recommend these semantically different but related co-occurring tags so that the recommended tags are fully informative and expressive to characterize a user.

Moreover, we can also find some synonyms or near-synonym from the expanded tags. For example, in Table 1, 'travel' is very semantically close to 'tour', so does 'media' to 'communication'. We next quantify the extent to which synonyms occur in the expanded list. To do this, we first selected 1000 most frequently used tags as the parent tags. For each of them, we summarized the proportion of the (near-)synonym tags that occur in its expanded tag list. We will introduce our approach to distinguish (near-)synonym tags in Sec. 3.3. We reported the average proportions over all parent tags under different top-k expanded tags. The results are shown in Fig. 4 where (near-)synonym tags account for 15%~20% in the expanded list. It shows that most of expanded tags are meaningful. On the other hand, it also implies that we still need to remove the semantic redundancy caused by the (near-)synonyms. This problem can be solved by our CKG (Chinese knowledge graph) based approach that will be discussed in the next section.

3   Tag Recommendation Algorithm

In this section, we elaborate our tag recommendation algorithm which contains three major steps, as shown in Alg. 1. For a user u, our algorithm generates k recommended tags ordered by a ranking score. In the first step (line 2), we generate candidate tags by homophily based recommendation scheme. In the second step (line 5 to 9), we expand the tag list by the co-occurrence based recommendation scheme. In the third step (line 11 to 18), we remove all semantically redundant tags by a CKG based method.

3.1   Step 1: Recommendation by Homophily

According to the empirical results of Sec. 2.1, i.e., close social relationships imply similar tags, we can profile a Weibo user by his/er collective tags. This strategy can solve the data sparsity problem of Weibo tags. Recall Eq. 1, we directly collect the tags from u's friends, i.e., the direct neighbors of u, to constitute CT_u. This naive approach has two weaknesses. First, it will fail if no direct neighbors have real tags. Second, it does not take into account the intimacy between two friends. Next, we will improve it by taking into account indirect neighbors' information and user intimacies. We use tag propagation to materialize the effects of these factors. To better explain our algorithm, we first give some preliminary definitions.

Definition 3 (Weibo Influence Graph).   The Weibo influence graph $G(V, E, w)$ is an edge-weighted directed graph, where V is user set and E is influence edge set. Each directed edge $e_{u \to v}$ indicates the social influence from user u to user v. Furthermore, we assign a weight w_{uv} to this edge to quantifies the extent to which u can influence v through it. In general, a followee has much more influence on his/er follower than the vice versa that is indicated by Fig. 3. Hence, for a better interpretation of our algorithm, we assume that only followee can influence his/er followers resulting in tag propagation from followees to followers only. Specifically, if and only if user v follows u, there is an edge $e_{u \to v}$ in the influence graph. We further set w_{uv} as the frequency that v retweets u in a given period[3].

Based on the Weibo influence graph, we further define *social influence* which characterizes the intimacy between two Weibo users. It is similar to the influence proposed by Mashiach et al. for optimizing PageRank algorithm [2].

Definition 4 (Social Influence).   For a directed path $p = (u_0, u_1, ..., u_r)$ in G, the *social influence* along p from u_0 to u_r equals to

$$si(p) = \prod_{i=0}^{r-1} \frac{w_{u_i u_{i+1}}}{\sum\limits_{u:u \to u_{i+1}} w_{uu_{i+1}}} \qquad (6)$$

where u is u_{i+1}'s in-neighbor in G. Let $P_r(v, u)$ be the set of all paths of length r from v to u, thus the social influence of v on u at radius r is

$$si_r(v, u) = \sum_{p \in P_r(v,u)} si(p). \qquad (7)$$

Furthermore, we define $si_0(u, u) = 1$ and $si_0(v, u) = 0$ for all $v \neq u$. Then, the total social influence of v on u is $si(v, u) = \sum\limits_{r=0}^{\infty} si_r(v, u)$.

---

[3] The frequency of mention (@username) and comment can also be used to quantify the influence weight between Weibo users. Our experimental results show that the selection of weighting scheme does not affect the performance of our algorithm.

Algorithm 1. Tag recommendation algorithm with three steps

Input: a Weibo user u; parameter k, q, λ, α;
Output: recommended tag list;

1: $C \leftarrow \phi$;
2: compute $\boldsymbol{S}_u$; //Step1: recommendation by homophily.
3: $i \leftarrow 1$;
4: **while** $|C| < k$ **do**
5:     $k' \leftarrow k \times i$; // begin Step 2: expansion by co-occurrence.
6:     $C \leftarrow C \bigcup \{\text{top-}k' \text{ tags ranked by } s(t)\}$;
7:     **for** each tag t in C in the descending order of $s(t)$ **do**
8:         $C \leftarrow C \bigcup \{\text{top-}q \text{ tags ranked by } s_t(t_i)\}$;
9:     **end for**
10:    set all newly added tags' parents;
11:    Rank tags in C by $\hat{s}(t)$ defined in Eq. 10; //begin Step 3.
12:    **for** each tag t in C in the descending order of $\hat{s}(t)$ **do**
13:        **for** each tag t' ordered after t **do**
14:            **if** $sim(t, t') \geq \alpha$ **then**
15:                remove t' from C;
16:            **end if**
17:        **end for**
18:    **end for**
19:    $i \leftarrow i + 1$;
20: **end while**
21: **return** the top-k tags in C;

Computation: Suppose there are overall N tags in G, the first step of our algorithm aims to calculate a *tag score vector* $\boldsymbol{S}_u = [s(1), ..., s(N)] \in \mathbb{R}^N$ for a user u, in which $s(j)$ $(1 \leq j \leq N)$ quantifies the extent to which tag j can profile u, i.e., the ranking score of candidate tag j. To consider the influence of indirect neighbors, we let the tags of indirect neighbors propagate along the path in the influence graph. Intuitively, *if a user v has a more significant influence on u (i.e., larger $si(v, u)$), u will be more tending to use v's tags to profile him/erself.* To reflect these facts, we define:

$$\boldsymbol{S}_u = \sum_{v \in V} si(v, u) \boldsymbol{T}_v = \sum_{v \in V} \sum_{j=0}^{r} si_j(v, u) \boldsymbol{T}_v \tag{8}$$

where $\boldsymbol{T}_v \in \mathbb{R}^N$ is v's real tag distribution vector and its entry $t_j = 1/n (1 \leq j \leq N)$ if user v originally uses tag j, otherwise $t_j = 0$. n is the number of user v's real tags and $\sum t_j = 1$. Refer to Eq. 6 and Eq. 7, we can recursively compute the social influence of user v on user u at radius r as

$$si_r(v, u) = \sum_{x:v \to x} \frac{w_{vx}}{\sum_{v':v' \to x} w_{v'x}} si_{r-1}(x, u) \tag{9}$$

where x is v's out-neighbor who has a path of $r-1$ length to u at least, and v' is x's in-neighbor. That is, the social influence of v on u at radius r equals to the weighted average influence of v's out-neighbors on u at radius $r-1$. This implies

Algorithm 2. Step1: Computing u's tag score vector $\boldsymbol{S}_u$.

Input: u, r;
Output: $\boldsymbol{S}_u$;
 1: $\boldsymbol{S}_u \leftarrow \phi$;
 2: $layer_0 \leftarrow u$;
 3: $si_0(u, u) \leftarrow 1$;
 4: **if** u has origin tags **then**
 5:     $\boldsymbol{S}_u \leftarrow \boldsymbol{T}_u$;
 6: **end if**
 7: **for** $i=1$ to r **do**
 8:     $layer_i \leftarrow \{$all in-neighbors of the nodes in $layer_{i-1}\}$;
 9:     **for** $\forall v \in layer_i$ **do**
10:         **if** v has real tags **then**
11:             **for** each v's out-neighbor x **do**
12:                 $si_i(v, u) \leftarrow \sum\limits_{x:v \to x} \frac{w_{vx}}{\sum\limits_{v':v' \to x} w_{v'x}} si_{i-1}(x, u)$;
13:             **end for**
14:             $\boldsymbol{S}_u \leftarrow \boldsymbol{S}_u + si_i(v, u) \times \boldsymbol{T}_v$;
15:         **end if**
16:     **end for**
17: **end for**
18: **return** $\boldsymbol{S}_u$;

that we can compute $\boldsymbol{S}_u$ iteratively as shown in Alg. 2. The computation starts from u. In the i-th iteration (line 7 to 17), for each user v that is i steps away from u and have real tags, we calculates its social influence on u by summing up the weighted social influences on u of each v's out-neighbor x at radius $i - 1$.

Optimization: Next, we optimize above computation from two aspects.
1. *Setting A Shorter r.* Obviously, the computation cost of Eq. 8 is unbearable if r is big. Refer to the observations on Twitter that more than 95% of information diffusion is less than the scope of 2 hops from the origin [15], we can set $r \leq 2$ in the real applications. We will present how to learn this upper bound of r in the experiment section.
2. *Suppressing General Tags.* Similar to co-occurrence tag expansion, we should suppress the tags that are too generally used by all users in order to find the specific and informative tags. Therefore, we also import an idf factor matrix D into Eq. 8. That is replacing $\boldsymbol{T}_v$ with $\boldsymbol{T}_v D$, where $D = diag[d_1, ..., d_N]$ is an $N \times N$ diagonal matrix and each non-zero entry $d_j (1 \leq j \leq N)$ is defined as Eq. 4. After $\boldsymbol{S}_u$ is computed, we rank all tags according to $s(j)$ and then select the top-k tags as the candidate set, namely C, that will be fed as the input of Step 2. k is the number of tags to profile a user.

3.2   Step 2: Expansion by Co-Occurrence

We have shown in Sec. 2.2 that co-occurrence is also an important tag recommendation mechanism. Therefore, we use this mechanism to enrich the recommended tags. The input of this step is the ranked tag list C generated in Step 1. The output is a new ranked list consisting of C and other expanded tags.

In Step 2, for each tag $t \in C$, in order to generate its *expansion list*, we select the top-q co-occurring tags, namely t_i, according to $s_t(t_i)$ value (refer to Eq. 5). If a co-occurring tag t_i can be found in more than one expansion list, t_i will only join the expansion list of the tag t having the maximal $s(t)$. We refer to such t as t_i's *parent tag*, namely $p(t_i)$. Thus, for each t_i, $p(t_i)$ is unique. If an expanded tag has existed in C, we just ignore it. As a result, at most $k \times q$ new tags can be discovered. Let C' be the new candidate tag list after expansion. Thus, $C' - C$ contains all newly expanded tags.

Re-ranking: After we generated the new recommendation tag set C', we need to re-rank each member of C'. The key of the new ranking is to ensure that the tags in $C' - C$ can fairly compete with those tags in C. To meet this requirement, we define a new ranking score $\hat{s}(t_i)$ for each tag $t_i \in C'$:

$$\hat{s}(t_i) = \begin{cases} s(t_i) & t_i \in C; \\ \lambda \times s(p(t_i)) \times \frac{s_{p(t_i)}(t_i)}{Z} & \text{otherwise} \end{cases} \tag{10}$$

where $\lambda \in (0,1)$ is a damping parameter, Z is used for normalization and set as the maximal $s_t(t_i)$ of all t_is that co-occur with t. If t_i is one of the original tag found in Step 1 (i.e., $t_i \in C$), we directly use the $s(t_i)$ as its new score. Otherwise, we inherit the score from $p(t_i)$'s ranking score $s(p(t_i))$ generated in Step 1 and use λ and $\frac{s_{p(t_i)}(t_i)}{Z}$ as two multiplicators to suppress it ($s_{p(t_i)}(t_i)$ is also defined according to Eq. 5). Since $p(t_i)$ is unique, $\hat{s}(t_i)$ is well defined.

The rationality of the new score is two-fold:
1. $\hat{s}(t_i)$ should be smaller than $s(p(t_i))$. The definition can ensure this because $\lambda \in (0,1)$ and $\frac{s_{p(t_i)}(t_i)}{Z} \in (0,1]$. On the other hand, to ensure t_i is competitive enough, we usually set $\lambda \geq 0.5$.
2. For any two tags t_i, t_j in one tag t's expansion list, $\hat{s}(t_i) < \hat{s}(t_j)$ should hold if $s_t(t_i) < s_t(t_j)$. It is not difficult to prove that $\hat{s}(t_i)$ satisfies the requirement.

3.3   Step 3: Removing Semantic Redundancy

As pointed out in [8], users often tag the same resource with different terms for their various habits or recognition. Similarly, Weibo users may use different terms to express the same or close semantics. As a result, many synonyms or near-synonyms tend to exist in Weibo tags. For example, tag 'tour' and 'travel' are both widely used in Weibo. Thus, the candidate tag set may have some tags of the same or similar semantics. These tags are redundant and should be avoided due to space limitation of a Weibo user's tags. For this purpose, we first construct a *Chinese Knowledge Graph* (CKG in short) and then use an *Explicit Semantic Analysis* (ESA in short) [7] based model to represent a tag's semantics through the concepts in CKG.

The CKG is a big graph constituted by millions of concepts and entities extracted from online encyclopedias such as Baike[4]. Each concept can be

---

[4] http://baike.baidu.com

classified into one or more categories and there exists a unique Web article to explain it. In each Web article, there are many hyperlinks referring to other concepts, namely *reference concept*. These hyperlinks constitute the edges of CKG (refer to Fig. 2). A concept can be referred to by more than one article. As well, a concept can also be referred to more than once in an article. Thus, for a reference concept, we can use the concepts whose articles refer to it, to represent its semantics. The number of referring also allows us to calculate a tf-idf score to select expressive concepts.

Based on above idea, we can quantify the semantics of a Weibo tag by first mapping it into Baike concepts, i.e., the concepts in CKG. Specifically, given a tag a and a Baike concept b, we map a to b if $s_a = s_b$ or s_b is the maximal substring of s_a, where s_a and s_b are the name strings of a and b, respectively. Under this mapping scheme, we can find an appropriate Baike concept for 88.7% of Weibo tags.

According to ESA, two tags are considered semantically related if their mapped concepts are co-referred to in the same article pages of CKG. The more such articles can be found, the more semantically related the two tags are. Based on it, we first formalize a tag's semantic representation as follows. Suppose CKG has L concepts in total, the semantic interpretation of a tag i can then be represented by a *concept vector* defined as $\boldsymbol{C} \in \mathbb{R}^L$ of which each entry, namely c_j, represents the semantic relatedness of concept j to tag i. c_j can be calculated as the tf-idf score of tag i in concept j's article. We notice that many concepts in CKG are quite general and cover a wide range of topics. These concepts in general have less semantic descriptiveness on a tag than those specific concepts. Hence we need to suppress these general tags. Intuitively, the concepts belonging to more categories are more general than the concepts belonging to less categories. Consequently, we further define

$$c_j = \frac{ts_j(i)}{|cat(j)|} \tag{11}$$

to punish general tags, where $ts_j(i)$ is the tf-idf score and $cat(j)$ is concept j's category set.

Then, given two tags i and j, we can measure their semantic similarity by computing the cosine similarity of $\boldsymbol{C}_i$ and $\boldsymbol{C}_j$, i.e., $sim(i,j) = cosine(\boldsymbol{C}_i, \boldsymbol{C}_j)$. According to the definition of concept vector, the larger the $sim(i,j)$ is, the more possible that i and j are (near-)synonyms. The detailed procedure of removing semantically redundant tags is shown in Alg. 1. We first sort the tags in C by the descending order of $\hat{s}(t)$ value. For each tag t in the ordered list, we start an inner loop to scan each tag t' ordered after t. If $sim(t,t')$ is larger than a threshold α, we remove t' from C. Finally, if C contains more than k tags, we just return the top-k tags ranked by $\hat{s}(\cdot)$ function (refer to Eq. 10).

3.4   Parameter Learning

There are several parameters in our tag recommendation algorithm, i.e., r in Step 1, q and λ in Step 2 and α in Step 3. In this subsection, we introduce how to set the best parameter values.

To find the best α, we used the synsets in Cilin[5] (a popular Chinese synonym database) as positive samples and manually labeled non-synonym paris as the negative sample. We use these samples as the training dataset to train a binary classification model. Then we found that $\alpha=0.007$ is the most effective threshold for distinguishing (near-)synonyms.

Next, we introduce how to learn the best value for q, r and λ. We first introduce how to evaluate the goodness of a recommended tag set. For a user with real tags, we can take his/er real tags as the ground truth. We can compare the recommended tags to the ground truth for the evaluation. In Sec. 2.1, we use the results of *exact match* for the comparison. But it is too strict for tag recommendation. For example, it is reasonable to recommend 'tour' to a user with a tag of 'travel' although the two tags are lexically different. To relax the match, we use the aforementioned cosine similarity between concept vectors to measure the match between two tag sets.

More formally, suppose our algorithm of the parameter setting θ recommends u with a tag set, namely $T(u, \theta)$. Let $\boldsymbol{C}_u(\theta)$ be the concept vector of $T(u, \theta)$'s. According to Eq. 11, each entry of $\boldsymbol{C}_u(\theta)$, namely c_j, can be defined as

$$c_j = \sum_{t \in T(u, \theta)} \frac{ts_j(t)}{|cat(j)|} \times \hat{s}(t, \theta) \tag{12}$$

where t is a tag in $T(u, \theta)$ and $\hat{s}(t, \theta)$ is t's score derived by our algorithm under the setting θ. For computing u's real tag set RT_u's concept vector, namely $\bar{\boldsymbol{C}}_u$, we set $\hat{s}(t) = 1/|RT_u|$ because we can not acquire u's extent to which s/he prefers to a real tag. Then, we propose an objective function $\mathcal{F}$ to measure the semantic similarity between $T(u, \theta)$ and RT_u as

$$\mathcal{F}(u, \theta) = sim(T(u, \theta), RT_u) = cosine(\boldsymbol{C}_u(\theta), \bar{\boldsymbol{C}}_u).$$

Thus, the best parameter setting (including q, r and λ) should be

$$\arg \max_{\theta \in \Theta} \mathbb{E}(\mathcal{F}(u, \theta)) = \arg \max_{\theta \in \Theta} \frac{\sum_{u \in U} \mathcal{F}(u, \theta)}{|U|}. \tag{13}$$

U is the training user set consisting of the seed users having real tags in our Weibo dataset. Finally, we found that $q = 1, r = 2, \lambda = 0.5$ are the best θ in our tag recommendation algorithm.

4   Evaluation

In this section, we evaluate the performance of our tag recommendation algorithm through the comparisons with some state-of-the-art methods. We not only present the match performance of our recommendation algorithm, but also display the effectiveness of the recommended tags on inferring user profiles.

---

[5] http://www.datatang.com/data/42306/

4.1 Experimental Settings

We first introduce the evaluation method and the competitors of our algorithm.
Human Assessments: One direct way to assess the recommended tags is comparing them with the real tags since the real tags are each Weibo user's preferences. However, nearly half of Weibo users have no real tags. So we have to resort to human assessments for evaluating the recommended tags. Specifically, we inquired each test Weibo users whether s/he will accept the recommended tags. Each user can select an option of *yes*, *no* and *unknown* for a tag. We only take the tag of *yes* as matched tag.
Baselines:
1. FREQ.: The first baseline is a naive method because it selects the recommended tags merely by ranking the frequency of candidate tags used by a user's followees, i.e., collective tags.
2. TF-IDF: This baseline recommends the tags according to the TF-IDF scheme.
3. CF: The CF approach has been proposed in [25] to recommend tags for a Flickr image based on tag co-occurrence mining. That is, for a user with real tags, we recommend to him/her with some tags that are co-used with his/er own tags by many other users. In fact, this method can be viewed as an item-based collaborative filtering approach when we regard a tag as an item and the tags co-used by a user as similar or related items. Clearly, this recommendation method can not be applied for the users without real tags.
4. TWEET: This approach is a content-based recommendation scheme which has been widely used in previous recommender systems [6,9]. This approach extracts some keywords from a user's tweets as the recommended tags since a user's tweets are direct indicators of users interests or preferences.

In our algorithm, the tags are generated from local neighbors within radius 2 ($r=2$). Hence, we name our algorithm as *Local Tag Propagation Algorithm* (LTPA in short). Besides r, the parameters q and λ of our algorithm were also set as the best values tuned by corresponding learning models (see Sec. 3.4) in the experiments.

4.2 Effectiveness

We justify our algorithm's effectiveness from two aspects. We first present the global match performance of our tag recommendation algorithm by comparing to the baselines. Then we justify the effectiveness of each step of our algorithm.

Global Performance: From the 3000 seed users in our dataset, we randomly selected 500 users as the test users in our experiments in which the spam users were excluded. Then, we designed two groups of experiments to recommend tags to these test users. In the first group, we compared all recommendation algorithms on the 268 test users having real tags since CF can only work on the users with tags. In the second group, we compared all competitors except for CF on all 500 test users. The human assessment results are shown in

Fig. 5 and Fig. 6, respectively. The results show that all algorithms perform the best when recommending top-5 tags. It proves that the algorithms can rank the best tags to a top position. The results also reveal that LTPA performs the best in all cases. The superiority of LTPA and TF-IDF over FREQ. justifies that the effectiveness of *idf* factor to discover informative and personalized tags for profiling a user. We will illustrate it by case studies in the next subsection. TWEET almost performs the worst in all cases, implying that *the keywords directly extracted from tweets are generally not appropriate for user profiling.* Further investigation on the tweet content reveals that, most tweet keywords are *colloquialisms* or *person name* of friends and newsmakers. For example, 'Diaos' is a new Internet vocabulary and is widely used by Chinese youngsters. These words produced due to the oral and informal language style in short tweets (less than 140 characters) can not accurately and completely characterize a user.

Effectiveness of Each Step: In Sec. 2.1, we have justified the rationality of homophily based recommendation. Next, we present the recommendation results after we add Step 2 and Step 3 incrementally into our algorithm to justify the co-occurrence based expansion and removing semantic redundancy. Since our algorithm has the best performance when $k=5$, we only evaluated our algorithm by recommending top-5 tags to the test users.

Step 2: To justify Step 2, we investigated the expanded tags generated by our algorithm consisting of Step1 and Step 2. We found that 75.11% of the expanded

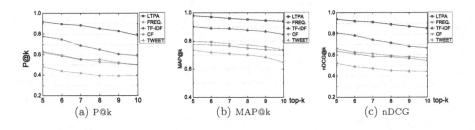

Fig. 5. Human assessment results of the recommended tags to the test users having real tags

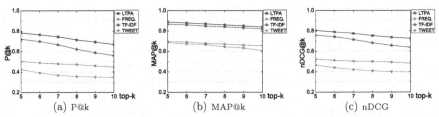

Fig. 6. Human assessment results of the recommended tags to all the test users

tags are newly discovered tags. In average, about 35.37% of these newly expanded tags were labeled as matched by the volunteers. These results imply that co-occurrence based expansion is necessary and effective in enriching the recommended tag list.

Step 3: Then, we ran the whole algorithm consisting of the three steps. We found that 14.55% of the tags after Step 2 were identified as (near-)synonyms of the previous tags. By surveying the volunteers' acceptance about the removed tags, we found that 74.7% of these (near-)synonyms identified by Step 3 are really redundant. These results justify the effectiveness of removing semantic redundancy.

Table 2. Inference accuracy in four profile categories

category	accuracy
location	94.64%
occupation	76.47%
education	95.24%
religion	99.21%

Table 3. Case studies to justify inference performance of the recommended tags

user	algorithm	tag list
userA	real tags	music, fashion
	CF	movie, food, listen to music, tour, 80s
	TWEET	Jehovah, Miss HongKong, beauty, child, good man
	FREQ.	Christian, food, movie, 80s, tour
	TF-IDF	Christian, Bible, Emmanuel, micro fashion, tide
	LTPA	Christian, Bible, faith, God's baby girl, God's child
userB	TWEET	Shantou (a Chinese city), WeChat, Internet, Shantou people, girl
	FREQ.	tour, food, movie, Internet, music
	TF-IDF	machine learning, Internet, data mining, Fudan University, technology
	LTPA	machine learning, IT, Internet, Fudan University, data mining

4.3   Inference of User Profiles

Many users are reluctant to publish their profiles, i.e., location, professions and religion, possibly due to the privacy concern. Hence, accurately inferring user profiles is very important for better understanding the users who have no tags or no informative tags. For the users who do not introduce themselves completely, the recommended tags can be used to infer the absent user characteristics. Identifying user profile characteristics can contribute to many real applications such as maintaining social cliques, search for target user and etc. To test the performance of inferring user profiles, we ran our algorithm on the test users to recommend top-5 tags. Then, we filtered out the test users whose recommended tags contain profile information and evaluated inference accuracy by inquiring the users. Table 2 lists the inference accuracy of tags generated by our algorithm w.r.t. four basic profile information: location, profession, education and religion. The results verify that our algorithm is effective on inferring user profiles.

Case Studies: Finally, we give two case studies to highlight our algorithm's effectiveness on recommending personalized and informative tags to enrich a user's profile.

Case 1: User A in Table 3 is a test user who has real tags. We can see that user A's real tags uncover nothing about her religion. CF can not recommend any tags indicating her religion either. In TWEET and FREQ., there is only one word, i.e., 'Jehovah' and 'Christian', implying user A's religion (Christianism). In contrast, TF-IDF and LTPA can recommend more than one tag that apparently reveal user A's religion. It is because these two algorithms can find more personalized and informative tags through idf factor. By investigating user A's tweets, we confirmed that she is really a Christian.

Case 2: Another test user B has no real tags. As a result, CF can not be applied on this user. From Table 3, we find that FREQ. only reveals general interests of youngsters. TWEET can only find keywords about his hometown ('Shantou'). In contrast, TF-IDF and LTPA can recommend more personalized and informative tags. From these tags we can confidently infer that user B is a university student (the university name is anonymous for blind review) who is interested in machine learning and data mining. In fact, user B is a student volunteer in the data mining laboratory of Fudan University.

5   Related Work

Tag Recommendation and Social Recommender: Most previous works of tag recommendation were employed on a triplet basis, i.e., user, tag and resource [11,29,18,13], instead of tagging a user. Xu et al. [29] proposed a set of criteria for a high quality tag. Based on these criteria, they further proposed a collaborative tag suggestion algorithm to discover the high-quality tags. Song et al. [26] recommended tags for a document according to the mutual information between words, documents and tags. In addition, Sigurbjornsson et al. [25] presented some recommendation strategies based on tag co-occurrence. Liu et al. [17] introduced a tag ranking scheme to automatically rank the tags associated with a given image according to tag relevance to the image content. All these methods were designed on the premise that each tagged object already has tags resulting in vulnerability to data sparsity towards tagging Weibo users. Similar to Step 1 in our algorithm, many scholars tried to improve recommendation performance by exploiting social context. These systems are generally called *social recommender* [3]. Ma et al. [18] used social relationships to solve the cold start problem of CF, but they mainly focused on rating objects instead of persons. Ben-Shimon et al. [4] explicitly quantified user similarity by computing their distances in the social graph without considering personality. Quijano-Sanchez et al. [22] resorted to a TKI survey upon users to acquire personality values which is not feasible to real on-line applications. Hotho et al. [13] also used a PageRank-based model to rank tags but they did not consider the semantic redundancy of tags.

Tag Semantics: One of the prerequisites to study the user tagging behavior is understanding the semantic of tags [1]. In general, to understand tag semantics, tags should be mapped into a thesauri or a knowledge base. E.g., mapping Flickr tags [25] and Del.icio.us tags [5] into WordNet, or mapping tags into Wikipedia categories by the content of tag-associated objects [21]. Moreover, some meta graphs are also constructed for understanding tags, such as a tag graph encoding co-occurrence relationships among tags [30,17]. Given that the low tag coverage of WordNet, we resort to Wikipedia-like encyclopedia, i.e., CKG in this paper. Furthermore, we improve ESA [7] by taking into account the categories of CKG concepts to improve the precision of a tag's semantic interpretation.

User Profile Inference: Sadilek et al. [23] presented a system to infer user locations and social ties between users. Mislove et al. [20] tried to infer user profile based on the open characteristics of a fraction of users. These mechanisms are not as flexible as our approach because they only work under the assumption that characteristics of some users have been uncovered in advance. The authors in [16] proposed an influence based model to infer home locations of Twitter users. Although their work also resorted to social relationships for an accurate inference, their model can only be used to infer location and needs expensive analysis of tremendous tweets. In contrast, our solution mainly depends on crawling and analyzing tags that are less costly than processing on tweets.

6   Conclusion

Motivated by many real applications built upon user profiles, we dedicate our efforts in this paper to tag recommendation for Weibo users. We conducted extensive empirical studies to unveil effective tag recommendation scheme based on which we proposed an integrated tag recommendation algorithm consisting of three steps, i.e., tag recommendation based on local tag propagation, tag expansion by co-occurrence and CKG-based elimination of semantically redundant tags. Extensive experiments validate that our algorithm can recommend more personalized and informative tags for profiling Weibo users than the state-of-the-art baselines.

References

1. Ames, M., Naaman, M.: Why we tag, motivations for annotation in mobile and online media. In: Proc. of CHI (2007)
2. Bar-Yossef, Z., Mashiach, L.T.: Local approximation of pagerank and reverse pagerank. In: Proc. of CIKM (2008)

3. Bellogin, A., Cantador, I., Diez, F., Castells, P., Chavarriaga, E.: An empirical comparison of social, collaborative filtering, and hybrid recommenders. ACM Transactions on Intelligent Systems and Technology **4** (2013)
4. Ben-Shimon, D., Tsikinovsky, A., Rokach, L., Meisles, A., Shani, G., Naamani, L.: Recommender system from personal social networks. In: Proc. of AWIC (2007)
5. Cattuto, C., Benz, D., Hotho, A., Stumme, G.: Semantic grounding of tag relatedness in social bookmarking systems (2008)
6. Chen, J., Geyer, W., Dugan, C., Muller, M., Guy, I.: Make new friends but keep the old, recommending people on social networking sites. In: Proc. of CHI (2009)
7. Gabrilovich, E., Markovitch, S.: Computing semantic relatedness using wikipedia-based explicit semantic analysis. In: Proc. of IJCAI (2007)
8. Gupta, M., Li, R., Yin, Z., Han, J.: Survey on social tagging techniques. In: Proc. of SIGKDD (2010)
9. Hannon, J., Bennett, M., Smyth, B.: Recommending twitter users to follow using content and collaborative filtering approaches. In: Proc. of RecSys (2010)
10. Hassanzadeh, O., Consens, M.: Linked movie data base. In: Proc. of LDOW (2009)
11. Heymann, P., Ramage, D., Garcia-Molina, H.: Social tag prediction. In: Proc. of SIGIR (2008)
12. Hofmann, T.: Latent semantic models for collaborative filtering. ACM Transactions on Information Systems **2**, 89–115 (2004)
13. Hotho, A., Jäschke, R., Schmitz, C., Stumme, G.: Information retrieval in folksonomies: search and ranking. In: Sure, Y., Domingue, J. (eds.) ESWC 2006. LNCS, vol. 4011, pp. 411–426. Springer, Heidelberg (2006)
14. Jarvelin, K., Kekalainen, J.: Cumulated gain-based evaluation of ir techniques. ACM Transactions on Information Systems **20**, 422–446 (2002)
15. Kwak, H., Lee, C., Park, H., Moon, S.: What is twitter, a social network or a news media? In: Proc. of WWW (2010)
16. Li, R., Wang, S., Deng, H., Wang, R., Chang, K.C.C.: Towards social user profiling: unified and discriminative influence model for inferring home locations. In: Proc. of SIGKDD (2012)
17. Liu, D., Hua, X.S., Yang, L., Wang, M., Zhang, H.J.: Tag ranking. In: Proc. of WWW (2009)
18. Ma, H., Yang, H., Lyu, M.R., King, I.: Sorec: social recommendation using probabilistic matrix factorization. In: Proc. of CIKM (2008)
19. McPherson, M., Smith-Lovin, L., Cook, J.: Birds of a feather: Homophily in social networks. Annual Review of Sociology **27**, 415–445 (2001)
20. Mislove, A., Viswanath, B., Gummadi, K.P., Druschel, P.: You are who you know: inferring user profiles in online social networks. In: Proc. of WSDM (2010)
21. Overell, S., Sigurbjornsson, B., van Zwol, R.: Classifying tags using open content resources. In: Proc. of WSDM (2009)
22. Quijano-Sanchez, L., Recio-Garcia, J.A., Diaz-Agudo, B., Jimenez-Diaz, G.: Social factors in group recommender systems. ACM Trans. on Intelligent Systems and Technology **4** (2013)
23. Sadilek, A., Kautz, H., Bigham, J.P.: Finding your friends and following them to where you are. In: Proc. of WSDM (2012)
24. Schafer, J., Konstan, J., Riedi, J.: Recommender systems in e-commerce. In: Proc. of EC (1999)
25. Sigurbjornsson, B., van Zwol, R.: Flickr tag recommendation based on collective knowledge. In: Proc. of WWW (2008)
26. Song, Y., Zhuang, Z., Li, H., Zhao, Q., Li, J., Lee, W.C., Giles, C.L.: Real-time automatic tag recommendation. In: Proc. of SIGIR (2008)

27. Wang, J., Hong, L., Davison, B.D.: Tag recommendation using keywords and association rules. In: Proc. of RSDC (2009)
28. Weng, J., Lim, E.P., Jiang, J., He, Q.: Twitterrank: finding topic-sensitive influential twitterers. In: Proc. of WSDM (2010)
29. Xu, Z., Fu, Y., Mao, J., Su, D.: Towards the semantic web: collaborative tag suggestions. In: Proc. of Collaborative Web Tagging Workshop in WWW (2006)
30. Zhou, T.C., Ma, H., Lyu, M.R., King, I.: Userrec: a user recommendation framework in social tagging systems. In: Proc. of AAAI (2010)

An Efficient Approach of Overlapping Communities Search

Jing Shan[✉], Derong Shen, Tiezheng Nie, Yue Kou, and Ge Yu

College of Information Science and Engineering, Northeastern University,
Liaoning 110004, China
mavisshan0129@gmail.com,
{shenderong,nietiezheng,kouyue,yuge}@ise.neu.edu.cn

Abstract. A great deal of research has been dedicated to discover overlapping communities, as in most real life networks such as social networks and biology networks, a node often involves in multiple overlapping communities. However, most work has focused on community detection, which takes the whole graph as input and derives all communities at one time. Community detection can only be used in offline analysis of networks and it is quite costly, not flexible and can not support dynamically evolving networks. Online community search which only finds overlapping communities containing given nodes is a flexible and light-weight solution, and also supports dynamic graphs very well. Thus, in this paper, we study an efficient solution for overlapping community search problem. We propose an exact algorithm whose performance is highly improved by considering boundary node limitation and avoiding duplicate computations of multiple input nodes, and we also propose three approximate strategies which trade off the efficiency and quality, and can be adopted in different requirements. Comprehensive experiments are conducted and demonstrate the efficiency and quality of the proposed algorithms.

1   Introduction

Community structure [9] is observed commonly existing in networks such as social media and biology. Nodes in one community are more highly connected with each other than with the rest of the network. Thus, community structure can provide rich information of the network. For example, community in social media reflects a group of people who interact with each other more frequently, so they may have common interest or background; community in protein-association network reflects a group of proteins perform one common cellular task. Therefore, finding communities is crucial for understanding the structural and functional properties of networks.

However, communities are often not separated in most real networks, they are often overlapped. In other words, one node often belongs to multiple communities. This phenomenon could be easily explained in social media: individuals could belong to numerous communities related to their social activities, hobbies, friends and so on. Thus, overlapping community detection (OCD) [13,10,8,11]

© Springer International Publishing Switzerland 2015
M. Renz et al. (Eds.): DASFAA 2015, Part I, LNCS 9049, pp. 374–388, 2015.
DOI: 10.1007/978-3-319-18120-2_22

has drawn a lot of attention in recent years. OCD dedicates to find all overlapping communities of the entire network, which has some shortcomings in some applications: First, it is time consuming when a network is quite large. Second, OCD uses a global criterion to find communities for all nodes in a network, which is inappropriate when the density of the network distributes quite unevenly. Third, OCD can not support dynamically evolving graphs, which is a typical characteristic for most real networks especially social network. Due to these reasons, overlapping community search (OCS) problem was proposed by Cui et al. [7].

OCS finds overlapping communities that a specific node belongs to. Thus, to support online query, OCS only needs to specify the query node and discover communities locally. Hence, OCS is light-weight, flexible, and can support dynamically evolving networks. In [7], an algorithm of finding overlapping communities of one query node was proposed, but it still has a large room for performance improvement. Besides, in some scenarios, the OCS query includes a set of nodes. For example, suppose a piece of news published on social network was read by a group of people, the service provider wants push the news to user communities in which people will be also interested in this news; or a product has been bought by a group of customers, and the producer wants to investigate the consumer groups in which people will also buy the product. In these scenarios, simply iterating the OCS algorithm for each query node could waste many computations and affect the efficiency. To this end, in this paper we propose an efficient approach for overlapping community search which not only highly improves the performance of single-node overlapping community search, but also includes an efficient framework for multiple-node query. In summary, we make the following contributions:

- We introduce the definition of boundary node, and use boundary node limitation to highly improve the performance of single-node overlapping community search algorithm.
- We propose a framework for multiple-node overlapping community search and try our best to avoid waste computations by strongly theoretical supports.
- We also propose a series of approximate strategies which trade off the efficiency and quality to suit different requirements.
- We conduct comprehensive experiments on real networks to demonstrate the efficiency and effectiveness of our algorithms and theories.

The rest of this paper is organized as follows. In Section 2 we review the related work. We formalize our problem in Section 3. In Section 4 we introduce both exact and approximate algorithms. We present our experimental results in Section 5. Finally, Section 6 concludes the paper.

2   Related Work

Our work is related to overlapping community detection problem, and local community detection problem, which can be also called community search problem.

Palla et al. first addressed overlapping community structure existing in most real networks [13], they proposed a clique percolation method (CPM), in which a community was defined as a k-clique component. Based on CPM, they developed a tool CFinder [1] to detect communities on biology networks. Besides structure based method like CPM, overlapping community detection could also be modeled as link-partition problem [8,2,11]. It first converts the original graph G into link graph $L(G)$, in which each node is a link of $L(G)$, and two nodes are adjacent if the two links they represent have common node in G. Then link partition of G can be mapped to node partition of $L(G)$, and by performing random walk [8], Jaccard-type similarity computation [2], or density-based clustering algorithm SCAN [11], node clusters of $L(G)$ are derived and then they can be converted to overlapping node communities of G. Label propagation method has been also widely used for OCD [10] [16], they propagate all nodes' labels to their neighbors for one step to make their community membership reach a consensus. Compared to OCD, OCS is more light-weight and flexible, it only needs to explore a part of the graph around query nodes, but not the whole graph, thus it is more appropriate for online query.

Considering the scalability problem of community detection, local community detection problem, also called community search, has also received a lot of attention [6,12,14,5]. These methods start from a seed node or node set, and then attach adjacent nodes to community as long as these nodes can increase some community quality metrics such as local modularity [6], subgraph modularity [12], or node outwardness [3]. In [15], the community is searched in an opposite way: they take the entire graph as input and delete one node which violates the condition such as minimum degree at each step, the procedure iterates until the query nodes are no longer connected or one of the query nodes has the minimum value of the condition. Although these community search methods are more flexible than OCD, none of these methods can discover overlapping communities, they can just find one community.

Our work is inspired by Cui et al [7], they proposed online overlapping community search problem. They defined a community as a k-clique component, and an algorithm which finds overlapping communities a given node belongs to was given. However, the algorithm of OCS still has a large room for performance improvement, and also, they did not consider the solution of overlapping community search for multiple nodes. Although simply iterating the algorithm in [7] could solve the problem, this method could produce a lot of waste computations and it is not an effective solution. Thus, we propose an efficient approach for OCS, considering both single query node and multiple query nodes situations.

3   Problem Definition

In this section, we define the problem of overlapping community search more formally, including OCS for both single query node and multiple query nodes.

Intuitively, a typical member of a community is linked with many but not necessarily all other nodes in the community, so we use k-clique as building

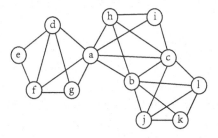

Fig. 1. A toy social network graph

blocks of a community to reflect this characteristic, just as Palla et al. [13]. Given a graph, we can derive a *k-clique graph* in which each node represents a *k*-clique, and if two cliques share $k - 1$ nodes, there exists an edge between them. A *community* is defined as a *k*-clique component, which is a union of all *k*-cliques sharing $k - 1$ nodes.

However, the definition of community above is too strict. Therefore, Cui et al. [7] proposed a less strict definition: two *k*-cliques are adjacent if they share at least α nodes, where $\alpha \leq k - 1$; and *k*-clique can be replaced by *γ-quasi-k-clique* [4] in which *k* nodes have at least $\lceil \gamma \frac{k(k-1)}{2} \rceil$ edges. Now, we give the problem definitions of OCS for single query node and multiple query nodes.

Problem 1 ((α, γ)-OCS). For a graph G, a query node v_0 and a positive integer k, the (α, γ)-OCS problem finds all *γ*-quasi-*k*-clique components containing v_0 and two *γ*-quasi-*k*-clique nodes of one component are adjacent if they share at least α nodes, where $\alpha \leq k - 1$.

Problem 2 ((α, γ)-OCS-M). For a graph G, a set of query nodes V_q and a positive integer k, the (α, γ)-OCS-M problem finds all *γ*-quasi-*k*-clique components containing at least one node in V_q and and two *γ*-quasi-*k*-clique nodes of one component are adjacent if they share at least α nodes, where $\alpha \leq k - 1$.

Example 1 ((α, γ)-OCS-M). For the graph in Fig. 1, given a set of query nodes $V_q = \{a,b,c\}$, let $k = 4$, consider $(3, 1)$-OCS-M, we get three communities $\{a, d, g, f\}$, $\{a, b, c, h, i\}$, $\{b, c, j, k, l\}$.

Apparently, both OCS problem and OCS-M problem are NP-hard, because they can be reduced from *k*-clique problem.

4   Overlapping Community Search Algorithms

We first propose a naive algorithm derived from OCS to solve OCS-M, then we propose optimized OCS and OCS-M algorithms based on a series of theorems with strong proofs. At last, we propose a series of approximate strategies to make the process more efficient.

4.1   Naive Algorithm

OCS algorithm searches overlapping communities of one single input node, intuitively, when given a set of nodes as input, we could iterate the OCS algorithm for each node, hence we get the naive algorithm of OCS-M as depicted in Alg. 1. For each node v_i in V_q, we first find a clique containing v_i, and then find the clique component which the clique belongs to. Notice that for a clique component (i.e. a community), we derived the same component no matter which clique it starts from. Thus, to avoid redundant enumeration, we only enumerate unvisited cliques for each round of iteration.

Algorithm 1. Naive OCS-M

Input: $G(V,E)$, V_q, α, γ, k;
Output: The overlapping communities containing $\forall v_i \in V_q$

1   $\mathcal{R} \leftarrow \emptyset$;
2   **foreach** $v_i \in V_q$ **do**
3 | **while** $C \leftarrow next\_clique(v_i), C \neq \emptyset$ **do**
4 | | $\mathbf{C} \leftarrow expand(C)$;
    | | // find the clique component $\mathbf{C}$ of C
5 | | $\mathcal{R} \leftarrow \mathcal{R} \cup \mathbf{C}$;

6   Return $\mathcal{R}$;

We adopt the same depth-first-search strategy for $next\_clique()$ and $expand()$ as in [7]. We omit the details and refer readers who are interested to [7]. Consider Example 1, when the input node is a, $next\_clique(a)$ may first return *Clique adfg*, and then $expand()$ on this clique gets *Community* $\{a,d,f,g\}$, the next call of $next\_clique(a)$ may return $abch$, and $expand()$ gets $acih$, thus we get *Community* $\{a,b,c,i,h\}$. Further call of $next\_clique(a)$ will not return any more cliques because all cliques containing a have been visited. When input node comes to b, we get $bclj$, and calling $expand()$ brings us $blkj$, thus we get *Community* $\{b,c,l,k,j\}$. Consider input node c, no further result is derived because all cliques containing c have been visited, and the procedure terminates.

4.2   Optimized OCS Algorithm

Though the DFS procedures of $next\_clique()$ and $expand()$ are pruned by checking the edge number of subgraph induced by current visiting node set in [7], we find a way which could further prune nodes to be checked. To optimize OCS algorithm, we first introduce two definitions, interior node and boundary node.

Definition 1 (Interior Node). *In the process of searching community* $\mathbf{C_m}$, *given a node i, if i and all its neighbors $neighbor(i)$ both exist in the currently found result set of* $\mathbf{C_m}$, *we say i is an interior node.*

Definition 2 (Boundary Node). *In the process of searching community* $\mathbf{C_m}$*, given a node b, if b exists in the result set of* $\mathbf{C_m}$*, and one or more neighbors of b do not exist in the result set, we say b is a boundary node.*

Nodes which are not interior nodes or boundary nodes are called exterior nodes. Considering the three types of node definitions, we propose a theorem which could be used to optimize the OCS algorithm.

Theorem 1. *For OCS algorithm, one community can be derived by only expanding boundary nodes or exterior nodes without losing completeness.*

Proof. At the beginning of searching community $\mathbf{C_m}$, every node is exterior node, thus a clique containing the query node can be found. In the procedure of expanding the clique, suppose set R is the result set including nodes of $\mathbf{C_m}$ that have already been found, node i is an interior node, if there exists an exterior node $n \in \mathbf{C_m} - R$, and it can be added into R from i, there must exist a clique C_l including i and n. Because n must be connected to at least one node b in $C_l - n$, thus the node b is a boundary node, and n can be added into $\mathbf{C_m}$ from b, therefore the theorem holds. $\square$

Algorithm 2. optimized next_clique(v_0)

Input: v_0: a query node
Output: C: next γ-quasi-k-clique
1 $U \leftarrow \{v_0\}$;
2 DFS (U, v_0);
3 **Procedure** DFS(U, u)
4 | **if** $|U| = k$ **then**
5 | | **if** U *is a γ-quasi-k-clique* **and** U *is unvisited* **then**
6 | | | **return** U;
7 | **else**
8 | | | **return;**
9 | **if** $g(U) < \gamma\frac{k(k-1)}{2}$ **then**
10 | | **return;**
11 | **foreach** $(u, v) \in E, v \notin U$ **do**
12 | | **if** $neighbor(v) \geq \lceil \gamma\binom{k}{2} - \binom{k-1}{2} \rceil$ **then**
13 | | | DFS$(U \cup \{v\}, v)$

According to Thm. 1, when expanding the current clique, if candidate nodes which are used to replace the current clique nodes are interior nodes, these nodes can be skipped. Besides, node degree could be taken into consideration as a pruning condition. For a γ-quasi-k-clique, the minimum degree of a node should be $\lceil \gamma\binom{k}{2} - \binom{k-1}{2} \rceil$. Base on the definition of community, if one node has less than $\lceil \gamma \binom{k}{2} - \binom{k-1}{2} \rceil$ edges, it is impossible to belong to a community. When $\gamma = 1$, the node should have at least $k - 1$ edges. Thus, utilizing interior node and node degree as pruning conditions, we could optimize $next\_clique()$ and $expand()$ of OCS as depicted in Alg. 2 and Alg. 3.

As mentioned before, we use DFS strategy to traverse nodes from the query node. Traversed nodes are iteratively added into set U, and check if a new valid clique is found (Alg.2 line 4-8), we use node degree condition to prune nodes (line 12), $neighbor(v)$ represents the degree of v. Besides, $g(U)$ is another pruning condition proposed in the original OCS Algorithm [7] (line 9), it represents the maximal number of edges that the resulting clique has, and

$$g(U) = |E(U)| + (k - |U|)|U| + \frac{(k - |U|)(k - |U| - 1)}{2} \tag{1}$$

where $|E(U)|$ is the number of edges in the subgraph induced by U.

Algorithm 3. optimized expand(C)

    Input: C: a γ-quasi-k-clique
    Output: A: the community of C
1 $A \leftarrow C$;
2 EXPAND_CLIQUE (C);
3 **return** A;
4 **Procedure** EXPAND_CLIQUE(C)
5   sort C by $d_{nc}(n)$, $n \in C$;
6   **foreach** $S_1 \in C$ **and** $|S_1| \geq \alpha$ **do**
7          $S_2 = C - S_1$;
8   **foreach** $u \in S_1$ **do**
9   **foreach** $v \in neighbor(u)$ **do**
10   **if** $d_{nc}(v) > 0$ *and* $neighbor(v) \geq \lceil \gamma \binom{k}{2} - \binom{k-1}{2} \rceil$ **then**
11                      $Cand \leftarrow Cand \cup v$;
12   **if** $|Cand| \leq |S_2|$ **or** $g(S_1) < \gamma \frac{k(k-1)}{2}$ **then**
13   Continue;
14   **foreach** $S_2' \in Cand, |S_2| = |S_2'|$ **do**
15             $C' \leftarrow S_1 \cup S_2'$;
16   **if** C' *is unvisited* **and** C' *is a γ-quasi-k-clique* **then**
17                $A \leftarrow A \cup S_2'$;
18   Update(A, S_2');
19   EXPAND_CLIQUE(C');

After find a clique C, we use $expand(C)$ to get the clique component of C, which can constitute a community. We adopt a DFS traversal on the clique graph. The key operation of the expanding procedure is to replace subset S_2 of C ($|S_2| \leq k - \alpha$) with the remaining subset S_1's ($|S_1| \geq \alpha$) neighbors S_2'

(line 15), where $|S_2'| = |S_2|$, and these neighbors should satisfy 1) they are not interior nodes, 2) degree should be not less than the lower bound (line 10). Note that $d_{nc}(v)$ denotes the number of v's neighbors which are not in the community, $d_{nc}(v) = 0$ means v is an interior node of the current explored community. Notice that d_{nc} is defined on the nodes which are already in the current community result set, if node v is not in the result set, its $d_{nc}(v)$ is unknown, and we initialize the value of $d_{nc}(v)$ with node degree at the beginning. For a new combination C', we check if it is a new valid clique (line 16). If so, S_2' is added into the result set A (line 17) and d_{nc} value of nodes in A need to be updated (line 18), then we expand C' (line 19). Note that at the beginning of expand procedure, we sort nodes of clique C by d_{nc} in ascending order (line 5), then we pick nodes of C by the order to form S_1. By doing this, we could guarantee that nodes with lower d_{nc} value change into interior nodes earlier, and we could get more interior nodes as early as possible.

Benefited from interior node and node degree pruning conditions, the enumerations of finding and expanding clique are sharply reduced. Thus the efficiency of OCS algorithm is highly improved, and this is shown by experiments in Sec. 5.

4.3 Optimized OCS-M Algorithm

When it comes to OCS-M problem, there is still room for efficiency improvement. Instead of simply iterating OCS, we try to avoid repeated computations by utilizing existing results. Note that there exists a consistency property for OCS problem:

Property 1 (Consistency). In (α, γ)-OCS, if $\mathbf{C_m}$ is a community that contains query node v_0, for any other node $v \in \mathbf{C_m}$ as query node, $\mathbf{C_m}$ is also returned as its community.

Consider Example 1, suppose we already finished the first round taking a as input node and got *Community* $\{a, d, f, g\}$, $\{a, b, c, i, h\}$, and now consider node b as input. Intuitively, since we already got $\{a, b, c, i, h\}$, according to Property 1, when we take b as input node, we will still get $\{a, b, c, i, h\}$. Thus, we wonder if we could omit some traversals related to $\{a, b, c, i, h\}$. The ideal situation is that all nodes in $\{a, b, c, i, h\}$ could be skipped, however, if we do that, we could only get $\{b, l, j, k\}$ as the result of the second round, and the exact result should be $\{b, c, l, j, k\}$. Apparently, node c is missing. So we try to find which nodes in the existing community can be skipped and which can not, and we get Thm. 2.

Theorem 2. *For $(k-1, 1)$-OCS-M, given a node v which is a member of existing community $\mathbf{C_m}$, and node v's degree $d(v) \leq k$, then v cannot exist in a new community $\mathbf{C_m'}$.*

Proof. Suppose $v \in \mathbf{C_m'}$, so there exists a clique C_l': $vn_1' \ldots n_{k-1}'$ which belongs to $\mathbf{C_m'}$, and the degree of v in C_l' is $d_{C_l'}(v) = |n_1' \ldots n_{k-1}'| = k - 1$, and we know that $v \in \mathbf{C_m}$, so there exists a clique C_l: $vn_1 \ldots n_{k-1}$ which belongs to $\mathbf{C_m}$ and $d_{C_l}(v) = |n_1 \ldots n_{k-1}| = k - 1$. Because C_l and C_l' are not in the same community, they are not adjacent, and satisfy $|C_l \cap C_l'| < k - 1$, so we have $|(C_l - v) \cap (C_l' - v)| < k - 2$. We know that $d_{min}(v) = |(C_l - v) \cup (C_l' - v)| = |C_l - v| + |C_l' - v| - |(C_l - v) \cap (C_l' - v)|$, so by computation we can derive $d(v) > k$, and this conflicts with the condition $d(v) \leq k$. Therefore, the theorem holds. □

According to Thm. 2, we could easily infer that for $(k - 1, 1)$-OCS-M, if a node already exists in a community and its degree is not larger than k, it can be skipped. Consider the example above, only $d(c)$ is larger than 4, it cannot be skipped, other nodes a, h, i can be skipped during DFS procedure taking b as input in the second round.

Now we discuss which nodes can be skipped for (α, γ)-OCS-M. For a γ-quasi-k-clique, the minimum degree of a node should be $d_{min}(v) = \lceil \gamma \binom{k}{2} - \binom{k-1}{2} \rceil$, and to keep the clique connected, $d_{min}(v) \geq 1$. Also, if two quasi cliques are not in the same community, they share less than α nodes. Thus, we replace the conditions in the proof of Thm. 2 and get Thm. 3.

Theorem 3. *For (α, γ)-OCS-M, given a node v which is a member of existing community $\mathbf{C_m}$, and node v's degree $d(v) \leq \max\{2\lceil \gamma \binom{k}{2} - \binom{k-1}{2} \rceil - (\alpha - 1), \lceil \gamma \binom{k}{2} - \binom{k-1}{2} \rceil\}$, then v cannot exist in a new community $\mathbf{C_m'}$.*

Example 2 (Optimized-(α, γ)-OCS-M). For the graph in Fig. 1, suppose input node set $V_q = \{a, b, c\}$, let $k = 5$, consider $(3, 0.9)$-OCS-M, after the first round of input node a, we get *Community* $\{a, b, c, h, i\}$, when taking input node b in the second round, according to Thm. 3, we only need to traverse nodes with $d(v) > 4$, thus h and i can be skipped during the DFS procedures of *next_clique()* and *expand()*.

From Example 2 we can see that utilizing Thm. 3, the performance of OCS-M Algorithm is remarkably improved. However, taking $(k - 1, 1)$-OCS-M as example, the lower bound of community node degree is k, which is not big enough for efficient pruning. Thus, we further discover other pruning rules. Base on the definitions of interior and boundary node, we have Thm. 4:

Theorem 4. *If one node i is an interior node of existing community $\mathbf{C_m}$, it cannot exist in a new community $\mathbf{C_m'}$.*

Proof. We know that node i exists in community $\mathbf{C_m}$, suppose it still exists in community $\mathbf{C_m'}$, then there exists a clique C_l': $in_1' \ldots n_{k-1}'$ which belongs to $\mathbf{C_m'}$, because community $\mathbf{C_m} \neq \mathbf{C_m'}$, thus there exists at least one node of $n_1' \ldots n_{k-1}'$ which is not in community $\mathbf{C_m}$, this conflicts with that node i is an interior node of $\mathbf{C_m}$, therefore the theorem holds. □

According to Thm. 4, after get the first community by expanding a clique, we could find the next clique of a query node by only traversing boundary nodes and exterior nodes. Utilizing Thm. 3 and Thm. 4, we could modify Alg. 2 by replacing line 12-13 with Alg. 4. When traversing to a node v, we first check if it is not an interior node (line 1), then check if it is already in an existing community (line 2), R represents the result set of OCS-M, if the node already exists in a community, we use the lower bound mentioned in Thm. 3 as pruning condition (line 3); if it does not exist in a community, we use the lower bound of node degree mentioned in Alg. 2 (line 6).

Algorithm 4. modify next_clique(v_0)

---

1 **if** $d_{nc}(v) > 0$ **then**
2   **if** $v \in R$ **then**
3   **if** $neighbor(v) > max\{2\lceil \gamma\binom{k}{2} - \binom{k-1}{2}\rceil - (\alpha - 1), \lceil \gamma\binom{k}{2} - \binom{k-1}{2}\rceil\}$ **then**
4        $\text{DFS}(U \cup v, v)$

5   **else**
6   **if** $neighbor(v) \geq \lceil \gamma\binom{k}{2} - \binom{k-1}{2}\rceil$ **then**
7        $\text{DFS}(U \cup v, v)$

Similarly, Alg. 3 could also be modified, we could replace line 10-11 with Alg. 4, in which line 4 and line 7 are changed into $Cand \leftarrow Cand \cup v$.

After modifying Alg. 2 and Alg. 3, we get the optimized algorithm of OCS-M, the improvement of performance will be shown through experiments in Sec. 5.

4.4 Approximate Strategies

Although the performance of the exact algorithm has been greatly improved, it is still an NP-hard problem. Thus, we propose a series of approximate strategies which could trade off the performance and quality of our OCS and OCS-M algorithms. We use two conditions boundary node and node degree to adjust the efficiency and quality of the algorithm.

Considering the search process of one community $\mathbf{C_m}$, it starts from a query node, and adjacent nodes are added into $\mathbf{C_m}$ as long as they satisfy that 1) they belong to a γ-quasi-k clique, 2) the clique they belong to can be reached from the start clique, 3) they are not in $\mathbf{C_m}$. We see the community as a growing circle with nodes scatter in it, if we traverse nodes out of the circle, but not wander in the circle, the entire community can be found more earlier. According to Thm. 1 and Thm. 4, interior nodes (i.e. nodes in the circle) can be omitted without losing the completeness, if we traverse the boundary nodes selectively or only traverse the exterior nodes, the search process could be terminated earlier with sacrificing result quality.

Besides, we also consider node degree as another traversing condition. Intuitively, nodes with higher degree have more possibility to belong to one or more cliques. Thus, if we want to prune traversed nodes during the process, we could raise the lower bound of $neighbor(v)$ in both $next\_clique()$ and $expand()$, the higher the lower bound is, the less the traversed nodes are, and the more the quality loss is.

With boundary node and node degree as traversing conditions, we could form three approximate strategies:

- *Strategy 1*: traverse boundary nodes with node degree restriction, and all exterior nodes. That means we partially traverse boundary nodes, and completely traverse exterior nodes.
- *Strategy 2*: traverse only exterior nodes. That means we skip all boundary nodes. By doing this, we could guarantee that if one node belongs to the community, it is only traversed once, no repetitive traversal is made.
- *Strategy 3*: traverse only exterior nodes with node degree restriction. That means we skip all boundary nodes, and partially traverse exterior nodes.

Note that we only apply our approximate strategies on the *expand*() procedure, the *next_clique*() procedure is still exact. Because compared to expanding seed cliques, the computations of finding a new clique as a seed occupy only a small portion of the whole procedure. But if we lose one seed clique, we may miss a bunch of cliques which could be reached from the seed clique. Thus, to guarantee the quality of approximate results, we only apply it on the *expand*() procedure.

Consider a k-clique community $\mathbf{C_m}$, the exact algorithm will explore $O(\binom{|\mathbf{C_m}|}{k})$ cliques. For Strategy 2, each time when a new node is found, a clique will be visited. Thus, it will only explore $O(|\mathbf{C_m}|)$ cliques. For Strategy 1, suppose the number of boundary nodes which exceed the lower bound of node degree is n, notice that $n \leq |\mathbf{C_m}|$, when the lower bound increases, n will decrease. Thus Strategy 1 will explore $O(\binom{n}{k} + |\mathbf{C_m}|)$ cliques. For Strategy 3, suppose the number of exterior nodes which exceed the lower bound of node degree is n, then it will explore $O(n)$ cliques. In this way, for Strategy 2 and 3, we reduce the exponential complexity to linear, and the efficiency is highly improved; for Strategy 1, the result is the most accurate of the three. Theoretically, the relationship of efficiency and quality of these three strategies is depicted in Lemma 1, we will demonstrate it by experiments in Sec. 5.

Lemma 1. *For the three approximate strategies of OCS and OCS-M, the efficiency of them is Strategy 1 < Strategy 2 < Strategy 3, and the quality of them is Strategy 1 > Strategry 2 > Strategy 3.*

5   Experimental Study

In this section, we present experimental study and demonstrate the efficiency and quality of our OCS and OCS-M algorithms.

5.1   Experiment Setup

We ran all the experiments on a PC with Intel Core2 at 2.67GHz, 4G memory running 32-bit Windows 7. All algorithms were implemented in C++. To intuitively show the performance of algorithms, we use $(k - 1, 1)$ OCS and OCS-M models to conduct our experiments.

We use three real-world networks as our experiment datasets, and the statistics are shown in Table 1. Amazon is a product co-purchasing network of Amazon

Table 1. Real-world Networks for Experiments

Dataset	# Nodes	# Edges	Average Degree
Amazon	334,863	925,872	5.53
DBLP	968,956	4,826,365	9.96
LiveJournal	3,997,962	34,681,189	17.4

website[1]. Nodes in Amazon represent products and if two products are frequently co-purchased, there exists an edge between them. DBLP is a scientific coauthor network extracted from a recent snapshot of the DBLP database[2]. Nodes in DBLP graph represent authors, and edges represent collaboration relationship. LiveJournal provides the LiveJournal friendship social network[3], it is a free online blogging community where users declare friendship.

5.2   Performance

We first compare the performance of exact algorithms of basic OCS, optimized OCS, and approximate Strategy 2. For each k, we randomly select 100 nodes (with degree not less than $k-1$) for queries, and compare the average answering time. Because exact algorithms have exponential complexity, we terminate them when the running time exeeds 60s. The results of the three algorithms on the three networks are shown in Fig. 2. We can see that our optimized OCS performs better than basic OCS, with about 20 times efficiency improvement, and the approximate strategy overwhelms the two exact algorithms on performance by about two or three orders of magnitudes respectively. Actually, the superiority is more significant than Fig. 2 shows. Because the maximal running time of exact algorithms is 60s in our setting. Especially for the biggest dataset LiveJournal, with millions of nodes, tens of millions of edges, and average degree 17.3, the executing time of approximate strategy is less than 100ms. This indicates that the approximate strategy can support online search on large real networks.

Then, we compare the performance of exact algorithms of basic OCS-M, optimized OCS-M, and approximate Strategy 2. We set $k = 5$ for Amazon, $k = 7$ for DBLP, $k = 9$ for LiveJournal, and change the query node number $|N|$ of query sets. For each $|N|$, we test 20 randomly selected query sets, and compare the average time cost. Also, we terminate the exact algorithms after 600s. The results are shown in Fig. 3. We can see that optimized OCS-M performs better than basic OCS-M, and as the query node number increases, the time cost of optimized OCS-M increases slowly than the basic algorithm. This indicates that our optimized algorithm avoiding duplicate computations works well on OCS-M problem. Also, the approximate Strategy 2 won on performance.

---

[1] http://snap.stanford.edu/data/com-Amazon.html
[2] http://dblp.uni-trier.de/xml/
[3] http://snap.stanford.edu/data/com-LiveJournal.html

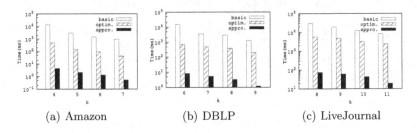

(a) Amazon   (b) DBLP   (c) LiveJournal

Fig. 2. Performance of basic OCS, optimized OCS, and approximate Strategy 2

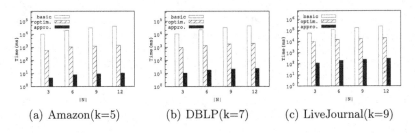

(a) Amazon(k=5)   (b) DBLP(k=7)   (c) LiveJournal(k=9)

Fig. 3. Performance of basic OCS-M, optimized OCS-M, and approximate Strategy 2

5.3 Quality

We compare the result quality of three approximate strategies of OCS problem, for OCS-M problem the situation is similar, thus we save the comparison for space limitation. We set the lower bound of node degree restriction at $2(k-1)$ for Strategy 1 and Strategy 3, and randomly select 100 valid query nodes for different k. Clearly, each community in the approximate result is smaller than its corresponding community in the exact result. Let $R'=\{\mathbf{C'_1}, \ldots, \mathbf{C'_m}\}$ be the approximate result, and $\mathbf{C_i}$ be the exact community containing $\mathbf{C'_i}$, thus the accuracy of the approximate result R' is defined as

$$Accuracy(R') = \frac{1}{m} \sum_{1 \leq i \leq m} \frac{\mathbf{C'_i}}{\mathbf{C_i}} \tag{2}$$

The average and variance accuracy of the three approximate strategies are shown in Fig. 4. It is clear that all the three strategies' accuracy is over 60%, and as our discussion in Lemma 1, Strategy 1 with more than 80% accuracy performs best on quality, and Strategy 2 is better than Strategy 3.

5.4 Influence of Node Degree Restriction

Now we investigate the influence of node degree restriction on approximate Strategy 1 and Strategy 3. For space limitation, we conduct experiments of OCS problem on DBLP dataset. We set $k = 7$, randomly select 100 valid query nodes for

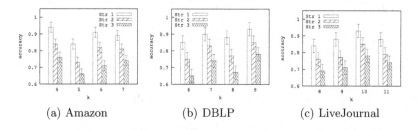

(a) Amazon   (b) DBLP   (c) LiveJournal

Fig. 4. Accuracy of approximate Strategy 1, 2, 3

different lower bounds of node degree , and compare the efficiency and quality of the algorithms, the results are shown in Table 2. We can see that for both of the two strategies, as the lower bound increases, the running time decreases sharply and the accuracy also decreases. However, the accuracy of Strategy 1 stays above 85%, and the accuracy of Strategy 3 stays above 50% with the efficiency improved 6 times. The results indicate that if the quality requirement is more important than the efficiency requirement, we could select Strategy 1, for the opposite situation, we could select Strategy 3, whose accuracy is also acceptable.

Table 2. Performance and Quality of Strategy 1 and Strategy 3

Lower Bound	8	11	14	17	Lower Bound	8	11	14	17
Time(ms)	14.8	10.2	7.5	6.9	Time(ms)	4.8	3.1	13	0.8
Accuracy	0.95	0.92	0.88	0.85	Accuracy	0.80	0.77	0.69	0.58

6   Conclusion

In this paper we studied an efficient solution for overlapping community search problem. We proposed an exact algorithm whose performance was highly improved for both single node overlapping community search and multiple nodes overlapping community search with strong theoretical supports. Besides, we proposed three approximate strategies which could satisfy different efficiency and quality requirements. Comprehensive experiments were conducted to evaluate the efficiency of the optimized exact algorithms, and the efficiency and quality difference of the three approximate strategies. Through the experiments we demonstrated that our solutions were effective and efficient to discover overlapping communities in real networks, and the approximate strategies are flexible for different requirements.

Acknowledgments. This work is supported by the National Basic Research 973 Program of China under Grant (2012CB316201) and the National Natural Science Foundation of China under Grant (61472070).

References

1. Adamcsek, B., Palla, G., Farkas, I.J., Derényi, I., Vicsek, T.: Cfinder: locating cliques and overlapping modules in biological networks. Bioinformatics **22**(8), 1021–1023 (2006)
2. Ahn, Y.Y., Bagrow, J.P., Lehmann, S.: Link communities reveal multiscale complexity in networks. Nature **466**(7307), 761–764 (2010)
3. Bagrow, J.P.: Evaluating local community methods in networks. Journal of Statistical Mechanics: Theory and Experiment **2008**(05), P05001 (2008)
4. Brunato, M., Hoos, H.H., Battiti, R.: On effectively finding maximal quasi-cliques in graphs. In: Maniezzo, V., Battiti, R., Watson, J.-P. (eds.) LION 2007 II. LNCS, vol. 5313, pp. 41–55. Springer, Heidelberg (2008)
5. Chen, J., Zaïane, O., Goebel, R.: Local community identification in social networks. In: International Conference on Advances in Social Network Analysis and Mining, ASONAM 2009, pp. 237–242. IEEE (2009)
6. Clauset, A.: Finding local community structure in networks. Physical Review E **72**(2), 026132 (2005)
7. Cui, W., Xiao, Y., Wang, H., Lu, Y., Wang, W.: Online search of overlapping communities. In: Proceedings of the 2013 International Conference on Management of Data, pp. 277–288. ACM (2013)
8. Evans, T., Lambiotte, R.: Line graphs, link partitions, and overlapping communities. Physical Review E **80**(1), 016105 (2009)
9. Girvan, M., Newman, M.E.: Community structure in social and biological networks. Proceedings of the National Academy of Sciences **99**(12), 7821–7826 (2002)
10. Gregory, S.: Finding overlapping communities in networks by label propagation. New Journal of Physics **12**(10), 103018 (2010)
11. Lim, S., Ryu, S., Kwon, S., Jung, K., Lee, J.G.: Linkscan*: Overlapping community detection using the link-space transformation. In: 2014 IEEE 30th International Conference on Data Engineering (ICDE), pp. 292–303. IEEE (2014)
12. Luo, F., Wang, J.Z., Promislow, E.: Exploring local community structures in large networks. Web Intelligence and Agent Systems **6**(4), 387–400 (2008)
13. Palla, G., Derényi, I., Farkas, I., Vicsek, T.: Uncovering the overlapping community structure of complex networks in nature and society. Nature **435**(7043), 814–818 (2005)
14. Papadopoulos, S., Skusa, A., Vakali, A., Kompatsiaris, Y., Wagner, N.: Bridge bounding: A local approach for efficient community discovery in complex networks. arXiv preprint arXiv:0902.0871 (2009)
15. Sozio, M., Gionis, A.: The community-search problem and how to plan a successful cocktail party. In: Proceedings of the 16th ACM SIGKDD International Conference on Knowledge Discovery and Data Mining, pp. 939–948. ACM (2010)
16. Šubelj, L., Bajec, M.: Unfolding communities in large complex networks: Combining defensive and offensive label propagation for core extraction. Physical Review E **83**(3), 036103 (2011)

A Comparative Study of Team Formation in Social Networks

Xinyu Wang[✉], Zhou Zhao, and Wilfred Ng

Department of Computer Science and Engineering,
The Hong Kong University of Science and Technology,
Clear Water Bay, Hong Kong, China
{xwangau,zhaozhou,wilfred}@cse.ust.hk

Abstract. Team formation in social networks is a fundamental problem in many database or web applications, such as community-based question answering and the collaborative development of software. It is also well-recognized that forming the right team in social networks is non-trivial. Although many algorithms have been proposed for resolving this problem, most of them are based on very different criteria, concerns and performance metrics, and their performance has not been empirically compared. In this paper, we first compare and contrast all the state-of-the-art team formation algorithms. Next, we propose a benchmark that enables fair comparison amongst these algorithms. We then implement these algorithms using a common platform and evaluate their performance using several real datasets. We also present our insights arising from the results of the comparison and uncover interesting issues for further research. Our experiments are repeatable, with the code and all the datasets publicly accessible in our website at the following url address: www.cse.ust.hk/~xwangau/TF.html.

Keywords: Team formation · Social networks · Benchmark · Comparative study

1 Introduction

With massive amount of collaboration experience and information from experts on the web, we have observed an explosive creation of large groups that communicate through social networks to deal with complex tasks, such as the GNU/Linux community and the collective efforts in Wikipedia.

How to form the right team of the right experts in a social network to accomplish a task at the least cost?

The above question leads to the study of the **Team Formation (TF)** problem. The TF problem has been attracting a lot of attention from the database, data mining and web research communities [1,2,6–11]. The problem is also a practical one, which is related to many real-life applications such as community-based question answering and project development in social networks. Examples of such social networks include LinkedIn[1], StackOverflow[2] and many others.

---

[1] http://www.linkedin.com
[2] http://stackoverflow.com/

© Springer International Publishing Switzerland 2015
M. Renz et al. (Eds.): DASFAA 2015, Part I, LNCS 9049, pp. 389–404, 2015.
DOI: 10.1007/978-3-319-18120-2_23

Basically, given a social network of experts and a collaborative task that requires a set of skills, the team formation problem is to find a team of experts who are able to cover all the required skills and communicate with one another in an effective manner. It could be that some experts may or may not be available in different periods of time. The communication costs between experts may also be very different. This problem was first formulated and proved to be NP-hard in [9].

In recent years, many team formation algorithms have been independently proposed, which are based on different communication cost functions such as *steiner distance*, *radius distance*, and so on. However, it is still unclear how to formulate an effective communication cost function for forming a team. We hereby give an example to illustrate the TF problem.

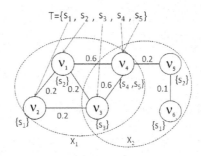

Fig. 1. A Social Network of Experts $V = \{v_1, \ldots, v_6\}$ and Their Communication Costs

Consider a social network of experts $V = \{v_1, \ldots, v_6\}$ in Figure 1, where each expert is associated with a specified set of skills. The communication cost between two adjacent experts v_i and v_j is represented as the weight of edge (v_i, v_j) while the communication cost between two non-adjacent experts is the shortest path between them. For example, expert v_1 is skilled in s_2 and the communication cost between experts v_1 and v_2 is 0.2.

Given a task T with the required skills $\{s_1, s_2, s_3, s_4, s_5\}$, we aim to find a team of experts X that collectively cover all the required skills with the least communication cost. However, team formation algorithms might return multiple teams with the lowest cost based on different communication cost functions. For example, both teams $X_1 = \{v_1, v_2, v_3, v_4\}$ and $X_2 = \{v_3, v_4, v_5, v_6\}$ cover all the required skills. The assignment of experts in the team covering the required skills is illustrated by red lines in Figure 1. We can observe that the communication cost of team X_1 on the radius distance is 0.8 (i.e. $0.2 + 0.6 = 0.8$) and their communication cost on steiner distance is 1.0 (i.e. $0.6 + 0.2 + 0.2 = 1.0$). The communication cost of team X_2 on both radius distance and steiner distance is 0.9 (i.e. $0.6 + 0.2 + 0.1 = 0.9$). Thus, team X_1 has the lower communication cost according to the steiner distance while team X_2 has the lower communication cost according to the radius distance. The performance of these two teams with respect to a task based on different cost functions is also difficult to compare. To the best of our knowledge, there is no existing work that evaluates the performance of the team formation algorithms based on different communication cost functions. To clarify this performance issue, in this work we have developed a unifying platform that enables such comparison.

Some of the existing work [2,7,8,10] considers even more cost factors, such as the workload and personal cost of experts to enrich the TF algorithm. However, the impact of these additional cost factors on team formation remain unclear. In this paper, we also study the benefits of these factors.

Another problem of existing team formation algorithms is that they have actually evaluated using different datasets and programming languages such as DBLP in [9], IMDB in [6] [7], Bibsonomy in [2] and Github in [10]. Thus, it makes the analysis of the algorithms' performance even more difficult. In order to clarify the performance analysis, we re-implement all popular team formation algorithms in the same programming language and study them by using the common basis of four real datasets.

Our contributions of this comparative study are summarized as follows:

- We study the differences between a comprehensive set of team formation algorithms, including RarestFirst, EnSteiner, MinSD, MinLD, MinDiaSol, MinAggrSol, MCC, ItRepace, LBRadius and LBSteiner. We categorize these algorithms into four groups based on their adopted communication cost functions: *radius distance*, *steiner distance*, *sum of distances*, and *leader distance*.
- We re-implement these team formation algorithms in C++ and develop a unifying platform on which to study them.
- We empirically compare the algorithms using four datasets *DBLP*, *IMDB*, *Bibsonomy* and *StackOverflow*. We conduct extensive experiments to evaluate their performance.

Organizations. The rest of the paper is organized as follows. Section 2 introduces the basic notation and formulates the problem of team formation in social networks. Section 3 surveys the related work. Section 4 presents details of popular team formation algorithms while Section 5 reports the experimental results. We conclude the work in Section 6.

2 Background

2.1 Notion and Notation

We now introduce the notion and notation for the problem of team formation in social networks. The summary of the notation is given in Table 1.

We assume that there is a collection of N candidate experts, $V = \{v_1, v_2, \ldots, v_N\}$ and a set of M specified skills $S = \{s_1, s_2, \ldots, s_M\}$. Each expert v_i is associated with a set of skills $s(v_i)$ (i.e. $s(v_i) \subseteq S$). We then denote the set of experts having skill s_k as $C(s_k)$ (i.e. $C(s_k) \subseteq V$). A task T is modeled by a set of required skills (i.e. $T = \{s_i, \ldots, s_j\} \subseteq S$). A team of experts X is said to

Table 1. Summary of Notation

Notation	Meaning
V	A set of experts
$G(V, E)$	A social network of experts
S	A set of specified skills
T	A task with required skills
X	A team of experts
$C(s_i)$	A set of experts skilled in s_i
$s(v_i)$	Skill of expert v_i
$l(v_i)$	Workload of expert v_i
$c(v_i)$	Packing constraint of expert v_i
$pc(v_i)$	Personal cost of expert v_i
$pc(X)$	Personal cost of a team of experts X
$d(v_i, v_j)$	Communication cost between experts v_i and v_j
$sp(v_i, v_j)$	Shortest path between experts v_i and v_j
$Cc\text{-}R(X)$	Communication cost on radius distance
$Cc\text{-}Steiner(X)$	Communication cost on steiner distance
$Cc\text{-}SD(X)$	Communication cost on sum of distances
$Cc\text{-}LD(X)$	Communication cost on distances to the leader

be feasible for task T, if and only if, the experts in X collectively cover all the required skills (i.e. $T \subseteq \bigcup_{v_i \in X} s(v_i)$).

We consider the workload of an expert v_i (i.e. $l(v_i)$) as the number of tasks for which expert v_i is employed (i.e. $l(v_i) = |\{X_j | v_i \in X_j\}|$). The packing constraint of expert v_i (i.e. $c(v_i)$) is the upper limit of the tasks that expert v_i can participate at the same time (i.e. $l(v_i) \leq c(v_i)$). We denote the personal cost of expert v_i as $pc(v_i)$. The personal cost of a team X is given by $\sum_{v_i \in X} pc(v_i)$.

The communication cost is the closeness of the experts in a social network $G(V, E)$. The communication cost of two adjacent experts v_i and v_j is given by the weight of the edge (v_i, v_j) while the communication cost of two non-adjacent experts v_i and v_j is given by the shortest path between them $sp(v_i, v_j)$. The four types of communication cost functions for team formation in social networks are defined as follows:

- $Cc\text{-}R(X)$. The communication cost of the team on radius distance (i.e. $Cc\text{-}R(X)$) is defined as the longest shortest path between any experts in team X (i.e. $Cc\text{-}R(X) = \arg\max_{v_i, v_j \in X} sp(v_i, v_j)$).
- $Cc\text{-}Steiner(X)$. The communication cost of team X on steiner distance (i.e. $Cc\text{-}Steiner(X)$) is defined as the weight cost of the minimum spanning tree for subgraph $G' \subseteq G$ formed by team X.
- $Cc\text{-}SD(X)$. The communication cost of team X on sum of distances (i.e. $Cc\text{-}SD(X)$) is defined as the sum of all shortest paths between any two experts in team X (i.e. $Cc\text{-}SD(X) = \sum_{v_i, v_j \in X} sp(v_i, v_j)$).
- $Cc\text{-}LD(X)$. Given a team leader v_L, the communication cost of the team on leader distance (i.e. $Cc\text{-}LD(X)$) is defined as the sum of the

Table 2. Summary of Team Formation Algorithms

Algorithm	Communication Cost				Personal Cost	Packing Constraint	Load Balancing
	Cc-R	Cc-Steiner	Cc-SD	Cc-LD			
RarestFirst [9]	✓						
EnSteiner [9]		✓					
MinSD [6]			✓				
MinLD [6]				✓			
MinDiaSol [10]	✓					✓	
MinAggrSol [10]		✓				✓	
MCC, ItReplace [7]			✓		✓		
LBRadius [2]	✓						✓
LBSteiner [2]		✓					✓

shortest paths from all the experts to the team leader (i.e. $Cc\text{-}LD(X) = \sum_{v_i \in X, v_i \neq v_L} sp(v_i, v_L)$).

2.2 Problem Definition

We now formally define the team formation problem in social networks as follows:

Problem 1. Given a social network of experts $G(V, E)$ where the experts are associated with specified skills in S, a task T, the problem of team formation is to find the team of experts X that can collectively cover all the required skills with the lowest communication cost $Cc(X)$.

We now summarize the popular TF algorithms based on their communication cost functions and some additional factors in Table 2. Some of the current works [2,7,10] also consider additional factors such as workload and personal cost of the experts for the TF problem. However, all of these algorithms are still based on the four types of communication cost functions listed in Table 2.

3 Related Work

The TF problem was well recognized by the database, data mining and web research communities in [1,2,6–10]. This problem was first formulated in [9], where the algorithms based on communication cost functions on steiner and radius were proposed. Kargar et al. [6] proposed the team formation algorithms whose communication cost functions are based on the sum of distances and leader distance. Majumder et al. [10] introduced the packing constraints such that the workload of the employed experts does not exceed his packing constraint. Kargar et al. [7,8] devised a bi-objective cost function for team formation that considers both communication cost function and personal cost of the team. Anagnostopoulos et al. [1,2] developed the algorithms to find the team of the experts where the workload is balanced. Avradeep Bhowmik et al. [5] developed the algorithms using the Submodularity method to find team of experts by relaxing the skill cover requirement, defining some skills are at the "must have" level while others are at the "should have" level.

However, it is still unclear how to determine which algorithm is the best to form the team, since the existing algorithms were proposed for different communication cost functions and many other factors. To the best of our knowledge, this work is the first approach to develop a common framework to compare the performance of the representative TF algorithms.

4   Implementation

We now give an overview of ten TF algorithms studied in this work and classify them into four categories in respective subsections. The full implementation details of these TF algorithms are given in Appendix [12].

4.1   R-TF Algorithm

The algorithms RarestFirst [9], MinDiaSol [10] and LBRadius [2] have been proposed for team formation based on Cc-R.

The **RarestFirst** algorithm is derived from the Multichoice algorithm [3]. For each skill s_i, the algorithm maintains a set $C(s_i)$ consisting of the experts associated with skill s_i. Given a task T, the algorithm first picks a skill $s \in T$ that has the lowest cardinality of $C(s)$, denoted as s_{rare}. Then, the algorithm enumerates all the experts in $C(s_{rare})$ and picks the one v_i^* that leads a subgraph with the smallest diameter. For other required skills $s_k \in T$, the algorithm collects an expert $v_j \in C(s_k)$ who has the lowest communication (shortest path) to v_i^* (i.e. $v_j = \arg\min_{v_l} d(v_i^*, v_l) \bigwedge v_l \in C(s_k)$) as well as the experts on the shortest path. The running time of RarestFirst is $O(|C(s_{rare})| \times N)$ while the worst-case can be $O(N^2)$.

The **MinDiaSol** algorithm aims to enroll the experts around the user who issues the task in social networks to form a team with packing constraints. The formation of a team is said to be feasible, if and only if, the total assignment of the skills to the experts does not exceed their packing constraints. Given a task with some skill requirement, the identification of feasible teams can be reduced to the problem of maximum flow on a bipartite graph, which can be computed in polynomial time. The worst case of this algorithm is of complexity $O((|V||T|(|V| + |T|) \log |V||T| \log h)$.

The **LBRadius** algorithm aims to find a team of experts such that the workload of the experts is balanced. The LBRadius algorithm also considers the distance constraint of the team, which means that the radius of the experts in the graph is bounded by threshold h. The LBRadius algorithm defines $l(v_i)$ the workload of an expert v_i by the number of teams in which the expert participates. (i.e. $l(v_i) = |\{T_j : v_i \in T_j\}|$). The LBRadius algorithm defines the workload cost of the team as the sum of the workload of each expert, given by

$$l(X) = \sum_{v_i \in X} (2N)^{\frac{l(v_i)}{\Lambda}},$$

where N is the number of experts in the graph and Λ is an appropriately chosen value.

4.2   Steiner-TF Algorithm

The algorithms EnSteiner [9], MinAggrSol [10], and LBSteiner [2] are proposed for team formation based on *Cc-Steiner*.

The **EnSteiner** algorithm chooses a set of experts based on *Cc-Steiner* on the graph. Given a task $T=\{s_1, ..., s_k\}$, the EnSteiner algorithm aims to find a team of experts to cover the required skills, where the experts are cohesively connected in the expert graph. For each required skill $s_i \in T$, a skill vertex v_{s_i} is created in the social network graph $G(V, E)$. Then, the EnSteiner algorithm adds the undirected edges between experts $v_j \in V$ and the corresponding skill vertex v_{s_i}, which results in the enhanced graph named H. The communication cost of the newly added edge (v_j, v_{s_i}) is set to the sum of pairwise communication costs of graph G (i.e. $d(v_j, v_{s_i}) = \sum_{v_m, v_n \in V} d(v_m, v_n)$).

The **MinAggrSol** algorithm is a user-oriented team formation algorithm that searches the experts within h hops from the user v_u. The expert set V' is created for the candidate experts within h hops from the user v_u. The cost of the experts in the set V' is the distance from the user v_u, denoted by $\lambda(v_i)$. It iteratively adds an expert v^* to the team X greedily until team X is feasible for the task (i.e. $maxflow(X, T) = k$). The selection of expert v^* is based on the ratio of an improvement on skill coverage to its cost. After searching a set of experts X covering the required skills in T, **MinAggrSol** connects experts X by enrolling additional experts in G (where $X \subseteq X'$). Finally, the algorithm returns X' as the team of experts.

The **LBSteiner** algorithm aims to search for a team in which the overall workload is fair (i.e. $\min_X l(X) = \min_X \sum_{v_i \in X} l(v_i)$) and the communication cost on Steiner is lower than a given bound B (i.e. $Cc\text{-}Steiner(X) \leq B$). Generally, it is difficult to jointly optimize the bi-criteria function. Thus, the parameter λ is employed to reduce the bi-criteria function to a single-criteria function, given by

$$f(\lambda) = \lambda l(X) + Cc\text{-}Steinter(X), \tag{1}$$

where the function $f(\lambda)$ is non-decreasing on λ. Given a fixed λ, we consider a team of experts X_λ that optimize Equation 1. We return the corresponding the team of experts X where the communication cost $Cc\text{-}Steiner(X) \leq B$ and the workload of the team $l(X)$ is minimized.

4.3   SD-TF Algorithm

The algorithms MinSD [6], MCC [7], MCCRare [7], and ItReplace [7] are proposed for team formation based on *Cc-SD*.

The **MinSD** algorithm aims to find a team of experts where the communication cost on the sum of distances is minimized ($\min_X SD(X) =$

$\min_X \sum_{i=1, v_i \in X, j=i+1, v_j \in X}^{|T|} d(v_{s_i}, v_{s_j}))$. The MinSD algorithm aims to enumerate all possible experts which have the required skills as the seed of the candidate team. Then it greedily adds the experts having other required skills until the experts in the team can cover all the required skills.

The **MCC** algorithm aims to find a team of experts having the lowest combined cost $ComCc(X)$ in Equation 2. The MCC algorithm is derived from the **MinSD** algorithm. The cost function λ in the MCC algorithm is replaced by Equation 2.

$$ComCc(X) = (1 - \lambda)pc(X) + \lambda Cc(X), \tag{2}$$

where $Cc(X)$ can be any type of the discussed communication cost functions.

The **ItReplace** algorithm is derived from the **MCC** algorithm, which ranks the experts $v_j \in C(s_i)$ by their personal cost $pc(v_j)$. Both MCC and ItReplace algorithms are similar with the MinLD algorithm. The details of the algorithms are omitted due to the space limitation.

4.4   LD-TF Algorithm

The **MinLD** [6] algorithm was proposed for team formation based on $Cc\text{-}LD$. To choose an appropriate leader of the team, the MinLD algorithm enumerates all possible experts in graph $G = (V, E)$ as the candidate leader expert to form N teams. Then, the MinLD algorithm returns a team of experts with the lowest communication cost in $Cc\text{-}LD$.

For each candidate leader expert v_i, MinLD finds the experts with the required skills (i.e. $v_k \in C(s_j)$ and $s_j \in T$) that have the lowest cost in $Cc\text{-}LD$ and form the team X'. Finally, the algorithm returns a team of experts which incur the lowest cost in $Cc\text{-}LD$ among N candidate teams.

5   Experiments

In this section, we evaluate the effectiveness and efficiency of the team formation algorithms detailed in Section 4. All the algorithms were implemented in C++ and tested on machines with Windows OS, Intel(R) Core(TM2) Quad CPU 3.40GHz, and 16GB of RAM.

Table 3. Summary Statistics of Datasets

Dataset	#Expert	#Skill	#Edge	Avg. #Skill per Expert	Avg. Distance	Avg. Degree	#CC	Largest CC
Bibsonomy	9269	36299	30557	30.862	7.028	6.623	339	8194
DBLP	7159	4355	15110	7.831	6.134	4.221	513	5880
IMDB	1021	27	11224	3.7	4.244	21.986	19	965
StackOverflow	8834	1603	62277	6.249	3.35	14.099	70	8688

5.1   Datasets

We collect four datasets to evaluate the performance of the team formation algorithms. Some statistics of the datasets are presented in Table 3.

DBLP. We restrict the DBLP dataset[3] to the following four fields of computer science: Database (DB), Theory (T), Data Mining (DM) and Artificial Intelligence (AI). Conferences that we consider for each field are given as follows. DB = {SIGMOD, VLDB, ICDE, ICDT, PODS}, T={SODA, FOCS, STOC, STACS, ICALP, ESA}, DM={WWW, KDD, SDM, PKDD, ICDM, WSDM}, AI = {IJCAI, NIPS, ICML, COLT, UAI, CVPR}. The expert set consists of authors that have at least three papers in DBLP. The skills of each expert are the set of keywords which appear in the titles of the authors' collaborative publications at least twice. Two experts are connected if they collaborate on at least two papers. The communication cost of expert v_i and v_j is estimated in the following way: Let P_{v_i} (P_{v_j}) denote the set of papers published by expert v_i (v_j), then their communication cost is estimated by $1 - \frac{|P_{v_i} \cap P_{v_j}|}{|P_{v_i} \cup P_{v_j}|}$. This weight of edges represents pairwise Jaccard distance between all pairs of experts. To compare the performance of different algorithms under this dataset, we use the above definitions such as skills of experts, connectivity between authors and communication cost exactly the same as that of the experimental setting in [9] for the DBLP dataset. This dataset is also used in [1],[6], [7] and [10] with similar settings. We extract the keyword from title of by using the tool-kit provide by NLTK[4]. First the title is segmented into several meaningful words or phrase with the tool. Then,it stemmed the words into a consistent form to extract words, which have different tenses or parts of speech but the same basic meaning. For example, for the title "Team Formation in Social Networks: An Experimental Evaluation", the set of extacted keywords will be {team, formation, Social Networks, Experimental, Evaluation}.

IMDB. The IMDB dataset[5] contains information about the actors and the set of movies that the actors have appeared in. The titles of the movies and the names of the actors are specified in the dataset. We consider the actors who have appeared in at least eight movies from the Year 2000 to the Year 2002 as experts. The skills of each expert is the set of terms that appear in the title of the movies. Two experts are connected if they collborate on at least four movies. The communication cost of two experts is determined in the same way as in the DBLP dataset. The IMDB dataset is used in [1], [2] and [6]. We choose the setting and definitions of the IMDB dataset be the same as DBLP because it is a more consistent to compare the performance of different algorithms under different datasets. We choose the same time period and quantitative limitation as [6]. From the statistics of the dataset it can be seen that the IMDB graph is

---

[3] http://www.informatik.uni-trier.de
[4] http://www.nltk.org
[5] http://www.imdb.com/interfaces

much denser and smaller than the others dataset, which can test the scalability of the algorithms using different datasets.

Bibsonomy. The Bibsonomy dataset [4] contains a large number of computer science related publications. The bibsonomy website is visited by a large community of users who use tags to annotate the publications such as theory, software and ontology. We consider the authors of these publications as experts. We also consider the set of tags associated with the papers as the skills. We specify the skills of the experts from the tags in their publications. The communication cost of two experts is set similarly as in the DBLP dataset. The Bibsonomy dataset is used in [1] and [2]. The definitions of skill, expertise, connectivity are same in this paper and [1],[2], but we have chosen Jaccard distance same as what we have for other datasets to be consistent.

StackOverflow. The StackOverflow dataset[6] contains a large amount of question and answer pairs. Each question is tagged by a set of keywords such as *Html, R* and *Python*. We consider the respondents in StackOverflow as the experts and specify these keywords as the skills of the experts. The StackOverflow dataset does not have an explicit notion of social graph, and therefore two experts are connected if they collaborate in solving at least three common questions. This collaboration is implied by their behavior of answering same question, hitting similar interests and expertise. The communication cost of two experts is determined in the same way as in the DBLP dataset. That is, by the Jaccard Distance, it show the common interest and expertise areas between two respondents. The more common they are, the easier it is for them to collaborate on a given task. To the best of our knowledge, the StackOverflow dataset has not been used in any other work as an experimental dataset. Thus we define the skill set of an expert as the same way we handle Bibsonomy dataset and others follow the DBLP dataset setting, which means that our setting is similar to those in [1], [2] and [6].

5.2   Performance Evaluation

We implemented the following ten team formation algorithms: RarestFirst, EnSteiner, MinSD, MinLD, MinDiaSol, MinAggrSol, MCC, ItReplace, LBRadius, and LBSteiner in C++. In most existing methods for solving the TF problem, the number of required skills is chosen as the parameter to compare the proposed method and baselines. This is because the generated tasks in these methods include defining the number of required skills and then randomly picking the corresponding number of skills [1][2][4][6][7][9]. Therefore, we also compare the performance of these algorithms with respect to the required task skills based on the following five commonly adopted metrics: (1) four types of communication costs of the team, (2) running time of team formation algorithms, (3) cardinality of the team, (4) workload of the team, and (5) personal cost of

---

[6] http://stackoverflow.com/

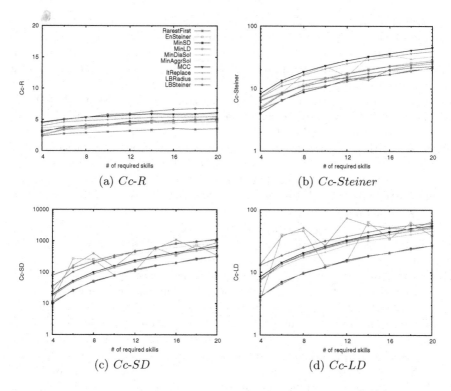

(a) *Cc-R*

(b) *Cc-Steiner*

(c) *Cc-SD*

(d) *Cc-LD*

Fig. 2. Effects by Number of Required Skills Based on Different Criteria

the team. We follow the experimental settings in [2],[4],[6],[7] and [9] for defining these metrics.

We now present the task generation procedure as follows: For each task T, we first set the number of required skills k. Similarly to [2],[4],[6],[7] and [9], we set $k \in \{4, 8, 12, 16, 20\}$ as used in the above work. [8] is similar to [7] work but returns multiple un-ranked teams. [5] relaxes the skill requirement which is different from other work settings, which makes them not comparable, so we did not evaluate the performance of the algorithms in these two. We then continue to sample a skill $s \in S$ and add s to the task T. After each sampling, we remove the sampled skill s from the skill set S in order to avoid repeated sampling. For each configuration, we repeat the task generation 100 times and report the average performance of the team formation algorithms on these 100 generated tasks.

Communication Cost of the Team Effect on *Cc-R*. We illustrate the effect of the number of required skills in the task on *Cc-R* in Figure 2(a). We investigate the communication cost of the team on radius distance formed by different TF algorithms on the datasets DBLP, IMDB, Bibsonomy, StackOverflow. We show the result on datasets DBLP in Figure 2(a) while results of other datasets are detailed in Appendix [12] due to the space limitation. We observe that the communication cost of all the TF algorithms first increases and then gradually converges. This is because, during the late stages of the process, some experts in the team may cover more than one skill, we can add less experts than existing team to cover remaining skills. Notably, the RarestFirst algorithm has the best performance of all the tested datasets. The algorithms EnSteiner, MinSD and MinLD also find the team with slightly larger communication cost on Cc-R. For other two algorithms based on Cc-R, MinDiaSol and LBRadius, the communication cost of the team formed by them is larger, since they also consider other factors, such as packing constraints and load balancing of the experts in the team. The performance of algorithms MinAggrSol, MCC and ItReplace are not as good as the previous algorithms, since they focus on more additional factors in different communication cost functions.

Effect on Cc-$Steiner$. We demonstrate the effect of the number of required skills in the task on Cc-$Steiner$ in Figure 2(b). We study the communication cost of the team on Steiner distance formed by different TF algorithms on the DBLP datasets in Figure 2(b). The communication cost on Cc-$Steiner$ of all the TF algorithms also first increases and then gradually converges. However, we also notice that the convergence of communication cost on Cc-$Steiner$ of all TF algorithms is slower than the convergence of communication cost of the algorithms on Cc-R. This is because Cc-$Steiner$ is more sensitive to the number of experts. Employing one more expert often increases the communication cost on Cc-$Steiner$ while it might not increase the communication cost on Cc-R. The MinSD and MinLD algorithms have the best performance of all the tested datasets. The algorithms RarestFirst, MinAggrsol and MinDiaSol have slightly larger communication cost in Cc-$Steiner$.

Effect on Cc-SD. We investigate the effect of the number of required skills in the task on Cc-SD in Figure 2(c). We illustrate the communication cost of the team on sum of distances formed by different TF algorithms on the DBLP datasets in Figure 2(c). The convergence of communication cost of all the algorithms on Cc-SD is faster than the convergence on Cc-$Steiner$ while slower than the convergence on Cc-R. This is because Cc-SD is not as sensitive as Cc-$Steiner$ to the number of experts. We notice that the algorithms, MinSD and MinLD, have the best performance of all the datasets tested. For other algorithms, the communication cost of the team found by them is larger.

Effect on Cc-LD. We study the effect of the number of required skills in the task on Cc-LD in Figure 2(d). We demonstrate the communication cost of the team on leader distance formed by different team formation algorithms on the DBLP datasets in Figure 2(d), respectively. We treat each candidate as a possible leader in the task. The convergence of communication cost is similar to the above analyzes. We find that MinSD and MinLD algorithms perform the best among all the tested datasets. The communication cost of the team returned by other algorithms is very large on Cc-LD.

There are some interesting observations for this evaluation. We find that the team formation algorithms based on Cc-SD usually have good performance on

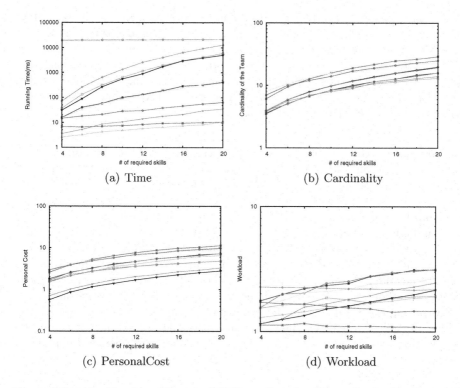

(a) Time   (b) Cardinality

(c) PersonalCost   (d) Workload

Fig. 3. Effects by Number of Required Skills on Running Time, Cardinality, Personal Cost and Workload

different communication cost functions. The results show that the Cc-SD is the best measurement for the communication cost of experts in the team. For other three communication cost functions Cc-R, Cc-$Steiner$ and Cc-LD, they do not include all the communications between the experts. The algorithms based on Cc-SD is more robust to the change in the social network but Cc-R, Cc-$Steiner$, and Cc-LD are very sensitive to adding or removing an expert. Clearly, Cc-R only measures the communication cost between the two experts that are furthest away from each other. Adding an expert in the social network may substantially reduce Cc-R.

We now study the performance of the team formation algorithms on the following issues.

Running Time of TF Algorithms. We present the running time of all team formation algorithms on the number of required skills and the task using the DBLP datasets in Figures 3(a). We observe that the running time of both MinLD and LBSteiner algorithms are costly and insensitive to the number of required skills in all the tested datasets. This is because the computational bottleneck of the MinLD algorithm is to enumerate all possible experts as the leader of the team and generate N candidate teams for all the tasks. For LBSteiner algorithm, it reduces the problem of balancing workload of team under the constraint of communication cost to a bi-objective optimization problem with parameter λ. The major computational cost of the LBSteiner algorithm is to estimate the proper value λ. The most efficient algorithms are MinDiaSol, RarestFirst and ItReplace. For other algorithms, the MinAggrSol and MinSD are also efficient while the EnSteiner algorithm is not scalable.

Cardinality of the Team. We present the cardinality of the team found by all the team formation algorithms on the number of required skills in the task using the DBLP datasets in Figure 3(b). We find that the algorithms based $Cc\text{-}SD$ such as the EnSteiner and MinSD algorithms often return the team of small cardinality for all datasets tested. This is because those algorithms based on $Cc\text{-}SD$ can find a team of experts who are "cohesive" in the social network. For the algorithms based on other communication cost functions, the cardinality of the team found by them is usually much larger. This is because the initially found experts may not communicate with one another. Thus, they have to employ more people to promote better expert communication, which make the cardinality of the team larger. Notably, the LBSteiner and LBRadius algorithms report the teams of small cardinality. This is because both algorithms do not count the people in the social network who make the communication between experts in the team.

Personal Cost of the Team. We present the personal cost of the team by all the team formation algorithms on the DBLP datasets in Figure 3(c). For all the datasets, we set the personal cost of experts to be proportional to the number of their skills, which is suggested in [7]. We notice that the MCC and ItReplace perform the best for all the datasets tested. This is because both algorithms aim to jointly optimize the bi-objective function of both communication cost and personal cost. Interestingly, we observe that the algorithms for $Cc\text{-}SD$ such as MinDiaSol and EnSeiner also find the teams with low personal cost. This is because these algorithms are able to find the teams of experts with small cardinality such that the personal cost of the team is normally lower. Generally, the team formation algorithms based on other communication cost functions may not be suitable for finding the teams with low personal cost, since the cardinality of the teams returned are usually large.

Workload of the Team. We present the workload of the team on the number of required skills in the task by all the team formation algorithms on four datasets in Figure 3(d). In this experiment, we carry out the team formation for 100 sequential given tasks and report the average workload of the teams returned by all the algorithms. We find that the LBRadius and LBSteiner algorithms return the teams with the lowest workload. This is because both algorithms aim to balance the workload of the team in the cost function. Interestingly, we notice that the rarest skill driven algorithms such as RarestFirst can also find the teams of experts where the workload is balanced. This is because the RarestFirst algorithm can find the team around the experts with rarest required skills. Thus, the found teams are diversified and the workload of the experts in the team is lower. However, the algorithms for personal cost such as MCC and ItReplace perform the worst, since these algorithms always choose the experts with large skill coverage and lower personal cost first.

6   Conclusion

In this paper, we present a comprehensive study of the team formation algorithms used in social networks. We survey and study a spectrum of algorithms: RarestFirst, EnSteiner, MinSD, MinLD, MinDiaSol, MinAggrSol, MCC, ItRepace, LBRadius and LBSteiner. We classify the algorithms into four categories based on their communication cost functions. We find that all the team formation algorithms are important but in different ways. For the algorithms in the first category, we can employ them to find a team of well-known experts to review the paper. For the team of reviewers found by algorithms in the second category, we are able to get more diversified reviews, since the background of the reviewers has more variations. The third category of algorithms are able to find a team of rising experts. The forth category of algorithms better balance the workload of the experts in the team, which is suitable for processing a large number of tasks.

We have conducted extensive experiments to evaluate the performance of the algorithms using four real datasets. Although there is no team formation algorithm that perform the best on all the communication cost functions, we find that the algorithms based on Cc-SD such as MinSD are able to find the teams with low cost in other communication cost functions like Cc-R, Cc-$Steiner$ and Cc-LD. We also show the benefits of additional factors such as packing constraints and personal cost for team formation. We observe that the team of diversified experts with the packing constraints can be found, since an expert cannot be employed for numerous tasks. By taking personal cost into consideration, we are able to find rising experts to be team members at lower personal cost.

Acknowledgments. We thank the reviewers for giving many constructive comments, with which we have significantly improve our paper. This research is supported in part by HKUST Grant No.FSGRF13EG22 and HKUST Grant No.FSGRF14EG31.

References

1. Anagnostopoulos, A., Becchetti, L., Castillo, C., Gionis, A., Leonardi, S.: Power in unity: forming teams in large-scale community systems. In: CIKM, pp. 599–608. ACM (2010)
2. Anagnostopoulos, A., Becchetti, L., Castillo, C., Gionis, A., Leonardi, S.: Online team formation in social networks. In: WWW, pp. 839–848. ACM (2012)

3. Arkiny, E.M., Hassinz, R.: Minimum diameter covering problems. Networks **36**(3), 147–155 (2000)
4. Benz, D., Hotho, A., Jäschke, R., Krause, B., Mitzlaff, F., Schmitz, C., Stumme, G.: The social bookmark and publication management system bibsonomy. VLDBJ **19**(6), 849–875 (2010)
5. Bhowmik, A., Borkar, V.S., Garg, D., Pallan, M.: Submodularity in team formation problem. In: Proceedings of the 2014 SIAM International Conference on Data Mining, Philadelphia, Pennsylvania, USA, April 24–26, 2014, pp. 893–901 (2014)
6. Kargar, M., An, A.: Discovering top-k teams of experts with/without a leader in social networks. In: CIKM, pp. 985–994. ACM (2011)
7. Kargar, M., An, A., Zihayat, M.: Efficient bi-objective team formation in social networks. In: Flach, P.A., De Bie, T., Cristianini, N. (eds.) ECML PKDD 2012, Part II. LNCS, vol. 7524, pp. 483–498. Springer, Heidelberg (2012)
8. Kargar, M., Zihayat, M., An, A.: Finding affordable and collaborative teams from a network of experts. In: SDM, pp. 587–595. SIAM (2013)
9. Lappas, T., Liu, K., Terzi, E.: Finding a team of experts in social networks. In: SIGKDD, pp. 467–476. ACM (2009)
10. Majumder, A., Datta, S., Naidu, K.: Capacitated team formation problem on social networks. In: SIGKDD, pp. 1005–1013. ACM (2012)
11. Rangapuram, S.S., Bühler, T., Hein, M.: Towards realistic team formation in social networks based on densest subgraphs. In: WWW, pp. 1077–1088. ACM (2013)
12. Xinyu, W., Zhou, Z., Wilfred, N.: Appendix: Implementation and experiments details. http://www.cse.ust.hk/~xwangau/appen.pdf

Inferring Diffusion Networks with Sparse Cascades by Structure Transfer

Senzhang Wang[1][(✉)], Honghui Zhang[2], Jiawei Zhang[3], Xiaoming Zhang[1], Philip S. Yu[3,4], and Zhoujun Li[1]

[1] State Key Laboratory of Software Development Environment, Beihang University, Beijing, China
{szwang,yolixs,zjli}@buaa.edu.cn
[2] Department of Computer Science and Technology, Tsinghua University, Beijing, China
zhh11@mails.tsinghua.edu.cn
[3] Department of Computer Science, University of Illinois at Chicago, Chicago, USA
[4] Institute for Data Science, Tsinghua University, Beijing, China
{jzhan9,psyu}@uic.edu

Abstract. Inferring diffusion networks from traces of cascades has been intensively studied to gain a better understanding of information diffusion. Traditional methods normally formulate a generative model to find the network that can generate the cascades with the maximum likelihood. The performance of such methods largely depends on sufficient cascades spreading in the network. In many real-world scenarios, however, the cascades may be rare. The very sparse data make accurately inferring the diffusion network extremely challenging. To address this issue, in this paper we study the problem of transferring structure knowledge from an external diffusion network with sufficient cascade data to help infer the hidden diffusion network with sparse cascades. To this end, we first consider the network inference problem from a new angle: link prediction. This transformation enables us to apply transfer learning techniques to predict the hidden links with the help of a large volume of cascades and observed links in the external network. Meanwhile, to integrate the structure and cascade knowledge of the two networks, we propose a unified optimization framework TrNetInf. We conduct extensive experiments on two real-world datasets: MemeTracker and Aminer. The results demonstrate the effectiveness of the proposed TrNetInf in addressing the network inference problem with insufficient cascades.

Keywords: Information diffusion · Network inference · Transfer learning

1 Introduction

Utilizing cascades to infer the diffusion network is an important research issue and has attracted a great deal of research attentions recently [17,20,22,23]. In many scenarios, we only have the traces of information spreading in a network

© Springer International Publishing Switzerland 2015
M. Renz et al. (Eds.): DASFAA 2015, Part I, LNCS 9049, pp. 405–421, 2015.
DOI: 10.1007/978-3-319-18120-2_24

without explicitly observing the network structure. For example, in virus propagation we only observe which people get sick at what time, but without knowing who infected them [18]; in viral marketing, viral marketers can track when customers buy products or subscribe to services, but it is hard to exactly know who influence the customers' decisions [25]. Inferring the underlying connectivity of diffusion networks is of outstanding interest in many applications, such as technological innovations spreading [16], word-of-mouth effect in viral marketing [26], and personalized recommendation in E-commerce websites [24].

Traditional approaches normally formulate a generative probability model to find the network which can generate all the cascades with the maximum likelihood, such as ConNIe [20], NETINF [23], NETRATE [21], and InfoPath [22]. Although these models can work well on synthetic datasets, their performance on real-world datasets is usually undesirable [3,21]. This is firstly due to the fact that information diffusion on real-world networks is too complex for existing information propagation models to handle. Secondly, the performance of generative models largely relies on a large volume of cascades, while in real-world scenarios the cascades may be rare or at least not sufficient [19].

To address above mentioned problems, in this paper we will study *how to borrow the structure knowledge from an external diffusion network whose links are known to help us infer a diffusion network whose links are hidden by transfer learning*. In many cases, although the cascades in the hidden diffusion network are sparse, a network related to the hidden diffusion network is known and may be helpful for our task [6]. For example, we want to infer the network of who influencing whom to buy some products based on the transaction logs of users' purchase history, such as iPhone 5S. The result might be quite inaccurate if we do not have enough such logs. However, if we know their following relationships and tweets about iPhone 5S in Twitter, the diffusion process of the tweets among them may potentially help us infer who influenced whom to buy an iPhone 5S.

Transfer learning has achieved significant success in many machine learning tasks including classification [14,15], regression [13], and clustering [12] to address the problem of lacking enough training data in the target domain. However, it is challenging to directly exploit transfer learning to our task. Traditional generative models formulate this task as an optimization problem, hence it is naturally hard for such models to extract and map feature spaces from one domain to another for knowledge transfer. Meanwhile, transfer learning normally can only capture and transfer knowledge from the source domain. In our task, we need to consider not only the structure knowledge transferred from an external diffusion network, but also the cascade information in the hidden network. How to integrate the knowledge from two different networks in a unified scheme to obtain a better network inference model also makes the problem challenging.

In this paper, we first formulate the network inference problem as a link prediction task by extracting various cascade related features. The advantages of the formulation are two-fold: 1) it paves the way of applying transfer learning techniques for structure transfer; and, 2) link prediction does not rely much on the particular information propagation model. As the links of the external diffusion network are known, we can use these labeled links to train a prediction model for predicting the links in the hidden network by transfer learning. To incorporate the transferred structure knowledge from the external network with the cascades in the hidden network, we next propose a unified optimization framework TrNetInf. TrNetInf jointly maximizes the likelihood of generating the cascades in the hidden diffusion network and minimizes the difference between the links inferred by traditional generative model and those predicted by transfer learning model simultaneously. We evaluate TrNetInf on two real-world datasets: Meme-Tracker dataset and AMiner citation network dataset. Experimental results on both datasets demonstrate the superior performance of TrNetInf, especially when the cascades are not sufficient. The main contributions of this paper are as follows:

- For the first time, to the best of knowledge, we study the network inference problem with the challenge of lacking enough cascade data (Section 2).
- To transfer structure knowledge from one diffusion network to another, we consider the network inference problem from a new angle: link prediction. Meanwhile, as the links of the hidden network is unknown and structure based features are hence not available, we propose to extract a set of cascade related features for learning (Section 3.2).
- We further propose a unified optimization framework TrNetInf. TrNetInf can efficiently integrate knowledge from source and target diffusion networks, and combine the results from the traditional generative model and the proposed link prediction model (Section 3.3).
- We evaluate the proposed approach on two real-world datasets by comparing it against various baselines. The results verify its effectiveness in addressing the network inference problem with very sparse cascades (Section 4).

The remainder of this paper is organized as follows. Section 2 formally defines the studied problem. Section 3 details the proposed model. Section 4 evaluates the model with two real-world datasets, followed by related work in Section 5. Section 6 concludes this research with directions for future work.

2   Problem Statement

In this section, we will give some terminologies to help us state the problem. Then we will formally define the studied problem. In information diffusion, a diffusion network is usually referred to a network with a set of information spreading in it [21]. Based on the diffusion network, we formally define a hidden diffusion network as follows.

Definition 1 *Hidden Diffusion Network* $G_{\mathcal{H}}$: We define a diffusion network $G_{\mathcal{H}} = (V, E_{\mathcal{H}})$ as a hidden diffusion network if only its nodes can be observed but the edges are hidden and need to be inferred. Here V denotes the set of node and $E_{\mathcal{H}}$ denotes the hidden edges.

There are usually many traces of information diffusion on a diffusion network. The traces are called cascades and can be formally defined as follows.

Definition 2 *Cascade*: *A cascade $\boldsymbol{t}^c$ associated with information c can be denoted as a N-dimensional vector $\boldsymbol{t}^c = (t_1^c, ..., t_N^c)^T$, where N is the number of nodes in the diffusion network. The i_{th} dimension of $\boldsymbol{t}^c$ records the time stamp when information c infects node i, and $t_i^c \in [0, T^c] \cup \{\infty\}$.*

The symbol ∞ labels nodes that are not infected during the observation window $[0, T^c]$. The time stamp is set to 0 at the start of each cascade. A cascade set $\mathbb{C}$ consists of a collection of cascades, i.e. $\mathbb{C} = \{\boldsymbol{t}^1, ..., \boldsymbol{t}^M\}$, where M is the number of cascades.

Based on above defined terminologies, the traditional network inference problem can be defined as follows [23].

Problem 1. *Given a hidden diffusion network $G_{\mathcal{H}} = (V, E_{\mathcal{H}})$ and a collection of cascades $\mathbb{C}$ on $G_{\mathcal{H}}$, the network inference problem aims to recover the network structure of $G_{\mathcal{H}}$, namely infer the hidden edges $E_{\mathcal{H}}$ based on the cascades $\mathbb{C}$.*

In our case, besides the hidden diffusion network we also have a related external diffusion network whose structure is known. Here we consider the hidden diffusion network as the target domain network and the related network as the source domain network. In traditional transfer learning setting, a *domain D* consists of two components: a feature space $\mathcal{X}$ and a marginal probability distribution $P(X)$, where $X = \{x_1, ..., x_n\} \in \mathcal{X}$ represent the features. Here we define a *domain $\hat{D}$* of information spreading in network G contains a cascade space $\mathcal{C}_G$ and also a marginal probability distribution $P(\mathbb{C}^{\mathbb{G}})$, where $\mathbb{C}^{\mathbb{G}} = \{c_1^G, ..., c_n^G\} \in \mathcal{C}^{\mathcal{G}}$. We will introduce how to compute $P(\mathbb{C}^{\mathbb{G}})$ later. Based on above definitions and terminologies, we formally define the studied problem as follows.

Problem 2. *Given the source domain diffusion network $G^s = (V^s, E^s)$ and the target domain diffusion network $G_{\mathcal{H}}^t = (V^t, E_{\mathcal{H}}^t)$ with corresponding cascades $\mathbb{C}^s \in \mathcal{C}^s$, $\mathbb{C}^t \in \mathcal{C}^t$, where the edges E^s of network G^s is known and the edges $E_{\mathcal{H}}^t$ of network $G_{\mathcal{H}}^t$ is hidden, the problem is how to transfer knowledge from G_s and $\mathbb{C}_s$ and incorporate it with $\mathbb{C}_t$ to better infer the edges $E_{\mathcal{H}}^t$ of $G_{\mathcal{H}}^t$.*

3   Methodology

In this section, we will first revisit some basic concepts and introduce some standard notations. Then we will introduce how to transform the network inference problem to a link prediction task, and how to apply transfer learning techniques to help predict links in the target diffusion network. Next, we will propose a unified scheme to incorporate the generative model on the target diffusion network and the knowledge transferred from the source domain network.

Before introducing the approach, we first give some basic concepts which are essential to model information diffusion. We define a nonnegative random variable T to be the time when an event happens, such as *user$_i$* adopting a piece of information. Let $f(t)$ be the probability density function of T, then the cumulative density function can be denoted as $F(t) = P(T \le t) = \int_0^t f(x)dx$.

Survival Function. The survival function $S(t)$ is the probability that a cascade $\mathbf{t}^c$ does not infect a node by time t:

$$S(t) = P(T \ge t) = 1 - F(t) = \int_t^\infty f(x)dx.$$

Hazard Function. Given functions $f(t)$ and $S(t)$, we can further define the *hazard function* $H(t)$, which represents the instantaneous rate that a cascade $\mathbf{t}^c$ infects a particular uninfected node within a small interval just after time t.

$$H(t) = \lim_{\Delta t \to 0} \frac{p(t \le T \le t + \Delta t | T \ge t)}{\Delta t} = \frac{f(t)}{S(t)}.$$

3.1 Network Inference Based on Generative Model

We define $g(\Delta_{ij}^c; \alpha_{ij})$ as the conditional likelihood of information transmission between node i and node j, where $\Delta_{ij}^c = t_j^c - t_i^c$ is the difference between the infecting time of the two nodes in cascade c and α_{ij} is the transmission rate from node i to j. Here we assume that within a cascade $\mathbf{t}^c$, a node j with a time stamp t_j^c can only be infected by the node i with an earlier time stamp, i.e. $t_j^c < t_i^c$. If $t_j^c > t_i^c$, we can refer node j as one of node i's child node and node i as one of node j's parent node.

Our goal is to infer the pair-wise transmission rate α_{ij}, and we consider that there exists an edge between two nodes if their transmission rate is larger than zero. Three models are used in most previous works to model the diffusion likelihood function $g(\Delta_{ij}^c; \alpha_{ij})$: Exponential model, Power law model, and Rayleigh model [21]. For brevity, we omit the description of the three models.

Likelihood of Node i Infecting j in Cascade $\mathbf{t}^c$. In a cascade, we assume 1) one node gets infected once the first parent infects it, and 2) all the parents infect their child nodes independently. Based on the two assumptions, the likelihood of the parent node i infecting the child node j in cascade $\mathbf{t}^c$ can be computed by

$$g(\Delta t_{ij}^c; \alpha_{ij}) \times \prod_{u \ne i, t_u^c < t_j^c} S(\Delta t_{uj}^c; \alpha_{uj}), \tag{1}$$

where $S(\Delta_{uj}^c; \alpha_{uj})$ is the survival function described before to denote the probability that node j has not been infected by node u before t_j^c under pairwise transmission rate α_{uj}^c between nodes u and j. In the cascade $\mathbf{t}^c$, the node j could be possibly infected by any one of its parent nodes. Hence the likelihood of j getting infected in the cascade $\mathbf{t}^c$ can be calculated by summing over the likelihoods of each potential parent being the first one to infect it:

$$\Gamma_j^+(\mathbf{t}^c) = \sum_{i:t_i^c < t_j^c} g(\Delta t_{ij}^c; \alpha_{ij}) \times \prod_{u \ne i, t_u^c < t_i^c} S(\Delta t_{uj}^c; \alpha_{uj}). \tag{2}$$

Likelihood of a Node j Survives from the Cascade $\mathbf{t}^c$. If node j survives from all the parents by the end time T^c of cascade $\mathbf{t}^c$, we say the node survives from the cascade $\mathbf{t}^c$. The likelihood that node j survives from the cascade $\mathbf{t}^c$ can be represented by the following product of survival function

$$\Gamma_j^-(\mathbf{t}^c) = \prod_{t_i^c < T^c} S(T^c - t_i^c; \alpha_{ij}). \tag{3}$$

Likelihood of the Cascade $\mathbf{t}^c$. Given a cascade $\mathbf{t}^c := (t_1^c, ..., t_N^c)$, its likelihood can be computed by multiplying the likelihoods of all the infected and survived nodes in the cascade. With Eq. (2), Eq. (3), and the hazard function $H(\Delta t_{ij}^c; \alpha_{ij}) = \frac{g(\Delta t_{ij}^c; \alpha_{ij})}{S(\Delta t_{ij}^c; \alpha_{ij})}$, the likelihood of cascade $\mathbf{t}^c$ can be represented as

$$g(\mathbf{t}^c; \mathbf{A}) = \prod_{t_j^c < T^c} \Gamma_j^+(\mathbf{t}^c) \times \prod_{t_j^c > T^c} \Gamma_j^-(\mathbf{t}^c)$$

$$= \prod_{t_j^c < T^c} \prod_{t_m^c > T^c} S(T^c - t_j^c; \alpha_{jm}) \times$$

$$\prod_{u:t_u^c < t_j^c} S(\Delta t_{uj}^c; \alpha_{uj}) \sum_{i:t_i^c < t_j^c} H(\Delta t_{ij}^c; \alpha_{ij}), \tag{4}$$

where $\mathbf{A}$ is a $N \times N$ matrix with each element $\mathbf{A}_{ij} = \alpha_{ij}$ denoting the link strength between node i and j.

Assuming the cascades spread independently in the network, the likelihood of a set of cascades $\mathbb{C} = \{\mathbf{t}^1, ..., \mathbf{t}^M\}$ can be represented as the product of the likelihoods of all the individual cascades,

$$\prod_{\mathbf{t}^c \in \mathbb{C}} g(\mathbf{t}^c; \mathbf{A}). \tag{5}$$

Network Inference Problem. The goal is to find the matrix $\mathbf{A}$ such that the network G with edge matrix $\mathbf{A}$ generates cascades $\mathbb{C}$ with the maximum likelihood. This can be achieved by solving the following optimization problem

$$min_{\mathbf{A}} - \sum_{c \in C} \log g(\mathbf{t}^c; \mathbf{A}). \tag{6}$$

$$\text{s. t.} \quad \alpha_{ij}^k \geq 0; i, j = 1, ..., N, i \neq j$$

3.2   Link Prediction in Diffusion Network with Structure Transfer

Fig.1. illustrates the framework of the proposed structure transfer scheme. The left part shows the source domain diffusion network with observed network structure and a large number of cascades. The right part is the target domain diffusion

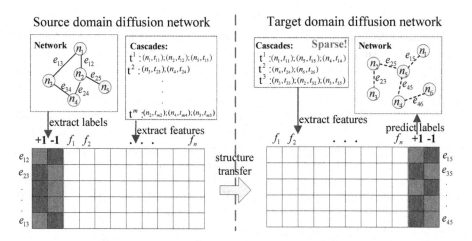

Fig. 1. An illustration of the proposed structure transfer scheme

network with sparse cascades and hidden network structure. In the network, n_i denotes node i, and e_{ij} denotes the edge between node i and j. In a cascade $\mathbf{t}^i$, we use (n_j, t_{ij}) to denote node j is infected at time t_{ij}. Given the two domain diffusion networks, our goal is to borrow the structure and cascade information of the source domain to help infer the network structure in the target domain. To this aim, we formulate it as a link prediction task. Specifically, we first extract features $\{f_1, f_2..., f_n\}$ from the cascades and extract link labels $l_{ij} \in \{+1, -1\}$ in the source domain network. $l_{ij} = +1$ means there exist an edge between node i and j, and otherwise $l_{ij} = -1$. In such a way we extract training samples from the source domain. Then we apply transfer learning technique to select training samples and use them to help predict the link labels in the target domain. Based on the brief description of the framework, next we will elaborate this scheme.

Traditionally, link prediction can be considered as a supervised classification task by constructing a set of features, such as neighborhood based features and path based features [4,10,11]. Motivated by this, we also formulate the network inference problem as a supervised classification problem since the links in the source domain network are known. However, the challenge is that the links in the target domain network are hidden and we cannot construct the features used in traditional link prediction setting. Alternatively, we can extract features from cascades. For example, if node i and j have never appeared in a cascade simultaneously, we can infer that there is probably no link between them; and if node i is the root node of a cascade with node j as the first child node, we can infer that there is definitely a link from i to j. In all we extract 16 cascade related features whose detailed descriptions are given in Table 1.

With the extracted features, we next utilize a popular transfer learning algorithm TrAdaBoost [9] to leverage the links of the source domain network to help us predict the links in the target domain network. TrAdaBoost is a transfer learning framework extended from AdaBoost. Given the limited number of training instances $\mathcal{M}_T^l$ and some test instances $\mathcal{T}$ in the target domain, TrAdaBoost aims to utilize the large volume of available labeled training instances $\mathcal{M}_S$ in the source domain to build a model $f : X \rightarrow Y$ such that the prediction error on $\mathcal{T}$ is minimized. Formally, let X_S be the instances in the source domain network, X_T be the instances in the target domain network, and $Y = \{-1, +1\}$ be the set of labels. Given the source domain network G_S whose edges are known and the target domain network G_T whose edges are hidden, we first assume that their label distribution is the same $\mathcal{Y}_S = \mathcal{Y}_T$, but the feature distribution is different $P_S(y|x) \neq P_T(y|x)$. To utilize TrAdaBoost, we further assume that a small number of labels of the instances X_T^l in the target domain network G_T is given. Therefore, the training data set $\mathcal{M} \subseteq \{X \times Y\}$ includes two parts: $\mathcal{M}_S$, and $\mathcal{M}_T^l$. $\mathcal{M}_S$ represents the source domain network data that $\mathcal{M}_S = \{(x_i^S, y_i^S)\}$, where $x_i^S \in X_S(i = 1, ..., n)$. $\mathcal{M}_T^l$ represents a small number of training data $\mathcal{M}_T^l = \{(x_j^T, y_j^T)\}$ in the target domain network, where $x_j^T \in X_T(j = 1, ..., m)$. n and m are the sizes of $\mathcal{M}_S$ and $\mathcal{M}_T^l$, respectively. By applying TrAdaBoost, we can finally obtain a label matrix $\mathbf{L}$ with $l_{ij} \in \{-1, +1\}$ denoting whether there exists a link from node i to j in the target domain network G_T.

Table 1. Cascade related features for structure transfer

feature	description
f_1	whether node i and j appear in at least one cascade simultaneously, and $t_i < t_j$
f_2	whether there exists a cascade with node i as the root node and node j as its first child node
f_3	the relative frequency of node i appearing before node j in all the cascades
f_4	the minimum time lag $min\Delta t_{ij}^c$ between node i and j in all the cascades
f_5	the average time lag $ave\Delta t_{ij}^c$ between node i and j in all the cascades
f_{6-8}	the maximum probability $maxf(\Delta_{ij}^c; 1)$ of node i infecting node j in all the cascades with three models
f_{9-11}	the average probability $avef(\Delta_{ij}^c; 1)$ of node i infecting node j in all the cascades with three models
f_{12}	for all the cascades that node i is before j, the minimum number of nodes $minN_{ij}^c$ between i and j
f_{13}	for all the cascades that node i is before j, the average number of nodes $aveN_{ij}^c$ between i and j
f_{14}	for all the cascades that node i is before j, the minimum number of nodes $minN_{ri}^c$ between root node r and i
f_{15}	for all the cascades that node i is before j, the minimum number of nodes $minN_{rj}^c$ between root node r and j
f_{16}	for all the cascades that node i is before j, the minimum sum of nodes $min(N_{ri}^c + N_{rj}^c)$ between root node r and i, j

3.3 TrNetInf: Network Inference Incorporating Structure Transfer

By solving the generative model in Eq. (6), we can infer a network matrix $\mathbf{A}$; while by structure transfer with TrAdaBoost, we can obtain a label matrix $\mathbf{L}$. In this section, we will describe how to combine the two parts.

Both methods can infer the connectivity of the target network independently, but the knowledge they used coming from different domains. The generative model only uses the cascades in the target network, and the link prediction based approach mainly relies on the structure knowledge transferred from the source domain network. The results of the two methods may be quite different, and their overlapping part is more likely to be accurate. Thus besides maximizing the probability of generating all the cascades in the target domain, we also want to minimize the difference between the inferred network links by the generative model and the predicted links by diffusion network transfer. We propose to achieve the two goals simultaneously by solving such an optimization problem

$$min_{\mathbf{A}} \quad -\sum_{c\in\mathbb{C}} \log g(\mathbf{t}^c; \mathbf{A}) + \gamma ||\mathbf{L} - \mathbf{A}||_2, \tag{7}$$

$$\text{s. t. } \alpha_{ij}^k \geq 0; \ l_{ij} = \{0,1\}; \ i,j = 1,...,N \text{ and } i \neq j$$

where $\mathbf{A} = \{\alpha_{ij}|i,j = 1,...,N, i \neq j\}$ are the variables and $\mathbf{L} = \{l_{ij}|i,j = 1,...,N, i \neq j\}$ contains the link labels from structure transfer.

Eq. (7) contains two parts. The first part computes the likelihood of the inferred network generating all the cascades, and we want it to be as high as possible. The second part incorporates the structure knowledge transferred from the source domain network. We expect the difference between the two results as small as possible by minimizing the L2 norm distance between $\mathbf{L}$ and $\mathbf{A}$. γ is a parameter used to control the importance of knowledge transferred from the source domain network. Smaller γ implies we trust more on the inferred network by generative model, while larger γ means we rely more on the transferred structure knowledge when available cascades are insufficient.

In addition, most networks are sparse in a sense that one node usually is connected to a small number of other nodes [1,20]. In order to encourage a sparse solution, we add a L2 norm penalty term $||\mathbf{A}||_2$. With the penalty term to control the sparsity of the network, we finally have such an optimization problem

$$min_{\mathbf{A}} \quad -\sum_{c\in\mathbb{C}} \log g(\mathbf{t}^c; \mathbf{A}) + \gamma_1 ||\mathbf{L} - \mathbf{A}||_2 + \gamma_2 ||\mathbf{A}||_2 \tag{8}$$

$$\text{s. t. } \alpha_{ij}^k \geq 0; \ l_{ij} = \{0,1\}; \ i,j = 1,...,N \text{ and } i \neq j$$

We have the following theorem to guarantee that the solution to the optimization problem in Eq. (8) is unique and consistent.

Theorem 1. *Given the optimization problem in Eq. (8), the following results hold:*

1. *Given the log-concave survival functions and concave hazard functions, the problem defined by Eq. (8) is strictly convex in **A** [21].*
2. *The optimization problem defined by Eq. (9) is convex for the proposed TrNetInf model with exponential, Rayleigh, and power law distributions.*
3. *The solution to Eq.(8) gives a unique and consistent maximum likelihood estimator.*

Proof Sketch. 1) Manuel et al. have proved that given the log-concave survival functions and hazard functions in the parameters of the pairwise transmission likelihoods by the exponential, power-law, and Rayleigh models, $\sum_{c \in \mathbb{C}} \log g(\mathbf{t}^c; \mathbf{A})$ is strictly convex in **A** [21]. 2) Due to the fact that all the norm functions are convex, we can further infer that both $||\mathbf{L} - \mathbf{A}||_2$ and $||\mathbf{A}||_2$ are convex. As the convex function follows from linearity and composition rules, the liner combination of the three convex functions is also a convex function. 3) For a strictly convex function, its global minimum is unique. Based on the criteria for consistency of identification, continuity and compactness defined by Newey and Mcfadden [27], we can further infer that the solutions to Eq.(8) is consistent. Due to space reason, we omit the proof here, and one can refer [21] for more details.

Solving TrNetInf. Since we have proved Eq. (8) is convex and the solution is unique, we can use a regular convex optimization algorithm to solve Eq. (8). Here we use CVX[1], a popular Matlab-based convex optimization package to solve this problem. We run the algorithm on a Dell PowerEdge T620 server with 32 cores Intel(R) Xeon(R) CPU E5-2670 2.60 GHz, and 64 GB main memory, running the Ubuntu 13.04 operating system.

4   Experimental Results

In this section we conduct a systematic empirical study on real datasets to verify the effectiveness of TrNetInf in inferring diffusion network with sparse cascades. We first introduce the experiment setup, including the used datasets and baselines. Next we give the parameter analysis to show how sensitive the proposed approach is to the parameters γ_1 and γ_2. Then we report the quantitive comparison results with baselines including state-of-the-art methods.

4.1   Experiment Setup

We use two real-world datasets to evaluate TrNetInf: MemeTracker dataset[2] [7] and AMiner citation network dataset[3] [5,8].

MemeTracker Dataset. The MemeTracker dataset contains more than 300 million blog posts and news articles collected from 3.3 million websites. *Memes*

---

[1] http://cvxr.com/cvx/

[2] http://www.memetracker.org/data.html

[3] http://arnetminer.org/citation

Table 2. Dataset statistics

MemeTracker Datasets			
phrase cluster	# of nodes	# of edges	# of cascades
"good morning America" (Target)	2,754	4,822	425
"put lipstick on a pig" (Source)	2,845	4,621	336
"I'm a mac I'm a pc" (Target)	1,766	2,303	207
"daily show Jon Stewart" (Source)	1,637	2,255	263
AMiner Citation Network Dataset			
research field	# of nodes	# of edges	# of cascades
Computer Theory (Target)	19,073	20,220	832
Graphic (Source)	16,469	21,705	707

are short textual phrases or quotes (like, "good morning America") that spread through the web. Each meme m can be considered as a piece of information, and all the time-stamped webpages which contain meme m forms a diffusion cascade. Memes related to the same topic are considered to be in a same cluster. With the aim of structure transfer, we consider memes in the same cluster coming from the same domain, and memes in different clusters coming from different domains. Given a meme cluster C_m, we first extracted the cascades collection $\mathbb{C}$, and all the websites containing one phrase in C_m as the nodes. For some memes with very long diffusion paths, we split it into several small cascades with length less than 30. The ground truth of the network is constructed by extracting the hyperlinks among all the extracted websites. If a site s_i publishes a phrase and uses a hyperlink to refer to another site s_j that also publishes a similar phrase, we think there exists a link from s_j to s_i.

AMiner Citation Network Dataset. The AMiner citation network dataset contains the citation relationships among papers extracted from DBLP, ACM, and other sources. The citation relationships among papers can be naturally considered as the ground truth of the diffusion network. Similar to MemeTracker dataset, we also consider some term pair phrases (like, "deep learning") extracted from the paper titles and abstracts as the information, and all the papers containing the same phrase can be considered as a cascade. To enable structure transfer, we distinguish the diffusion networks of different domains based on the research fields such as database and computer theory. For example, papers published in the field of database can be considered coming from a domain and those published in computer theory can be consider coming from another domain.

In our experiment, we extract four meme clusters from the MemeTracker dataset forming two groups of datasets for evaluation. For each group of dataset, we use one as the source domain data donated by "Source" and the other as the target domain data denoted by "Target". Similarly, we select the papers published in the venues of two research fields: computer theory and graphic from the AMiner dataset forming another group of dataset. Statistics of the datasets is given in Table 2. We compare TrNetInf with the following baselines.

- **NETRATE**[4] [21]. NETRATE is a representative model to infer both the connectivity of the network and the transmission rates over each edge. As the most relevant work to the proposed model, we choose it as a baseline.
- **NetInf**[5] [23]. Another type of network inference model only aims to infer the network connectivity, such as NetInf. To compare with such kind of methods, we choose the representative approach NetInf as the second baseline.
- **TrNetInf without Sparsity Penalty (TrNetInf-SP).** To study whether and to what extent the sparsity penalty can affect algorithm performance, we use TrNetInf without sparsity penalty as a baseline. For this baseline, we simply set the parameter $\gamma_2 = 0$.
- **TrNetInf without Structure Transfer (TrNetInf-ST)** Similarity, we also use the TrNetInf without structure transfer as a baseline to study how much improvement can be achieved by incorporating structure transfer. In this case, we set the parameter $\gamma_1 = 0$.
- **Link Prediction with Structure Transfer (LPST).** As the proposed TrNetInf combines the information from link prediction model, we use this baseline to study how well the pure link prediction model can perform on the network inference problem and how much achievement can be achieved by TrNetInf. For the LPST baseline, we use TrAdaBoost as the classifier.

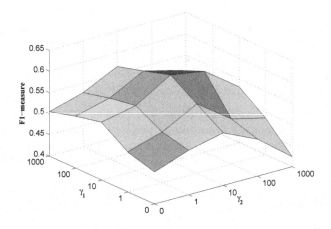

Fig. 2. F1-measure on "good morning America" dataset with various γ_1 and γ_2

4.2   Parameter Analysis

We first study the effect of parameters γ_1 and γ_2 on the performance of TrNetInf. Due to space limitation, we only report the result of the first group of Meme-Tracker dataset. The results of the other datasets are similar.

---

[4] http://people.tuebingen.mpg.de/manuelgr/netrate/
[5] http://snap.stanford.edu/netinf/

Fig. 2. shows the F1-measure of the "good morning America" dataset with "put lipstick on a pig" as the source domain network over various γ_1 and γ_2. One can see that with the increase of γ_1, the performance first increases, and then decreases, and finally becomes stable. It implies that structure transfer does help our task as the F1-measure are mostly higher than non-transfer with $\gamma_1 = 0$. From $\gamma_1 = 100$ on, the performance tends to be stable, which means the transferred structure knowledge dominates the final results when γ_1 is large. One can also see the F1-measure further increases if we add the sparsity penalty weighted by γ_2, but too large a γ_2 will also hurt the performance. How to choose a proper γ_2 may largely depend on the prior knowledge on the network. A denser network prefers a smaller γ_2, and a larger γ_2 means we may want to infer a less dense network. Fig. 2. suggests that $\gamma_1 = 10$, $\gamma_2 = 10$ seem a good choice of the two parameters for the MemeTracker dataset, and in the following experiments we choose $\gamma_1 = 10$, $\gamma_2 = 10$ as our default parameter settings.

4.3   Quantitive Comparison with Baselines

We quantitively evaluate the performance of TrNetInf via three measures: precision, recall, and F1-measure. We first study the effectiveness of TrNetInf with insufficient cascades by comparing with two state-of-the-art network inference approaches NETRATE and NetInf. To utilize TrAdaBoost for knowledge transfer, some link labels in the target domain network need to be available. In our experiment, we assume 1% links in the target domain network are given.

Comparison Against Network Inference Models. Fig. 3. shows the precision-recall curves of three approaches: NETRATE, NetInf, and TrNetInf over the three datasets. One can see that TrNetInf outperforms NETRATE and NefInf on the three datasets in terms of precision-recall. It implies that the performance can be improved if the structure knowledge is properly transferred. The result also shows that the AMiner dataset seems easier to infer than the two MemeTracker datasets.

Evaluation with Sparse Cascade Data. To study the effectiveness of TrNetInf with insufficient cascades, we compare the F1-measure achieved by TrNetInf against NETRATE and NetInf by sampling different numbers of cascades in the target domain network. Fig. 4. shows the F1-measures of the three approaches over various numbers of cascades. One can observe that TrNetInf achieves significantly higher F1-measure than NETRATE and NetInf when the number of cascades is relatively small. With the increase of the number of cascades, the performance of the three methods tends to be similar. It implies that structure transfer is especially helpful when the cascade data are very sparse. The performance of the two baselines becomes closer to TrNetInf when more and more cascades are available. It implies that the improvement by structure transfer becomes less significant when a large volume of cascades are available.

Comparison Against Two Variations and Link Prediction Models. Next, we conduct experiment to study whether transfer learning and sparsity

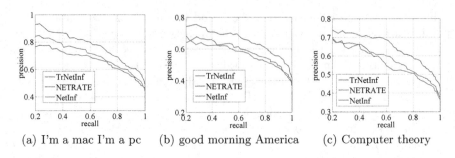

Fig. 3. The precision-recall curves of the three approaches on three groups of datasets

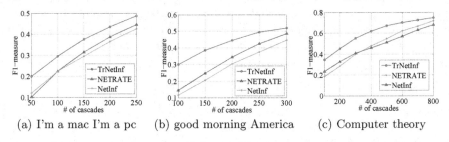

(a) I'm a mac I'm a pc   (b) good morning America   (c) Computer theory

Fig. 4. The F1-measure of the three approaches with various numbers of cascades

penalty can both help the network inference task. To this aim, we compare TrNetInf with two variations: TrNetInf without sparsity penalty (TrNetInf-SP) and TrNetInf without structure transfer (TrNetInf-ST). We report precision, recall, and F1-measure for each method on each dataset in Table 3. The figures in bold show the best results. One can see that TrNetInfer is consistently better than TrNetInf-SP and TrNetInf-ST. On average, the F1-measure has improved by about 4% compared with TrNetInf-SP on the three groups of datasets. Compared with TrNetInf-ST, the improvement is more significant, more than 13%. The result leads us to conclude that 1) sparsity penalty do help the studied task, and 2) transfer learning can significantly improve the performance. We also report the performance of link prediction with structure transfer model LPST. One can see that although slightly worth than TrNetInf-ST, LPST model still

Table 3. Experimental result by comparing TrNetInf against two variations and LPST

Method	"I'm a mac I'm a pc"			"good morning America"			Computer theory		
	precision	recall	F1	precision	recall	F1	precision	recall	F1
TrNetInf	0.575	**0.611**	**0.593**	0.621	**0.635**	**0.628**	**0.651**	0.700	**0.675**
TrNetInf-SP	0.557	0.598	0.576	0.601	0.598	0.600	0.622	0.657	0.640
TrNetInf-ST	0.534	0.515	0.524	0.546	0.526	0.536	0.540	**0.704**	0.611
LPST	**0.579**	0.379	0.458	0.515	0.534	0.524	0.534	0.598	0.564

gives rather good prediction results. It means that properly structure transfer can provide us useful information for better inferring the diffusion network.

5   Related Work

The problem of inferring the diffusion networks and estimating the diffusion probabilities has been extensively studied in many domains, such as the hyperlink network of on-line new articles [21–23], the coloration network of scientist [20], and the following network in social media [2,17,28]. Previous related works on this topic can be roughly divided into inferring the network structure [23] and inferring both the network structure and the transmission rates between nodes [21]. The representative work on inferring the network structure is NetInf [23]. NetInf formulates this problem as a submodular function maximization problem. NETRATE is a representative approach to infer the diffusion network through estimating the pairwise transmission rates between two nodes. Based on the general inference models, some fine-grained models are proposed. [17] and [19] studied the topic-level diffusion network inference problem.

A related research topic to the network inference problem is link prediction. Link prediction aims to predict the likelihood of a future association between nodes, knowing that there is no association between the nodes in the current state of the graph [4,11]. One of the earliest link prediction models is proposed by Liben-Nowell and Kleinberg [29]. Their proposed approach typically extracts the similarity between a pair of vet ices by various graph-based similarity metrics. Then they use the ranking on the similarity scores to predict the link between two vertices. Besides similarity ranking based approach, another popular approach is to model the link prediction problem as a supervised classification problem [4,10,11]. Such methods normally learn a prediction model by constructing a set of features, such as neighborhood based features [4] and path based features [10]. The main difference between link prediction and network inference is that link prediction aims to predict the future potential connections between nodes based on their current states. In the network inference setting, the network structure is totally hidden and needs to be inferred from traces of information diffusion.

6   Conclusion

To address the problem that traditional inference models may not be effective when lacking enough cascade data, in this paper we proposed a structure transfer scheme to infer the diffusion network with the help of an external diffusion network. We first formulated the network inference problem as a link prediction task by extracting cascade related features. This formulation thus enabled us effectively transfer the cascades and links of the external diffusion network to help predict the hidden links of the target domain network. We also proposed a unified optimization framework to integrate the traditional generative model and the proposed transfer learning model. Evaluations on two real-world datasets demonstrated the effectiveness of the proposed scheme.

In the future, we are particularly interested in further investigating: 1) How to extend one source domain to many source domains. Currently we only consider one source domain diffusion network, but multiple source domains may be more helpful as more information are available [30]. 2) Given multiple source domain diffusion networks, how to select the source domains that are most relevant to the target domain. Currently we only use the domain data which are highly relevant to the target domain. A domain that are irrelevant may also hurt the performance. Source domain diffusion network selection is an interesting and challenging research issue we will focus on in the future.

Acknowledgments. This work is supported in part by NSFC (Nos. 61170189, 61370126, 61202239), the Fund of the State Key Laboratory of Software Development Environment (No. SKLSDE- 2013ZX-19), the Fundamental Research Funds for the Central Universities (YWF-14-JSJXY-16), the Innovation Foundation of BUAA for PhD Graduates (No. YWF-14-YJSY-021), Microsoft Research Asia Fund (No. FY14-RES-OPP-105), NSF through grants CNS-1115234, and OISE-1129076.

References

1. Gomez-Rodriguez, M., Leskovec, J., Scholkopf, B.: Modeling information propagation with survival theory. In: ICML (2013)
2. Wang, S.Z., Yan, Z., Hu, X., Yu, P.S., Li, Z.J.: Burst time prediction in cascades. In: AAAI (2015)
3. Wang, L., Ermon, S., Hopcroft, J.E.: Feature-enhanced probabilistic models for diffusion network inference. In: Flach, P.A., De Bie, T., Cristianini, N. (eds.) ECML PKDD 2012, Part II. LNCS, vol. 7524, pp. 499–514. Springer, Heidelberg (2012)
4. Lu, L.Y., Zhou, T.: Link Prediction in Complex Networks: A Survey. Physica A: Statistical Mechanics and its Applications **390**(6), 1150–1170 (2011)
5. Tang, J., Zhang, D., Yao, L.M.: Social network extraction of academic researchers. In: ICDM (2007)
6. Zhang, J.W., Yu, P.S., Zhou, Z.H.: Meta-path based multi-network collective link prediction. In: KDD (2014)
7. Leskovec, J., Backstrom, L., Kleinberg, J.: Meme-tracking and the dynamics of the news cycle. In: KDD (2009)
8. Tang, J., Zhang, J., Yao, L.M., Li, J.Z., Zhang, L., Su, Z.: Arnetminer: extraction and mining of academic social networks. In: KDD (2008)
9. Dai, W.Y., Yang, Q., Xue, G.R., Yu, Y.: Boosting for transfer learning. In: ICML (2009)
10. Hasan, M.A., Chaoji, V., Salem, S., Zaki, M.: Link prediction using supervised learning. In: SDM (2006)
11. Lichtenwalter, R.N., Lussier, J.T., Chawla, N.V.: New perspective and methods in link prediction. In: KDD (2010)
12. Jiang, W., Chung, F.: Transfer spectral clustering. In: Flach, P.A., De Bie, T., Cristianini, N. (eds.) ECML PKDD 2012, Part II. LNCS, vol. 7524, pp. 789–803. Springer, Heidelberg (2012)
13. Pardoe, D., Stone, P.: Boosting for regression transfer. In: ICML (2010)
14. Zhu, Y., Chen, Y.Q., Lu, Z.Q., Pan, S.J., Xue, G.R., Yu, Y., Yang, Q.: Heterogeneous transfer learning for image classification. In: AAAI (2011)

15. Pan, S.J., Yang, Q.: A Survey on Transfer Learning. IEEE Trans. on Knowl. and Data Eng. **22**(10), 1345–1359 (2010)
16. Herlihy, M.: Diffusion in Organizations and Social Movements: From Hybrid Corn to Poison Pills. Annual Review of Sociology **24**, 265–290 (1998)
17. Wang, S.Z., Hu, X., Yu, P.S., Li, Z.J.: MMRate: inferring multi-aspect diffusion networks with multi-pattern cascades. In: KDD (2014)
18. Erdman, D.D.: Propagation and Identification of Viruses. Topley and Wilson's Microblology and Microblal Infections (2010)
19. Du, N., Song L., Woo, H., Zha, H.Y.: Uncover topic-sensitive information diffusion networks. In: AISTATS (2013)
20. Myers, S.A., Leskovec, J.: On the convexity of latent social network inference. In: NIPS (2010)
21. Gomez-Rodriguez, M., Balduzzi, D., Scholkopf, B.: Uncovering the temporal dynamics of diffusion networks. In: ICML (2011)
22. Gomez-Rodriguez, M., Leskovec, J., Scholkopf, B.: Structure and dynamics of information pathways in online media. In: WSDM (2013)
23. Gomez-Rodriguez, M., Leskovec, J., Krause, A.: Inferring networks of diffusion and influence. In: KDD (2010)
24. Leskovec, J., Singh, A., Kleinberg, J.M.: Patterns of influence in a recommendation network. In: Ng, W.-K., Kitsuregawa, M., Li, J., Chang, K. (eds.) PAKDD 2006. LNCS (LNAI), vol. 3918, pp. 380–389. Springer, Heidelberg (2006)
25. Chen, W., Wang, C., Wang, Y.J.: Scalable influence maximization for prevalent viral marketing in large-scale social networks. In: KDD (2010)
26. Kempe, D., Kleinberg, J., Tardos, E.:Maximizing the spread of influence through a social network. In: KDD (2003)
27. Newey, W.K., McFadden, D.: Large sample estimation and hypothesis testing. In: Handbook of Econometrics, pp. 2111–2245 (1994)
28. Wang, L., Ermon, S., Hopcroft, J.E.: Feature-enhanced probabilistic models for diffusion network inference. In: Flach, P.A., De Bie, T., Cristianini, N. (eds.) ECML PKDD 2012, Part II. LNCS, vol. 7524, pp. 499–514. Springer, Heidelberg (2012)
29. Liben-Nowell, D., Kleinberg, J.: The Link Prediction Problem for Social Networks. Journal of the American Society for Information Science and Technology **58**(7), 1019–1031 (2007)
30. Chen, Z.Y., Liu, B.: Topic modeling using topics form many domains, lifelong learning and big data. In: ICML (2014)

Information Integration
and Data Quality

Scalable Inclusion Dependency Discovery

Nuhad Shaabani$^{(\boxtimes)}$ and Christoph Meinel

Hasso-Plattner-Institut, University of Potsdam,
Prof.-Dr.-Helmert-Str. 2-3, 14482 Potsdam, Germany
{nuhad.shaabani,christoph.meinel}@hpi.de
http://www.hpi.de

Abstract. Inclusion dependencies within and across databases are an important relationship for many applications in anomaly detection, schema (re-)design, query optimization or data integration. When such dependencies are not available as explicit metadata, scalable and efficient algorithms have to discover them from a given data instance.

We introduce a new idea for clustering the attributes of database relations. Based on this idea we have developed S-INDD, an efficient and scalable algorithm for discovering all unary inclusion dependencies in large datasets. S-INDD is scalable both in the number of attributes and in the number of rows. We show that previous approaches reveal themselves as special cases of S-INDD. We exhaustively evaluate S-INDD's scalability using many datasets with several thousands attributes and rows up to one million. The experiments show that S-INDD is up to 11x faster than previous approaches.

Keywords: Inclusion dependency · Data integration · Data profiling

1 Introduction

Dependencies are metadata that describe relationships between relational attributes. Dependencies play very important roles in database design, data quality management, and knowledge representation. In the case that they are modeled as part of the application requirements, they are then used in database normalization and are implemented in the designed database to ensure data quality. In contrast, dependencies in knowledge discovery are extracted from the existing data of the database. The extraction process is called dependency discovery and aims to find dependencies satisfied by existing data. A typical type of dependency is inclusion dependencies (INDs), which represent value reference relationships between two sets of attributes. Together with functional dependencies, they represent an important part of database semantics.

In the context of data integration, the discovery of inclusion dependencies can help to solve a very common and difficult problem: discovering foreign key constraints. There are many reasons for an absence of foreign key constraints in databases. These include a simple lack of domain knowledge within the development team during the design and development time, the worry that checking

© Springer International Publishing Switzerland 2015
M. Renz et al. (Eds.): DASFAA 2015, Part I, LNCS 9049, pp. 425–440, 2015.
DOI: 10.1007/978-3-319-18120-2_25

such constraints by the hosted system would hamper database performance, or the lack of support for checking foreign key constraints in the host system.

The manual search for INDs by domain experts is usually not feasible due to the large number of data sources, a widespread lack of reliable metadata about legacy databases, and the possibility of a high number of attributes in real-world relations. Therefore, efficient and scalable algorithms to detect INDs enable easy integration of new data sources that previously would not have been used, because their relationships with existing data was not known.

N-ary INDs cover pairs of n attributes, while unary INDs (uINDs) cover only pairs of single attributes (formal definitions are in Sec. 2). All known algorithms for detecting high-dimensional INDs require the discovery of all unary inclusion dependencies (single-column INDs) [10–13]. This is because any valid IND of a size greater than one implies that all unary INDs derivable from it have to be valid in the same database. This means that the reliability of the algorithms for detecting high-dimensional INDs is dependent on a scalable and efficient discovery of unary INDs.

There are three approaches in related work focused on exhaustive detecting single-column inclusion dependencies: Bell and Brockhausen [3], De Marchi et al. [11,13], and Bauckmann et al. [1,2] (see Sec. 6).

The algorithm proposed in [11,13] for discovering unary INDs uses an inverted index associating every value in the database with the set of all attributes having this value. Because for every attribute A the intersection of all attribute sets containing A is the set of all attributes including A, the algorithm runs through all values in order to compute such an intersection for every attribute. However, this approach is inefficient because an attribute set in the index can be associated with many different values. This means, the algorithm executes a lot of redundant intersection operations. These operations are very costly if the dataset has a large number of attributes sharing a lot of values.

The first research question addressed in this paper is how we can eliminate such redundant operations caused by using the inverted index. We tackle this problem by introducing the concept of *attribute clustering* (see Sec. 3).

SPIDER [1,2] is an external algorithm that writes the values of every attribute to a file after sorting them and removing duplicate values. Then it opens all files at once and starts comparing the values in parallel and in the same way in which the merge-sort algorithm does. During this process, SPIDER applies an efficient method for discarding unsatisfied unary INDs (see Sec. 6 for more details). SPIDER outperforms the approach proposed in [11] up to orders of magnitude [1,2]. However, the drawback in SPIDER's approach is its dependency on the number of attributes. This means, that by increasing the number of attributes, SPIDER's scalability decreases: the number of I/O- operations increases because the size of buffers allocated for the opened files becomes smaller.

The second research challenge addressed in this paper is how we can make SPIDER independent from the number of attributes in order to improve its scalability in two dimensions: in the number of attributes and the number of rows.

Table 1. Running example

A	B	C	D
1	1	5	1
2	2	5	1
2	3	6	3
4	4	7	3

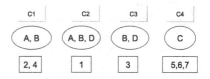

Fig. 1. Attribute clustering based on the data of table 1

We tackle this challenge by devising S-INDD, a scalable approach for computing the *attribute clustering* (see Sec. 4).

Every cluster in the *attribute clustering* is a subset of attributes sharing a subset of values that can not be shared by the attributes of any different cluster in the *attribute clustering*. E.g., $\{\{A,B\},\{A,B,D\},\{B,D\},\{C\}\}$ shown in figure 1 is the *attribute clustering* over the values of table 1. Attributes A and B shape cluster $\mathcal{C}_1$ because both share the values $\{2,4\}$ that can not be shared by the attributes of $\mathcal{C}_2$, $\mathcal{C}_3$, or $\mathcal{C}_4$. Every attribute of the attribute set, denoted by $\mathcal{A}$, must be contained in at least one cluster.

Clustering the attributes in this way allows us to derive the following inference rule[1]: Attribute X is included in attribute Y if and only if every cluster containing X contains Y. E.g., the set of D's values in table 1 is included in the set of B's values because the clusters $C2$ and $C3$ that both contain D also contain B, but B's values are not included in D's values because cluster $C1$ contains B and does not contain D.

For every attribute A, S-INDD stores all elements of the set $\mathcal{V}_A \times \{\{A\}\}$ ($\mathcal{V}_A$ denotes the value set of A) as a sorted list in an external repository. Then, for every value $v \in \mathcal{V}$ ($\mathcal{V}$ denotes the whole set of values in the dataset), S-INDD computes incrementally the set of attributes, denoted by $\mathcal{A}^v$, whose value sets contain v. The incremental computing of the sets $\mathcal{A}^v$ is achieved by executing a sequence of merging operations. Every merging operation merges simultaneously k lists from the repository ($k > 1$ is a given number) and replaces them with a new list. The new list contains the union of all sets $\mathcal{A}^v$ contained in the k lists read previously. In this way the sets $\mathcal{A}^v$ are incrementally computed. Such merging continues until the repository contains less than k lists. After finishing merging, S-INDD generates the clusters from the remaining lists by processing them in parallel. The possibility that S-INDD can control the number of lists to be merged makes its scalability independent from the number of attributes.

To handle a large dataset with a very large number of rows S-INDD partitions the whole dataset and computes the attribute clustering of every partition. The whole attribute clustering is then the union of all attribute clusterings of all partitions (see Sec. 4.2). This method makes the S-INDD's scalability independent from the number of rows.

---

[1] This rule is a generalization of property 1 formulated in [11]

Contributions. (1) We introduce the concept of *attribute clustering*, a new concept for inferencing all unary inclusion dependencies much more efficiently than using the inverted index introduced in [11].

(2) We devise S-INDD, a scalable algorithm for computing the *attribute clustering* in large datasets. Its scalability neither dependents on the number of attributes nor on the number of rows.

(3) We experimentally validate S-INDD on real and synthetic datasets and compare it with SPIDER [1,2]. The results show that S-INDD is up to 11x faster than Spider. Furthermore, we show that SPIDER is a special case of S-INDD.

2   Preliminaries

Let $\mathcal{A}$ be a finite set of attributes. Each attribute $A \in \mathcal{A}$ has an associated domain $dom(A)$, which defines the set of all its possible values. For $A_1, A_2, \ldots A_n \in \mathcal{A}$ and for a symbol R, $R[A_1, A_2, \ldots, A_n]$ is called a relational schema over $A_1, A_2, \ldots, A_n$ and R is the relation name. A tuple t over R is an element from $dom(A_1) \times dom(A_2) \times \cdots \times dom(A_n)$. For a tuple t over R and $X \subseteq \mathcal{A}$, we use $t[X]$ to denote the projection of t to X. A finite set r of tuples over R is called an instance of R. For an instance r of R and for a sequence X of attributes in R, the projection of r onto X, denoted by $\pi_X(r)$, is defined as $\pi_X(r) = \{t[X] \mid t \in r\}$.

A set $\mathcal{R}$ of relational schemata $R_i[A_{i,1}, \ldots, A_{i,n_i}]$, where $A_{i,1}, \ldots, A_{i,n_i} \in \mathcal{A}$, $1 \le n_i \le |\mathcal{A}|$ and $1 \le i \le m = |\mathcal{R}|$, is called a database schema. A relational database instance $\mathcal{D}$ over $\mathcal{R}$ is a set of instances r_i over each $R_i \in \mathcal{R}$.

Definition 1. *(Inclusion dependency) Let $R_i[A_{i,1}, \ldots, A_{i,n_i}]$ and $R_j[A_{j,1}, \ldots, A_{j,n_j}]$ be two relational schemata. Let X be a set of k distinct attributes from R_i and Y a set of k distinct attributes from R_j, with $1 \le k \le \min(n_i, n_j)$. An inclusion dependency (IND) is an assertion of the form $R_i[X] \subseteq R_j[Y]$ where k is the size of the IND. For $k = 1$ the inclusion dependency is called a unary inclusion dependency (uIND).*

Definition 2. *(IND satisfaction) Let $\mathcal{D}$ be a database over a database schema $\mathcal{R}$. An inclusion dependency $R_i[X] \subseteq R_j[Y]$ over $R_i, R_j \in \mathcal{R}$ is satisfied or valid in $\mathcal{D}$ iff $\forall u \in r_i, \exists v \in r_j$ such that $u[X] = v[Y]$.*

Thus, a satisfied IND $R_i[X] \subseteq R_j[Y]$ states that every value combination for attribute set X in relation R_i is also present as a value combination of attribute set Y in R_j. INDs are a prerequisite for foreign keys, and their discovery is particularly helpful to understand how records of two relations might be joined.

To simplify the formulation of the algorithm, we assume without loss of generality that attribute names are unique across all relations. Under this assumption, we can denote a unary inclusion dependency $R_i[A] \subseteq R_j[B]$ by $A \subseteq B$. We also define the two sets $\mathcal{V}_A$ and $\mathcal{V}$ to ease notation:

$\mathcal{V}_A$ is the set of A's values occurring in the corresponding instance of the relation schema in which A occurs:

$$\mathcal{V}_A = \{v \in dom(A) \mid \exists R \in \mathcal{R} : A \in R \land v \in \pi_A(r)\}$$

Then $\mathcal{V}$ is the set of all values of all attributes occurring in the database instance.

$$\mathcal{V} = \cup_{R_i \in \mathcal{R}} \cup_{A \in R_i} \mathcal{V}_A$$

It is now obvious that a unary inclusion dependency $A \subseteq B$ is valid if and only if $\mathcal{V}_A \subseteq \mathcal{V}_B$. Accordingly, the discovery of all valid unary inclusion dependencies in a database over a database schema $\mathcal{R}$ is equivalent to the computation of the following set: $\mathcal{I} = \{A \subseteq B \mid A, B \in \mathcal{A} \land \mathcal{V}_A \subseteq \mathcal{V}_B\}$

3   Attribute Clustering

We now formally introduce the concept of *attribute clustering*.

Definition 3. *(Attribute Clustering) The set $\mathcal{AC} \subseteq 2^{\mathcal{A}}$, where $\mathcal{AC} \neq \emptyset$ and $2^{\mathcal{A}}$ is the power set of $\mathcal{A}$, is an attribute clustering over $\mathcal{V}$ if there is a surjective function that maps every value $v \in \mathcal{V}$ to a $\mathcal{C} \in \mathcal{AC}$ that contains all attributes $A \in \mathcal{A}$ with $v \in \mathcal{V}_A$. In other words, $\mathcal{AC}$ is an attribute clustering if there is $f : \mathcal{V} \to \mathcal{AC}$ satisfying the following condition:*

1. *$(\forall \mathcal{C} \in \mathcal{AC})(\exists v \in \mathcal{V}) : f(v) = \mathcal{C}$ (i.e., f is surjective).*
2. *$(\forall v \in \mathcal{V})(\neg \exists A \in \mathcal{A}) : v \in \mathcal{V}_A \land A \notin f(v)$ (i.e., $f(v)$ is the maximal set of attributes $A \in \mathcal{A}$ with $v \in \mathcal{V}_A$).*

Each $\mathcal{C} \in \mathcal{AC}$ is called a cluster. *Clusters need not be mutually disjoint.*

The next lemma shows the relationship between the clusters and the values of the dataset.

Lemma 1. *An attribute clustering $\mathcal{AC} = \{\mathcal{C}_1, \mathcal{C}_2, \dots, \mathcal{C}_c\}$ divides the set $\mathcal{V}$ into $|\mathcal{AC}|$ disjoint partitions $\mathcal{P}_1, \mathcal{P}_2 \dots, \mathcal{P}_c$ so that for every cluster $\mathcal{C}_i \in \mathcal{AC}$ there is a partition $\mathcal{P}_i$ with $\mathcal{P}_i \subseteq \cap_{A \in \mathcal{C}_i} \mathcal{V}_A$.*

Proof. According to definition 3, there is a surjective function $f : \mathcal{V} \to \mathcal{AC}$ where $f(v) = \mathcal{C}$ is the set of the all attributes A with $v \in \mathcal{V}_A$. For each cluster $\mathcal{C}_i(1 \leq i \leq c)$, we can define the set

$$\mathcal{P}_i = f^{-1}(\mathcal{C}_i) = \{v \in \mathcal{V} \mid f(v) = \mathcal{C}_i\} \subseteq \cap_{A \in \mathcal{C}_i} \mathcal{V}_A \tag{1}$$

because f is surjective.

Because any $v \in \mathcal{V}$ can not be mapped to two different clusters, we have

$$\mathcal{P}_i \cap \mathcal{P}_j = \emptyset \text{ for } i \neq j (1 \leq i, j \leq c) \tag{2}$$

Because there is $\mathcal{P}_i(1, \leq i \leq c)$ for any $v \in \mathcal{V}$, we have

$$\cup_{1 \leq i \leq c} \mathcal{P}_i = \mathcal{V} \tag{3}$$

According to (2) and (3), the sets $\mathcal{P}_1, \mathcal{P}_2 \dots \mathcal{P}_c$ are disjoint partitions of $\mathcal{V}$. $\square$

The next lemma states that for each two different attributes A, B, the set of A's values is included in the set of B's values if and only if the intersection of all clusters containing A contains B. In other words, we have the following inference rule: for any attribute A, the set of all attributes including A is the intersection of all clusters containing A.

Lemma 2. *Let $\mathcal{AC} = \{\mathcal{C}_1, \ldots, \mathcal{C}_c\}$ be an attribute clustering over $\mathcal{V}$. Then the following holds:*

$$\forall A, B \in \mathcal{A} : \mathcal{V}_A \subseteq \mathcal{V}_B \Leftrightarrow B \in \cap_{\mathcal{C} \in \mathcal{AC}, A \in \mathcal{C}} \mathcal{C}$$

Proof. 1) "$\Rightarrow$": We assume $B \notin \cap_{A \in \mathcal{C}} \mathcal{C}$. This means, there is $\mathcal{C}$ with $A \in \mathcal{C}$ and $B \notin \mathcal{C}$. According to definition 3, there is at least $v \in \mathcal{V}$ mapped to $\mathcal{C}$ with $v \in \mathcal{V}_A$ and $v \notin \mathcal{V}_B$ because $A \in \mathcal{C}$ and $B \notin \mathcal{C}$. This means, $\mathcal{V}_A \nsubseteq \mathcal{V}_B$, contradicting $\mathcal{V}_A \subseteq \mathcal{V}_B$.

2) "$\Leftarrow$": We assume $\mathcal{V}_A \nsubseteq \mathcal{V}_B$. This means, there is at least $v \in \mathcal{V}$ with $v \in \mathcal{V}_A$ and $v \notin \mathcal{V}_B$. According to definition 3, v can only be mapped to a cluster $\mathcal{C}$ containing all attributes whose value sets contain v. This means, $A \in \mathcal{C}$ and $B \notin \mathcal{C}$ because $v \in \mathcal{V}_A$ and $v \notin \mathcal{V}_B$. This means, $B \notin \cap_{A \in \mathcal{C}} \mathcal{C}$, contradicting $B \in \cap_{A \in \mathcal{C}} \mathcal{C}$ □

We can now formulate the motivation for the introduction of the concept of *attribute clustering* as the answer of the following question.

Why is the deriving of all unary INDs from the *attribute clustering* much more efficient than deriving them from the inverted index?

Let $\mathcal{AC} = \{\mathcal{C}_1, \ldots, \mathcal{C}_c\}$ be an *attribute clustering* and let $\mathcal{P} = \{\mathcal{P}_1, \ldots, \mathcal{P}_c\}$ be the partitions defined by its clusters (see lemma 1). The inverted index defined in [11] can now be formulated as $\mathbb{B} = \cup_{1 \le i \le c}(\mathcal{P}_i \times \{\mathcal{C}_i\})$. Furthermore, let $\mathcal{I}_A$ be the set[2] of all attributes including A. $\mathcal{I}_A$ is initially initialized with $\mathcal{A}$ in [11]. For every subset $\mathbb{B}_i = \mathcal{P}_i \times \{\mathcal{C}_i\} \subseteq \mathbb{B}$, the algorithm in [11] must run through $|\mathcal{P}_i|$ iterations in order to compute the set $\cap_{(v, \mathcal{C}_i) \in \mathbb{B}_i} \mathcal{C}_i \cap \mathcal{I}_A$. However, from all $|\mathcal{P}_i|$ intersections we need only to compute one intersection because the result of the remaining $|\mathcal{P}_i| - 1$ intersections is known, namely the set $\mathcal{C}_i$ itself. This means, using the clusters allows us to save $\Sigma_{1 \le i \le c}|\mathcal{P}_i| - |\mathcal{AC}| = |\mathcal{V}| - |\mathcal{AC}|$ redundant intersection operations compared to using the inverted index. Such intersection operations are very costly if we have a large dataset with a large number of attributes sharing a lot of values.

In fact, the runtime for computing the set $\mathcal{I}$ by using the inverted index is $\mathcal{O}(|\mathcal{V}| \times |\mathcal{A}|^2)$ while it is $\mathcal{O}(|\mathcal{AC}| \times |\mathcal{A}|^2)$ by using the *attribute clustering* (see line 5 in algorithm 1 in Sec. 4.1).

Furthermore, the way in which the inverted index has to be computed and presented has a big impact on the efficiency and the scalability of the algorithm in [11]. However, there is no explicit method suggested in [11] for computing the inverted index (one can only assume that it is computed in [11] as a kind of dictionary data structure presented in the main memory).

---

[2] This set is denoted as $rhs(A)$ in [11]

The scalable computing of the *attribute clustering* is the main objective of S-INDD's development.

The following lemma shows that the *attribute clustering* exists for every database instance $\mathcal{D}$. Its proof can be considered as the proof of S-INDD's correctness because S-INDD incrementally computes the sets $\mathcal{A}^v$ ($v \in \mathcal{V}$) defined in the proof and then generates the set $\mathcal{AC}$ (see Sec. 4.1).

Lemma 3. *For any database instance $\mathcal{D}$ over a database schema $\mathcal{R}$, there always exists an attribute clustering to satisfy Definition 3.*

Proof. For every value $v \in \mathcal{V}$, let $\mathcal{A}^v$ be the set of all attributes A whose values sets contain v. I.e.,

$$\forall A \in \mathcal{A}^v : v \in \mathcal{V}_A \text{ and } \neg \exists A' \in \mathcal{A} : v \in \mathcal{V}_{A'} \wedge A' \notin \mathcal{A}^v \tag{4}$$

For all values $v_{i_1}, v_{i_2}, \ldots, v_{i_j}$ $(1 \le i, j \le |\mathcal{V}|)$ with $\mathcal{A}^{v_{i_1}} = \mathcal{A}^{v_{i_2}} = \cdots = \mathcal{A}^{v_{i_j}}$, we replace the sets $\mathcal{A}^{v_{i_1}}, \mathcal{A}^{v_{i_2}}, \ldots, \mathcal{A}^{v_{i_j}}$ with a set $\mathcal{C}_i$, i.e. $\mathcal{C}_i = \mathcal{A}^{v_{i_1}} = \cdots = \mathcal{A}^{v_{i_j}}$. We show now that the set

$$\mathcal{AC} = \{\mathcal{C}_1, \ldots, \mathcal{C}_c\} = \{\mathcal{C} \mid \exists v \in \mathcal{V} : \mathcal{C} = \mathcal{A}^v\}$$

is an *attribute clustering*:

Assuming, for a $v \in \mathcal{V}$, there are two different sets $\mathcal{C}_i$ and $\mathcal{C}_j$ with at least a common attribute A satisfying $v \in \mathcal{V}_A$. That contradicts (4) and consequently, the construction of the sets $\mathcal{C}_i (1 \le i \le c)$. This means, our assumption is wrong. This means, the function

$$f : \mathcal{V} \to \{\mathcal{C}_1, \ldots, \mathcal{C}_c\} \text{ with } f(v) = \mathcal{C} \text{ where } \mathcal{C} = \mathcal{A}^v$$

satisfies definition 3.   □

The next lemma allows us to increase the scalability of S-INDD in the case of having datasets with a large number of rows (see Sec. 4.2).

Lemma 4. *Let $\mathcal{V}_1, \ldots, \mathcal{V}_n$ be disjoint partitions of the set $\mathcal{V}$ and let $\mathcal{AC}_1, \ldots, \mathcal{AC}_n$ be the corresponding attribute clusterings. Then $\cup_{1 \le i \le n} \mathcal{AC}_i$ is an attribute clustering over $\mathcal{V}$.*

Proof. For any $\mathcal{AC}_i$ $(1 \le i \le n)$ we can define a function $f_i : \mathcal{V}_i \to \mathcal{AC}_i$ satisfying definition 3 because $\mathcal{AC}_i$ is an attribute clustering over $\mathcal{V}_i$. Based on these functions and on the fact that the sets $\mathcal{V}_i$ $(1 \le i \le n)$ are disjoint partitions of $\mathcal{V}$, we define the function:

$$f : \mathcal{V} \to \cup_{1 \le i \le n} \mathcal{AC}_i \text{ with } \forall v \in \mathcal{V} : f(v) = f_i(v) \text{ iff } v \in \mathcal{V}_i$$

Obviously, f satisfies definition 3. This means, $\mathcal{AC} = \cup_{1 \le i \le n} \mathcal{AC}_i$ is an attribute clustering over $\mathcal{V}$.   □

4 Algorithm

4.1 S-indd

Overall Idea. As an external algorithm (see algorithm 1), S-INDD uses a repository on a hard drive (as an external memory) in order to store temporary computation results. The input parameter $\mathcal{L}$ denotes the name of the repository. $\mathcal{L}$ contains initially the lists $L_1, L_2, \ldots, L_{|\mathcal{A}|}$ where every list $L \in \mathcal{L}$ relates to a different attribute $A \in \mathcal{A}$ and its elements are all elements of the set $\mathcal{V}_A \times \{\{A\}\}$ sorted according to the values in $\mathcal{V}_A$. Example 1 illustrates these data structures.

Algorithm 1.

```
Input    : L, A, k
Output   : I
1 while (L contains k or more than k
         lists) do
    ⌊ mergeLists(L, k)

3 AC ← computeAttClustering(L)

4 I ← ∅
5 foreach A ∈ A do
    ⌊ I_A ←      ⋂       C
            C∈AC∧A∈C

7 foreach A ∈ A do
    foreach B ∈ I_A do
      ⌊ ⌊ I ← I ∪ {A ⊆ B}
```

Algorithm 1. S-indd

Algorithm 2.

```
Input    : L, k
1 L_1, L_2, ..., L_k ← selectLists(L, k)
2 Queue ← createPriorityQueue(
                      L_1, L_2, ..., L_k, L)
  L ← [ ]
5 while Queue·size() ≠ 0 do
    (v, AS) ← readNextAttSets()
    C ←    ⋃     A^v
        A^v∈AS
    ⌊ L ← L + [(v, C)]

9 remove(L_1, L2, ..., L_k, L)
10 write(L, L)
```

Algorithm 2. mergeLists

Example 1. Using the data of table 1, repository $\mathcal{L}$ will be initialized with the following four lists:

$$L_1 = [(1, \{A\}), (2, \{A\}), (4, \{A\})], \quad L_2 = [(1, \{B\}), (2, \{B\}), (3, \{B\}), (4, \{B\})]$$
$$L_3 = [(5, \{C\}), (6, \{C\}), (7, \{C\})], \quad L_4 = [(1, \{D\}), (3, \{D\})]$$

The purpose of these data structures is to compute the sets $\mathcal{A}^v$ ($v \in \mathcal{V}$) incrementally, where $\mathcal{A}^v$ is the set of all attributes $A \in \mathcal{A}$ whose values sets contain v (i.e., $v \in \mathcal{V}_A$). After computing the sets $\mathcal{A}^v$, S-INDD generates the set $\{\mathcal{C} \mid \exists v \in \mathcal{V} : \mathcal{C} = \mathcal{A}^v\}$ which is, according to the constructive proof of lemma 3, an *attribute clustering*. Having the *attribute clustering*, S-INDD computes for every attribute the intersection of all clusters containing it (line 5). The set $\mathcal{I}$, the set of all uINDs, is then computed based on lemma 2 (line 7).

The incremental computing of the sets $\mathcal{A}^v$ is achieved in two stages. The first stage (line 1) consists of a sequence of merging operations. The second stage (line 3) implicitly completes the computation of the sets $\mathcal{A}^v$ and generates the *attribute clustering*.

Merging. The merging operation reads k ($2 \le k \le |\mathcal{A}|$) lists

$$L_1 = [(v_{11}, \mathcal{A}^{v_{11}}), \ldots, (v_{1l_1}, \mathcal{A}^{v_{1l_1}})], \ldots, L_k = [(v_{k1}, \mathcal{A}^{v_{k1}}), \ldots, (v_{kl_k}, \mathcal{A}^{v_{kl_k}})]$$

from $\mathcal{L}$ and then replaces them with the new list

$$L = [(v_1, \mathcal{A}^{v_1}), (v_2, \mathcal{A}^{v_2}), \ldots, (v_n, \mathcal{A}^{v_n})]$$

that satisfies the following condition:

$$v_1 = \min_{\substack{1 \leq i \leq k \\ 1 \leq j \leq l_i}} \{v_{il_j}\}, \qquad \mathcal{A}^{v_1} = \bigcup_{\substack{v_{il_j} = v_1 \\ 1 \leq i \leq k \\ 1 \leq j \leq l_i}} \mathcal{A}^{v_{il_j}}$$

$$\vdots$$

$$v_s = \min_{\substack{1 \leq i \leq k \\ 1 \leq j \leq l_i}} \{v_{il_j}\} \setminus \{v_1, \ldots, v_{s-1}\}, \qquad \mathcal{A}^{v_s} = \bigcup_{\substack{v_{il_j} = v_s \\ 1 \leq i \leq k \\ 1 \leq j \leq l_i}} \mathcal{A}^{v_{il_j}}$$

with $s = 2, \ldots, n$

In other words, the new list L is sorted according to the values $v_s \in \{v_{il_j} \mid 1 \leq i \leq k, 1 \leq j \leq l_i\}$ $(1 \leq s \leq n)$ and every set $\mathcal{A}^s$ is the union of all sets $\mathcal{A}^{v_{il_j}}$ identified by the value v_s in the k lists.

S-INDD repeats the merging operation (line 1) until the repository $\mathcal{L}$ has less than k lists where every new list generated by the merging operation has to be stored as a temporary result in the repository $\mathcal{L}$ (line 10 in algorithm 2). Example 2 illustrates the merging operation.

Example 2. According to example 1 and for $\underline{k = 3}$, S-INDD has to execute only one merging operation.
If the first three lists L_1, L_2, and L_3 (see line 1 in algorithm 2) are selected for merging, the following list

$$L_{1,2,3} = [(1, \{A, B\}), (2, \{A, B\}), (3, \{B\}), (4, \{A, B\}), (5, \{C\}), (6, \{C\}), (7, \{C\})]$$

will be generated and the repository $\mathcal{L}$ will be changed to contain only the lists: $L_{1,2,3}$ and L_4.

For an efficient implementation of the merging operation and for managing a simultaneous reading of k lists (files) from the repository $\mathcal{L}$, a priority queue is used by algorithm 2 (and also by algorithm 3 - see below). The queue manages k readers (sequential file readers). Every reader is associated with a list and points to the entry that can currently be read from the list. For every two readers r, r', reader r has a higher priority than r' if and only if the value v in $(v, \mathcal{A}^v)$ is smaller than or equal to the value v' in $(v', \mathcal{A}^{v'})$ where $(v, \mathcal{A}^v)$ is the entry that r can currently read and $(v', \mathcal{A}^{v'})$ is the entry that r' can currently read.

The purpose of using a priority queue is to enable an efficient collecting of all sets $\mathcal{A}_1^v, \ldots, \mathcal{A}_{l_v}^v$ $(1 \leq l_v \leq k)$ by a simultaneous and sequential reading of k lists where v is the smallest value among all values that have not been read from the k lists in the queue yet. That is possible in a simultaneous sequential reading because the lists are sorted according to the values $v \in \mathcal{V}$ and the priority in

Input   : $\mathcal{L}$
Output   : $\mathcal{AC}$

$Queue \leftarrow$ `createPriorityQueue`$(\mathcal{L})$
$\mathcal{AC} \leftarrow \emptyset$
while $Queue\cdot$`size`$() \neq 0$ **do**
  $(v, \mathcal{AS}) \leftarrow$ `readNextAttSets`$()$
  $\mathcal{C} \leftarrow \bigcup_{\mathcal{A}^v \in \mathcal{AS}} \mathcal{A}^v$
  $\mathcal{AC} \leftarrow \mathcal{AC} \cup \{\mathcal{C}\}$

Algorithm 3. computeAttClustering

Input   : $Queue$
Output   : $(v, \mathcal{AS})$

$\mathcal{AS} \leftarrow \emptyset$
repeat
  $r \leftarrow Queue \cdot$`pull`$()$
  $(v, \mathcal{A}^v) \leftarrow r\cdot$`readCurrent`$()$
  $(v, \mathcal{AS}) \leftarrow (v, \mathcal{AS} \cup \{\mathcal{A}^v\})$
  if $r\cdot$`hasNext`$()$ **then**
    $r\cdot$`readNext`$()$
    $Queue\cdot$`add`(r)
  $r' \leftarrow Queue \cdot$`peek`$()$
  $(v', \mathcal{A}^{v'}) \leftarrow r'\cdot$`readCurrent`$()$
until $(Queue\cdot$`size`$() = 0) \vee (v' \neq v)$

Algorithm 4. readNextAttSets

the queue is defined according to the ascending order of the values. This kind of applying the priority queue is well-known by external merge-sort algorithms.

Clusters Computing. After finishing the merging, algorithm 3 will generate the clusters of the *attribute clustering* $\mathcal{AC}$ by processing all remaining k' ($1 \leq k' < k$) lists simultaneously. For every value v, there are still l_v ($1 \leq l_v < k'$) lists containing entries of the form $(v, \mathcal{A}_i^v)$ ($1 \leq i \leq l_v$). Algorithm 3 collects all these entries, computes the set $\mathcal{C} = \cup_{1 \leq i \leq l_v} \mathcal{A}_i^v$, and adds $\mathcal{C}$ as a cluster to the set $\mathcal{AC}$. Example 3 illustrates the computing of the clusters.

Example 3. According to example 2 and for $\underline{k = 3}$, $\mathcal{L}$ will contain the lists

$$L_{1,2,3} = [(1, \{A, B\}), (2, \{A, B\}), (3, \{B\}), (4, \{A, B\}), (5, \{C\}), (6, \{C\}), (7, \{C\})]$$
$$L_4 = [(1, \{D\}), (3, \{D\})]$$

after finishing the merging.
For the value $v = 1$ there are two entries: $(1, \{A, B\})$ in $L_{1,2,3}$ and $(1, \{D\})$ in L_4. Therefore, algorithm 3 collects the two sets $\{A, B\}$ and $\{D\}$ by calling algorithm 4 in the first run of the *while*-loop which delivers the tuple: $(1, \{\{A, B\}, \{D\}\})$. The first cluster is then $\mathcal{C}_1 = \{A, B\} \cup \{D\} = \{A, B, D\}$ and consequently $\mathcal{AC} = \{\{A, B, D\}\}$. After a second run of the *while*-loop we have $\mathcal{AC} = \{\{A, B, D\}, \{A, B\}\}$. Calling algorithm 4 in the third run of the *while*-loop delivers the tuple: $(3, \{\{B\}, \{D\}\})$. Consequently, $\mathcal{AC}$ will be extended to $\mathcal{AC} = \{\{A, B, D\}, \{A, B\}, \{B, D\}\}$. Computing $\mathcal{AC}$ will be finished after the seventh run of the *while*-loop resulting in $\mathcal{AC} = \{\{A, B, D\}, \{A, B\}, \{B, D\}, \{C\}\}$.

Repository Size. During the whole process of computing the *Attribute Clustering*, the repository size remains almost constant. This is because (i) the selected k lists in every merging operation will not be needed any more after merging them, which allows algorithm 2 to remove them from the repository after merging them (see line 9), and (ii) the size of the new list that results from merging the selected k lists can not exceed the total size of these k lists.

We can now answer the following question.

Why Spider [1] is a special case of S-indd? SPIDER can only process the whole set of the attributes at once. That means, SPIDER is only a form of algorithm 3. To let S-INDD process all attributes at once we need only to put $k = |\mathcal{A}| + 1$.

```
Input    : ℒ₁,…,ℒₚ,𝒜,k
Output   : ℐ
AC ← ∅
for i ← 1 to p do
   while (ℒᵢ contains k or more
   than k lists) do
      mergeLists(ℒᵢ, k)

   ACᵢ ← computeAttClustering(ℒᵢ)
   AC ← AC ∪ ACᵢ

foreach A ∈ 𝒜 do
   ℐ_A ←      ⋂      C
          C∈AC∧A∈C

foreach A ∈ 𝒜 do
   foreach B ∈ ℐ_A do
      ℐ ← ℐ ∪ {A ⊆ B}
```

(with line labels **2** at the `for i` line and **6** at the `AC ← AC ∪ ACᵢ` line)

Algorithm 5. Extended S-indd

```
Input    : L_A for every A ∈ 𝒜
Input    : m₁,…,m_{p−1}
Output   : ℒ₁,…,ℒₚ
for i ← 1 to p − 1 do
   ℒᵢ ← ∅
   foreach L_A with L_A ≠ ∅ do
      L_A^i ← getSubList(L_A, mᵢ)
      ℒᵢ ← ℒᵢ ∪ {L_A^i}
      L_A ← L_A \ L_A^i

ℒₚ ← ∅
foreach L_A with L_A ≠ ∅ do
   ℒₚ ← ℒₚ ∪ {L_A}
```

(with line label **4** at the `L_A^i ← getSubList` line)

Algorithm 6. Partition

4.2 Extending S-indd

In the case that the dataset is very large and its values are shared among a lot of attributes, many temporary lists generated by the merging operation in subsequent iterations will have a relatively large size. Processing such large lists by algorithm 2 or algorithm 3 may demand more I/O-operations.

To avoid generating large temporary lists in this case, the dataset can be partitioned into disjoint partitions, and the *attribute clustering* will then be, according to lemma 4, the union of all clusters computed for all partitions.

Algorithm 5 is an implementation of this idea and consists of computing iterations whose number equals the number of the partitions of the dataset. Every iteration is an instance of S-INDD applied for computing the *attribute clustering* over a different partition. The *attribute clustering* over the whole dataset is computed based on lemma 4 in line 6. The input of algorithm 5 contains the names of p repositories $\mathcal{L}_i$ ($1 \leq i \leq p$) where every repository corresponds to a different partition and contains the initial data structures (lists) generated from the corresponding partition.

The disjunction of the partitions has an important computational advantage. It avoids redundant computation of the set $\mathcal{A}^v$ of any value $v \in \mathcal{V}$. However, the important question arising now is how can we partition a dataset to meet the requirement of the extended version of S-INDD (algorithm 5)? The answer to this question is given in algorithm 6.

The main idea applied by algorithm 6 to partition the dataset is to choose p values $m_1,\ldots,m_p$ with $m_1 < m_2 < \cdots < m_p$ (to ease notation and formulation, we put $m_p = \infty$) and then to divide every initial list L_A into disjoint sublists

L_A^i ($1 \leq i \leq p$) where every sublist L_A^i has to satisfy the following condition:

$$\max\{v \mid (v, \{A\}) \in L_A^i\} \leq m_i$$

In other words, the maximal value in partition i does not exceed the value m_i.

The disjunction of the partitions is guaranteed by algorithm 6 because (i) the lists L_A are sorted, (ii) every sublist L_A^i is generated from L_A by processing L_A from the first element until all elements from it have been obtained that are less or equal to m_i (line 4 in algorithm 6), and (iii) after its generating and adding to the repository $\mathcal{L}_i$, L_A^i will be removed from L_A. Example 4 illustrates the extended version of S-INDD.

Example 4. Using the lists in example 1 and for $m_1 = 3$, algorithm 6 produces two partitions. The first partition $\mathcal{L}_1$ contains the lists:

$$L_A^1 = [(1, \{A\}), (2, \{A\})], \quad L_B^1 = [(1, B), (2, B), (3, B)], \quad L_D^1 = [(1, \{D\}), (3, \{D\})]$$

The second partition $\mathcal{L}_2$ contains the lists:

$$L_A^2 = [(4, \{A\})], \quad L_B^2 = [(4, B)], \quad L_C^2 = [(5, \{C\}), (6, \{C\}), (7, \{C\})]$$

Based on these partitions algorithm 5 will be provided with two repositories $\mathcal{L}_1$ and $\mathcal{L}_2$. It generates $\mathcal{AC}_1 = \{\{A, B, D\}, \{A, B\}, \{B, D\}\}$ from the first repository and $\mathcal{AC}_2 = \{\{A, B\}, \{C\}\}$ from the second repository. The whole *attribute clustering* is then $\mathcal{AC} = \mathcal{AC}_1 \cup \mathcal{AC}_2 = \{\{A, B\}, \{A, B, D\}, \{B, D\}, \{C\}\}$.

5   Experiments

The main aim of our experiments is to compare the performance of S-INDD with that of SPIDER. This is our focus because SPIDER is reported to be the current leading algorithm for unary INDs discovery [1,2]. SPIDER already significantly outperforms other approaches, in particular [3] and [11].

Experimental Conditions. We implemented both algorithms in Java 7 and performed the experiments on the Windows 7 Enterprise system with an Intel Core i5-3470 (Quad Core, 3.20 GHz CPU) and 8 GB RAM. We used an external 500 GB hard drive as external memory. We set the minimum Java heap size to 4 GB and the maximum to 6 GB for all our experiments.

Datasets. Two groups of synthetic datasets are generated for conducting two different groups of experiments. The purpose of the first group is to evaluate and compare the scalability of both algorithms by varying the number of attributes and fixing the number of rows, while in the second group of datasets the number of rows is varied and the number of attributes is fixed.

Experiments with real-word datasets are conducted using datasets from the life science domain (see below).

Scaling the Number of Attributes. In these experiments, we generate thirteen synthetic datasets with the same number of rows, namely 200,000 rows.

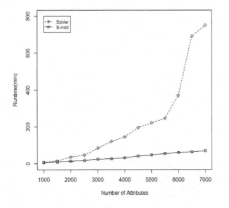

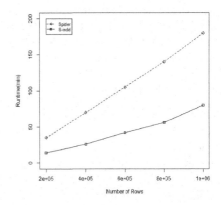

Fig. 2. Comparing scalability by scaling the number of attributes and fixing the number of rows to 200,000

Fig. 3. Comparing scalability by scaling the number of rows and fixing the number of attributes to 2,000

Starting with 1,000 different attributes and ten unary INDs in the first dataset, the attributes set in the next dataset consists of the attributes set in the previous dataset plus 500 new different attributes and ten new different INDs so that the thirteenth dataset has 7,000 different attributes and 130 unary INDs. For all these datasets , S-INND is configured to merge 200 lists ($k = 200$) simultaneously.

Figure 2 shows the results of these experiments. (i) For every dataset S-INDD is faster than SPIDER. For example, for the dataset with 7,000 attributes and 36.2 GB size, S-INDD needs one hour and ten minutes while SPIDER needs twelve hours and thirty minutes. This means, S-INDD is about 11x faster than SPIDER. (ii) By increasing the number of attributes, SPIDER's runtime grows much faster than S-INDD's runtime. For example, by increasing the number of attributes from 6,000 to 7,000, SPIDER'runtime increases by 38 % while S-INDD'runtime increases by 1 % (for the dataset with 6,000 attributes and 31 GB SPIDER needs six hours and 10 minutes while S-INDD needs only about one hour).

Scaling the Number of Rows. In these experiments, we generate 5 synthetic datasets with the same number of attributes, namely 2,000 attributes. Starting with 200,000 rows in the first dataset, the next dataset contains all rows in the previous dataset plus 200,000 new different rows so that the fifth dataset has 1,000,000 rows and 48 GB size. Every dataset has the same number of INDs, namely 15 unary INDs. For all these datasets, we applied the extended version of S-INDD configured to merge 200 lists and to partition the datasets so that every partition had a maximum of 200,000 rows. For example, the dataset with 1,000,000 rows was divided into 5 disjoint partitions, i.e., algorithm 5 had to execute the *For*-loop 5 times (line 2).

Figure 3 shows the results of these experiments. These results also show that S-INDD is faster than SPIDER for every dataset. For example, for the dataset with 1,000,000 rows, S-INDD needs one hour and twenty-minutes while SPIDER needs about three hours. Furthermore, by increasing the number of rows, SPIDER's runtime grows faster than S-INDD's runtime. For example, by increasing the number of rows from 200,000 to 1,000,000, SPIDER'runtime increases by 0.18 per thousand rows while S-INDD'runtime increases by 0.08 per thousand rows (for the dataset with 200,000 rows and 10 GB SPIDER needs 35 minutes while S-INDD needs only about 14 minutes).

INDs discovery in life science datasets. As real-word datasets we used SCOP[3], BIOSQL[4], CATH[5], and PDB[6] from the life science domain. To discover dependencies inside every dataset and between the datasets we processed the four datasets as a whole dataset. Their complete size is about 46 GB. Together they have total of 1,262 attributes. Life science databases are an example of the unreliability of the data type of the attributes. This means, we can not apply restriction on the data type of the attributes but rather, must assume that all attributes have the same data type (e.g. string). For this test, S-INDD needed about 9 minutes while SPIDER needed about 17 minutes.

6   Related Work

Bell and Brockhausen [3] generate all unary IND candidates from previously collected statistics, such as min-max values and data types. Then they validate them using SQL join-statements. The transitivity of INDs is exploited to reduce the number of untested candidates. However, SQL-based validation is very costly because it is accesses the database for every candidate.

De Marchi et al. [11,13] propose an algorithm for unary INDs discovery that generates an inverted index associating every value to the attributes having the value. Because for every attribute A the intersection of all attribute sets containing A is the set of all attributes including A, the algorithm runs through all values in order to compute such an intersection for every attribute. However, this approach is inefficient because an attribute set in the index can be associated with many values. This means, the algorithm executes a lot of redundant intersection operations. The concept of *attribute clustering* we introduced in this paper solves this problem.

Bauckmann et al. propose SPIDER [1,2]. SPIDER is an external algorithm that writes the sorted values of every attribute to a file. Then it opens all files at once and starts comparing the values in parallel and in the same way in which the merge-sort algorithm does. SPIDER prunes IND candidates as follows: for each two attributes A and B, A is not included in B (i) if there is an iteration i in which the current A's value is greater than the current B's value and in the subsequent iteration $i + 1$, B does not have value equal to the A's value in iteration i, or (ii) if there is an iteration in which the current A's value is less

---

[3] http://scop.mrc-lmb.cam.ac.uk/scop
[4] http://obda.open-bio.org
[5] http://www.biochem.ucl.ac.uk/bsm/cath_new
[6] http://www.rcsb.org/pdb

than the current B's value and in the subsequent iteration $i + 1$, A does not have value equal to the B's value in iteration i. This technique makes SPIDER the most efficient algorithm for unary IND detection in related work. However, SPIDER's scalability decreases by increasing the number of attributes. To solve this problem we developed S-INDD in this paper.

Dasu et al. [5] compute a summary of data from which they calculate a "rate of similitude" between attributes. Based on this "rate of similitude" unary INDs can be found approximately. This means, some discovered unary INDs aren't satisfied, but also satisfied unary INDs can be missed.

Mannila and Toivonen [9] suggest the first known approach for an exhaustive search of N-ary INDs. They point out that this problem can fit in the framework of level-wise algorithms and is representable as sets; algorithms and implementations are proposed in [7,10–13]

Rostin et al. [14] propose rule-based discovery technique based on machine learning to derive foreign keys from INDs.

Zhang et al. [15] propose an approximate techniques to discover foreign keys. They assume that the value sets of foreign keys and the value sets of corresponding primary keys obey the same probability distribution. They premise availability of primary keys. Furthermore, Their approach may produce unsatisfied references and may miss satisfied references. For this reason, they focus on precision and recall rather than on runtime. The specialization on foreign key discovery also makes their approach inapplicable to other IND use cases, such as schema matching [8], query optimization [6], or integrity checking [4].

7 Conclusion

We introduced a new idea for clustering the attributes of database relations. We showed that the inferencing of all unary inclusion dependencies from the *attribute clustering* is much more efficient than inferencing them from the inverted index introduced in [11,13]. We then devised S-INDD for computing the *attribute clustering* in large datasets. S-INDD computes such clusters incrementally by extending the idea of sort-merge-join approach. S-INDD is a composite of configurable computing iterations. In each iteration, it can control the number of rows and the number of attributes having to be processed. This flexibility makes each iteration efficiently executable. We showed how to parametrize S-INDD to present SPIDER [1,2] as a special case of this algorithm. Therefore, S-INDD is much more faster and scalable than SPIDER.

Acknowledgments. Discussions and collaboration with Felix Naumann and Thorsten Papenbrock supported this paper.

References

1. Bauckmann, J.: Dependency Discovery for Data Integration. Ph.D. thesis, Hasso Plattner Institute at the University of Potsdam (2013). http://opus.kobv.de/ubp/volltexte/2013/6664/

2. Bauckmann, J., Leser, U., Naumann, F.: Efficiently computing inclusion dependencies for schema discovery. In: Proceedings of the International Workshop on Database Interoperability (InterDB) (2006)
3. Bell, S., Brockhausen, P.: Discovery of Data Dependencies in Relational Databases. Tech. rep., Universitat Dortmund (1995)
4. Casanova, M.A., Tucherman, L., Furtado, A.L.: Enforcing inclusion dependencies and referencial integrity. In: Proceedings of the 14th International Conference on Very Large Data Bases, VLDB 1988, pp. 38–49. Morgan Kaufmann Publishers Inc., San Francisco (1988). http://dl.acm.org/citation.cfm?id=645915.671795
5. Dasu, T., Johnson, T., Muthukrishnan, S., Shkapenyuk, V.: Mining database structure; or, how to build a data quality browser. In: Proceedings of the International Conference on Management of Data (SIGMOD), pp. 240–251 (2002). http://doi.acm.org/10.1145/564691.564719
6. Gryz, J.: Query folding with inclusion dependencies. In: In Proc. of the 14th IEEE Int. Conf. on Data Engineering (ICDE 1998), pp. 126–133 (1998)
7. Koeller, A., Rundensteiner, E.: Discovery of high-dimensional inclusion dependencies. In: Proceedings of the International Conference on Data Engineering (ICDE), pp. 683–685 (2003)
8. Levene, M., Vincent, M.W.: Justification for inclusion dependency normal form. IEEE Transactions on Knowledge and Data Engineering 12 (2000)
9. Mannila, H., Toivonen, H.: Levelwise search and borders of theories in knowledgediscovery. Data Min. Knowl. Discov. 1(3), 241–258 (1997). http://dx.doi.org/10.1023/A:1009796218281
10. De Marchi, F., Flouvat, F., Petit, J.-M.: Adaptive strategies for mining the positive border of interesting patterns: application to inclusion dependencies in databases. In: Boulicaut, J.-F., De Raedt, L., Mannila, H. (eds.) Constraint-Based Mining and Inductive Databases. LNCS (LNAI), vol. 3848, pp. 81–101. Springer, Heidelberg (2006). http://www.dx.doi.org/10.1007/11615576_5
11. De Marchi, F., Lopes, S., Petit, J.-M.: Efficient algorithms for mining inclusion dependencies. In: Jensen, C.S., Jeffery, K., Pokorný, J., Šaltenis, S., Bertino, E., Böhm, K., Jarke, M. (eds.) EDBT 2002. LNCS, vol. 2287, pp. 464–476. Springer, Heidelberg (2002). http://www.dl.acm.org/citation.cfm?id=645340.650245
12. Marchi, F.D., Petit, J.M.: Zigzag: a new algorithm for mining large inclusion dependencies in databases. In: Proceedings of the Third IEEE International Conference on Data Mining, ICDM, pp. 27–34 (2003). http://dl.acm.org/citation.cfm?id=951949.952179
13. Marchi, F., Lopes, S., Petit, J.M.: Unary and n-ary inclusion dependency discovery in relational databases. Journal of Intelligent Information Systems 32(1), 53–73 (2009). http://dx.doi.org/10.1007/s10844-007-0048-x
14. Rostin, A., Albrecht, O., Bauckmann, J., Naumann, F., Leser, U.: A machine learning approach to foreign key discovery. In: Proceedings of the ACM SIGMOD Workshop on the Web and Databases (WebDB), Providence, RI (2009)
15. Zhang, M., Hadjieleftheriou, M., Ooi, B.C., Procopiuc, C.M., Srivastava, D.: On multi-column foreign key discovery. Proc. VLDB Endow. 3(1–2), 805–814 (2010). http://dx.doi.org/10.14778/1920841.1920944

Repairing Functional Dependency Violations in Distributed Data

Qing Chen, Zijing Tan$^{(\boxtimes)}$, Chu He, Chaofeng Sha, and Wei Wang

School of Computer Science, Shanghai Key Laboratory of Data Science,
Fudan University, Shanghai, China
{13210240082,zjtan,12210240018,cfsha,weiwang1}@fudan.edu.cn

Abstract. One of the problems central to data consistency is data repairing. Given a database D violating a set Σ of data dependencies as data quality rules, it aims to modify D for a new relation D' satisfying Σ. When D is a centralized database, a host of methods have been provided to address this problem. In practice, a database may be fragmented and distributed to multiple sites, which is advocated by distributed systems for better scalability and is readily supported by commercial systems. This paper makes a first effort to develop techniques for repairing functional dependency violations in a horizontally partitioned database. (1) Based on a message-passing distributed computing model and two complexity measures (parallel time and data shipment) for distributed algorithms, we study data repairing with equivalence classes in the distributed setting. We show that it is NP-complete to build equivalence classes when the data is horizontally partitioned, and when we aim to minimize either data shipment or parallel computation time. (2) Despite the intractability, we propose efficient distributed algorithms and optimization techniques for data repairing based on equivalence classes. (3) We experimentally verify the effectiveness and efficiency of our algorithms, using both real-life and synthetic data.

1 Introduction

Functional dependencies (FDs) are constraints that data values in a relation are required to satisfy. In practice, however, we often encounter relations that violate a predefined set of FDs and hence are inconsistent. Among techniques for resolving FD violations, optimal repair computation is well studied. It aims to repair an inconsistent relation by minimally modifying it *w.r.t.* some cost measure, so as to get a new relation satisfying constraints (*a.k.a.* a *repair* of the inconsistent relation). Despite the intractability of optimal repair computation for FD violations, several heuristic or approximation algorithms [2–5,15] are presented to repair a centralized database.

In this paper, we contend that it is necessary to develop algorithms for repairing *distributed* data. (1) In practice, a relation is often fragmented and distributed across different machines, *e.g.*, horizontal or vertical partition supported by commercial systems. With this comes the need for repairing distributed data. (2)

© Springer International Publishing Switzerland 2015
M. Renz et al. (Eds.): DASFAA 2015, Part I, LNCS 9049, pp. 441–457, 2015.
DOI: 10.1007/978-3-319-18120-2_26

Existing algorithms for optimal repair computation are typically quadratic, or even cubic in the data size, and are hence too costly on real-life large data set. Several optimizations are then provided to improve the scalability, while these techniques necessarily have a negative impact on the repair quality. To overcome the limitation of scalability, another way is to partition and distribute the large data to multiple machines, so as to leverage more resources, as advocated by distributed systems. Distributed repairing problem necessarily introduces new challenges that we do not encounter in the centralized setting, and makes our lives much harder. To our best knowledge, no such algorithms are in place yet.

Example 1. Fig. 1 gives an EMP relation D (Fig. 1(a)); each tuple specifies an employee's name, job (title, level, salary) and contact info (phn, street, city, zip). The following functional dependencies (FDs) are defined on this relation:

φ_1 : $title, level \rightarrow salary$

φ_2 : $phn \rightarrow street, zip$

φ_3 : $zip \rightarrow city$

It is easy to see that D is inconsistent, since it violates given FDs. When D is a centralized relation, we can employ existing repairing techniques to repair D. We give one possible repair D' (Fig. 1(b)), by modifying some attribute values.

Now suppose D is horizontally fragmented into three fragments (Fig. 1(c)), and each fragment D_i resides at site S_i. Then to repair FD violations, data shipments between different sites are generally required. We present some shipping schedules for illustration. (1) The baseline approach is to collect all tuples at a single site and employ a centralized data repairing algorithm. Even this simple idea has some variants. For example, Collecting all tuples at site S_1 is better than collecting all tuples at S_2, in terms of communication cost. (2) By analyzing the FD set, we see that data modifications on an employee's job are independent of those on his contact info. In light of this, we can ship employees' title, level and salary attributes to one site and ship phn, street, city and zip attributes to another site. Then, computations at these two sites can be done simultaneously, and hence enjoy parallelism for better parallel computation time. (3) If we decide not to ship data from/to site S_3, we may introduce distinct new values to attribute level, phn and zip of tuple t_4 and t_5. As will be seen in Sec. 3, this guarantees no FD violation. This approach favors communication cost and parallel computation time, however, possibly at the cost of poor repair quality.

Putting these together, we know that strategies of data shipment have a great impact on the communication cost, effectiveness of parallel computation, and even the repair quality as well. □

Contributions. We make a first effort to investigate the problem of repairing functional dependency violations in horizontally partitioned data.

(1) Based on a message-passing distributed computing model and two complexity measures: parallel time and data shipment, for the analyses of distributed algorithms (Section 3), we study the distributed version of equivalence class technique, to develop distributed repairing algorithms with good repair quality (Section 4). We show that it is NP-complete to build equivalence classes when

	name	title	level	salary	phn	street	city	zip
t_1:	Daisy	VP	1	350K	021-11111111	Meiyuan	SH	200070
t_2:	Jack	staff	1	80K	021-11116666	Qingyun	BJ	200070
D t_3:	Bob	staff	2	50K	021-11111111	Meiyuan	SH	200070
t_4:	Joe	staff	1	60K	025-22222222	Zhujiang	NJ	210008
t_5:	Mike	staff	1	80K	025-22221111	Hankou	NJ	210008

(a) An EMP relation D.

	name	title	level	salary	phn	street	city	zip
t_1:	Daisy	VP	1	350K	021-11111111	Meiyuan	SH	200070
t_2:	Jack	staff	1	80K	021-11116666	Qingyun	SH	200070
D' t_3:	Bob	staff	2	50K	021-11111111	Meiyuan	SH	200070
t_4:	Joe	staff	1	80k	025-22222222	Zhujiang	NJ	210008
t_5:	Mike	staff	1	80K	025-22221111	Hankou	NJ	210008

(b) One possible repair D' of D.

	name	title	level	salary	phn	street	city	zip
D_1 t_1:	Daisy	VP	1	350K	021-11111111	Meiyuan	SH	200070
t_2:	Jack	staff	1	80K	021-11116666	Qingyun	BJ	200070

	name	title	level	salary	phn	street	city	zip
D_2 t_3:	Bob	staff	2	50K	021-11111111	Meiyuan	SH	200070

	name	title	level	salary	phn	street	city	zip
D_3 t_4:	Joe	staff	1	60K	025-22222222	Zhujiang	NJ	210008
t_5:	Mike	staff	1	80K	025-22221111	Hankou	NJ	210008

(c) A horizontal partition of D.

Fig. 1. A relation D, one possible repair D' and D's horizontal partitions

the data is horizontally partitioned, and when the complexity is measured by either data shipment or parallel computation time.

(2) Despite the intractability, we present efficient distributed algorithms and optimization techniques for data repairing based on equivalence classes (Section 5). Our work is built upon an implementation of equivalence classes that are distributed to multiple sites.

(3) Using both real-life and synthetic data, we conduct an extensive experimental study to verify the effectiveness and efficiency of our algorithms (Section 6).

Related Work. In the field of data consistency management, repair computation [2–6, 8, 9, 13, 15, 18–20] is the most well studied. There are different versions of this problem, by considering various settings of constraint, repair primitive and cost model, among other things. To our best knowledge, former works deal with a centralized data set. This paper presents *distributed* algorithms for repairing FD violations in distributed data. It is easy to see that the centralized setting is a special case of the distributed one when only one site is available. In addition, different complexity measures, *e.g.,* data shipment or parallel time, are employed to evaluate the performance of distributed algorithms and guide the design of such algorithms. In light of this, the framework and techniques for distributed repair computation are necessarily much more intricate.

[12] studies the problem of conditional FD violation detection in fragmented and distributed relations, and [14] further provides algorithms for incrementally detecting violations of conditional FDs in fragmented data. Note that violation detection is to identify tuples violating FDs, while repairing aims at resolving violations to obtain a consistent data set, and is hence more complicated. Indeed, repairing one FD can break another, and simple heuristics could even fail to terminate in the presence of interrelated FDs. In contrast, violation detection can deal with FDs one by one and in any order. When it comes to distributed computation, data repairing requires to balance the repair quality and the efficiency of parallel computation, since there are possibly exponential number of repairs. In contrast, FD validation has a deterministic result.

One solution for our problem is to employ existing frameworks, *e.g.*, MapReduce [10], and delegate most work to the system. However, a good solution for distributed data repairing must exploit the nature of data repairing itself; existing systems fall short of these abilities. For example, recursive computation is typically required in data repairing due to complicate interactions between FDs, while MapReduce is generally not fit for this setting, which needs a series of chained MapReduce invocations.

2   Preliminaries

In this section, we review some basic notations.

Data Repair for a Relation. We consider an instance D of relation schema $R(A_1, \ldots, A_m)$. $t[A]$ denotes the projection of tuple t onto attribute A, referred to as a *cell*. We assume each tuple t is associated with a distinct identifier (*id*) $t.id$, which is not subject to updates.

We consider functional dependency (FD) of the form $X \to A$, where $X \subseteq A_1, \ldots, A_m$. Any FD can be converted to this form by splitting right hand side (RHS) attributes. For a given FD $\varphi = X \to A$ and an instance D, D satisfies φ, denoted $D \models \varphi$, when there does not exist two tuples t_1, t_2 in D such that $t_1[B] = t_2[B]$ for all $B \in X$ and $t_1[A] \neq t_2[A]$. D satisfies a set Σ of FDs, denoted $D \models \Sigma$, when $D \models \varphi$ for $\forall \varphi \in \Sigma$.

When there exist FD violations in D w.r.t. Σ, we say D' is a *repair* of D, if (1) D' is an instance of R, having the same tuple *ids* as D; and (2) $D' \models \Sigma$. Note that in this definition of repair, cell modification is used as the only repair operation, similar to [3,5,15]. There are generally a large or even infinite number of repairs. To this end, *optimal repair computation* aims to find one single repair that minimizes some *cost* measure among all repairs. Recall that optimal repair computation with cell modifications is proved to be NP-complete, even when the cost of a repair is computed as the number of modified cells [15].

Example 2. Recall repair D' presented in Fig. 1. When the number of modified cells is taken as the repair cost, D' has a cost of 2. □

For space limitation, in this paper we consider relation D that is horizontally partitioned (fragmented) and distributed to multiple sites.

Horizontal Partition [1,17]. Relation D may be partitioned into a disjoint set of fragments $D_1, \ldots, D_n$ that share the same schema R as D. Specifically, $D_i = \sigma_{F_i}(D)$, $D = \bigcup_{i \in [1,n]} D_i$: (1) F_i is a predicate such that the selection $\sigma_{F_i}(D)$ identifies fragment D_i; and (2) D can be reconstructed by the union of these fragments. W. l. o. g., we assume fragment D_i is placed at site S_i, *i.e.*, one fragment at each site. We also extend tuple *ids* by adding site number as a prefix; therefore, the site at which a tuple resides can be identified by its *id*.

Repairing FD Violations in a Horizontally Fragmented Relation. Given a horizontally fragmented relation $D = \bigcup_{i \in [1,n]} D_i$ of schema R and an FD set Σ, the problem of *repairing* D *w.r.t.* Σ is to find another fragmented relation $D' = \bigcup_{i \in [1,n]} D'_i$ of R, such that (1) D'_i has the same tuple *ids* as D_i, possibly with modified cell values; and (2) $D' \models \Sigma$.

3   Analyses of Distributed Data Repairing

In this section, we first present a message-passing computational model and two complexity measures for distributed algorithms, and then investigate the complexities of distributed data repairing based on the given model.

Model of Distributed Computation. We consider a pure message passing model, which is flexible enough to express a large class of distributed algorithms [16], and is fit for the problem of distributed data repairing. There are several identical sites that can directly send arbitrary number of messages to each other, and those sites work together by message-passing and local computations. Specifically, messages sent from a site S_i to another site S_j only consist of the local data available at S_i. Local computations executed on S_i utilize only data at S_i, *i.e.*, local data and messages received at S_i.

Complexity Measures for Distributed Algorithms. We use two measures to evaluate distributed algorithms: (a) parallel computation time, the time measuring the completion time at different sites in parallel, and (b) total data shipment, the size of total messages among sites during the computation.

It is worth mentioning that repair quality is not considered in the complexity measures for distributed algorithms, while any meaningful distributed repairing algorithms should produce a repair with good quality. Indeed, if only the efficiency of distributed computation is concerned, we next present a simple distributed repairing algorithm, referred to as NaiveLocal. NaiveLocal is optimal in data shipment and when the relation is evenly fragmented and distributed to all sites, it is also optimal in parallel computation time. NaiveLocal resolves all FD violations locally, based on an adaption of the notion of *core implicant* [15]. Specifically, in parallel at site S_i, NaiveLocal first computes a set Z of attributes that intersects with at least one left hand side (LHS) attribute of each FD $\varphi \in \Sigma$,

(a) ECs of relation D (b) Data structure of an EC

Fig. 2. Example equivalence class

and then introduces a distinct new value to each attribute in Z for each tuple t in D_i . Here "distinct new value" implies a value not used in that attribute in D.

Example 3. {level, phn, zip } is a set of attributes that intersects with at least one left hand side attribute of each FD given in Example 1. Then, a repair is obtained by introducing new values to these attributes of all tuples. □

To avoid values used in other fragments, we generally have to introduce meaningless values in NaiveLocal, just as placeholders. Therefore, the repair produced by NaiveLocal is of low quality and is not acceptable in practice. We stress that an effective distributed repairing algorithm should be developed based on some repairing technique with good repair quality. As will be seen shortly, this makes the optimization of distributed algorithms much harder.

4   Distributed Equivalence Classes

Since it is beyond reach to find the optimal repair, there is no available "best" repairing technique that we can follow in the distributed setting. In this section, we first review the notion of *equivalence class*, which is an effective heuristic repairing technique, and then discuss the complexities of its distributed version.

Equivalence class. Equivalence class (EC) is a technique used in data repairing [2,3,5,9,14], for keeping track of cells having a same value in the generated repair. We use the following notations: (1) an EC e^A on attribute A is a set of cells of the form $t_i[A]$; (2) any cell c belongs to exactly one EC at any time, denoted by $ec(c)$; and (3) ξ denotes the set of all equivalence classes (ECs).

Given a relation D and a set Σ of FDs as input, ECs are built as follows.

(1) Initialization. Each cell c is in EC $\{c\}$, *i.e.*, a singleton set containing itself.

(2) Merge equivalence classes: merging two ECs in ξ means replacing them by a new EC that is equal to their union. Two distinct ECs e^C, e'^C are merged when (*i*) there exist $t_i[C] \in e^C$, $t_j[C] \in e'^C$, such that $t_i[C] = t_j[C]$, or (*ii*) there exists $X \to C \in \Sigma$, $t_i[C] \in e^C$, $t_j[C] \in e'^C$ such that $\forall D \in X$, $ec(t_i[D]) = ec(t_j[D])$. Note that merging ECs needs recursive computations and terminates when no change to ξ is possible. Also note that building ECs reaches a deterministic result, *i.e.*, a unique fixpoint, in finite steps.

(3) Assign a target value to each EC. We get a repair of D by providing all cells in EC e^A with a same value, referred to as the target value of e^A. This value is

typically set to minimize the total cost of value modifications from cell values in e^A to the target value.

Example 4. In Fig. 2(a), we show in dashed boxes ECs of D that have multiple cells. We get a repair by ensuring all cells in the same EC have a same value. Consider the EC that $t_1[city], t_2[city], t_3[city]$ belongs to; choosing "SH" as the target value of this EC incurs one cell modification.   □

Remark. As stated in former works, EC technique delays the choice of target value as late as possible, to avoid poor local decisions. Also, EC avoids introducing values that are not meaningful, in contrast to NaiveLocal.

Equivalence Classes on a Fragmented Relation. This paper considers a relation that is fragmented and distributed to multiple sites. Then to repair FD violations using EC, we need to develop distributed algorithms that can build ECs upon a fragmented relation. We find the improvement in repair quality introduced by EC comes at a cost: it is intractable to build ECs on horizontally partitioned data, for the optimization of distributed algorithms.

Theorem 1. *On a horizontally partitioned relation, it is NP-complete to build ECs with either minimum data shipment or minimum parallel time.*   □

One may want to minimize data shipment and parallel computation time at the same time. However, these two measures may be controversial with each other, even in FD violation detection [12]. Since parallel time is typically the dominating factor of algorithm design, in the rest of paper, we present algorithms to optimize parallel time. Note that data shipment time is part of the parallel time, and hence is considered in our algorithms as well.

5 Distributed Data Repairing Based on Equivalence Class

We present algorithms for repairing FD violations in horizontally partitioned data, based on equivalence class (EC). In light of the intractability, our algorithms are heuristic. We first provide an efficient implementation of EC to facilitate the design of distributed algorithms, and then give repairing algorithms and optimization techniques that distribute ECs to multiple sites for parallelism.

5.1 Implementation of Equivalence Class

We aim to give an implementation of EC that can effectively support basic operations on EC, and that can be extended to handle ECs distributed to multiple sites. To this end, we implement EC in Algorithm 1 by combining the disjoint-set forest data structure [7] with the linked list technique. Each EC is denoted by a tree, whose nodes are cells in this EC. Since an EC is associated with a specific attribute, each cell c is denoted by the tuple *id* of c in the tree. Slightly abusing notation, we use *c.id* to denote the related tuple *id* of c.

(1) Initialization (Procedure Init). For each cell c, we build as its initial EC a

---

Algorithm 1. BuildEC

Procedure Init /* initialize EC for each cell c */
1   **foreach** *cell c* **do**
2        $T[c]$.parent:= $c.id$; $T[c]$.rank:= 0; $T[c]$.next:= NULL; $T[c]$.tail:= $c.id$;
3   insert $(c,1)$ into $T[c]$.HTab;

Procedure Merge(c_1, c_2) /* merge two trees rooted at c_1, c_2 */
1   **if** $T[c_1].rank < T[c_2].rank$ **then** $T[c_1]$.parent := $c_2.id$;
2   **else if** $T[c_1].rank > T[c_2].rank$ **then** $T[c_2]$.parent := $c_1.id$;
3   **else** $T[c_1]$.parent := $c_2.id$; $T[c_2]$.rank := $T[c_2]$.rank + 1;
     /* W. l. o. g., below we assume the tree rooted at c_1 is attached to c_2. */
4   **foreach** (v,cnt_1) in $T[c_1]$.HTab **do**
5   **if** $T[c_2]$.HTab *has an entry* (v,cnt_2) **then** update it as $(v,cnt_2 + cnt_1)$; ;
6   **else** insert (v,cnt_1) into $T[c_2]$.HTab;;
7    $T[T[c_2]$.tail].next:= $c_1.id$; $T[c_2]$.tail:= $T[c_1]$.tail;

Procedure Chase(T, T')/* When two ECs T, T' on D are merged, deal with the possible mergence of ECs on C, via $X \rightarrow C$ ($D \in X$). */
1   initialize set l (resp. l') $\leftarrow$ all cells (tuple *ids*) in T (resp. T'); $L := \{(l,l')\}$;
2   **foreach** $B \in X \setminus D$ **do**
3        $List := L$; $L := \emptyset$;
4   **foreach** $(l,l') \in List$ **do** /* join tuples from l, l' on their ECs of B */
5   split it into set $M := \{(l_1, l_1'), (l_2, l_2'), \dots\}$ such that $l_1, l_2, \dots$ (resp. l_1', $l_2', \dots$) are non-emtpy disjoint subsets of l (resp. l'), and $\forall t \in l_i$, $\forall t' \in l_i'$, $Find(t[B]) = Find(t'[B])$;
6            $L := L \cup M$;
7   **foreach** $(l,l') \in L$ **do**
8   **foreach** $t \in l$, $t' \in l'$ **do** /* t, t' agree on all ECs of X*/
9   **if** $Find(t[C]) \neq Find(t'[C])$ **then** $Merge(Find(t[C]), Find(t'[C]))$;

Function Find(c)/* find the root of the tree that $T[c]$ belongs to */
1   **if** $T[c].parent \neq c.id$ **then** $T[c]$.parent := $Find(T[c]$.parent);
2   **return** $T[c]$.parent;

---

single-node tree $T[c]$ with five fields: parent, rank, HTab, next and tail. parent and rank are initialized to be $c.id$ and 0 respectively, to be used by the union-by-rank heuristics [7]. HTab is a Hash table, keeping distinct values and their related counts, *i.e.*, the number of cells having that value in this EC. Initially, we insert an entry $(c, 1)$ into HTab, with c as the key field of the hash table; here c denotes the value of cell c. next and tail are initialized to be NULL and $c.id$, respectively; they link to the next cell following c and the last cell in the linked list.

Complexity. It takes $O(|D| \times m)$ time for the initialization phase, where $|D|$ is the number of tuples, and m is the number of attributes involved in FDs.

(2) Merge equivalence classes (Procedure Merge). (*i*) Following [7], the union-by-rank heuristics is applied to union two trees rooted at c_1 and c_2 (lines 1-3). Intuitively, it aims to always attach the tree with a smaller rank to the root of the tree with a larger rank. When two trees have equal rank, we arbitrarily choose one of them as the parent and increase its rank by 1. (*ii*) We maintain HTab tables when merging two ECs (lines 4-6). (*iii*) Finally, we maintain the

linked list by attaching the list starting from c_1 to the end of the list starting from c_2 (assuming the tree rooted at c_1 is attached to c_2). Note that we maintain fields parent and next for all cells, but maintain other fields only for the root cell.

Complexity. It takes $O(1)$ for (i) and (iii), and at most $O(max(i,j))$ for (ii), where i, j is the number of entries in $T[c_1].$HTab and $T[c_2].$HTab respectively.

(3) When to merge ECs? As stated in Section 4, there are two cases: (i) two ECs are merged when having same values in their HTab; or (ii) the mergence of ECs on attribute D may lead to mergence of ECs on attribute C, when there exists an FD $X \rightarrow C$ and $D \in X$. Case (i) requires similar operations on HTab as $(2)(ii)$. Case (ii) is much more subtle, since it involves FD reasonings. Procedure Chase is provided for this case. Chase first enumerates all cells (ids) in given ECs (line 1); this can be efficiently done by following the next field from the root cell. Chase then joins tuples (ids) from two ECs based on their ECs on attribute $B \in X \backslash D$ one by one (lines 2-6). This requires to find the EC that a given tuple belongs to (Procedure Find). Here, Find uses the path compression heuristics [7] to shorten path to the root. Finally, Merge is called for each pair of t, t' that agrees on all ECs of X, and that does not agree on ECs of C (lines 7-9).

Complexity. We study the complexity of Chase. (a) It takes linear time in the number of cells for line 1. (b) Hash join of set l with i tuples and set l' with j tuples on $|X|$-1 attributes takes at most $(|X|-1)(i+j)$ Find operations. Note that k find operations on a tree of N nodes, can be performed on a disjoint-set forest with "union by rank" and "path compression" heuristics in its worst-case time $O(k\alpha(N))$ [7]. Here $\alpha(N)$ is the inverse Ackermann function, which is incredibly slowly growing and is less than 5 for all remotely practical values of N. Hence, $\alpha(N)$ can be regarded as a constant. (c) It requires in its worst case $i + j$ Find and $i \times j$ Merge for lines 7-9, but quite rare in practice.

Example 5. Consider the EC on attribute city that $t_1[city], t_2[city], t_3[city]$ belongs to. The data structure of this EC is shown in Fig. 2(b), with *valid* fields, *i.e.,* parent, next for all cells, and rank, tail, HTab for the root cell.   □

5.2   Distributed Equivalence Class for Data Repairing

We come to the distributed setting and start with the baseline algorithm, referred to as DisBuild. In DisBuild, at each site partial ECs are built on the fragmented relation, upon which *global* ECs are then built at some *coordinator* sites.

DisBuild follows the distributed computation model stated in Section 3. To simplify presentation, we use *remote function* as a wrapper of some message passings. At site S_i, algorithm may call a remote function of the form $S_j : f(p_1, \ldots, p_n)$, to be executed at another site S_j. Technically, to do so, algorithm needs to send messages to site S_j by encoding $f(p_1, \ldots, p_n)$, and receives answers via messages from S_j. There are two basic remote functions supported by all sites. (1) $r\_list(root)$ is to list all cells in the tree (EC) rooted at $root$; and (2) $r\_find(cell)$ is to find the EC that $cell$ belongs to. Since $root, cell$ are tuple ids, the site at which remote function is to be conducted, can be readily identified.

Algorithm. Algorithm DisBuild takes as input a set Σ of FDs and partial relation $D_i = \sigma_{F_i}(D)$ at site S_i. It finds a repair of D using ECs in four stages. Without loss of generality, we suppose data are evenly distributed to all sites.

Stage 1: ECs are built on D_i at site S_i in parallel, by following Algorithm 1.

Stage 2: DisBuild merges ECs on the same attribute at different sites when they have same values in their HTab tables. To do so, (1) DisBuild heuristically picks a *coordinator* site for each attribute A involved in Σ, denoted by S^A. If possible (the number of sites is larger than the number of attributes), DisBuild assigns a coordinator to each attribute. Otherwise, DisBuild prefers to assign a coordinator to each of LHS attributes of FDs, and shares coordinators among attributes when necessary. (2) Site S_i identifies ECs at S_i on attribute A. For each such EC tree T rooted at cell c, S_i sends $(c.id, T[c].\mathsf{HTab})$ to S^A. (3) For each received $(id, table)$ at S^A, DisBuild builds as an EC a single-node tree $T[c]$, with $(id, 0, table, \mathrm{NULL}, id)$ as values for fields (parent, rank, HTab, next, tail), respectively. We refer to ECs built in Stage 2 at S^A as *global* ECs, while refer to ECs built in Stage 1 as *local* ECs. Intuitively, DisBuild builds *global* ECs upon roots of *local* ECs. (4) DisBuild identifies *global* ECs with same values in HTab at S^A, and merges them following Merge in Algorithm 1.

Data shipment. For each EC, only its root cell (id) is shipped. The number of entries of all HTab tables shipped from site S_i to coordinator S^A equals the number of distinct $t[A]$ values in fragment D_i.

Stage 3: Triggered by mergence of *global* ECs at S^A in Stage 2 and iteratively in Stage 3, DisBuild conducts EC computations for all FDs of the form $X \to C$ ($A \in X$) at S^A, and informs coordinator S^C (by message passing) to merge its *global* ECs on C when required. This repeats until no change happens at any site. To do so, DisBuild extends Chase in Algorithm 1, by obtaining data via message passings (including remote functions). Specifically, (1) To fetch all cells in a *global* EC rooted at r, DisBuild first at the corresponding coordinator lists all cells in this *global* EC, and then for each listed cell c, calls remote function $r\_list(c)$ to list all cells in the *local* EC rooted at c. (2) To find the EC that cell c belongs to, DisBuild first calls remote function $r\_find(c)$ to fetch the root r of the *local* EC containing c, and then at the coordinator identifies the root of the *global* EC containing r. (3) Note that all cells in a *local* EC are at the same site, so are related tuples. Therefore, DisBuild introduces a single message protocol to fetch ECs that $t_i[B]$ belongs to, for all tuples t_i containing cells in a *local* EC on A and for all attributes $B \in \{C\} \cup X \setminus A$, when handling $X \to C$ ($A \in X$). Although this may incur more data shipment compared to the approach that fetches data when necessary, this avoids the overhead of multiple rounds of communication and can be partly done in parallel with Chase.

Data shipment. All messages consist of only tuple *ids*. For a local EC with k cells and an FD $X \to C$, it requires to fetch at most $k \times |X|$ *ids*, with a single round of communication. Note that for an FD $A \to C$, *i.e.,* FD with only one LHS attribute, all tuples in the same *local* EC on A must be in the same *local* EC on C; in this case, only one *id* is required to be obtained for C.

Stage 4: For each *global* EC, DisBuild identifies its target value based on HTab

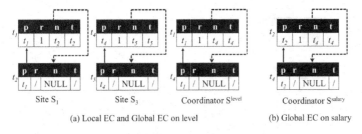

(a) Local EC and Global EC on level   (b) Global EC on salary

Fig. 3. Example of DisBuild

of the root cell, and in this *global* EC, identifies *local* ECs with value(s) other than the target value by their HTab collected in Stage 2. DisBuild informs sites containing those ECs to modify cell values accordingly, to produce a repair of D.
Data Shipment. We need to ship one value for each *local* EC with value(s) different from the target value.

Example 6. (1) In Fig. 3(a), we show *local* ECs at site S_1, S_3 and *global* ECs on level after Stage 2. (2) For EC computation via FD *title, level* $\rightarrow$ *salary* in Stage 3, DisBuild lists all cells in the *global* EC, and identifies ECs that $t_i[B]$ belongs to, for $i \in [1, 2, 4, 5]$ and $B \in \{title, salary\}$, by local computations and message passings. This causes mergence of *global* ECs at the coordinator site for salary, shown in Fig. 3(b). Here we suppose t_4 is the root of the *local* EC containing t_4, t_5 at site S_3. (3) Finally, a target value is selected for this EC in Stage 4. Suppose 80K is the target value, site S_3 is required to be informed of this. □

Remark. (1) DisBuild distributes to multiple sites computations for (a) different fragments in Stage 1, (b) different attributes in Stage 2 and 4, and (c) FDs with different LHS attributes in Stage 3. (2) DisBuild is a distributed implementation of data repairing technique with EC. Therefore, DisBuild is guaranteed to terminate in finite steps and correctly find a repair.

5.3   Optimization Strategies

We next introduce two optimization strategies.

Fully Distributed Mode. A limitation of DisBuild is that it requires to visit coordinators for most operations. Alternatively, we present another approach that fully distributes computations to all sites, denoted as FullDis. FullDis is based on fully distributed ECs: upon mergence of ECs, related fields of ECs are modified (excluding HTab) following Merge of Algorithm 1, with trivial data shipment. After that, some cell may have cell at other site as its parent (similarly for tail and next); this enables FullDis to tune basic operations. Specifically, by following parent, when $r\_find(c)$ executed at site S_i reaches a cell c' at other site, say S_j, FullDis in turn calls $r\_find(c')$ at S_j; this continues until reaching the site containing the root of EC that c belongs to. In this way, FullDis distributes computations to more sites other than coordinators. Better, the path compression heuristics in Find helps reduce the number of sites to be visited. Similarly, $r\_list(r)$ is conducted by following next, possibly through multiple sites.

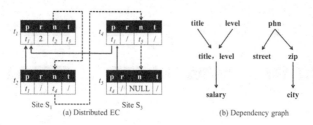

Fig. 4. Example of optimizations

Example 7. As shown in Fig. 4(a), ECs on level at site S_1, S_3 are merged to form a *distributed* EC rooted at t_1. One can find the root of this EC and list cells in this EC by following fields parent, next, respectively.   □

More specifically, FullDis is also conducted in four stages, but differs from DisBuild in Stages 2-4, as follows.

Stage 2: When there are abundant sites, FullDis may employ multiple coordinators for attribute A, whose values are from an ordered domain. Given k as the number of coordinators for A, FullDis works as follows. (1) As a preprocessing step, FullDis mines some D_i for values $v_1 < v_2 < \cdots < v_{k-1}$ as boundary values to partition the domain of A. It then identifies k coordinators $S^A(i)(i \in [1, k])$, and informs all sites of boundary values and coordinators. (2) At site S_j in parallel, for each EC on A rooted at cell r, FullDis ships $r.id$ and $\sigma_{G(i)}(T[r].\mathsf{HTab})$ to $S^A(i)$ when $\sigma_{G(i)}(T[r].\mathsf{HTab})$ is not empty. Here $\sigma_{G(i)}(T[r].\mathsf{HTab})$ is a horizontally partitioned fragment of $T[r].\mathsf{HTab}$: $(v, cnt) \in \sigma_{G(i)}(T[r].\mathsf{HTab})$ if (i) $i=1$ and $v < v_1$, or (ii) $i = k$ and $v \geq v_{k-1}$, or (iii) $i \in [2, k - 1]$, and $v_{i-1} \leq v < v_i$. Note that roots of some ECs may be shipped to multiple coordinators, but each with disjoint partial HTab tables. (3) Coordinators in parallel, merge collected ECs based on HTab values, using the aforementioned distributed ECs.

Stage 3: FullDis employs sites other than coordinators, for EC computations via FD. FullDis can delegate such tasks to any *idle* site, by sending it a message with root *ids* of ECs that are merged. This site then identifies involved FDs based on LHS attributes, and fetches required data by visiting the distributed ECs.

Stage 4: FullDis employs more sites to determine target values for ECs. (1) Each site S_i in parallel identifies ECs at S_i whose root cell r is at other site, and ships HTab table to the site at which r resides (HTab tables of ECs with the same root are firstly merged locally). Recall that FullDis does not modify HTab when merging ECs in Stage 2 and 3. (2) For ECs rooted at S_i, S_i determines target values for them by considering local HTab and HTab received from other sites.

Remark. As verified by our experiments (Section 6), FullDis allows a higher degree of parallelism than DisBuild in Stages 2-4. Indeed, even when we cannot afford multiple coordinators for one attribute in Stage 2, distributing EC computations to more sites in Stage 3 is proved to be very effective by itself.

Build EC Following Dependency Graph. Former approaches are *eager* in that they perform mergence of ECs via FD as early as possible. This maximizes parallelism but may incur unnecessary computation in certain cases. Consider Example 1. t_2 from site S_1 and t_4, t_5 from site S_3 will be put into the same EC on salary via FD *title, level $\rightarrow$ salary*. However, this is doable only when t_2, t_4, t_5 are already in the same EC on title (level); the aforementioned EC computation via FD may fail if it is conducted before ECs on title or level are merged. This *false negative* is possible because DisBuild deals with ECs on title and level at different sites in parallel. Although DisBuild will successfully conduct mergence of ECs eventually, we see it may incur unnecessary computations in the process.

We present an approach that may avoid some unnecessary computations, denoted as SerBuild. SerBuild *serializes* some EC computations via FD, by following the *dependency graph*. As a preprocessing step, SerBuild builds dependency graph at a selected *master* site S_m. In the graph, (a) each attribute or (b) each set of attributes that are LHS attributes of a same FD, is treated as a (*composite*) vertex, and there is an edge from LHS attribute(s) to RHS attribute for each FD, and an edge from attribute A to each composite vertex containing A. As an example, we show the dependency graph for Example 1 in Fig. 4(b).

At master site S_m, SerBuild identifies edges (FDs) that start from composite vertex, and that are not part of strongly connected components in linear time [7]. SerBuild differs from DisBuild in Stage 3 when handling these FDs. Specifically, for each of these FDs in the form of $X \rightarrow A$, SerBuild performs mergence of ECs on A via this FD only after mergence of all attributes $B \in X$. To do so, master site S_m communicates with coordinators sites, to monitor the progress of EC computation at those sites, and to guide some sites for the next step.

Remark. Through experiments (Section 6), we find SerBuild avoids some unnecessary computations, without affecting parallelism.

6 Experimental Study

Experimental Setting. We use 8 machines (sites), each with 2.53GHz Intel Xeon X3440 CPU, 4GB memory and Windows 7, connected by a local area network. Each experiment was run 5 times and the average is reported here.

As noted earlier, our algorithms provide a distributed implementation of EC technique and produce the same repair as the centralized approach. Since the effectiveness of EC in terms of repair quality is well demonstrated by former works, we omit the results concerning repair quality, *e.g.,* precision, here.

Data. (1) Real-life HOSP data is taken from US Department of Health & Human Services. We obtain a relation having more than 200K tuples with 16 attributes (https:// data.medicare.gov/data/hospital-compare) and design 8 FDs for it. (2) Synthetic Person data combines the schema of Fig. 1 with that of the UIS Database generator [3,21]. We create a relation with 10 attributes, and populate it using a modified version of the UIS Database generator.

Algorithms. We implement the following algorithms in Java: distributed repairing algorithm DisBuild, and optimizations FullDis and SerBuild. For comparison, we also implement a naive approach Naive, which collects all tuples at a single coordinator site, and then repairs data using centralized equivalence class.

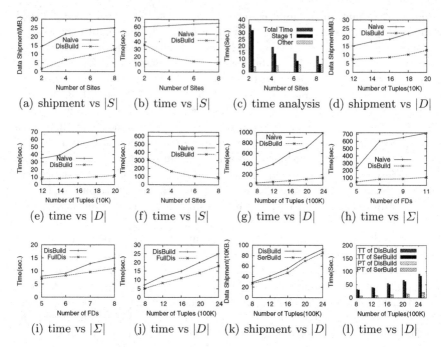

Fig. 5. Experimental Results

All experiments are controlled by two parameters: (a) $|D|$: the number of tuples; and (b) $|S|$: the number of (fragments) sites. We uniformly distribute $|D|$ tuples to $|S|$ sites in all experiments.

Exp-1. Using HOSP data, we compare the performance of DisBuild against Naive. *Varying* $|S|$. By fixing $|D| = 200$K, varying $|S|$ from 2 to 8, Fig. 5(a) shows results of total data shipment. This comparison favors Naive, since shipments in DisBuild are distributed among sites. We see the following. (1) Data shipments of all algorithms increase with larger $|S|$, as expected. (2) DisBuild consistently outperforms Naive. As stated earlier, most shipments in DisBuild consist of only tuple *ids*, and for the most expensive part of shipment conducted in Stage 2, DisBuild ships only distinct values in each fragment.

Fig. 5(b) shows the parallel time of all algorithms. DisBuild consistently outperforms Naive, and the gap increases as $|S|$ increases; it takes less time for DisBuild but more time for Naive with larger $|S|$. Note that Naive always takes the same time for EC computation at its coordinator, while more time for data shipment with the increase of $|S|$. To further analyze the results of DisBuild, in Fig. 5(c), we decompose its time into two parts: time for Stage 1, and time for other stages. We see that the former time decreases while the latter one increases as $|S|$ increases, as expected. Specifically, DisBuild leverages more sites to significantly reduce the time for Stage 1 from 32 Sec. to 6 Sec., and the time for other stages slightly increases from 4 Seconds to 6 Seconds.

Varying $|D|$. We then evaluate the scalability of algorithms with $|D|$. By fixing $|S| = 8$, varying $|D|$ from 120K to 200K in 20k increment, Fig. 5(d), 5(e) show the total data shipment and parallel time. As expected, the required shipments and times of all algorithms increase as the data size increases. Compared to Naive, DisBuild scales better with $|D|$, especially in the parallel time.

Exp-2. Using Person data, we compare DisBuild against Naive on large data sets, in terms of parallel time. We use one more parameter $|\Sigma|$ to vary the number of FDs $\in \Sigma$. We set $|S| = 8$, $|D| = 1,600K$, $|\Sigma| = 7$ by default, and vary one parameter in each of Fig. 5(f), 5(g) and 5(h), respectively.

Varying $|S|$. By varying $|S|$ from 2 to 8, Fig. 5(f) confirms our observations on HOSP data. DisBuild outperforms Naive in reducing parallel time by 48% to 85%, as $|S|$ increases. We find in DisBuild, the time for Stage 1 decreases from 301 Sec. to 68 Sec., and the time for other stages slightly increases from 8 Sec. to 14 Sec. (not shown in figures). The time for Stage 1 remains the dominant factor, and can be effectively optimized with the increase of $|S|$.

Varying $|D|$. Fig. 5(g) shows experimental results when $|D|$ increases from 800k to 2,400k, in 400k increment. We see that DisBuild scales well with $|D|$: its parallel time increases from 38 Sec. to 135 Sec., as $|D|$ triples.

Varying $|\Sigma|$. We increase $|\Sigma|$ from 5 to 11, and report results in Fig. 5(h). As expected, the increase of $|\Sigma|$ has a negative impact on the running time. Compared to Naive, DisBuild scales better. This is because DisBuild distributes computations concerning Σ to multiple sites, both in Stage 1 and in Stage 3.

Exp-3. We compare FullDis against DisBuild. We use 8 machines and get a vertically fragmented Person data with 7 attributes. We assign a coordinator to each attribute, and use one additional machine as a *worker* site, to be used by FullDis in Stage 3 for EC computation via FD. In this experiment, we report parallel time of Stages 2-4; FullDis differs from DisBuild in these states.

Varying $|\Sigma|$. We fix $|D| = 1,600K$, increase $|\Sigma|$ from 5 to 8, and report results in Fig. 5(i). We see that FullDis outperforms DisBuild, and the gap widens when $|\Sigma|$ increases. Indeed, the time of FullDis is about [73%, 90%] of the time taken by DisBuild. With the increase of $|\Sigma|$, we find several sites become bottlenecks in DisBuild: each of these sites is used as the coordinator for an attribute that is LHS attribute of multiple FDs. These sites are required to conduct EC computations for all related FDs in DisBuild, and hence take longer time than other sites. FullDis avoids this by distributing such computations to other idle sites, *e.g.*, the worker site, or sites as coordinators only for RHS attributes of FDs. Combining these with the fully distributed ECs, FullDis further improves parallelism.

Varying $|D|$. Fig. 5(j) shows experimental results when $|D|$ increases from 800k to 2,400k and $|\Sigma| = 8$. FullDis is faster than DisBuild and scales well with $|D|$.

Exp-4. We compare SerBuild against DisBuild using Person, by fixing $|S| = 8$, $|\Sigma| = 7$ and varying $|D|$ from 800k to 2,400k. Among the 7 FDs, two of them have multiple LHS attributes. We report results of Stage 3, since SerBuild differs from DisBuild in this stage. Fig. 5(k) shows that SerBuild requires less shipment compared to DisBuild. All shipments in Stage 3 are conducted to fetch data for EC computation via FD; this implies that SerBuild avoids some of the unnecessary computations. Also note that there are small data shipments in Stage 3, since only tuple *ids* are shipped. Fig. 5(l) shows parallel time (PT), and in addition, shows *total* computation time (TT), which is the sum of computation times at all sites. We find SerBuild has similar PT as DisBuild, and more evidently, improves TT by 5% to 8%. SerBuild avoids unnecessary computations at some sites, and those sites proceed to other computations, without affecting parallelism.

7   Conclusions

We have studied the complexity of distributed data repairing (with equivalence class), presented algorithms and optimizations for distributed repairing based on EC, and experimentally verified our approach. We are currently experimenting with more real-life datasets, extending algorithms to support vertically partitioned relations, developing distributed repairing techniques for more constraints, *e.g.,* conditional functional dependencies [11].

Acknowledgments. This paper is supported by Shanghai technology innovation project 14511107403.

References

1. Abiteboul, S., Hull, R., Vianu, V.: Foundations of Databases. Addison-Wesley (1995)
2. Bohannon, P., Fan, W., Flaster, M., Rastogi, R.: A cost based model and effective heuristic for repairing constraints by value modification. In: SIGMOD (2005)
3. Beskales, G., Ilyas, I., Golab, L., Galiullin, A.: Sampling from repairs of conditional functional dependency violations. VLDB Journal **23**(1), 103–128 (2014)
4. Beskales, G., Ilyas, I., Golab, L., Galiullin, A.: On the relative trust between inconsistent data and inaccurate constraints. In: ICDE (2013)
5. Cong, G., Fan, W., Geerts, F., Jia, X., Ma, S.: Improving data quality: Consistency and accuracy. In: VLDB (2007)
6. Chu, X., Ilyas, I., Papotti, P.: Holistic data cleaning: Putting violations into context. In: ICDE (2013)
7. Cormen, T., Leiserson, C., Rivest, R., Stein, C.: Introduction to Algorithms. MIT Press (2009)
8. Chiang, F., Miller, R.: A unified model for data and constraint repair. In: ICDE (2011)
9. Dallachiesa, M., Ebaid, A., Eldawy, A., Elmagarmid, A., Ilyas, I., Ouzzani, M., Tang, N.: NADEEF: a commodity data cleaning system. In: SIGMOD (2013)
10. Dean, J., Ghemawat, S.: MapReduce: Simplified data processing on large clusters. In: OSDI (2004)
11. Fan, W., Geerts, F., Jia, X., Kementsietsidis, A.: Conditional functional dependencies for capturing data inconsistencies. In: TODS **33**(2) (2008)

12. Fan, W., Geerts, F., Ma, S., Muller, H.: Detecting inconsistencies in distributed data. In: ICDE (2010)
13. Fan, W., Li, J., Ma, S., Tang, N., Yu, W.: Towards certain fixes with editing rules and master data. VLDB Journal **21**(2), 213–238 (2012)
14. Fan, W., Li, J., Tang, N., Yu, W.: Incremental detection of inconsistencies in distributed data. TKDE **26**(6), 1367–1383 (2014)
15. Kolahi, S., Lakshmanan, L.: On approximating optimum repairs for functional dependency violations. In: ICDT (2009)
16. Lynch, N.: Distributed Algorithms. Morgan Kaufmann (1996)
17. Ozsu, M., Valduriez, P.: Principles of Distributed Database Systems (2nd edition). Prentice-Hall (1999)
18. Song, S., Cheng, H., Yu, J., Chen, L.: Repairing vertex labels under neighborhood constraints. In: VLDB (2014)
19. Wang, J., Tang, N.: Towards dependable data repairing with fixing rules. In: SIGMOD (2014)
20. Yakout, M., Elmagarmid, A., Neville, J., Ouzzani, M., Ilyas, I.: Guided data repair. In: VLDB (2011)
21. UIS data generator. http://www.cs.utexas.edu/users/ml/riddle/data.html

GB-JER: A Graph-Based Model for Joint Entity Resolution

Chenchen Sun$^{(\boxtimes)}$, Derong Shen, Yue Kou, Tiezheng Nie, and Ge Yu

College of Information Science and Engineering,
Northeastern University, Shenyang, China
dustinchenchen_sun@163.com,
{shenderong,kouyue,nietiezheng,yuge}@ise.neu.edu.cn

Abstract. To resolve multiple classes of related entity representations jointly promotes accuracy of entity resolution. We propose a graph-based joint entity resolution model: GB-JER, who exploits a dynamic entity representation relationship graph. It contracts the neighborhood of the matched pair, where enrichment of semantics provides new evidences for subsequent entity resolution iteratively. Also GB-JER is an incremental approach. The experimental evaluation shows that GB-JER outperforms existing the state-of-the-art joint entity resolution approach in accuracy.

Keywords: Joint entity resolution · Similarity propagation · Structure-based similarity · Entity representation relationship graph

1 Introduction

Entity resolution (ER) is a key aspect of data quality and is very important to data integration and data mining [1,6,17]. ER identifies which entity representations correspond to the same real-world entity in databases. Traditional ER approaches rely on attribute-based similarity (ABS), focusing on a single class of representations [1,6]. In the big data era, data are related mostly, such as citation data including papers, authors & venues and movie data including movies, actors & directors.

Example 1. In the toy database of Fig. 1., (a) shows the schema corresponding to a set of representations in (b). The schema includes three classes: paper, author and venue, each with a set of attributes. The paper class includes a simple attribute "name" and two reference attributes "writtenBy" and "publishedIn", linking to author and venue respectively. In Fig. 1. (b), the representation set contains 6 paper representations, 13 author representations and 6 venue representations. For example, c_1 has a title "Incremental entity resolution", which is written by a_1 & a_2 and is published in v_1. Fig. 1. (c) shows the ground truth of ER result of the set in (b). Given (a) and (b), how to get (c) is a joint ER problem.

© Springer International Publishing Switzerland 2015
M. Renz et al. (Eds.): DASFAA 2015, Part I, LNCS 9049, pp. 458–473, 2015.
DOI: 10.1007/978-3-319-18120-2_27

The accuracy of result over related data by traditional ER approaches is limited for such approaches don't exploit relationships among different representations. In Fig. 1., traditional ER approaches resolve papers, authors and venues separately, but can't utilize relationships to promote accuracy. They (1) can't resolve a_1 & a_3 because a_3 is an incomplete representation and (2) can't decide whether a_5 match a_1 or a_{12} because a_5 is an ambiguous representation for a_1 & a_{12}. There exist a few ER approaches utilizing relationships, some of which resolve multiple classes of representations jointly. However, existing joint ER approaches can't handle the two problems above at the same time. Getoor et al. [2] utilize co-occurrences of authors in citation dataset to resolve authors, which handles problems like (1) but not (2). Dong et al. [5] build a dependent graph with candidate representation pairs to do joint ER, which handles problems like (1) but not (2). These two approaches utilize direct relations between representations. Kalashnikov et al. [10,14] exploits entity relationship graph to resolve a single class of representations, which handles problems like (2) but not (1). Besides, nowadays data updates faster than past so that incremental ER approaches are necessary.

paper {title, writtenBy, publishedIn};   author {name};   venue {name}.
(a) Schema

Paper:
p_1 {"Incremental entity resolution", $\{a_1, a_2\}$, v_1 }; p_2 { "Incremental ER", $\{a_3, a_4\}$, v_2 };
p_3 { "Graph partition", $\{a_5, a_6\}$, v_3 }; p_4 {"Big graph partition", $\{a_7, a_8\}$, v_4 };
p_5 {"Graph data management", $\{a_9, a_{10}\}$, v_5 }; p_6 {"Cliques in graph", $\{a_{11}, a_{12}, a_{13}\}$, v_6}.

Author:
a_1 {"Josh Doe"}; a_2 {"Steven Pelley"}; a_3 {"J. "}; a_4 {"S. Pelley"}; a_5 {"J. Doe"};
a_6 {"A. Widom"}; a_7 {"Alex Widom"}; a_8 {"Jenny Brown"}; a_9 {"Jen. Brown"};
a_{10} {"M. Jacob"}; a_{11} {"Jennifer Brown"}; a_{12} {"Jonathan Doe"}; a_{13} {"Marie Jacob "}.

Venue:
v_1 {"ICDE"}; v_2 {"International cof on data engineering"}; v_3 {"CIKM"};
v_4 {"EDBT"}; v_5 {"dasfaa"}; v_6 {"kdd"}.
(b) Representation set

$\{<p_1, p_2>,$ $<p_3>,$ $<p_4>,$ $<p_5>,$ $<p_6>,$ $<a_1, a_3>,$ $<a_2, a_4>,$ $<a_5, a_{12}>,$ $<a_6, a_7>,$
$<a_8, a_9, a_{11}>,$ $<a_{10}, a_{13}>,$ $<v_1, v_2>,$ $<v_3>,$ $<v_4>,$ $<v_5>,$ $<v_6>\}$
(c) Ground truth

Fig. 1. Motivational example

A relational dataset consisted of multiple classes of related representations can be mapped to an entity representation relationship graph (ERRG), which retains corresponding relational schema and holds topology. With an ERRG, it's convenient to fully exploit semantic relationships among representations. The similarity between two representations can be measured via structure in addition to attributes that it helps solve problems like (1) and (2). The structure similarity can be measured by combining the paths between representations and schemata of links on paths. When two representations match, they are merged and the neighborhood is contracted, which enriches semantics. In Fig. 1. after p_1 and p_2 are resolved, the two pairs a_1 & a_3 and a_2 & a_4 are more likely to match. This is called similarity propagation, which helps solve problems like (1). With joint

ER going on, the ERRG evolves and semantics of it become richer and richer that promotes subsequent ER. However, such an iterative process results in a higher complexity. A pair of representations may be compared a few times until they match or neighborhood around them no longer changes. To save cost, the resolution order should be optimized so that the most probably duplicate pairs should be resolved preferentially. Getoor's and Dong's approaches are also faced with high complexities but they don't handle it well. Besides, when increment comes, the existing ERRG can be utilized to do incremental ER with less time cost but no accuracy loss. This paper makes contributions as following:

- We propose a graph-based, iterative joint entity resolution model: GB-JER. It fully exploits a gradually converged ERRG and iteratively utilizes dynamic relationships among representations to jointly resolve multiple classes of representations con-currently.
- Experimental evaluation demonstrates that GR-JER can improve accuracy of joint ER.

In section 2, we present the problem formalization and entity representation relationship graph; as the core, section 3 specifies GB-JER model including joint match, joint merge & similarity propagation and also we discuss two extra aspects of GB-JER; section 4 evaluates the proposed model; section 5 introduces related work; section 6 concludes.

2   Preliminaries

2.1   Problem Formalization

In our model, a schema contains a set of classes, each of which includes a set of attributes. Each instance of a class is an entity representation. The attributes are categorized as: simple attributes, whose values are simple types like string and integer; and reference attributes, whose values are references to other representations. Like example 1, given a set of multiple classes of representations and its corresponding schema , how to work out the ER result is a joint ER problem. Specifically, we try to solve three types of problems categorized as following:

- type 1, incomplete representations such as problem (1) in example 1;
- type 2, ambiguous representations such as problem (2) in example 1;
- type 3, resolution order optimization that the most probable duplicate pairs should be resolved preferentially and results in lower cost.

Unlike traditional ER, joint ER takes relationships into consideration. Joint ER can be solved by a pair of joint match operation and joint merge operation. A joint match operation dependently decides whether two representations match. The match of two representations is affected both by attribute-based similarities and the ER results of representations related to them. A joint merge operation merges two matched representations into a complex representation (denoted as $\langle *, * \rangle$) and let the new representation inherit two original ones' relationships. A pair of joint match and joint merge operations iteratively work until the joint ER is done.

2.2   Entity Representation Relationship Graph

Definition 1 (Entity Representation Relationship Graph (ERRG)).
Given a schema $S = (N, \Gamma)$, where N is a set of classes and Γ is a set of relations, an entity representation relationship graph is a directed graph $G = (R, L)$, with a representation class mapping function $\varphi : R \to N$ and a link type mapping function $\psi : L \to \Gamma$. For each representation $r \in R$, $\varphi(r) \in N$; for each link denoted as $l = l(r_i, r_j)$ and $l \in L$, $\psi(l) \in \Gamma$.

As a template for an ERRG [17,18], the schema graph depicts the classes and relations among classes. For a relation γ from class μ_1 to class μ_2, denoted as $\mu_1 \xrightarrow{\gamma} \mu_2$, μ_1 are source class and μ_2 are target class, denoted as $\gamma.S$ and $\gamma.T$. γ can be also denoted as $\mu_1 \to \mu_2$ or $\gamma(\mu_1, \mu_2)$ or (μ_1, μ_2). The inverse relation naturally exists as $\mu_1 \xleftarrow{\gamma} \mu_2$. Most of times, γ and γ^{-1} are not equal unless μ_1 and μ_2 are the same class and γ is symmetric. Citation network is an ERRG and its schema graph exists. In Fig. 1. (a), there are paper (P), author (A) and venue (V); relations are "write" & "writtenBy" between author and paper, and "publishedIn" & "publish" between paper and venue. Its corresponding schema graph is showed in In Fig. 2. Since usually γ and γ^{-1} are not equal, different weights should be given to them when measuring representations' similarity. For example, in average a venue corresponds to 56.8 papers and a paper corresponds to one venue in a database. If we know two papers' venues are the same, the probability of they matching is denoted $prob(papers|venues)$; the opposite situation is denoted as $prob(venues|papers)$. Intuitively, $prob(papers|venues) \leq prob(venues|papers)$. We propose *unidirectional related weight* to reflect such differences, denoted as $urw(\gamma)$, implying γ's importance to similarity computation. The unidirectional related weight can be set by domain experts or by experiments. In Fig. 2., $urw(venue \to paper) = 1/56.8$ and $urw(paper \to venue) = 1/1$.

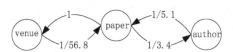

Fig. 2. Examplary citation schema graph

Definition 2 (Schema Path). *Given a schema $S = (N, \Gamma)$, a schema path ρ is denoted as $\rho(\mu_0, \mu_k) = \mu_0 \xrightarrow{\gamma_0} \mu_1 \xrightarrow{\gamma_1} \mu_2 \cdots \mu_{k-1} \xrightarrow{\gamma_{k-1}} \mu_k$, which defines a composite relation $\gamma = \gamma_0 \circ \gamma_1 \cdots \circ \gamma_{k-1}$ between μ_0 and μ_k where $\circ$ denotes the composition operator on relations. The length of ρ is number of relations: k.*

A schema path can be denoted with classes if there is no multiple relations between two classes, such as $\rho = (\mu_0 \mu_1 \cdots \mu_k)$. In Fig. 2., schema path from author to venue with length two is $A \xrightarrow{write} P \xrightarrow{publishedIn} V$ or APV. The reverse path of ρ is denoted as ρ^{-1}, which defines a reverse relation of γ. A representation path $\mathfrak{p} = (r_0 r_1 \cdots r_k)$ is a path instance of ρ, where $\varphi(r_i) = \mu_i$ and $\psi(l_i(r_i, r_{i+1})) = \gamma_i$. It's denoted as $\rho(\mathfrak{p}) = (\varphi(r_0), \varphi(r_1), \cdots, \varphi(r_k))$ and $\mathfrak{p} \in \rho$.

Definition 3 (Relatedness of a schema path). *Given a schema path* $\rho = (\mu_0\mu_1\cdots\mu_k)$, *its relatedness is,*

$$rel(\rho) = \prod_{i=0}^{k-1} urw(\mu_i, \mu_{i+1}) \tag{1}$$

3   The Graph-Based Joint Entity Resolution Model

3.1   Overview

The graph-based joint entity resolution model (GB-JER) iteratively exploits a gradually converged ERRG, fully utilizing both direct and indirect relationships among representations, to jointly resolve multiple classes of related representations. It consists of three core modules (joint match, joint merge and similarity propagation) and an initialization module; it includes an ERRG and a candidate queue Q. It takes a dirty related dataset as input and outputs a clean one.

The initialization includes ERRG initialization and candidate queue initialization. A dirty ERRG is built from the input dataset and consists of multiple separate sub-ERRGs. The Canopy blocking technology [12] is used to produce blocks, each of which contains similar representations from a single class. In each block initial candidate representation pairs are worked out by Cartesian product, called *type I candidate pairs*. A candidate pair is denoted as $(*, *)$. A candidate queue Q is used to keep the order of candidate pairs. Q is a priority queue, in which each node contains a candidate pair and a priority. The initial candidate pairs are inserted into Q with the same but low priories in a partially random order, where candidate pairs from related representations are put closely so that GB-JER resolves related representations together as soon as possible. During the joint ER procedure, priorities of candidate pairs may be switched because of similarity propagation, which is an optimization of resolution order.

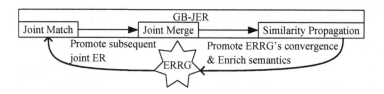

Fig. 3. Iterative process of GB-JER

GB-JER includes three iterative steps: (a) The first pair from Q is sent to the joint match. If the pair matches, go to next step; otherwise, repeat the same procedure. The joint match hires a hybrid similarity, combining an attribute-based similarity (ABS) and a structure-based similarity (SBS), to measure the pair's similarity. We propose a schema path based similarity computation algorithm to compute SBS. (b) The joint merge merges the matched pair and contracts the neighborhood. (c) The contraction change the neighborhood's topology and increases SBSes of some pairs, triggering re-computation. As shown in Fig. 3., the above three steps form an iterative procedure. With the procedure going on, the separate sub-ERRGs are merged together gradually. Finally, the Q is empty and the ERRG is stable and clean. GB-JER output the clean ERRG as the result of joint ER.

3.2   Joint Match

Joint Match Operator. Given a dirty ERRG $G(R, L)$, where $R = \{R_t\}_{t=1}^T$ is a representation set including multiple classes of related representations, R_t is t-th subset of a single class and $L \subseteq R * R$ is the link set of R, $r_i, r_j \in R$, the *hybrid similarity* between them is,

$$sim_{hyb}(r_i, r_j) = (1 - \delta) * sim_{abs}(r_i, r_j) + \delta * sim_{sbs}(r_i, r_j) \tag{2}$$

sim_{abs} is the attribute-based similarity; sim_{sbs} is the strucutre-based similarity; δ is the balance weight, which is set by users upon the importance of each similarity. sim_{abs} is chosen from existing ABS algorithms [3]; we propose a schema path based similarity computation algorithm to compute sim_{sbs}.

The joint match operator computes $sim_{hyb}(r_i, r_j)$ and compares it with a given match threshold θ_m, where if $sim_{hyb}(r_i, r_j) \geq \theta_m$, they match; otherwise, they fail to match. The match threshold θ_m is set by domain experts or by experiments.

A Schema Path Based Similarity Computation Algorithm. There are two types of algorithms to compute similarity of two nodes via structure on a graph: common neighbors based algorithms and paths based algorithms [11]. The proposed schema path based similarity computation algorithm belongs to the second type that takes both direct and indirect relationships into consideration. Since a schema path is unidirectional, we use a variant of random walk [13] that walks along the schema path and its reverse path to get a symmetric relatedness of two nodes.

Definition 4 (Relatedness of two representations along a representation path). *Given $r_i, r_j \in R, i < j$ and a representation path $\mathfrak{p} = (r_i r_{i+1} \cdots r_j)$, then the relatedness of r_i, r_j along $\mathfrak{p}$ is,*

$$rel(r_i, r_j)_{\mathfrak{p}} = rel(r_i, r_j)_{\mathfrak{p}^{-1}} = 0.5 * (rel(\mathfrak{p}) + rel(\mathfrak{p}^{-1})) \tag{3}$$

$$rel(\mathfrak{p}) = prob_{rw}(\mathfrak{p}) * rel(\rho(\mathfrak{p})) \tag{4}$$

Equation 3 shows that the relatedness of r_i, r_j along $\mathfrak{p}$ or $\mathfrak{p}^{-1}$ is symmetric. In equation 4, $rel(\mathfrak{p})$ is the relatedness of the representation path $\mathfrak{p}$, which combines the random walk probability along $\mathfrak{p}$ on G, $prob_{rw}(\mathfrak{p})$ and relatedness of its schema path $\rho(\mathfrak{p})$, $rel(\rho(\mathfrak{p}))$.

In order to compute the SBS of two representations in an ERRG, we sum up relatedness along all paths between them. Notice that too long paths contribute little to the SBS but cost much. To balance between cost and preciseness, we limit lengths of paths in consideration. The path set $\mathfrak{P}_{len}(r_i, r_j)$ includes all paths between r_i and r_j whose length is no longer than len. In this paper, $len = 8$.

Definition 5 (Schema path based similarity). *Given $r_i, r_j \in R, i < j$, then the schema path based similarity between them is,*

$$sim_{path}(r_i, r_j) = \sum_{\mathfrak{p} \in \mathfrak{P}_{len}(r_i, r_j)} rel(r_i, r_j)_{\mathfrak{p}} \tag{5}$$

Problem (2) in section 1 is whether $a_5\{$"J. Doe"$\}$ matches $a_1\{$"Josh Doe"$\}$ or $a_{12}\{$"Jonathan Doe"$\}$. ABS ER approaches and Dong's approach can't solve it. We show how to solve problem (2) with schema path based similarity, which demonstrate its usefulness in solving type 2 problems. At start time, since ABSes of (a_5, a_1) and (a_5, a_{12}) are almost the same and the three representations are in three separate sub-ERRGs so their SBSes both are 0, it's impossible to decide which pair matches. However, when the joint ER procedure runs into a certain phase that the neighbors around the three representations are resolved and the neighborhoods become denser, it's highly possible to solve the problem via schema path based similarity. As Fig. 4. shows, two paths between a_5 and a_{12} exist: $(a_5, c_3, \langle a_6, a_7 \rangle, c_4, \langle a_8, a_9, a_{11} \rangle, c_6, a_{12})$ and $(a_5, c_3, \langle a_6, a_7 \rangle, c_4, \langle a_8, a_9, a_{11} \rangle, c_5, \langle a_{10}, a_{13} \rangle, c_6, a_{12})$; a_1 and a_3 are merged into a complex representation $\langle a_1, a_3 \rangle$ and no path exists between a_5 and $\langle a_1, a_3 \rangle$. Now GB-JER combines both ABS and schema path based similarity to decide that a_5 and a_{12} match but a_5 and $\langle a_1, a_3 \rangle$ do not.

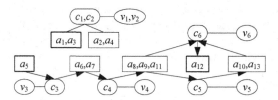

Fig. 4. A Joint ER example with schema path based similarity

3.3   Joint Merge

After a pair of representations match, the joint merge operator merges them as a complex representation and contracts the neighborhood around them, which makes it denser and enriches semantics. Re-computations then are triggered around the neighborhood, called similarity propagation. Joint merge and similarity propagation are closely related that they together make the ERRG a dynamic one.

Principle of representation merge: information amount should never change during representation merge. We propose a representative attribute value based merge method that it chooses the most representative value among multiple values from different original representations per attribute. This method saves both time cost and storage cost compared to the common method that keep all values per attribute [1].

In addition to representation merge, the neighborhood around the merged pair is contacted.

Definition 6 (Neighborhood contraction). *Given $r_i, r_j \in R$, they match and are merged as $\langle r_i, r_j \rangle$, then connect all links from r_i and r_j to $\langle r_i, r_j \rangle$ and remove duplicate links.*

As shown in Fig. 5., when c_1 and c_2 are merged as $\langle c_1, c_2 \rangle$, who inherits all links $(l_1, l_2, l_3, l_4, l_5, l_6)$ connected to c_1 and c_2, there exists two duplicate links l_5 and l_6 between $\langle c_1, c_2 \rangle$ and $\langle v_1, v_2 \rangle$. When duplicate links exist, randomly remove the redundant ones but keep one. In Fig. 5., l_5 is kept.

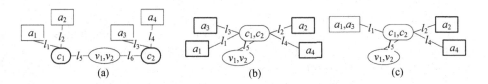

$$(a) \qquad\qquad (b) \qquad\qquad (c)$$

Fig. 5. An example of joint merge & similarity propagation

3.4   Similarity Propagation

When the joint merge of a matched pair is done, semantics enrichment provides positive evidences for ER in the neighborhood. As shown in Fig. 5., after joint merge of the pair (c_1, c_2), SBSes of the pairs (a_1, a_3) and (a_2, a_4) increase so that they should be resolved preferentially later. Besides links, paths also propagate similarity, such that co-author APA is a strong relation.

Definition 7 (Similarity propagation). *Given that $r_i, r_j \in R_x, r_g, r_h \in R_y$, $1 \leq x, y \leq T$, r_i and r_j match and are merged as $\langle r_i, r_j \rangle$, if $\rho(\mathfrak{p}(r_i, r_g)) = \rho(\mathfrak{p}(r_j, r_h))$, $rel(\rho(\mathfrak{p}(r_i, r_g))) = rel(\rho(\mathfrak{p}(r_j, r_h))) \geq \theta_{sp}$ and r_g, r_h satisfy the Canopy block requirement, then (r_g, r_h) is a type II candidate pair; specially, if (r_g, r_h) fail to satisfy the Canopy blocking requirement because of attribute values incompletion, also set them as a type II candidate pair.*

In this paper, the upper bound of the length of schema path in definition 7 is set to 2. θ_{sp} is the similarity propagation threshold. The special setting in definition 7 ensures that duplicate representations can match via structure even if their attribute values are incomplete, which solves type 1 problems in subsection 2.1. Although in problem (1) of section 1 a_1 and a_3 are not a type I candidate pair because of incomplete name values, a_1 and a_3 are generated as a type II candidate pair and match via SBS, as shown in Fig. 5. (b),(c).

It's easy to find out that the type I candidate pair set and the type II candidate pair set overlap but it's not a meaningless repetition. When type II candidate pairs are generated, they are inserted into proper positions if they don't exist in Q; otherwise their positions may be switched. GB-JER expect to first resolve candidate pairs that match with high probabilities, which are estimated by a weighted combination of the pairs' approximate ABS and relatedness of the schema path in the similarity propagation. Here the approximate ABS (denoted as $sim_{aprox-abs}$) is computed by the same low-cost approximate similarity methods used in Canopy blocking. GB-JER decides the order of candidate pairs in Q according to scores generated by the priority score function below.

Definition 8 (Priority score function). *Given that $r_i, r_j \in R_x, r_g, r_h \in R_y$, $1 \le x, y \le T$, the type II candidate pair (r_g, r_h) are generated by $\langle r_i, r_j \rangle$ and $\rho(\mathfrak{p}(r_i, r_g)) = \rho(\mathfrak{p}(r_j, r_h))$, then the priority of (r_g, r_h) in Q is,*

$$score(r_g, r_h) = (1 - \eta) * sim_{aprox-abs}(r_g, r_h) + \eta * rel(\rho(\mathfrak{p}(r_i, r_g))) \quad (6)$$

η is the balance weight, which is set by users upon the importance of each factor.

When a new type II candidate pair already exists in Q, switch it to a new position if its new priority is higher than the old one; otherwise, skip. There are other options. *Head insertion* inserts the pair into the head of Q if it's not in the head; otherwise, skip. *Tail insertion* inserts the pair into the tail of Q if it's not in Q; otherwise, skip. *Random insertion* inserts the pair into a random position of Q if it's not in Q; otherwise, it insert the pair into a random position between its current position and the head. The evaluation shows that priority based insertion method performs best.

Similarity propagation does not only improve the accuracy of joint ER but also optimizes the resolution order, which solves type 3 problems in subsection 2.1. For example in Fig. 5., similarity propagation generates two type II candidate pairs (a_1, a_3) and (a_2, a_4), whose probabilities of matching just increase because (c_1, c_2) are merged. In such way, the two candidate pairs will be resolved soon. This is important to pay-as-you-go ER applications [19], who desire most of ER results as soon as possible.

3.5   Discussion

In this subsection we briefly discuss two useful aspects of GB-JER without steady theoretic evidences: the knowledge graph output by the model and the incremental property.

Knowledge Graph. The output of GB-JER is a clean ERRG that naturally is a knowledge graph. Such a knowledge graph is the input of link mining and heterogeneous information network mining [7,17]. Mining studies assume that a knowledge graph of high quality exists without unambiguous nodes and links. Unfortunately, the real world graph data are dirty. Our work fills the gap between dirty graph data and knowledge graphs of high quality. Here "high quality" means clean and with complete semantic relationships. As presented before, duplicate nodes are merged and their neighborhoods are contracted, during which all semantic relationships are discovered and set up.

ER with knowledge base. On the other hand, since the produced knowledge graph is clean, it's trustworthy and is a knowledge base for new data. The knowledge base can be utilized when new dirty data of the same domain come, which promotes new joint ER tasks. Here new dirty data do not have to be from the same source. For instance, a clean ERRG of dblp may be helpful to resolution of data from citeseer if they refer to similar groups of related entities in the real world.

Incremental Property. In the big data era, the volume of data is large and data updates faster than past, which call for incremental ER approaches. Gruenheid et al. [8] propose an incremental ER approach for single classes of entities with clustering algorithms. However, the approach does not suit joint ER. Data increment include addition, delete and modification. In this paper, we consider increment as data addition but leave delete and modification as future work.

Definition 9 (Incremental Joint ER). *Gvien R is a set of multiple classes of representations, ΔR is an increment to R, and $jer(R)$ is the output of joint ER on R, incremental joint ER resolves representations from $(R + \Delta R)$ with $jer(R)$, denoted as $incjer(R, \Delta R, jer(R))$.*

Generally, ΔR is much smaller than R so it costs much if $(R + \Delta R)$ are recomputed from scratch. GB-JER iteratively evolves an ERRG that the more relationships the easier to resolve data. Semantics in the ERRG provide positive evidence for joint ER. This characteristic makes the model good at incremental joint ER. GB-JER does not requires a cold start when increment comes. The old ERRG produced previously can be directly re-used when incremental joint ER. The old ERRG with rich semantics is exploited to compute SBS and propagate similarity for the new data who will be gradually integrated into the ERRG. Finally, the old ERRG evolves into a new and clean one. Appraentlly, for GB-JER, incremental joint ER is a special case of ER with knowledge base. The evaluation testifies our model's incremental property.

4   Experimental Evaluation

4.1   Experiment Preliminaries

Experiment Setting. CPU: Intel(R) Core(TM) i7-2600, 3.4GHz, 8 cores; Main memory: 8 G; OS: Ms Windows 7 Ultimate 64 bits.

Dataset. We extract $100,000$ citation items from Citeseer[1], whose schema is {title, authors, address, venue, date}. The schema is partitioned into paper =

---

[1] http://citeseerx.ist.psu.edu/

{title, writtenBy, publishedIn, date}, author = {name, address} and venue = {name, date}. All representations of paper, author and venue are manually labeled, which is ground truth for ER evaluation. A sub-set Citeseer-1 containing $50,000$ citation items are randomly generated, which is used later by default if no specially setting.

Measure. We use F index to evaluate GB-JER. F index is the harmonic mean of precision and recall,

$$F = 2 * prec * recall/(prec + recall) \tag{7}$$

Parameters. We manually set the same parameters for three types of representations. The balance weight of hybrid similarity $\delta = 0.45$, the match threshold $\theta_m = 0.85$, the similarity propagation threshold $\theta_{sp} = 1/50$ and the balance weight in priority score function $\eta = 0.6$. These settings don't have to be optimized but experiment results in such settings verify advantages of GB-JER.

Unidirectional related weights setting. $10,000$ items are sampled from Citeseer-1 in this way: randomly select an item at first and then select all other items that contain similar paper title, author names or venue name in the item; repeat the process until the size of items reach $10,000$. This training dataset is called Citerseer-train, whose ER result is known. Build a clean ERRG (denoted as ERRG-train)according to the ER result of Citerseer-train. In equation 8, the numerator is size of representations belonging to the relation's source class and the denominator is size of representations belonging to the relation's target class. According to equation 8 and ERRG-train, $urw(venue, paper) = 1/56.8$, $urw(paper, venue) = 1/1$, $urw(author, paper) = 1/5.1$ and $urw(paper, author) = 1/3.4$.

$$urw(\mu_i, \mu_j) = \sum_{\varphi(r_k)=\mu_i} |r_k| \left/ \sum_{\varphi(r_k)=\mu_i} \sum_{\varphi(r_m)=\mu_j, r_k \to r_m} |r_m| \right. \tag{8}$$

4.2 General Test

Comparisons with Existing Approaches. Traditional ER approaches only consider ABS, denoted as ABS ER. Dong's joint ER approach builds a dependent graph of candidate pairs and considers direct neighbors when computing SBS, denoted as DepGraph. Kalashnikov proposed a single class ER approach based on connections between representations on a graph, denoted as RelDC. As shown in Fig. 6., accuracies of GB-JER is higher than other three approaches because it fully exploits semantic relationships in a dynamic ERRG, where SBS is more precise leading to higher precision and similarity propagation contributes to higher recall. DepGraph perform better than RelDC. Both DepGraph and RelDC perform better than ABS ER.

Scalability. In order to test scalability of GB-JER, we increase data size by $10,000$ each time, starting from $10,000$. As shown in Fig. 7., time cost increases almost linearly with data size.

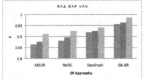

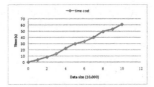

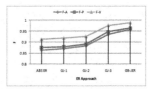

Fig. 6. Accuracy comparisons with existing approaches

Fig. 7. Scalability

Fig. 8. Component contribution

4.3 Components Test

This sub-section tests contribution of each key component. Table 1. clearly shows all components of different ER approaches in the sub-section. '+' means that this component exists; '−' means that this component doesn't exist; an abbreviation stands for a specific component, which is specified below.

SPBS: schema path based similarity; RWBS: random walk based similarity; SRWBS: symmetric random walk based similarity; C-mg: common representation merge method; RAB-mg: representative attribute value based representation merge method; PB-in: priority based insertion; H-in: head insertion; R-in: random insertion; T-in: tail insertion.

Table 1. Different ER approaches with components

ER Approaches	ABS	SBS	Merge	Neighborhood contraction	similarity propagation	Insertion
ABS ER	+	−	S-mg	−	−	−
GJ-1	+	SPBS	S-mg	−	−	−
GJ-2	+	SPBS	RAB-mg	−	−	−
GJ-3	+	SPBS	RAB-mg	+	−	−
GB-JER	+	SPBS	RAB-mg	+	+	PB-in
GJ-RW	+	RWBS	RAB-mg	+	+	PB-in
GJ-SRW	+	SRWBS	RAB-mg	+	+	PB-in
GJ-Cmerge	+	SPBS	C-mg	+	+	PB-in
GJ-InQH	+	SPBS	RAB-mg	+	+	H-in
GJ-InR	+	SPBS	RAB-mg	+	+	R-in
GJ-InQT	+	SPBS	RAB-mg	+	+	T-in

Component Contribution. We compare ABS ER, GJ-1, GJ-2, GJ-3 and GB-JER, whose details are in Table 1. As shown in Fig. 8., GJ-1 performs better than ABS ER, because schema path based similarity help representations match, solving type 2 problems in subsection 2.1. GJ-1 and GJ-2 perform almost the same, which testifies that the simple merge method and the representative attribute value base merge method affect the ER results almost the same. GJ-3 performs much better than GJ-2, which shows that neighborhood contraction is very important to joint ER. When the neighborhood is contracted after representations match, schema path based similarities among representations in the neighborhood increase, which promote subsequent joint ER. GB-JER performs better than GJ-3, showing that similarity propagation increases the precision of joint ER. Similarity propagation helps solve type 1 problems in subsection 2.1.

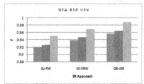
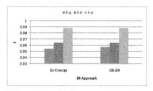
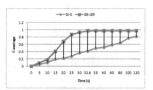

Fig. 9. Structure based similarities

Fig. 10. Representation merge methods' comparisons

Fig. 11. Similarity propagation's influence on efficiency

Schema Path Based Similarity. We switch the SBS component as Table 1., GJ-RW is with random walk based similarity and GJ-SRW is with symmetric random walk based similarity. In Fig. 9., GB-JER performs better than both GJ-RW and GJ-SRW, because schema path based similarity combines schema and topology to measure SBS, which is more precise than random walk and symmetric random walk.

Representative Attribute Value based Representation Merge Method. As Table 1., GJ-Cmerge is with common representation merge method. In Fig. 10., GJ-Cmerge and GB-JER perform almost the same. GB-JER gets a little higher accuracy in authors' resolution. Authors' names have many variants or may be incomplete but the representative value based merge method selects most representative value each time so it performs a little better. On the other hand, GJ-Cmerge costs 39.4s but GB-JER costs 32.6s that GJ-Cmerge cost 21.5% more time than GB-JER. Considering both accuracy and time cost, the proposed representative attribute value based merge method is better.

Similarity Propagation's Influence on Efficiency. We compare average F indexes (denoted as F-$average$) of GJ-3 and GB-JER with time increment. As shown in Fig. 11., in range $[0, 32.6s]$, F-$average$ of GB-JER is bigger than that of GJ-3 and increases faster, which is easy to be known by slopes of two curves. F-$average$ of GB-JER reaches the peak (0.97) at 32.6s but F-$average$ of GJ-3 is only about 0.41, far from its own peak. At the largest value of time axis, F-$average$ of GJ-3 is about 0.82, still far from its own peak. GJ-3 need much more time to reach its own peak of F-$average$. Above all, similarity propagation helps resolve duplicate representations faster.

Type II Candidate Pair Insertion. How to insert type II candidate pairs doesn't influence accuracy of joint ER, but it influences time cost. We compare time costs of GJ-InQH, GJ-InR, GJ-InQT and GB-JER, whose details are described in Table 1. As shown in Fig. 12., GB-JER cost least time because the priority based insertion method optimizes the resolution order by estimating probabilities of candidate pairs matching. Head insertion method inserts all type II candidate pairs into the head of Q, which helps resolve some duplicate representations fast. Some pairs' neighborhood is not dense and its semantics is not rich enough that they can't be resolved as duplicates but they are repeatedly inserted into head of Q by similarity propagation. These pairs experience evolving of their neighborhoods and finally are resolved as duplicates or not duplicates. Such repetitions cause much time costs. Random insertion cost longer time than head insertion because it does not give type II candidate pairs any higher priority. Tail insertion costs longest time because it always put new type II candidate pairs into tail of Q, which blocks resolution of pairs with higher duplicate probabilities.

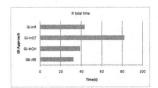 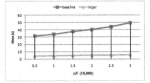 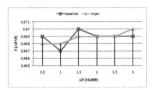

Fig. 12. Insertion methods comparison **Fig. 13.** Incremental joint ER time test **Fig. 14.** Incremental joint ER accuracy test

4.4 Incremental Joint ER Test

We run GB-JER over Citeseer-1 first and then add 5000 to it as increment each time. The baseline approach re-compute $(R+\Delta R)$ from scratch when increments come. The incjer approach do it incrementally: re-compute ΔR with R and the old ERRG. Fig. 13 shows time costs of two approaches with ΔR. The baseline's incremental time cost increases apparently. The incjer's incremental time cost increases little. The gap between the two time costs is very large at beginning and is enlarged with increments. At beginning the baseline costs 5 times longer than the incjer and finally the former costs 9 times longer than the incjer. Fig. 14 shows that accuracies of two approaches with increments are close. Although the incjer costs much less time than the baseline, its accuracy is almost the same as the baseline.

5 Related Work

ER has a long history, attracting researchers from research fields like database, data mining and AI. It's also called entity match, record linkage, de-duplicate, merge & purge [1,4,6,16]. Traditional ER approaches rely on ABS, focusing on single classes of representations. However, most data are related in the real world, such as citation data and movie data. In order to resolve representations in related dataset quickly and precisely, relationships among representations should be utilized and multiple classes of representations should be resolved concurrently.

There are some ER approaches based on relationships in literature. Dong et al. [5] analysis relationships among representations such as authors, papers, email et al. and build a dependent graph of candidate pairs, where each pair is a node. This approach considers common neighbors to compute SBS. When two representations are resolved as duplicate, the result is propagated to their direct neighbors via dependent links, which increases SBS. Such an iterative process is formed. This approach focuses on personal data management. Getoor et al. [2] utilize co-occurrences to do ER via relational clustering. Their approach focuses on authors' resolution in citation data. The above two approaches utilize direct relationships, called context ER approaches. Kalashnikov et al. [10,14] build a relationship graph assuming only one class of representations are not resolved, such as authors in citation data. Their approach computes connections among representations without schema to do ER, which helps resolve different entities with the same names. In AI, there are a few ER approaches based on relationships [4,16], who learn a global detailed probabilistic model from training data that it guides ER process. Such approaches are suitable for data with few relations and little heterogeneity. For complex data from multiple classes, they don't guarantee accuracy and efficiency. A few researches focus on scalability of joint ER [9,15]. Nowadays new data are updated more frequently that efficient incremental approach is needed. Existing incremental approaches [8] focus on single classes.

The proposed GB-JER model fully exploits a dynamic ERRG, where it combines schema and topology to compute SBS and propagates similarities among representations. It guarantees precision and efficiency for joint ER. Dong's, Getoor's and Kalashnikov's approaches are subsets of GB-JER. It efficiently processes increments without loss of accuracy.

6   Conclusion

ER is an key aspect of data quality. Nowadays most data are related that it's an important problem to resolve multiple classes of representations jointly with high precision and efficiency. We propose a graph based joint ER model who iteratively exploits a dynamic ERRG; on the other hand, the ERRG gradually converges to be a clean one. The evaluation verifies its advantages. This model can process data addition efficiently. In the future, we will focus on other increments such as deletion and modification.

Acknowledgments. This work is supported by the National Basic Research 973 Program of China under Grant (2012CB316201) and the National Natural Science Foundation of China under Grant (61472070).

References

1. Benjelloun, O., Garcia-Molina, H., Menestrina, D., Su, Q., Whang, S.E., Widom, J.: Swoosh: a generic approach to entity resolution. VLDB J. **18**(1), 255–276 (2009). http://www.dx.doi.org/10.1007/s00778-008-0098-x

2. Bhattacharya, I., Getoor, L.: Collective entity resolution in relational data. ACM Transactions on Knowledge Discovery from Data **1**(1), 1–36 (2007)
3. Cohen, W., Ravikumar, P., Fienberg, S.: A comparison of string metrics for matching names and records. In: KDD Workshop on Data Cleaning and Object Consolidation, vol. 3, pp. 73–78 (2003)
4. Culotta, A., McCallum, A.: Joint deduplication of multiple record types in relational data. In: Proceedings of the 14th ACM International Conference on Information and Knowledge Management, pp. 257–258. ACM (2005)
5. Dong, X., Halevy, A., Madhavan, J.: Reference reconciliation in complex information spaces. In: Proceedings of the 2005 ACM SIGMOD International Conference on Management of Data, pp. 85–96. ACM (2005)
6. Elmagarmid, A.K., Ipeirotis, P.G., Verykios, V.S.: Duplicate record detection: a survey. IEEE Transactions on Knowledge and Data Engineering **19**(1), 1–16 (2007)
7. Getoor, L., Diehl, C.P.: Link mining: a survey. SIGKDD Explor. Newsl. **7**(2), 3–12 (2005). http://www.doi.acm.org/10.1145/1117454.1117456
8. Gruenheid, A., Dong, X.L., Srivastava, D.: Incremental record linkage. Proceedings of the VLDB Endowment **7**(9) (2014)
9. Herschel, M., Naumann, F., Szott, S., Taubert, M.: Scalable iterative graph duplicate detection. IEEE Transactions on Knowledge and Data Engineering **24**(11), 2094–2108 (2012)
10. Kalashnikov, D.V., Mehrotra, S., Chen, Z.: Exploiting relationships for domain-independent data cleaning. In: SDM, pp. 262–273. SIAM (2005)
11. Liben-Nowell, D., Kleinberg, J.: The link-prediction problem for social networks. Journal of the American Society for Information Science and Technology **58**(7), 1019–1031 (2007)
12. McCallum, A., Nigam, K., Ungar, L.H.: Efficient clustering of high-dimensional data sets with application to reference matching. In: Proceedings of the Sixth ACM SIGKDD International Conference on Knowledge Discovery and Data Mining, KDD 2000, pp. 169–178. ACM (2000)
13. Motwani, R., Raghavan, P.: Randomized Algorithms. Cambridge University Press (1995)
14. Nuray-Turan, R., Kalashnikov, D.V., Mehrotra, S.: Adaptive connection strength models for relationship-based entity resolution. Journal of Data and Information Quality (JDIQ) **4**(2), 8 (2013). http://www.doi.acm.org/10.1145/2435221.2435224
15. Rastogi, V., Dalvi, N., Garofalakis, M.: Large-scale collective entity matching. Proceedings of the VLDB Endowment **4**(4), 208–218 (2011)
16. Singla, P., Domingos, P.: Entity resolution with markov logic. In: Sixth International Conference on Data Mining, ICDM 2006, pp. 572–582. IEEE (2006)
17. Sun, Y., Han, J.: Mining heterogeneous information networks: a structural analysis approach. SIGKDD Explorations **14**(2), 20–28 (2012). http://www.doi.acm.org/10.1145/2481244.2481248
18. Sun, Y., Han, J., Yan, X., Yu, P.S., Wu, T.: Pathsim: meta path-based top-K similarity search in heterogeneous information networks. Proceedings of the VLDB Endowment **4**(11), 992–1003 (2011). http://www.vldb.org/pvldb/vol4/p992-sun.pdf
19. Whang, S.E., Marmaros, D., Garcia-Molina, H.: Pay-as-you-go entity resolution. IEEE Transactions on Knowledge and Data Engineering **25**(5), 1111–1124 (2013)

Provenance-Aware Entity Resolution: Leveraging Provenance to Improve Quality

Qing Wang[1]([✉]), Klaus-Dieter Schewe[2], and Woods Wang[3]

[1] Research School of Computer Science,
Australian National University, Canberra, Australia
qing.wang@anu.edu.au
[2] Software Competence Center Hagenberg
and Johannes-Kepler-University Linz, Linz, Austria
kd.schewe@scch.at
[3] Alcatel-Lucent Beijing, Beijing, China
woods.wang@alcatel-lucent.com

Abstract. Entity resolution (ER) - the process of identifying records that refer to the same real-world entity - pervasively exists in many application areas. Nevertheless, resolving entities is hardly ever completely accurate. In this paper, we investigate a provenance-aware framework for ER. We first propose an indexing structure that can be efficiently built for provenance storage in support of an ER process. Then a generic repairing strategy, called *coordinate-split-merge* (CSM), is developed to control the interaction between repairs driven by must-link and cannot-link constraints. Our experimental results show that the proposed indexing structure is efficient for capturing the provenance of ER both in time and space, which is also linearly scalable over the number of matches. Our repairing algorithms can significantly reduce human efforts in leveraging the provenance of ER for identifying erroneous matches.

Keywords: Entity resolution · Data matching · Record linkage · Deduplication · Data provenance · Repair · Indexing structure

1 Introduction

Studies on entity resolution (ER) have been carried out over the last 50 years [11,15]. Numerous ER techniques have been proposed under a variety of perspectives such as probabilistic [15,18], cost-based [22], ruled-based [13], supervised [16], active learning [4,19], and collective approaches [7]. Nevertheless, ER results in real-world applications are still largely *imprecise*. Take Scopus for example, searching authors named "Qing Wang" or "Q. Wang" yields over 69,000 publication records, in which the publications of "Qing Wang" authoring this paper are mixed with publications of other people who have similar names over different author entities in Scopus. There are various reasons for having such imprecise ER results, e.g., dirty data, limitations of ER techniques, or the dynamic nature

© Springer International Publishing Switzerland 2015
M. Renz et al. (Eds.): DASFAA 2015, Part I, LNCS 9049, pp. 474–490, 2015.
DOI: 10.1007/978-3-319-18120-2_28

of an ER process. These factors together make it particularly challenging to identify, understand, and repair errors in imprecise ER results.

In this paper we aim to develop provenance-aware ER for improving the quality of ER results. The following example illustrates why the provenance of ER is needed. For simplicity, we use $\langle a_1, \ldots, a_n \rangle$ to denote the set $\{a_1, \ldots, a_n\}$ of records that are resolved to the same entity.

Example 1. Fig. 1.(a)-(b) contain several sample records of AUTHOR *and* PUB-LICATION. *The records 1, 3, 5, 7 and 8 in* AUTHOR *are resolved to represent the same entity e_1, as shown in Fig. 1.(c). Suppose that e_1 is inconsistent because 1 and 7 in* AUTHOR *indeed refer to two different persons. Then there are many possible ways of splitting 1 and 7, such as $\{\langle 7 \rangle, \langle 1, 3, 5, 8 \rangle\}$ or $\{\langle 1 \rangle, \langle 3 \rangle, \langle 5 \rangle, \langle 7 \rangle, \langle 8 \rangle\}$, but it is not known a priori which way can repair such an inconsistency. Naturally, we would ask: (a) why does an inconsistency happen? (b) what is the implication of an inconsistency for entities in the ER result?*

aid	name	affiliation	email
1	Qing Wang		qw@gmail.com
2	Mike Lee	Curtin University	
3	Qinqin Wang	Curtin University	
4	Jan Smith		jan@gmail.com
5	Q. Wang	University of Otago	qw@gmail.com
6	Jan V. Smith	RMIT	jan@gmail.com
7	Q. Q. Wang		
8	Wang, Qing	University of Otago	

(a) AUTHOR

pid	title	authors	year
p_1	Knowledge-Aware Services	1, 2	2012
p_2	Schema Evolution	3, 4	2009
p_3	Data Migration	5	2010
p_4	XML Data Exchange	6, 7	2010
p_5	Cloud Computing	8	2011

(b) PUBLICATION

eid	aids
e_1	$\langle 1, 3, 5, 7, 8 \rangle$
e_2	$\langle 2 \rangle$
e_3	$\langle 4, 6 \rangle$

(c) AN ER RESULT

Fig. 1. Sample records

Assume that we have the provenance information about e_1 as follows: (1) 1 and 3 were resolved, resulting in $\langle 1, 3 \rangle$; (2) 5 and 1 were resolved, resulting in $\langle 1, 3, 5 \rangle$; (3) 3 and 7 were resolved, resulting in $\langle 1, 3, 5, 7 \rangle$ and (4) 5 and 8 were resolved, resulting in $\langle 1, 3, 5, 7, 8 \rangle$. Based on these, we can identify two matches – 3 and 7, and 1 and 3 – which are relevant to the inconsistency of e_1. If the match between 1 and 3 is erroneous, the correct repair in this case would be $\{\langle 1, 5, 8 \rangle, \langle 3, 7 \rangle\}$. Other options would generate new inconsistencies, e.g., 3 and 7 in $\{\langle 7 \rangle, \langle 1, 3, 5, 8 \rangle\}$ may indeed refer to the same entity. Therefore, to find accurate repairs *for an inconsistent ER result, we need to store the provenance information about the ER process, which allows us to analyze the interaction among ER decisions (i.e., matches or non-matches) that have occurred in the ER process. Although a domain expert may be able to repair such inconsistencies manually, the level of human efforts has been proven to be very expensive in many real-life applications [11]. Ideally, whenever an inconsistency is detected in an ER result, we want to minimize human efforts in repairing the errors that cause the inconsistencies.*

Contributions. We first develop an ER indexing method, called ERI, for storing the provenance information of ER. An ERI index maps each entity in an ER result to an ER tree in which each node represents a match. This point of view on the ER process, together with several nice properties of the proposed indexing structure, empowers us to analyze inconsistent ER results. Whenever an inconsistency is found, relevant matches can be pinpointed, providing a unified view to understand such an inconsistency. In doing so, the provenance information serves as a ground on which inconsistent ER results can be repaired in a meaningful and efficient way. Our ER indexing method can thus make an ER system provenance-aware through capturing matches, which is independent of the chosen ER techniques.

We study algorithms that construct ERI indexing to capture the provenance information. We show that there is an efficient linear-time algorithm to construct such ERI indices. We also develop algorithms for two important operations – merge and split – which can eliminate inconsistent entities violating must-link and cannot-link constraints, respectively. Our algorithm for the split operation traverses an ER tree to identify and remove erroneous matches in order to eliminate inconsistencies. Such traversals downward and upward an ER tree are both efficient (see Theorems 2 and 3). We further propose a generic strategy, called CSM, to improve the effectiveness of ER repairs by taking into account how must-link and cannot-link constraints interact, in connection with the provenance information represented by ERI indices.

To evaluate the efficiency and effectiveness of our proposed method, we have conducted three sets of experiments over two real-world data sets. The experimental results show that: (1) The ERI construction method for capturing the provenance is efficient in time and space, and has a linear scalability over the number of matches; (2) The algorithms for repairing ER results using the provenance information stored in ERI indices can consistently improve precision with significantly reduced human efforts; (3) Under the CSM strategy, the algorithms can achieve both high precision and high recall after certain number of cannot-link constraints are identified over time.

Outline. The remainder of the paper is structured as follows. We introduce the related work in Section 2 and basic definitions in Section 3. Then we present the ERI indexing method in Section 4. Section 5 is devoted to discuss the algorithms and strategies that leverage the provenance information for solving inconsistent ER results. Our experimental results are presented in Section 6. The paper is concluded in Section 7.

2   Related Work

Up till now, research efforts have been focused on the areas of entity resolution [12] and data provenance [21] separately, although both areas are related to the quality of data. In the following, we present the works related to these two areas.

Studies on ER have been heavily carried out over the last 50 years [11,15]. Traditional ER techniques analyse attributes between records by using various similarity measures [12]. Recently, a number of works have incorporated constraints into such similarity-based ER techniques to improve the quality of ER [5,10,20,24]. These works mostly focused on preventing an ER result from violating certain types of constraints, rather than repairing an ER result that violates constraints. For instance, the authors of [24] developed strategies for resolving inconsistent before an ER result is produced, and did not consider backtracking for splitting records of inconsistent entities. In contrast, our ER repairing method provides backtracking search for analyzing why errors occur in the ER process as well as for identifying all propagated errors, based on leveraging the provenance information. Both early approach and late approach (i.e., inconsistencies are resolved before and after an ER result is produced) can be well supported in our framework.

ER indexing is an important technique used for improving computational performance and scalability of the ER process (i.e., also called *ER blocking*) [12]. It groups records into a set of possibly overlapping but small blocks, and only records that potentially represent the same entity are placed into the same block. To our best knowledge, no studies of using indexing techniques to represent the provenance information of ER have been previously reported. This paper shows, for the first time, how to use an indexing structure to efficiently store ER decisions in a simple tree structure, and how to exploit ER decisions to accurately repair inconsistent entities in an ER result.

Data provenance [6,8,9,17,21] provides a detailed view on the derivation of single pieces of data. Our work relates to data provenance in the sense that we use an indexing structure to capture the provenance information of ER, such as: which matches were generated by a match rule, when matches were generated and how matches were involved in resolving entities. Such provenance information provides a rich source for analyzing and improving the quality of ER, but not much attention has been paid so far in this respect. Previously, a research prototype DBMS, called Trio, has been developed at the Stanford InfoLab [3], which can track lineage information of a merged result for entity resolution. Nevertheless, we are not aware of any existing studies that integrate provenance-based resolution into the ER process for improving the quality.

Database repair has been previously investigated in many works [2,14,25]. Database repair only deals with data errors existing in a database, while ER repair focuses on handling errors that are caused by the problematic ER process. Nonetheless, although database repair and ER repair are studied under different purposes, their techniques can interact with and complement each other. It is desirable to establish a mechanism that combines database and ER repair tools for repairing errors *collectively*, i.e., different types of errors are repaired jointly rather than independently.

3   Preliminaries

A *(database) instance* is a finite, non-empty set of relations, each having a finite set of records. Each record is *uniquely identifiable* by a (record) key in the instance. We use $key(I)$ to denote the set of all record keys in an instance I. A *(record) cluster* c is a set of records in the form of $\langle k_1, \ldots, k_n \rangle$. We use $|c|$ to refer to the *size* of a cluster c. A *match* is a pair (k_1, k_2) of records that represent the same entity. A *(match) rule* r is a function that, given an instance I as the input, generates a set $r(I)$ of matches. Every ER algorithm that generates a set of matches can be viewed as a match rule in our work, including machine-learning algorithms. We assume that matches are symmetric in $r(I)$, e.g., if $(k_1, k_2) \in r(I)$, then $(k_2, k_1) \in r(I)$, and use $r(k_1, k_2)$ to denote that (k_1, k_2) is generated by r. An *ER model* M is constituted by a (finite) number of match rules. The *ER result* of applying M over I is a set E^* of clusters determined by $E = \bigcup_{r \in M} r(I)$ such that: (1) E^* is a partition of $key(I)$; (2) If $(k_1, k_2) \in E$, then $\exists c \in E^*.\{k_1, k_2\} \subseteq c$; (3) If $\exists c \in E^*.\{k_0, k_n\} \subseteq c$, then $\exists k_1, \ldots, k_{n-1}.(k_0, k_1), (k_1, k_2) \ldots, (k_{n-1}, k_n) \in E$. Conceptually, each cluster in an ER result represents an entity. We use $rec(e)$ to denote the set of record keys in the cluster that represents an entity e.

In this paper, we focus on two kinds of ER constraints, so-called *must-link* and *cannot-link* constraints [23]. Let k_1 and k_2 be two record keys. Then a *must-link constraint* $k_1 \asymp k_2$ means that k_1 and k_2 must be matched to the same entity, and a *cannot-link constraint* $k_1 \not\asymp k_2$ means that k_1 and k_2 cannot be matched to the same entity. Both must-link and cannot-link constraints are instance-level constraints [23], and symmetric in the sense that if $k_1 \asymp k_2$ (resp. $k_1 \not\asymp k_2$) is satisfied, then $k_2 \asymp k_1$ (resp. $k_2 \not\asymp k_1$) is also satisfied. Although must-link and cannot-link constraints look simple, they serve as the building blocks of expressing the integrity of an ER result. In real-life applications, such constraints can be gathered from either user feedback provided by the user, or integrity constraints [5,20] specified by the domain expert. Consider a constraint φ over AUTHOR in Example 1, which states that two author records refer to the same person if they have the same email address, i.e., $x \asymp x' \Leftarrow \text{AUTHOR}(x, y, z, w) \wedge \text{AUTHOR}(x', y', z', w)$. Then, we would obtain a set of must-link constraints w.r.t. a database instance by evaluating φ.

Let Σ be a set of constraints. An entity is *consistent* w.r.t. Σ if it does not violate any constraint in Σ. More precisely, an entity e *violates* $k_1 \asymp k_2$ if $k_1 \in rec(e)$ but $k_2 \notin rec(e)$, and e *violates* $k_1 \not\asymp k_2$ if $k_1 \in rec(e)$ and $k_2 \in rec(e)$. It is possible that an entity violates must-link and cannot-link constraints at the same time. An ER result E^* is *consistent* w.r.t. Σ if every entity represented in E^* is consistent w.r.t. Σ.

4   ER Indexing

An *entity resolution index* (ERI) is a data structure that maps each entity to an ER tree made of record keys. An index entry in ERI has the form:

(entity e, ER tree t_e),

where t_e keeps track of the matches that are relevant to the entity e. Given an ER tree t, we use $V(t)$ to denote the set of nodes in t, and $V(t, k)$ to denote the set of nodes in t with k child nodes. A *subtree* of t at a node v is also an ER tree that consists of v and all of its descendants in t under the preservation of the labels on nodes and edges. We use $leaf(t)$ to refer to the set of labels on the leaves of t, $parent(v)$ to the parent node of a node v, and $ledge(v)$ and $redge(v)$ to the labels of the edges to the left and right child nodes of v, respectively. Formally, an *ER tree* of e is a binary tree t together with a labelling function θ such that:

- θ assigns to each leaf v of t a label $\ell \in rec(e)$, and $\theta(v_1) \neq \theta(v_2)$ for any different leaves v_1 and v_2;
- θ assigns to each edge $(v, parent(v))$ of t a label $\ell \in rec(e)$, and $\ell \in leaf(t')$ for the subtree t' of t at v.

Each leaf labelled by ℓ in an ER tree represents a record whose key is ℓ. For clarity we use $v_{(k_1, k_2)}$ to indicate an internal node, representing a match between two records with keys k_1 and k_2. The *height* of t is the number of edges in the longest path to a leaf from the root.

Example 2. Fig. 2 depicts two possible ER trees t_{e_1} and t'_{e_1} for the entity e_1 in Example 1, in which each contains four matches $(1, 3), (5, 1), (3, 7)$ and $(5, 8)$ corresponding to the internal nodes $v_{(1,3)}$, $v_{(5,1)}$, $v_{(3,7)}$ and $v_{(5,8)}$, respectively. The cluster that associates with t_{e_1} and t'_{e_1} is $\langle 1, 3, 5, 7, 8 \rangle$.

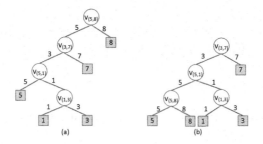

Fig. 2. (a) an ER tree t_{e_1}; (b) an ER tree t'_{e_1}

Given an ER model M, an ERI index can be initially constructed when applying M over an instance I, and then be dynamically maintained. In the following we elaborate the key steps involved in constructing such an ERI index. Let E be the set of matches generated by applying M over I. We use T to represent the set of ER trees constructed during the computation, and K to represent the set of record keys that have been processed.

1. Firstly, initialize T and K to be $\emptyset$, and then build a family T of ER trees gradually by doing the following until there is no more change on T.

1.1. If $r(k_1, k_2) \in E$, then $K_{tmp} := K \cap \{k_1, k_2\}$.

(a) If $K_{tmp} = \emptyset$, then construct a new ER tree t with one internal node v, and label the edges to two child nodes of v with k_1 and k_2, respectively. We have $T := T \cup \{t\}$.

(b) If $K_{tmp} = \{k_1\}$ (resp. $K_{tmp} = \{k_2\}$), then extend the ER tree t that satisfies $k_1 \in leaf(t)$ (resp. $k_2 \in leaf(t)$) to another ER tree t' by adding a new node v such that t is the subtree rooted at one child node v_1 of v and the other child node of v is a leaf v_2 labelled by k_2 (resp. k_1). The edges from v to v_1 and v_2 are labelled as k_1 and k_2 as appropriate. We then have $T := T \cup \{t'\} - \{t\}$.

(c) If $K_{tmp} = \{k_1, k_2\}$, then check if $\exists t_1, t_2 \in T.t_1 \neq t_2$, $k_1 \in leaf(t_1)$ and $k_2 \in leaf(t_2)$ hold. If yes, then construct a new ER tree t' rooted at a node that has t_1 and t_2 as the subtrees at its child nodes, and the edges to its child nodes are labelled as k_1 and k_2, respectively. We then have $T := T \cup \{t'\} - \{t_1, t_2\}$.

1.2. $K := K \cup \{k_1, k_2\}$ and $E := E - \{r(k_1, k_2)\}$.

2. For each $k \in key(I)$ but $k \notin K$, create a singleton tree with one leaf labelled by k, and add it into T.

Example 3. Consider the ER-trees in Fig. 2. For t_{e_1}, it starts with one node $v_{(1,3)}$ (by Step 1.1.a), then is extended to have the nodes $v_{(5,1)}$, $v_{(3,7)}$ and $v_{(5,8)}$ one by one (all by Step 1.1.b). For t'_{e_1}, it starts with two ER trees that have only one node $v_{(1,3)}$ and $v_{(5,8)}$ (by Step 1.1.a), respectively. Then they are merged into one ER tree with the root node $v_{(5,1)}$ (by Step 1.1.c). The resulting ER tree is finally extended to t'_{e_1} by adding the node $v_{(3,7)}$ as the root (by Step 1.1.b).

Theorem 1. *Given a finite set E of matches, the construction of an ERI index over E is in linear time $O(n)$, where $n = |E|$.*

5   ER Result Repairs

We now discuss how an ERI index can support repairs of inconsistent ER results.

5.1   Must-link: Merging

Suppose that two entities e_1 and e_2 violate a must-link constraint $\sigma = k_1 \backsimeq k_2$. Let t_1 and t_2 be the ER trees of e_1 and e_2, respectively, $k_1 \in leaf(t_1)$ and $k_2 \in leaf(t_2)$. Then merging t_1 and t_2 w.r.t. σ is an operation that yields an ER tree t satisfying σ, i.e., $\{k_1, k_2\} \subseteq leaf(t)$, written as $\mathtt{merge}(t_1, t_2, \sigma) = t$. A well-defined merge operation $\mathtt{merge}(t_1, t_2, \sigma)$ should satisfy the following conditions: (1) $leaf(t_1) \cup leaf(t_2) = leaf(t)$, (2) there is an injective mapping $f_m : V(t_1, 2) \cup V(t_2, 2) \mapsto V(t, 2)$ that preserves nodes and their edge labels, and (3) there is a node $v \in V(t, 2)$ with $redge(v) \cup ledge(v) = \{k_1, k_2\}$. A merge operation can be implemented in different ways as long as it is well-defined. Algorithm 1 describes a simple implementation of a well-defined merge operation.

> *Input:* two ER trees t_1 and t_2, and a constraint $\sigma = k_1 \asymp k_2$
> *Output:* an ER tree t
>
> 1. Create a new ER tree t with the root node v
> 2. Extend t by adding t_1 and t_2 as the subtrees at the child nodes v_l and v_r of v, respectively
> 3. Label the edges (v, v_l) with k_1 and (v, v_r) with k_2
> 4. Return t

Algorithm 1. mergeTrees

5.2 Cannot-link: Splitting

Given an entity e that violates a cannot-link constraint $\sigma = k_1 \not\asymp k_2$, splitting the ER tree t of e w.r.t. σ is an operation that yields a set of ER trees, written as $\texttt{splitTrees}(t, \sigma) = T$, where $T = \{t_1, \ldots, t_n\}$. A split operation is *well-defined* if it satisfies three conditions: (1) $\{k_1, k_2\} \subseteq leaf(t)$ but $\bigwedge_{t_i \in T} \{k_1, k_2\} \not\subseteq leaf(t_i)$, (2) $\{leaf(t_1), \ldots, leaf(t_n)\}$ is a partition of $leaf(t)$, and (3) there is an injective mapping $f : \bigcup_{t_i \in T} V(t_i, 2) \mapsto V(t, 2)$ that preserves nodes and their edge labels. In the following we propose an efficient algorithm for a well-defined split operation.

Traversing Downward. Let t be an ER tree and $\sigma = k_1 \not\asymp k_2$. A node v in t is called a *guard-node* of σ if $\{k_1, k_2\} \subseteq leaf(t_1)$ for the subtree t_1 at v, but $\{k_1, k_2\} \not\subseteq leaf(t_2)$ for every smaller subtree t_2 of t_1. A node v in t is *critical* w.r.t. σ if removing the match represented by v from the set of matches represented by t would lead to a set T of ER trees satisfying σ, i.e., $\forall t' \in T.k_1 \notin V(t', 0) \vee k_2 \notin V(t', 0)$.

Example 4. Consider the ER tree t_{e_1} in Fig. 2. The guard-node of $1 \not\asymp 7$ in t_{e_1} is $v_{(3,7)}$, which is also a critical node w.r.t. $1 \not\asymp 7$, i.e., removing the match represented by $v_{(3,7)}$ would lead to two ER trees t_1 and t_2 with $V(t_1, 0) = \{1, 3, 5, 8\}$ and $V(t_2, 0) = \{7\}$ which both satisfy $1 \not\asymp 7$.

By the definition of an ER tree, we have the following proposition.

Proposition 1. *Let t be an ER tree with $\{k_1, k_2\} \subseteq leaf(t)$ and $\sigma = k_1 \not\asymp k_2$. Then there exists exactly one guard-node of σ in t.*

Since removing the guard-node of σ in t can always yield two ER trees each having either k_1 or k_2, satisfying σ, we also have the following proposition.

Proposition 2. *If v is the guard-node of σ in an ER tree t, then v is also a critical node in t w.r.t. σ.*

We further observe that all critical nodes of t w.r.t. σ must occur in the subtree rooted at the guard-node v of σ, i.e., in the subtree "guarded" by v. Based on this, we develop Algorithm 2 to identify critical nodes by first finding the guard-node of σ and then traversing downward the ER tree. Each node in the ER tree only needs to be navigated *at most once*.

Input: an ER tree t, and a constraint $\sigma = k_1 \nleq k_2$
Output: a set V of critical nodes

1. $V := \emptyset$ and $P := \{(k_1, k_2)\}$
2. Do the following iteratively if there is $(k_1', k_2') \in P$ with $k_1' \neq k_2'$:
   (a) Find the guard-node v of $k_1' \nleq k_2'$ in t
   (b) $V := V \cup \{v\}$
   (c) If $k_1' \in leaf(t_l)$ and $k_2' \in leaf(t_r)$ for t_l and t_r at the child nodes of v, then $P' := \{(k_1', ledge(v)), (k_2', redge(v))\}$; otherwise $P' := \{(k_1', redge(v)), (k_2', ledge(v))\}$
   (d) $P := P - \{(k_1', k_2')\} \cup P'$
3. Return V

Algorithm 2. findCiticalNodes

Example 5. Consider the ER tree t_{e_1} in Fig. 2 and $1 \nleq 7$ again. We show how to find all critical nodes using Algorithm 2. First, Step 1 starts with $V = \emptyset$ and $P = \{(1,7)\}$. Then, by Steps 2(a)-(b), we have $V = \{v_{(3,7)}\}$ where $v_{(3,7)}$ is the guard-node of $1 \nleq 7$. By Step 2(c), we have $P' = \{(1,3), (7,7)\}$. By Step 2(d), $P = \{(1,3), (7,7)\}$. Since $1 \neq 3$, we repeat Steps 2(a)-(d) to find the guard-node $v_{(1,3)}$ of $1 \nleq 3$. Consequently, $V = \{v_{(3,7)}, v_{(1,3)}\}$, $P' = \{(1,1), (3,3)\}$ and $P = \{(1,1), (3,3), (7,7)\}$, and the computation terminates because all matches in P are trivial. Hence, all critical nodes in t_{e_1} w.r.t. $1 \nleq 7$ is $v_{(3,7)}$ and $v_{(1,3)}$.

Once critical nodes have been identified in an ER tree, we can obtain all records occurring in the edge labels from a critical node to its child nodes. Such records are *critical* because matches represented by critical nodes are generated based on these records. Following Algorithm 2, we have the following theorem.

Theorem 2. *Finding all critical nodes of an ER tree t w.r.t. σ is in logarithmic time $O(logn)$ where $n = |leaf(t)|$.*

Human Decision. Given an ER tree t that violates a cannot-link constraint σ, every critical node identified is relevant to the violation of σ, but does not necessarily represent an erroneous match. To decide whether or not a critical node represents an erroneous match, human guidance is required. For this, we consider that human decisions are made to classify critical records into disjoint groups - each group contains the records referring to the same entity, and records in different groups refer to different entities. Then human decisions are sent to a *cut-solver* that can automatically identify all erroneous matches. The nodes that correspond to erroneous matches in an ER tree are called *cut nodes*. Each cut node must be a critical node, but not vice versa. Algorithm 3 describes the process of finding cut nodes in an ER Tree.

Input: a set V of critical nodes
Output: a set V_{cut} of cut nodes

1. $V_{cut} := \emptyset$
2. $K := \bigcup_{v \in V} \{ledge(v), redge(v)\}$
3. Given a human decision $\{K_1, \ldots, K_n\}$ of K, check every $v \in V$:
   (a) $K_v := \{ledge(v), redge(v)\}$
   (b) If $K_v \cap K_i \neq \emptyset$ and $K_v \cap K_j \neq \emptyset$ for $1 \leq i \neq j \leq n$, then $V_{cut} := V_{cut} \cup \{v\}$
4. Return V_{cut}

Algorithm 3. findCutNodes

Example 6. Recall that the ER tree t_{e_1} has two critical nodes $v_{(3,7)}$ and $v_{(1,3)}$ w.r.t. $1 \not\le 7$. Thus, the critical records are $\{1,3,7\}$. Suppose that the human decision on classifying $\{1,3,7\}$ is $\{\{1\},\{7,3\}\}$, then by Algorithm 3, the cut-solver would identify $v_{(1,3)}$ as a cut node.

In the case of not using an ERI index, all records in $leaf(t)$ must be manually reviewed, which is generally much larger than the number of critical records, e.g., for an ER tree with 200 - 300 leaves in the Cora data set, the percentage of critical records is 5%-7% (see Section 6 for more details). In such cases, identifying critical records helps improve the efficiency of identifying erroneous matches dramatically. By narrowing down manually reviewed records to be only critical records, we say that the level of human efforts is *minimized*.

Input: an ER tree t and a set V_{cut} of cut nodes
Output: a set T of ER trees

1. Choose $v \in V_{cut}$ where the subtree at v does not contain other cut nodes
2. $T := \{t_l, t_r\}$ where t_l and t_r are the subtrees at the child nodes of v
3. Do the following until v is the root node of t:
   (a) Find $parent(v)$ and the subtree t_3 at the other child node of $parent(v)$. If t_3 contains $V' \subseteq V_{cut}$, then $T' := \text{removeCutNodes}(t_3, V')$ and $T := T \cup T'$; otherwise, $T := T \cup \{t_3\}$.
   (b) If $parent(v) \notin V_{cut}$, then check the edge label ℓ from $parent(v)$ to each of its child nodes. If $\exists t' \in T.\ell \in leaf(t')$, then replace the subtree at the child node by t' and label the edge from $parent(v)$ to $root(t')$ as ℓ.
   (c) $T := (T \cup \{t'\}) - \{t_l', t_r'\}$ where t_l' and t_r' are at the child nodes of $parent(v)$
   (d) $v := parent(v)$
4. Return T

Algorithm 4. `removeCutNodes`

Traversing Upward. Given a set of cut nodes, the question left is how we should remove these cut nodes and restructure the ER tree? This needs to be considered in terms of the conditions stipulated on a well-defined split operation. To satisfy Condition (1), the subtrees at the child nodes of each cut node must be split into different ER trees; otherwise, the errors caused by cut nodes cannot be eliminated. To satisfy Condition (3), we are only allowed to remove (not add) nodes from the original ER tree t. Together with Condition (2), this leads to decomposing t into a number of smaller trees. However, there are two additional things to note. First, in order to avoid false negatives caused by over-splitting, removing cut nodes from an ER tree should yield a minimal number of smaller ER trees that satisfy Conditions (1)-(3). Moreover, since edge labels between each node v and its two child nodes indicate records relevant to the match represented by v, removing a cut node should also remove nodes that have records from both subtrees rooted at the child nodes of the cut node. In doing so, we can also repair erroneous matches that are hidden. Algorithm 4 describes the process of removing cut nodes from an ER tree.

Theorem 3. *Let V_{cut} be a set of cut nodes in an ER tree t. Then splitting t by removing the nodes in V_{cut} is in time $O(n\log n)$ where n is the height of t.*

Example 7. Using Algorithm 4 to cut $v_{(1,3)}$ from the ER tree in Fig. 3.(a) would result in two smaller ER trees in Fig. 3.(b). Step 3 is executed iteratively by traversing upward the path from $v_{(1,3)}$ to $v_{(5,8)}$.

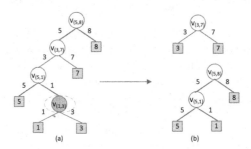

Fig. 3. ER trees after removing $v_{(1,3)}$

5.3   Repair Strategies

A naive strategy of repairing ER results is to apply a merge operation whenever a must-link constraint is violated, and apply a split operation whenever a cannot-link constraint is violated. However, this naive strategy does not work well when must-link and cannot-link constraints coexist and interweave. Merging ER trees to solve inconsistencies w.r.t. must-link constraints may yield an ER result violating cannot-link constraints, and vice versa. Hence, we propose a strategy, called *coordinate-split-merge* (CSM), to repair ER results by taking into account the interaction between must-link and cannot-link constraints.

Let $\Sigma = \Sigma_{\simeq} \cup \Sigma_{\not\simeq}$ be a set of constraints where $\Sigma_{\simeq}$ and $\Sigma_{\not\simeq}$ refer to the must-link and cannot-link constraints in Σ, respectively, and E^* be an inconsistent ER result w.r.t. Σ. In practice, we can build a knowledge base for Σ to store all must-link and cannot-link constraints, and incrementally update it whenever new constraints are identified. We assume that Σ are consistent (i.e., the consistency checking can be handled using existing techniques [1]), and use $T(E)$ to refer to the set of ER trees constructed from the set E of matches that determine E^*. Then we repair E^* using the CSM strategy:

- Start with $T' := T(E)$
- Check each $\sigma \in \Sigma$: if $\sigma \in \Sigma_{\not\simeq}$ and t violates σ, then $T := \text{splitTrees}(t, \sigma)$, and $T' := T' \cup T - \{t\}$; if $\sigma \in \Sigma_{\simeq}$, then do the following:
  (1). *Coordinating*: Check T' w.r.t. Σ to identify conflicts among constraints that involve σ (will be defined soon), then expand Σ to $\Sigma := \Sigma \cup \Delta$ where Δ is a set of constraints newly discovered from solving such conflicts (will be discussed).
  (2). *Splitting*: Split the ER trees in T' w.r.t. each $\sigma' \in \Delta_{\not\simeq}$, where $\Delta_{\not\simeq}$ denotes the subset of all cannot-link constraints in Δ. For each split $T := \text{splitTrees}(t', \sigma')$, set $T' := T' \cup T - \{t'\}$.
  (3). *Merging*: Merge the ER trees in T' w.r.t. σ, i.e., $t := \text{mergeTrees}(t_1, t_2, \sigma)$ where t_1 and t_2 each contain a distinct key in σ, and $T' := T' \cup \{t\} - \{t_1, t_2\}$.

– Return T', i.e., ER trees being repaired from $T(E)$.

The key idea is to govern merge operations using cannot-link constraints such that each merge operation is performed in a way that does not violate any cannot-link constraints. The coordinating stage is to determine whether there is a conflict among constraints in $\Sigma_{\not\simeq} \cup \{\sigma\}$ w.r.t. a set E of matches. A set of constraints contains a *conflict* in E if there does not exist a partition of ER trees in $T(E)$ such that all constraints can be satisfied. In Fig. 4.(a), the ER trees t_1 and t_2 satisfy $k_1 \not\simeq k_2$ but violate $k_1' \simeq k_2'$. Merging t_1 and t_2 would yield an ER tree satisfying $k_1' \simeq k_2'$ but violating $k_1 \not\simeq k_2$. Such conflicts can be systematically discovered by representing the set $T(E)$ of ER trees and their constraints Σ as a graph $G_{(E,\Sigma)}$ - each ER tree $t \in T(E)$ is a vertex, and each constraint $k_1 \simeq k_2 \in \Sigma$ (resp. $k_1 \not\simeq k_2 \in \Sigma$) is an edge between the vertices t_1 and t_2 if $k_1 \in leaf(t_1)$ and $k_2 \in leaf(t_2)$, labelled as $+$ (resp. $-$).

Proposition 3. *A set Σ of constraints contains a* conflict *among E if the graph $G_{(E,\Sigma)}$ contains at least one cycle that has exactly one edge labelled as $-$.*

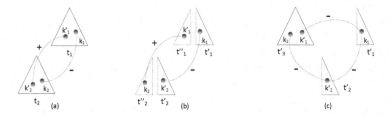

Fig. 4. ER trees with conflicts

In Fig. 4.(a), the conflict indicates that $k_1 \not\simeq k_1'$ or $k_2 \not\simeq k_2'$ may possibly happen. If both $k_1 \not\simeq k_1'$ and $k_2 \not\simeq k_2'$ are identified as being true, then $\Delta = \{k_1 \not\simeq k_1', k_2 \not\simeq k_2'\}$. Fig. 4.(b) shows that the ER trees t_1 and t_2 thus need to be split w.r.t. $k_1 \not\simeq k_1'$ and $k_2 \not\simeq k_2'$, respectively. This yields four ER trees t_1', t_1'', t_2' and t_2'', and by $\sigma = k_1' \simeq k_2$, a merge operation is applied to merge t_1'' and t_2 into the ER tree t_3', as illustrated in Fig. 4.(c).

5.4   Bootstrapping

In the CSM strategy, each merge operation is governed by a set of cannot-link constraints that is iteratively expanded during the ER repair process. This is because the split operation can accurately remove erroneous matches, whereas the merge operation behaves unpredictably, i.e., by merging two ER trees, both false positives and true positives may increase, which could yield an unexpected result. Therefore, the CSM strategy utilizes cannot-link constraints to control the behaviors of each merge operation. If merging two ER trees would violate a cannot-link constraint, then it can only be conducted on the subtrees of these two ER trees after splitting them properly. Hence, to improve the efficiency, we can bootstrap the ER repair process using the existing cannot-link constraints.

6   Experiments

We have evaluated our provenance-aware ER method from three aspects: (1) Resource requirements for building and maintaining an ERI index (i.e., time and space); (2) Efficiency of repairing ER results (i.e., reduced human efforts); (3) Effectiveness of repairing ER results (i.e., improved ER quality).

6.1   Experimental Setup

Our experiments were performed on a Windows 7 machine with Intel(R) Core(TM) i3-2330M 2.20 GHz CPU, 8GB main memory and 64 bit operating system.

We used two data sets[1] in our experiments: one is from Scopus and the other is Cora. The Cora data set contains 1879 publications, and its "gold standard" is available for the public. Scopus is a bibliographic database containing millions of records for publications, authors and affiliations. We downloaded 10784 publication records from the querying API provided by Scopus. A "gold standard" for 4865 publication and 19527 author records in this Scopus data set was established by domain experts through manually checking.

We implemented the ER models: one for each data set, in Java with JDBC access to a PostgeSQL database. For the Cora data set, two publications are matched if: (1) their titles are similar, (2) they were published in the same year, and their pages and booktitles are similar, (3) their authors and pages are similar, or (4) their authors and book titles are similar. In these rules, the similarity of attributes was compared by using q-gram Jaccard similarity and the thresholds were set to 0.85. The ER result of applying this ER model is determined by the matches in the union of the sets of matches generated by these rules. For the Scopus data set we only used one rule – two author records are matched if their names are similar, e.g., "Baker M." and "Baker M.G.". Note that we can certainly use more sophisticated ER algorithms to obtain an initial ER result with better quality. Nevertheless, an initial ER result only serves as a baseline for illustrating ER repair techniques. The better quality an initial ER result has, the less repairs we would need for achieving a high-quality ER result. For this reason, these simple ER models were used in our experiments. We implemented ERI indices in a PostgeSQL database, as part of metadata managed by the database system.

6.2   Time and Space Requirements

To test the time efficiency, we implemented our ERI construction algorithm. The ER model over the Cora data set generates 61453 matches, while the ER model over the Scopus data set generates 75447 matches. For each data set, we ran the ERI construction algorithm on these matches to build an ERI index 5 times. In Fig. 5, (a) and (c) illustrate that the scalability of our ER index construction

---

[1]   http://www.scopus.com/home.url;http://www.cs.umass.edu/~mccallum/

method is in linear time w.r.t. the number of matches, which empirically verified Theorem 1. Although the numbers of matches in two data sets are similar, the time of building an ERI index in Scopus was doubled than in Cora. This is because they have different data characteristics. The ERI index for Scopus contains 2969 non-singleton ER trees that contain at least one match, and 36003 singleton ER trees containing only one leaf (i.e., no match with other records). The percentage of author records contained in non-singleton ER trees in terms of all author records in Scopus is 24%. In contrast, the ERI index for Cora has 117 non-singleton ER trees which contain 1674 publication records in total. Thus, the percentage of publication records contained in non-singleton ER trees in terms of all publication records in Cora is 89%. These results are presented in Fig. 6. We also tested the space efficiency of our ERI construction method. For this we computed the space rates of ER trees in ERI, i.e., for an ER tree t_e of an entity e,

$$\text{space rate} = \frac{\text{the number of internal nodes in } t_e}{\text{the number of matches for } e}.$$

In Fig. 5, (b) and (d) show that the space rates drop dramatically when the total number of matches associated with an entity increases, i.e., nearly %40 for entities with about 10 matches, %20 for entities with about 50 matches, %10 for entities with about 200 matches, and %4 for entities with about 1000 matches. The results on the Cora and Scopus data sets are highly consistent.

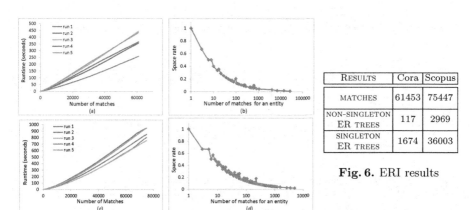

Fig. 5. (a)-(b) Time and space requirements (Cora); (c)-(d)
Time and space requirements (Scopus)

RESULTS	Cora	Scopus
MATCHES	61453	75447
NON-SINGLETON ER TREES	117	2969
SINGLETON ER TREES	1674	36003

Fig. 6. ERI results

6.3 Human Effort

In our experiments we measured the level of human efforts reduced by using ERI for repairing inconsistent ER results. In the case that no ERI index is available, all the records in the cluster of an inconsistent entity have to be manually checked. However, by using an ERI index, we only need to check critical records.

Therefore, for each inconsistent entity, we first compare the total number of records in its cluster (i.e., the size of cluster) with the number of critical records identified for repairing the inconsistency in this cluster. Fig. 7 shows that about 30% - 50% records are critical when the size of a cluster is small (i.e., less than 10), whereas only 5% - 9% records are critical when the size of a cluster is over 100. As indicated by the power trendline for critical records in Fig. 7, human efforts are significantly reduced when the size of cluster increases. We also compared the total number of records in the cluster of an inconsistent entity with the number of cut nodes required for repairing the inconsistency. Each cut node represents an erroneous match involved in producing the inconsistency. Fig. 7 shows that the percentage of the number of cut nodes in terms of the size of cluster remains quite stable (between 1% - 2%) when the size of cluster is greater than 50. In addition to this, the number of cut nodes is much smaller than the number of critical records in our experiments. This difference indicates the quality of the ER result generated by the chosen ER model.

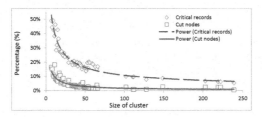

Fig. 7. Human effort: critical and cut nodes for different sizes of clusters in Cora

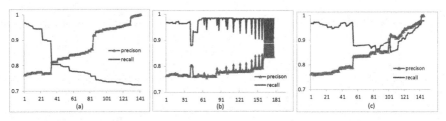

Fig. 8. Repair E_{cora} using: (a) only cannot-link constraints; (b) the naive strategy; (c) the CSM strategy, where the x-axis plots the number of constraints, and the y-axis plots precision and recall of the repaired results

6.4   ER Quality

We evaluated how effectively the proposed ERI can improve the quality of an ER result. Our experiments all start with the same initial ER result E_{cora}, which was generated by applying the ER model of the Cora data set. E_{cora} has the precision 76.35% and the recall 96.80%. We thus use E_{cora} as a base line to compare the changes of precision and recall in the three settings as described in Fig. 8. Fig. 8.(a) shows that the precision can be consistently improved if applying the splitting operation to repair E_{cora} in terms of cannot-link constraints. However, the recall decreases correspondingly. Fig. 8.(b) shows that there is no clear evidence on the improvement of ER quality when using the native strategy to repair E_{cora}. However, there is an inverse correlation between the precision and recall. Applying the splitting operation to repair E_{cora} sometimes gives a sharp jump in increasing precision but decreasing recall. But over time, precision and recall converge to certain range, reaching some fix-points. Fig. 8.(c) shows that the CSM strategy can control the negative effects of merging operations on precision so that the precision can be consistency improved over time. After reaching certain point, both precision and recall increase towards achieving high-quality ER results. To simulate constraints that should be identified by humans, we developed a human decision simulator that can randomly generate cannot-link and must-link constraints based on ground truth. A constraint that is violated in an ER result may not necessarily be violated after repairing the ER result with other constraints. For this reason, constraints used in our experiments need to be dynamically identified according to an up-to-date ER result. These experiments illustrate that the user can interactively provide feedback to an ERI index and ask for repairing an ER result in real time.

7   Conclusions

We studied entity resolution in terms of provenance, which is largely unexplored in the literature. An indexing method was proposed, which can efficiently manage the provenance information of the ER process. Together with a knowledge base capturing must-link and cannot-link constraints, the ERI indexing structure enables us to repair inconsistent ER results. Our experimental results confirmed that the ERI indexing method not only exhibits good scalability properties for building a provenance structure, but also supports efficient algorithms to repair erroneous ER matches with reduced human efforts.

References

1. Abiteboul, S., Hull, R., Vianu, V.: Foundations of Databases. Addison-Wesley (1995)
2. Afrati, F.N., Kolaitis, P.G.: Repair checking in inconsistent databases: algorithms and complexity. In: ICDT, pp. 31–41 (2009)
3. Agrawal, P., Ikeda, R., Park, H., Widom, J.: Trio-ER: The Trio system as a workbench for entity-resolution. Technical report, Stanford InfoLab (2009)
4. Arasu, A., Götz, M., Kaushik, R.: On active learning of record matching packages. In: SIGMOD, pp. 783–794 (2010)
5. Arasu, A., Ré, C., Suciu, D.: Large-scale deduplication with constraints using dedupalog. In: ICDE, pp. 952–963 (2009)
6. Benjelloun, O., Sarma, A.D., Halevy, A., Theobald, M., Widom, J.: Databases with uncertainty and lineage. The VLDB Journal 17(2), 243–264 (2008)
7. Bhattacharya, I., Getoor, L.: Collective entity resolution in relational data. TKDD 1(1), 5 (2007)

8. Buneman, P., Khanna, S., Tan, W.-C.: Why and where: a characterization of data provenance. In: Van den Bussche, J., Vianu, V. (eds.) ICDT 2001. LNCS, vol. 1973, pp. 316–330. Springer, Heidelberg (2000)

9. Buneman, P., Tan, W.-C.: Provenance in databases. In: SIGMOD, pp. 1171–1173 (2007)

10. Chaudhuri, S., Das Sarma, A., Ganti, V., Kaushik, R.: Leveraging aggregate constraints for deduplication. In: SIGMOD, pp. 437–448 (2007)

11. Christen, P.: Data Matching. Springer (2012)

12. Christen, P.: A survey of indexing techniques for scalable record linkage and deduplication. IEEE TKDE **24**(9), 1537–1555 (2012)

13. Cohen, W.: Data integration using similarity joins and a word-based information representation language. TOIS **18**(3), 288–321 (2000)

14. Cong, G., Fan, W., Geerts, F., Jia, X., Ma, S.: Improving data quality: consistency and accuracy. In: PVLDB, pp. 315–326 (2007)

15. Fellegi, I., Sunter, A.: A theory for record linkage. J. Amer. Statistical Assoc. **64**(328), 1183–1210 (1969)

16. Han, J., Kamber, M.: Data mining: concepts and techniques. Morgan Kaufmann (2006)

17. Karvounarakis, G., Ives, Z.G., Tannen, V.: Querying data provenance. In: SIGMOD, pp. 951–962 (2010)

18. Newcombe, H., Kennedy, J.: Record linkage: making maximum use of the discriminating power of identifying information. Comm. of the ACM **5**(11)

19. Sarawagi, S., Bhamidipaty, A.: Interactive deduplication using active learning. In: KDD (2002)

20. Shen, W., Li, X., Doan, A.: Constraint-based entity matching. In: AAAI, pp. 862–867 (2005)

21. Simmhan, Y.L., Plale, B., Gannon, D.: A survey of data provenance in e-science. ACM SIGMOD Record **34**(3), 31–36 (2005)

22. Verykios, V., Moustakides, G., Elfeky, M.: A Bayesian decision model for cost optimal record matching. The VLDB Journal **12**(1), 28–40 (2003)

23. Wagstaff, K., Cardie, C.: Clustering with instance-level constraints. In: AAAI, pp. 1097 (2000)

24. Whang, S.E., Benjelloun, O., Garcia-Molina, H.: Generic entity resolution with negative rules. The VLDB Journal **18**(6), 1261–1277 (2009)

25. Wijsen, J.: Database repairing using updates. TODS **30**(3), 722–768 (2005)

Information Retrieval
and Summarization

A Chip Off the Old Block – Extracting Typical Attributes for Entities Based on Family Resemblance

Silviu Homoceanu[✉] and Wolf-Tilo Balke

IFIS TU Braunschweig, Mühlenpfordstraße 23, 38106 Braunschweig, Germany
{silviu,balke}@ifis.cs.tu-bs.de

Abstract. Google's Knowledge Graph offers structured summaries for entity searches. This provides a better user experience by focusing on the main aspects of the query entity only. But to do this Google relies on curated knowledge bases. In consequence, only entities included in such knowledge bases can benefit from such a feature. In this paper, we propose ARES, a system that automatically discovers a manageable number of attributes well-suited for high precision entity summarization. With any entity-centric query and exploiting diverse facts from Web documents, ARES derives a common structure (or schema) comprising attributes *typical* for entities of the same or similar entity type. To do this, we extend the concept of typicality from cognitive psychology and define a practical measure for *attribute typicality*. We evaluate the quality of derived structures for various entities and entity types in terms of precision and recall. ARES achieves results superior to Google's Knowledge Graph or to frequency-based statistical approaches for structure extraction.

Keywords: Entity summarization · Schema extraction · Knowledge graph

1   Introduction

Entity-centric searches account for more than 50% of the queries on the Web [13, 14]. There are two main types of entity search queries: those focused on finding entities and those returning properties of given entities. The first type has been extensively researched: many systems performing query by example for Related Entity Finding (REF) or for Entity List Completion (ELC) have been published [5, 18]. Also belonging to the first category, searching for entities by means of properties e.g. "President of USA" known as "Web-based Question Answering", received significant attention [8, 17, 27]. Systems like the well-known IBM Watson [23] stand as a proof of their success. However, most queries are of the second type, popular entities according to user search behavior (identified in [14]) being celebrities, organizations, or health concerning issues like medical conditions. Unfortunately not much has been done to accommodate such queries. The problem of entity-centric search focused on returning information for given entities is the central topic of this paper.

Google's Knowledge Graph represents the state of the art for such entity queries. It summarizes knowledge of common interest using some fixed schema to provide an overview. After typing some entity name into Google's search field, an entity

© Springer International Publishing Switzerland 2015
M. Renz et al. (Eds.): DASFAA 2015, Part I, LNCS 9049, pp. 493–509, 2015.
DOI: 10.1007/978-3-319-18120-2_29

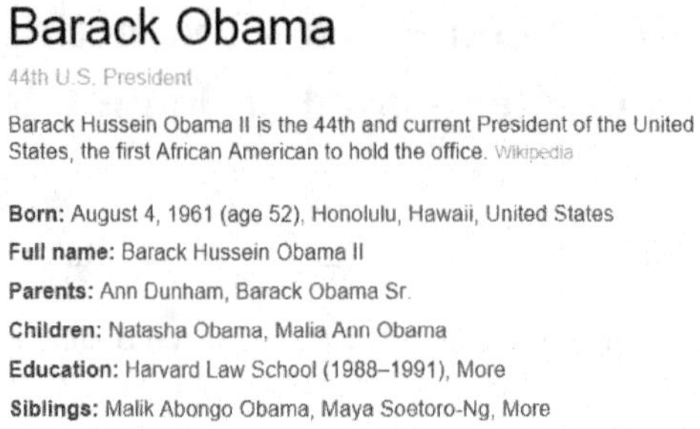

Barack Obama

44th U.S. President

Barack Hussein Obama II is the 44th and current President of the United States, the first African American to hold the office. Wikipedia

Born: August 4, 1961 (age 52), Honolulu, Hawaii, United States
Full name: Barack Hussein Obama II
Parents: Ann Dunham, Barack Obama Sr.
Children: Natasha Obama, Malia Ann Obama
Education: Harvard Law School (1988–1991), More
Siblings: Malik Abongo Obama, Maya Soetoro-Ng, More

Fig. 1. Knowledge Graph - result for "Barack Obama"

summary is provided on the right hand side of the search results, if the Knowledge Graph contains the entity. A sample entity summary for 'Barack Obama' is shown in Figure 1. According to Google's official blog (http://www.googleblog.blogspot.de/20 12/05/introducing-knowledge-graph-things-not.html), the Graph mainly relies on manually curated data sources like Wikipedia Infoboxes, Google's Freebase, and schema.org annotations on the Web. But with this, the Knowledge Graph has a major shortcoming: it only provides information on well-known entities already having a Wikipedia article, Freebase record or sufficient schema.org annotations. Our extensive evaluation presented in Section 2 shows that this is indeed rather limited.

In this paper we argue that a *data-driven approach* of building entity summaries directly from unstructured data on the Web is more suitable for entity-centric search. Inspecting the popular ClueWeb09 (lemurprject.org/clueweb09/) data set consisting of 500 million documents from the Web, we extracted approximately 11,000 statements regarding Barack Obama. The statements are structured as triples of the form (*subject, predicate, object*), with predicates representing *attributes* and objects represent the corresponding *values*. But the volume of information is huge. With the information needs of the majority of users in mind, when browsing through the variety of *attribute: value* pairs for 'Barack Obama', we found that many of them, e.g., visit: Israel, love: Broccoli or spent_vacation_in: Hawaii, are irrelevant. Can such attributes be recognized as irrelevant and pruned to obtain a suitable, yet concise structure?

The first idea that comes to mind is a frequency-based solution. Approaches like the count of witnesses as a measure for the importance of attributes have often proven efficient [8, 19]. Together with the Knowledge Graph, they will serve as baseline for evaluating the approach presented in this paper. But browsing through the triples for Barack Obama, one can observe that some of the information, like the year of election, term in office, being member of some party, etc., is common to all American presidents. Intuitively, a data-driven entity summary for "Barack Obama" as an "American president" would comprise a few, good descriptive properties selected from these shared characteristics. Taking a closer look at how the attributes extracted for Barack Obama are actually shared among the 44 American presidents (see Figure 2) a typical power law distribution can be observed. While the attributes that are common and important for this small world of presidents fall into the head of the distribution, the tail mostly comprises trivia about individual presidents. That means, by simply chopping off the tail, one might already identify common attributes of good quality. Is such a distribution valid for *all types* of entities, i.e. can the lessons learned from the small and homogeneous set of 44 American presidents be generalized? And, how can this distribution *efficiently* be derived and pruned, i.e. can this also be performed for classes with thousands of entities or heterogeneous entity types with only limited similarity?

Approach and contribution: Motivated by these observations we present *ARES* (AttRibute selector for Entity Summaries) a system for extracting data-driven structure for entity summarization. Starting from a query comprising an entity and the category of interest (like SCAD [2] we use categories for query disambiguation), e.g., "Barack Obama: American President", ARES delivers highly typical attributes for the query entity in the context of the provided category. To implement our approach we extend the concept of *typicality* from cognitive psychology and define *attribute typicality* together with a novel and practical rule for actually calculating it. We evaluate the quality of extracted attributes in terms of *precision* and *recall* deploying the basic structure from matching Wikipedia articles as *ground truth* together with human assessment. As a baseline, we employ the well-known ReVerb Open Information Extraction (OpenIE) system [10] enhanced with frequency-based scoring and state-of-the-art paraphrase discovery [4] and the Google Knowledge Graph. All measurements have been performed over *sufficiently large real-world datasets* namely the freely accessible part of PubMedCentral (www.ncbi.nlm.nih.gov/pmc) and ClueWeb09.

2   A Brief Glance at Google's Knowledge Graph

According to Google the Knowledge Graph mainly relies on the Wikipedia Infoboxes, Freebase and schema.org annotations. Schema.org was launched in early 2011 as joint initiative of major search engine providers. It provides a unified set of vocabularies for semantically annotating data published on the Web. But as we have shown in [12] schema.org did not gain traction. One year after schema.org was introduced, only about 1.56% of the websites comprised annotated data. In consequence

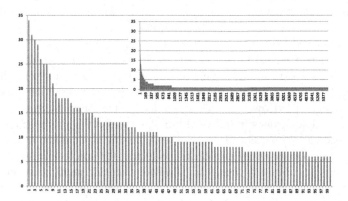

Fig. 2. Distribution of extracted attributes (x-axis) sorted by how many American presidents (y-axis) share each attribute (with zoom-in on the first 100 attributes).

schema.org can't significantly contribute to the Knowledge Graph. It seems that the Knowledge Graph is mostly limited to entities from Wikipedia and Freebase.

Wikipedia and Freebase largely overlap in terms of the entities and entity structure. Freebase is mainly focused on providing structured information (same as Wikipedia Infoboxes but more extensive). Wikipedia additionally provides a textual description, each entity being presented in a comprising article. Infoboxes are fixed-format tables built on one or more hierarchical Infobox templates. The purpose of Infoboxes is to consistently present a summary of some unifying aspects that articles share. The idea is that articles of entities of a similar kind share the same Infobox structure. This way, similar entities share the same structure that should flow into the corresponding Knowledge Graph summaries. But the Infoboxes can be quite extensive, having much more attributes than the Knowledge Graph snippet should comprise. Choosing the "right" attributes to build the entity summary is vital for the whole system. For instance, the snippet for Barack Obama (Figure 1) comprises 6 attributes taken from the Infobox Person template (en.wikipedia.org/wiki/Template:Infobox_person). However, nothing really specific regarding his activity as a president is mentioned, other than the first sentence being copied from the Wikipedia article. The information presented in the Knowledge Graph is quite general, common to any person be it a politician, writer, actor, etc. In fact, the same snippet structure is provided also for actor Kevin Bacon. But relevant information like the year Obama took office or which political party he belongs to, are not being included here despite being present in the corresponding Infobox. It seems that for this entity, the Knowledge Graph only presents the first few attributes of the broader Infobox template the entity is associated with – in this case the *Person* Infobox. Instead, we believe that finding a sweet-spot between too broad and too specific information, like for example a subset of the Office holder template (en.wikipedia.org/wiki/Template:Infobox_officeholder), is sensible.

But is this only a problem of choosing the right attributes from Infoboxes, i.e. do Infoboxes include the right attributes for entity summaries? Manually inspecting different entities with same Infobox templates it can be observed that for entities of common type, the structure of Wikipedia articles is often very similar with nearly identical first-level headings. Encouraged by this observation we analyzed a larger number of entities. Starting from the list of 3,000 diseases featuring an article on Wikipedia we extracted the headings of all articles (structural headings of Wikipedia like "References" and "External Links" were pruned). Indeed, even over large samples of entities, a common structure can be extracted (see Figure 3). A similar result holds in the case of American presidents, yet with lower percentages. Both entity types form homogeneous groups. However, this is not always the case: the same experiment performed on all companies from the S&P 500 list shows that, with the exception of only two headings (Products and Acquisitions) there is no common article structure for this category. Going a step further and inspecting the article headings for companies from the same business field, the structure becomes more homogeneous. For instance, articles for automotive companies often cover topics like 'Alliances' or 'Motorsport', in contrast to articles for pharmaceutical companies, where topics like 'Clinical Trials' and 'Litigation' are more common. Despite articles for companies being highly heterogeneous, all their Infoboxes follow the same template (the

Company Infobox template - en.wikipedia.org/wiki/Template:Infobox_company), summarizing information with 41 generic attributes. Does this general structure provide suitable selections of attributes that reflect article differences?

Focusing on the structure provided by the Infoboxes we conducted an experiment to investigate two aspects: The *number of expected attributes* (i.e., how many an entity summary should feature) and the *suitability of generic attributes* for building knowledge snippet structures reflecting the heterogeneous Wikipedia articles. We selected 50 companies, split into 10 groups, each group corresponding to a major business field (e.g. Automotive, Energy, Financial, IT, Retail, etc.). Each company and its corresponding attributes and values have then been presented to 25 human subjects with the task to select those few relevant properties they would like to see in a short description of the company. The experiment was conducted through a crowdsourcing platform (CrowdFlower). In total we collected 1250 judgments. Companies were presented in random order and attributes were shuffled for each task.

The number of selected attributes over all judgments on all companies ranges from 1 to 18, with a clear focus between 3 and 7, an average of 5.3 and a standard deviation of 3.12. This behavior is consistent for all companies: averages of the selected number of attributes per company range between 5.1 and 6.0. Also in terms of attribute relevance there is large consensus: the same few typical attributes are considered relevant by most subjects for all companies. The histogram presented in Figure 4 shows companies from the financial sector (histograms for all other companies are very similar). In fact, low standard deviation values for each attribute on all companies, show subjects selected the same attributes over and over, regardless of the company. As a consequence, histogram based similarity metrics like the Minkowski distance measured pairwise between all companies, can't really differentiate between various business sectors or other semantically meaningful criteria.

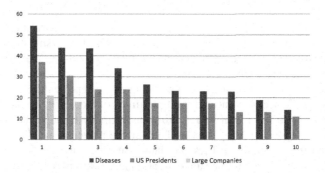

Fig. 3. Wikipedia Article Structure – % of entities (y-axis) belonging to the same category and sharing a certain heading (x-axis, values shared by less than 10% omitted)

Of course, since the Infobox structure has to cover all kinds of companies, the respective attributes were general. But the fact that popular attributes selected from this structure are not correlated to the different topics presented in the articles suggests more sophisticated measures have to be taken when categories are heterogeneous. Our experiments show a clear tendency regarding the number of attributes an entity

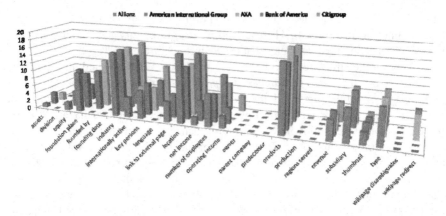

Fig. 4. The number of subjects (y-axis) that have selected an attribute (x-axis) for a certain company (z-axis). For companies from the financial sector only.

summary should feature and a surprisingly high consensus about what attributes are considered important. A good entity summary structure highlights *between 3 and 7 attributes*, and focuses on *typical properties* of the entity. But overall, seeing the respective articles' richness, considering just generic properties may poorly reflect the real world. If the entity is part of a homogeneous category, properties are usually typical for the entire category. But, if categories are heterogeneous, good structures have to be derived in a data-driven fashion with properties typical for a more homogeneous semantic subgroup.

3   Related Work

Knowledge graphs have been used in the past for entity summarization. In [22] the authors present a greedy algorithm that adapts the idea of diversification from information retrieval [1] to extract entity summaries from *subject-predicate-object* triples. The authors argue that a diversity unaware system is likely to present only certain aspects of an entity. For instance in a knowledge base representing movie information, entity Tom_Cruise is connected multiple times to movies by the acted_in predicate but just once to the literal representing his birthday through the born_on_date predicate. In this case, the many movies Tom Cruise played in would be more likely to be included in the summary. Personal data would be ignored. To incorporate the concept of diversification into the summarization algorithm they rely on a knowledge graph built form the triples. Added to the edges, weights should represent the "importance" of the attributes of the entity. These weights are assumed to be provided as input. We consider that the concept of attribute typicality introduced in this paper is perfect for establishing such weights. Furthermore, we separate between predicates and values. Each predicate is considered only once. In consequence, acted_in has same chance of making it into the summary as born_on_date has. The decision which attributes to include in the summary is made based on the attribute typicality, value controlled by the user through the entity category.

Related to our work, in [16] the authors propose a probabilistic approach to compute attribute typicality. But there is a fundamental difference: the authors ignore the difference between entities and sub-concepts representing instances of a concept. This way for any concept, say company, both IT company and Toyota are instances of company. This simplifying assumption doesn't consider data heterogeneity: for concepts comprising heterogeneous entities the extracted attributes only loosely represent the corresponding entities. In contrast, our approach distinguishes between concepts and sub-concepts. It follows a data-driven approach with attributes typical for each sufficiently homogeneous semantic subgroup.

From a broader perspective, our work is related to the field of schema matching and mapping. Such systems use various matching techniques and data properties to overcome syntactic, structural or semantic heterogeneity. But most approaches focus on data from relational databases [20]. Systems like WEBTABLES [7] or OCTOPUS [6] rely on semi-structured data like html lists and tables on the Web to extract data structure. In contrast, our system may use all extractions from text without any restrictions increasing the number of supported entities.

4   The Concept of Typicality

Leading the quest for defining the psychological concept of typicality [21], Eleanor Rosch showed that the more similar an item was to all other items in a domain, the more typical the item was for that domain. Her experiments show that typicality strongly correlates (Spearman rhos from 0.84 to 0.95 for six domains) with family resemblance a philosophical idea made popular by Ludwig Wittgenstein in [25]. Wittgenstein postulates that the way in which family members resemble each other is not defined by a (finite set of) specific property(-ies), but through a variety of properties that are shared by some, but not necessarily all members of a family. Based on this insight, Wittgenstein defines a simple family-member similarity measure based on property sharing:

$$S(X_1, X_2) = |X_1 \cap X_2| \qquad (1)$$

where X_1 and X_2 are the property sets of two members of the same family. However, this measure of family resemblance assumes that a larger number of common properties increase the perceived typicality, while larger numbers of distinct properties do not decrease it. To overcome this problem, the model proposed by Tversky in [24] suggests that typicality increases with the number of shared properties, but to some degree is negatively affected by the distinctive properties:

$$S(X_1, X_2) = \frac{|X_1 \cap X_2|}{|X_1 \cap X_2| + \alpha|X_1 - X_2| + \beta|X_2 - X_1|} \qquad (2)$$

where α and $\beta \geq 0$ are parameters regulating the negative influence of distinctive properties. In particular, when measuring the similarity of a family member X_2 to the family prototype X_1, a choice of $\alpha \geq \beta$ poses the same or more weight to the properties of the prototype itself. For $\alpha = \beta = 1$ this measure becomes the well-known

Jaccard coefficient. For $\alpha+\beta \leq 1$ more weight is given to shared features, while for $\alpha+\beta > 1$ diverse properties are emphasized, which is useful for heterogeneous families.

4.1   Attribute Typicality

Applying Tversky's family resemblance model enables the selection of a *most typical family member* or entity. However, our main goal is to find a common structure, i.e. a *most typical set of attributes* for some entity and its respective entity type. Hence, we need to find out which of the attributes occurring in a family actually are typical with respect to this family. Since the family definition relies on the measure of members' similarity, we adapt Tversky's measure as follows: assume we can determine some family F consisting of n entities $E_1, \dots, E_n$ and a total of k distinct attributes given by predicates $p_1, \dots, p_k$ are observed for family F. Let X_i and X_j represent the respective attribute sets for two members E_i and E_j, then:

$$|X_i \cap X_j| = 1_{X_i \cap X_j}(p_1) + 1_{X_i \cap X_j}(p_2) + \cdots + 1_{X_i \cap X_j}(p_k) \tag{3}$$

where $1_X(p) = \begin{cases} 1 \; if \; p \in X \\ 0 \; if \; p \notin X \end{cases}$ is a simple indicator function.

Now we can rewrite Tversky's shared similarity measure to make all attributes explicit:

$$S(X_i, X_j) = \frac{\sum_{l=1}^{k} 1_{X_i \cap X_j}(p_l)}{|X_i \cap X_j| + \alpha|X_i - X_j| + \beta|X_j - X_i|} \tag{4}$$

where the same conditions as above apply to α and β.

According to Tversky, each attribute shared by X_i and X_j contributes evenly to the similarity score between X_i and X_j. This allows us to calculate the *contribution score* of each attribute to the similarity of each pair of members:

Let p be an attribute of a member from F. The *contribution score* of p to the similarity of any two attribute sets X_i and X_j, denoted by $C_{X_i,X_j}(p)$, is:

$$C_{X_i,X_j}(p) = \frac{1_{X_i \cap X_j}(p)}{|X_i \cap X_j| + \alpha|X_i - X_j| + \beta|X_j - X_i|} \tag{5}$$

where $\alpha = \beta \geq 0$. (Additionally further normalization could be applied to avoid small values.) The contribution of some attribute towards the similarity of two family members is this way dependent on the degree of similarity between the two members. This is a fundamental difference to simply performing property set intersections (like in Figure 2). In particular, this enables us to cope even with difficult cases where entity collections are heterogeneous. Building on the *contribution score* we are now ready to introduce the notion of *attribute typicality*.

Definition 1: Attribute Typicality. *Let F be a set of n entities $E_1, \dots, E_n$ of similar kind represented by their respective attribute sets $X_1, \dots, X_n$. Let U be the set of all distinct attributes of all entities from F. The typicality $T_F(p)$ of an attribute/predicate*

$p \in U$ w.r.t. F is the average contribution of p to the pairwise similarity of all entities in F:

$$T_F(p) = \frac{1}{C_2^n} \cdot \sum_{i=1}^{n-1} \sum_{j=i+1}^{n} C_{X_i X_j}(p) \tag{6}$$

where $C_{X_i X_j}(p)$ is the contribution score of attribute p regarding the similarity between X_i and X_j (eq. 5) and C_2^n is the number of all combinations of entities from F.

5 Designing the Retrieval System

Building on state-of-the-art information extraction, our prototype system discovers all those attributes that are typical for the entity and entity type provided by the user. Figure 5 shows an overview of the system. In brief, the system works as follows:

5.1 Information Extraction

Documents from the Web, are processed with Open Information Extraction (OpenIE) methods. This results in a large number of (subject, predicate, object) triples. Since the same entity can be expressed in multiple forms, an *Entity Dictionary* listing unique entities and their possible string representations is kept and updated. Two problems have to be discussed regarding the entity dictionary: *synonymy*, i.e. every entity can have more than just one string representation form, e.g. "Barack Obama", "B. H. Obama", etc. and *ambiguity*, i.e. every string can refer to different entities e.g. "Clinton" may refer either to "Bill Clinton" or to "Hillary Clinton". Since synonymy is not our main focus, our prototype uses thesauri like WordNet, Mesh and entity string representations from Wikipedia. The problem of *ambiguity* is known as Entity Disambiguation. In order to solve this, the assumption is made that any ambiguous reference to some entity, say "Clinton", is preceded in the document by some clear entity reference like "President Clinton" or "Mrs. Clinton". If no such reference is found, we relax our assumption like in presented [19] and assume that each entity string is uniquely addressing exactly one entity within a document.

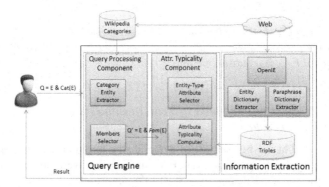

Fig. 5. ARES – System Architecture

Predicates may also have synonym terms e.g. president_of, won_elections_in, was_elected_president_of, etc. Also in this case we keep listing unique predicates – the *Paraphrase Dictionary*. However, for predicates there are no acceptable thesauri. The field of paraphrase discovery is concerned with this problem [4]. State-of-the-art methods rely on a class of metrics called *distributional similarity metrics* [15]. In the context of paraphrase discovery, this hypothesis is applied as: two predicates are paraphrases of each other, if they are similarly distributed over a set of pairs of entity-types. However, in contrast to the entity ambiguity problem, a simplifying assumption is made: predicates can't have multiple meanings (single-sense assumption [26]). Following on these insights and similar to the method presented in [11], we applied hierarchical clustering to the predicate/entity-type pairs distributions. As a similarity measure we have used the well-known cosine metric with mean linkage as criteria. Still, despite experimenting with different similarity thresholds, the success of the paraphrasing process is rather limited. While on manual inspection the clusters prove good precision, just about 7% (for 0.9 similarity threshold) actually build clusters. The rest of the predicates build single node clusters although a substantial number of cases show obvious paraphrases. This is consistent with results from the literature [26], where, the recall barely reaches 35%.

All extracted facts are cleaned based on these dictionaries. Then, they are stored in a knowledge base (we use a Virtuoso RDF database in our prototype).

5.2   Query Engine

The query engine module is responsible for extracting the entity structure for user queries comprising the entity and corresponding type. The first step in this direction is to identify all entities that belong to the same category as the query entity. To do this, a mapping between the entities and the corresponding categories is needed. Such mappings can be extracted in the preprocessing phase directly from text with lexico-syntactic patterns, like "…an X such as Y…" or "… all X, including Y…" expressing "is-a" hierarchies between entity category X and entity Y.

Experiments in section 2 show that categories may comprise heterogeneous entities. Our approach relies on Tversky's similarity measure (eq. 2) to find those k-nearest neighbors to the query entity. These entities not only belong to the same category as the query entity, but they also share similar structure. We call this special collection of entities, the *family* of the query entity.

Definition 2: Family. *Let X be the query entity and C be the set of entities of the same category as the category given by the user to represent X. The family of X w.r.t. category C, denoted $F_{X,C}$, is a subset of entities from C, with:*

$$F_{X,C} = \{Y | Y \in C \wedge S(X,Y) > \theta\}$$

where $S(X,Y)$ represents the similarity between entities X and Y (see eq. 2) and θ is a family specific threshold.

The value of θ has to be established dynamically, based on the start entity and the entities falling into the same category. For this purpose, we employ automatic thresholding methods, in particular the ISODATA algorithm. Applied to the entities falling into the same category as the query, this method identifies the similarity threshold that splits the entities in two groups: one comprising homogeneous entities with high similarity to the query entity and one containing all the less similar entities.

With the query rewritten from "entity plus type" to "entity plus family", we can now proceed to extract the attributes that are central for the entity types' structure. Following the definition of *attribute typicality*, introduced in section 4.1, for each attribute, we calculate its contribution to defining the family of the query entity. For better overview of how the quantification of attribute typicality is performed we present the pseudo-code of our system's algorithm in Algorithm 1. Being the online part of the system, in the following we present an analysis of the systems' efficiency.

Runtime Analysis: Even for broad categories with thousands of entities our system requires about 40 seconds per query. For instance, in the case of diseases, 3,513

Algorithm 1: Extraction algorithm for typical attributes.

Input: X - query entity, C - set of entities of same category as X, ϕ - attribute quality threshold, RDF triple collection

Output: T - set of typical attributes

```
1:  F ← FAMILY(X, C)
2:  T ← {}; U ← {}
3:  foreach X in F do
4:      x_attr ← ATTRIBUTES(X, RDF)
        // all attributes from triples where X is the subject or the object
        // stored in memory for heavy reuse (*)
5:      U ← U ∪ x_attr
6:  end for
7:  foreach a in U do
8:      a_typ ← 0
9:      for Xᵢ ∈ F do
10:         for Xⱼ ∈ F −{Xₖ|1 ≤ k ≤ i ∧ Xₖ ∈ F} do
11:             xᵢ_attr ← ATTRIBUTES(Xᵢ, RDF)    // (*)
12:             xⱼ_attr ← ATTRIBUTES(Xⱼ, RDF)    // (*)
13:             contr←0
14:             if  a ∈ xᵢ_attr ∧ a ∈ xⱼ_attr then
```

$$contr \leftarrow \frac{1}{|x_i\_attr \cap x_j\_attr| + \alpha|x_i\_attr - x_j\_attr| + \beta|x_j\_attr - x_i\_attr|}$$

```
                // contribution of p to similarity between Xᵢ and Xⱼ (eq.5)
16:             end if
17:             a_typ ← a_typ + contr
18:         end for
19:     end for
```

$$20:\quad a\_typ \leftarrow 2 \cdot \frac{a\_typ}{|F|\cdot(|F|-1)} \qquad \text{// the number of pairwise comparisons } (C_2^{|F|})$$

```
21:     if a_typ > φ then
22:         T ← T ∪ a
23:     end if
24: end for
25: return T

26: function FAMILY(X, C)
27:     F ← {X}; x_attr ← ATTRIBUTES(X, RDF)
28:     foreach Y in C do
29:         y_attr ← ATTRIBUTES(Y, RDF)
30:         sim←similarity(x_attr, y_attr)        // Tversky's similarity, eq. 2
            // computed only once then stored in memory for later use (**)
31:         S←S ∪ sim
32:     end for
33:     θ ← ISODATA(S)
34:     foreach Y in C do
35:         y_attr ← ATTRIBUTES(Y, RDF)
36:         sim←similarity(x_attr, y_attr)        // (**)
37:         if sim ≥ θ then
38:             F ← F ∪ Y
39:         end if
40:     end for
41:     return F
```

---

entities in the disease category have articles on Wikipedia. For entity "hypertension", ARES needs 42.977 seconds to extract typical attributes on commodity hardware. The time required is broken down as follows: computing the family of the query (lines 26 to 41 in Alg. 1) takes 22.472 seconds to complete. This covers the following parts: extracting all attributes for all entities (13.350 seconds – an average of 3.8 milliseconds per entity); pairwise comparing the 1,329 diseases that also appear in PubMed-Central statements (8.917 seconds – an average of 6.7 milliseconds per comparison); computing the family threshold (21 milliseconds). All other operations (assignments, logical, arithmetical operators) for the family computation require 184 milliseconds. The computation of typicality values for all 2,711 attributes (lines 2 to 25 in Alg. 1) takes 20.505 seconds to compute (about 7.5 milliseconds per attribute).

All tests have been performed single threaded. But since all major operations allow for parallelization, ARES should run in real-time on a cluster with up to 100 nodes.

6   Qualitative Evaluation

6.1   Experimental Setup

Dataset. For our tests, we used ClueWeb09 a 500 million Web documents corpus and PubMedCentral comprising about 250,000 biomedicine and life sciences research papers. All documents are processed offline by our IE module. For ClueWeb09 billions of noisy triples are extracted. After filtering out triples that are infrequent only approx. 15 million triples remain. With the same process on the PubMedCentral corpus, we extracted about 23 million triples. The IE module needs about 1 minute to process 8,000 sentences. On commodity hardware, the process took about 11 days.

Queries. We experiment with persons (in the sense of American presidents), organizations (in the sense of companies) and medical conditions, three types of entities identified in [14] as most popular entity-centric queries on the Web.

Measures. Our goal is to extract a high quality structure with limited, yet precise attributes. Therefore the success of all algorithms is measured in terms of precision (also in aggregated form as mean average precision MAP). Given the small number of attributes in an entity summary, recall is less important but still relevant for our task.

The frequency-based baseline algorithm. Drawing on the literature, we assume that attributes frequently appearing together with either the query entity or with entities of the same type, build a good structure for the query. Relying on the same infrastructure as described in Section 5 we thus implemented the frequency-based baseline approach (in the following called Frequency-based Entity Summarization - short FES). Another reference system is of course Google Knowledge Graph, the attributes from the knowledge snippet to be specific.

Establishing Ground Truth. For entity-centric search falling into the categories of REF or ELC, TREC [3] and INEX [9] provide data samples and gold standards. Unfortunately, no such data is available for problem presented in this paper. We rely on the basic structure from matching Wikipedia articles as ground truth together with human assessment. For the evaluation based on human assessment all attributes provided by ARES, FES and the Knowledge Graph were mixed together and ordered alphabetically for each query. We presented the resulting lists to subjects and provided the same instructions as in Section 2. We selected relevant attributes based on the 'majority rule'. All assessments showed *substantial agreement* [19] showing Fleiss' Kappa agreement levels between 0.71 and 0.76.

6.2   Experiments

In the following we present two sets of experiments: one set focused on evaluating different query forms to proper disambiguate query entities and one for testing our approach on large categories with either heterogeneous or homogeneous structure.

6.2.1 Disambiguation of Queries

A query consists of two parts: an entity and the corresponding entity category. This is because a single entity is not enough to capture the user intent. Still, allowing users to give examples might help disambiguation. For instance, with "Ronald Reagan" as a query entity and "Clint Eastwood" as additional example, users will be referring to American actors rather than American presidents. However, they might also have other categories in mind like Western actors, actors from California, etc. The more examples, the better the disambiguation, however increasing query complexity. On the other hand, a user provided category leads to easy and high quality disambiguation [13]. To establish which query form is better, we performed three experiments:

a)   Users provide an entity. Since our approach relies on shared attributes, a single entity cannot be disambiguated. This experiment is however interesting for FES.

b)   Users provide five similar entities. The cognitive burden increases heavily beyond that. FES cumulates the frequency of the attributes over the five entities and selects the most frequent ones. We modified ARES for this experiment to use the five entities as a family.

c)   Users provide an entity together with the category of interest. The frequency-based method cumulates the frequency of the attributes for all entities of the category and selects the most frequent ones. ARES works as described in Section 5.

In Figure 6 (a, b and c) we present the top 10 attributes for our running example: American presidents. For just one entity (Fig. 6.a), the precision of FES proves really poor. No disambiguation can be performed and thus, all kinds of attributes are considered. Barack Obama has proven to be an unlucky choice for FES: averaging the precision over multiple American presidents shows better results. In the case of five examples (Fig. 6.b), both the frequency-based method and the modified attribute-sharing method achieve average results. Again, the reason is proper disambiguation: only a small part of the personal aspects are evened out. Finally when the category of American presidents is also provided (see Fig. 6.c), both methods achieve quite good precision. Hence we can state that a query consisting of some entity and a respective type leads to better disambiguation and allows us to extract better structure.

Experiment c) shows the normal functionality of both ARES and FES. The precision and recall values obtained by the systems are presented in Table 1. For the case of American presidents, ARES is with a MAP of 0.75 superior to the other systems. The Knowledge Graph focuses in this case on family and education, elements considered irrelevant by the assessors. This severely affects its recall.

Table 1. Precision & Recall by system and query category

	Precision			Recall		
	ARES	FES	Knowledge Graph	ARES	FES	Knowledge Graph
American Presidents	0.75	0.33	0.37	0.5	0.3	0.2
Companies	0.73	0.44	0.49	0.6	0.3	0.25
Diseases	0.87	0.37	0.52	0.7	0.4	0.38

6.2.2 The Structure of Categories

While categories are perfect for disambiguation, some categories may prove hetero-geneous. Fortunately, our approach features a self-tuning resemblance measure able to automatically refine categories: all queries are focused on a *family* of entities with sufficiently homogeneous structure, while keeping the focus on the query entity. For instance, out of the S&P 500 list of companies, for queries like "Toyota Motor Corpo-ration", "Renault S.A.", or "Volkswagen A.G.", our systems builds families with 17 to 24 entities, clearly focusing on car companies (30% - 50% of the selected family members are car makers). For queries like "Apple Inc", "Google Inc" or "Microsoft" the results are similar, with 40% - 60% of IT companies in the selected family.

Indeed the self-tuning works well. In contrast to the results observed for generic attributes from Wikipedia Infoboxes (Section 2), where Minkowski similarity metrics (Manhattan distance) showed no difference between companies in different fields, for the attributes extracted by our system these differences are much more expressive. The average histogram distance for the attributes from Wikipedia Infoboxes selected by the crowd for car companies is of 69.8. The same distance for attributes selected by ARES for car companies is of 6.7. For IT companies the respective average values are 56.4 for attributes from the Wikipedia Infoboxes vs. 1.4 for the ARES selection. However, the average distance between different sectors stays large also for ARES: 78.16 vs. 65.3 for the selection from the Wikipedia Infoboxes and ARES respectively, for car vs. IT companies. This shows that ARES is able to extract attributes particular to homogeneous entities (small histogram distances for similar companies) while keeping heterogeneous entities apart. It's remarkable that, when presented with data-driven attributes, assessors picked attributes differentiating between business sectors.

In terms of precision, our approach achieves 0.73 MAP for all company queries, superior to both baselines. We present the results, averaged by sector, in Fig. 6.d and Fig. 6.e. For car makers, 8 attributes proved relevant according to the majority of human assessors. There is an important difference regarding the precision achieved by FES in the two sectors. Deeper inspection showed that there is more information about IT companies than about car makers in ClueWeb09. In fact most information on automotive topics actually refers to cars and not to the respective companies. Thus, it is understandable that a frequency based method shows such poor results.

Motivated by these findings, we repeated the experiments on PubMedCentral with diseases as query entities. Unlike in the case of companies, here we can't recognize any particular (especially taxonomically motivated) patterns regarding the members in our automatically derived families. Cardiovascular diseases are mixed together with infectious diseases, skin conditions and forms of cancer. We evaluated the systems over five well-known medical conditions ("cancer", "diabetes mellitus", "hepatitis", "hypertension" and "tuberculosis"). Boosted by the high quality information (Fig. 6.f.), our method achieves an impressive MAP of 0.87. The Knowledge Graph returns some of the National Library of Medicine headings obtaining fair average precision of 0.52. We would have expected that FES also performs better given the large amount of relevant information. But it seems that the broad coverage of various subjects leads to the frequency of attributes being spread rather evenly. Also in terms of recall ARES is consistently superior, showing its overall practical usefulness.

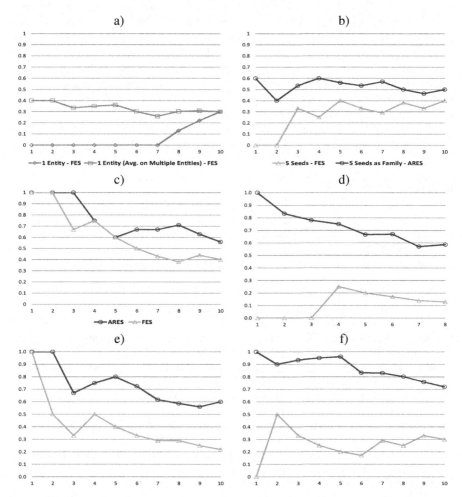

Fig. 6. Precision@k - for experiments a) and b) and c), for automotive companies (d), IT companies (e) and diseases (f)

7   Conclusions and Future Work

Google's Knowledge Graph represents the state of the art for entity summarization. However, our experiments show that even simple, frequency-based approaches already reach similar or even better quality results. But for good user experience, especially for heterogeneous collections of entities even more is needed. Therefore, tuning the data-driven schema selection for entities over the vast variety of facts from the Web is a major contribution of the ARES approach presented in this paper. ARES relies on the concept of family resemblance introduced by cognitive psychology and intelligently blends the homogeneity/heterogeneity of entity families with schema integration techniques in the light of all extracted facts. ARES is self-tuning in the sense that after family selection, entities within families show high intra-family similarity, while entities from heterogeneous categories show low inter-family similarity. Given the current advances in OpenIE, that allow to work directly on text, any entity being described somewhere on the Web, can thus be summarized appropriately. Our experiments on real-world entity classes representing different degrees of class homogeneity show that ARES is indeed superior to both, frequency-based statistical approaches and the Knowledge Graph, in terms of precision and recall. Moreover, also the run-time performance is already quite practical.

In future work, the improvement of response times by advanced indexing techniques, parallelization, and the exploitation of specialized hardware will be addressed.

References

1. Agrawal, R., et al.: Diversifying search results. In: Proceedings of the Second ACM International Conference on Web Search and Data Mining (WSDM), p. 5. ACM Press (2009)
2. Bakalov, A., Fuxman, A.: SCAD: collective discovery of attribute values categories and subject descriptors. In: Procedings of the 20th International World Wide Web Conference (WWW), Hyderabad, India, pp. 447–456 (2011)
3. Balog, K., et al.: Overview of the trec 2011 entity track. In: TREC (2011)
4. Barzilay, R., Lee, L.: Learning to paraphrase: an unsupervised approach using multiple-sequence alignment. In: Proceedings of the Conference of the North American Chapter of the Association for Computational Linguistics on Human Language Technology (NAACL), Edmonton, Canada, pp. 16–23 (2003)
5. Bron, M., Balog, K., de Rijke, M.: Example based entity search in the web of data. In: Serdyukov, P., Braslavski, P., Kuznetsov, S.O., Kamps, J., Rüger, S., Agichtein, E., Segalovich, I., Yilmaz, E. (eds.) ECIR 2013. LNCS, vol. 7814, pp. 392–403. Springer, Heidelberg (2013)
6. Cafarella, M.J., et al.: Data integration for the relational web. In: Proceedings of the Very Large Database Endowment (PVLDB), Lyon, France (2009)
7. Cafarella, M.J., et al.: WebTables: exploring the power of tables on the web. In: Proceedings of the Very Large Database Endowment (PVLDB), Auckland, New Zealand, pp. 538–549 (2008)
8. Cheng, T., et al.: EntityRank: searching entities directly and holistically. In: Proceedings of the 33rd International Conference on Very Large Databases. (VLDB), pp. 387–398 (2007)
9. Demartini, G., Iofciu, T., de Vries, A.P.: Overview of the INEX 2009 entity ranking track. In: Geva, S., Kamps, J., Trotman, A. (eds.) INEX 2009. LNCS, vol. 6203, pp. 254–264. Springer, Heidelberg (2010)
10. Fader, A., et al.: Identifying relations for open information extraction. In: Proceedings of the Conference on Empirical Methods in Natural Language Processing (EMNLP), Edinburgh, Scotland, UK, pp. 1535–1545 (2011)

11. Hasegawa, T., et al.: Discovering relations among named entities from large corpora. In: Proceedings of the 42th Annual Meeting of the Association for Computational Linguistics (ACL), Barcelona, Spain (2004)
12. Homoceanu, S., Geilert, F., Pek, C., Balke, W.-T.: Any suggestions? active schema support for structuring web information. In: Bhowmick, S.S., Dyreson, C.E., Jensen, C.S., Lee, M.L., Muliantara, A., Thalheim, B. (eds.) DASFAA 2014, Part II. LNCS, vol. 8422, pp. 251–265. Springer, Heidelberg (2014)
13. Homoceanu, S., Balke, W.: What makes a phone a business phone. In: Proceedings of the International Conference on Web Intelligence (WI), Lyon, France (2011)
14. Kumar, R., Tomkins, A.: A characterization of online search behavior. Proc. IEEE Data Eng. Bull. **32**(2), 1–9 (2009)
15. Lee, L.: Measures of distributional similarity. In: Proceedings of the 37th Annual Meeting of the Association for Computational Linguistics (ACL), pp. 25–32 (1999)
16. Lee, T., et al.: Attribute extraction and scoring: a probabilistic approach. In: Proc. of. ICDE (2013)
17. Lin, J., Katz, B.: Question answering from the web using knowledge annotation and knowledge mining techniques. In: Proceedings of the International Conference on Information and Knowledge Management (CIKM), p. 116. ACM Press (2003)
18. Metzger, S., et al.: QBEES: query by entity examples. In: Proc. of the 22nd ACM Int. Conf. on Information and Knowledge Management (CIKM), pp. 1829–1832. ACM, New York (2013)
19. Metzger, S., Schenkel, R.: S3 K: seeking statement-supporting top-k witnesses. In: Proceedings of the 20th Conference on Information and Knowledge Management (CIKM), Glasgow, Scotland, UK, pp. 37–46 (2011)
20. Qian, L., et al.: Sample-driven schema mapping. In: Proceedings of the International Conference on Management of Data (SIGMOD). ACM Press, Scottsdale (2012)
21. Rosch, E.: Cognitive representations of semantic categories. J. Exp. Psychol. Gen. **104**(3), 192–233 (1975)
22. Sydow, M., et al.: DIVERSUM: towards diversified summarisation of entities in knowledge graphs. In: Proceedings of the International Conference on Data Engineering Workshop (ICDEW), pp. 221–226 (2010)
23. Tesauro, G., et al.: Analysis of Watson's Strategies for Playing Jeopardy! J. Artif. Intell. Res. **21**, 205–251 (2013)
24. Tversky, A.: Features of similarity. Psychol. Rev. **84**(4), 327–352 (1977)
25. Wittgenstein, L.: Philosophical Investigations. The MacMillan Company, New York (1953)
26. Yates, A., Etzioni, O.: Unsupervised methods for determining object and relation synonyms on the web. J. Artif. Intell. Res. **34**, 255–296 (2009)
27. Zhou, M., et al.: Learning to rank from distant supervision: exploiting noisy redundancy for relational entity search. In: Proceedings of the International Conference on Data Engineering (ICDE) (2013)

Tag-Based Paper Retrieval: Minimizing User Effort with Diversity Awareness

Quoc Viet Hung Nguyen$^{(\boxtimes)}$, Son Thanh Do,
Thanh Tam Nguyen, and Karl Aberer

École Polytechnique Fédérale de Lausanne, Lausanne, Switzerland
{quocviethung.nguyen,sonthanh.do,tam.nguyenthanh,karl.aberer}@epfl.ch

Abstract. As the number of scientific papers getting published is likely to soar, most of modern paper management systems (e.g. ScienceWise, Mendeley, CiteULike) support tag-based retrieval. In that, each paper is associated with a set of *tags*, allowing user to search for relevant papers by formulating tag-based queries against the system. One of the most critical issues in tag-based retrieval is that user often has difficulties in precisely formulating his information need. Addressing this issue, our paper tackles the problem of automatically suggesting new tags for user when he formulates a query. The set of tags are selected in such a way that resolves query ambiguity in two aspects: *informativeness* and *diversity*. While the former reduces user effort in finding the desired papers, the latter enhances the variety of information shown to user. Through studying theoretical properties of this problem, we propose a heuristic-based algorithm with several salient performance guarantees. We also demonstrate the efficiency of our approach through extensive experimentation using real-world datasets.

1 Introduction

With the rapid advances in science and technology, large collections of papers have been published every year. To manage such paper collections efficiently, many tag-based systems such as ScienceWise [1], Mendeley [3], and CiteULike [2] have been developed and received spectacular attentions. In these systems, each paper is associated with multiple tags, which often represent the domains it belongs to, the concepts it is related to, or the terms it contains. All associated tags in the repository are essential to enable tag-based retrieval that allows users to represent their search intents by choosing from a suggested list of tags and returns the relevant papers. For example, a user wants to retrieve the paper that he read before, but does not remember its name. He only has partial information about the paper (e.g. its domain and terms). By using the suggested tags, the user can easily figure out what he is exactly searching. As an another example, consider a user searching for papers of relevance to the research proposal he is working on. While the user is eventually interested in one or few papers, at the beginning he may have a lot of search queries in mind; thus a search with useful suggestion of tags is necessary to narrow down the choices. Motivated by these

© Springer International Publishing Switzerland 2015
M. Renz et al. (Eds.): DASFAA 2015, Part I, LNCS 9049, pp. 510–528, 2015.
DOI: 10.1007/978-3-319-18120-2_30

examples, we argue that tags can better help users specify their search intents rather than letting them issue the queries by themselves, especially if they do not know important keywords in the field.

In this work, we study the problem of minimizing user's effort in finding his expected paper(s) through an effective tag suggestion. More precisely, our goal is to minimize the expected number of tags which user need to put into the query. To the best of our knowledge, the closest work to ours is the research on query reformulation. In general, users are often not be able to state their search intents clearly when formulating a search query. The purpose of query reformulation is to provide additional information via query terms for users to reformulate their search intents. The terms are often ranked by different criteria such as co-occurrence patterns [20], latent topic model [5], and via knowledge bases [26]. The main difference between our work and the previous ones is that we rank the tags by their potential information towards reducing user effort.

The problem is challenging for several reasons. First, the dependencies between tags dynamically change according to the search context (i.e. current user query). Hence, it is necessary to develop a suggestion model that takes into account both the currently retrieved papers and the tags which were previously chosen into user query. Second, since the user's intent is not known until he is satisfied with the search, the problem of minimizing user effort cannot be solved in advance. As such, the suggestion needs to look-ahead possible choices by user when he formulates the next query, so that the user can reach the desired paper(s) with minimal (expected) number of querying steps. Third, there is a trade-off between information and diversity of the tags. Although suggesting the tags with high amount of information might improve the chances of reducing the search results quickly, user is also prevented from having a broad view of different domains on top of the suggestion.

Addressing these challenges via a unified model of tag-based paper retrieval, this paper makes the following contributions.

- Section 2: We first provide a generic user interaction scheme for tag-based retrieval. Further, we introduce a formal model of the retrieval process. Then we motivate the requirements of tag suggestion.
- Section 3: We propose a goodness function that quantifies the quality of a tag suggestion solution by combining the two dimensions *informativeness* and *diversity* mentioned above. We also show that our function satisfies a set of useful properties.
- Section 4: We formulate the problem of finding a tag set with maximal goodness value. We prove that this problem is NP-hard. And thus, we propose a greedy algorithm with several salient performance guarantees to approximate the solution.

The remaining sections are structured as follows. Section 5 presents the experimental evaluations. Section 6 summarizes related work, before Section 7 concludes the paper.

2   Tag-Based Paper Retrieval

User Interaction Scheme. Our tag-based paper retrieval framework implements a user interaction scheme as illustrated in Figure 1. Given a list of available tags in (2), a user chooses one of them to put into the query box (1). For this tag-based query, the system returns results as a set of papers in (3). Using tags as a query for retrieving papers helps user to narrow down the scope of research topics and quickly obtain the papers of interest. Moreover, he is also given an overview of all research topics, without spending any effort to rediscover these topics by manually reading the papers. In general, the result quality of tag-based search depends on how well the papers in the repository are annotated by tags. Our work is based on existing paper repositories, such as ScienceWise [1], Mendeley [3], and CiteULike [2], in which each paper is well-annotated with many meaningful tags by the experts in the field.

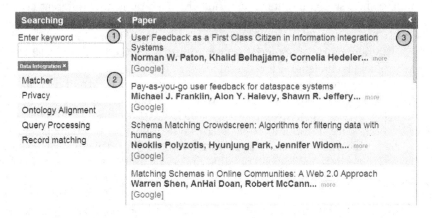

Fig. 1. Tag-based Exploration User Interface

Tag-Based Retrieval. We denote a repository of papers by $\mathcal{D}$, in which each paper $d \in \mathcal{D}$ is annotated by a set of tags T_d. We also denote $\mathcal{T} = \bigcup_{d \in \mathcal{D}} T_d$ as the set of all tags available in the system. Tag-based paper retrieval is the process of finding a paper (or papers) of interest through dynamically suggesting the tags of this paper (or these papers) to user. We assume that user does not know in advance exactly which paper he is looking for and which tags he should choose. Instead, he explores the repository by sequentially selecting the tags suggested by the system until he is satisfied with the search result. More precisely, we model the retrieval process as an interactive process, where in each step three actions are performed: (I) the system suggests a list of tags to user, (II) user chooses one of these suggested tags into the query, and (III) the system updates the set of retrieved papers of relevance to the chosen tags. In general, the retrieval set of papers is reduced after each step and the process ends once the retrieval goal is reached (e.g., user is satisfied). The main focus of our work is to suggest a good set of tags in each step such that the number of retrieved papers is reduced as fast as possible.

Technically, each user interaction step is characterized by a specific index i. Then $Q_i = \langle Q_i^+, Q_i^- \rangle$ denotes the tag-based query formulated by users in step i, where Q_i^+ contains a set of inclusive tags and Q_i^- contains a set of exclusive tags; i.e. $Q_i^+ \cap Q_i^- = \emptyset$, $Q_i^+, Q_i^- \subseteq \mathcal{T}$. For convenience, we denote the size of user query as $|Q_i| = |Q_i^+| + |Q_i^-|$. In the beginning, we have $Q_0 = \langle \emptyset, \emptyset \rangle$. Based on the query Q_{i-1} and the repository $\mathcal{P}$, the system suggests a list of tags $T_i \subseteq \mathcal{T} \setminus (Q_{i-1}^+ \cup Q_{i-1}^-)$, $|T_i| = k$ (action I). Among the suggested tags T_i, user chooses a particular tag t as either inclusion or exclusion into Q_i (action II). That is, $Q_i = \langle Q_{i-1}^+ \cup \{t\}, Q_{i-1}^- \rangle$ or $Q_i = \langle Q_{i-1}^-, Q_{i-1}^- \cup \{t\} \rangle$. In action III, the set of retrieved papers relevant to Q_i is denoted as D_i. A paper is considered relevant to a tag-based query if it contains all inclusive tags and does not contain any exclusive tags; i.e., $D_i = \{d \in \mathcal{D} \mid Q_i^+ \subseteq T_d \wedge Q_i^- \cap T_d = \emptyset\}$. A possible retrieval goal is that there remains only one paper or a set of papers sharing the same tags; i.e. $|D_i| = 1$ or $\forall d, d' \in D_i$, $T_d = T_d'$. Note that for brevity sake, we overload set notation for the suggestion list of tags T_i (or T), meaning that set operators applied to the list are evaluated based on the set of list elements.

Minimal User Effort with Diversity Awareness. In this work, we study the question of how to design a tag suggestion method that minimizes user effort with diversity awareness. In other words, the tags are ranked for suggestion along two dimensions:

- *Informativeness*: The tags are not independent; some tags always appear together in common papers while some others never go along with each other. Therefore, each tag has a distinguished amount of potential information. Suggesting the tags with higher potential information would provide more chances of minimizing the number of user interaction steps for retrieving the papers that truly match user intent.
- *Diversity*: The tags with high potential information might belong to the same domains, since they often have similar dependencies with the others. As such, only focusing on the *informativeness* dimension might prevent user from having a broad view of different domains. Therefore, there is a need of diversifying the list of tags suggested to user. In the absence of explicit knowledge about user intent, increasing the diversity (i.e. the number of domains) of the suggested tags would increase the probability of retrieving some papers that truly match the user's expectation.

To provide a unified quality measurement of tag suggestion, we propose a single comprehensive goodness function that combines both the informativeness degree and the diversity of the tags. The details are given in the next section.

3   Tag Suggestion Quality

In this section, we propose a quality measurement for tag suggestion. Given a user query Q_i and the set of retrieved papers D_i at step i, the quality of a tag set T_i is measured by a goodness function $g : 2^T \to \mathbb{R}$, where 2^T denotes the domain of possible tag sets. For brevity sake, we hereby omit the step index i of the notations (Q_i, T_i, etc.). The goodness value is composed of two notions: informativeness and diversity penalty. While the former reflects the degree of saving user effort of a tag when it is chosen by user, the latter addresses the diversity aspect by penalizing tags that are similar to each other.

3.1   Informativeness

As described above, user expresses his search intent by formulating a query from available tags. Based on the formulated query, our system retrieves a set of relevant papers. However, since user cannot often provide a concrete query that truly describes his search intent, the retrieved papers might not satisfy user expectation. In other words, there are always some degrees of uncertainty about matching user search intent with the retrieved papers. At the beginning of the retrieval process, this uncertainty is high since the query only has few tags and thus the set of retrieved papers is still broadened. During the course of the process, user incrementally refines his search intent by adding more tags into the query. When more tags are added, the set of retrieved papers is narrowed downed. Its uncertainty is continuously reduced until the query is specific enough to reflect user search intent.

Therefore, to minimize user effort (i.e. the number of tags needed to put into the query), we have to suggest the tags with the highest uncertainty reduction. For example, we have two currently retrieved papers p_1 and p_2, which are associated with the tag sets $\{t_1, t_2\}$ and $\{t_1, t_3\}$ respectively. User has three tags t_1, t_2, t_3 as options to formulate the next query. Consider two cases:

(i) User chooses t_1: the set of retrieved papers does not change since both p_1 and p_2 contain t_1. In other words, the uncertainty of the retrieved papers does not change. Suggesting t_1 has no benefit of reducing the uncertainty.

(ii) User chooses t_2 (or t_3): the number of retrieved papers reduces to only one (p_1 or p_2, for both inclusive and exclusive options). In other words, only one set of associated tags remains; i.e. the retrieved papers become certain or there is no uncertainty. Suggesting t_2 reduces the uncertainty.

Based on this observation, we introduce the concept of *informativeness*, which measures the amount of uncertainty reduction of a tag when it is chosen into user query (e.g. informativeness of t_1 is 0, of t_2 is > 0). Suggesting the tags with low informativeness (low information gain) like t_1 requires user to choose many tags, while suggesting the tags with high informativeness (high information gain) like t_2 or t_3 makes the retrieval process faster. Hence, to minimize user effort, we should suggest the tags with high informativeness. In the following, we propose a probabilistic formulation to compute the informativeness of a tag.

Probability of a Tag. As mentioned earlier, we denote D as the set of retrieved papers given a user query Q. The probability that a particular tag t is used in D then becomes:

$$p_t = \frac{|\{d \in D | t \in T_d\}|}{|D|} \tag{1}$$

Recall that T_d is the set of tags annotated with the paper d. The probability distribution of all tags available in the retrieval is thus denoted as $\Omega(D) = \{p_t | t \in \mathcal{T}\}$. Intuitively, tags that appear in all papers have probability of 1; whereas, tags that do not appear in the same papers with the tags in user query have probability of 0.

Uncertainty of Matching User Intent. We compute the uncertainty of matching user intent of a set of retrieved papers D as the Shannon entropy over the probability distribution of the tags:

$$H(D) = - \sum_{p_t \in \Omega(D)} [p_t \log p_t + (1 - p_t) \log(1 - p_t)] \tag{2}$$

where $H(D) \geq 0$. A set of papers in which each paper is annotated with different sets of tags implies a high uncertainty and vice-versa. The more user effort (i.e. more tags are added to the query), the lower value of the uncertainty. As a consequence, the retrieval process ends when the uncertainty reaches zero. Indeed, $H(D) = 0$ means that all the associated tags have probability equal to either 0 or 1. In other words, all the retrieved papers are annotated with an identical set of tags, which converges to user search intent.

Conditional Uncertainty. We now compute the uncertainty of the retrieved papers if user chooses a particular tag. Since the choice of regarding t as inclusive tag or exclusive tag in the query is not known before-hand, the conditional uncertainty should be measured as the expected amount across both cases. Formally, we define the conditional uncertainty w.r.t a particular tag as the entropy conditioned on that tag:

$$H(D|t) = p_t \times H(D^{+t}) + (1 - p_t) \times H(D^{-t}) \tag{3}$$

where $p_t \in \Omega(D)$ is the probability that t is used in D as aforementioned. $D^{+t} = \{d \in D | t \in T_d\}$ and $D^{-t} = \{d \in D | t \notin T_d\}$ are respectively the set of retrieved papers after the inclusiveness and exclusiveness of t in user query.

Informativeness Computation. We compute the informativeness of a tag t following a decision theoretic approach, cf. [28]. More precisely, we measure the amount of uncertainty reduction obtaining by the decision that t is selected; i.e. this reduction is computed as the difference between the ambiguity of the retrieved papers before and after user selects t. Formally, we have:

$$IG(t) = H(D) - H(D|t) \tag{4}$$

With a normalized form ($\in [0, 1]$) as:

$$h(t) = \frac{IG(t)}{\max_{t' \in T_D} IG(t')} \tag{5}$$

Any tag with informativeness equal to zero would have no contribution to reduce the uncertainty. The more informativeness of the tag, the more chances of the uncertainty being improved. In the sense of user effort, we should suggest the high informative tags to reduce the number of user interaction steps. Moreover, it is worth noting that at the beginning of the retrieval process, the query is empty. In this case, we consider $h(t) = 1$ for every tag, implying that all of them are initially considered equal for the suggestion.

3.2 Diversity Penalty

Computing informativeness helps us select more minimum-effort driven tags into the suggestion list. However, the most informative tags often belong to many common papers, resulting in a redundant suggestion. To increase the variety of the suggestion list, we penalize the tags that are similar to each other. The similarity between two tags reflects the amount of information that is shared in their common papers. While the computation of tag similarity is given at the end of this section, we first formulate the notion of diversity penalty. The idea is that the more similar between the tags, the higher amount of penalty is applied. Technically, the diversity penalty of a set of suggested tags T is calculated in terms of the pair-wise similarity between the tags weighted by their informativeness:

$$\phi(T) = \sum_{t,t' \in T} h(t)S(t,t')h(t') \tag{6}$$

where $S(t,t') \in [0,1]$ is the similarity score between any two tags t and t' (the more similar, the higher value). We weight the tag similarity by informativeness of the tags to penalize similar tags with high informativeness more than those with low informativeness. This is motivated by the need to allow more chances of selecting dissimilar tags (despite of lower informativeness) to increase the diversity of the suggestion.

Similarity Computation. The similarity between tags should depend on user query and thus be computed dynamically during the paper retrieval process. For example, we consider two scenarios: (i) user query is "data mining", (ii) user query contains "data mining" and "clustering". In the first scenario, the two tags "DBSCAN" and "k-means" are similar since they are one of many well-known techniques in the field of data mining. In the second scenario, since "DBSCAN" and "k-mean" are the name of two different approaches in the clustering topic, they are dissimilar. Therefore, we propose a query-based probabilistic measurement for tag similarity (or dissimilarity) as follows.

Given a user query Q, the dissimilarity between two tags t_1, t_2 can be measured by the KL divergence [35] of two probability distributions when user chooses either t_1 or t_2:

$$\xi(t_1||t_2) = \sum_t p_t^1 \log \frac{p_t^1}{p_t^2} \tag{7}$$

where $p_t^1 \in \Omega(D_1)$ and $p_t^2 \in \Omega(D_2)$ are the probability of a tag t when either t_1 or t_2 is chosen. That is, $D_1 = \{d \in D | t_1 \in T_d\}$ and $D_2 = \{d \in D | t_2 \in T_d\}$. Since there is no meaningful notion of order in similarity, we use a commonly used symmetric variation:

$$\xi'(t_1,t_2) = \xi(t_1||t_2) + \xi(t_2||t_1) \tag{8}$$

However, the KL divergence still does not take into account the relationship between the two tags and user query Q. For example, for the tag set $T = \{$"data mining"$\}$, we could add $t_1 = $ "shared memory" and $t_2 = $ "message passing" whose meanings are not related to "data mining". To improve this, we weight the KL divergence by the conditional probabilities of the two tags and therefore discount additional tags that have no real relation with the query. As a result, the tag dissimilarity can be defined as:

$$\xi''(t_1, t_2) = p_{t_1} p_{t_2} \xi'(t_1, t_2) \tag{9}$$

where $p_{t_1}, p_{t_2} \in \Omega(D)$. With a further normalization (into $[0,1]$) and inversion of dissimilarity, we have the final form of tag similarity:

$$S(t_1, t_2) = 1 - \frac{\xi''(t_1, t_2)}{\max_{i,j} \xi''(t_i, t_j)} \tag{10}$$

In general, the larger similarity between two given tags, the higher penalty they receive (i.e. the higher chance they are not selected). The aim of diversifying the tag suggestion becomes the selection of the tags that are sufficiently dissimilar with each other.

3.3 Put It Altogether

To balance informativeness and diversity in a top-k selection of tags, we design a quality measure for such a selection. On the one hand, the goodness measure should incorporate given informative scores of tags in a fine-grained level, by weighting the importance of tags unequally. The idea behind is that tags stemming from a large group of similar tags are often associated with popular papers, which implies a high chance to satisfy user information needs. On the other hand, the goodness measure should penalize similar tags. This is motivated by the need to increase diversity in the suggestion.

Our goodness measure for a selection of tags T is based on the overall, weighted informativeness of a selected tag, which is reduced by the diversity penalty of the tags that have also been selected. Intuitively, this approach favors tags from big clusters of similar tags, but penalizes the selection of multiple informative tags that are very similar to each other. Technically, given T_D as the set of tags associated with the current set of retrieved papers D, we define $q(t) = \sum_{t' \in T_D} S(t, t') \cdot h(t')$ as the importance of tag $t \in T_D$. With $w \in \mathbb{R}^+$ as a positive weight parameter, our goodness measure is defined as follows:

$$g(T) = w \sum_{t \in T} q(t)h(t) - \phi(T) \tag{11}$$

The proposed notion of goodness satisfies the following properties [10], whose proofs can be found in the appendix. First, our notion of goodness shows monotonicity; i.e., when adding more tags to an existing selection, the goodness of the overall selection will increase.

Proposition 1 (Monotonicity). *Let T_D be a set of tags associated with a particular paper set D. For any $w \geq 2$, $\forall T_1, T_2 \subseteq T_D, T_1 \cap T_2 = \emptyset$, we have $g(T_1 \cup T_2) \geq g(T_1)$.*

Second, our goodness measure shows submodularity, which refers to the property that marginal gains in goodness start to diminish due to saturation of the objective. That is, the marginal benefit of adding tags to the selection decreases w.r.t. the size of the selection.

Proposition 2 (Submodularity). *Let T_D be a set of tags associated with a particular paper set D. For any $w > 0$, $\forall T \subseteq T_D, t_1, t_2 \in T_D \setminus T$, we have $g(T \cup \{t_1\}) + g(T \cup \{t_2\}) \geq g(T \cup \{t_1, t_2\}) + g(T)$.*

4   Efficient Tag Suggestion

In this section, we first formulate our tag suggestion problem. Due to the NP-hardness of the problem, we then propose a greedy algorithm. After that, we prove various performance guarantees for the proposed algorithm.

4.1   Problem Definition

Using the notion of goodness, we define tag suggestion as an optimization problem. That is, we are interested in finding a selection of top-k tags that maximize the goodness measure:

Problem 1 (Tag Suggestion). *Let T_D be a set of tags associated with the retrieved papers and k be a threshold for the number of tags. Then, the tag suggestion problem is defined to be:*

$$\underset{T \subseteq T_D, |T| = k}{\operatorname{argmax}} \ g(T) \tag{12}$$

Here, selection of the top-k tags is of particular practical relevance for information retrieval, cf., [23]. An appropriate value for k depends on the user and the application context. In general, the problem of tag suggestion turns out to be NP-Complete, whose proof can be found in the appendix.

Theorem 1. *The k-tag suggestion problem is NP-Complete.*

4.2   Algorithm

Given the complexity of the tag suggestion problem, we now present a heuristic algorithm to approximate its optimal solution [10]. The main idea of our algorithm is to start from the null set and add one element at a time, taking at each step the element which increases the goodness of the suggestion list most. To achieve a provably near-optimal solution, our algorithm exploits the two aforementioned properties of the goodness function g, i.e., monotonicity ad submodularity. In essence, the algorithm iteratively expands the selection of tags by adding the tag that maximizes the goodness value, thus it can be bounded. Solving the problem requires k iterations.

Algorithm 1. Heuristic algorithm for tag suggestion.

---

input : A set of tags T_D associated with the retrieved papers, a weight factor $w \geq 2$, and a threshold for the number of tags k.

output: A selection of tags $T^* = \langle t_1, \ldots, t_k \rangle, t_i \in T_D, 1 \leq i \leq k$.

1 $T^* \leftarrow \emptyset$;
  // Compute ranking score for each tag
2 Let $r : T_D \rightarrow \mathbb{R}, r(t) \mapsto w \cdot h(t) \cdot \sum_{t' \in T_D} S(t, t') h(t')$;
3 **while** $|T^*| < k$ **do**
4       $t_m \leftarrow \text{argmax}_{t \in T_D, t \notin T^*} r(t)$;
5       $T^* \leftarrow T^* \cap \{t_m\}$;
       // Update ranking score for the remaining tags
6       $r' \leftarrow r$;
7   Let $r : T_D \rightarrow \mathbb{R}, r(t) \mapsto r'(t) - 2 \cdot h(t_m) \cdot S(t, t_m) \cdot h(t)$;
8 **return** T^*

The details of our heuristic are given in Algorithm 1. It takes a set of tags T_D, a weight factor w, and a threshold for the number of tags k as input and returns a selection T^* of k tags. We begin by computing a ranking score for each tag $t \in T_D$ that is based on the weight factor, the tag informativeness, and the tag importance (line 2). In the actual greedy selection step, we select k tags. In each iteration, we add the tag with the highest ranking score (lines 4 and 5), before the ranking score is updated for the remaining tags (line 7). The latter avoids re-computation of the ranking scores from scratch in each iteration. As mentioned above, we overload set notation for the suggestion list of tags T^* for brevity sake. When presented in user interface, the tags are listed top-down in the decreasing order of ranking score (i.e. from left to right of the sequence representation).

4.3 Algorithm Analysis

The proposed algorithm shows several desirable properties. First, the approximation error is bounded.

Guarantee 1 (Near-Optimality). *Algorithm 1 is a (1- 1/e)-approximation for the tag suggestion problem.*

Proof. For any monotone, submodular function f with $f(\emptyset) = 0$, it is known that an iterative algorithm selecting the element e with maximal value of $f(I \cup \{e\}) - f(I)$ with I as the set of elements selected so far has a performance guarantee of $(1 - 1/e) \approx 0.63$ [24]. This result is applicable to line 1, since our goodness function g is monotonic (proposition 1) and submodular (proposition 2), it holds $g(\emptyset) = 0$ (eq. (11)), and the ranking score is defined as $r(t) = g(T^* \cup \{t\}) - g(T^*)$ (lines 2 and 7).

Next, we consider the complexity of our heuristic.

Guarantee 2 (Complexity). *The time complexity and the space complexity of Algorithm 1 are $\mathcal{O}(|T_D|^2 + k|T_D|)$ and $\mathcal{O}(|T_D|)$, respectively.*

Proof. Time complexity: The quadratic term $|T_D|^2$ stems from the computation of the ranking score. The linear term $k|T_D|$ is explained by k iterations, in each of which we iterate over all remaining tags, for selection of t_{max} and for updating the ranking score. Space complexity: Storing tag similarities requires $\frac{|T_D||T_D-1|}{2}$ space since S is symmetric and $S(t,t)$ is fixed.

Further, our algorithm shows stability in the selection, which is important to support multi-resolution (i.e. in cases user wants to see more tags in the suggestion list). For example, if a user is first presented with the top-10 tags, but then extends the suggestion list to the top-20, the expectation is clearly that the top-10 remain unchanged.

Guarantee 3 (Stability). *For T^* as returned by line 1, let $T^*_{k_1} = \langle t_1, \ldots, t_{k_1} \rangle$, $T^*_{k_2} = \langle t'_1, \ldots, t'_{k_2} \rangle$ be selections with $t_i \in T^*$, $1 \leq i \leq k_1$, $t'_j \in T^*$, $1 \leq j \leq k_2$, and $0 < k_1 \leq k_2$. Then, it holds that $t_i = t'_i$ for $1 \leq i \leq k_1$.*

Proof. In Algorithm 1, the construction of T^* is performed stepwise and elements are never removed from T^*. Moreover, the selection is deterministic: we always add a new tag with the highest ranking score (line 4). Thus, the larger selection sequence comprises the smaller selection sequence as a prefix.

5   Experiments

This section presents a comprehensive experimental evaluation to verify the effectiveness of our tag-based paper retrieval framework. In particular, we first discuss the experimental setup including datasets and evaluation measures. Then, we proceed to report the following experiments: (i) evaluations on informativeness, and (ii) evaluations on diversity. The results highlight that the proposed tag suggestion algorithm performs well in terms of both user effort and diversity aspect.

5.1   Experimental Settings

Dataset. Our prototype is developed on top of the ScienceWise platform since it supports API to retrieve data and has a rich tag collection. The ScienceWise's data contains 16725 scientific papers and 15083 tags. Each paper has 70 tags in average. The ScienceWise platform itself has not supported tag suggestion in the search results yet.

Evaluation Measures. For comparative evaluation, we study the following measures.

Domain Coverage. This metric measures the diversity of a top-k list of tags in terms of coverage of domains. It indicates the proportion of possible domains (which might be of interest to user) the tag list can capture. Formally, we run k-meloids clustering to divide the set of all available tags T_D into k clusters, based on the tag similarity proposed in Section 3.2. The domain $dom(t)$ of a tag t is the cluster it belongs to. The domain coverage ($\in [0,1]$) of top-k tag suggestion T^* is defined as the number of domains of its tags over the total number of domains:

$$DC(T^*) = \frac{|\bigcup_{t \in T^*} dom(t)|}{k} \tag{13}$$

Normalized Informativeness. This metric measures the informativeness of the tag suggestion list with respect to the top-k tags with highest informativeness; i.e., it indicates how well the informativeness of the tags is preserved when diversity is taken into account. Formally, the normalized informativeness ($\in [0,1]$) of top-k tag suggestion T^* from the set of candidate tags T_D is defined as the sum of their informativeness scores over the sum of the k highest informativeness scores:

$$nH(T_D, T^*) = \frac{\sum_{t \in T^*} h(t)}{\max_{T \subseteq T_D, |T| = |T^*|} \sum_{t \in T} h(t)} \tag{14}$$

User Effort. To quantify the amount of time user spends to retrieve the desired papers, we compute the user effort as the number of interaction steps of the retrieval process described in Section 2. Each interactive step is counted when user selects a new tag to be added into the query. Formally, we have:

$$E = |Q^+| + |Q^-| \tag{15}$$

5.2 Evaluations on Informativeness

The goal of this evaluation is to verify the soundness of the proposed informativeness function of a tag. To this end, we will study the informativeness in two aspects: (i) paper amount reduction – how many retrieved papers are reduced after user chooses a tag, and (ii) user effort – how many tags user need to choose in the retrieval.

Informativeness vs. Paper Amount Reduction. In this experiment, we only consider one user interaction step of the retrieval process. We assume that user is interested in a particular paper, which is associated with a set of tags. The user query is simulated by randomly choosing some of these tags. Given a simulated query, we rank the tags by the decreasing order of informativeness. For each of the top-10 tags, we put it into the query as inclusive if it is contained in the tag set and exclusive otherwise. Then we retrieve the papers of the new query and measure the amount reduction of retrieved papers.

Figure 2 and Figure 3 illustrate the results for different query sizes (size = 0 and size = 10). The report numbers are averaged over 100 different targeted papers (these papers have more than 15 tags). The X-axis is the rank of tags in terms of informativeness. The Y-axis is the relative reduction of the amount of retrieved papers. An interesting finding is that the higher rank of tags, the more papers are reduced. For example, the tag with highest informativeness (rank 1) gives about 50% reduction, whereas the tag with lowest informativeness (rank 10) gives less than 5% reduction. This supports the soundness of our

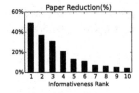 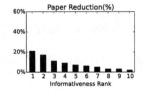 

Fig. 2. Query Size = 0   **Fig. 3.** Query Size = 10   **Fig. 4.** Informativeness vs. User Effort

informativeness function in capturing user effort. Another noticeable observation is that as more tags are selected into the query, the number reduction of retrieved papers is smaller. For example, with query size = 0, the reduction of the rank-1 tag is about 50%, while this number is only about 20% with query size = 10. This is reasonable because after each user interaction step, the set of retrieved papers and the set of their associated tags are narrowed down. As such, the percentage of papers sharing the common tags is higher and selecting these tags would return mostly the same papers.

Informativeness vs. User Effort. In this experiment, we simulate the whole retrieval process. Like the previous experiment, we assume that user is interested in a particular paper, which is associated with a set of tags. At the beginning, we initialize user query by randomly choosing one of these tags. In each interaction step, user receives a suggested tag and put it into the query (the tag is regarded as inclusive or exclusive based on the target paper). The process stops when only one paper remains or all the remaining papers share the same set of tags. Three tag suggestion strategies are studied: (i) *1st Rank* – suggest the tag with highest informativeness, (ii) *3rd Rank* – suggest the tag of rank-3 in the decreasing order of informativeness, (iii) *Random* – suggests a random tag to user.

Figure 4 depicts the result, which is averaged over 100 simulations (i.e. 100 different target paper). The X-axis is the tag suggestion strategy. The Y-axis is the percentage of user effort over the number of tags contained in the target paper. A key finding is that the *Random* strategy incurs most user effort (95.73%). This can be explained by the fact that with *Random* strategy, user has to go through many redundant tags, which do not (or rarely) reduce the number of papers. Another interesting observation is that the more informativeness of the tag, the more user effort is reduced. Indeed, the *1st Rank* strategy takes the least user effort (67.38%), whereas the *3rd Rank* strategy requires more user effort (74.94%). This supports that user effort can be reflected through the informativeness of a tag. Suggesting the tag with the highest informativeness does indeed reduce user effort the most.

5.3   Evaluations on Diversity

In this experiment, we would like to verify the soundness of the diversity aspect of tag suggestion. More precisely, we will compare two tag suggestion strategies: (1) with diversity: the suggestion list of tags is computed by the proposed algorithm,

(2) without diversity: the suggestion list of tags is computed by returning the top tags with highest informativeness values. For the strategy 1, we randomly set the tunning parameter w (trade-off between informativeness and diversity) according to uniform distributions $\mathscr{U}(0,1)$ and $\mathscr{U}(1,2)$, respectively. The final numbers are computed as the average over 100 runs. We vary the number of suggested tags k from 5 to 55, and compare the two strategies according to different aspects as follows.

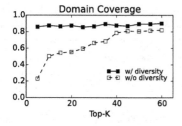

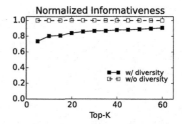

Fig. 5. Diversity   **Fig. 6.** Informativeness

Figure 5 illustrates the results on the diversity aspect. A key finding is that strategy 1 is always better than strategy 2 in terms of domain coverage. For example, while the domain coverage of strategy 1 is always greater than 0.8, the domain coverage of strategy 2 is only about 0.2 with k = 5. This supports the fact that our proposed algorithm performs well in producing a diverse list of suggested tag. Another noticeable observation is that the difference of domain coverage between the two strategies is smaller when k increases. For instance, with $k = 60$, the domain coverage of strategy 2 is nearly 0.8. This is because when k is higher, strategy 2 will include more tags with lower informativeness, up to the point that all tags in the list are dissimilar enough among themselves, resulting in high domain coverage.

Figure 6 presents the result on the informativeness aspect. By definition, the normalized informativeness of strategy 2 is always equal to 1. An interesting finding is that the normalized informativeness of strategy 1 is not much lower than strategy 2 in comparison with the domain coverage. For example, with $k = 5$, the difference of domain coverage is more than 0.6 while the difference of normalized informativeness is less than 0.25. This implies that in spite of producing a diverse list of suggested tags, our proposed algorithm still keeps most of the informativeness amount of the tags. In other words, the tags with high informativeness values are preserved, which goes beyond the trade-off between diversity and informativeness. Another important observation is that the normalized informativeness of strategy 1 increases when the suggestion size is higher. This is reasonable since the tags that are diverse often have different values of informativeness. When the number of suggested tags increases, our algorithm will add both the tags with high informativeness and the tags with low informativeness, up to the point that the two strategies share most of common tags with each other.

6 Related Work

Our work aims to reduce user effort for retrieving relevant papers in tag-based paper management platforms. It is mainly related to tag-based retrieval, query suggestion, and diversification, which are briefly reviewed as follows.

Tag-based Retrieval. In the last decades, there has been an increasing development of tag-based retrieval systems, which allow to add tags (manually or automatically) to existing resources such as images and videos. The research efforts in tag-based retrieval can be broadly categorized into three types, namely *annotating, ranking,* and *presenting. Annotating* involves determining the set of tags best describing a resource [16,34–36,39]. *Ranking* aims to compute a relevance score between a query and a resource [13,19,22]. *Presenting* focuses on improving user satisfaction by effectively presenting the tags or search results to users [18,33,37].

Tag-based retrieval for scientific papers is a distinguished and recognized direction. This is because using textual search on research articles has some limitations, for example, full-text access is not always available [6] and OCR errors are inherently found [30]. Moreover, different from other resources (web pages, image, videos), scientific papers are associated with much more tags since there is a lot of scientific concepts across different domains (e.g. in our dataset, each paper is associated with 70 tags in average). This distinct characteristic opens up an opportunity to design more complex mechanisms by exploiting potential information of the tag collection. A wide range of tag-based paper retrieval systems have been developed with reliable and high-quality tags such as ScienceWise [1], Mendeley [3], and CiteULike [2]. Moreover, there is also a considerable number of research outcomes on this direction, including tag-based search engine [14], semantic-based framework [27], and collaborative tagging [25].

Query Suggestion. Query suggestion (a.k.a. query reformulation, query expansion, query completion) is a supportive method to improve search productivity. In general, users are often not be able to state their search intents clearly when formulating a search query. The purpose of query suggestion is to provide additional information for users to help them reformulate their queries. In the literature, query suggestion has been studied in different contexts. In [5], the authors exploited query log of the search engine to suggest new query terms for the current user query. Instead of using query log, the authors of [20] made use of existing keywords provided by social annotation services to generate and rank the new queries for suggestion. In the same line, the authors of [26] extracted candidate query terms from existing Wikipedia articles related to user query. In the context of image search, the work in [37] uses representative images for user to look ahead the search results of query terms.

Diversification. The diversification problem has been long acknowledged in information retrieval [9,18]. It aims to improve user satisfaction by providing a diverse view of information, thereby increasing the probability of returning some information that truly matches the user's expectation. Various applications that have benefited from diversification include sentiment analysis [4], web search [12], database search [7], large-scale visualization [31], social network [40] and recommender systems [11]. In our case, since users cannot often precisely and exhaustively describe their queries, increasing diversity of tag-query suggestion will provide users more chances to find the desired papers quickly. We propose a *function-based* approach [17] for tag diversification, which is "less heuristic" than the *threshold-based* [32] and the *graph-based* [38] approaches.

To summarize, our work differs from previous research in the following aspects: (1) we do not aim to provide an "auto-complete" feature like the previous works. Rather, we study a different aspect of query suggestion with the goal of minimizing user's effort in retrieving the information that truly matches his search intent. (2) we jointly consider user effort minimization and diversification by designing a comprehensive goodness function, which guides the on-the-fly computation of suggested tags according to the current user query. Moreover, it is worth noting that although the proposed algorithm is demonstrated on the context of paper retrieval, it can be applied for other domains such as document retrieval and image retrieval. It should be also emphasized that our work is not about tagging online contents [16] (i.e. *Annotating*. Instead, we leverage the generated tags to better support the retrieval of these contents (which cannot be accessed via textual search).

7 Conclusions and Future Work

This work proposes a novel approach that enables tag-based retrieval in online archives of scientific papers. To make these archives searchable, each paper is associated with a set of pre-defined descriptive keywords, so-called tags. We study the problem of how to efficient suggest new tags for user to formulate his query intent. The goal is to not only reduce the efforts of user in reaching his search intent, but also increase the diversity of the suggested tags. In particular, we define the notion of goodness measure that captures both the informativeness and diversity aspects of the tags. Based on this measure, we formulate the tag suggestion problem as the identification of a set of k tags with maximal goodness value. Through studying theoretical properties of this problem, we propose a heuristic-based algorithm with several salient performance guarantees. Finally, we present a comprehensive experimental evaluation indicating that the approach allows for effective and efficient retrieval of real-world scientific data.

Our work opens up several future research directions. First, the proposed quality measurement can be used to evaluate existing query suggestion methods, especially the user-effort aspect. Second, we can investigate other dimensions to be considered in the quality measurement. Third, this paper focuses on searching for scientific papers, yet, our tag-based retrieval framework (in particular the tag suggestion algorithm) can be applied for a variety of domains, such as business documents and social medias. Fourth, although the suggested tags are presented as a list in our context, we can also study other presentation options such as hierarchical and categorical-like structures. Fifth, our work could be tailored to take into account the meta-data (e.g. citation [21]), if available, of the scientific papers to further refine their relevance (not only based on tags). When each paper has multiple search dimensions, we can develop more sophisticated cost models [15] as well. Moreover, one can also improve the retrieval performance by relevance feedback [29], which is out of the scope of this paper.

Acknowledgments. The research has received funding from the EU-FP7 EINS project (grant number 288021) and the ScienceWise project.

Appendix - Proofs

NP-Complete. We prove Theorem 1 by reduction to the Densest k-Subgraph problem, which is known to be NP-Complete [8,10]. Let $G = (V, E)$ be an undirected graph with vertices V and edges E. Let W be the $|V| \times |V|$ binary connectivity matrix (symmetric), i.e., $W_{i,j} = 1$ if $\{i, j\} \in E$, and $W_{i,j} = 0$ otherwise. Then, the Densest k-Subgraph problem requires identifying a subgraph of k vertices with a maximal number of edges:

$$\operatorname*{argmax}_{\hat{V} \subseteq V, |\hat{V}| = k} \sum_{i,j \in \hat{V}} W_{i,j}$$

which is equivalent to

$$\operatorname*{argmax}_{I = (V \setminus \hat{V}), |\hat{V}| = k} 2 \sum_{i \in \hat{V}, j \in I} W'_{i,j} + \sum_{i,j \in I} W'_{i,j} \tag{16}$$

where $W'_{i,j} = 1 - W_{i,j}$. Now we will show that eq. (16) can be viewed as an instance of the optimization problem in eq. (12). To this end, let all informative scores be one ($h(t) = 1$ for all $t \in T_D$) and choose $w = 2$. Then, our objective function $g(T)$ becomes:

$$g(T) = 2 \sum_{t \in T} q(t) - \sum_{t_1, t_2 \in T} S(t_1, t_2) = 2 \sum_{t_1 \in T} \sum_{t_2 \in T_D} S(t_1, t_2) - 2 \sum_{t_1, t_2 \in T} S(t_1, t_2) + \sum_{t_1, t_2 \in T} S(t_1, t_2)$$

$$= 2 \sum_{t_1 \in (T_D \setminus T)} \sum_{t_2 \in T} S(t_1, t_2) + \sum_{t_1, t_2 \in T} S(t_1, t_2) \tag{17}$$

The latter is equivalent to the objective function in eq. (16), so that selection of k tags corresponds to the finding the densest subgraph of $(|V| - k)$ nodes.

Monotonicity. With $w \geq 2$, we have:

$$g(T_1 \cup T_2) - g(T_1) = w \sum_{t \in T_2} q(t)h(t) - \left(\sum_{t \in T_2, t' \in T_1} h(t)S(t, t')h(t') + \sum_{t \in T_1, t' \in T_2} h(t)S(t, t')h(t') \right.$$

$$+ \sum_{t, t' \in T_2} h(t)S(t, t')h(t')) = w \sum_{t \in T_2} h(t) \sum_{t' \in T_D} S(t, t')h(t') - (2 \sum_{t \in T_1, t' \in T_2} h(t)S(t, t')h(t')$$

$$+ \sum_{t, t' \in T_2} h(t)S(t, t')h(t')) \geq 2 \sum_{t \in T_2} h(t) \sum_{t' \in T_D} S(t, t')h(t') - (2 \sum_{t \in T_1, t' \in T_2} h(t)S(t, t')h(t')$$

$$+ \sum_{t, t' \in T_2} h(t)S(t, t')h(t')) = 2 \sum_{t \in T_2} (\sum_{t' \in T_D} S(t, t')h(t') - \sum_{t' \in T_1 \cup T_2} S(t, t')h(t'))$$

$$= 2 \sum_{t \in T_2} \sum_{t' \notin T_1 \cup T_2} S(t, t')h(t') \geq 0$$

which completes the proof of monotonicity.

Submodularity. From eq. (11), we have:

$$g(T \cup \{x\}) - g(T) = wq(x)h(x) - 2h(x) \sum_{t \in T} S(x,t)h(t) + h^2(x) \qquad (18)$$

Following eq. (18), we have:

$$g(T \cup \{t_1\}) + g(T \cup \{t_2\}) \geq g(T \cup \{t_1, t_2\}) + g(T) \Leftrightarrow g(T \cup \{t_1\}) - g(T) \geq g(T \cup \{t_2\} \cup \{t_1\}) - g(T \cup \{t_2\})$$

$$\Leftrightarrow wq(t_1)h(t_1) - 2h(t_1) \sum_{t \in T} h(t)S(t, t_1) + h^2(t_1) \geq wq(t_1)h(t_1) - 2h(t_1) \sum_{t \in T \cup \{t_2\}} h(t)S(t, t_1) + h^2(t_1)$$

$$\Leftrightarrow 2h(t_1)h(t_2)S(t_1, t_2) \geq 0$$

which completes the proof of submodularity.

References

1. http://sciencewise.info
2. http://www.citeulike.org/
3. http://www.mendeley.com/
4. Aktolga, E., Allan, J.: Sentiment diversification with different biases. In: SIGIR, pp. 593–602 (2013)
5. Bing, L., Lam, W., Wong, T.L.: Using query log and social tagging to refine queries based on latent topics. In: CIKM, pp. 583–592 (2011)
6. Cohen, A.M., Hersh, W.R.: A survey of current work in biomedical text mining. Briefings in Bioinformatics, 57–71 (2005)
7. Drosou, M., Pitoura, E.: Disc diversity: result diversification based on dissimilarity and coverage. In: PVLDB, pp. 13–24 (2012)
8. Feige, U., Peleg, D., Kortsarz, G.: The dense k-subgraph problem. Algorithmica, 410–421 (2001)
9. Goffman, W.: A searching procedure for information retrieval. ISR, 73–78 (1964)
10. He, J., Tong, H., Mei, Q., Szymanski, B.: Gender: a generic diversified ranking algorithm. In: NIPS, pp. 1142–1150 (2012)
11. Hurley, N., Zhang, M.: Novelty and diversity in top-n recommendation - analysis and evaluation. TOIT, 1–30 (2011)
12. Iwata, M., Sakai, T., Yamamoto, T., Chen, Y., Liu, Y., Wen, J.R., Nishio, S.: Aspectiles: tile-based visualization of diversified web search results. In: SIGIR, pp. 85–94 (2012)
13. Jain, V., Varma, M.: Learning to re-rank: query-dependent image re-ranking using click data. In: WWW, pp. 277–286 (2011)
14. Jomsri, P., Sanguansintukul, S., Choochaiwattana, W.: A comparison of search engine using "tag title and abstract" with citeulike - an initial evaluation. In: ICITST, pp. 1–5 (2009)
15. Kashyap, A., Hristidis, V., Petropoulos, M.: Facetor: cost-driven exploration of faceted query results. In: CIKM, pp. 719–728 (2010)
16. Kim, J.W., Candan, K.S., Tatemura, J.: Organization and tagging of blog and news entries based on content reuse. J. Sign. Process. Syst., 407–421 (2010)
17. Küçüktunç, O., Saule, E., Kaya, K., Çatalyürek, U.V.: Diversified recommendation on graphs: pitfalls, measures, and algorithms. In: WWW, pp. 715–726 (2013)
18. van Leuken, R.H., Garcia, L., Olivares, X., van Zwol, R.: Visual diversification of image search results. In: WWW, pp. 341–350 (2009)

19. Li, X., Snoek, C.G.M., Worring, M.: Learning social tag relevance by neighbor voting. In: TMM, pp. 1310–1322 (2009)
20. Lin, Y., Lin, H., Jin, S., Ye, Z.: Social annotation in query expansion: a machine learning approach. In: SIGIR, pp. 405–414 (2011)
21. MacRoberts, M.H., MacRoberts, B.R.: Problems of citation analysis: a critical review. JASIST, 342–349 (1989)
22. Maniu, S., Cautis, B.: Network-aware search in social tagging applications: instance optimality versus efficiency. In: CIKM, pp. 939–948 (2013)
23. Manning, C.D., Raghavan, P., Schütze, H.: Introduction to Information Retrieval, vol. 1. Cambridge University Press (2008)
24. Nemhauser, G., Wolsey, L., Fisher, M.: An analysis of approximations for maximizing submodular set functions-i. MP, 265–294 (1978)
25. Noël, S., Beale, R.: Sharing vocabularies: tag usage in citeulike. In: BCS-HCI, pp. 71–74 (2008)
26. Oliveira, V., Gomes, G., Belém, F., Brandão, W., Almeida, J., Ziviani, N., Gonçalves, M.: Automatic query expansion based on tag recommendation. In: CIKM, pp. 1985–1989 (2012)
27. Prokofyev, R., Boyarsky, A., Ruchayskiy, O., Aberer, K., Demartini, G., Cudré-Mauroux, P.: Tag recommendation for large-scale ontology-based information systems. In: Cudré-Mauroux, P., Heflin, J., Sirin, E., Tudorache, T., Euzenat, J., Hauswirth, M., Parreira, J.X., Hendler, J., Schreiber, G., Bernstein, A., Blomqvist, E. (eds.) ISWC 2012, Part II. LNCS, vol. 7650, pp. 325–336. Springer, Heidelberg (2012)
28. Russell, S.J., Norvig, P., Canny, J.F., Malik, J.M., Edwards, D.D.: Artificial Intelligence: A Modern Approach, vol. 74. Prentice Hall Englewood Cliffs (1995)
29. Salton, G., Buckley, C.: Improving retrieval performance by relevance feedback. JASIST (1997)
30. Sebastiani, F.: Machine learning in automated text categorization. CSUR, 1–47 (2002)
31. Skoutas, D., Alrifai, M.: Tag clouds revisited. In: CIKM, pp. 221–230 (2011)
32. Vieira, M.R., Razente, H.L., Barioni, M.C.N., Hadjieleftheriou, M., Srivastava, D., Traina, C., Tsotras, V.J.: On query result diversification. In: ICDE, pp. 1163–1174 (2011)
33. Wang, M., Yang, K., Hua, X.S., Zhang, H.J.: Towards a relevant and diverse search of social images. In: TMM, pp. 829–842 (2010)
34. Wang, Q., Ruan, L., Zhang, Z., Si, L.: Learning compact hashing codes for efficient tag completion and prediction. In: CIKM, pp. 1789–1794 (2013)
35. Weinberger, K.Q., Slaney, M., Van Zwol, R.: Resolving tag ambiguity. In: MM, pp. 111–120 (2008)
36. Xie, L., He, X.: Picture tags and world knowledge: learning tag relations from visual semantic sources. In: MM, pp. 967–976 (2013)
37. Zha, Z.J., Yang, L., Mei, T., Wang, M., Wang, Z.: Visual query suggestion. In: MM, pp. 15–24 (2009)
38. Zhang, B., Li, H., Liu, Y., Ji, L., Xi, W., Fan, W., Chen, Z., Ma, W.Y.: Improving web search results using affinity graph. In: SIGIR, pp. 504–511 (2005)
39. Zhu, G., Yan, S., Ma, Y.: Image tag refinement towards low-rank, content-tag prior and error sparsity. In: MM, pp. 461–470 (2010)
40. Zhu, X., Goldberg, A.B., Van Gael, J., Andrzejewski, D.: Improving diversity in ranking using absorbing random walks. In: HLT-NAACL, pp. 97–104 (2007)

Feedback Model for Microblog Retrieval

Ziqi Wang and Ming Zhang[✉]

School of EECS, Peking University, Beijing 100871, China
wangziqi@pku.edu.cn, mzhang@net.pku.edu.cn

Abstract. Information searching in microblog services has become common and necessary for social networking. However, microblog retrieval is particularly challenging compared to web page retrieval because of serious vocabulary mismatch problem and non-uniform temporal distribution of relevant documents. In this paper, we propose a feedback model, which includes a feedback language model and a query expansion model considering both lexical expansions and temporal expansions. Experiments on TREC data sets have shown that our proposed model improves search effectiveness over standard baselines, lexical only expansion model and temporal only retrieval model.

Keywords: Microblog retrieval · Feedback model · Query expansion · Pseudo-relevance feedback

1 Introduction

Microblog services, such as Twitter, have become new sources of information. To get relevant information of trends or breaking news, users submit queries on the microblog sites instead of web search engines. However, microblog retrieval differs from general information retrieval (IR) due to the following reasons: (1) Tweets are short. Vocabulary mismatch problem is extremely significant in microblog retrieval. (2) Time plays an important role. Temporal distribution of relevance documents is not uniform. In this paper, we propose a novel feedback model incorporating a feedback language model and a query expansion model to tackle these challenges.

Query document vocabulary mismatch happens when user and authors of documents use different terms to represent the same concept. Vocabulary mismatch has always been a critical challenge in information retrieval. For web search, documents are relatively long and authors usually use keywords repeatedly to describe the topic. Term frequency is heavily relied on in most of retrieval models such as query likelihood model. However, in microblog retrieval, tweets have fewer terms (no more than 140 characters). Most terms, especially key concepts, only appear once in documents, which makes statistical method less reliable. Vocabulary mismatch problem becomes worse in microblog retrieval.

Since temporal distribution of relevant documents in microblog retrieval is not uniform, some work has been focusing on incorporating time information

© Springer International Publishing Switzerland 2015
M. Renz et al. (Eds.): DASFAA 2015, Part I, LNCS 9049, pp. 529–544, 2015.
DOI: 10.1007/978-3-319-18120-2_31

into the retrieval model. Many researchers proposed various methods of using temporal information to improve term selection in query expansion model [1] [2] [3]. Temporal evidence has also been explored under the language model framework to improve document ranking [4] [5]. It is very important to make use of temporal information.

In this paper, we propose a feedback model for microblog retrieval. Our model includes a feedback language model and a query expansion model. The feedback language model is built on the search results from the initial retrieval. Document relevance scores are adjusted based on the feedback language model. The query expansion model expands the query by using both lexical expansions and temporal expansions.

We evaluated the proposed model using the TREC 2011 and 2012 Microblog data set. The experiment results have shown that our proposed feedback language model outperforms the query likelihood baseline and our proposed query expansion model performs better than the relevance model. Overall, the proposed feedback model improves microblog retrieval effectiveness over previously proposed baselines.

The rest of the paper is organized as follows. In Section 2, we review related work including temporal information retrieval, microblog retrieval and pseudo-relevance feedback. In Section 3, we present motivation of this study. Our proposed feedback language model, query expansion model and feedback model are presented in Section 4. Experiments and analysis of results are shown in Section 5. Finally, we conclude this paper in Section 6.

2   Related Work

Related work can be found in three areas. The first is temporal information retrieval. Time plays a very important role in microblog retrieval. The second is general microblog retrieval. There has been some research focusing on other aspects besides using temporal information to improve search performance. The third is pseudo-relevance feedback via query expansion.

2.1   Temporal Information Retrieval

Previous researches incorporate recency into retrieval. Newly published documents are assumed to have a larger probability to be relevant than older documents. Li and Croft proposed a time-based language model by adding document prior based on recency [4]. Efron and Golovchinsky proposed an extension by using query-specific information to estimate parameters [1]. Massoudi et al. expanded queries by using terms in the most recent documents [6].

Instead of focusing on recency queries, some works have been trying to deal with more general time-sensitive queries. Jones and Diaz proposed a temporal query model and an approach to distinguish different types of temporal queries [7]. Dakka et al. proposed a general framework to combine lexical and temporal evidence together [5]. Liang et al. detected burst and aggregated ranking results from different retrieval methods [8].

Pseudo-relevance feedback via query expansion has been widely used in temporal retrieval. Liang et al. proposed a two-stage pseudo-relevance feedback query expansion method [3]. Whiting et al. proposed a pseudo-relevance feedback model using the correlation between temporal profiles of n-grams obtained from query and feedback documents [9]. Whiting et al. built a graph using temporal and TF evidence and selected n-gram using PageRank [10]. Keikha et al. proposed a time-based relevance model using temporal distribution of retweets [2]. Miyanishi extended latent concept expansion model based on the temporal relevance model for query expansion [11]. Metzler et al. proposed a temporal query expansion model for event retrieval based on temporal co-occurrence of terms in a timespan [12].

The major difference between previous work and our work is how we use temporal information. Most previous work used temporal information to select lexical expansions. Our proposed model identifies bursts and conducts temporal expansions for the query. Temporal expansions and lexical expansions are then combined together in a query expansion model. Besides, our proposed model can deal with both temporally unambiguous and ambiguous queries while previous temporal models could only handle temporally unambiguous queries.

2.2   Microblog Retrieval

Query document vocabulary mismatch problem is one of the critical challenges of information retrieval, especially microblog retrieval due to the short length of documents. Methods such as query expansion and document expansion have been studied to address the query document vocabulary mismatch problem.

For query expansion methods, besides using temporal information as we discussed above, external sources can also be useful. Chen et al. used external knowledge including Google and Wikipedia to conduct query expansion [13]. Bandyopadhyay et al. proposed a query expansion model using Google API and BBC site [14].

Efforts have also been made to explore document expansion methods. Efron et al. proposed an aggressive document expansion based on pseudo-relevance feedback [15]. Han et al. proposed a document expansion by using nearest neighbors of documents [16].

One of the most important differences between microblog retrieval and web page retrieval is that many tweets are low-quality and contain a lot of noise. Choi et al. proposed a quality model to demote uninformative content [17]. Gurini and Gasparetti proposed an effective real time ranking algorithm using noise features [18].

2.3   Pseudo-Relevance Feedback via Query Expansion

Pseudo-relevance feedback techniques, represented as the relevance model [19], have been widely studied in information retrieval. Relevance model has been improved by some researchers. Lv and Zhai proposed a model that optimizes the balance of the query and feedback information, and automatically learns the parameters of relevance model [20]. Tao and Zhai proposed a probabilistic mixture model using different parameters to each document and integrating the original query with feedback documents [21], and then this model was modified by Dillon and Collins-Thompson [22].

There have been work on term selection and document selection of relevance model. Cao et al. used SVM to classify good and bad terms [23]. Lv and Zhai extended relevance model to exploit term positions in the feedback documents [24]. Raman et al. chose terms that discriminate pseudo-relevant documents from pseudo-irrelevant documents [25]. Huang et al. proposed an approach to determine the optimal number of feedback documents with clarity score and cumulative gain [26]. He and Ounis used classification model to select good documents [27].

3   Motivation

3.1   Language Model in Microblog Retrieval

A statistical language model assigns a probability to a sequence of words by means of a probability distribution. In information retrieval, language model is used in the query likelihood model. Each document in the collection is represented as a language model. Documents are ranked based on the probability of query $Q = q_1, q_2, \ldots, q_n$ given document's language model $P(Q \mid M_D)$. Since authors usually use topic words repeatedly, keywords of document is expected to have large probabilities in the corresponding language model. The unigram language model is commonly used to achieve this.

$$P(Q \mid M_D) = \prod_{i=1}^{n} P(q_i \mid M_D) \tag{1}$$

$$P(q_i \mid M_D) = \frac{f_{q_i,D} + \mu \frac{c_{q_i}}{|C|}}{|D| + \mu} \tag{2}$$

When language model is used in microblog retrieval, one of the biggest challenge we have is that documents are too short. Most of the terms only appear once in one document. Keywords of document cannot be differentiated from other words in language model. Table 1 shows an example query from TREC Microblog Track 2011.

All the listed non-relevant documents and relevant documents contain two query terms. For example, relevant documents 2, non-relevant documents 1 and 2 all have "British" and "politician", but topic of non-relevant documents is not about the query. When we are calculating query likelihood as in Eq. (2), they all get 1 in $f_{q_i,D}$, and only difference will be at smoothing and document length. Therefore, language model does not work well in microblog retrieval.

Table 1. Examples of Retrieval Results

Query MB008: **phone hacking British politicians**
Non-Relevant Documents:
1. Boris Johnson has to be my favourite British politician of all time. He is an absolute LEGEND.
2. Politicians may be too nervous to address Britain's increasing irrelevance on the world stage, but they must
Relevant Documents:
1. British Tabloid Dismisses Editor Over Hacking Scandal
2. To Spy Politicians, British Aide to Prime Minister Resigns:
3. Ex-PM Brown feared voicemail hacking amid scandal: Former British Prime Minister Gordon Brown wrote to the police last summer to ask ...

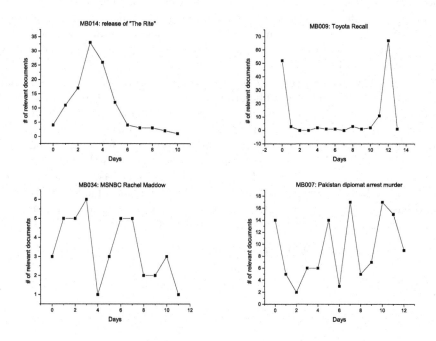

Fig. 1. Temporal in microblog retrieval

3.2 Temporal in Microblog Retrieval

Previous studies have shown that temporal distribution of relevant tweets is not uniform and should be considered in the ranking. Fig. 1 shows a visualization of four different types of query from TREC Microblog Track 2011. X-axis shows time prior to the query time, in days.

Query MB014 "release of 'The Rite' " has a single burst, which happens at the day of movie "The Rite" premiere. Query MB009 "Toyota Recall" has two bursts, which are the day that Toyota initiated vehicle recalls and the day government announced the investigation report. Query MB007 "Pakistan diplomat arrest murder" has more than one burst. Although recency is an important fact, documents do not always cluster right before the query time.

4   Models

4.1   Feedback Language Model

The reason that language model does not perform well in microblog retrieval is that documents are short and most of query terms only show once. We need a better language model to describe topic of query. We adopted the idea of using document likelihood and relevance model [19]. Relevance model is a language model that represents the topic covered by relevant documents. Query Q can be seen as a small sample generated by the relevance model, and relevant document can be seen as a big sample generated by the same model. Document likelihood model use $P(D|R)$ as the probability of document generated by the given relevance model. In web page retrieval, document likelihood is very difficult to estimate since the variation of document length can be very large. One document may contains 10 terms while other may contains 10,000 terms. However, in microblog retrieval, we notice that lengths of tweets are not varied largely.

Here we use language model generated by the pseudo-relevance feedback documents to estimate document likelihood. Pseudo-relevance feedback documents $F = D_1, D_2, \ldots, D_k$ are the search results that returned from the first retrieval of the original query. Based on the idea of pseudo-relevance feedback, we assume that top k ranked documents are relevant. Feedback language model M_F is generated based on all the documents in F. Therefore, we can estimate probability of document generated by feedback language model.

$$P(D \mid M_F) = \prod_{w_i \in D} P(w_i \mid M_F) \tag{3}$$

$$P(w_i \mid M_F) = \frac{\sum_{D_j \in F} f_{w_i, D_j}}{\sum_{D_j \in F} |D_j|} \tag{4}$$

where f_{w_i, D_j} is the frequency of term w_i in document D_j.

Although lengths of tweets are not varied much, we apply normalization to Eq. (3) so that affects caused by different document length can be reduced [28].

$$P_{norm}(D \mid M_F) = APW^d \cdot P(D \mid M_F) \tag{5}$$

where APW^d denotes the penalty factor depending on document length. d equals to average document length subtracts length of D, and APW is average probability weight.

$$APW_{P(D|M_F)} = \frac{1}{|D|} \sum_{w_i \in D} P(w_i \mid M_F) \tag{6}$$

As described above, we generate document probability based on feedback language model. We now define the new score of document as a linear combination of scores produced by query likelihood model and feedback language model.

$$P'(D \mid Q) = \lambda P(D \mid Q) + (1 - \lambda)P(D \mid M_F) \qquad (7)$$
$$= \lambda \frac{P(Q \mid D)P(D)}{P(Q)} + (1 - \lambda)P(D \mid M_F)$$
$$= c\lambda P(Q \mid D) + (1 - \lambda)P(D \mid M_F)$$

where λ determines the weights of two models which is trained in the experiments, and constant $c = \frac{P(D)}{P(Q)}$ since $P(D)$ is usually assumed to be uniform.

4.2   Query Expansion Model

Query expansion is a well-studied technique to overcome the vocabulary mismatch problem in information retrieval. Several query expansion techniques have been developed. Pseudo-relevance feedback technique has been proven useful in previous work for improving retrieval performance.

Here we proposed a query expansion model that conducts both lexical expansions and temporal expansions. Our query expansion model is based on pseudo-relevance feedback technique. The idea of the proposed model is to expand original query with terms and times based on top-ranked documents from initial retrieval. For query $Q = q_1, q_2, \ldots, q_n$, we expand Q with:

1) Lexical expansion: expand original query with terms $Q_{lex} = w_1, w_2, \ldots, w_{lex}$.
2) Temporal expansion: expand original query with times $Q_{tem} = t_1, t_2, \ldots, t_{tem}$.

Here we have the new query $Q' = \{Q, Q_{lex}, Q_{tem}\}$.

We adopted framework proposed by Dakka et al. [5]. The framework assumed that document D can be split into a content component c_D and a temporal component t_D, and content relevance and temporal relevance are independent. The ranking function can be written as:

$$P(D \mid Q) = P(c_D, t_D \mid Q) \qquad (8)$$
$$= P(c_D \mid Q)P(t_D \mid Q)$$

c_D can been considered as D in language models.

Lexical Expansion. Relevance model generated expansion terms using pseudo-relevance feedback documents.

$$P(w \mid R) \propto \sum_{D \in R} P(w \mid D)P(D) \prod_{i=1}^{n} P(q_i \mid D) \qquad (9)$$

Every terms from feedback documents are extracted and ranked according to Eq. (9). Top terms are chosen to expand the query. We interpolate the lexical expansion model with the retrieval model.

$$P(c_D \mid Q) \propto \alpha \sum_{w \in V} P(w \mid Q) \log P(w \mid D) \qquad (10)$$
$$+ (1 - \alpha) \sum_{w \in V} P(w \mid R) \log P(w \mid D)$$

Temporal Expansion. The idea of picking several times for temporal expansion is to build temporal profile for query and identify bursts in it. We take following steps to generate temporal expansion.

1. For query Q, get the top ranked documents $F_t = D_1, D_2, \ldots, D_t$.
2. Each document D has an associated time stamp, and we partition them into bins. Each bin corresponds to a time, for example days, hours, minutes. Number of bins depends on the time span of the document F_t.
3. Bins can be scored in two ways. The first way is to count the number of the documents in the bin. The second one is to add query likelihood scores of the documents in the bin.

$$score_{count}(bin(t)) = |D \in bin(t)| \qquad (11)$$
$$score_{ql}(bin(t)) = \sum_{D \in bin(t)} P_{QL}(Q \mid D) \qquad (12)$$

4. Rank bins based on their scores and expand the query using corresponding times of the top ranked bins.

After temporal expansions of query are generated, we can have temporal relevance $P(t_d \mid Q)$ as follows:

$$P(t_D \mid Q) = P(t_D \mid Q_{tem}) \qquad (13)$$
$$= P(t_D \mid t_1, t_2, \ldots, t_{tem})$$

Given a serious of times, we use two ways to get the probability. The first one is to assume that the probability of document depends on the time that has the biggest impact of the document. The second one is to take the sum of all the impact of all the times. Thus we have:

$$P_{max}(t_D \mid t_1, t_2, \ldots, t_{tem}) = \max_{t_i \in Q_{tem}} P(t_D \mid t_i) \qquad (14)$$
$$P_{sum}(t_D \mid t_1, t_2, \ldots, t_{tem}) = \sum_{t_i \in Q_{tem}} P(t_D \mid t_i) \qquad (15)$$

To estimate the impact of temporal evidence, we use two ways to get $P(t_D \mid t_i)$.

1. Exponential function

$$P_{exp}(t_D \mid t_i) = e^{-\beta|t_i - t_D|} \tag{16}$$

2. Gaussian function

$$P_{gauss}(t_D \mid t_i) = e^{-\frac{|t_i - t_D|^2}{2\sigma^2}} \tag{17}$$

4.3 Feedback Model

We proposed our feedback model by combining our proposed feedback document model and query expansion model together. More specifically, we take the following steps:

1. Get initial retrieval results returned by original query.
2. Apply the proposed feedback language model to rerank the initial results.
3. Apply the proposed query expansion model to generate the new query with lexical expansions and temporal expansions.
4. Get retrieval results returned by the new query.
5. Apply the proposed feedback language model again to rerank the retrieval results.

5 Experiments

We have experimentally evaluated our model on TREC Microblog data. In Section 5.1, we first describe our experiment setup. Then we show the evaluation results of our proposed feedback language model in Section 5.2 and query expansion model in Section 5.3. In Section 5.4, we report the evaluation results of the feedback model, which is a combination of the feedback language model and the query expansion model. Finally, we conduct temporal query analysis by looking into different temporal types of queries in Section 5.5.

5.1 Setup

The experiments are conducted on TREC Microblog Track 2011 and 2012 data sets. The Track 2011 and 2012 evaluations are based on Track 2011 collection. The collection consists of an approximately 16 million tweets (1% sample of tweets from January 23, 2011 to February 7, 2011). There are 49 topics in Track 2011 and 59 topics in Track 2012 (MB050 topic and MB076 are deleted because of the absence of relevant documents). Each topic consists a query and its corresponding time stamp. Relevance judgements were based on a standard pooling strategy, and 3-point scale were used ("not relevant", "relevant", "highly relevant"). We removed all retweets and non-English tweets since TREC judged them as non-relevant. We indexed tweets posted before the time stamp associated with each topic using the Indri search engine[1]. No more than 1000 results are retrieved per topic.

---

[1] http://www.lemurproject.org/indri/

Table 2. Retrieval performance among non-expansion models

(a) TREC 2011

Methods	MAP	P@30	NDCG@30
QL	0.3082	0.3483	0.4254
SDM	0.2981	0.3463	0.4169
Recency Prior	$0.3112^\dagger$	0.3483	0.4330
FLM	$0.3202^{\dagger\ddagger}$	$0.3626^{\dagger\ddagger}$	$0.4417^{\dagger\ddagger}$

(b) TREC 2012

Methods	MAP	P@30	NDCG@30
QL	0.1868	0.2955	0.2836
SDM	0.1860	0.2955	0.2903
Recency Prior	0.1870	0.3006	0.2856
FLM	$0.1937^{\dagger\ddagger}$	0.3051	$0.2919^\dagger$

Each of our test models requires training data, we employ 2-fold cross-validation within each test collection. Parameters were trained with respect to precision at rank 30. We report mean average precision (MAP), precision at rank 30 (P@30) and NDCG at rank 30 (NDCG@30), which were the primary metrics used in the TREC Microblog evaluation. Statistical differences in our experiments are tested using a two-tailed paired t-test with level $\alpha = 5\%$.

5.2   Evaluation of Feedback Language Model

First we discuss the performance of our proposed feedback language model, referred to as **FLM**. Since this model doesn't involve query expansion, we picked several retrieval baselines without using query expansion techniques.

- **QL:** Standard query likelihood approach with Dirichlet smoothing ($\mu = 1500$).
- **SDM:** Sequential dependence model proposed by Metzler and Croft [29]. The model uses the original query words and bigrams extracted from the original query. We took default parameter settings, which are 0.85 for original query words, 0.15 for unwindowed bigrams, and 0.1 for windowed bigrams.
- **Recency Prior:** Recency prior for document is used in query likelihood model. It is one part of the time-based language model proposed by Li and Croft [4]. In $P(D|Q) \propto P(Q|D)P(D)$, $P(D)$ is assigned as a recency prior instead of being uniform. The recency prior is defined as $P(D) = \lambda e^{-\lambda(t_c - t_D)}$, where t_c is the query issued time and t_D is the time of the document.

Experiment results are shown in Table 2. Please note that † means performance of the method improves statistical significantly over QL baseline, and ‡ means performance of the method improve statistical significantly over both QL

Table 3. Retrieval performance on TREC 2011 among different variations of query expansion models

Methods	Description	MAP	P@30	NDCG@30
QEL	only lexical expansions	0.3217	0.3571	0.4431
QELX_ECM	exp + count + max	$0.3396^\dagger$	$0.3823^\dagger$	$0.4643^\dagger$
QELX_ECS	exp + count + sum	0.3348	0.3803	0.4595
QELX_EQM	exp + ql + max	$0.3391^\dagger$	$0.3830^\dagger$	$0.4668^\dagger$
QELX_EQS	exp + ql + sum	0.3360	0.3810	0.4645
QELX_GCM	gauss + count + max	$0.3376^\dagger$	$0.3789^\dagger$	$0.4649^\dagger$
QELX_GCS	gauss + count + sum	0.3367	0.3769	$0.4644^\dagger$
QELX_GQM	gauss + ql + max	$0.3373^\dagger$	0.3789	$0.4646^\dagger$
QELX_GQS	gauss + ql + sum	0.3373	0.3776	$0.4653^\dagger$

and *Recency Prior* baselines. Although *SDM* model has shown effectiveness in previous research of information retrieval, it fails in microblog retrieval. As we discussed above, documents are very short in microblog retrieval so that query bigrams are not likely to be seen in a certain window size. In TREC 2011, we can see that *Recency Prior* method helps the performance, but the improvements are not significant for all the metrics. Our proposed model outperforms both *QL* and *Recency Prior* baselines significantly. However, in TREC 2012, the performance of initial retrieval is not very effective so that feedback documents cannot provide much useful information. Although our proposed model has shown effectiveness on the performance, the improvements are not significant for all the metrics. *Recency Prior* method does not perform very well at this data because the temporal distribution of the query is not always clustered before the query time.

5.3   Evaluation of Query Expansion Model

In Section 4.2, we suggest different ways of getting three functions, which are $score(bin(t))$, $P(t_D \mid t_1, t_2, \ldots, t_{tem})$ and $P(t_D \mid t_i)$. To explore the effectiveness of our proposed different functions, we tested all the combinations of the functions. Experiment results are shown in Table 3. Descriptions of each abbreviation are also listed in the table. Please note that *QEL* denotes query expansion method with only lexical expansions, which is equivalent to the relevance model [19]. Notation † means performance of the method improves statistical significantly over *QEL*. Due to the limitation of space, we only demonstrate the results from TREC 2011.

We can see from the results that all eight combinations outperform query expansion method without temporal expansions. Some of them get significant improvements, while some of them do not. We think different combinations working with different types of queries. For example, methods with "sum" component work well with temporally unambiguous queries and methods with "max" component work well with temporally ambiguous queries. "exp" and "gauss"

Table 4. Retrieval performance among proposed feedback model and baseline models

(a) TREC 2011

Methods	MAP	P@30	NDCG@30
RM3	0.3261	0.3653	0.4504
Recency	0.3376	$0.3769^{\dagger}$	0.4632
QELX	$0.3391^{\dagger}$	$0.3830^{\dagger}$	$0.4668^{\dagger}$
MFM	$0.3520^{\dagger\dagger}$	$0.4027^{\dagger\ddagger}$	$0.4820^{\dagger\ddagger}$

(b) TREC 2012

Methods	MAP	P@30	NDCG@30
RM3	0.2006	0.3062	0.2902
Recency	0.2042	0.3051	0.2951
QELX	$0.2116^{\dagger\ddagger}$	$0.3232^{\dagger\ddagger}$	$0.3084^{\dagger\ddagger}$
MFM	$0.2122^{\dagger\ddagger}$	$0.3260^{\dagger\ddagger}$	$0.3107^{\dagger\ddagger}$

functions work similarly. We think more reasonable distance functions can be explored in the future. In the following experiments, we use *QELX* as a short for *QELX_EQM* to represent the query expansion model.

5.4   Evaluation of Feedback Model

We combine our proposed feedback document model and query expansion model together as the feedback model, referred to as **MFM**. To conduct the comparison experiments, we picked two retrieval baselines.

- **RM:** Relevance model proposed by Lavrenko and Croft [19].
- **Recency:** Time-based language models proposed by Li and Croft [4]. In *Recency Prior* method, recency prior is only used in the query likelihood model. Here, recency prior for document $P(D)$ is also used in relevance model $P(w \mid R) \propto \sum_{D \in R} P(w \mid D) P(D) \prod_{i=1}^{n} P(q_i \mid D)$.

We display the performance of the baselines and our proposed models in Table 4. Please note that † and ‡ mean the performance of the method is statistically significant over *RM* and *Recency* respectively. In TREC 2011, both of our proposed models *QELX* and *MFM* significantly outperform *RM3* baseline . The performance of *MFM* model is also significantly better than *Recency* baseline. We observe some improvements of *QELX* over *Recency* as well. In TREC 2012, the results show that our proposed models perform significantly better than both of the baselines.

5.5   Temporal Query Analysis

To explore how our proposed model performs on different types of temporal queries, we identify the number of bursts for each query using the relevant judgments. For each document D, we partition them into bins using their associated

Table 5. Examples of Retrieval Results

Track	#Bursts	#Topics	Topic
TREC 2011	0	1	33
	1	37	1-6, 10-18, 21, 22, 24, 27, 28, 30-33, 34-36, 38-45, 47-49
	2	9	8, 9, 19, 20, 23, 25, 26, 29, 37
	>2	2	7, 46
TREC 2012	1	33	52-54, 56, 57, 60, 62, 63, 65, 66, 68, 70-73, 75, 77, 80-82, 85, 86, 89, 91-94, 96, 99, 100, 104, 106, 108
	2	21	51, 58, 59, 61, 64, 67, 69, 74, 79, 83, 87, 88, 90, 95, 98, 101-103, 105, 109, 110
	>2	5	55, 78, 84, 97, 107

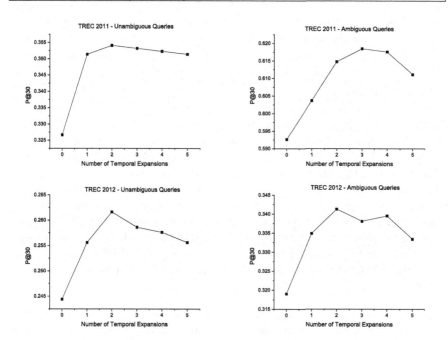

Fig. 2. Performance of queries with different temporal types with different numbers of bins

time stamps. Here "day" is used as time span of the bins. We use $B = \{b\}$ to represent set of bins. We define score of bins as $score(b) = |D \in b|$. Then we build the set of bursts U:

$$U = \{u \in B | \forall b \in (B - U), score(u) - score(b) >= \sigma\} \qquad (18)$$

σ represents the standard variation of all the bins' scores.

Queries from TREC 2011 and 2012 classified by their number of bursts are listed in Table 5. We consider queries with zero or one burst as temporally unambiguous queries and queries with more than one bursts as temporally ambiguous queries. For different types of queries, we tested how the performance changes according to different numbers of the bins. The experiment results are shown in Fig. 2. In TREC 2011, the performance of both query types get a big improvement by using the first temporal expansion. After that, the performance of temporally unambiguous queries begin to steady and then slightly decline, which is consistent with the ground truth that these queries only have one burst. For temporal ambiguous queries, the performance keeps growing and stops at the third expansion and then drops. In TREC 2012, the performance reaches the maximum value when adding two temporal expansions in both cases, then they slowly go down. From all the lines, we can see that when the number of temporal expansion are less than the number of bursts in the ground truth, adding more temporal expansions improves the performance. When the model using the same number of temporal expansion as the number of bursts, adding more temporal expansions also helps the performance in same cases. Overall, we can conclude that our proposed model involving temporal expansion improves the performance of not only temporally ambiguous queries but also temporally unambiguous queries.

6   Conclusion

In this paper, we proposed a feedback model for microblog retrieval. The feedback model includes a feedback language model and a query expansion model considering both lexical expansions and temporal expansions. Experiment results on TREC Microblog Track 2011 and 2012 data sets show that our proposed models improve upon existing baselines. Researchers have been paying growing attention to temporal evidence in information retrieval area. We think there is a lot more benefit that we can get from it. As future work, we will investigate more interesting and effective ways to use temporal information. It is also worth mentioning that there are many other directions in microblog retrieval that can be followed including using URLs information and modeling the noise and quality of tweets.

Acknowledgments. This paper is partially supported by the National Natural Science Foundation of China (NSFC Grant No. 61472006), the Doctoral Program of Higher Education of China (Grant No. 20130001110032) and the National Basic Research Program (973 Program No. 2014CB340405).

References

1. Efron, M., Golovchinsky, G.: Estimation methods for ranking recent information. In: Proceedings of the 34th International ACM SIGIR Conference on Research and Development in Information Retrieval, SIGIR 2011, pp. 495–504. ACM, New York (2011)
2. Keikha, M., Gerani, S., Crestani, F.: Time-based relevance models. In: Proceedings of the 34th International ACM SIGIR Conference on Research and Development in Information Retrieval, SIGIR 2011, pp. 1087–1088. ACM, New York (2011)

3. Liang, F., Qiang, R., Yang, J.: Exploiting real-time information retrieval in the microblogosphere. In: Proceedings of the 12th ACM/IEEE-CS Joint Conference on Digital Libraries, JCDL 2012, pp. 267–276. ACM, New York (2012)
4. Li, X., Bruce Croft, W.: Time-based language models. In: Proceedings of the Twelfth International Conference on Information and Knowledge Management, CIKM 2003, pp. 469–475. ACM, New York (2003)
5. Dakka, W., Gravano, L., Ipeirotis, P.G.: Answering general time sensitive queries. IEEE Transactions on Knowledge and Data Engineering **24**(2) (2012)
6. Massoudi, K., Tsagkias, M., de Rijke, M., Weerkamp, W.: Incorporating query expansion and quality indicators in searching microblog posts. In: Clough, P., Foley, C., Gurrin, C., Jones, G.J.F., Kraaij, W., Lee, H., Mudoch, V. (eds.) ECIR 2011. LNCS, vol. 6611, pp. 362–367. Springer, Heidelberg (2011)
7. Jones, R., Diaz, F.: Temporal profiles of queries. ACM Trans. Inf. Syst. **25**(3), Article 14 (2007)
8. Liang, S., Ren, Z., Weerkamp, W., Meij, E., de, Rijke, M.: Time-Aware rank aggregation for microblog search. In: Proceedings of the Twelfth International Conference on Information and Knowledge Management, CIKM 2014. ACM, Shanghai (2014)
9. Whiting, S., Moshfeghi, Y., Jose, J.M.: Exploring term temporality for pseudo-relevance feedback. In: Proceedings of the 34th International ACM SIGIR Conference on Research and Development in Information Retrieval, SIGIR 2011, pp. 1245–1246. ACM, New York (2011)
10. Whiting, S., Klampanos, I.A., Jose, J.M.: Temporal pseudo-relevance feedback in microblog retrieval. In: Baeza-Yates, R., de Vries, A.P., Zaragoza, H., Cambazoglu, B.B., Murdock, V., Lempel, R., Silvestri, F. (eds.) ECIR 2012. LNCS, vol. 7224, pp. 522–526. Springer, Heidelberg (2012)
11. Miyanishi, T., Seki, K., Uehara, K.: Time-aware latent concept expansion for microblog search. In: Proceedings of the Eighth International Conference on Weblogs and Social Media, ICWSM, pp. 1–4. Ann Arbor, Michigan (2014)
12. Metzler, D., Cai, C., Hovy, E.: Structured event retrieval over microblog archives. In: Proceedings of the 2012 Conference of the North American Chapter of the Association for Computational Linguistics: Human Language Technologies, NAACL HLT 2012, pp. 646–655. Association for Computational Linguistics, Stroudsburg (2012)
13. Chen, L., Chun, L., Ziyu, L., Quan, Z.: Hybrid pseudo-relevance feedback for microblog retrieval. J. Inf. Sci. **39**(6), 773–788 (2013)
14. Bandyopadhyay, A., Ghosh, K., Majumder, P., Mitra, M.: Query expansion for microblog retrieval. IJWS **1**(4), 368–380 (2012)
15. Efron, M., Organisciak, P., Fenlon, K.: Improving retrieval of short texts through document expansion. In: Proceedings of the 35th International ACM SIGIR Conference on Research and Development in Information Retrieval, SIGIR 2012, pp. 911–920. ACM, New York (2012)
16. Han, Z., Li, X., Yang, M., Qi, H., Li, S., Zhao, T.: HIT at TREC 2012 microblog track. In: Proceedings of Text Retrieval Conference (2012)
17. Choi, J., Bruce Croft, W., Kim, J.Y: Quality models for microblog retrieval. In: Proceedings of the 21st ACM International Conference on Information and Knowledge Management, CIKM 2012, pp. 1834–1838. ACM, New York (2012)
18. Gurini, D.F., Gasparetti, F.: Real-time algorithm for microblog ranking systems. In: Proceedings of The Twentyfirst Text Retrieval Conference, TREC 2012, Gaithersburg, pp. 6–9 (November 2012)

19. Lavrenko, V., Bruce Croft, W.: Relevance based language models. In: Proceedings of the 24th Annual International ACM SIGIR Conference on Research and Development in Information Retrieval, SIGIR 2001, pp. 120–127. ACM, New York (2001)

20. Lv, Y., Zhai, C.X.: Adaptive relevance feedback in information retrieval. In: Proceedings of the 18th ACM Conference on Information and Knowledge Management, CIKM 2009, pp. 255–264. ACM, New York (2009)

21. Tao, T., Zhai, C.X.: Regularized estimation of mixture models for robust pseudo-relevance feedback. In: Proceedings of the 29th Annual International ACM SIGIR Conference on Research and Development in Information Retrieval, SIGIR 2006, pp. 162–169. ACM, New York (2006)

22. Dillon, J.V., Collins-Thompson, K.: A unified optimization framework for robust pseudo-relevance feedback algorithms. In: Proceedings of the 19th ACM International Conference on Information and Knowledge Management, CIKM 2010, pp. 1069–1078. ACM, New York (2010)

23. Cao, G., Nie, J.-Y. Gao, J., Robertson, S.: Selecting good expansion terms for pseudo-relevance feedback. In: Proceedings of the 31st Annual International ACM SIGIR Conference on Research and Development in Information Retrieval, SIGIR 2008, pp. 243–250. ACM, New York (2008)

24. Lv, Y., Zhai, C.-X.: Positional relevance model for pseudo-relevance feedback. In: Proceedings of the 33rd International ACM SIGIR Conference on Research and Development in Information Retrieval, SIGIR 2010, pp. 579–586. ACM, New York (2010)

25. Raman, K., Udupa, R., Bhattacharya, P., Bhole, A.: On improving pseudo-relevance feedback using pseudo-irrelevant documents. In: Gurrin, C., He, Y., Kazai, G., Kruschwitz, U., Little, S., Roelleke, T., Rüger, S., van Rijsbergen, K. (eds.) ECIR 2010. LNCS, vol. 5993, pp. 573–576. Springer, Heidelberg (2010)

26. Huang, Q., Song, D., Rüger, S.M.: Robust query-specific pseudo feedback document selection for query expansion. In: Macdonald, C., Ounis, I., Plachouras, V., Ruthven, I., White, R.W. (eds.) ECIR 2008. LNCS, vol. 4956, pp. 547–554. Springer, Heidelberg (2008)

27. He, B., Ounis, I.: Finding good feedback documents. In: Proceedings of the 18th ACM Conference on Information and Knowledge Management, CIKM 2009, pp. 2011–2014. ACM, New York (2009)

28. Maier, V.: Facing the problem of combining the language model with the acoustic model in speech recognition. Master Degree Thesis. University of Sheffield (2003)

29. Metzler, D., Croft, W.B.: A Markov random field model for term dependencies. In: Proceedings of the 28th Annual International ACM SIGIR Conference on Research and Development in Information Retrieval, SIGIR 2005, pp. 472–479. ACM, New York (2005)

Efficient String Similarity Search:
A Cross Pivotal Based Approach

Fei Bi[1]([✉]), Lijun Chang[1], Wenjie Zhang[1], and Xuemin Lin[1,2]

[1] University of New South Wales, Sydney, Australia
f.bi@student.unsw.edu.au, {ljchang,zhangw,lxue}@cse.unsw.edu.au
[2] East China Normal University, Shanghai, China

Abstract. In this paper, we study the problem of string similarity search with edit distance constraint; it retrieves all strings in a string database that are similar to a query string. The state-of-the-art approaches employ the concept of pivotal set, which is a set of non-overlapping signatures, for indexing and query processing. However, they do not fully exploit the pruning power potential of the pivotal sets by using only the pivotal set of the query string or the data strings. To remedy this issue, in this paper we propose a cross pivotal based approach to fully exploiting the pruning power of multiple pivotal sets. We prove theoretically that our cross pivotal filter has stronger pruning power than state-of-the-art filters. We also propose a more efficient algorithm with better time complexity for pivotal selection. Moreover, we further develop two advanced filters to prune unpromising single-match candidates which are the set of candidates introduced by one and only one of the probing signatures. Our experimental results on real datasets demonstrate that our cross pivotal based approach significantly outperforms the state-of-the-art approaches.

1 Introduction

The problem of string similarity search that finds similar strings in a string database to a query string has attracted a great deal of attentions recently. It is an important operation in data cleaning and data integration, and has many applications, such as duplicate detecting [1,6], spelling checking [8,21], and sequence alignment comparison in bioinformatics [9,16]. Edit distance [13] is a well-known metric to measure the similarity between two strings; two strings are similar if and only if their edit distance is no larger than a user-given threshold.

The existing studies for string similarity search with edit distance constraint in [2,4,5,8,11,12,14,15,21] have adopted the signature-based filter-verification framework. The framework consists of two phases: indexing phase and query processing phase. In the indexing phase, a set of signature based inverted indexes are constructed offline. During the query processing phase, firstly, a set of candidate strings are obtained by probing the inverted indexes with filtering conditions, and then each candidate string is verified to compute the similarity to the query string.

© Springer International Publishing Switzerland 2015
M. Renz et al. (Eds.): DASFAA 2015, Part I, LNCS 9049, pp. 545–564, 2015.
DOI: 10.1007/978-3-319-18120-2_32

In the literature, a variety of signature-based filters, such as the counter filter [11] and the prefix filter [3], are proposed to extract candidates. The number of signatures plays an important role on both the pruning power and the filtering cost of a filter, because it directly links to the number of probings to the inverted index as well as the number of candidates retrieved. It has been proven in [14] that the minimum number of signatures is $\tau + 1$, where τ is the user-given similarity threshold. Based on this result, the concept of pivotal set has been proposed for indexing, which indexes a set of $\tau + 1$ non-overlapping signatures. Several pivotal set based techniques [4,14,15] have recently been proposed to improve the pruning power and reduce the filtering cost and shown to outperform other existing works. Nevertheless, these techniques only exploit the single-side pivotal set for query processing; that is, they either use pivotal set of data string or query string [14,15], or choose dynamically during query processing [4].

In this paper, we propose a cross pivotal set based approach to further exploit the pruning power of multiple pivotal sets. Thus, we can achieve significantly less number of candidates than the state-of-the-art approaches; we also prove theoretically that the pruning power of the proposed cross pivotal filter is stronger than the state-of-the-art pivotal based filters. Through our empirical studies, we observe that a large portion of the candidates obtained by the pivotal based approaches are introduced by only one probing signature, and they are very unlikely to be the results; we call these candidates as single-match candidates. Due to the unique feature of our cross pivotal filter, these single-match candidates can be identified, and furthermore we propose two advanced filters to reduce the number of non-results in the set of single-match candidates. Our experimental results demonstrate that the proposed filters can significantly reduce the number of candidates with very little extra filtering cost.

Contributions. The contributions of this paper are summarized as follow:

- We propose a novel cross pivotal filter, which achieves much smaller number of candidates compared to the state-of-the-art approaches. We also prove the superiority of the proposed filter theoretically.
- We propose an efficient dynamic programming approach for pivotal selection, which runs in $O(q\tau^2)$ time in contrast to the $O(q^2\tau^3)$ time of the existing approach (where q is the length of signatures(q-grams)).
- We develop two advanced filters to further reduce the number of candidates.

We conduct extensive experimental studies and demonstrate that our cross pivotal based approach significantly outperforms the state-of-the-art approaches in terms of both candidate number and query processing time.

Organization. The rest of the paper is organized as follows. A brief overview of related work is given below. We formally define the problem in Section 2 and present our cross pivotal filter and the framework in Section 3. The two advanced filters are developed in Section 4. We show the extension of our approach in Section 5, and present the experimental results in Section 6. Finally, we conclude the paper in Section 7.

Related Work. The problem of string similarity search with edit distance constraint has been extensively studied [2,4,5,8,10–12,14,15,17,20,21], and most of the existing studies employ the signature-based filter-verification framework.

There are various representations of signatures, among which q-gram [7] is a very popular and effective one. Based on q-grams, the count filter [11] was first proposed as an effective method of pruning unpromising candidates, which requires that two similar strings must share a certain amount of q-grams. By ordering q-grams of each string according to an universal order, the prefix filter [3] ensures that two similar strings must share one common q-gram in their prefixes. Due to the simplicity of prefix filter, several effective methods such as the position prefix filter [19] and the mismatch filter [18], were designed to further improve the pruning power of the prefix filter.

Recently, several pivotal based filters [4,14,15] were proposed to reduce the signature size and gain more pruning power. The Q-Chunk method [14] employs the fixed-position q-chunks as signatures to perform filtering, and it devises two filters, IndexGram and IndexChunk, which utilize $\tau + 1$ q-chunks in the querying phase and in the indexing phase, respectively. IndexGram-Turbo and IndexChunk-Turbo [15] are designed to optimize Q-Chunk with floating q-chunks and to reduce filtering cost. Pivotal prefix filter [4] improves the prefix filter by probing the inverted index with a pivotal set. In this paper, we propose a new filter named cross pivotal filter which outperforms all the existing signature-based approaches.

2 Problem Definition

In this paper, we focus on a string database S which consists of a set of strings. For a string $s \in S$, we let $|s|$ denote the length of s and let $s.id$ denote the unique identifier of s in S.

Definition 2.1. (Edit Distance) The edit distance between two strings s and r, is the minimum number of edit operations required to transform one string to the other, denoted as $\mathsf{ed}(s,r)$. There exists three types of edit operations: insertion, deletion, and substitution.   □

Example 2.1. Given two strings $s = $ "$datalearningx$" and $r = $ "$datacleanings$", the edit distance between s and r is $\mathsf{ed}(s,r) = 3$, since the optimal sequence of edit operations to transform s to r is: 1) insert 'c' before 'l'; 2) delete 'r'; 3) substitute 'x' with 's'.   □

Problem Statement. Given a string database S, a query string r, and a threshold τ, we study the problem of *string similarity search*; that is, finding all strings $s \in S$ such that $\mathsf{ed}(s,r) \leq \tau$.

Notations. Frequently used notations are summarized in Table 1.

Table 1. Notation Table

Notation	Description
s / S / r	the data string / string database / query string
τ	the query threshold
$\mathsf{ed}(s,r)$	the edit distance between s and r
$\mathsf{q}(s)$ / $\mathsf{q}(r)$	the q-gram set of s / r
$\mathsf{pre}_x(s)$ / $\mathsf{pre}_x(r)$	the x-prefix set of s / r
$\mathsf{piv}(s)$ / $\mathsf{piv}(r)$	the pivotal set of s / r
$\mathsf{piv}_{\tau+1}(s)$ / $\mathsf{piv}_{\tau+1}(r)$	the $(\tau+1)$-th pivotal q-gram in $\mathsf{piv}(s)$ / $\mathsf{piv}(r)$, according to the universal order
lp_s / lp_r	the position of $\mathsf{piv}_{\tau+1}(s)$ / $\mathsf{piv}_{\tau+1}(r)$ in ordered $\mathsf{q}(s)$ / $\mathsf{q}(r)$
g / $g.pos$ / $g.order$	a q-gram / its start position / its universal order number

3　A Cross Pivotal Based Approach

In this section, we develop a new filter called cross pivotal filter, based on which we propose efficient query processing algorithms for string similarity search. In the following, we first present our cross pivotal filter in Section 3.1, then give our algorithm in Section 3.2, while an efficient algorithm for pivotal set selection is presented in Section 3.3.

3.1　Cross Pivotal Filter

Given two strings s and r, let $\mathsf{q}(s)$ and $\mathsf{q}(r)$ denote the sets of *q-grams* of s and r, respectively. We sort all q-grams in $\mathsf{q}(s)$ and $\mathsf{q}(r)$ by an universal order (e.g., in q-gram frequency ascending order), and denote the $(q\tau+1)$-*prefix* sets of $\mathsf{q}(s)$ and $\mathsf{q}(r)$ by $\mathsf{pre}_{q\tau+1}(s)$ and $\mathsf{pre}_{q\tau+1}(r)$, respectively. From $\mathsf{q}(s)$ and $\mathsf{q}(r)$, we respectively choose sets of $\tau+1$ disjoint q-grams as the *pivotal* sets of s and r, denoted as $\mathsf{piv}(s)$ and $\mathsf{piv}(r)$; two q-grams are defined to be disjoint if and only if they have no overlap (i.e., the difference between their start positions is not smaller than q). Then, we have the theorem below.

Theorem 3.1. *If two strings s and r are similar (i.e., $\mathsf{ed}(s,r) \le \tau$), then* $\mathsf{piv}(s) \cap \mathsf{q}(r) \ne \emptyset$ *and* $\mathsf{piv}(r) \cap \mathsf{q}(s) \ne \emptyset$. □

Proof Sketch: We first prove by contradiction that if s and r are similar, then $\mathsf{piv}(s) \cap \mathsf{q}(r) \ne \emptyset$. Suppose $\mathsf{piv}(s) \cap \mathsf{q}(r) = \emptyset$, then none of the $\tau+1$ disjoint q-grams in $\mathsf{piv}(s)$ appears in $\mathsf{q}(r)$. Consequently, at least $\tau+1$ edit operations are required to transform $\mathsf{q}(s)$ into $\mathsf{q}(r)$ (i.e., one for each q-gram in $\mathsf{piv}(s)$); thus, $\mathsf{ed}(s,r) \ge \tau+1$ which is a contradiction. Therefore, $\mathsf{piv}(s) \cap \mathsf{q}(r) \ne \emptyset$. Similarly, we can prove that $\mathsf{piv}(r) \cap \mathsf{q}(s) \ne \emptyset$. □

Let $\mathsf{piv}_{\tau+1}(s)$ denote the $(\tau+1)$-th q-gram in $\mathsf{piv}(s)$ according to the universal order, and lp_s denote its *position* in the ordered $\mathsf{q}(s)$ (i.e., $\mathsf{piv}_{\tau+1}(s)$ is the lp_s-th q-gram in $\mathsf{q}(s)$ according to the universal order); lp_r is defined similarly. Then, we can further reduce $q(r)$ and $q(s)$ in Theorem 3.1 to be their prefix sets, as shown in the lemma below.

Lemma 3.1. *If two strings s and r are similar (i.e., $\text{ed}(s, r) \leq \tau$), then we have* $\text{piv}(s) \cap \text{pre}_{lp_s+(q-1)\tau}(r) \neq \emptyset$ *and* $\text{piv}(r) \cap \text{pre}_{lp_r+(q-1)\tau}(s) \neq \emptyset$. $\qquad\square$

Proof Sketch: If s is similar to r, then to transform s to r, at least one q-gram in $\text{piv}(s)$ remains unchanged, and τ edit operations will introduce at most $q\tau$ new q-grams. Let g be the q-gram in $\text{piv}(s)$ that has the smallest position in the ordered $\text{q}(s)$ among all q-grams that remain unchanged, let x be its position in the ordered $\text{q}(s)$, and let y be the number of q-grams before g in $\text{piv}(s)$. Then, g must be in $\text{pre}_{x-y+q\tau}(r)$, since the τ edit operations will introduce at most $q\tau$ new q-grams and also destroy at least y q-grams. Note that $x + (\tau - y) \leq lp_s$, where $(\tau - y)$ is the number of q-grams after g in $\text{piv}(s)$. Therefore, $\text{piv}(s) \cap \text{pre}_{lp_s+(q-1)\tau}(r) \neq \emptyset$. Similarly, we can prove that $\text{piv}(r) \cap \text{pre}_{lp_r+(q-1)\tau}(s) \neq \emptyset$. $\qquad\square$

If we select the pivotal set $\text{piv}(s)$ from $\text{pre}_{q\tau+1}(s)$, then $lp_s \leq q\tau + 1$; note that, the existence of such a selection is proven in [4]. Then, following from Lemma 3.1, we have the corollary below.

Corollary 3.1. *Given any two similar strings s and r, if the pivotal set is selected from the $(q\tau + 1)$-prefix, then* $\text{piv}(s) \cap \text{pre}_{(2q-1)\tau+1}(r) \neq \emptyset$ *and* $\text{piv}(r) \cap \text{pre}_{(2q-1)\tau+1}(s) \neq \emptyset$. $\qquad\square$

Cross Pivotal Filter. Given any two strings s and r, if $\text{piv}(s) \cap \text{pre}_{lp_s+(q-1)\tau}(r) = \emptyset$ or $\text{piv}(r) \cap \text{pre}_{lp_r+(q-1)\tau}(s) = \emptyset$, then s and r cannot be similar. Lemma 3.1 above proves the correctness of this filter.

Table 2. Dataset S and query string r $(q = 2)$

id	string	q-gram set ordered by global order
s_1	datamining	$\{\langle am, 4\rangle\langle mi, 5\rangle\langle da, 1\rangle\langle ni, 7\rangle\langle ng, 9\rangle\langle ta, 3\rangle\langle at, 2\rangle\langle in, 6\rangle\langle in, 8\rangle\}$
s_2	datalearning	$\{\langle ar, 7\rangle\langle rn, 8\rangle\langle le, 5\rangle\langle al, 4\rangle\langle da, 1\rangle\langle ea, 6\rangle\langle ni, 9\rangle\langle ng, 11\rangle\langle ta, 3\rangle\langle at, 2\rangle\langle in, 10\rangle\}$
s_3	dutaleatings	$\{\langle du, 1\rangle\langle ut, 2\rangle\langle ti, 8\rangle\langle le, 5\rangle\langle gs, 11\rangle\langle al, 4\rangle\langle ea, 6\rangle\langle ng, 10\rangle\langle ta, 3\rangle\langle at, 7\rangle\langle in, 9\rangle\}$
s_4	datalearnings	$\{\langle ar, 7\rangle\langle rn, 8\rangle\langle le, 5\rangle\langle gs, 12\rangle\langle al, 4\rangle\langle da, 1\rangle\langle ea, 6\rangle\langle ni, 9\rangle\langle ng, 11\rangle\langle ta, 3\rangle\langle at, 2\rangle\langle in, 10\rangle\}$
s_5	datalweatings	$\{\langle lw, 5\rangle\langle we, 6\rangle\langle ti, 9\rangle\langle gs, 12\rangle\langle al, 4\rangle\langle da, 1\rangle\langle ea, 7\rangle\langle at, 2\rangle\langle ng, 11\rangle\langle ta, 3\rangle\langle at, 8\rangle\langle in, 10\rangle\}$
r	datacleaning	$\{\langle ac, 4\rangle\langle cl, 5\rangle\langle an, 8\rangle\langle le, 6\rangle\langle da, 1\rangle\langle ea, 7\rangle\langle ni, 9\rangle\langle ng, 11\rangle\langle ta, 3\rangle\langle at, 2\rangle\langle in, 10\rangle\}$

Table 3. Universal order of q-grams (increasing frequency order)

Frequency	q-grams
1	$\langle 1 : am\rangle\langle 2 : an\rangle\langle 3 : aw\rangle\langle 4 : li\rangle\langle 5 : lw\rangle\langle 6 : mi\rangle\langle 7 : we\rangle$
2	$\langle 8 : ar\rangle\langle 9 : du\rangle\langle 10 : ut\rangle\langle 11 : rn\rangle\langle 12 : ti\rangle$
3	$\langle 13 : le\rangle\langle 14 : gs\rangle$
4	$\langle 15 : al\rangle\langle 16 : da\rangle\langle 17 : ea\rangle\langle 18 : ni\rangle$
5	$\langle 19 : ng\rangle\langle 20 : ta\rangle$
6	$\langle 21 : at\rangle\langle 22 : in\rangle$

Example 3.2. Consider string s_1 and string r in Table 2. Here, $q = 2$, $\tau = 2$, and the universal order of q-grams is shown in Table 3; for example, $\langle 1 : am\rangle$

in Table 3 indicates that am is the first q-gram in the universal order, and $\langle am, 4\rangle$ in Table 2 denotes that the start position of am in s_1 is 4. Assume that $\{am, da, ni\}$ is selected as $\mathsf{piv}(s_1)$ and $\{ac, an, le\}$ is selected as $\mathsf{piv}(r)$; then $lp_s = 4$, $lp_r = 4$. The prefix lengths are $lp_s + (q-1)\tau = 6$, and $lp_r + (q-1)\tau = 6$. Therefore, $\mathsf{pre}_{lp_r+(q-1)\tau}(s_1) = \{am, mi, da, ni, ng, ta\}$ and $\mathsf{pre}_{lp_s+(q-1)\tau}(r) = \{ac, cl, an, le, da, ea\}$. Then, we have $\mathsf{piv}(s_1) \cap \mathsf{pre}_{lp_s+(q-1)\tau}(r) = \{da\}$ and $\mathsf{piv}(r) \cap \mathsf{pre}_{lp_r+(q-1)\tau}(s_1) = \emptyset$. Thus, s_1 and r cannot be similar according to the cross pivotal filter.   □

Compared with Existing Filters. In the literature, there are other filters studied, such as IndexChunk-Turbo [15], IndexGram-Turbo [15], and pivotal prefix filter [4]. Given a data string s and a query string r, the pruning condition (i.e., the condition that s and r cannot be similar) for IndexChunk-Turbo is $\mathsf{piv}(s) \cap \mathsf{pre}_{(2q-1)\tau+1}(r) = \emptyset$, and for IndexGram-Turbo is $\mathsf{piv}(r) \cap \mathsf{pre}_{(2q-1)\tau+1}(s) = \emptyset$; while for pivotal prefix filter it is $\mathsf{piv}(r) \cap \mathsf{pre}_{q\tau+1}(s) = \emptyset$ if $last(\mathsf{pre}_{q\tau+1}(r)) < last(\mathsf{pre}_{q\tau+1}(s))$, otherwise it is $\mathsf{piv}(s) \cap \mathsf{pre}_{q\tau+1}(r) = \emptyset$, where $last(\mathsf{pre}_{q\tau+1}(s))$ and $last(\mathsf{pre}_{q\tau+1}(r))$ denote the universal order of the last q-gram in $\mathsf{pre}_{q\tau+1}(s)$ and in $\mathsf{pre}_{q\tau+1}(r)$, respectively. We summarize the pruning conditions of these filters in Table 4.

Table 4. Pruning Conditions of Filters

Filter	Pruning condition
IndexChunk-Turbo	$\mathsf{piv}(s) \cap \mathsf{pre}_{(2q-1)\tau+1}(r) = \emptyset$
IndexGram-Turbo	$\mathsf{piv}(r) \cap \mathsf{pre}_{(2q-1)\tau+1}(s) = \emptyset$
Pivotal Prefix filter	$\mathsf{piv}(r) \cap \mathsf{pre}_{q\tau+1}(s) = \emptyset$, if $last(\mathsf{pre}_{q\tau+1}(r)) < last(\mathsf{pre}_{q\tau+1}(s))$
	$\mathsf{piv}(s) \cap \mathsf{pre}_{q\tau+1}(r) = \emptyset$, if $last(\mathsf{pre}_{q\tau+1}(r)) \geq last(\mathsf{pre}_{q\tau+1}(s))$
Cross Pivotal filter	$\mathsf{piv}(r) \cap \mathsf{pre}_{lp_r+(q-1)\tau}(s) = \emptyset$ or $\mathsf{piv}(s) \cap \mathsf{pre}_{lp_s+(q-1)\tau}(r) = \emptyset$

We prove that our cross pivotal filter has the best pruning power among all filters in Table 4 in the theorem below.

Theorem 3.2. *Our cross pivotal filter has the best pruning power among all filters in Table 4.*   □

Proof Sketch: Firstly, we have $lp_s \leq q\tau + 1$ and $lp_r \leq q\tau + 1$ if the pivotal sets are selected from the $(q\tau + 1)$-prefix sets. Thus, given a query string r, any data string s that is pruned by IndexChunk-Turbo or pruned by IndexGram-Turbo must also be pruned by our cross pivotal filter.

Secondly, for pivotal prefix filter, if $last(\mathsf{pre}_{q\tau+1}(r)) < last(\mathsf{pre}_{q\tau+1}(s))$, then $\mathsf{piv}(r) \cap \mathsf{pre}_{q\tau+1}(s) = \emptyset$ is equivalent to $\mathsf{piv}(r) \cap \mathsf{q}(s) = \emptyset$, which can be transferred into $\mathsf{piv}(r) \cap \mathsf{pre}_{lp_r+(q-1)\tau}(s) = \emptyset$ according to Corollary 3.1. Similarly, if $last(\mathsf{pre}_{q\tau+1}(r)) \geq last(\mathsf{pre}_{q\tau+1}(s))$, then $\mathsf{piv}(s) \cap \mathsf{pre}_{q\tau+1}(r) = \emptyset$ is equivalent to $\mathsf{piv}(s) \cap \mathsf{pre}_{lp_s+(q-1)\tau}(r) = \emptyset$. Thus, given a query string r, any data string s that is pruned by pivotal prefix filter must also be pruned by our cross pivotal filter.

Thus, the theorem holds.   □

3.2   Cross Pivotal Based Approach

In this subsection, based on the proposed cross pivotal filter, we present our app-roach for string similarity search, which consists of two phases: Phase-I, indexing q-grams; and Phase-II, query processing.

Indexing. Given a query string r, every data string $s \in S$ that survives the cross pivotal filter has $\mathsf{piv}(s) \cap \mathsf{pre}_{lp_s+(q-1)\tau}(r) \neq \emptyset$ and $\mathsf{piv}(r) \cap \mathsf{pre}_{lp_r+(q-1)\tau}(s) \neq \emptyset$. Therefore, in order to efficiently retrieve all the candidate strings in S that pass the cross pivotal filter, we construct inverted index on the pivotal set and the prefix set for each data string in S. However, the query string r is not given at the time of indexing, so lp_r is unknown; moreover, lp_s may vary for different data strings $s \in S$. Therefore, we select $\mathsf{piv}(s)$ and $\mathsf{piv}(r)$ from $\mathsf{pre}_{q\tau+1}(s)$ and $\mathsf{pre}_{q\tau+1}(r)$, respectively, and set lp_s and lp_r to be $q\tau+1$ which is their upper bound.

We denote the inverted index constructed for the pivotal sets of all data strings as I_{piv}, and denote the inverted index constructed for the $((2q-1)\tau+1)$-prefix sets of all data strings as I_{pre}. In the inverted indexes, for each q-gram g, we store not only the ids of strings that contain g but also the start positions of g in the corresponding strings; the start positions are stored to enable position filtering during query processing.

The pseudocode is shown in Algorithm 1, denoted INDEXING. We first gen-erate all the q-grams $\mathsf{Q}(s)$ for each string $s \in S$ and count the frequency of the generated q-grams as well (Line 1). Then, we sort the set of all generated q-grams for all strings in S in increasing frequency order, which defines the universal order of q-grams (Line 2). The inverted indexes I_{piv} and I_{pre} are ini-tialized to be empty (Line 3). Then, we process each string s in S (Lines 4-13). For string s, we first obtain and store the length $|s|$ of s (Line 5), and then sort $\mathsf{q}(s)$ according to the universal order (Line 6). The $((2q-1)\tau+1)$-prefix of s is selected and indexed by I_{pre} (Lines 7-9). The pivotal set $\mathsf{piv}(s)$ of s is chosen from the $(q\tau+1)$-prefix of $\mathsf{q}(s)$ by algorithm PIVOTALSELECTION which will be discussed in Section 3.3 (Lines 10-11), and is indexed by I_{piv} (Lines 12-13).

Example 3.3. Consider the string database $S = \{s_1, \ldots, s_5\}$ in Table 2 with $q = 2$ and $\tau = 2$; then $(2q-1)\tau+1 = 7$. For string s_1, $\mathsf{pre}_{(2q-1)\tau+1}(s_1) = \{am, mi, da, ni, ng, ta, at\}$ and assume $\mathsf{piv}(s_1) = \{am, da, ni\}$. Then, the $\langle id,$ start position$\rangle$ pairs $\langle s_1, 4\rangle$, $\langle s_1, 5\rangle$, $\langle s_1, 1\rangle$, $\langle s_1, 7\rangle$, $\langle s_1, 9\rangle$, $\langle s_1, 3\rangle$, and $\langle s_1, 2\rangle$ are put into $I_{pre}[am]$, $I_{pre}[mi]$, $I_{pre}[da]$, $I_{pre}[ni]$, $I_{pre}[ng]$, $I_{pre}[ta]$, and $I_{pre}[at]$, respectively. Similarly, for the inverted index of pivotal sets, $\langle s_1, 4\rangle$, $\langle s_1, 1\rangle$ and $\langle s_1, 7\rangle$ are put into $I_{piv}[am]$, $I_{piv}[da]$ and $I_{piv}[ni]$, respectively. The final inverted index I_{pre} and I_{piv} are shown in Table 5.   □

Time Complexity. The time complexity of Algorithm 1 is $O(\sum_{s \in S} |s| \log \sum_{s \in S} |s|)$, if we exclude the pivotal selecting time consumed by PIVOTALSELECTION. The reason is that putting a q-gram into I_{pre} or I_{piv} takes constant time, and Line 2 of Algorithm 1 is the dominating cost.

Algorithm 1. INDEXING

Input: String database S

Output: Inverted index I_{piv} for pivotal sets, and I_{pre} for $((2q-1)\tau+1)$-prefix sets

1 Generate q-grams $\mathsf{q}(s)$ for all strings $s \in S$;

2 Sort all q-grams in S in increasing frequency order, which defines the universal order;

3 $I_{piv} \leftarrow \emptyset, I_{pre} \leftarrow \emptyset$;

4 **for each** $s \in S$ **do**

5   Obtain and store the length of s;

6   Sort $\mathsf{q}(s)$ according to the universal order;

    /* Lines 7-9: Index prefix set   */

7      $\mathsf{pre}_{(2q-1)\tau+1}(s) \leftarrow$ {the first $(2q-1)\tau+1$ q-grams in $\mathsf{q}(s)$};

8   **for each** q-gram $g \in \mathsf{pre}_{(2q-1)\tau+1}(s)$ **do**

9          $I_{pre}[g] \leftarrow I_{pre}[g] \cup \langle s.id, g.pos \rangle$;

    /* Lines 10-13: Index pivotal set   */

10     $\mathsf{pre}_{q\tau+1}(s) \leftarrow$ {the first $q\tau+1$ q-grams in $\mathsf{q}(s)$};

11     $\mathsf{piv}(s) \leftarrow$ PIVOTALSELECTION($\mathsf{pre}_{q\tau+1}(s)$);

12   **for each** q-gram $g \in \mathsf{piv}(s)$ **do**

13         $I_{piv}[g] \leftarrow I_{piv}[g] \cup \langle s.id, g.pos \rangle$;

Table 5. Example of Inverted Index for String Database S

I_{pre} for dataset S	I_{piv} for dataset S
$am \to \langle s_1, 4 \rangle$; $at \to \langle s_1, 2 \rangle$; $du \to \langle s_3, 1 \rangle$; $lw \to \langle s_5, 5 \rangle$; $mi \to \langle s_1, 5 \rangle$;	$am \to \langle s_1, 4 \rangle$; $du \to \langle s_3, 1 \rangle$;
$ng \to \langle s_1, 9 \rangle$; $ta \to \langle s_3, 3 \rangle$; $ut \to \langle s_3, 2 \rangle$; $we \to \langle s_5, 6 \rangle$;	$lw \to \langle s_5, 5 \rangle$; $ni \to \langle s_1, 7 \rangle$;
$ar \to \langle s_2, 7 \rangle, \langle s_4, 7 \rangle$; $ni \to \langle s_1, 7 \rangle, \langle s_2, 9 \rangle$; $rn \to \langle s_2, 8 \rangle, \langle s_4, 8 \rangle$;	$ar \to \langle s_2, 7 \rangle, \langle s_4, 7 \rangle$;
$ti \to \langle s_3, 8 \rangle, \langle s_5, 9 \rangle$;	$da \to \langle s_1, 1 \rangle, \langle s_2, 1 \rangle$;
$gs \to \langle s_3, 11 \rangle, \langle s_4, 12 \rangle, \langle s_5, 12 \rangle$; $le \to \langle s_2, 5 \rangle, \langle s_3, 5 \rangle, \langle s_4, 5 \rangle$;	$gs \to \langle s_4, 12 \rangle, \langle s_5, 12 \rangle$;
$al \to \langle s_2, 4 \rangle, \langle s_3, 4 \rangle, \langle s_4, 4 \rangle, \langle s_5, 4 \rangle$; $da \to \langle s_1, 1 \rangle, \langle s_2, 1 \rangle, \langle s_4, 1 \rangle, \langle s_5, 1 \rangle$;	$ti \to \langle s_3, 8 \rangle, \langle s_5, 9 \rangle$;
$ea \to \langle s_2, 6 \rangle, \langle s_3, 6 \rangle, \langle s_4, 6 \rangle, \langle s_5, 7 \rangle$;	$le \to \langle s_2, 5 \rangle, \langle s_3, 5 \rangle, \langle s_4, 5 \rangle$;

Query Processing. Given the inverted indexes I_{piv} and I_{pre} constructed for a string database S, for any query string r, our query processing algorithm consists of two stages: stage-I, generating candidate sets, and stage-II, verifying each string in the candidate set to find the true similar strings. The candidate set is defined as the set of strings in S that pass the cross pivotal filter as proposed in Section 3.1; that is, $\{s \in S \mid \mathsf{piv}(s) \cap \mathsf{pre}_{(2q-1)\tau+1}(r) \neq \emptyset, \mathsf{piv}(r) \cap \mathsf{pre}_{(2q-1)\tau+1}(s) \neq \emptyset\}$.

To obtain the candidate set, we first use the q-grams in the prefix of $\mathsf{q}(r)$, $\mathsf{pre}_{(2q-1)\tau+1}(r)$, to query the inverted index I_{piv} to generate an initial set of candidates. Here, we also apply the *length filtering* and the *position filtering*. For each q-gram $g \in \mathsf{pre}_{(2q-1)\tau+1}(r)$, and each entry in the inverted list of g (i.e., $\langle sid, pos \rangle \in I_{piv}[g]$), the length filtering requires that the length of the string with id sid must be within the range $[|r| - \tau, |r| + \tau]$, and the position filtering requires that the position difference must satisfy $|g.pos - pos| \leq \tau$. Then, we use the pivotal q-grams in $\mathsf{piv}(r)$ to query the inverted index I_{pre} to further refine the candidate set using the position filtering again.

The pseudocode is shown in Algorithm 2, denoted as SEARCH. Firstly, we generate the q-gram set $\mathsf{q}(r)$ of the query string r, and sort it according to the universal order (Line 1). Then, through Lines 2-8, we generate the initial candidate set by using the prefix of $\mathsf{q}(r)$, $\mathsf{pre}_{(2q-1)\tau+1}(r)$, to probe the inverted index I_{piv}. That is, for each q-gram $g \in \mathsf{pre}_{(2q-1)\tau+1}(r)$ (Line 4), and each pair $\langle sid, pos \rangle \in I_{pre}[g]$ (Line 5), we add sid to the candidate set if its corresponding string passes both the length filtering and the position filtering (Lines 7-8). Next, through Lines 9-15, we refine the candidate set by using the pivotal set $\mathsf{piv}(r)$ to probe the inverted index I_{pre}. Finally, for each string s in the candidate set, we verify it by checking whether $\mathsf{ed}(s, r)$ is larger than τ or not (Lines 16-20); note that, here we use the same verification algorithm as that was used in the existing works [4, 15].

Note that, we can also apply the tight cross pivotal filter as presented in Lemma 3.1 at Lines 7,14, by storing the lp_s for each sid in I_{piv} and the position of each q-gram g in the prefix of the string with id sid in I_{pre}. For presentation briefness, we omit the details.

Example 3.4. Consider the inverted indexes I_{piv} and I_{pre} built in Example 3.3 and the query string r in Table 2. $\mathsf{pre}_{(2q-1)\tau+1}(r) = \{\langle ac, 4 \rangle, \langle cl, 5 \rangle, \langle an, 8 \rangle, \langle le, 6 \rangle, \langle da, 1 \rangle, \langle ea, 7 \rangle, \langle ni, 9 \rangle\}$ where the numbers indicate the start position of the corresponding q-gram in r, and assume $\mathsf{piv}(r)$ is chosen as $\{\langle ac, 4 \rangle, \langle an, 8 \rangle, \langle le, 6 \rangle\}$. The algorithm starts by using $\mathsf{pre}_{(2q-1)\tau+1}(r)$ to probe the inverted index I_{piv} with both the length filtering and the position filtering. Taking $\langle da, 1 \rangle$ in $\mathsf{pre}_{(2q-1)\tau+1}(r)$ as an example, we have $I_{piv}[da] = \{s_1, s_2\}$ and both of them pass the length filtering and the position filtering, thus, the candidates obtained by querying I_{piv} with $\langle da, 1 \rangle$, denoted as $cand_{piv}(da, 1)$, are $\{s_1, s_2\}$. Similarly, we have $cand_{piv}(ac, 4) = cand_{piv}(cl, 5) = cand_{piv}(an, 8) = cand_{piv}(ea, 7) = \emptyset$, $cand_{piv}(le, 6) = \{s_2, s_3, s_4\}$ and $cand_{piv}(ni, 9) = \{s_1\}$. Merging them altogether, we obtain the candidate set obtained by querying I_{pre}, denoted as $cand_{piv}$, which is $\{s_1, s_2, s_3, s_4\}$. Then $\mathsf{piv}(r)$ is used to probe I_{pre} to extract candidates. We denote the candidates obtained by $\langle ac, 4 \rangle$ as $cand_{pre}(ac, 4)$; then, we have $cand_{pre}(ac, 4) = cand_{pre}(an, 8) = \emptyset$ and $cand_{pre}(le, 6) = \{s_2, s_3, s_4\}$. Therefore, the candidate set obtained by querying I_{Pre}, denoted as $cand_{pre}$, is $\{s_2, s_3, s_4\}$. Finally, we get the candidate set $cand = cand_{piv} \cap cand_{pre} = \{s_2, s_4, s_5\}$, with s_1 in $cand_{piv}$ being pruned. □

Time Complexity. Obviously, the two inverted-index probing processes (i.e., Lines 4-8 and Lines 12-15) dominate the time complexity of Algorithm 2. For each q-gram g in $\mathsf{pre}_{(2q-1)\tau+1}$, it needs $O(|I_{piv}[g]|)$ time to probe the inverted list. Therefore the time complexity for the first probing (i.e., Lines 4-8) is $O(\sum_{g \in \mathsf{pre}_{(2q-1)\tau+1}(r)} |I_{piv}[g]|)$. Similar, we obtain that the time complexity for the second probing (i.e., Lines 12-15) is $O(\sum_{g \in \mathsf{piv}(r)} |I_{pre}[g]|)$. Consequently, the total time complexity of Algorithm 2 (excluding the time for verification at Lines 17-19) is $O(\sum_{g \in \mathsf{pre}_{(2q-1)\tau+1}(r)} |I_{piv}[g]| + \sum_{g \in \mathsf{piv}(r)} |I_{pre}[g]|)$.

Algorithm 2. SEARCH

Input: String dataset S, inverted indexes I_{piv} and I_{pre}, and query string r

Output: All the similar strings $Res \subseteq S$ to r, (i.e., $Res = \{s \in S \mid ed(s, r) \leq \tau\}$)

1   $q(r) \leftarrow \{$the q-grams of r sorted according to the universal order$\}$;

    /* Lines 2-8: $Candidate_{temp} \leftarrow \{s \in S \mid piv(s) \cap pre_{(2q-1)\tau+1}(r) \neq \emptyset\}$ */

2   $pre_{(2q-1)\tau+1}(r) \leftarrow \{$the first $(2q-1)\tau+1$ q-grams in $q(r)\}$;

3   $Candidate_{temp} \leftarrow \emptyset$;

4   **for** *each q-gram $g \in pre_{(2q-1)\tau+1}(r)$* **do**

5   **for** *each pair $\langle sid, pos \rangle \in I_{piv}[g]$* **do**

6   Let s be the string with id sid;

7   **if** $|s| \in [|r| - \tau, |r| + \tau]$ *and* $pos \in [g.pos - \tau, g.pos + \tau]$ **then**

8            $Candidate_{temp} \leftarrow Candidate_{temp} \cup \{sid\}$;

    /* Lines 9-15: $Candidate \leftarrow \{s \in Candidate_{temp} \mid piv(r) \cap pre_{(2q-1)\tau+1}(s) \neq \emptyset\}$ */

9   $pre_{q\tau+1}(r) \leftarrow \{$the first $q\tau+1$ q-grams in $q(r)\}$;

10   $piv(r) \leftarrow$ PIVOTALSELECTION $(pre_{q\tau+1}(r))$;

11   $Candidate \leftarrow \emptyset$;

12   **for** *each q-gram $g \in piv(r)$* **do**

13   **for** *each pair $\langle sid, pos \rangle \in I_{pre}[g]$* **do**

14   **if** $sid \in Candidate_{temp}$ *and* $pos \in [g.pos - \tau, g.pos + \tau]$ **then**

15            $Candidate \leftarrow Candidate \cup \{sid\}$;

    /* Lines 16-19: Verification   */

16   $Res \leftarrow \emptyset$;

17   **for** *each $sid \in Candidate$* **do**

18   Let s be the string with id sid;

19   **if** VERIFY (s, r) **then** $Res \leftarrow Res \cup \{s\}$;

20   **return** Res;

3.3   Pivotal Set Selection

In this section, we propose a new efficient dynamic programming algorithm for selecting the pivotal set in $O(q\tau^2)$ time.[1]

Objective Function. In Table 4, we have shown that the pruning condition of cross pivotal filter is $piv(r) \cap pre_{lp_r+(q-1)\tau}(s) = \emptyset$ or $piv(s) \cap pre_{lp_s+(q-1)\tau}(r) = \emptyset$. Intuitively, the smaller lp_s and lp_r are, the shorter the prefixes of s and r in the cross pivotal filter, and thus the more powerful the pruning condition. Hence, we compute the pivotal set with the aim of minimizing lp_s and lp_r, which is equivalent to minimizing the maximum order of the selected pivotal q-grams. Note that, each q-gram has an universal order number, denoted $g.order$, which is the position of g in the sorted set of all q-grams in the string database (i.e., in the universal order).

---

[1] Note that, our method is different from the algorithm in [4] which runs in $O(q^2\tau^3)$ time.

Algorithm. We develop a dynamic programming algorithm to find the pivotal set (i.e., $\tau + 1$ disjoint q-grams) with the minimum max-order from the q-grams q of a string. It has been proven in [4] that there always exists a pivotal set in the $(q\tau + 1)$-prefix of q; thus, it is sufficient for us to take the $(q\tau + 1)$-prefix $pre_{q\tau+1}$ as input. That is, the optimal pivotal set will be in $pre_{q\tau+1}$.

We sort q-grams in $pre_{q\tau+1}(r)$ by their start positions, and let g_i denote the i-th q-gram in this order with $1 \leq i \leq q\tau + 1$. For each q-gram g_i in $pre_{q\tau+1}(r)$, we let $ld(i)$ denote the last non-overlapping q-gram that appears before g_i; that is, $ld(i) = \max\{1 \leq j < i \mid g_i.pos - g_j.pos \geq q\}$. If all q-grams before g_i overlap with g_i, then $ld(i)$ is set to 0.

Now, we let $mmo(i,j)$ denote the optimal solution (i.e., with minimum max-order) of selecting j disjoint q-grams from $\{g_1, \ldots, g_i\}$. Then, $mmo(i,j)$ can be obtained from two cases: 1) including g_i, thus the other part of the solution is $mmo(ld(i), j-1)$; and 2) not including g_i, thus, it is the same as $mmo(i-1, j)$ by greedy selection. Therefore, we can compare the two cases, and choose the one with smaller max-order.

Let $mmo(i,j).maxOrder$ denote the max-order of the q-grams in $mmo(i,j)$, let $order_{in}$ denote $\max\{mmo(ld(i), j-1).maxOrder, g_i.order\}$, and let $order_{not\_in}$ denote $mmo(i-1, j).maxOrder$. Then we can compute $mmo(i,j)$ recursively as follows,

$$\begin{cases} mmo(i,j) \leftarrow mmo(i-1,j), & If \quad order_{not\_in} < order_{in} \\ mmo(i,j) \leftarrow mmo(ld(i), j-1) \cup g_i, & If \quad order_{not\_in} \geq order_{in} \end{cases} \quad (1)$$

Lemma 3.2. *Equation (1) correctly computes $mmo(i,j)$ for all $1 \leq i \leq q\tau + 1$ and $1 \leq j \leq \tau + 1$.* □

Proof Sketch: Obviously, the optimal solution $mmo(i,j)$ must be one of the two cases: 1) including g_i; and 2) not including g_i. For case 1), since g_i is included in the solution, then the other part of the solution must be $mmo(ld(i), j-1)$ by greedy selection due to the fact that $mmo(i', j') \geq mmo(i'+1, j')$ for all i' and j'. For case 2), as g_i is not selected, the optimal solution must be $mmo(i-1, j)$ by greedy selection. Thus, $mmo(i,j)$ is the better one between the above two cases. Thus, the lemma holds. □

Following from the above, the pseudocode is shown in Algorithm 3, denoted as PIVOTALSELECTION. It starts by sorting the set of q-grams by their start positions (Line 1). $ld(i)$s are computed at Lines 3-6, while Lines 7-12 compute $mmo(i,j)$ following Equation (1). Note that, for efficiency concerns, in Algorithm 3, $mmo(i,j)$ actually stores $mmo(i,j).maxOrder$; that is, $mmo(i,j)$ stores the max-order of the set of selected q-grams instead of the actual q-grams. Finally, we obtain the optimal pivotal set by backtracking on $mmo(i,j)$ (Lines 13-16); the observation is that, the q-gram at position ri of p is not selected in the pivotal set if and only if $mmo(ri, i)$ equals $mmo(ri-1, i)$, and we start the construction from $mmo(q\tau + 1, \tau)$.

Time Complexity. The time complexity of Algorithm 3 is $O(q\tau^2)$. It is easy to verify that the $ld(i)$s are constructed in $O(q\tau)$ time at Lines 3-6, the $mmo(i,j)$s

Algorithm 3. PIVOTALSELECTION

Input: A set of q-grams p
Output: The pivotal set piv of p with minimum max-order
1 Sort q-grams in p by their start positions;
2 Allocate arrays $ld[q\tau + 1]$ and $mmo[q\tau + 1][\tau + 1]$;
    /* Lines 3-6: Compute $ld[1],\dots,ld[q\tau + 1]$ */
3 $ld[1] \leftarrow 0$;
4 **for** $i \leftarrow 2$ **to** $q\tau + 1$ **do**
5     $ld[i] \leftarrow ld[i - 1]$;
6   **while** $g_i.pos - g_{ld[i]+1}.pos \geq q$ **do** $ld[i] \leftarrow ld[i] + 1$;

    /* Lines 7-12: Compute $mmo(i, j)$ */
7 $mmo[1][1] \leftarrow g_1.order$;
8 **for** $i \leftarrow 2$ **to** $q\tau + 1$ **do**
9     $mmo[i][1] \leftarrow \min\{mmo[i - 1][1], g_i.order\}$;
10   **for** $j \leftarrow 2$ **to** $\tau + 1$ **do**
11         $maxOrder \leftarrow \max\{mmo[ld[i]][j - 1], g_i.order\}$;
12         $mmo[i][j] \leftarrow \min\{maxOrder, mmo[i - 1][j]\}$;

    /* Lines 13-16: Construct the pivotal set   */
13 piv $\leftarrow \emptyset$; $ri \leftarrow q\tau + 1$;
14 **for** $i \leftarrow \tau + 1$ to 1 **do**
15   **while** $mmo[ri][i] = mmo[ri - 1][i]$ **do** $ri \leftarrow ri - 1$;
16   piv$[i] \leftarrow$ the q-gram at position ri of p; $ri \leftarrow ld[ri]$;
17 **return** piv;

---

are computed in $O(q\tau^2)$ at Lines 7-12, and Lines 13-16 build the pivotal set in $O(q\tau)$ time. Note that, a similar but different dynamic programming approach has been proposed in [4] for selecting weight-based pivotal set in $O(q^2\tau^3)$ time. Therefore, our pivotal selection algorithm is more efficient.

4 Advanced Filters

As it is time-consuming to verify each candidate by checking whether the edit distance between the candidate string and the query string is larger than τ, in this section we propose two advanced filters to further refine the candidate set, thus reduce the number of candidates to be verified. Before that, we first introduce the concept of single-match candidate.

Definition 4.2. (Single-Match Candidates) A candidate string s is called a single-match candidate if s contains only one q-gram of the pivotal set of the query string r and moreover, that pivotal q-gram has only one copy in r.   □

Through our experiments, we observe that a large portion (i.e., 50% to 95%) of the candidates are single-match candidates; that is, they are introduced by only one pivotal q-gram of r. However, many of the single-match candidates are not included in the final result. For example, in **DNA** dataset with τ being 10, out of 1.4 million candidates, there are 1 million single-match candidates, none of which contributes to the actual result; in **Title** and **URL** with τ being 10 and 3 respectively, the number of all candidates, single-match candidates, and actual results that are single-match candidates are 9.5 millions, 8.1 millions, 0 and 18.8 millions, 17.4 millions, 9030 respectively.

Example 4.5. Consider the three candidates s_2, s_3 and s_4 computed by SEARCH in Example 3.4. Obviously, they are single-match candidates, all of which are solely introduced by the pivotal q-gram $\langle le, 6 \rangle$ that has only one matched position (i.e., start position 5) in each of the three candidate strings.   □

Motivated by the above, in the following we propose two advanced filters, pivotal substitution filter and position match filter, to refine the single-match candidates.

4.1 Pivotal Substitution Filter

For our cross pivotal filter, a natural extension of Theorem 3.1 is that, let $\mathsf{piv}(r)$ and $\mathsf{piv}'(r)$ denote two different pivotal sets computed from $\mathsf{q}(r)$, if s and r are similar, then $\mathsf{piv}(s) \cap \mathsf{q}(r) \neq \emptyset$ and $\mathsf{piv}(r) \cap \mathsf{q}(s) \neq \emptyset$ and $\mathsf{piv}'(r) \cap \mathsf{q}(s) \neq \emptyset$. Moreover, this can be extended to many more pivotal sets. However, the query with multiple pivotal sets would involve significant filtering cost, as it increases the number of strings required to be processed. To address this issue, based on the above observation of single-match candidates, we propose the pivotal substitution filter to perform multiple pivotal queries efficiently.

Definition 4.3. (Pivotal Substitution) Given a pivotal set $\mathsf{piv}(r)$ and a q-gram $g_i \in \mathsf{piv}(r)$, a pivotal substitution of g_i is a q-gram from $\mathsf{q}(r)$ that is disjoint with the other τ pivotal q-grams of $\mathsf{piv}(r)$.   □

Pivotal Substitution Filter. For a single-match candidate s introduced by the i-th pivotal q-gram g_i, if there exists such a pivotal substitution of g_i in $\mathsf{q}(r)$ that is not in $\mathsf{q}(s)$, then s and r cannot be similar.

The intuition of the pivotal substitution filter is as follows. Let g' be such a pivotal substitution, and let $\mathsf{piv}'(r)$ be the result of substituting g_i with g' in $\mathsf{piv}(r)$. Then, $\mathsf{piv}'(r) \cap \mathsf{q}(s) = \emptyset$; thus, s and r are not similar.

Example 4.6. Consider, the candidate s_3 in Example 4.5. For s_3, the matched pivotal q-gram is $\langle le, 6 \rangle$. Assume we substitute it with $\langle da, 1 \rangle$, but $\langle da, 1 \rangle$ is not in $\mathsf{q}(s_3)$; thus, s_3 is pruned by the pivotal substitution filter.   □

Implementation and Time Complexity. To minimize the filtering cost, when selecting a pivotal substitution for a pivotal q-gram g_i, we select the one with minimum universal order among all q-grams in $\mathsf{q}(r)$ that satisfy the non-overlapping constraint. Once a pivotal substitution g' is selected, the candidate set is refined by conducting an intersection with the inverted list of g'. Note that, we perform the above pivotal substitution filter for each pivotal in $\mathsf{piv}(r)$. Thus, the time complexity of applying each pivotal substitution filter linear to the size of the inverted list of the pivotal substitution (i.e., $O(|I_{pre}[g']|)$).

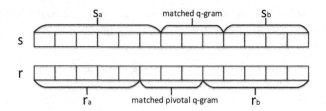

Fig. 1. A Demonstration of Pivotal Match Filter

4.2   Position Match Filter

In this subsection, we propose the position match filter for single-match candidates based on their matched positions.

Position Match Filter. For a single-match candidate s that is introduced by the i-th pivotal q-gram g_i in $\mathsf{piv}(r)$ and has a matched position pos, if $|pos - g_i.pos| > i-1$ or $|(|s| - pos) - (|r| - g_i.pos)| > \tau+1-i$, then s cannot be similar to the query string r.

We prove the correctness of the position match filter by showing that, assuming s and r are similar, then $|pos - g_i.pos| \leq i-1$ and $|(|s|-pos)-(|r|-g_i.pos)| \leq \tau + 1 - i$. Since s is a single-match candidate and only matches the pivotal q-gram g_i, then g_i must be matched to the q-gram of s with start position pos and remains unchanged in the optimal sequence of edit operations. As illustrated in Figure 1, let s_a and s_b denote the left substring and the right substring in s separated by the matched q-gram respectively, and r_a and r_b are defined similarly. Then, we have $\mathsf{ed}(s,r) = \mathsf{ed}(s_a,r_a) + \mathsf{ed}(s_b,r_b)$.

Moreover, we have $\mathsf{ed}(s_a,r_a) \geq i - 1$ and $\mathsf{ed}(s_b,r_b) \geq \tau + 1 - i$, since there are $i - 1$ ($\tau + 1 - i$) unmatched pivotal q-grams between s_a and r_a (s_b and r_b), respectively. For s and r being similar, $ed(s,r) \leq \tau$. Thus, $\mathsf{ed}(s_a,r_a) \leq \mathsf{ed}(s,r) - \mathsf{ed}(s_b,r_b) \leq i-1$, and $\mathsf{ed}(s_b,r_b) \leq \tau+1-i$. Therefore, $\mathsf{ed}(s_a,r_a) = i-1$ and $\mathsf{ed}(s_b,r_b) = \tau + 1 - i$.

Note that $\mathsf{ed}(s_a,r_a) \geq |pos - g_i.pos|$ and $\mathsf{ed}(s_b,r_b) \geq |(|s| - pos - q + 1) - (|r| - g_i.pos - q + 1)| = |(|s| - pos) - (|r| - g_i.pos)|$, due to lower bounding by length difference. Therefore, the position match filter is correct.

Example 4.7. As mentioned in Example 4.5, s_4 is a single-match candidate introduced by $\langle le, 6 \rangle$ with matched position $pos = 5$. Now we apply the position match filter to s_4. The length difference of the two substrings on the left is $|pos - g_i.pos| = |5 - 6| = 1$ and the length difference of the two on the right is $|(|s| - pos) - (|r| - g_i.pos)| = |(13 - 5) - (12 - 6)| = 2$. Also, we have $i = 2$. Therefore, $|pos - g_i.pos| = 1 = i - 1$ and $|(|s| - pos) - (|r| - g_i.pos - q)| = 2 > \tau + 1 - i = 1$. Consequently, s_4 is pruned by the position match filter. □

Time Complexity. Obviously, the running time for applying position match filter on each single-match candidate is constant. Thus, this filter can be applied very efficiently.

5   Extension for Dynamic Thresholds

In this section, we extend our techniques to support dynamic thresholds. Given a maximum threshold τ_{max}, we compute the pivotal set piv for τ_{max}, and then for each $1 \leq \tau \leq \tau_{max}$, the pivotal set for τ is selected from piv. In the indexing phase, we construct two sets of inverted indexes $I_{pre} = \{I_{pre}^0, I_{pre}^1 \ldots I_{pre}^{\tau_{max}}\}$ and $I_{piv} = \{I_{piv}^0, I_{piv}^1 \ldots I_{piv}^{\tau_{max}}\}$ where I_{pre}^i and I_{piv}^i ($0 \leq i \leq \tau_{max}$) are subsidiary indexes. I_{pre}^i is the inverted index built for the q-grams from the $((2q-1)(i-1)+1)$-th position (excluded) to the $((2q-1)i+1)$-th position in the prefix of each string in S, and I_{piv}^i is the inverted index for the $(i+1)$-th pivotal q-gram in the pivotal set for each string in S. In query processing phase, given a query string r and a threshold τ_q, the candidate set is computed as $(\bigcup_{i=0}^{i=\tau_q}(\mathsf{pre}_{(2q-1)*\tau_q+1}(r) \cap I_{piv}^i)) \cap (\bigcup_{i=0}^{i=\tau_q}(\mathsf{piv}(r) \cap I_{pre}^i))$.

6   Experiments

In this section, we evaluate the performance of our proposed approach, CROSSPIV-OTALSEARCH. We compare it with three state-of-the-art approaches, PIVOTAL-PREFIX [4], INDEXGRAM-TURBO [15], and INDEXCHUNK-TURBO [15], and we obtained the source code of the three approaches from the authors. All approaches are implemented in C++ and complied using g++ 4.8.2 with -o3 flag. All experiments are conducted on a machine with an Intel Quad-Core 3.20G CPU and 16 GB memory running 64bit Ubuntu. We use three real datasets in our experiments: the medical publication title dataset **Title**, a DNA sequence dataset **DNA** and a hyperlink dataset **URL**. Statistics of the datasets are shown in Table 6. For each dataset, we generate a set of query strings randomly selected from the dataset.

Table 6. Statistics of Datasets

Dataset	# of strings	Avg Length	Size (MB)	Query Size
DNA	2,476,276	108	269	2,305
Title	4,000,000	100.6	402	4,000
URL	1,000,000	28.03	28	1,000

In our approach, CROSSPIVOTALSEARCH, we first generate candidates by using the cross pivotal filter, and then apply the position match filter and the pivotal substitution filter to further refine the candidate set; for verification, we first use the alignment filter [4] to perform the final pruning for each candidate string, and then verify the candidates with length-aware verification method [5]. The q-gram lengths are tuned for the best performance and shown as follow: for **DNA** dataset, q = 12, 12, 12, 12, 11, 11, 11, 10, 9, 9, 8, 7, 6, 6 for τ varying from 2 to 15; for **Title** dataset, q = 8,8,8,8,6,6,6,5,4,4,4,3,3,3 for τ varying from 2 to 15; and for **URL**, q = 6,3,3,2,2,2,2,2 for τ varying from 1 to 8. Thus, we have both small τ (i.e., $1 \leq \tau \leq 10$) and large τ (i.e., $10 \leq \tau \leq 15$).

6.1   Comparison with State-of-the-art Approaches

In this subsection, we evaluate the approaches by two metrics: candidate number and query processing time.

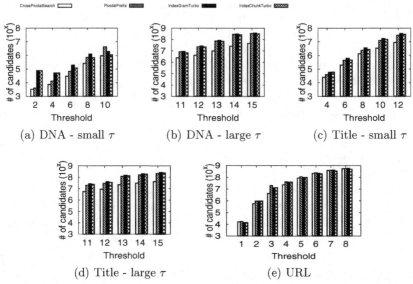

Fig. 2. Comparison with state-of-the-art Approaches - Candidate Number

Candidate Number. The results of candidate number are shown in Figure 2. We can see that CROSSPIVOTALSEARCH achieves the least candidates among the four algorithms. For example, consider Figure 2(c) for the **Title** dataset with $\tau = 10$, CROSSPIVOTALSEARCH extracts 3.3 million candidates, while the candidate number for PIVOTALPREFIX, INDEXGRAM-TURBO and INDEXCHUNK-TURBO are 12 millions, 17 millions and 15 millions respectively. CROSSPIVOTALSEARCH has the smallest number of candidates due to the stronger pruning power of cross pivotal filter and the further pruning power of the two advanced filters. For the best case scenario, CROSSPIVOTALSEARCH only extracts 13.4%, 11.9% and 14.1% of the candidates of PIVOTALPREFIX, INDEXGRAM-TURBO and INDEXCHUNK-TURBO respectively for the **DNA** dataset; those numbers for the **Title** and **URL** datasets, are 18.3%, 14.7%, 15.9% and 21.5%, 33.5%, 32.4%.

Processing Time. Figure 3 shows the average query processing time for the four approaches. CROSSPIVOTALSEARCH and PIVOTALPREFIX run significantly faster than INDEXGRAM-TURBO and INDEXCHUNK-TURBO, with CROSSPIVOTALSEARCH being the best. For instance, in **DNA** dataset with $\tau = 13$, CROSSPIVOTALSEARCH only takes 10.8 milliseconds, while the average processing time for PIVOTALPREFIX, INDEXGRAM-TURBO and INDEXCHUNK-TURBO are 36.3 milliseconds, 159.9 milliseconds and 135.9 milliseconds, respectively.

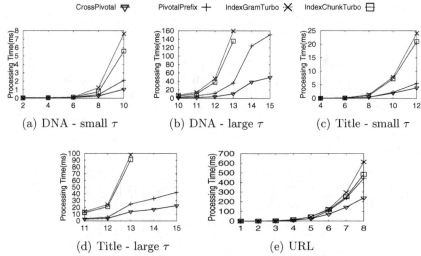

Fig. 3. Comparison with state-of-the-art Approaches- Query Time

Note that, in Figure 3(b) and Figure 3(d), we omit the result for $\tau = 14$ and $\tau = 15$ for INDEXGRAM-TURBO and INDEXCHUNK-TURBO, due to the excessive processing time compared to that of CROSSPIVOTALSEARCH and PIVOTALPREFIX. For the best case scenario, CROSSPIVOTALSEARCH only achieves 31.2%, 5.6% and 5.7% of processing time of PIVOTALPREFIX, INDEXGRAM-TURBO and INDEXCHUNK-TURBO respectively for the **DNA** dataset; those numbers for the **Title** and **URL** datasets, are 51.8%, 10.2%, 10.5% and 53.7%, 38.9%, 49.2%.

We also compare the index size of the approaches and conclude that CROSSPIVOTALSEARCH has the smallest index size. For example, given a small scale medical title dataset (75.5MB), with $\tau = 8$ and $q = 6$, the index size of CROSSPIVOTALSEARCH is 1.4GB, while that of CROSSPIVOTALSEARCH, INDEXGRAM-TURBO and INDEXCHUNK-TURBO are 1.9GB, 3.1GB and 2.3GB, respectively.

6.2   Evaluation on Advanced Filters

In this subsection, we evaluate the two advanced filters: position match filter and pivotal substitution filter. We compare CROSSPIVOTALSEARCH with CROSSPIVOTALBASIC, which is the same as CROSSPIVOTALSEARCH except without the two advanced filters.

Figures 4 and 5 depict the candidate numbers and the average query processing time of the two approaches. We can observe that CROSSPIVOTALSEARCH only has 50% to 80% of the candidates of CROSSPIVOTALBASIC, and consequently CROSSPIVOTALSEARCH runs faster than CROSSPIVOTALBASIC. This is due to that single-match candidates account for a large portion of the candidates obtained by the cross pivotal filter, while the two advanced filters can prune out a large amount of single-match candidates.

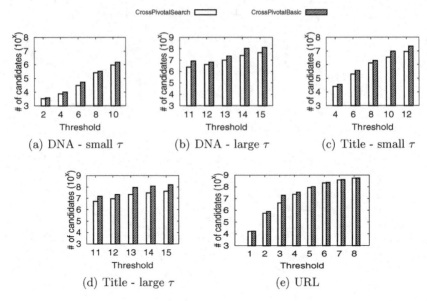

Fig. 4. Evaluation of Advanced Filters - Candidate

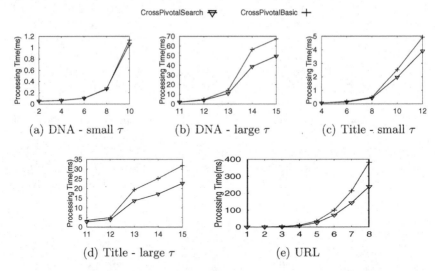

Fig. 5. Evaluation of Advanced Filters - Query Time

7   Conclusion

In this paper, we studied the problem of string similarity search with edit distance constraint. We proposed an efficient cross pivotal based approach, and proved its strongest pruning power compared with the state-of-the-art approaches. We then devised two advanced filters, position match filter and pivotal substitution filter to further reduce the number of candidates. Finally, we compared our cross pivotal based approach with other three state-of-the-art pivotal based approaches by performing a comprehensive experimental study, and empirical evaluations on real datasets demonstrate the superiority of our approach in terms of both candidate number and query processing time.

Acknowledgments. Lijun Chang is supported by ARC DE150100563. Wenjie Zhang is supported by ARC DE120102144, DP120104168, ARC DP150103071 and DP150102728. Xuemin Lin is supported by NSFC61232006, ARC DP120104168, ARC DP140103578, and ARC DP150102728.

References

1. Arasu, A., Ganti, V., Kaushik, R.: Efficient exact set-similarity joins. In: VLDB (2006)
2. Chaudhuri, S., Ganjam, K., Ganti, V., Motwani, R.: Robust and efficient fuzzy match for online data cleaning. In: SIGMOD (2003)
3. Chaudhuri, S., Ganti, V., Kaushik, R.: A primitive operator for similarity joins in data cleaning. In: ICDE (2006)
4. Deng, D., Li, G., Feng, J.: A pivotal prefix based filtering algorithm for string similarity search. In: SIGMOD (2014)
5. Deng, D., Li, G., Feng, J., Li, W.: Top-k string similarity search with edit-distance constraints. In: ICDE (2013)
6. Forman, G., Eshghi, K., Chiocchetti, S.: Finding similar files in large document repositories. In: SIGKDD (2005)
7. Gravano, L., Ipeirotis, P.G., Jagadish, H.V., Koudas, N., Muthukrishnan, S., Srivastava, D.: Approximate string joins in a database (almost) for free. In: VLDB (2001)
8. Hadjieleftheriou, M., Koudas, N., Srivastava, D.: Incremental maintenance of length normalized indexes for approximate string matching. In: SIGMOD (2009)
9. Kahveci, T., Singh, A.K.: Efficient index structures for string databases. In: VLDB (2001)
10. Kim, Y., Shim, K.: Efficient top-k algorithms for approximate substring matching. In: SIGMOD (2013)
11. Li, C., Lu, J., Lu, Y.: Efficient merging and filtering algorithms for approximate string searches. In: ICDE (2008)
12. Li, C., Wang, B., Yang, X.: VGRAM: improving performance of approximate queries on string collections using variable-length grams. In: VLDB (2007)
13. Navarro, G.: A guided tour to approximate string matching. ACM Comput. Surv. **33**(1) (2001)
14. Qin, J., Wang, W., Lu, Y., Xiao, C., Lin, X.: Efficient exact edit similarity query processing with the asymmetric signature scheme. In: SIGMOD (2011)
15. Qin, J., Wang, W., Xiao, C., Lu, Y., Lin, X., Wang, H.: Asymmetric signature schemes for efficient exact edit similarity query processing. ACM Trans. Database Syst. **38**(3) (2013)
16. Sokol, D., Benson, G., Tojeira, J.: Tandem repeats over the edit distance. Bioinformatics **23**(2) (2007)
17. Wang, J., Li, G., Feng, J.: Can we beat the prefix filtering?: an adaptive framework for similarity join and search. In: SIGMOD (2012)

18. Xiao, C., Wang, W., Lin, X.: Ed-join: an efficient algorithm for similarity joins with edit distance constraints. PVLDB **1**(1) (2008)
19. Xiao, C., Wang, W., Lin, X., Yu, J.X.: Efficient similarity joins for near duplicate detection. In: WWW (2008)
20. Yang, X., Wang, B., Li, C.: Cost-based variable-length-gram selection for string collections to support approximate queries efficiently. In: SIGMOD (2008)
21. Zhang, Z., Hadjieleftheriou, M., Ooi, B.C., Srivastava, D.: Bed-tree: an all-purpose index structure for string similarity search based on edit distance. In: SIGMOD (2010)

Security and Privacy

Authentication of Top-k Spatial Keyword Queries in Outsourced Databases

Sen Su[1]($\boxtimes$), Han Yan[1], Xiang Cheng[1],
Peng Tang[1], Peng Xu[1], and Jianliang Xu[2]

[1] State Key Laboratory of Networking and Switching Technology,
Beijing University of Posts and Telecommunications, Beijing, China
{susen,yanh,chengxiang,tangpeng,xupeng}@bupt.edu.cn
[2] Department of Computer Science, Hong Kong Baptist University,
Kowloon Tong, Hong Kong
xujl@comp.hkbu.edu.hk

Abstract. In this paper, we study the authentication of top-k spatial keyword queries in outsourced databases. We first present a scheme based on tree-forest indexes, which consist of an MR-tree (which is the state-of-the-art authenticated data structure for the authentication of spatial queries) and a collection of Merkle term trees (MT-trees). The tree-forest indexes can support efficient top-k spatial keyword query (kSKQ) processing and authentication. To derive a small verification object (VO) to be returned to the user, we put forward an entry pruning based scheme, where an MT*-tree is presented. The entries in each node of MT*-tree are ordered and an embedded Merkle hash tree (embedded-MHT) is constructed over them. By employing a novel pruning strategy, the redundant entries in each node of MT*-trees can be eliminated from VO. Our extensive experiments verify the effectiveness, efficiency and scalability of our proposed schemes on several performance metrics, including the index construction time, index size, running time, VO size and authentication time.

1 Introduction

Owing to the popularization of the positioning-enabled devices (e.g., smart phones) and booming of the mobile internet, location-based services (LBSs) have become a vital part in our daily activities in recent years. Such services provide users with location-aware query experiences based on their locations. Since a great many of real-world applications have requirements to support the top-k spatial keyword query (kSKQ), it has attracted considerable attention from both academia and industry communities. In a spatial-textual database, each object has two attributes: one is the location, the other is the textual description (or called document). Given such a database and a query request with a location, a set of keywords and a positive integer k, kSKQ finds k objects which are relatively nearer to the query location and whose documents are comparatively more similar to the

© Springer International Publishing Switzerland 2015
M. Renz et al. (Eds.): DASFAA 2015, Part I, LNCS 9049, pp. 567–588, 2015.
DOI: 10.1007/978-3-319-18120-2_33

keywords in the query. For instance, a user may want to find a "restaurant that serves good beer and barbecue" and close to the user's current location.

If the data owner (DO) of a massive spatial-textual database wants to provision kSKQ services, he/she needs to build up the basic IT infrastructure and hire specialized personnel. However, as such cost might be unaffordable for small-to-medium businesses, database outsourcing to a third-party location-based services provider (LBSP) has been an appealing option for better making use of the spatial-textual data. Yet, this outsourcing model brings a great challenge that the query results returned by the LBSP might be incomplete or incorrect. There are a variety of reasons for this. Firstly, the LBSP may return tailored results for profit purposes (e.g., tampering with the ranking of top-k results in favor of sponsors). Secondly, even if the LBSP is trustworthy, it is still likely that its server is intruded by attackers. If an attacker takes control of the server, he/she may forge the results for his/her own interest.

The aforementioned reasons necessitate the development of mechanisms that will allow users to authenticate the kSKQ results that the LBSP returns. The users need to verify the *soundness* and *completeness* of query results through a proof, called verification object (VO) returned by the LBSP. In particular, the *soundness* means that the original spatial-textual data in the result set is not tampered with, while the *completeness* implies that no valid result is missing.

In this paper, to make one step closer towards practical deployment of LBSs in outsourced databases, we study the authentication of top-k spatial keyword queries (AkSKQ) problem. Most of the existing work related to this problem considered either spatial queries [21,22] or textual searchings [14], but not a combination of them. However, as we described above, each object in the spatial-textual database has a composite of both spatial and textual attributes and the definition of the kSKQ involves the computation of both spatial proximity and textual similarity. Therefore, the authentication techniques proposed in these previous studies cannot be applied to solve our problem. Most recently, Wu *et al.* [18] studied a similar problem which is the authentication of moving top-k spatial keyword queries, where they used a special ranking function (see Section 2). In contrast, we focus on the authentication of snapshot top-k spatial keyword queries with a more widely adopted ranking function [4].

A basic approach for tackling this problem is to simply combine IR-tree (which is the state-of-the-art index to answer the kSKQ) [4,19] and Merkle hash tree (MHT) [13] to form an MIR-tree. Based on MIR-tree, a best-first traversal algorithm can be employed to process the kSKQ. Meanwhile, VO is generated based on the nodes which have been visited. After receiving query results and VO, the hash value of the root of MIR-tree is reconstructed by the user in a bottom-up manner to verify the *soundness* of query results. As for verifying the *completeness*, the user re-computes the ranking score of each returned object. However, this approach is not very practical as large inverted files (which are used to index the textual information in MIR-tree) are included in VO. Thus, it will result in a tremendous communication overhead between the LBSP and the user. In addition, it also leads to excessive computation cost at the user-side. Therefore, the AkSKQ still remains a very challenging problem.

To reduce the VO size and make VO more suitable for the authentication, we propose a scheme based on tree-forest indexes, where MIR-tree is split into an MR-tree and a collection of Merkle term trees, denoted by MT-trees (each term corresponds to an MT-tree). MR-tree is the state-of-the-art authenticated data structure (ADS) for the authentication of spatial queries [21,22], while the structure of each MT-tree is similar to that of MR-tree. In MT-trees, only the textual information of each object is stored. Based on tree-forest indexes, we introduce an *extensional priority queue* to assist the query processing. Since the query processing only involves the access to the MT-trees associated with the keywords the user inputs, VO does not include inverted files any more. Therefore, the VO size and authentication time can be sharply reduced. Moreover, to further optimize VO, we present an entry pruning based scheme, where an MT*-tree is developed. Specifically, the entries in each node of MT*-tree are ordered and organized by an embedded Merkle hash tree (embedded-MHT). Based on MT*-trees, we present a novel pruning strategy to avoid, as much as possible, returning the entries in each node of MT*-trees which are irrelevant to the authentication of query results. A thorough experimental study on real datasets is conducted over a wide range of parameter settings to evaluate the effectiveness, efficiency and scalability of our proposed authentication schemes on several performance metrics, including the index construction time, index size, running time, VO size and authentication time.

The rest of this paper is organized as follows. Section 2 introduces the necessary background and presents the problem formulation. Section 3 presents a basic approach. A tree-forest indexes based scheme is proposed in Section 4, followed by Section 5, where an entry pruning based scheme is put forward. Section 6 presents the experimental evaluation results. Related work in spatial keyword query processing and query authentication is surveyed in Section 7. In the end, we conclude the paper in Section 8.

2   Background and Problem Formulation

2.1   Background

Cryptographic Primitives. In this section, we review the cryptographic primitives that underlie our proposed schemes.

One-Way Hash Function. A one-way hash function $H(\cdot)$ maps a message m with arbitrary length to a fixed-length output $H(m)$. It works in one direction. It is easy to compute $H(m)$ for a message m. However, it is computationally infeasible to find a message m that maps to a given $H(\cdot)$.

Cryptographic Signature. A cryptographic signature (or simply *signature*) is a mathematical scheme for demonstrating the authenticity of a digital message. In particular, a signer creates a pair of a private key and a public key. The former is kept by the signer secretly and the latter is publicly distributed. A digital message can be signed by its owner using his/her private key. The authenticity of the message can be verified by anyone who receives this message using the owner's public key. RSA [17] is the most widely used signature algorithm.

Merkle Hash Tree. The Merkle hash tree (MHT) [13] is an authenticated data structure (ADS) used for collectively authenticating a set of messages. The Merkle hash tree is a binary tree and built in a bottom-up manner, by first computing the hash values of the messages in leaf nodes. The hash value of each internal node is derived from its two child nodes. Finally, the hash value of the root is signed by the owner of the messages. Moreover, MHT can be used to authenticate any subset of messages, in conjunction with a *proof*. The *proof* consists of the signed root and sibling nodes (auxiliary hash values) on the path from the root down to the messages which need to be authenticated. Its idea can be extended to multi-way trees.

Top-k Spatial Keyword Query. Due to the space limitations, we refer readers to [4,19] for details of the spatial-textual database, IR-tree and top-k spatial keyword query. Here we just show some notations used in this paper. ① O: an object in the database; ② R: a minimum bounding rectangle (MBR) of the objects which are relatively close to each other; ③ $O.d$: the document of O; ④ $R.d$: the pseudo document of R. Besides, a collection of all the distinct terms constitutes a *dictionary* and its size is denoted by M. In addition, the ranking function used to compute the ranking score of an object O with respect to the query Q is defined as follows (note that the smaller is the ranking score, the better)

$$RS(Q,O) = \alpha \frac{D(Q,O)}{D_{\max}} + (1-\alpha)(1 - \frac{S(Q,O)}{S_{\max}}), \tag{1}$$

where $\alpha \in [0,1]$ is a parameter used to balance the spatial proximity and textual similarity. The Euclidian distance between the query location and the object O is computed by the function $D(Q,O)$ (can also be called the *distance score*). Besides, the function $S(Q,O)$ is the Okapi formulation [28] which is a textual similarity function and is effective in practice (can also be called the *textual score*), i.e.,

$$S(Q,O) = \sum_{t \in Q} w_{Q,t} \cdot w_{O,t}, \tag{2}$$

where $w_{Q,t}$ and $w_{O,t}$ are the associated weights of the term t in the query Q and the document of the object O, respectively. $w_{Q,t} \cdot w_{O,t}$ can be called the *term score* of the term t.

Note that our ranking function is widely adopted in existing studies [4]. Although Wu *et al.* [18] studied a similar problem which is the authentication of moving top-k spatial keyword queries, they used a special ranking function as defined below

$$RS(Q,O) = \frac{D(Q,O)}{S(Q,O)}. \tag{3}$$

Our proposed schemes are irrelevant to the selection of the ranking functions. Therefore, they can also be extended to support the Eq. (3). We will experimentally compare with the method proposed in [18] for snapshot queries in Section 6.

2.2   Problem Formulation

System Model. Our system involves three entities: the data owner (DO), the location-based services provider (LBSP) and the users.

Before outsourcing a spatial-textual database to the LBSP, the DO builds an authenticated data structure (ADS) on the database. To support efficient kSKQ processing at the LBSP-side, the ADS is often a tree-like index structure. To ensure the integrity of the original spatial-textual database, the DO signs the root of the ADS using his/her private key. Besides, algorithms for processing the kSKQ are also designed by the DO. The LBSP provides the storage resources for the spatial-textual database, ADS, root's signature of the ADS and algorithms. After processing a user's kSKQ using the algorithms designed by the DO, the LBSP returns query results as well as a verification object (VO) which is generated based on the ADS. The *soundness* and *completeness* of query results can be verified by users using the returned VO, root's signature of the ADS and DO's public key.

Threat Model. Among three entities in our system model, we consider the DO is unconditionally trusted but the third-party LBSP is the potential adversary. The LBSP might be under the control of malicious attackers who might return incorrect results to users intentionally. It is also possible in the case of out-of-control of the DO, the LBSP modifies the original spatial-textual database and its index, executes the query processing algorithm incorrectly or tampers with query results in favor of sponsors.

Problem Statement. In this paper, we study the **A**uthentication of top-k **S**patial **K**eyword **Q**ueries (AkSKQ) in outsourced databases. That is, given a top-k spatial keyword query, the LBSP not only needs to efficiently retrieve a ranking list of k objects according to their ranking scores but also needs to generate VO for users to verify the *soundness* and *completeness* of query results. VO should be generated as small as possible for minimizing the communication cost between the LBSP and the user. At the mean time, VO should be suitable for user's authentication of query results.

3   A Basic Approach

In this section, we present a basic approach to solve the AkSKQ problem. The main idea is to integrate Merkle hash tree (MHT) into IR-tree [4,19] to form a new index: MIR-tree which underlies the kSKQ processing and authentication of query results.

3.1   Index Structure

MIR-tree combines the concepts of Merkle hash tree (MHT) and IR-tree. As Fig. 1 shows, the leaf nodes of MIR-tree are identical to those of IR-tree. Each entry in the leaf nodes of MIR-tree corresponds to an object, i.e., O, while each entry in the internal nodes of MIR-tree is represented by a tuple, i.e., $(R, H(R))$, where R is an MBR and $H(R)$ is the hash value of R.

The hash value of each entry in the internal nodes of MIR-tree is computed by the binary concatenation of the entries and inverted file included in its child node. Specifically, if the child of R is a leaf node, $H(R)$ is given by: $H(R) = H(O_1|...|O_i|IF)$, where $O_1, ..., O_i$ and IF are the entries and inverted file included in this leaf node. Otherwise, if the child of R is an internal node, $H(R)$ is given by: $H(R) = H((R_1|H(R_1))|...|((R_i|H(R_i))|IF'))$, where $(R_1, H(R_1)), ..., (R_i, H(R_i))$ and IF' are the entries and inverted file included in this internal node. The hash value of the root of MIR-tree is signed by the DO using his/her private key and stored with the MIR-tree.

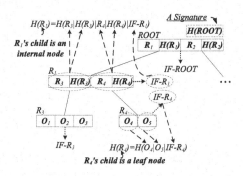

Fig. 1. MIR-tree

3.2   Query Processing

To process the kSKQ over MIR-tree, a best-first traversal algorithm is employed to retrieve the ranking list of k objects. Meanwhile, VO is generated based on the entries and inverted files included in the nodes which have been visited. After the query processing, VO is returned to the user together with the query results. Algorithm 1 shows the pseudocode of the kSKQ processing and VO generation over MIR-tree.

As for the kSKQ processing, similar to IR-tree, we start from the root of MIR-tree (line 1) and traverse the tree in a best-first manner. When deciding which entry to be visited next, we pick the entry with the smallest ranking score in a priority queue which includes all the candidate entries that have yet to be visited (line 3). Recall that the smaller is the ranking score, the better. When an MBR is picked from the priority queue (line 4), we compute the ranking score of each entry in its child node and put them into the priority queue again

---

Algorithm 1. kSKQ Processing and VO Generation over MIR-tree

Input: A kSKQ request, MIR-tree
Output: The result set and VO

1 Put each entry R_i in the root into a priority queue and initialize VO with ① '['; ②
   each R_i and $H(R_i)$; ③ the inverted file associated with the root; ④ ']';
2 **while** k *objects have not been found* **do**
3 | Pick the entry with the smallest ranking score from the priority queue;
4 | **if** *the picked entry is an MBR R_i* **then**
5 | | Put each R_j (or O_j) in R_i's child node into the priority queue again;
6 | | Replace R_i and $H(R_i)$ with ① '['; ② each R_j and $H(R_j)$ (or O_j) in R_i's
   | | child node; ③ the inverted file associated with R_i's child node; ④ ']' in VO;
   |
7 | **else**
8 | | // the picked entry is an object O_i
9 | | Put O_i into the result set;
   |
10 Return the result set and VO;

(line 5). Otherwise, when an object is picked (lines 7-8), we put it into the result set (line 9). As for the VO generation, we use a pair of tokens '[' and ']' to indicate the scope of the entries in a node. In the beginning, we initialize VO with following four parts: ① '['; ② each R_i and $H(R_i)$ in the root; ③ the inverted file associated with the root; ④ ']' (line 1). Once an MBR R_i is picked from the priority queue (line 4), we adopt a replacement strategy for R_i and its hash value $H(R_i)$. Specifically, we replace R_i and $H(R_i)$ with ① '['; ② each R_j and $H(R_j)$ (or O_j) in R_i's child node; ③ the inverted file associated with R_i's child node; ④ ']' (line 6). Otherwise, if an object is picked (lines 7-8), it has been included in VO with certainty since we have adopted the replacement strategy for its parent MBR before. In this case, VO remains unchanged. The above procedure is repeated until k objects have been found (line 2).

3.3 Authenticating Query Results

To authenticate the *soundness* of query results, the user needs to scan VO to reconstruct the hash value of the root of MIR-tree and compare it against the root signature using the DO's public key. Since VO includes the entries which have been visited during the query processing, the user can simulate the procedure of MIR-tree traversal and recursively reconstruct each MBR and compute its hash value in a bottom-up manner. Specifically, each MBR and its hash value can be computed from the entries and inverted file in its child node which are indicated by '[' and ']'.

To authenticate the *completeness* of query results, the user first needs to check each object in the result set is indeed present in VO and the ranking scores of them are smaller than those of other entries returned in VO.

3.4 Limitations of the Basic Approach

The basic approach discussed above, however, is not very practical. The major problem faced by this approach is that the VO size is too large. VO includes the nodes which have been visited during the query processing and large inverted files are associated with these nodes. Therefore, it will incur high communication cost when VO is returned to the user. Moreover, it will also impose excessive computation cost for the authentication process at the user-side. This is because when the user verifies the *completeness* of query results, he/she needs to traverse the corresponding inverted files to find the weights of the keywords he/she inputs to re-compute the ranking score of each entry in VO.

4   Tree-Forest Indexes Based Scheme

An optimized scheme, tree-forest indexes based scheme, is designed to overcome the drawbacks of the basic approach. The motivation of this scheme is based on the following observation. In the basic approach, when the user re-computes the ranking score of each entry in VO, he/she needs to retrieve the weights of the keywords from the corresponding inverted files. However, only the keywords that he/she inputs are involved in the computation of ranking scores. The number of these keywords is rather smaller than that of terms in any of those inverted files. Therefore, we decouple the spatial and textual information in MIR-tree by splitting MIR-tree into tree-forest indexes which include an MR-tree and a collection of Merkle term trees (MT-trees). Specifically, we use the state-of-the-art authenticated data structure (ADS) MR-tree to authenticate the spatial attribute of each object and compute the spatial proximity between the query location and that object. In addition, for each term in the *dictionary*, we build an MT-tree and use it to authenticate the textual attribute of each object and compute the textual similarity of the keywords in the query with respect to the document of that object. Since the query can be processed just based on MR-tree and a few MT-trees, we can avoid returning VO with large inverted files. Therefore, the VO size and authentication time can be both dramatically reduced.

4.1 Index Structure

The structure of tree-forest indexes is shown in Fig. 2. In total, the data owner (DO) needs to build one MR-tree and M MT-trees. Recall that M is the size of the *dictionary*. Each MT-tree is associated with a term in the *dictionary*.

Structural Inconsistency. When the ranking score of each entry is computed, the *distance score* and *textual score* need to be computed simultaneously from MR-tree and corresponding MT-trees. Hence, the access to entries in MR-tree and MT-trees needs to be synchronized. We refer to this kind of access as *synchronous access*. Thus, to guarantee the synchronization of the access, we let the structure of each MT-tree be **completely** the same as MR-tree. The structural consistency results in a fact that there is a one-to-one correspondence between each entry of MR-tree and MT-tree. Therefore, we introduce the notion of the corresponding entry.

Definition 1. *(**Corresponding Entry***) Given an entry e_L (which includes an object, i.e., O) in a leaf node of MR-tree, its corresponding entry (denoted by ce_L) is the entry with the same position as e_L in an MT-tree. In a similar way, R is an MBR included in an entry e_I in an internal node of MR-tree, and its corresponding entry (denoted by ce_I) is the entry with the same position as e_I in an MT-tree.*

Lemma 1. *In total, each e_L (e_I) has M corresponding entries.*

Lemma 2. *Each ce_L (ce_I) in an MT-tree associated with a term t only stores the weight of the term t in the document $O.d$, i.e., $O.d.t.w$ (pseudo document $R.d$, i.e., $R.d.t.w$).*

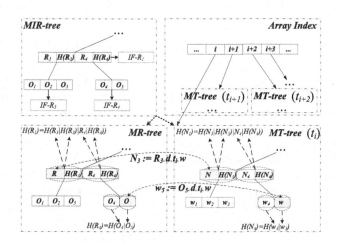

Fig. 2. Tree-Forest Indexes

Note that in the implementation, we use a same ID to indicate e_L (e_I) and each of its corresponding entries ce_L (ce_I). If the document of an object (or the pseudo document of an MBR) does not include the term t, we can store a '0' in the entry in the corresponding MT-tree for guaranteeing the structural consistency between MR-tree and MT-trees. Although the structural consistency can facilitate the synchronization of the access to MR-tree and MT-trees, this method will incur excessive extra storage cost because the average number of the objects whose documents include the term t is far less than that of the objects in the database. Thus, there will be too many '0's in the MT-tree associated with the term t and so do other MT-trees. Therefore, to save the storage cost, we only store non-zero weights in MT-trees. However, removing the redundant '0's in MT-trees results in the structural inconsistency between MR-tree and MT-trees which will make the synchronization of the access unavailable and we will discuss it later in Section 4.2.

Hashing Operations. Hash values stored in internal nodes of MR-tree and MT-trees are computed in a similar way to those of MIR-tree.

High Level Index. We use an array index to index all the MT-trees. As Fig. 2 shows, each MT-tree corresponds to a position in the array through the term ID and each position in the array includes a pointer which points to the root of the corresponding MT-tree. Contrast to MR-tree and all the MT-trees, this array index can be stored in memory. Therefore, extra I/O operations when processing the kSKQ over tree-forest indexes can be reduced.

4.2 Query Processing

The procedure of the kSKQ processing over tree-forest indexes is similar to that of the basic approach. However, there are two main differences between them, which are detailed below.

1). Multiple Trees are Visited during the Query Processing
When the ranking score of each entry is computed, the Euclidian distance between the query location and the entry is computed through MR-tree, while the textual similarity of the keywords in the query with respect to the document (or pseudo document) of this entry is computed through corresponding MT-trees.

As we discussed above, removing the redundant '0's in MT-trees will benefit saving the storage cost. However, this will result in the inconsistency between the structure of MR-tree and MT-trees. In particular, the spatial information of each entry always exists in MR-tree, while each of its corresponding entries is not necessarily present in the corresponding MT-tree. Therefore, when the ranking score of an entry is computed, we need to use MR-tree as the benchmark and find the weights stored in its corresponding entries in MT-trees. This kind of access to the entries in MR-tree and MT-trees is referred to as *asynchronous access*. For having a better adaptation to the *asynchronous access*, we use an *extensional priority queue*, where we assign a vector for each entry in it. Each vector has $k+1$ elements and each element is a pointer. One points to the entry in MR-tree and other k pointers point to the entry's k corresponding entries in MT-trees. Thus, if an entry is not present in MT-tree, its corresponding pointer in the vector is *null*. For instance, if $k = 2$ and R_3 is in the *extensional priority queue*, we assign a vector $< p_{R_3}, p_{R_3,t_1}, null >$ for R_3 in which p_{R_3} points to R_3 in MR-tree and p_{R_3,t_1} points to R_3's corresponding entry in the MT-tree associated with the term t_1. We assume R_3's corresponding entry in the MT-tree associated with the term t_2 is not present, thus, p_{R_3,t_2} is *null*. Note that in the basic approach, each element in the priority queue is only one pointer which points to the entry in MIR-tree.

The procedure of the query processing using the *extensional priority queue* as assistance is described as follows. If we have picked R_3 from the *extensional priority queue*, we first need to use its vector's first element p_{R_3} to access the entries in its child node (e.g., O_1, O_2) and assign each of them a vector with $k+1$ elements. For instance, for O_1, the vector is initialized with $< p_{O_1}, null, null >$ and for O_2, the initial vector is $< p_{O_2}, null, null >$, where p_{O_1} and p_{O_2} point to O_1 and O_2 in MR-tree. Then, we use other elements of R_3's vector to access the entries in their child nodes in MT-trees. For instance, through p_{R_3,t_1}, we can access O_1 and O_2's corresponding entries in the MT-tree associated with the term t_1. Then, we fill in O_1 and O_2's vectors with the addresses of O_1 and O_2's corresponding entries if they are present. For instance, for O_1, the vector is $< p_{O_1}, p_{O_1,t_1}, null >$ and for O_2, the vector is $< p_{O_2}, null, null >$, where we assume O_2's corresponding entry is not present in the MT-tree associated with the term t_1. Since p_{R_3,t_2} is *null*, we need not access the entries in R_3's child nodes in the MT-tree associated with the term t_2 and the vectors for O_1 and O_2 remain the same. After we have found the weights in R_3's child nodes in MT-trees, we can compute their ranking scores and put their vectors into the *extensional priority queue* again.

Based on tree-forest indexes, there is no need for the LBSP to retrieve and process the entire inverted file associated with the node being visited, which is necessary when using MIR-tree as the index. Therefore, it also improves the efficiency of the query processing.

2). Multiple Items are Included in VO

One of items in VO is generated through MR-tree, which is denoted by *VO(MR)*. Other items in VO are generated through the MT-trees associated with the keywords the user inputs. Let *VO(MT-t)* denote each of these items. Since the access to MR-tree and each MT-tree is asynchronous, *VO(MR)* contains the spatial information of entries which have been visited, while their textual information are not necessarily included in the corresponding *VO(MT-t)*s. Therefore, the structure of *VO(MR)* and each *VO(MT-t)* are not exactly the same. *VO(MR)* and each *VO(MT-t)* together make up VO.

4.3 Authenticating Query Results

The *soundness* of query results can be verified by reconstructing the hash values of roots of MR-tree and each MT-tree through *VO(MR)* and each *VO(MT-t)* in VO, respectively and judging whether they can match the hash values restored by DO's signatures or not.

The *completeness* of query results can be verified by re-computing the ranking score of each entry returned in VO. The spatial proximity between the query location and that entry is computed from *VO(MR)* while the textual similarity of the keywords in the query with respect to the document (or pseudo document) of that entry is computed through *VO(MT-t)*s. However, as we discussed above, we do not know whether the textual information (i.e., the weight) of each entry in *VO(MR)* is included in *VO(MT-t)*s or not. An intuitive method to solve this problem is that when we compute the ranking score of an entry in *VO(MR)*, we traverse every *VO(MT-t)* to find its textual information. However, it results in a high time complexity of $O(n^{k+1})$, where n is the number of entries in *VO(MR)* and k is the number of the keywords the user inputs. To better solve this problem, we compute the ranking score of each entry returned in VO in the following manner. Firstly, for *VO(MR)* and each *VO(MT-t)* in VO, we set pointers to indicate the positions of the entries being visited. At the beginning, the pointers all point to the first entry of each item. Secondly, we use *VO(MR)* as the benchmark and scan the item from the front to back. Each time we meet '[', ']' or the hash value, we just skip it. When we meet the first entry in *VO(MR)*, we begin to scan *VO(MT-t)*s using the corresponding pointers to find the weights stored in the corresponding entries of this entry. If the first weight we meet in *VO(MT-t)* does not belong to the document (or pseudo document) of this entry, we stop scanning this item and consider the textual information of this entry is not included in this *VO(MT-t)*. Otherwise, we fetch the weight from the item. Finally, until all the scannings are finished, we compute the entry's ranking score. This procedure is repeated until all the ranking scores of the entries in *VO(MR)* have been computed. Therefore, the number of traversal to each *VO(MT-t)* is significantly reduced and the time complexity of this procedure is only $O(n(k + 1))$.

5   Entry Pruning Based Scheme

Recall that in MT-trees, each node consists of a simple list of entries. During the query processing, the authentication information (i.e., weights or auxiliary hash values) of each pruned entry is included in VO. In practice, a node may include a large number of entries while only a small fraction of them will be relevant to query results. Consequently, many of them are pruned but their contents are inserted into VO. Based on this observation, to further reduce the VO size, we present an entry pruning based scheme. The main idea is to avoid, as much as possible, returning the redundant entries in each node of MT-trees by employing a novel pruning strategy. The redundant entries that we want to prune satisfy two conditions. The first is, obviously, that they are irrelevant to query results. The second condition is that they must not affect the authentication of query results.

It has been shown that some terms in the *dictionary* are included in most objects' documents in the database because these terms are always common terms in people's daily lives [14]. However, for some uncommon terms, there are only a few objects' documents that include them. Therefore, the number of entries in each MT-tree follows a highly skewed distribution. In particular, most of terms have only a few entries in their corresponding MT-trees (these MT-trees are referred to as *sparse MT-trees*), whereas a small minority of MT-trees include several orders of magnitude more entries (we refer to these MT-trees as *dense MT-trees*). Although the number of *dense MT-trees* is relatively small, most queries may still involve a mix of *dense MT-trees* and *sparse MT-trees* since *dense MT-trees* correspond to common terms. Therefore, in practice, our scheme is effective in helping further reducing the VO size although some entries may not be present in each node of MT-trees.

5.1 Index Structure

To better utilize the entry pruning strategy, some modifications are made to the structure of each MT-tree. In particular, we sort the entries in each node N of an MT-tree in descending order by their weights. Moreover, we adopt an embedded-MHT technique to compute the hash value of each N. Each embedded-MHT is a binary tree. The hash value of the root of the embedded-MHT summarizes the authentication information about N and is computed in a bottom-up manner, where the leaves of the embedded-MHT are the entries included in this node. We refer to this modified index structure as MT*-tree which is shown in Fig. 3. Note that each entry in the internal node of MT*-tree includes a weight and its hash value, thus we first use the concatenation of them to compute the hash value of this entry, while each entry in the leaf node of MT*-tree only includes a weight, which can be simply used to compute the entry's hash value. For example, $h_1 = H(H(N_5|H(N_5))|H(N_3|H(N_3)))$, whereas $h'_1 = H(w_3|w_4)$.

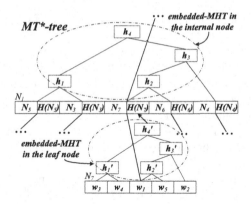

Fig. 3. MT*-tree

5.2 Query Processing

The procedure of the kSKQ processing can be divided into two independent processes. The first is to answer the kSKQ to retrieve top-k results. Based on the query results, the second process is to prune the entries in the nodes which have been visited in MT*-trees and generate VO simultaneously.

In the first process, we no longer generate VO immediately. Instead, we record the traversal processes of MR-tree and each MT*-tree. These processes will form several spanning trees. Each of these spanning trees is a subtree either of MR-tree (denoted by sMR-tree) or of MT*-trees (denoted by sMT*-tree) and it only records the nodes which have been visited during the query processing. Therefore, the entries in the internal nodes of these spanning trees can be divided into two categories: one is that query results are included in these entries and the other is that query results are **not** included in these entries. The former can be called *result entries* and the latter can be referred to as *non-result entries*. Similarly, the entries in a leaf node can also be divided into *result entries* and *non-result entries*.

In the second process, we first retrieve the collection of entries in the root of sMR-tree (denoted by l_s) and their corresponding entries in each sMT*-tree (denoted by l_t). Using l_s and the ranking score of the kth result (denoted by v_{max}) which is retrieved from the first process, we can prune the redundant entries in each l_t which will not affect the authentication of *completeness* of query results with certainty. Thus, we do not have to include these pruned entries in VO. Then, this procedure is repeated on the entries in each node of sMR-tree and their corresponding entries in sMT*-trees. At last, we use all the entries in each node of sMR-tree, the entries which are not pruned and the auxiliary hash values of the pruned entries in each node of sMT*-trees to generate VO.

The entry pruning strategy is shown in Algorithm 2. We first compute the minimum distance (d_{min}) between the query location and the *non-result entries* in l_s (line 1). Then, we use d_{min} and the ranking score of the kth result v_{max} to compute a textual threshold t using Eq. (1), which is used to bound the entries that will not affect the authentication of *completeness* of query results (line 2). Next, we compute a dummy *textual score* s using the largest *term score* in each l_t (lines 3-7). If $s > t$, among the *term scores* involved in computing the dummy *textual score*, we find the largest one and denote l_t (which it belongs to) by l (lines 8-13). Then, we mark this entry and at the same time, mark all the entries in other l_ts which have the same ID (lines 14-18). Then, we visit the next entry in l and update the dummy *textual score* s (lines 19-20). The procedure above is repeated until $s \leq t$. Thus, the un-marked entries that are yet to be visited are pruned and need not be returned. Then, we mark the boundary in each l_t (which is the last entry that has been visited in each l_t). We proceed to use the marked entries and the auxiliary hash values of the un-marked entries to update each l_t (lines 21-23). Finally, we return all these updated l_ts (line 24).

5.3   Authenticating Query Results

To authenticate the *soundness* of query results, the user needs to reconstruct the hash values of the roots of MR-tree and each MT*-tree. In the tree-forest indexes based scheme, the hash value of a node in MT-tree is computed by the concatenation of the entries included in this node. In this entry pruning based scheme, the hash value of a node in MT*-tree is the hash value of the root of the embedded-MHT in this node, which can be computed using the entries which are not pruned and the auxiliary hash values.

To authenticate the *completeness* of query results, the user first needs to verify the procedure of the pruning is correct. Therefore, for the entries (l_s) between each pair of '[' and ']' in *VO(MR)*, the user first computes the distance

(d_{min}) between the query location and the closest *non-result entry*. Then, among the corresponding entries (l_t) in each *VO(MT-t)*, the user finds the entry at the boundary. Then, the entries at the boundaries in all l_ts are coupled with the d_{min} to compute a dummy ranking score v. If v is lager than the ranking score of the kth object v_{max}, the user can claim that the procedure of the pruning is correct. Moreover, the user needs to compute the ranking score of each entry in *VO(MR)*. If its corresponding entry is not returned in *VO(MT-t)*, it may not exist or be pruned and this entry will not affect the authentication of *completeness* of query results with certainty.

6   Experimental Evaluation

In this section, we proceed to experimentally evaluate the performance of our authentication schemes.

6.1   Experiment Settings

Datasets. In total, seven datasets are used in our experiments: one for effectiveness and efficiency and six others for scalability. Firstly, a real spatial dataset named as LA, which contains a total of 131,461 objects located in Los Angeles streets, California[1] is used. We use a real document dataset of 20 Newsgroups[2], which consists of short user-generated content, to resemble the textual information attached to the spatial objects. The document attached to an object is selected randomly from 20 Newsgroups. Specifically, the average number of terms in the document attached to each object is 108; the total number of unique terms in the document dataset is 29476 and the total number of terms in the document dataset is 145,308. Secondly, to evaluate the scalability of our proposed schemes, we generate six datasets including 1, 2, 3, 5, 8, 10 million objects, respectively, where locations are selected randomly from LA and documents selected at random from 20 Newsgroups are attached to the locations. Besides, we generate five query sets, which consist of randomly selected locations from LA and the number of keywords ranges from 1 to 5. For each query set, 300 queries are included, and the average cost in each query set is reported.

System Configuration. All the experiments were run on a server with Intel(R) Pentium(R) CPU G640 @2.4GHz (Dual Processor) and 4 GB RAM. The programs were implemented in Java and the Java Virtual Machine Heap is set to 3GB.

Performance Metrics. The performance metrics for performance evaluation include: (i) index construction time, (ii) index size, (iii) running time, (iv) VO size and (v) authentication time.

---

[1] http://www.rtreeportal.org/

[2] http://qwone.com/~jason/20Newsgroups/

Algorithm 2. Entry Pruning

Input: l_s, $\{l_t\}$ and v_{max}
Output: $\{$updated $l_t\}$

1 Compute the minimum distance (d_{min}) between the query location and the *non-result entries* in l_s;
2 $t =$ THRESHOLD(d_{min}, v_{max});
3 $s = 0$;
4 **for** *each* l_t **do**
5   Set a pointer p that points to the first entry e in l_t;
6     $l_t.s' = w_{Q,t} \cdot l_t.p.e$;
7     $s+ = l_t.s'$;

8 **while** $s > t$ **do**
9     $s'_{max} = -\infty$;
10   **for** *each* l_t **do**
11   **if** $l_t.s' > s'_{max}$ **then**
12          $s'_{max} = l_t.s'$;
13          $l = l_t$;

14   **if** $l.p.e.flag = F$ **then**
15       $l.p.e.flag = T$;
16   **for** *each* $l_t! = l$ **do**
17   **if** $\exists p'.e \in l_t \&\& p'.e = l.p.e$ **then**
18             $p'.e.flag = T$;

19    $l.p++$;
20   Update s; // lines 4,6 and 7

21 **for** *each* l_t **do**
22    $l_t.p.e.flag = T$;
23   Use the entries with $flag = T$ and auxiliary hash values to update l_t;

24 Return each updated l_t;

Algorithms. Algorithms to be evaluated in our experiments include: ① SK (kSKQ over IR-tree which is proposed in [4] without any functionality of authentication); ② BA (the basic approach); ③ TFI (the tree-forest indexes based scheme) and ④ EP (the entry pruning based scheme). Moreover, we compare our proposed TFI and EP with the method proposed in [18] for snapshot queries on the same ranking function (Eq. (3)). Their method is denoted by AMSK.

6.2 Cost at the DO/LBSP

In this section, we evaluate the performance metrics (i) and (ii), which mainly burden the DO and LBSP. Figs. 4(a) and 4(b) plot the index construction time and index size of SK, BA, TFI and EP as a function of the number of objects. We can see that the index construction time and index size of SK, BA, TFI and EP increase linearly with the number of objects. Regarding the index construction time, EP is more expensive than the others because it involves not only the building of the embedded-MHT over the entries in each node of MT-trees but also the process of sorting them in an descending order. Constructing the index of TFI costs more time than that of BA because it needs to compute more hash values. As for the index size, EP is the same as TFI because only the leaf nodes of each embedded-MHT are stored in each node of MT*-trees. Since the index of TFI includes more hash values, it is larger than that of BA. Both the index construction time and index size of SK are the smallest because it does not involve any computation of hash values. Since the construction of the index is a one-time cost, it is reasonable for the DO to build the index in an off-line manner.

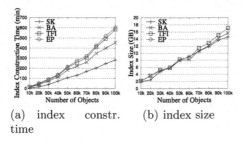

(a) index constr. time

(b) index size

Fig. 4. Varying the number of objects

6.3 Cost at the LBSP

Next, the performance metric (iii) is evaluated. The running time of SK, TFI and EP as a function of the number of keywords, k and α is shown in Figs. 5(a), 5(b) and 5(c), respectively. In all cases tested, the running time of EP is slightly larger than that of TFI mainly because it involves the process of pruning but the loss is not more than 20ms. The running time of TFI is larger than that of SK due to the procedure of VO's generation and the access to multiple trees. Nevertheless, for EP and TFI, their running time is at most about 100 ms longer than that of SK. Therefore, our proposed schemes can be considered efficient.

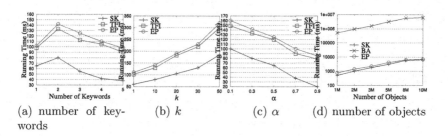

(a) number of keywords

(b) k

(c) α

(d) number of objects

Fig. 5. Evaluation of running time

6.4 Cost Between the LBSP and the User

Another important performance metric is (iv), i.e., VO size, which will affect the communication overhead between the LBSP and user. Figs. 6(a), 6(b) and 6(c) illustrate the VO size of TFI and EP under the experimental settings with varying the number of keywords, k and α, respectively. From these figures it can be seen that the VO size of EP is about 30% smaller than that of TFI since the redundant entries that will not affect the authentication of *completeness* of query results are pruned. In all cases tested, the VO size of our proposed schemes are all less than 100 KB.

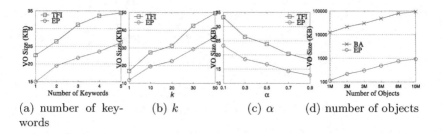

(a) number of key-   (b) k (c) α (d) number of objects
words

Fig. 6. Evaluation of VO size

6.5   Cost at the User

The last performance metric is the authentication time at the user-side. The cost of the authentication consists of two aspects. One is hashing operations, and the other is re-computing the ranking scores of returned entries in VO. From Figs. 7(a), 7(b) and 7(c), we can see that the authentication time of EP is slightly larger than that of TFI because the hash value of the root of each embedded-MHT needs to be reconstructed first. The authentication time of both TFI and EP is always less than 1s under the experimental settings with varying the number of keywords, k and α, which is believed reasonable as a cost for the user.

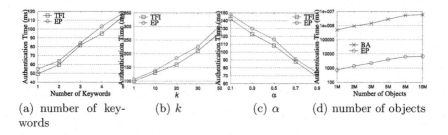

(a) number of key-   (b) k (c) α (d) number of objects
words

Fig. 7. Evaluation of authentication time

6.6   Scalability

To evaluate the scalability of our proposed schemes, we run an experiment on the six datasets containing 1, 2, 3, 5, 8, 10 million objects, respectively. From Figs. 5(d), 6(d) and 7(d), we can see that the running time, VO size and authentication time all increase with the increase of the dataset. In terms of all these three metrics, EP outperforms BA by 2-3 orders of magnitude because the latter needs to process large inverted files. Our scheme scales well for large datasets. When the number of objects comes to 10 million, the running time, VO size and authentication time of our proposed schemes are only about 10 s, 1 MB and 10 s, respectively.

6.7   Comparison with AMSK

As AMSK proposed in [18] only supports the special ranking function as discussed in Section 2, we extend TFI and EP to support the same ranking function and compare with it. Fig. 8 shows the results by varying k. TFI outperforms AMSK because the latter needs to traverse the index twice while TFI needs to traverse the index only once (Fig. 8(a)). Although EP also traverses the index twice, the VO size of EP is smaller than that of AMSK, which can be seen from Fig. 8(b). The authentication time of TFI and EP is almost the same as that of AMSK (Fig. 8(c)).

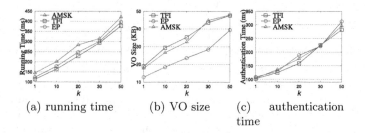

(a) running time   (b) VO size   (c)   authentication
                                                 time

Fig. 8. Varying k

7   Related Work

In this section, we retrospect the related work on spatial keyword query processing and query authentication.

7.1   Spatial Keyword Query

There have been many studies on spatial keyword query due to its importance to the commercial search engines. Felipe et al. [5] tackled a kNN searching problem which finds k objects that include the keywords in the query and are near to the query location. They proposed an IR^2-tree by combining an R-tree with superimposed text signatures. Cong et al. [4] and Wu et al. [19] presented the IR-tree which takes advantage of the inverted file for the textual similarity retrieval and the R-tree for the spatial proximity searching. Zhang et al. [26] studied the collective keyword searching (mCK), which finds the spatially closest objects that match m user-specified keywords. Different from the mCK, Cao et al. [2] studied the collective spatial keyword searching, which finds a group of spatial objects such that the group's keywords cover the keywords in the query and objects are nearest to the query location and have the lowest inter-object distances. Huang et al. [8] proposed a model to represent the safe region and devised searching algorithms to compute the safe region to support moving top-k spatial keyword queries. However, these studies do not support the authentication of top-k spatial keyword queries.

7.2   Query Authentication

Authenticated query processing has been studied extensively. Most studies on query authentication are based on an authenticated data structure (ADS) called Merkle hash tree (MHT) [13] as introduced in Section 2. Following the concept of MHT, the query authentication problem has been studied for relational databases [9,20], data streams [10,15,16], and textual search engines [14]. Yang et al. [21,22] first introduced this problem to the domain of spatial databases and studied the authentication of spatial range queries. They proposed an authenticated index structure called MR-tree, which combines the ideas of MB-tree [9] and R*-tree [1]. Yiu et al. investigated how to efficiently authenticate moving kNN [24], range [25] queries and shortest-path queries [23]. Hu et al. [7] proposed a novel approach that authenticates spatial queries based on neighborhood information. More recently, Hu et al. [6] and Chen et al. [3] developed new schemes for range and top-k query authentication that preserves the location privacy of queried objects. Besides, Lin et al. [11,12] investigated the authentication of location-based skyline queries in subspaces. A new authenticated index structure called MR-Sky-tree was proposed. Zhang et al. [27] studied the authentication of the location-based top-k queries which ask for the POIs in a certain region and with the highest k ratings for an interested POI attribute. However, these problems are different from ours studied in this paper since they considered either the authentication of spatial queries or the authentication of textual searchings. Moreover, their methods cannot be applied to our problem.

Wu et al. [18] studied the authentication of moving top-k spatial keyword queries. However, they only supported a special ranking function (Section 2). Instead, our problem focuses on the snapshot top-k spatial keyword queries and we adopt a general ranking function. Moreover, our schemes can also support other spatial-textual ranking functions. When we use their ranking function, our schemes also outperform their method for snapshot queries in terms of both efficiency and communication cost.

8   Conclusion

In this paper, we have studied the AkSKQ problem in outsourced databases. We propose a tree-forest indexes based scheme, where the indexes include an MR-tree and a collection of MT-trees. Based on tree-forest indexes, the kSKQ is processed by employing an *extensional priority queue* as assistance. Both the VO size and authentication time of the tree-forest indexes based scheme are much smaller than those of the basic approach. To further optimize VO, we propose an entry pruning based scheme, where we present an MT*-tree. The redundant entries in each node of MT*-trees are pruned and will not be included in VO. Our experiments show the effectiveness, efficiency and scalability of our proposed schemes. Moreover, the schemes proposed in this paper can also be applied to a broader set of query types, such as collective spatial keyword query, etc. This work may open up many promising directions for future work. Firstly, it is worth supporting incremental updates on the authenticated data structures proposed in this paper. Secondly, it is worth studying the authentication of top-k spatial keyword queries without compromising the location and keyword privacy.

Acknowledgments. We would like to thank all reviewers for their valuable comments. This work was supported in part by the following funding agencies of China: National Natural Science Foundation under grant 61170274, National Key Basic Research Program (973 Program) under grant 2011CB302506 and Fundamental Research Funds for the Central Universities under grant 2014RC1103. This work was also partially supported by HK-RGC Grants HKBU12200114 and HKBU12202414.

References

1. Beckmann, N., Kriegel, H.-P., Schneider, R., Seeger, B.: The r*-tree: An efficient and robust access method for points and rectangles. In: SIGMOD, pp. 322–331 (1990)
2. Cao, X., Cong, G., Jensen, C.S., Ooi, B.C.: Collective spatial keyword querying. In: SIGMOD, pp. 373–384 (2011)
3. Chen, Q., Hu, H., Xu, J.: Authenticating top-k queries in location-based services with confidentiality. In: VLDB, pP. 49–60 (2013)
4. Cong, G., Jensen, C.S., Wu, D.: Efficient retrieval of the top-k most relevant spatial web objects. In: VLDB, pp. 337–348 (2009)
5. Felipe, I.D., Hristidis, V., Rishe, N.: Keyword search on spatial databases. In: ICDE, pp. 656–665 (2008)
6. Hu, H., Xu, J., Chen, Q., Yang, Z.: Authenticating location-based services without compromising location privacy. In: SIGMOD, pp. 301–312 (2012)
7. Hu, L., Ku, W.-S., Bakiras, S., Shahabi, C.: Spatial query integrity with voronoi neighbors. IEEE TKDE **25**(4), 863–876 (2013)
8. Huang, W., Li, G., Tan, K.-L., Feng, J.: Efficient safe-region construction for moving top-k spatial keyword queries. In: CIKM, pp. 932–941 (2012)
9. Li, F., Hadjieleftheriou, M., Kollios, G., Reyzin, L.: Dynamic authenticated index structures for outsourced databases. In: SIGMOD, pp. 121–132 (2006)
10. Li, F., Yi, K., Hadjieleftheriou, M., Kollios, G.: Proof-infused streams: Enabling authentication of sliding window queries on streams. In: VLDB, pp. 147–158 (2007)
11. Lin, X., Xu, J., Hu, H.: Authentication of location-based skyline queries. In: CIKM, pp. 1583–1588 (2011)
12. Lin, X., Xu, J., Hu, H., Lee, W.-C.: Authenticating location-based skyline queries in arbitrary subspaces. IEEE TKDE **26**(6), 1479–1493 (2014)
13. Merkle, R.C.: A Certified Digital Signature. In: Brassard, G. (ed.) CRYPTO 1989. LNCS, vol. 435, pp. 218–238. Springer, Heidelberg (1990)
14. Pang, H., Mouratidis, K.: Authenticating the query results of text search engines. In: VLDB, pp. 126–137 (2008)
15. Papadopoulos, S., Yang, Y., Papadias, D.: Cads: Continuous authentication on data streams. In: VLDB, pp. 135–146 (2007)
16. Papadopoulos, S., Yang, Y., Papadias, D.: Continuous authentication on relational streams. VLDB J. **19**(2), 161–180 (2010)
17. Rivest, R.L., Shamir, A., Adleman, L.M.: A method for obtaining digital signatures and public-key cryptosystems. Commun. ACM **21**(2), 120–126 (1978)

18. Wu, D., Choi, B., Xu, J., Jensen, C.S.: Authentication of moving top-k spatial keyword queries. IEEE TKDE **27**(4), 922–935 (2015)
19. Wu, D., Cong, G., Jensen, C.S.: A framework for efficient spatial web object retrieval. VLDB J. **21**(6), 797–822 (2012)
20. Yang, Y., Papadias, D., Papadopoulos, S., Kalnis, P.: Authenticated join processing in outsourced databases. In: SIGMOD, pp. 5–18 (2009)
21. Yang, Y., Papadopoulos, S., Papadias, D., Kollios, G.: Spatial outsourcing for location-based services. In: ICDE, pp. 1082–1091 (2008)
22. Yang, Y., Papadopoulos, S., Papadias, D., Kollios, G.: Authenticated indexing for outsourced spatial databases. VLDB J. **18**(3), 631–648 (2009)
23. Yiu, M.L., Lin, Y., Mouratidis, K.: Efficient verification of shortest path search via authenticated hints. In: ICDE, pp. 237–248 (2010)
24. Yiu, M.L., Lo, E., Yung, D.: Authentication of moving knn queries. In: ICDE, pp. 565–576 (2011)
25. Yung, D., Lo, E., Yiu, M.L.: Authentication of moving range queries. In: CIKM, pp. 1372–1381 (2012)
26. Zhang, D., Chee, Y.M., Mondal, A., Tung, A.K.H., Kitsuregawa, M.: Keyword search in spatial databases: Towards searching by document. In: ICDE, pp. 688–699 (2009)
27. Zhang, R., Zhang, Y., Zhang, C.: Secure top-k query processing via untrusted location-based service providers. In: INFOCOM, pp. 1170–1178 (2012)
28. Zobel, J., Moffat, A.: Inverted files for text search engines. ACM Comput. Surv. **38**(2) (2006)

Privacy-Preserving Top-k Spatial Keyword Queries over Outsourced Database

Sen Su[1]([✉]), Yiping Teng[1], Xiang Cheng[1], Yulong Wang[1], and Guoliang Li[2]

[1] State Key Laboratory of Networking and Switching Technology,
Beijing University of Posts and Telecommunications, Beijing, China
{susen,typ,chengxiang,wyl}@bupt.edu.cn
[2] Department of Computer Science and Technology,
Tsinghua University, Beijing, China
liguoliang@tsinghua.edu.cn

Abstract. In this paper, we study the privacy-preserving top-k spatial keyword query problem in outsourced environments. Existing studies primarily focus on the design of privacy-preserving schemes for either spatial or keyword queries, and they cannot be applied to solve the privacy-preserving spatial keyword query problem. To address this problem, we present a novel privacy-preserving top-k spatial keyword query scheme. In particular, we build an encrypted tree index to facilitate privacy-preserving top-k spatial keyword queries, where spatial and textual data are encrypted in a unified way. To search with the encrypted tree index, we propose two effective techniques for the similarity computations between queries and tree nodes under encryption. Thorough analysis shows the validity and security of our scheme. Extensive experimental results on real datasets demonstrate our scheme achieves high efficiency and good scalability.

1 Introduction

With the increasing popularity of location-based services in mobile Internet, spatial keyword queries have drawn growing interest from both the industrial and academic communities in recent years. Given a set of spatio-textual objects (e.g., points of interest) and a query with a location and a set of keywords, a top-k spatial keyword query finds k objects that are most relevant to the query in terms of both spatial proximity and textual relevancy [5], which has been widely used in real-life applications such as Google Maps and Foursquare. To realize the great flexibility and cost savings, more and more data owners are motivated to outsource their data service (including data, indices, querying algorithms, etc.) to the cloud.

However, directly outsourcing such service to cloud may arise serious privacy concerns. On one hand, the spatio-textual database may involve some private data objects whose locations or textual descriptions cannot be learned by any third parties including the cloud provider. Moreover, it requires human and financial resources to collect the spatio-textual objects, which can be regarded

© Springer International Publishing Switzerland 2015
M. Renz et al. (Eds.): DASFAA 2015, Part I, LNCS 9049, pp. 589–608, 2015.
DOI: 10.1007/978-3-319-18120-2_34

as business secrets to competitors, and it prohibits any unauthorized parties to grab the data. On the other hand, if the locations and the query keywords in the spatial keyword queries of data users are acquired illegally by the untrusted third parties, the travel habits or the query manners will be analyzed or even utilized by some potential attackers [12]. Thus, it is of great significance to study the privacy-preserving scheme for top-k spatial keyword queries in outsourced environments.

Existing studies primarily focus on the design of privacy-preserving schemes for either spatial or keyword queries. They cannot be applied to solve the privacy-preserving top-k spatial keyword query problem. Even though the spatial keyword queries are performed by simply combining such separate schemes, available queries cannot be provided due to the efficiency and validity, since both text relevancy and spatial proximity are exploited for search space pruning and results ranking [7,25]. Therefore, it calls for effective methods to efficiently process privacy-preserving top-k spatial keyword queries.

To this end, we first define the problem of the top-k spatial keyword query over outsourced spatio-textual databases in cloud. We then present a brand new scheme for achieving privacy-preserving top-k spatial keyword queries (PkSKQ). Specifically, in our scheme, a secure index based on existing tree-based index [7] is built to facilitate PkSKQ. In this index, to achieve a unified encryption, the spatial and textual data (i.e., coordinates and keyword weights) are converted into vectors and encrypted by an enhanced version of Asymmetric Scalar-product-Preserving Encryption (ASPE) [24], namely ASPE with Noise (ASPEN). We prove that ASPEN is resilient to chosen-plaintext attack and known-plaintext attack.

To search with the secure index, a basic operation is to compute the similarity between a query point and a tree node in the secure index. However, since the coordinates and keywords of the query point and the tree node are encrypted, we cannot directly compute such similarity. To solve this problem, we develop two techniques, anchor-based position determination and position-distinguished trapdoor generation. In particular, by adding auxiliary points into the query point and each tree node, the anchor-based position determination method allows to determine the positional relation between them under encryption; and for the position-distinguished trapdoor generation method, to facilitate the similarity computations between the query point and tree nodes, it generates query vectors corresponding to all the possible positional relations between the query point and tree nodes in the trapdoor generation process. In this way, the similarity computations between the query point and tree nodes can be performed without privacy breaches.

To summarize, our contributions are as follows:

- To the best of our knowledge, this is the first attempt to define and solve the privacy-preserving top-k spatial keyword query problem in outsourced environments.
- We propose a new privacy-preserving scheme for top-k spatial keyword queries. In particular, we devise a secure index to facilitate the privacy-

preserving top-k query, where spatial and textual data are encrypted in a unified way.

- To search with the secure index, we propose two techniques, anchor-based position determination and position-distinguished trapdoor generation, for the similarity computations between the query point and tree nodes under encryption.

- Thorough analysis shows the validity and security of our scheme, where it is proven to be resilient to chosen-plaintext attack and known-plaintext attack. Extensive experimental results on real datasets further demonstrate our scheme can achieve high efficiency and good scalability.

The rest of the paper is organized as follows. Section 2 first describes the privacy-preserving top-k spatial keyword query problem over outsourced database in cloud, and then introduces our encryption method. Section 3 presents our privacy-preserving top-k spatial keyword query scheme, followed by Section 4, which analyzes the validity and security of the proposed scheme. In Section 5, experimental evaluation is presented. Section 6 reviews related works on spatial keyword queries and privacy-preserving schemes for spatial and keyword queries. We conclude in Section 7.

2 Problem Formulation

In this section, we first introduce preliminaries on spatial keyword queries, and then describe the system and threat model of privacy-preserving top-k spatial keyword queries. At last, the encryption method used in our scheme is presented.

2.1 Top-k Spatial Keyword Query

Let D be a spatio-textual database owned by the data owner. Each object O in D is defined as a tuple $(O.l, O.t)$, where $O.l$ represents the location descriptor in multidimensional space and $O.t$ denotes the text that describes the object. A top-k spatial keyword query Q is also defined as a tuple $(Q.l, Q.t)$, where $Q.l$ is a location descriptor and $Q.t$ is a set of keywords. To evaluate the relevancy between object O and query Q, we require to combine their spatial proximity and textual relevancy [7,25], which is usually defined as below,

$$
\begin{aligned}
D_R(Q, O) &= \alpha \frac{D(Q.l, O.l)}{D_{max}} + (1 - \alpha)(1 - \frac{R(Q.t|O.t)}{R_{max}}) \\
&= \alpha \frac{(x_o - x_q)^2 + (y_o - y_q)^2}{D_{max}^2} + (1 - \alpha)(1 - \frac{\sum_{i=1}^{n} \xi_i q_i}{R_{max}}),
\end{aligned}
\tag{1}
$$

where $\alpha \in [0, 1]$ is a parameter to balance distance proximity and text relevancy. $\frac{D(Q.l, O.l)}{D_{max}}$ represents the normalized Euclidian distance between Q and O, where D_{max} can be the maximum distance between objects in D, and it is defined as $\frac{(x_o - x_q)^2 + (y_o - y_q)^2}{D_{max}^2}$ in 2-dimensional space. The textual relevancy score $\frac{R(Q.t|O.t)}{R_{max}}$

between the querying keywords and the terms of the object is defined as $\frac{\sum\limits_{i=1}^{n} \xi_i q_i}{R_{max}}$, where R_{max} can be the sum of maximum weights of keywords. ξ_i is the weight of term w_i, q_i is a boolean parameter representing whether w_i is in the query and n is the total number of distinct keywords. $D_R(Q, O)$ is called *spatio-textual similarity* which is computed the lower, the better. The top-k spatial keyword query is to find the k objects which are with the lowest *spatio-textual similarities* in the database.

To efficiently support spatial keyword queries, we should utilize spatio-textual indices. For simplicity, in this paper we use the IR-tree [7,18] as an example. IR-tree exploits the Minimum Bounding Rectangle (MBR) and inverted files to index spatio-textual objects. Using IR-tree, the *spatio-textual similarity* between the query point Q and an MBR N is defined as $\text{MIND}_R(Q, N)$,

$$
\begin{aligned}
\text{MIND}_R(Q,N) &= \alpha \frac{\text{MINDIST}(Q.l, N.r)}{D_{max}} + (1-\alpha)(1 - \frac{R(Q.t|N.t)}{R_{max}}) \\
&= \alpha \frac{(x_N - x_q)^2 + (y_N - y_q)^2}{D_{max}^2} + (1-\alpha)(1 - \frac{\sum\limits_{i=1}^{n} \xi_i q_i}{R_{max}}),
\end{aligned}
\tag{2}
$$

where $\text{MINDIST}(Q.l, N.r)$ is the minimal Euclidean distance from query point $Q.l$ to region $N.r$ (the MBR N). In two-dimensional space, it is defined as $\frac{(x_N - x_q)^2 + (y_N - y_q)^2}{D_{max}^2}$, the square of the minimum distance between the query point and arbitrary points on the rectangle whose coordinate is denoted as (x_N, y_N). $\frac{R(Q.t|N.t)}{R_{max}}$ is computed by Equation 1 replacing $O.t$ by $N.t$ (the pseudo text of N).

To process top-k spatial keyword queries with the IR-tree index, a best-first traversal algorithm is applied to retrieving the top-k objects. We refer the readers to [7,25] for details of the spatial keyword query algorithm. We assume a two-dimensional geographical space composed of latitude and longitude, but the scheme proposed in this paper can be generalized to other multidimensional spaces of low dimensionality.

2.2   System and Threat Model

The system model consists of three roles: the data owner, the data user and the cloud server (see Fig. 1). The data owner has a spatio-textual database to be outsourced to the cloud server. To improve the efficiency of spatial keyword query, the data owner builds an index for the database. For the privacy issue, the data owner encrypts the database and its index before outsourcing. Authorized data users can access the query processing service from the cloud server. When the data users want to start spatial keyword queries with their locations and query keywords, each of them first acquires a

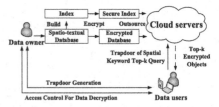

Fig. 1. The system of spatial keyword query in cloud

corresponding trapdoor from the data owner and submits it to the cloud server along with the optional number k for requesting top-k best objects. After receiving the trapdoor, the cloud server then executes the top-k spatial keyword query and returns the top-k results which are most relevant to the query. At last, the data user asks for the data decryption permission from the data owner though access control.

We assume that the cloud server is "honest but curious" in our model as in [2, 17]. It means that the cloud sever will correctly follow the designed protocols, but acting in a "curious" fashion, it may try to collect and analyze the meaningful information such as the location or textual information of the database, the contents of the index and the users' queries. We consider the proposed scheme in our system is under *chosen-plaintext attack* model and *known-plaintext attack* model [9]. In *chosen-plaintext attack* model, assume that the cloud server can derive ciphertext for plaintext of chosen objects in the database and attempt to recover the ciphertext for which it does not have the plaintext. In *known-plaintext attack* model, assume that the cloud server can obtain plaintext-ciphertext pairs of objects in the database. Using the information of these pairs, it attempts to solve the secret key in order to decrypt more ciphertext.

In our model, we assume that the authorization and the access control are well performed, which are not focused on in this paper, and we also do not assume that the cloud server will collude with data users. The access pattern, which is the sequence of results, is not considered to be protected in our scheme. Although existing techniques such as private information retrieval (PIR) [6] and Oblivious RAM [13,19] can be utilized to hide the access pattern, they are not efficient to be applied in our model.

2.3   Asymmetric Scalar-Product-Preserving Encryption with Noise

Asymmetric Scalar-product-Preserving Encryption (ASPE) proposed in [24] is a secure scheme for kNN queries over encrypted data. In ASPE, Euclidean distance between a database record p and a query point q is calculated as scalar product of vectors. The secret key in ASPE consists of two $l' \times l'$ invertible matrix $\{M_1, M_2\}$, one l'-bit string S and $l' - (l+1)$ random numbers $\{w_{l+2}, w_{l+3}, ..., w_{l'}\}$, where l is the dimension of the data vector p. First, each data vector p and query vector q are extended to $(l+1)$-dimensional vectors by adding $-0.5\|p\|^2$ and 1 as the $(l+1)$-th dimension, respectively. Besides, the query vector q is still scaled by a positive random number r as $r(q,1)$. Then, both the data vector and the query vector, denoted as $\hat{p}$ and $\hat{q}$, are extended from $l+1$ dimensions to l' dimensions with artificial values according to S. Particularly, for i from $l+2$ to l', if $S[i] = 1$, set $\hat{p}[i]$ to w_i; otherwise, set $\hat{p}[i]$ to a random number. Correspondingly, if $S[i] = 0$, set $\hat{q}[i]$ to w_i; otherwise, set $\hat{q}[i]$ to a random number. For the last dimension with which $S[i] = 0$ (resp., $S[i] = 1$), $\hat{p}[i]$ (resp., $\hat{q}[i]$) is given a value so that the scalar product over the artificial dimensions is equal to 0. After extension, $\hat{p}$ and $\hat{q}$ are split into two vectors as $\{\hat{p}', \hat{p}''\}$ and $\{\hat{q}', \hat{q}''\}$ according to each value of S. In particular, if $S[i] = 1$, the value of $\hat{p}[i]$ is split into $\hat{p}'[i]$ and $\hat{p}''[i]$ $(\hat{p}'[i] + \hat{p}''[i] = \hat{p}[i])$; otherwise, $\hat{p}'[i]$ and $\hat{p}''[i]$ are both set to $\hat{p}[i]$. In the contrary, the value of $\hat{q}[i]$ is split into $\hat{q}'[i]$ and $\hat{q}''[i]$ ($\hat{q}'[i] + \hat{q}''[i] = \hat{q}[i]$), when $S[i] = 0$. And if $S[i] = 1$, we set both $\hat{q}'[i]$ and $\hat{q}''[i]$ with the value of $\hat{q}[i]$. Finally, the split data vectors $\{\hat{p}', \hat{p}''\}$ are encrypted as $\{M_1^T \hat{p}', M_2^T \hat{p}''\}$, and the split query vectors $\{\hat{q}', \hat{q}''\}$ are encrypted as $\{M_1^{-1}\hat{q}', M_2^{-1}\hat{q}''\}$.

In doing so, the scalar products of the encrypted data vectors and query vectors can be preserved as the indicator of Euclidean distance to find k nearest neighbors during query processing. ASPE is proven to be secure against known plaintext attack so that neither data vector nor query vector can be recovered by analyzing the plaintext-ciphertext pairs. However, as indicated in [26], existing ASPE cannot be resilient to the chosen-plaintext attack. To address this problem, we propose an enhanced version of ASPE, namely *ASPE with noise* (ASPEN). In particular, we insert an additional dimension into each data vector and set the noise to it. The noise is a random number δ, which is drawn from a Laplacian distribution, $\delta \sim Laplace(\mu, b)$. Since the scalar products are disturbed by the added noise, the query result may not be exactly accurate. The scale parameter b is considered as a trade-off parameter between query accuracy and security. We will show a detailed evaluation of how the added noise affects the query accuracy in experiments.

3 Privacy-Preserving Top-K Spatial Keyword Query Scheme

In this section, we first give an overview of the privacy-preserving top-k spatial keyword query (PkSKQ) scheme, and then describe this scheme in details.

3.1 Overview

As the system model described in Section 2, the data owner first builds a regular IR-tree for indexing spatio-textual database. To protect the privacy of the IR-tree, it is encrypted by a unified encryption method. In particular, for the entries (i.e., MBRs and objects) in the IR-tree nodes, the coordinates and weights of keywords are converted into one vector and encrypted using ASPEN, while the parent-children relationships are preserved. The details of the unified encryption for the secure index are presented in Section 3.2.

To enable the similarity computations between the query point and MBRs under encryption, some auxiliary points need to be generated by the data owner before oursourcing the index. Specifically, to search with the IR-tree, as a basic operation, the computation of the distance between a query point and an MBR relies on the positional relation between them, which cannot be directly determined under encryption. To this end, an anchor-based position determination method is proposed to help determine the relations, where auxiliary points, called anchors, are generated for each MBR and outsourced as a part of index by the data owner. The details of this method is introduced in Section 3.3.

To issue a top-k spatial keyword query, the data user first needs to acquire a trapdoor from the data owner, which is the encrypted form of the user's top-k spatial keyword query. Due to the different positional relations between the query point and MBRs, different coordinates of the query point are used in the distance computations, which cannot be chosen in ciphertext. To address this problem, we present the position-distinguished trapdoor generation method, where the trapdoor is generated with all possible coordinates for different positional relations. To compute the distance between the query and MBRs, the cloud server can choose the proper coordinates in trapdoors based on the determined positional relations using anchors. The detailed method is proposed in Section 3.4.

In the query processing of PkSKQ, based on the techniques above, the cloud server computes and compares the spatio-textual similarities between queries and MBRs (objects) to retrieve the top-k objects by the best-first traversal algorithm.

3.2   Unified Encryption for Secure Index Construction

To enable top-k spatial keyword queries over encrypted spatio-textual database in the cloud, we design a unified encryption over the spatio-textual data to facilitate the computations of the similarities between query points and entries in the tree nodes of the secure index. To achieve the unified encryption, for the IR-tree, the coordinates and the keyword weights of MBRs and objects are transformed into one vector and encrypted using ASPEN, while the parent-children relationships are not encrypted. Exploiting the scalar-product preserving property of ASPEN, the similarities can be computed in the way of the scalar product between vectors.

Particularly, for MBRs and objects, the inverted files are first transformed and extended into textual vectors of the same length to facilitate the computation of the textual relevancy. For an MBR or object, the textual vector, $E.tv$ or $O.tv$, is denoted by $(\xi'_1, \xi'_2, ..., \xi'_n)$, where n is the total number of distinct keywords in the database; $\xi_i{}'$ is equal to $\frac{\xi_i}{R_{max}}$, if w_i is contained in its inverted files; otherwise, ξ'_i is set to 0.

Then, the coordinates of MBRs or spatial objects are converted and extended into spatial vectors by adding additional dimensions to enable the computation of the spatial proximity. For an MBR $(x'_{min}, y'_{min}, x'_{max}, y'_{max})$, its spatial vector is denoted by

$$E.lv = (x'_{min}, x'^2_{min}, 1|y'_{min}, y'^2_{min}, 1|x'_{max}, x'^2_{max}, 1|y'_{max}, y'^2_{max}, 1),$$

and for an object (x', y'), its spatial vector is denoted by

$$O.lv = (x', x'^2, 1|y', y'^2, 1|0, 0, 0|0, 0, 0),$$

where all coordinates mentioned are normalized by D_{max}.

Finally, for each MBR or object, the spatial vector and the textual vector are combined into one data vector and appended with an additional dimension of 1. Considering the security issue, before encryption with ASPEN, each the data vector is further padded with an additional random number δ. The final data vector is denoted as $E.v = (E.lv|E.tv|1|\delta)$ (resp. $O.v = (O.lv|O.tv|1|\delta)$). To achieve the encryption, the specific encryption is that each data vector is first extended into a d-dimensional vector by adding artificial dimensions and split into two random vectors. Then, each pair of the split vectors are encrypted with two invertible $d \times d$ matrices M_1 and M_2. The encrypted data vectors of the MBR and the object are denoted as $\overline{E.v} = \{M_1^T E'.v, M_2^T E''.v\}$ and $\overline{O.v} = \{M_1^T O'.v, M_2^T O''.v\}$ respectively. After all the entries in the IR-tree nodes

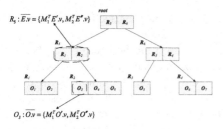

Fig. 2. Secure index based on IR-tree

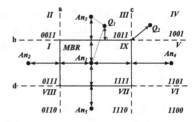

Fig. 3. Anchors and region codes for MBR

are encrypted, the encrypted IR-tree is regarded as the secure index I, which is shown in Fig. 2 as an example.

3.3 Anchor-Based Position Determination

To compute the similarities, a basic operation is to compute the distance between a query point and an MBR. According to definition of Euclidean distance, the distance from a point to a rectangle is the shortest distance between the point and a point on the rectangle. In this case, such distance can be computed with the vertices or the points on the edges of the rectangle, which relies on the positional relation of the query point around the rectangle. For example, in Fig. 3, the distance between Q_1 and the MBR is the perpendicular distance from Q_1 to the line b, while that for Q_2 is the distance from Q_2 to the vertex of the MBR. As shown in Fig. 3, four edges of an MBR can divide the plane into nine regions. To determine the positional relation between the query point and the MBR, we only need to find the region where the query point is located.

Since each region lies on the deterministic sides of the four edges of an MBR, each region can be coded as 4 binary bits, each of which represents one edge of the MBR. The coding strategy can be customized. For ease of presentation, we use the strategy shown in Fig. 3 as an example. In Fig. 3, if the region is on the right of line a or lies below line b, the corresponding bit is set to 1. If the region is on the left of line c or lies above line d, the corresponding bit is also set to 1. Otherwise, it is set to 0. To obtain the region of the query point is just to determine which sides of the four edges it is on. This determination can be easily performed with their coordinates in plaintext, but considering the privacy of the query point and the MBR, it can hardly performed in ciphertext. Therefore, with ASPEN allowing to find the nearer point in ciphertext, we add auxiliary points for each MBR, called *anchors*, to facilitate such determination.

An observation is that if the query point is closer to one of two symmetry points about a line, the query point is on the same side of the line with the closer point (an example see Q_1, An_3 and An_1 in Fig. 3). Accordingly, for each edge, a pair of symmetry anchors is generated randomly, totally at least five anchors generated for an MBR. The first anchor is picked as a random point in the MBR and other four are picked in an axial symmetric way with the first one about each edge of the MBR (see Fig. 3).

Formally, for an anchor (x_i, y_i), its vector is generated and extended as $An_i = (x_i, y_i, -0.5(x_i + y_i)^2)$, and $\{An_i\}$, i from 1 to 5, are in the position of inside, left, upside, right and downside of the MBR respectively. All these vectors of the anchors are encrypted using ASPEN. They are first extended and split into two random d'-dimensional vectors An'_i and An''_i. Then, the split vectors are encrypted as $\overline{An_i} = \{m_1^T An'_i, m_2^T An''_i\}$, where m_1 and m_2 are invertible $d' \times d'$ matrices.

Correspondingly, for the query point (x_q, y_q), an auxiliary query vector is created as $Q.a = (x_i, y_i, 1)$. According to Section 2, encrypted using ASPEN, the vector is first scaled by a positive random number r. Then, $Q'.a$ and $Q''.a$, extended and split from $Q.a$, are encrypted as $\overline{Q.a} = \{m_1^{-1} Q'.a, m_2^{-1} Q''.a\}$.

To perform the determination of positional relation, given the anchors $\{An_i\}$ of the MBR and the auxiliary query vector $\overline{Q.a}$, the cloud server computes the following equation with corresponding vectors in order from $\overline{An_2}$ to $\overline{An_5}$,

$$PD(\overline{An_1}, \overline{An_i}, \overline{Q.a}) = (m_1^T An'_1 - m_1^T An'_i) \cdot m_1^{-1} Q'.a + (m_2^T An''_1 - m_2^T An''_i) \cdot m_2^{-1} Q''.a, \quad (3)$$

where $2 \le i \le 5$. If $PD(\overline{An_1}, \overline{An_i}, \overline{Q.a})$, where $2 \le i \le 5$, is positive, return $h_i = 1$. Otherwise, return $h_i = 0$. Thus, a 4-bit binary code $H = h_2|h_3|h_4|h_5$ combined from these four bits indicates the query point is located in the region of code H.

3.4   Position-Distinguished Trapdoor Generation

As shown in Fig. 3, according to the positional relation between the query point and the MBR after determination, the distance computation can be classified into three categories: a) the distance is equal to 0, when the query point is inside an MBR; b) the distance is calculated with the vertices of an MBR, when the query point is in the corner regions (Region II, IV, VI and VIII); c) the distance is calculated with the edges of an MBR, when the query point is in Region I, III, V and VII. Except Region IX, the distance computation needs to exploit the different combinations of the coordinates of an MBR. For example, in Fig. 3, the distance from Q_1 to the MBR is calculated only with the coordinate of line b (i.e., y_{max}), while the distance from Q_2 requires the vertex (x_{max}, y_{max}). However, as the coordinates of the MBR and the query point are encrypted, the distance between them cannot be calculated either, even if the positional relation is determined. To this end, we design a method to generate the trapdoor, where all the possible combinations of the coordinates used in the distance computation are preprocessed by adding the additional dimensions in query vectors.

Specifically, a query point $Q(x_q, y_q)$ with the querying keywords is transformed into query vectors $\{Q_i.v | 1 \le i \le 9\}$, each of which is corresponding to region I to IX. The one-to-one mapping between the query vectors and the region codes is shown in Table 1 so that the proper query vector can be selected by the code. Each query

Table 1. Binary codes mapping to query vectors

$Q_i.v$	code	$Q_i.v$	code	$Q_i.v$	code
$Q_1.v$	0111	$Q_4.v$	1001	$Q_7.v$	1110
$Q_2.v$	0011	$Q_5.v$	1101	$Q_8.v$	0110
$Q_3.v$	1011	$Q_6.v$	1100	$Q_9.v$	1111

vector $Q_i.v$ is generated with two parts: the spatial vector and the textual vector.

For the spatial vector $Q_i.lv$, the coordinates (x_q, y_q) are extended to 12 dimensions by adding their proper forms used in the distance computation, which are set in corresponding dimensions to $e.lv$. Other dimensions in the spatial vector of the query vector are set to 0. In summary, $Q_i.lv$ (i from 1 to 9) are denoted as

$Q_1.lv = (-2x'_q, 1, x'^2_q | 0, 0, 0 | 0, 0, 0 | 0, 0, 0)$, $Q_5.lv = (0, 0, 0 | 0, 0, 0 | -2x'_q, 1, x'^2_q | 0, 0, 0)$,

$Q_2.lv = (-2x'_q, 1, x'^2_q | 0, 0, 0 | 0, 0, 0 | -2y'_q, 1, y'^2_q)$, $Q_6.lv = (0, 0, 0 | -2x'_q, 1, x'^2_q | -2y'_q, 1, y'^2_q | 0, 0, 0)$,

$Q_3.lv = (0, 0, 0 | 0, 0, 0 | 0, 0, 0 | -2y'_q, 1, y'^2_q)$, $Q_7.lv = (0, 0, 0 | -2y'_q, 1, y'^2_q | 0, 0, 0 | 0, 0, 0)$,

$Q_4.lv = (0, 0, 0 | 0, 0, 0 | -2x'_q, 1, x'^2_q | -2y'_q, 1, y'^2_q)$, $Q_8.lv = (-2x'_q, 1, x'^2_q | -2y'_q, 1, y'^2_q | 0, 0, 0 | 0, 0, 0)$,

$Q_9.lv = (0, 0, 0 | 0, 0, 0 | 0, 0, 0 | 0, 0, 0)$,

where x'_q and y'_q are normalized by D_{max}.

Correspondingly, the textual vector $Q.tv$ is created as $Q.tv = \{q_1, q_2, ..., q_n\}$, where q_i is set to -1, if the query contains the keyword w_i; otherwise, q_i is set to 0. To achieve the balance of spatial proximity and textual relevancy, $Q_i.lv$ and $Q.tv$ are multiplied with α and $1 - \alpha$ respectively, and the query vector $Q_i.v$ is combined and extended as $Q_i.v = (\alpha Q_i.lv | (1 - \alpha) Q.tv | (1 - \alpha) | 1)$.

For security, the query vectors are first scaled by a positive random number r as $\{rQ_i.v | 1 \le i \le 9\}$. Notice that the random number r is different for different queries. Then, using ASPEN, they are split and encrypted with M_1^{-1} and M_2^{-1}, denoted as $\{\overline{Q_i.v} = M_1^{-1}Q'_i.v, M_2^{-1}Q''_i.v | 1 \le i \le 9\}$.

In query processing, after the position determination of the query point, the cloud server matches the code H with the codes in Table 1, and choose the corresponding vectors $\overline{Q_i.v}$. To queue nodes into priority queue, for an MBR, the cloud server computes the scalar product of $\overline{E.v}$ and $\overline{Q_i.v}$, shown as:

$$SP(\overline{E_j.v}, \overline{Q_i.v}) = M_1^T E'.v \cdot M_1^{-1}Q'_i.v + M_2^T E''.v \cdot M_2^{-1}Q''_i.v.$$

When the visiting an object, the cloud server computes the scalar product of $\overline{O.v}$ and $\overline{Q_8.v}$, shown as:

$$SP(\overline{O.v}, \overline{Q_8.v}) = M_1^T O'.v \cdot M_1^{-1}Q'_8.v + M_2^T O''.v \cdot M_2^{-1}Q''_8.v.$$

The results of these scalar products are used as the keys of MBRs and objects in the priority queue, and the best-first traversal algorithm can be executed. If k objects have been found, the cloud server terminates the algorithm and returns them to the data user.

4   Analysis

In this section, we present the analysis of our PkSKQ scheme including the validity and security, respectively.

4.1   Validity Analysis

In PkSKQ scheme, to retrieve the top-k objects, the cloud server determines the positional relations for each MBR using anchor-based method, and then computes the similarities between the query and MBRs (objects). We first analyze the validity of the anchor-based position determination method. In this method, given anchors $\{\overline{An_i} | 1 \le i \le 5\}$ of the MBR and auxiliary query vector

$\overline{Q.a}$, such determination is performed by computing the following equation, for i from 2 to 5,

$$
\begin{aligned}
& PD(\overline{An_1}, \overline{An_i}, \overline{Q.a}) \\
&= (m_1^T An_1' - m_1^T An_i') \cdot m_1^{-1} Q'.a + (m_2^T An_1'' - m_2^T An_i'') \cdot m_2^{-1} Q''.a \\
&= (m_1^T An_1' - m_1^T An_i')^T m_1^{-1} Q'.a + (m_2^T An_1'' - m_2^T An_i'')^T m_2^{-1} Q''.a \\
&= (An_1' - An_i')^T Q'.a + (An_1'' - An_i'')^T Q''.a \qquad (4) \\
&= (An_1 - An_i)^T r Q.a \\
&= 0.5r[[(x_i - x_q)^2 + (y_i - y_q)^2 - \delta_i] - [(x_1 - x_q)^2 + (y_1 - y_q)^2 - \delta_1]] \\
&= 0.5r[d^2(An_i, Q) - d^2(An_1, Q)].
\end{aligned}
$$

Note that in equation 4, r is the positive random number generated for the query. If the result is positive, it means Q is closer to An_1 than An_i, so that it can be determined the query point lies in which region. Assuming the query point lies in the region "0011", according to Table 1, the encrypted query vector $\overline{Q_2.v}$ should be selected to compute the similarity between Q and the MBR, which is shown as follows:

$$
\begin{aligned}
& SP(\overline{E.v}, \overline{Q_2.v}) \\
&= M_1^T E'.v \cdot M_1^{-1} Q_2'.v + M_2^T E''.v \cdot M_2^{-1} Q_2''.v \\
&= (M_1^T E'.v)^T M_1^{-1} Q_2'.v + (M_2^T E''.v)^T M_2^{-1} Q_2''.v \\
&= E'.v^T Q_2'.v + E''.v^T Q_2''.v \\
&= E.v^T Q_2.v \qquad (5) \\
&= r[\alpha E.lv^T Q_2.lv + (1-\alpha) E.tv^T Q_2.tv + \delta] \\
&= r[\alpha(x_{min}' - x_q')^2 + \alpha(y_{max}' - y_q')^2 + (1-\alpha)(1 - \sum_{i=0}^{n} \xi_i' q_i) + \delta] \\
&= r[\alpha \frac{(x_{min} - x_q)^2 + (y_{max} - y_q)^2}{D_{max}^2} + (1-\alpha)(1 - \frac{\sum_{i=1}^{n} \xi_i q_i}{R_{max}}) + \delta].
\end{aligned}
$$

Similarly, if an object is visited, the similarity between the query Q and the object O can be computed as $SP(\overline{O.v}, \overline{Q_8.v})$. During a given query since the random number r is particular and positive, the comparison of the similarity between query and MBRs or objects can be achieved. Therefore, the best-first traversal algorithm in PkSKQ can be executed correctly by the cloud server in the "honest but curious" model.

Note that since the added noise δ will lead to the inexactness of the query results, we give a quantificational evaluation for the query accuracy in Section 5.

4.2   Security Analysis

We analyze our PkSKQ scheme is secure under the attack model mentioned in Section 2. To achieve the spatial keyword queries in cloud, the proposed scheme ensures that scalar products between encrypted vectors can be calculated and compared. To prove the security guarantee of the proposed scheme, without loss of generality, we assume that the query Q (i.e., $Q_i.v$) and the data object P (i.e., $E.v$ or $O.v$) are d-dimensional, and their ciphertexts, $\overline{Q}$ and $\overline{P}$, are $(d+1)$-dimensional including the random numbers δ and r.

Theorem 1. *The proposed scheme is resilient to the chosen-plaintext attack, if the random number r for each query and δ for each object cannot be known by the adversary.*

Proof. To launch the chosen-plaintext attack, assume that the adversary (i.e. the cloud server) can derive a set of queries and their ciphertexts. For each query, the adversary would have one encrypted pair $\overline{Q} = \{M_1^{-1}Q', M_2^{-1}Q''\}$ used in the query processing. In the proposed scheme, the scalar product between encrypted query $\overline{Q}$ and any encrypted data $\overline{P} = \{M_1^T P', M_2^T P''\}$ can be calculated based on ASPEN as follows,

$$
\begin{aligned}
SP(\overline{P}, \overline{Q}) \\
&= M_1^T P' \cdot M_1^{-1}Q' + M_2^T P'' \cdot M_2^{-1}Q'' \\
&= (M_1^T P')^T M_1^{-1}Q' + (M_2^T P'')^T M_2^{-1}Q'' \\
&= (P')^T Q' + (P'')^T Q'' \\
&= r(P^T Q + \delta).
\end{aligned}
\tag{6}
$$

As described in [26], since the adversary can derive the plaintext of query Q and the corresponding scalar product, Equation 6 contains only $d + 2$ variables unknown, i.e., the d dimensions of P, the random number r and δ. If the random number r is the same for each query, the adversary only needs to collect the plaintext-ciphertext pairs of $d + 2$ query points and constructs $d + 2$ equations like Equation 6 to solve the $d + 1$ unknowns in P and δ. Thus, the attack in [26] can work in this case.

However, in the proposed scheme, the random number r are generated differently for different users. In this case, the equation set is constructed by collecting plaintext-ciphertext pairs of d query points as follows:

$$
SP(\overline{P}, \overline{Q_i}) = r_i(P^T Q_i + \delta), \quad i \in [1, d].
\tag{7}
$$

In Equation 7, there are $2d + 1$ variables unknown, i.e. d dimensions of P, d random numbers $\{r_1, r_2, ...r_d\}$ and δ. Since there are only d equations, which are less than the number of unknowns, the adversary does not have sufficient information to solve P, even if d queries and corresponding scalar products are known by the adversary.

Similarly, assume that the adversary derives a set of objects in the database with their ciphertexts. For one certain encrypted query $\overline{Q}$, scalar products can be calculated with $\{\overline{P_j} | 1 \le j \le d\}$ as follow,

$$
SP(\overline{P_j}, \overline{Q}) = r(P_j^T Q + \delta_j), \quad j \in [1, d]
\tag{8}
$$

Since the random number δ_i for each object is not known by the adversary, there are $2d + 1$ unknowns, i.e., the d dimensions of Q, the random number $\{\delta_1, \delta_2, ...\delta_d\}$ and r, in d equations. It is not sufficient for the adversary to solve the unknowns, because the number of equations is less than that of the unknowns. Hence, the proposed scheme is resilient against the chosen-plaintext attack.

In fact, since $\delta \sim Laplace(\mu, b)$, to improve security, the parameter b is expected to be larger so that it has higher probability to produce disturbance. In our scheme, since such disturbance is introduced to the computations of scalar products (i.e., the computations of similarities between queries and MBRs or objects), it only impacts the accuracy of top-k queries. The plaintext of objects can be still recovered by data users, because the random number δ is set to an additional dimension of each data vector. Thus, even if the adversary ignores the disturbance introduced to similarities and constructs $d+1$ equations to solve the unknowns in P, he/she will only obtain the disturbed plaintext of objects. It is meaningless for the adversary to use this disturbed database for providing other services, as the true values of objects cannot be recovered.

Theorem 2. *The proposed scheme is resilient to the known-plaintext attack, if the bit string used for splitting (i.e. S of the key) cannot be known by the adversary.*

Proof. Assume that the adversary knows the data object P with its corresponding encryption $\overline{P} = \{M_1^T P', M_2^T P''\}$. Without loss of generality, no artificial attributes are added. For any data object, if the adversary does not know the bit string used for splitting, $\overline{P}$ has to be modeled as two unknown $(d+1)$-dimensional vectors. The equations for solving the secret matrices can be constructed with $\overline{P}$ and P. Notice that there are $2(d+1)|P|$ unknown variables in P' and P'', where $|P|$ is the number of data objects in the database. There are also $2(d+1)^2$ unknowns in the secret matrices, however only $2(d+1)|P|$ equations constructed. Similarly, using the queries, there are $2d|Q|$ equations constructed which contain $2(d+1)|Q|+2(d+1)^2+1$ unknown variables, where $|Q|$ is the number of obtained queries. Therefore, the information to solve the unknowns is insufficient for the adversary, and the scheme is resilient to the known-plaintext attack.

It is noteworthy that in the anchor-based position determination, the coding strategy of the 4-bit binary code, which is used to label the regions around an MBR, can be customized by the data owner. This strategy should be kept confidential against the cloud server. In fact, even through such strategy can be learned by the cloud server, the locations of objects or users cannot be estimated precisely, since the spatio-textual database, indices, and spatial keyword queries are secure based on the analysis above.

As for the attack based on the order statistics in [23], in our scheme, since the noise disturbs the similarities of queries, the ordering information of encrypted objects cannot be derived precisely. Thus, even though the adversary can obtain such ordering information, the exact distribution of the database cannot be estimated to infer the plaintext.

5   Experimental Evaluation

In this section, we evaluate the performance of our PkSKQ scheme. Our experimental goal is to evaluate the index construction, trapdoor generation, query processing, query accuracy and scalability of our scheme.

5.1   Setup

We use a server served as the cloud with Intel(R) Xeon(R) CPU L5638 @2.00GHz
Dual and 40.0GB RAM, and a PC served as the data owner with Intel(R)
Core(TM) i7-3610QM CPU @ 2.30GHz 2.30GHz and 6.0GB RAM for the experiments. Both of them run at 64-bit Windows 7 operating system. All the programs
are implemented in Java and the Java Virtual Machine Heap is set to 3GB.

Table 2. Datasets in Experiments

Name	# of objects	# of unique keyword	total # of keywords	Descriptions
WU	10,493	1,000	145,308	Popular places in the west part of US
LA	131,461	10,000	14,760,856	Streets of Los Angeles

In the experiments, we use real spatial datasets whose objects are chosen
from the locations in parts of America [1]. A real document dataset of 20 Newsgroups[2] consists of short user-generated documents which aim to resemble text
attached to spatial objects. We randomly select keywords from documents of 20
Newsgroups attached to an object and limit the total number of the unique keywords. The detailed settings of datasets are shown in Table 2. For the generation
of spatial keyword queries, the locations and keywords are generated randomly
in the datasets and the number of keywords is selected from 1 to 10. For each
experiment, 200 queries are compromised, and average costs of the queries are
reported. The schemes for comparison with our PkSKQ in the experiments are
shown in Table 3.

Table 3. Compared Schemes in Experiments

Schemes	Descriptions
Plaintext	The top-k spatial keyword query in [7] performed with the IR-tree index in plaintext.
PkSKQ	The privacy-preserving top-k spatial keyword query scheme proposed in this paper.
Baseline	The objects in the spatio-textual database and user queries are encrypted using ASPEN. The top-k spatial keyword query is processed via a linear scan of the encrypted database without any indices.

5.2   Index Construction

In Fig 4(a), the size of the secure index is shown. The storage cost mainly results
from the encrypted vectors for MBRs and the added anchors which are additional
data to build the secure index. When the number of objects rises, the size of the
secure index will closely increase in a linear fashion. When the fanout is set to
either 100 or 200, the tend of the size is almost the same. With the scale of
objects rising to 100,000, the size of secure index is going up to over 300MB.
In any case, since the storage in cloud is an inexpensive resource, to provide

---

[1] http://www.chorochronos.org/

[2] http://qwone.com/~jason/20Newsgroups/

PkSKQ in cloud, it deserves to take reasonable space to construct the secure index.

Fig. 4(b) shows the construction time of the secure index. This part of time is mainly incurred for two reasons. The one is that to construct a secure index, a regular IR-tree is needed to be built first, and still needed to add anchors into each MBR. Another is that all the MBRs together with their anchors should be encrypted using ASPEN, in which there are multiplications of $d \times d$ matrices and d-dimensional vectors for MBRs and $d' \times d'$ matrices and d'-dimensional vectors for anchors. When the number of objects goes up, there are more nodes to be generated in the secure index. The construction time will almost increase linearly following the rising number of objects. For different fanouts, there is no obvious difference in the construction time. When the number of objects contained in the dataset reaches 100,000, the construction time comes to the minute level. Since the construction of the secure index is needed only once and can be done offline, its time is considered reasonable for the data owner.

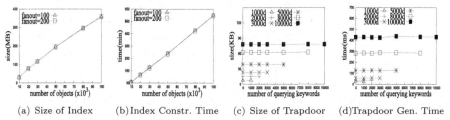

| (a) Size of Index | (b) Index Constr. Time | (c) Size of Trapdoor | (d) Trapdoor Gen. Time |

Fig. 4. Evaluation of Secure Index Construction and Trapdoor Generation

5.3 Trapdoor Generation

Fig. 4(c) shows the size of the trapdoor which is evaluated for all candidate query vectors. Since the encrypted query vectors in the trapdoor have constant length, the size of trapdoor is always a fixed value for a given dimensionality following the increasing number of querying keywords. As shown in this figure, the maximum size of the trapdoor with 10,000 dimensions equals to around 80KB. As the size of the trapdoor is still within kilobytes, it will not cause too much storage burden and can be considered affordable for the data user.

As shown in Fig. 4(d), given a certain dimensionality, the generation time of the trapdoor presents as a constant nearly, which is not affected by the number of querying keywords. If the dimensions are varied from 1,000 to 10,000, the generation time increases to over 400ms at most. Similarly, the main computation cost in the trapdoor generation is the encryption of query vectors, which contains two multiplications of a $d \times d$ matrix and a d-dimensional vector for each query vector. As the trapdoor generation time of trapdoor is still in milliseconds, it can be considered efficient enough for the data owner.

5.4 Query Processing

In the evaluation of the query response time for the PkSKQ scheme, we perform experiments in both memory-resident setting and disk-resident setting

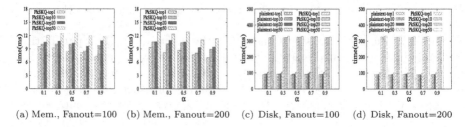

(a) Mem., Fanout=100 (b) Mem., Fanout=200 (c) Disk, Fanout=100 (d) Disk, Fanout=200

Fig. 5. Query Response Time of PkSKQ in Memory-resident and Disk-resident Setting

respectively. The dataset WU is used in the memory setting, which can be loaded totally into main memory, and LA is used in the disk setting.

Memory-Resident Setting. Fig. 5(a) and Fig. 5(b) show the query response time in the memory setting, when the fanout is set to 100 and 200 respectively. In both these figures, when the number k of the required results goes up to 50, there is an obvious increasing in the query response time compared with other three bars for each value of α. When varying α, a large α means that the spatial distance is more important, while a small α means the keywords are more important. Our secure index performs better for large α, but there is no apparent change due to the efficiency of processing in main memory. When the fanout is set to 100 or 200, the query response time is still under 13ms. Such performance comes from the efficient computations of the similarities between the query point and MBRs (objects). In these computations, for an MBR, there is one pair of scalar products of d-dimensional vectors and 4 pairs of scalar products of d'-dimensional anchors, whose complexity is only $O(d)$ and $O(d')$ for each scalar product respectively. For an object, there is only one pair of scalar products of d-dimensional vectors. From the results, our scheme can be considered efficient under the memory-resident setting.

Disk-Resident Setting. The query response time in the disk setting is shown in Fig. 5(c) and Fig. 5(d), when the fanout is set to 100 and 200 respectively. In this set of experiments, we take the query response time of the plaintext scheme for comparison. In Fig. 5(c) and 5(d), it can be observed that in either the PkSKQ scheme or the plaintext scheme, when the parameter k increases, it will cost more time to perform queries. For any values of α, the query response time of our PkSKQ scheme is under 400ms. Compared with the plaintext scheme, the query response time of PkSKQ is longer. This is because the larger I/O costs and the computations of the scale products of encrypted vectors, whose dimensionality is over 10,000, cost the majority of the query response time. Nevertheless, for our PkSKQ scheme, the query response time is still within milliseconds. Therefore, our scheme can be considered efficient on query response time under the disk-resident setting.

5.5 Query Accuracy

As analyzed in Section 4, since the added noise (i.e., the random numbers δ) will introduce disturbance to the similarity between MBRs (objects) and queries, the query results based on the similarity will be no longer exactly accurate. In other words, some of real top-k objects are not in the result set returned to the data user. This is because the similarities may be computed increased or decreased due to the addition of the noise. To evaluate the accuracy of our PkSKQ scheme, we define a measure as the accuracy $A = \overline{k}/k$, where $\overline{k}$ is the number of real top-k objects returned. Recall that the added noise is drawn from a Laplacian distribution and b is its scale parameter. As shown in Fig 6, the accuracy A is obviously impacted by b. Considering the availability of our scheme, the parameter b is expected to be smaller to acquire a high accuracy. Since the parameter b is a trade-off parameter between query accuracy and security, it can be provided as a balance factor for the data owner to satisfy different requirements about the query accuracy and security.

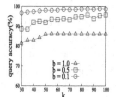

Fig. 6. Query Accuracy

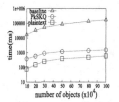

Fig. 7. Query Response Time, Scalability

5.6 Scalability

To evaluate the scalability of our scheme, similar with [7], we generate groups of datasets containing from 100,000 to 1,000,000 data points which are selected randomly from LA. In these experiments, the parameter k is set to 10, and fanout is set to 200. We take the plaintext scheme and the baseline scheme for comparisons. In the baseline scheme, all the encrypted objects are visited and ranked in a linear scan fashion to retrieve the top-k objects, thus the whole dataset should be accessed. As shown in Fig. 7, the query response time increases with the size of the dataset in these three schemes. Compared with the plaintext scheme, the query response time of our PkSKQ scheme is slightly slower, but is faster than the baseline scheme by two orders of magnitude.

6   Related Work

Spatial Keyword Query. Spatial keyword queries have been studied for several years with the increasing popularity of location-based services. Zhou et al. [29] use a hybrid index structure that integrates inverted files and R*-trees for computing both textual and location aware queries. The IR²-tree [11] is another hybrid index that combines an R-tree with superimposed text signatures for only Boolean keywords filter. In the IR-tree [7,18,25], each node in the R-tree is augmented with inverted files. It supports the ranking of objects based on a weighted sum of the spatial distance and text relevancy. Cao et al. [4] study a variant of the spatial keyword query that retrieves a group of spatial objects such that the retrieved objects have the lowest inter-object distances. More introduction and evaluation of spatial keyword queries can refer to [3,5]. However,when spatial keyword queries are performed over the outsourced database, none of the above techniques takes into account the security problem of protecting the privacy.

Secure Spatial Query. The issue of secure spatial query over outsourced data has been addressed in some recent works. Yiu et al.[27,28] propose a spatial transformation that re-distributes the locations in space before outsourcing them and a cryptographic-based transformation. Hore et al. [14] propose a solution based on bucketization to support multidimensional range queries on encrypted data, which prevents the server from learning exact values. In [23], Wang et al. present the $\hat{R}$-tree that can be securely placed in the cloud and searched efficiently, which is designed for encrypted halfspace range queries using ASPE. Yao et al. [26] design a new secure kNN method based on partition-based secure Voronoi diagram, where the relevant encrypted partitions are retrieved for the encrypted query so that the partition is guaranteed to contain the k-nearest neighbors of the query. Hu et al. [15] proposed a method based on Privacy Homomorphism to address the secure kNN problem on the R-tree index, in which the encryption function and the decryption function are conducted in the client and the server respectively. Elmehdwi et al. [10] present a secure kNN protocol based Paillier cryptosystem that protects the confidentiality of the data, user's input query, and data access patterns. These works focus on the security problem on spatial queries, however, considering the text relevancy leveraged in spatial keyword queries, none of which can be applied to protect the privacy of textual information.

Secure Textual Query. Studies on secure textual queries in recent years can be classified into two types based on the number of keywords in the query: 1) Secure Single Keyword Search. Curtmola et al. give the formal definition of the searchable encryption and propose an index scheme based on the inverted list in [8]. In [21], Wang et al. solve the result ranking problem utilizing the keyword frequency and order-preserving encryption. Boneh et al. [1] propose the first searchable encryption scheme using the asymmetric encryption scheme. Li et al. proposed a wildcard based fuzzy search over encrypted data in [17]. Then Wang et al. [22] improved the scheme by constructing a trie-based index. In [16], the LSH functions are used to generate file index. 2) Secure Multiple Keywords Search. Cao et al. [2] propose a privacy preserving multi-keyword ranked search scheme using symmetric encryption. Sun et al. [20] propose an efficient privacy preserving multi-keyword supporting cosine similarity measurement. These studies can only support part of the secure spatial keyword queries by keyword filtering. However, when the keywords are attached to the spatial points, none of them can conduct spatial keyword queries without privacy leakage.

7   Conclusion

In this paper, we have studied the privacy-preserving top-k spatial keyword query problem in outsourced environments. We proposed a privacy-preserving top-k spatial keyword query scheme. In this scheme, we built a secure index based on a state-of-the-art index for spatio-textual database (ie., IR-tree), where the spatial and textual data are encrypted in a unified way. To search with such index, we developed anchor-based position determination and position-distinguished trapdoor generation for the similarity computations between query points and tree nodes under encryption. Thorough security analysis and extensive experimental results illustrate the validity and security of our scheme. This work may open up many promising directions for future work. First, it is worth supporting incremental updates on the encrypted IR-tree index. Second, it is worth studying the case that the cloud cannot be trusted.

Acknowledgments. We thank the reviewers for their valuable comments. The work was supported in part by the following funding agencies of China: National Natural Science Foundation under grant 61170274, National Key Basic Research Program (973 Program) under grant 2011CB302506 and Fundamental Research Funds for the Central Universities under grant 2014RC1103.

References

1. Boneh, D., Di Crescenzo, G., Ostrovsky, R., Persiano, G.: Public Key Encryption with Keyword Search. In: Cachin, C., Camenisch, J.L. (eds.) EUROCRYPT 2004. LNCS, vol. 3027, pp. 506–522. Springer, Heidelberg (2004)
2. Cao, N., Wang, C., Li, M., Ren, K., Lou, W.: Privacy-preserving multi-keyword ranked search over encrypted cloud data. In: INFOCOM, pp. 829–837 (2011)
3. Cao, X., Chen, L., Cong, G., Jensen, C.S., Qu, Q., Skovsgaard, A., Wu, D., Yiu, M.L.: Spatial keyword querying. In: ER, pp. 16–29 (2012)
4. Cao, X., Cong, G., Jensen, C.S., Ooi, B.C.: Collective spatial keyword querying. In: SIGMOD Conference, pp. 373–384 (2011)
5. Chen, L., Cong, G., Jensen, C.S., Wu, D.: Spatial keyword query processing: An experimental evaluation. PVLDB **6**(3), 217–228 (2013)
6. Chor, B., Kushilevitz, E., Goldreich, O., Sudan, M.: Private information retrieval. J. ACM **45**(6), 965–981 (1998)
7. Cong, G., Jensen, C.S., Wu, D.: Efficient retrieval of the top-k most relevant spatial web objects. PVLDB **2**(1), 337–348 (2009)
8. Curtmola, R., Garay, J.A., Kamara, S., Ostrovsky, R.: Searchable symmetric encryption: improved definitions and efficient constructions. In: ACM Conference on Computer and Communications Security, pp. 79–88 (2006)
9. Delfs, H., Knebl, H.: Introduction to Cryptography - Principles and Applications. Information Security and Cryptography, Springer (2007)
10. Elmehdwi, Y., Samanthula, B.K., Jiang, W.: Secure k-nearest neighbor query over encrypted data in outsourced environments. In: ICDE, pp. 664–675 (2014)
11. Felipe, I.D., Hristidis, V., Rishe, N.: Keyword search on spatial databases. In: ICDE, pp. 656–665 (2008)

12. Ghinita, G.: Privacy for Location-based Services. Synthesis Lectures on Information Security, Privacy, and Trust, Morgan & Claypool Publishers (2013)
13. Goldreich, O., Ostrovsky, R.: Software protection and simulation on oblivious rams. J. ACM **43**(3), 431–473 (1996)
14. Hore, B., Mehrotra, S., Canim, M., Kantarcioglu, M.: Secure multidimensional range queries over outsourced data. VLDB J. **21**(3), 333–358 (2012)
15. Hu, H., Xu, J., Ren, C., Choi, B.: Processing private queries over untrusted data cloud through privacy homomorphism. In: ICDE, pp. 601–612 (2011)
16. Kuzu, M., Islam, M.S., Kantarcioglu, M.: Efficient similarity search over encrypted data. In: ICDE, pp. 1156–1167 (2012)
17. Li, J., Wang, Q., Wang, C., Cao, N., Ren, K., Lou, W.: Fuzzy keyword search over encrypted data in cloud computing. In: INFOCOM, pp. 441–445 (2010)
18. Li, Z., Lee, K.C.K., Zheng, B., Lee, W.C., Lee, D.L., Wang, X.: Ir-tree: An efficient index for geographic document search. IEEE Trans. Knowl. Data Eng. **23**(4), 585–599 (2011)
19. Stefanov, E., Shi, E., Song, D.X.: Towards practical oblivious RAM. In: NDSS, pp. 5–8 (2012)
20. Sun, W., Wang, B., Cao, N., Li, M., Lou, W., Hou, Y.T., Li, H.: Privacy-preserving multi-keyword text search in the cloud supporting similarity-based ranking. In: ASIACCS, pp. 71–82 (2013)
21. Wang, C., Cao, N., Li, J., Ren, K., Lou, W.: Secure ranked keyword search over-encrypted cloud data. In: ICDCS, pp. 253–262 (2010)
22. Wang, C., Ren, K., Yu, S., Urs, K.M.R.: Achieving usable and privacy-assured similarity search over outsourced cloud data. In: INFOCOM, pp. 451–459 (2012)
23. Wang, P., Ravishankar, C.V.: Secure and efficient range queries on outsourced databases using R̂-trees. In: ICDE, pp. 314–325 (2013)
24. Wong, W.K., Cheung, D.W.L., Kao, B., Mamoulis, N.: Secure knn computation on encrypted databases. In: SIGMOD Conference, pp. 139–152 (2009)
25. Wu, D., Cong, G., Jensen, C.S.: A framework for efficient spatial web object retrieval. VLDB J. **21**(6), 797–822 (2012)
26. Yao, B., Li, F., Xiao, X.: Secure nearest neighbor revisited. In: ICDE, pp. 733–744 (2013)
27. Yiu, M.L., Ghinita, G., Jensen, C.S., Kalnis, P.: Outsourcing search services on private spatial data. In: ICDE, pp. 1140–1143 (2009)
28. Yiu, M.L., Ghinita, G., Jensen, C.S., Kalnis, P.: Enabling search services on outsourced private spatial data. VLDB J. **19**(3), 363–384 (2010)
29. Zhou, Y., Xie, X., Wang, C., Gong, Y., Ma, W.Y.: Hybrid index structures for location-based web search. In: CIKM, pp. 155–162 (2005)

Bichromatic Reverse Nearest Neighbor Query without Information Leakage

Lu Wang[1], Xiaofeng Meng[1]([✉]), Haibo Hu[2], and Jianliang Xu[2]

[1] School of Information, Renmin University of China, Beijing, China
{luwang,xfmeng}@ruc.edu.cn
[2] Department of Computer Science, Hong Kong Baptist University,
Hong Kong, China
{haibo,xujl}@comp.hkbu.edu.hk

Abstract. Bichromatic Reverse Nearest Neighbor (BRNN) Query is an important query type in location-based services (LBS) and has many real life applications, such as site selection and resource allocation. However, such query requires the client to disclose sensitive location information to the LBS. The only existing method for privacy-preserving BRNN query adopts the cloaking-region paradigm, which blurs the location into a spatial region. However, the LBS can still deduce some information (albeit not exact) about the location. In this paper, we aim at strong privacy wherein the LBS learns nothing about the query location. To this end, we employ private information retrieval (PIR) technique, which accesses data pages anonymously from a database. Based on PIR, we propose a secure query processing framework together with various indexing and optimization techniques. To the best knowledge, this is the first research that preserves strong location privacy in BRNN query. Extensive experiments under real world and synthetic datasets demonstrate the practicality of our approach.

Keywords: Privacy preservation · Location privacy · Private information retrieval · Bichromatic RNN

1 Introduction

Given two point sets S (the servers) and R (the objects), and a server $q \in S$, a bichromatic reverse nearest neighbor (BRNN) query finds the set of objects whose nearest server is q. BRNN has been receiving increasing attention since the boom of mobile computing and location-based services (LBS). It has numerous

X. Meng – This research was partially supported by the grants from the Natural Science Foundation of China (No. 61379050, 91224008); the National 863 High-tech Program (No. 2013AA013204); Specialized Research Fund for the Doctoral Program of Higher Education(No. 20130004130001); the Fundamental Research Funds for the Central Universities, and the Research Funds of Renmin University(No. 11XNL010);HK-RGC GRF grants HKBU12200914 and HKBU 210612;HK-RGC Grants HKBU211512 and HKBU12200114.

M. Renz et al. (Eds.): DASFAA 2015, Part I, LNCS 9049, pp. 609–624, 2015.
DOI: 10.1007/978-3-319-18120-2_35

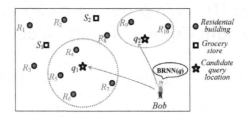

Fig. 1. Example of BRNN query

applications in map search, resource allocation, emergency service dispatching, military planning, and mobile reality games [3]. Figure 1 illustrates two sets of points of interest (POIs) from an online map service, where red circles are residences R_i and black squares are grocery stores S_i. Bob has a few candidate locations q_i (the black star) to open up a new store, so he wants to know which location can attract the most residences from existing stores based on distance. By issuing a BRNN query at each candidate location, he is able to tell q_1 is the best location to open a new grocery store, as it leads to the largest BRNN results — four residences R_4, R_5, R_6 and R_7.

However, the query location as well as Bob's business intention has been disclosed to the server during this process. Such privacy disclosure also occurs in other BRNN application scenarios. For example, in taxi dispatching, a taxi driver has to report the cab's current location in order to know the customers to whom he/she is the nearest to serve. However, such location can reveal sensitive information about the passenger, such as his/her medical or financial condition, as well as political affiliations [2]. Therefore, protecting the query input of a BRNN query against the LBS is indispensable. In the literature, the only existing approach for privacy-preserving BRNN query adopts the cloaking-region paradigm [3], which sends to the LBS a spatial region that contains the query point. Based on this region, the LBS returns a superset of the genuine reverse nearest neighbors, from which the client user will refine the true result. Unfortunately, this approach still reveals to the LBS substantial information about the location.

To guarantee strong location privacy, a promising cryptography tool is private information retrieval (PIR) [16]. PIR allows a data item (e.g., a disk page) to be retrieved from a server without leaving any clue of the item being retrieved. PIR was considered to be resource-intensive, but thanks to the recent progress in cryptography, practical software or hardware PIR solutions have been proposed [4]. Since then it has been successfully applied to database problems, such as kNN and shortest path search [6] [8].

In this paper, we investigate privacy-preserving BRNN query without the LBS deducing any information about the query point. To this end, we adopt practical PIR techniques that retrieve a single data page as the building block. The challenges of a PIR-based BRNN solution lies in the following aspects: (1) although PIR guarantees secure access of a single page from the server, the variation of the number of page accesses from different queries may reveal information about the query point, and (2) as the database contains voluminous points, directly applying PIR for the BRNN query is inefficient, thus calling for an integration with spatial index, such as KD-tree. To address these challenges, we first propose a PIR-based BRNN query processing framework that guarantees strong privacy. We then apply to this framework two indexing schemes, whose performance varies with the data distribution. An orthogonal optimization technique is also proposed to further enhance the performance. To summarize, our main contributions are:

(1) To the best knowledge, this is the first work on BRNN query processing with no information leakage.

(2) We propose a framework for PIR-based BRNN query and prove its security.

(3) We design two indexing schemes for different data distributions, and propose an optimization to further bring down the transformation cost.

(4) We conduct extensive experiments under real-world and synthetic datasets, which shows our proposed approach is practical.

The rest of the paper is organized as follows. Section 2 reviews the related work. In Section 3, we formalize the system model and problem definition. In Section 4, we present the framework for the PIR-based BRNN query processing, followed by two indexing schemes based basic methods, namely KD-tree based method and Adaptive grid based method in Section 5. We then propose an optimization in Section 6. The solutions are evaluated by experiments in Section 7.

2 Related Work

In this section, we review existing literature on bichromatic reverse nearest neighbor query and private information retrieval.

2.1 Bichromatic Reverse Nearest Neighbor

In light of its critical applications ranging from social life domain such as location selection to military activities such as the placement of food [9], bichromatic reverse nearest neighbor query (BRNN query) attracts considerable attention since its first seminal work [12]. To efficiently find the BRNNs for a query point, Voronoi polygon is widely used under various circumstances such as static or continuous query processing [14] [13]. In these works, the Voronoi polygon determines candidate or accurate region for the query point, within which object points are the query point's BRNN result. However, all these queries have not considered the privacy issue of disclosing the plaintext query location to the LBS. There is only one recent work addressing this issue [3]. In this work, the client issues a query region instead of a point to the server, and the server returns object points that are BRNNs to every point in the region. The client then refines the actual BRNNs based on his/her actual location point. While this solution still exposes a query region, our work supersedes it by revealing no location information of the query point.

2.2 Private Information Retrieval

Prior to PIR-based methods, data transformation based methods are considered to provide strong location privacy. [11] presents a model that adopts Hilbert mapping to transform the location data. Such transformation encrypts coordinates in a way that preserves distance proximity and thus can be applied for approximate nearest neighbor query or range query. [10] proposes a secure transformation to guarantee the approximate distances of POIs to the query point and answers kNN query. However, these methods are vulnerable to exposing relative distance [1] or access pattern attack[8].

Thanks to the advances of modern hardware and distributed/cloud computing, PIR has become a viable solution to oblivious data page access in malicious server [5]. However, it is not trivial to apply it to privacy-preserving location queries, because the processing for different queries incurs different numbers of PIR access, which may be exploited by adversaries to induce the query location [7]. Existing PIR-based methods includes PIR-based NN query[7], kNN query[6] and shortest path computation[8]. To guarantee equal number of PIR access for any query point, all these methods imposes the maximum number of PIR access on the dataset. Their main objective, therefore, is to design elaborate data structures (e.g., grid file or KD-tree) that decompose the space to reduce this number. Nonetheless, no existing work has been on applying PIR-based method to BRNN query.

3 Problem Definition

In this section we present the system model and formally define the problem as well as the security model.

3.1 System Model

Figure 2 illustrates the system model in this paper. The server (LBS) owns two POI datasets, namely, the server points S and object points R. The client issues a bichromatic reverse nearest neighbor (BRNN) query $q \in S$, for which the server returns the set of objects whose nearest server is q. Formally, $BRNN(q) = \{r \in R | \forall s \in S : dist(r, q) \leq dist(r, s)\}$. As for the privacy requirement, the server should not learn any information about q.

To enable privacy protection, a naive solution is to ship both S and R datasets to the client for processing. However, due to their large volume and dynamic nature, this solution cannot scale well. Thanks to the recent advances in private information retrieval (PIR), we adopt the state-of-the-art hardware-based PIR as follows. The server installs a secure co-processor ($SCOP$), which offers unobservable and unmolested computation inside an untrusted hosting device. The $SCOP$ performs a hardware-based PIR protocol with the client, and offers the latter oblivious access of a data page [4]. With the $SCOP$, we propose a general secure processing framework for spatial queries, which is composed of

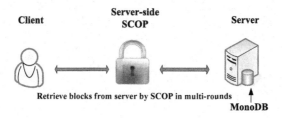

Fig. 2. System architecture

multi-round PIR access to a database $MonoDB$. It is a monolithic database that integrates both the datasets and indexes. In each round, the client retrieves a specific page of $MonoDB$ through $SCOP$. The fetched data helps the client determine the next page to retrieve, and the procedure repeats until the BRNN query is answered. Therefore, the secure query processing problem is reduced to the efficient design of the $MonoDB$ and the associated retrieval plan.

3.2   Adversary and Security Model

Adversary. The adversary in our problem is the LBS server. As a common assumption in private information retrieval, the computational power of the adversary is polynomially bounded.

Security Model. Our objective is to develop practical protocol for processing BRNN query without the LBS deducing any location information about the queries. Similar to [6], we assert that every BRNN query follows the same retrieval plan, which is necessary in order to achieve our privacy goal. Specifically, we ensure that every query (1) executes in the same number of rounds in the same order and (2) in each round it retrieves the same number of data pages. The retrieval plan is determined by the processing protocol and is publicly available. For example, if the protocol states that in the second round, 5 pages are fetched from the database, then every query must fetch 5 pages from database in the second round. If a query needs fewer than 5 pages, the protocol pads with dummy page requests. Since each invocation of PIR is secure, we can naturally reach the following theorem regarding the security of our proposed framework.

Theorem 1. *The BRNN query processing framework leaks no information to the adversary about query location. Equivalently, from the adversary's perspective, every query is indistinguishable from any other.*

4   Private BRNN Processing Framework on MonoDB

In this section, we overview private BRNN query processing in the proposed $MonoDB$ framework. Recall that in this framework, any query processing is equivalent to a multi-round retrieval of data pages of the $MonoDB$. The $MonoDB$ can be logically split into n databases $DB_1, DB_2, ..., DB_n$, where DB_i $(1 \le i \le n-1)$ are indexes and DB_n is the object database R. The retrieval sequence of the query can also be split accordingly as $[c_1, c_2, ..., c_n]$, where the client fetches a set of pages c_i from DB_i $(1 \le i \le n)$. In this section, we first present a baseline BRNN processing algorithm, based on a key observation that reduces the number of servers and objects to retrieve for the query evaluation. Based on this algorithm, we describe the detailed $MonoDB$ design and retrieval plan.

4.1   Baseline BRNN Processing

There are several existing BRNN query processing methods in the literature. In the Voronoi Diagram-based method, a Voronoi Diagram is constructed for all server points and the query point q, and the object points that are in q's Voronoi cell are the BRNN results. However, this method cannot be directly applied in our framework as the query point q is dynamic and cannot be learnt by the LBS. Nonetheless, we observe that the Voronoi Diagram of the server points gives a nice bound of the result objects.

Figure 3(a) illustrates the Voronoi Diagram for all server points (they are called "seeds" in some literatures). For each seed, any object point in its corresponding Voronoi cell is closer to it than to any other seed. When a query q is issued, q is added to the set of seeds, and the Voronoi Diagram is updated as in Figure 3(b). Compared with these two diagrams, we observe that changes only occur in Voronoi cells v_5, v_6, v_7, v_8, and v_{10}, where q is in v_7. In other words, the Voronoi cell for the query point q only depends on the seed v_7 and its neighboring seeds in the original Voronoi diagram. As a result, only object points in these Voronoi cells can reside in the new Voronoi cell of q. Therefore, the BRNN results can be bounded by those object points in the Voronoi cells that contains q and its neighboring cells. This observation can be formally stated as below.

Theorem 2. *Given query point q and the Voronoi diagram of all server points, any object point M outside of the Voronoi cell that contains q and its neighboring cells cannot be q's BRNN.*

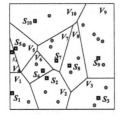
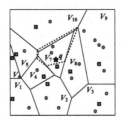
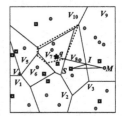

(a) Voronoi Diagram by server points

(b) Voronoi Diagram by server points plus query point

(c) Proof for the observation

Fig. 3. Voronoi Diagram for server points and query point

Proof. As Figure 3(c) illustrates, the segment between q and M must cross some Voronoi cell that neighbors with q at point I. Suppose that the seed of this cell is S. The distance between q and M equals to the segment length between q and M, which is equal to $dist(M, I) + dist(I, q)$. Since I is S's BRNN, it must hold that $dist(I, S) < dist(I, q)$. Further, according to the triangle inequality, $dist(M, S) < dist(M, I) + dist(I, S)$. Therefore, $dist(q, M) = dist(M, I) + dist(I, q) > dist(M, I) + dist(I, S) > dist(M, S)$. As such, the distance between q and M is larger than the distance between M and S, which means M cannot be q's BRNN.

By Theorem 2, the BRNN processing protocol between the client and LBS is as follows. The LBS first computes the Voronoi Diagram of all server points offline. When query q arrives, it sends to the client (1) the servers (i.e., seeds) whose Voronoi cell contains q or is a neighbor of it; and (2) all object points in the corresponding cells of these seeds. The client then refines the BRNN results by verifying among these seeds if q is the nearest neighbor to each object point.

4.2 MonoDB Design and Retrieval Plan

Three Databases. Based on the above baseline BRNN query processing algorithm, the *MonoDB* can be split into three logical databases as illustrated in Figure 4: DB_1 stores all the Voronoi cells, DB_2 records the Voronoi neighbors of each cell, and DB_3 stores object points of each cell. Note that DB_1 implies a space partition, from which only the relevant Voronoi cells need to be retrieved. The partition is non-overlapping so that only one record in DB_1 will be retrieved for any query q. The detailed partition algorithms are discussed in the next section.

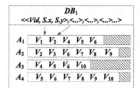

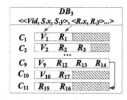

(a) Example for DB_1 (b) Example for DB_2 (c) Example for DB_3

Fig. 4. Examples for the three index structures

Retrieval Plan. Given *MonoDB* and a BRNN query q, the PIR retrieval plan is as follows. The client first accesses DB_1 for the record of the partition where q is located. This record stores the coordinates of all seeds in this partition, so that the client can compute their distances to q and finds the seed i that is the closest to q. Then the client accesses DB_2 for the record of i, which stores i's Voronoi neighbors. According to Theorem 2, the Voronoi cell of q can be derived from i and its Voronoi neighbors. So the client accesses DB_3 for the records of q and q's Voronoi neighbors. These records store all object points that are in these Voronoi cells. A final refinement step is needed to remove from the results those objects outside of the Voronoi cell of q.

Figure 4 illustrates the $MonoDB$ for the whole space and the server points of Figure 3(c). If the query q is issued at the star point in Figure 3(c), it will be located in record A_4 in DB_1, where we can find v_7 is the closest to q. So we obtain v_7's Voronoi neighbors from record B_7 of DB_2, i.e., server points $v_2, v_5, v_6, v_8, v_{10}$. We then access records $C_2, C_5, C_6, C_7, C_8, C_{10}$ from DB_3. These records give us the candidate result objects such as R_2, R_3 that are further refined by the client.

Rationale of Three Databases. Splitting the $MonoDB$ into three logical databases has a variety of benefits: (1) it decouples the server and object points so that the update in one dataset will not significantly change the $MonoDB$; (2) it removes redundancy information and thus enhances the PIR performance. For example, if DB_1 and DB_2 were merged into DB_1', there would be a lot of common neighboring seeds in different records of DB_1'.

Overflow and Underflow Handling. Normally a record spans a single page of a database. If it is not full, it will be padded with dummy data. On the other hand, if a record overflows in any database DB_i, the LBS creates extra pages and appends them at the end of DB_i. These pages are chained up by the overflow pointer at the end of each page, e.g., B_2 with B_{11} in DB_2 and C_9 with C_{11} in DB_3. In what follows, we use cnt_i to denote the maximum number of pages for a single record in DB_i.

5   Spatial Partition

In our $MonoDB$ framework, while DB_2 and DB_3 depend only on the two datasets, DB_1 also depends on the space partition scheme. In this section, we present two space partition algorithms, which leads to two different indexes for DB_1.

5.1   KD-tree Partition

KD-tree is a widely adopted method for space partition due to its at least 50% space utilization. In what follows, we show how to construct the DB_1 based on a KD-tree.

Figure 5(a) illustrates that a KD-tree partition of the space over server points, which produces four node: N_1, N_2, N_3 and N_4. Each node contains a minimum of 2 and a maximum of 2*2-1=3 seeds. As such, DB_1 has four records N_1 through N_4, each of which stores the seeds whose Voronoi cell overlaps with this node. Note that we set each data page can hold 4 seeds, so each record spans two pages in Figure 5(a).

Algorithm 1 illustrates the BRNN query evaluation routine. First, the partition that covers the query point is obtained (Line 2) and the corresponding

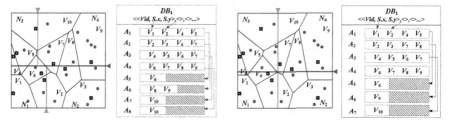

(a) KD-tree based partitioning   (b) Adaptive grid based partitioning

Fig. 5. The two partition schemes

Algorithm 1 KD-Tree Based Method

Require: Three databases DB_1, DB_2 and DB_3, the index of KD-tree, query point q, QP

Ensure: Reverse nearest neighboring object points of q

1: $result = \emptyset$
2: $leafnode = kd.search(q)$
3: Fetch the seeds Cs of record $leafnode$ from DB_1 by cnt_1 PIR accesses
4: $c = argmin_{i \in Cs}dist(i,q)$
5: Fetch all neighboring Voronoi cells Nc of record c from DB_2 by cnt_2 PIR accesses
6: Construct Voronoi Diagram VD according to $Nc \cup q \cup c$
7: $region = VD.q$
8: **for** each Voronoi Cell $i \in VD$ **do**
9:     Fetch all object points O belonging to the record i from DB_3
10:    **for** each object point $o \in O$ **do**
11:       **if** o is contained in $region$ **then**
12:          $result = result \cup o$
13: return $result$

---

record in DB_1 is fetched (Line 3). The client then finds out the server point c whose Voronoi cell contains the query point (Line 4). Next, we compute the Voronoi cell of q (Line 5-7). Finally, those candidate object points that fall in the Voronoi cell of q are the BRNN results (Line 8-13).

5.2   Adaptive Grid Partition

The disadvantage of KD-tree partition for DB_1 is the non-uniform distribution of record size. Although the number of server points in each partition is almost uniform, the number of points whose Voronoi cells overlap each partition is not. As such, a record in DB_1 may span too many pages and thus degrades the PIR retrieval performance. In what follows, we present an adaptive-grid based method that addresses this issue.

The motivation is to have fine-granularity partition over regions with dense Voronoi cells and coarse-granularity partition over regions with sparse Voronoi cells. In this way, it can avoid a single record in DB_1 that hold too many seeds.

Specifically, this method partitions the whole space into an $n \times n$ grid in an adaptive manner as follows. It first finds $n - 1$ vertical lines one by one that partition the space into n grid cells. Each time, a vertical line is found to minimize the difference of number of Voronoi cells overlapping with both cells. This is achieved by using the standard plane sweep algorithm. Then, it similarly finds $n - 1$ horizontal lines that further partition each one of the n cells into n sub-cells. In Figure 5(b), there are only 2×2 grid cells, so only one grid line is needed for each dimension.

6   Optimization

In this section, we present an orthogonal general optimization to the two basic indexing methods by packing small records in DB_3 into one page. Note that the default placement for records in DB_3 assigns every record a different page, which suffers from low utilization and leads to inefficient PIR access that only fetches very few useful data from a page. Therefore, we propose to pack those records of DB_3 if they correspond to the same record in DB_2, as these records are always retrieved altogether. As a result, the PIR access of both DB_2 and DB_3 will be more efficient. Figure 6(a) illustrates the default placement that requires 2 PIR accesses to fetch object points for any query, whereas Figures 6(b) and 6(c) illustrate two example packing results. In Figure 6(b), only 1 PIR access is needed to fetch object points of DB_3 for any query. By contrast, 2 PIR accesses are still needed for any query if the records are packed as in Figure 6(c).

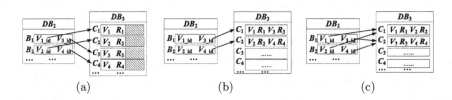

(a)   (b)   (c)

Fig. 6. Example for packing records of DB_3

Let N_{DB_2}, N_{DB_3} denote the number of records in DB_2 and DB_3, respectively. Let $e_i^{DB_2}$ denote the i-th record in DB_2, and $\{e_1^{DB_3}, e_2^{DB_3}, ..., e_m^{DB_3}\}$ the m records that corresponds to $e_i^{DB_2}$. Further, let B_m denote the size of $e_m^{DB_3}$; since it might span multiple pages, and only the last page requires packing, the actual packing size b_m is the fraction part of B_m, i.e., $b_m = B_m \% Page\_Size$. Then, the problem of record packing can be formalized as follows:

Definition 1. Record Packing Problem. *To pack records in DB_3 into data pages, so that $max_i \sum_{j=1}^{m} B_j$ for $\forall e_i^{DB_2}$ is minimized.*

The following theorem shows that this problem is NP-hard.

Theorem 3. *Record packing problem is NP-hard.*

Proof. This problem can be reduced from the "bin packing" problem. The aim of the latter is to find the fewest number of pages to accommodate a total of m items, each of which is smaller than a page. We reduce a bin packing problem to our problem as follows. We create a single record in DB_2 which contains all server points. Then each item in the bin packing problem is mapped to a server point and the size of the item equals to the number of object points for this server point. It is obvious that this straightforward mapping is polynomial, thus completing the proof.

To design an approximation algorithm, in what follows we first present an integer programming solution to the problem, and then relax it to a linear programming problem.

Let variable $y_{m,j} \in \{0,1\}$ denote whether record $e_m^{DB_3}$ is stored in page j of DB_3, and $x_{i,j} \in \{0,1\}$ denote whether any record $e_m^{DB_3} \in e_i^{DB_2}$ is stored in page j. Formally, we have $\forall e_m^{DB_3} \in e_i^{DB_2}$, $x_{i,j} \geq y_{m,j}$. And $\sum_{j=1}^{P} y_{m,j} = 1$, where P is the number of data pages in the default placement for DB_3.

With these variables defined, the number of PIR accesses for object points for a record in DB_2 is the number of full data pages of corresponding object points in DB_3 plus the packed size for this record. That is,

$$e_i^{DB_2} = \sum_{e_m^{DB_3} \in e_i^{DB_2}} \lfloor B_m/Page\_Size \rfloor + \sum_{j=1}^{P} x_{i,j}$$

Finally, the total number of object points in a page should not exceed the page capacity. That is, $\sum_{m=1}^{N_{DB_3}} b_m y_{m,j} \leq Page\_Size$. Let K be the maximum number of PIR accesses for any record in DB_2. Therefore, we reach the following integer programming problem for K as follows:

$$\begin{aligned}
\text{minimize}\quad & K \\
\text{subject to}\quad & \sum_{e_m^{DB_3} \in e_i^{DB_2}} \lfloor \tfrac{B_m}{Page\_Size} \rfloor + \sum_{j=1}^{P} x_{i,j} \leq K, \forall 1 \leq i \leq N_{DB_2} \\
& \sum_{m=1}^{N_{DB_3}} b_m y_{m,j} \leq Page\_Size, && \forall 1 \leq j \leq P \\
& x_{i,j} \geq y_{m,j}, \forall 1 \leq i \leq N_{DB_2}, && \forall e_m^{DB_3} \in e_i^{DB_2} \\
& \sum_{j=1}^{P} y_{m,j} = 1, && \forall 1 \leq m \leq N_{DB_3} \\
& && x_{i,j}, y_{m,j} \in \{0,1\}
\end{aligned}$$

The above integer programming problem can be approximately solved in polynomial time in two steps. First, one can solve a linear relaxation of the problem, where $x_{i,j}$ and $y_{m,j}$ is a fraction in $[0,1]$. In this regard, $y_{m,j}$ serves as the probability of placing record $e_m^{DB_3}$ into data page j, and $x_{i,j}$ serves as the probability of records corresponding to $e_i^{DB_2}$ being placed into data page j. As the second step, we adopt the randomized rounding strategy to obtain a feasible solution as follows. We assign object points in the m-th record of DB_3 to the j-th page with probability $y_{m,j}$. If the page overflows, we will assign new empty pages until all object points in this record can be accommodated.

7   Experimental Evaluation

In this section, we conduct experiments under real world and synthetic datasets to demonstrate the effectiveness of our PIR-based BRNN algorithm. We also compare the performance with a weaker location privacy preservation approach — the cloaking-based PARNN method [3] and show our algorithm is of great practical value. We also carry out experiments to analyze the effect of our optimization approach.

7.1   Experiment Settings

The real world dataset is collected from Open Street Map[1], with location data from Boston and New York, respectively. Both datasets have relatively uniform distribution, while there are more points in New York than in Boston.

As for the synthetic dataset, we vary the number of server points from 10^5 to 10^6, and object points are 10 times those of server points. To emulate a skewed distribution, these points are generated by a widely adopted benchmark defined by Chen et al. [15]. In this benchmark, a portion $f \in (0, 1]$ of points are generated in a skewed way to capture object clusters while the rest $1 - f$ portion of points are uniformly generated. Specifically, the portion f of the points are controlled by another skewed parameter s and are generate not far from one of the s randomly selected server points. Table 1 summarizes the detailed parameters of the datasets.

The two indexing methods are implemented with the optimization in Section 6 in place. All codes are written in C# and run on a machine with an Intel Core2 Quad CPU 2.53Ghz and 4 GByte of RAM. We also adopt the open source GNU Linear Programming Kit[2] as the solver the record packing problem in Section 6. As with previous hardware-based PIR methods, we assume the IBM 4764 PCI-X Cryptographc Coprocessor as the SCOP and strictly simulate its performance. The client communicates with the LBS using a link with round trip time of 700ms and bandwidth 384 Kbit/s, which emulates a moving client connected via a 3G network.

Table 1. Summary of Experimental Settings

Dataset	The number of server points	The number of object points
Boston	8381	146207
New York	126900	1462057
Synthetic	10000-1000000	100000-10000000

---

[1] www.openstreetmap.org
[2] https://www.gnu.org/software/glpk/

7.2   Performance Comparison

In this section, we compare the performance of our PIR-BRNN method with the PARNN method under both real world and synthetic datasets. The latter method fetches all Voronoi cells that overlap with the client-issued cloaking region, and then returns to the client all object points that are covered by these cells. Note that the performance of PARNN is plotted only for reference, as it still discloses a cloaking region to the LBS.

Figure 7(a) illustrates that PIR-BRNN method outperforms PARNN method with different cloaking region size by a factor of $2-4$ in terms of execution time. We can see that when the cloaking region size shrinks from $100 \times 100m^2$ to $10 \times 10m^2$ (in practice, from a plaza to a road crossing), the execution time for PARNN can improve by about 50%, because fewer server points and object points will be accessed by PARNN. Nonetheless, it still takes more than 2 times the execution time than our proposed PIR-BRNN method.

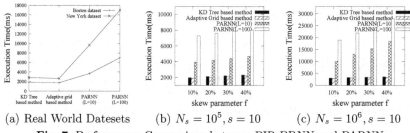

(a) Real World Datesets   (b) $N_s = 10^5, s = 10$ (c) $N_s = 10^6, s = 10$

Fig. 7. Performance Comparison between PIR-BRNN and PARNN

In synthetic datasets, the performance of PIR-BRNN approach deteriorates as more dummy PIR accesses need to be carried out due to the skewed data distribution. However, Figures 7(b) and 7(c) illustrate that the performance of PIR-BRNN approach is still better than PARNN. The experimental results show that our PIR-BRNN approach is superior to the PARNN method by providing stronger privacy guarantee as well as faster query result time.

7.3   Effect of Space Partitions

In this experiment, we evaluate the effect of partition scheme without any optimization[3]. Figure 8(e) illustrates the performance under the real world dataset, where two methods have similar performance. However, as the point distribution becomes more and more skewed in Figures 8(a)-(d), the adaptive grid based method significantly outperforms KD-tree based method. This result coincides with our analysis in Section 5.2 that while the KD-tree keeps each partition

---

[3] The real SCOP only has 32MB of main memory and can only support up to 2.5GB addressable space. To enable the experiment in this subsection, however, we simply assume there are enough memory buffer in the SCOP emulator.

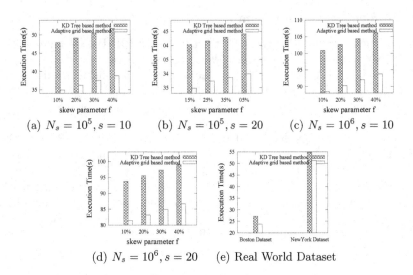

(a) $N_s = 10^5, s = 10$ (b) $N_s = 10^5, s = 20$ (c) $N_s = 10^6, s = 10$

(d) $N_s = 10^6, s = 20$ (e) Real World Dataset

Fig. 8. Performance under two basic methods

approximately equal number of points, it fails to keep each partition approximately equal number of overlapped Voronoi cells. Therefore, as the server points become skew, it suffers from more overflow pages for dense records in DB_1, and thus incurs unnecessary PIR accesses for these extra data pages.

7.4   Effectiveness of Optimizations

In this subsection, we evaluate the effect of the optimization proposed in Section 6. First, we show that the number of PIR accesses in both indexing methods is reduced significantly by the optimization. Then we show that although the linear programming (LP) based optimization does not yield the optimal packing of records, it runs much faster than the integer programming (IP) based optimization while still leads to reasonable performance.

Figure 9 illustrates that under real world dataset, both IP and LP based optimization reduce the number of PIR accesses by more than 70% on average. In particular, the effect of our optimization is most significant for skewed data distribution. This is because many server points have a lot of corresponding records in DB_3 with only very few object points; as they are packed together, the maximum number of PIR accesses is greatly reduced.

It is worth noting that although the LP based optimization dose not yield the optimal packing of records, it achieves comparable performance as the optimal IP-based optimization. On the other hand, Figure 10 illustrates that the running time of the former is much faster and is thus more practical than the latter. In fact, in our experiment we cannot complete the IP based optimization on the New York dataset or any synthetic dataset with more than 10^6 server points.

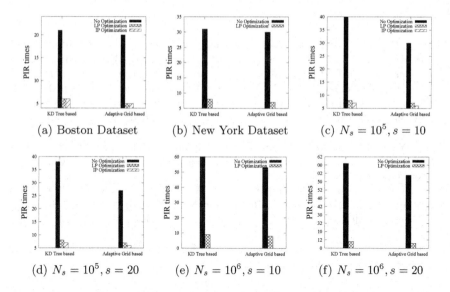

Fig. 9. Effect of Optimization on Various Datasets

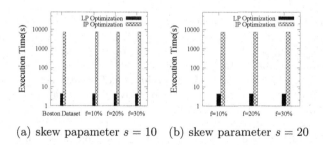

Fig. 10. Performance Comparison between LP and IP Optimization

8   Conclusion

In this paper we introduce the novel problem of BRNN query with strong privacy guarantee, where an adversary cannot distinguish a query point from any other point in the space. This is the first work that applies PIR to BRNN query. Further, we show that it is NP-hard to minimize the number of PIR accesses given any partition scheme over the whole space, and therefore propose a linear programming approximation to the optimal packing problem in our proposed *MonoDB*. Finally, we evaluate our methods on real world dataset and synthetic dataset. Extensive experiments demonstrate the practicality of our method.

References

1. Tian, F., Gui, X., Yang, P., Zhang, X., Yang, J.: Security analysis for hilbert curve based spatial data privacy-preserving method. In: 10th IEEE International Conference on High Performance Computing and Communications and 2013 IEEE International Conference on Embedded and Ubiquitous Computing, pp. 929–934, Zhangjiajie, China (2013)
2. Wicker, S.B.: The loss of location privacy in the cellular age. Commun. ACM **55**(8), 60–68 (2012)
3. Du, Y.: Privacy-aware RNN query processing on location-based services. In: 8th International Conference on Mobile Data Management, pp. 253–257, Mannheim, Germany (2007)
4. Williams, P., Sion, R.: Usable PIR. In: NDSS, San Diego, California, USA (2008)
5. Kushilevitz, E., Ostrovsky, R.: Replication is NOT needed: SINGLE database, computationally-private information retrieval. In: FOCS, pp. 364–373, Florida, USA (1997)
6. Papadopoulos, S., Bakiras, S., Papadias, D.: Nearest neighbor search with strong location privacy. In: VLDB, pp. 619–629 (2010)
7. Ghinita, G., Kalnis, P., Khoshgozaran, A., Shahabi, C., Tan, K.L.: Private queries in location based services: anonymizers are not necessary. In: SIGMOD, pp. 121–132, Vancouver, BC, Canada (2008)
8. Mouratidis, K., Yiu, M.L.: Shortest path computation with no information leakage. In: VLDB, pp. 692–703, Istanbul, Turkey (2012)
9. Stanoi, I., Riedewald, M., Agrawal, D. and Abbadi, A.E.: Discovery of influence sets in frequently updated databases. In: VLDB, pp. 99–108, Roma, Italy (2001)
10. Wong, W.K., Cheung, D.W.L., Kao, B., Mamoulis, N.: Secure kNN computation on encrypted databases. In: SIGMOD, pp. 139–152, Rhode Island, USA (2009)
11. Khoshgozaran, A., Shahabi, C.: Blind evaluation of nearest neighbor queries using space transformation to preserve location privacy. In: Papadias, D., Zhang, D., Kollios, G. (eds.) SSTD 2007. LNCS, vol. 4605, pp. 239–257. Springer, Heidelberg (2007)
12. Korn, F., Muthukrishnan, S.: Influence sets based on reverse nearest neighbor queries. In: SIGMOD, pp. 201–212, Dallas, Texas, USA (2000)
13. Tran, Q.T., Taniar, D., Safar, M.: Bichromatic Reverse Nearest-Neighbor Search in Mobile Systems. J. IEEE SYSTEMS **4**(2), 230–242 (2010)
14. Kang, J.M., Mokbel, M.F., Shekhar, S., Xia, T. and Zhang, D.: Continuous evaluation of monochromatic and bichromatic reverse nearest neighbors. In: ICDE, pp. 806–815, Istanbul, Turkey (2007)
15. Chen, S., Jensen, C.S., Lin, D.: A benchmark for evaluating moving object indexes. In: VLDB, pp. 1574–1585, New Zealand (2008)
16. Chor, B., Kushilevitz, E., Goldreich, O., Sudan, M.: Private information retrieval. J. ACM **45**(6), 965–981 (1998)

Authentication of Reverse k Nearest Neighbor Query

Guohui Li, Changyin Luo[⊠], and Jianjun Li

School of Computer Science and Technology,
Huazhong University of Science and Technology, Wuhan, China
{Guohuili,luochangyin,jianjunli}@hust.edu.cn

Abstract. In outsourced spatial databases, the LBS provides query services to the clients on behalf of the data owner. However, the LBS provider is not always trustworthy and it may send incomplete or incorrect query results to the clients. Therefore, ensuring spatial query integrity is critical. In this paper, we propose efficient RkNN query verification techniques which utilize the influence zone to check the integrity of query results. The methods in this work aim to verify both monochromatic and bichromatic RkNN queries results. Specifically, our methods can gain efficient performance on verifying bichromatic RkNN query results. Extensive experiments on both real and synthetic datasets demonstrate the efficiency of our proposed authenticating methods.

Keywords: Authentication · RkNN queries · Influence zone

1 Introduction

With the ever-increasing use of mobile handset devices (e.g.,smartphones and tablet devices) and rapid development of wireless communication technologies, location-based services (LBSs) have experienced explosive growth over the past decade. Users carrying location-aware mobile devices are able to query their interests at any time and anywhere. Among the many types of location-based queries, one important class is location-based reverse k nearest neighbor (RkNN) query [1,2,6,11,14], which has various applications in location based services, marketing and decision support systems. Consider the example of a gas station, the drivers for which this gas station is one of the k nearest gas stations are its potential customers. The drivers are the RkNN of the gas station and can be monitored so that the gas station can provide better services for these drivers. In this paper, the objects that provide a facility or service (e.g., gas stations) are called *facilities* and the objects (e.g., the drivers) that use the facility are called *users*. The facility can be considered as a query object (q) when it makes a RkNN query.

To scale up LBSs along with their ever-growing popularity, there has been a rising trend of outsourcing of relational databases [10,15–17] to the LBS, which provides query services to clients on behalf of data owners (DO). While such

© Springer International Publishing Switzerland 2015
M. Renz et al. (Eds.): DASFAA 2015, Part I, LNCS 9049, pp. 625–640, 2015.
DOI: 10.1007/978-3-319-18120-2_36

Fig. 1. Database outsourcing framework

an outscourcing model has its benefits in performance, cost and flexibility in resource management, it also brings a great challenge to *query integrity assurance* [7,8,10,15]. As the LBS provider is not the real owner of the data, it may return incorrect or incomplete query results to clients. Furthermore, query results might be tampered with by malicious attackers. Therefore, in the data-outscouring model, the client must be able to verify that 1) all data returned from the LBS provider originated at the DO and 2) the result set is correct and complete.

Figure 1 shows a framework of authenticated query processing, which is based on digital signatures utilizing a public-key cryptosystem (e.g., RSA). Initially, the DO obtains a private and a public key through a certificate authority. Before delegating a spatial dataset to the LBS provider, DO builds an authenticated data structure (ADS) of the dataset. To support efficient query processing, the ADS is often a tree-like index structure where the root is signed by the DO using its private key. The LBS provider keeps the spatial dataset, ADS and its root signature. Upon receiving a query from the client, the LBS provider returns the root signature and the verification object ($\mathcal{VO}$) that contains the corresponding authentication information. By using the $\mathcal{VO}$, root signature and public key, the client can verify the result.

In spatial queries processing, in order to significantly reduces the communication frequency between the user and LBS provider, the LBS provider often returns query objects an *influence zone* [1] in addition to the query results. Consider a set of facilities $F = \{f_1, f_2, ..., f_n\}$ where f_i represents a point in Euclidean space and denotes the location of the ith facility. Given a query $q \in F$, the influence zone Z_k is the area such that for every point $p \in Z_k$, q is one of its k closet facilities, and for every point $p' \notin Z_k$, q is not one of its k closet facilities.

Recently, techniques for authenticating query results have received a lot of attention. For example, [18,19] study the authentication of moving kNN queries and rang queries; [3] addresses authenticating top-k queries with confidentiality; [8] study authenticating location-Based skyline queries in arbitrary subspaces. Yang et al. [16,17] introduce the MR- and MR*-tree, which are space-efficient authenticating data structures supporting fast query processing and verification. [5] studies the authentication of k nearest neighbor query on road networks. Hu et al. [4] address query integrity assurance with voronoi neighbors, they only focus on *monochromatic* RkNN authentication. Whereas our work aims to authenticate both *monochromatic and bichromatic* RkNN queries, which are based on the *influence zone* verification. That is, once the influence zone is verified, both types of RkNN results can be easily authenticated.

Our contributions can be summarized as follows.

- A compact construction method for the $\mathcal{VO}$ is designed, and the query processing method at server side is proposed.
- Based on the authentication of influence zone Z_k, the verification for the RkNN query at client is designed.
- A set of experiments on both real data are conducted to study the efficiency of our proposed methods.

2   Related Work and Background

RkNN query processing was first studied by Korn et al. [6], and then, some classic literatures [2,11,13,14] have done further research on it. Specifically, Wu et al. [14] propose a method named FINCH. Instead of using bisectors to prune the objects, they use a convex polygon that approximates the unpruned area. Continuous monitoring RkNN is first proposed by Wu et al. [13]. Cheema et al. [2] also present Lazy Updates algorithm to continuously monitor RkNN queries. Based on the TPL [11] and FINCH methods, the recent work [1] designs an innovative *Influence Zone* that can be used to process snapshot and continuous RkNN queries.

A lot of researches have been conducted to investigate data integrity verification for years in area of database outsourcing [7,10,15]. Most authentication techniques [8,16–19] are based on *Merkle tree* [9], which is an authenticated data structure (ADS). The current state-of-the-art ADS for authenticating spatial queries is the Merkle R-tree(MR-tree) [16] based on Merkle tree [9] and R-tree. An extension of MR-tree is called MR*-tree [17], which can reduce the number of entries in the $\mathcal{VO}$ but it has higher construction cost. In this paper, we use the MR-tree because of its popularity and low construction cost.

Yiu et al. [18] and Yung et al. [19] examine how to authenticate the *safe region* of moving kNN queries and moving range queries respectively. The *safe region* for kNN (e.g., convex polygons) and range queries (e.g., set unions/differences of circles) are different from influence zone for RkNN queries (e.g., differences area of k bisectors combination) in our methods. Thus their methods are inapplicable to the problem in this paper. Wu et al [12] first study the authentication of spatial keyword queries, Issues related to keyword authentication are beyond the scope of this paper.

The most relative work [4] proposes an approach named VN-Auth to verify monochromatic RkNN queries. While VN-Auth cannot be used to verify bichromatic RkNN query. However, our approaches can authenticate both monochromatic and bichromatic RkNN queries in the same framework.

3   Problem Definition and Preliminaries

RkNN Authentication Problem. Given a RkNN query, the authentication problem studies how the client can verify the correctness of the query result returned by the LBS provider. It involves two correlated issues: i) server query processing and the $\mathcal{VO}$ construction for each RkNN query on the LBS provider; ii) result verification based on the received $\mathcal{VO}$ and signature at the client. In order to study this problem, we first present some definitions as follows.

Given two facilities a and q in a 2-dimensional data space, a perpendicular bisector $B_{a:q}$ between a and q divides the space into two halves as shown in Fig. 2. The half plane containing a is denoted as $H_{a:q}$ and the half plane containing q is denoted as $H_{q:a}$. Any point p (depicted by a star in Fig. 2) that lies in $H_{a:q}$ is closer to a than q (i.e., $dist(p, a) \leq dist(p, q)$) and any point y that lies in $H_{q:a}$ is closer to q than a (i.e., $dist(y, q) \leq dist(y, a)$). Hence, q cannot be the closet facility of any point p that lies in $H_{a:q}$. The point p can be pruned by bisector $B_{a:q}$ if p lies in $H_{a:q}$. Alternatively, the point a can prune the point p.

Definition 1. Monochromatic RkNN query: *Given a set of data points P of the same type and a query point $q \in P$, a monochromatic RkNN query returns every point $p \in P$ s.t. $dist(p, q) \leq dist(p, p_k)$ where $dist()$ is a distance function, and p_k is the kth nearest point to p according to the distance function $dist()$.*

Definition 2. Bichromatic RkNN query: *Given a set of facilities $F = \{f_1, f_2, ..., f_n\}$ and a users set $U = \{u_1, u_2, ..., u_n\}$, a bichromatic R$k$NN query for a point $q \in F$ is to retrieve every object $u \in U$ s.t. $dist(u, q) \leq dist(u, f_k)$ where f_k is the kth nearest point of u in F according to the distance function $dist()$.*

A special circle C_p for a point p is defined as follows.

Definition 3. *Given a point p, for a query point $q \in F$, C_p denotes a circle centered at p with radius equalling to $dist(p, q)$, where $dist(p, q)$ is the distance between p and q.*

Fig. 2 shows a C_p for facility p. $|C_p|$ denotes the number of facilities that lie within the circle C_p. Both types of RkNN queries can be described as follows: given a set of users U, a set of facilities F and a query $q \in F$, a bichromatic RkNN query is to retrieve every user $u \in U$ for which $|C_u| < k$. Analogically, a monochromatic RkNN for a query $q \in F$ is to retrieve every facility $f \in F$ for which $|C_f| < k + 1$.

We mainly focus on authenticating the bichromatic RkNN queries. The solution can be adapted for monochromatic RkNN queries verification through subtle modification. That is, the inequality of $|C_f| < k$ for bichromatic RkNN verification should be changed to $|C_f| < k + 1$. Our verification methods for RkNN are based on *influence zone*. Its definition for bichromatic RkNN queries is described as follows.

Definition 4. Influence zone Z_k *(Ref. [1]).* *Given a query $q \in F$, the influence zone Z_k is the area such that for every point $p \in Z_k$, $|C_p| < k$ and for every point $p' \notin Z_k$, $|C_{p'}| \geq k$.*

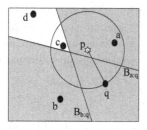
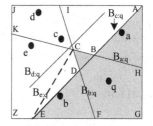

Fig. 2. Unpruned area is not influence   **Fig. 3.** Influence zone Z_k (k=2)

Definition 5. *Counter*. *Given a data point v, the **counter** of v is a number which counts the number of perpendicular bisectors that prune it in the data space. Note that, the perpendicular bisectors are constructed between the facilities and query point q.*

In Fig. 3, the counter of intersection point C is 1 because it is only pruned by $B_{c:q}$.

Definition 6. *Generator*. *Given a data space D, a facility p is defined to be a generator of the influence zone Z_k if the perpendicular bisector between p and query facility q ($B_{p:q}$) contributes to Z_k. Let $G(q, D)$ be the set of all generators of Z_k.*

Suppose $k = 2$, the influence zone Z_2 is the shaded area in Fig. 3, which is formed by considering the perpendicular bisectors of q with every facility point in the data space. However, only the perpendicular bisectors of facilities a, b and c contribute to Z_2, while the perpendicular bisectors of other facilities (e.g., d) do not affect Z_2. Therefore, $G(q, D) = \{a, b, c\}$.

Based on the definition of influence zone, once the influence zone Z_k is computed, the users in it are the RkNN result. Thus, at client side, the authentication methods designed for RkNN query mainly focus on the verification of influence zone Z_k. However, there are some challenges in verifying the correctness of influence zone Z_k. Let's take an example to illustrate them in Fig. 3. Suppose that the LBS provider needs to compute RkNN of q and its influence zone. Based on Definition 6, the LBS provider can represent influence zone Z_2 by query facility q and its generator set: $G(q, D) = \{a, b, c\}$, which can be used to reconstruct Z_2 and verify the RkNN query result at a client.

Suppose the LBS provider intentionally reports a bigger influence zone $ABCEGA$ by adding $\triangle CDE$ into real Z_2 in Fig. 3, then the facility e is falsely considered as a generator. In this case, $G(q, D) = \{a, c, e\}$. The problem here is that the client cannot verify the correctness of this fake generator set. Because all points in $G(q, D)$ originate from the dataset D, they can pass the data correctness checking by examining the signature of $\mathcal{VO}$. However, the client cannot determine whether those facilities belong to the real generator set. Similarly, the LBS provider may intentionally omit $\triangle BCD$ and only sends a fake $Z_2 = AEGA$ to the client. In this case, the client will get a smaller influence zone, which faces the similar authentication problem. Obviously, only performing signature verification is not feasible. In order to authenticate the soundness and completeness of query results, the client must do *geometric verification*, which can check correctness of the results based on the geometric properties of the influence zone.

4   Solution

Section 4.1 describes the query processing mechanism at the LBS provider. Section 4.2 presents authenticating procedures at a client.

4.1   Query Processing at the LBS Provider Side

The MR-tree is adopted in this paper, which is an index based on the R*-tree and is capable of authenticating arbitrary spatial queries. The LBS provider is required to return a verification object $\mathcal{VO}$ that contains three types of data [16]: (i) all objects in each leaf node visited during query processing, (ii) the MBR (minimum bounding rectangle) for each internal node and hash values of pruned nodes, and (iii) special tokens that mark the scope of a node. With this information, the client can reconstruct the root digest and compare it against the one that was signed by the owner to make signature verification. However, as discussed in the literatures [4,5,18,19], the construction of $\mathcal{VO}$ in [16] is not compact enough. So we should design algorithm to construct the $\mathcal{VO}$ for verifying the correctness of the influence zone, while minimizing the size of $\mathcal{VO}$.

Algorithm 1 is the pseudo-code of the LBS provider algorithm. Upon receiving the query facility q and the number k of required RkNN, it computes Z_k from the MR-tree T_D (Line 1). Then, it defines a *Verification Region (VR)* so as to identify facilities that are useful for authenticating Z_k and put them into the $\mathcal{VO}$. Specifically, the VR is defined as the union of C_{v_i} for each vertex v_i of Z_k (Line 2, 3), i.e., $VR = \cup_{v_i \in V} C_{v_i}$, where V denotes the set of vertices of Z_k.

Algorithm 1. Query processing at the LBS provider

Input: MR-tree T_D (on dataset D), Query facility $q \in F$, k
Output: influence zone, $\mathcal{VO}$
  1: compute Influence Zone Z_k from the MR-tree T_D(using the method in [1])
  2: V=collect the set of vertex v_i of Z_k
  3: $VR = \cup_{v_i \in V} C_{v_i}$
  4: $\mathcal{VO}$=DepthFirstRangeSearch($T_D.root$, VR)
  5: **send** the $\mathcal{VO}$ to the client.

Fig. 4 illustrates the VR for LBS provider algorithm. Z_k (i.e., ABCD) is in gray color, which has four vertices. For each vertex v_i, there is corresponding C_{v_i}. $VR = \cup_{v_i \in V} C_{v_i}$, so VR is the region formed by the union of these C_{v_i} (i.e., $VR = C_A \cup C_B \cup C_C \cup C_D$).

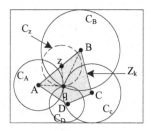

Fig. 4. Illustration of VR

The facilities in the VR are necessary to construct the influence zone at the client side, while the facilities outside the VR have no influence on the construction of influence zone Z_k as stated in the following lemma.

Lemma 1. *If a facility f lies outside C_p for every facility p in the influence zone Z_k, it cannot prune any facility in Z_k and f can be ignored for constructing Z_k*

Proof. Given facilities p, f, q, if $dist(p, f) < dist(p, q)$ which means $f \in C_p$ and p is pruned by the perpendicular bisector $B_{f:q}$. In other words, p can be pruned by the facility f. Therefore, if f lies outside C_p (i.e., $f \notin C_p$), it cannot prune p. Analogically, if f lies outside C_p for every facility p in the influence zone Z_k, it cannot prune any facility in Z_k. The Lemma 1 is proved.

Apparently, the facilities in Z_k cannot be ignored. The relationship between Z_k and VR is described in following lemma.

Lemma 2. *The influence zone Z_k is contained in VR.*

The lemma 2 is obvious so we omit its proof. Based on Lemma 1, we know parts of facilities outside Z_k can be omitted when constructing Z_k. We focus on the borders of Z_k. If f lies outside C_p for each facility p on the borders, it cannot prune any point on the borders and can be ignored. In order to check whether f can be ignored, *a baseline method* is to check whether $f \in C_p$ for every point p on the borders. Obviously, this can result in high cost. Now, we describe an important lemma as follows which can guide us to introduce more efficient methods.

Lemma 3. *Given a line segment AB and a point z on AB. The circle C_z is contained by $C_A \cup C_B$. i.e., every point in the circle C_z is either contained by C_A or by C_B (see Fig. 4) (Ref. [1]).*

Based on the Lemma 3, we only need to check two endpoints in each edge of Z_k. If for $\forall v_i \in Z_k$, $f \notin C_{v_i}$, f cannot prune any facilities in Z_k and can be ignored. Otherwise, f can be used to construct influence zone and should be included in the $\mathcal{VO}$. Hence, the facilities in area $\cup_{v_i \in V} C_{v_i} = VR$ have to be taken into consideration. In other words, the facilities outside the VR cannot alter the influence zone. which is stated in the following theorem.

Theorem 1. *The facilities in the VR are necessary to construct the influence zone Z_k. While the facilities outside the VR cannot alter the influence zone Z_k.*

Proof. Based on the Lemma 1 and 2 and 3, Theorem 1 can be easily proved.

Based on the above theorem, the $\mathcal{VO}$ is computed by a depth-first traversal of the MR-tree (Line 4). When accessing a node n_i in the MR-tree, all facilities in n_i intersecting with VR are put into the $\mathcal{VO}$. Note that the $\mathcal{VO}$ can include some objects which lie outside VR and this will not violate the correctness of query verification at a client side. For a non-leaf entry e which does not intersect VR, e is added to the $\mathcal{VO}$ and its subtree is not visited further.

The $\mathcal{VO}$ constructed by algorithm 1 provides all the information that the client needs to verify the correctness of the influence zone and RkNN results. In the next section, we describe how to authenticate RkNN results at the client.

4.2   Authentication Processing at the Client

The LBS provider sends the $\mathcal{VO}$ to the client. Once Z_k is verified based the $\mathcal{VO}$, it is easy to obtain the correct RkNN. To verify the query results, the client should take two sequential steps in a authentication process. First, the signature of $\mathcal{VO}$ is examined by the client to ensure that all returned objects originated at the DO. Concretely, at the beginning of the query, the client downloads the root signature from the LBS provider and the public key from a certificate authority. Upon receiving the $\mathcal{VO}$, the client first checks the correctness of $\mathcal{VO}$ by constructing the digest of the root of the MR-tree from the $\mathcal{VO}$, and then verifies it against the root signature using the public key of the DO. If the $\mathcal{VO}$ fails the signature authentication process, the result is considered as corrupted and the authentication process should terminate. Otherwise, the client should take the second important step to conduct the geometric verification.

The influence zone verification can be implemented by vertex verification and edge verification. The following Lemma describes the features of the edges of influence zone.

Lemma 4. *The edges of influence zone Z_k originate from perpendicular bisectors drawn between the generators and the query q, or from the boundaries of the data space D.*

Proof. This lemma is obvious so we omit its proof.

For any point p, if we draw C_p, two sets of objects can be obtained: $S_1(p) = \{o_i | o_i$ *is on circle of* $C_p\}$, and $S_2(p) = \{o_j | o_j \in C_p\}$. Note that bisector $B_{o_i:q}$ for all objects o_i in S_1 can intersect at p. If p is a vertex of a real influence zone and it dose not lie on a boundary of data space, we are sure that p is the intersection point of two perpendicular bisectors, that is $p = B_{o_1:q} \cap B_{o_2:q}$. Now for a vertex v_i in the influence zone sent from the LBS provider, if we can not get two perpendicular bisectors $B_{o_1:q}$ and $B_{o_2:q}$ where $v_i = B_{o_1:q} \cap B_{o_2:q}$ and $o_1 \in S_1(v_i)$ and $o_2 \in S_1(v_i)$, we can infer that v_i is a fake vertex. Furthermore, the bisector $B_{o_j:q}$ for all objects o_j in $S_2(p)$ can prune p, so $counter(p) = |S_2(p)|$.

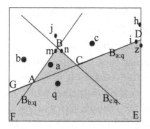

Fig. 5. Verifying the vertex

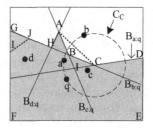

Fig. 6. Fake influence zone

Based on the definition of Z_k, for each edge e_i of Z_k, suppose its associated two endpoints are v_i and v_{i+1}, if e_i originates from a perpendicular bisector, it meets the condition: $\forall p \in e_i \wedge p \neq v_i \wedge p \neq v_{i+1}$, $counter(p) = k - 1$. On the other hand, if e_i lies on a boundary of D, we have $\forall p \in e_i \wedge p \neq v_i \wedge p \neq v_{i+1}$, $counter(p) \leq k-1$. Below, Lemma 5 depicts how to authenticate a vertex in the influence zone.

Lemma 5. *Given a vertex v_i in Z_k, there are two edges, e_{i_1} and e_{i_2}, associated with v_i in the influence zone. If these two edges do not lie on the boundaries of data space D, two points, m and n, which are infinitely close to v_i, are taken from e_{i_1} and e_{i_2} respectively, and an additional point j is also taken which lies outside Z_k and is infinitely close to v_i. If $counter(m) = counter(n) = k - 1$ and $counter(j) \geq k$, then v_i is a legal vertex in Z_k.*

Proof. As shown in Fig. 5, the points m and n that are infinitely close to B are taken from the edge AB and BC respectively. Note that AB is on the $B_{b:q}$ while BC lies on the $B_{c:q}$. We take the point j that is infinitely close to the vertex **B** and lies outside Z_k. All the perpendicular bisectors which prune m can also prune j. Furthermore, j is pruned by an additional perpendicular bisector $B_{b:q}$. So we have $counter(j) \geq counter(m) + 1$. It is also the case for the point n and j, i.e., $counter(j) \geq counter(n) + 1$. Now since we have $counter(m) = counter(n) = k - 1$ and $counter(j) \geq k$, we can infer $j \notin Z_k$ and vertex **B** is a legal vertex in Z_k.

Note that, if one of the associated edges e_i lies on the boundary of data space, for each point $p \in e_i$ except the two endpoints of e_i, we have $counter(p) \leq k-1$, for example, in Fig. 5, $counter(z) \leq k - 1$. The vertices on the boundaries of data space can be verified in the similar way as described in Lemma 5.

There are two possible fake situations in the influence zone Z_k^{out} accepted by a mobile client: (i) $Z_k^{out} > Z_k^{real}$; (ii) $Z_k^{out} < Z_k^{real}$ where Z_k^{real} is the real Z_k. For instance, in Fig. 6, the shaded area is Z_3^{real}. If $Z_3^{out} = ACDEFGHA$, it falsely adds $\triangle ABC$ into Z_3^{real}. If $Z_3^{out} = ABCDEFIJHA$, $\triangle GIJ$ is falsely omitted from Z_3^{real}. However, these two types fake Z_3^{out} can be easily detected by vertex verification. Specifically, suppose it starts to make verification on the vertex A in $Z_3^{out} = ACDEFGHA$, we assume two points p_1 and p_2 are taken from the edge HA and AC respectively, it is easy to find $counter(p_2) = 3$,

($p_2 \in AC$ can be pruned by $B_{a:q}$, $B_{b:q}$, $B_{c:q}$), which indicates AC is a fake edge. So the verification process should terminate. If we make authentication on $Z_3^{out} = ABCDEFIJHA$, suppose it is time to make verification on vertex J (or I), because the edge IJ does not lie on any perpendicular bisector and boundary of data space, thus, IJ can be judged as a fake edge based on Lemma 4.

In the above discussion, some kinds of fake Z_k can be detected when we verify the vertices. However, there are other kinds of fake Z_k which can not be detected by vertex verification. For example, in Fig. 6, if $Z_3^{out} = GCDEFG$, $\triangle AHB$ is omitted from Z_3^{real} in fact. But this error cannot be detected in the procedure of vertex verification. The reason is that, each edge in Z_3^{out} originates from the perpendicular bisectors, or from the boundaries of the data space. Thus, it can pass examination on the edges based on Lemma 4. Furthermore, each vertex can also be verified successfully based on Lemma 5. Therefore, only making verification on the vertices is not enough and it needs to make authentication on the edges. Below, we describe the feature of the counter values of the points on a line segment in a perpendicular bisector or a boundary of data space.

Theorem 2. *Given a line segment AB in a two dimensional data space D, which originates from a perpendicular bisector $B_{o_i:q}$ or from a boundary L_i of data space, if there is no other bisector $B_{o_j:q}$ intersecting with $B_{o_i:q}$ (L_i) in AB, the counter values for all points except two endpoints A and B are equal.*

Proof. Given a line segment AB in a two dimensional data space D, suppose it lies on $B_{o_i:q}$, if the other bisector $B_{o_j:q}$ has no intersection with $B_{o_i:q}$, we can infer $B_{o_i:q}$ and $B_{o_j:q}$ are parallel. Otherwise, these two bisectors must intersect with each other. There are only two cases that the other bisectors and $B_{o_i:q}$ are parallel, as shown in Fig. 7: (i) $B_{o_j:q}//B_{o_i:q}$, (ii) $B_{o_z:q}//B_{o_i:q}$. In both cases, for two arbitrary points $o_i \in AB$ and $o_j \in AB$, we have $counter(o_1) = counter(o_2)$.

If the other $B_{o_j:q}$ intersects with $B_{o_i:q}$, suppose $p = B_{o_j:q} \cap B_{o_i:q}$, and p is not in AB, or it may be at two endpoints, s.t. $p \notin AB$, or $p = A$ or $p = B$, there are four intersecting cases, which are depicted in Fig. 8. For case (i), suppose $p = B_{o_h:q} \cap B_{o_i:q}$, we find $p \notin AB$. Because $B_{o_h:q}$ cannot prune any point on AB, for $o_1 \in AB$ and $o_2 \in AB$, we have $counter(o_1) = counter(o_2)$. For case (ii), we assume $p' = B_{o_z:q} \cap B_{o_i:q}$, and find $p' \notin AB$, although $B_{o_z:q}$ can

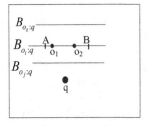

Fig. 7. Parallelism

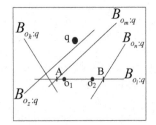

Fig. 8. Intersecting with AB

prune two arbitrary points o_1 and o_2 on AB, s.t. $counter(o_1) = counter(o_1) + 1$ and $counter(o_2) = counter(o_2) + 1$, $counter(o_1) = counter(o_2)$ is still held. For case (iii), because of $B_{o_m:q} \cap B_{o_i:q} = A$, $B_{o_m:q}$ can prune all the points **except** endpoint A so $counter(o_1) = counter(o_2)$. For case (iv), if $B_{o_n:q}$ is considered, it cannot prune any point on AB, thus, $counter(o_1) = counter(o_2)$ is also guaranteed. Note that, if AB lies on a boundary (L_i) of data space, the verification can be discussed in the similar way. Therefore, the theorem is proved.

The following lemma provides hints on how to verify the edge in influence zone Z_k.

Lemma 6. *Given an edge AB in the output influence zone with its two endpoints A and B verified to be legal vertices, if there is no intersection point on AB except two endpoints A and B, AB is a legal edge.*

Proof. Note that in order to verify vertex A and B, two points m and n, which are infinitely close to A and B respectively, are taken from the edge AB. Since A and B pass the vertex verification, based on Lemma 5, we have $counter(m) = counter(n) = k - 1$ if AB originates from a perpendicular bisector. Furthermore, there is no intersection point on AB except two endpoints A and B , which satisfies the condition required in Theorem 2, therefore, we can infer that $\forall o_i \in AB \wedge o_i \neq A \wedge o_i \neq B$, $counter(o_i) = k - 1$. Note that, if AB originates from a boundary of data space, $\forall o_i \in AB \wedge o_i \neq A \wedge o_i \neq B$, $counter(o_i) \leq k - 1$. Thus, AB is a legal edge. $\qquad \square$

Furthermore, we can check whether AB is legal based on the number of bisectors intersecting with it, which is described in following lemma.

Lemma 7. *Given an edge AB in the output influence zone in a two dimensional data space, suppose its two endpoints A and B are verified as legal vertices, if there are an* odd *number of perpendicular bisectors intersecting with AB, and each intersection point p_i satisfies $p_i \in AB \wedge p_i \neq A \wedge p_i \neq B$, AB is an illegal edge.*

Proof. Given an edge AB of an influence zone, if there is a perpendicular bisector B_i intersecting with AB, and their intersection point p_i satisfies: $p_i = AB \cap B_i \wedge p_i \neq A \wedge p_i \neq B$, $\forall o_i \in Ap_i$ and $\forall o_j \in p_iB$, $|counter(o_i) - counter(o_j)| = 1$, which means if B_i prunes points on Ap_i, it does not prune points on p_iB. In order to ensure the counter values for points on both sides of p_i on AB are equal, there should be an other bisector B_j passing p_i in the opposite pruning direction of B_i. However, since there are an *odd* number of perpendicular bisectors intersecting with AB, wherever they intersect with AB, the counter values of the points on AB except two endpoints A and B can not be equal, which contradicts the characteristics of a legal edge. Therefore, the lemma is proved. $\qquad \square$

Based on above discussion, if there are an *even* number of perpendicular bisectors intersecting with e_i, it has an opportunity to become a legal edge. So it needs to do further verification and the detailed steps is described in algorithm 2.

Algorithm 2. Authenticate Influence Zone (q,$\mathcal{VO}$,k)

Result: Influence Zone, RkNN

1 $Z_k = \mathcal{VO}.Result()$;

2 check $q \in Z_k$;

3 h'_{root} = reconstruct the root digest from $\mathcal{VO}$;

4 authenticate h'_{root} against the MR-tree root signature;

5 **if** h'_{root} *is correct* **then**

6   D' = the set of data points extracted from $\mathcal{VO}$;

7   R' = the set of non-leaf entries extracted from $\mathcal{VO}$;

8   **if** $\forall e \in R', e \cap (\forall v_i \in V, \odot(v_i, |v_i, q|)) = \phi$ **then** /* V is the vertices of Z_k */

9   **if** *each vertex of Z_k is verified* **then**   /* Lemma 5 */

10   **for** *each edge $e_i \in Z_k$* **do**

11   **if** *draw C_{v_i} and $C_{v_{i+1}}$ to collect BS* **then**

12        $P = \{p_i | p_i = B_i \cap e_i \wedge p_i \neq v_i \wedge p_i \neq v_{i+1} \wedge B_i \in BS\}$;

13   **if** $|P| \neq 0$ **then**   /* Lemma 6 */

14         $IB = \{B_i | p_i = B_i \cap e_i \wedge p_i \neq v_i \wedge p_i \neq v_{i+1} \wedge B_i \in BS\}$

        if *the number of $|IB|$ is even* **then** /* Lemma 7 */

15          $P = \{p_i | p_i = B_i \cap e_i \wedge B_i \in IB\}$;

16   **if** $\exists p_i \in P \wedge p_i$ *is a legal vertex* **then**

17   return e_i error

18   **else**

19   return e_i error

20   **else**

21   return vertex error

22   return $RkNN = \{o_i | o_i \in Z_k\}$

23   return authentication failed;

The pseudo-code of the verification method at a client side is shown in Algorithm 2, which avoids computing Z_k from scratch. Upon receiving the $\mathcal{VO}$ from the server, the algorithm first retrieves Z_k by calling $\mathcal{VO}.Result()$. It then judges whether q is in Z_k. If q is not in Z_k, the algorithm reports an error and terminates. Next, it reconstructs the root digest from the $\mathcal{VO}$ and verifies it against the MR-tree root signature signed by the data owner (Lines 4). If this verification is successful, the $\mathcal{VO}$ is guaranteed to contain only entries from the original MR-tree. Next, it starts to conduct the geometric verification for Z_k. Particularly, it extracts from the $\mathcal{VO}$: (i) a set D' of data points, and (ii) a set R' of non-leaf entries (Lines 6-7). As presented in Section 4.1, every non-leaf entry of R' should not intersect $\odot(v_i, |v_i, q|)$ where v_i is a vertex of Z_k (Line 8). If this geometric condition is satisfied, it begins to verify each vertex and edge one by one in clockwise direction. We first conduct vertices verification and some kinds of fake Z_k can be detected at this step. And then, we make verification on each edge (Line 10). In order to verify e_i, we draw C_{v_i} and $C_{v_{i+1}}$ at two vertex v_i and v_{i+1} associated with e_i to collect *object set* $OS = \{o_i | o_i \in C_{v_i} \vee o_i \in C_{v_{i+1}}\}$ and compute *perpendicular bisector set* $BS = \{B_{o_i:q} | o_i \in OS\}$ (Line 11). The algorithm should check whether the perpendicular bisector in BS can intersect with e_i, thus, the algorithm needs to compute the intersection set P (Line 12). If there is no intersection point in e_i, i.e., $|P| = 0$, based on Lemma 6, e_i is a legal edge. Otherwise, we continue to conduct further verification. Particularly, the number of *intersection perpendicular bisector IB* is computed (Line 14). Based on Lemma 7, if the number of IB, $|IB|$, is odd, it indicates e_i is a fake edge, the algorithm should report an error (Line 19). If $|IB|$ is even, it has to do further verification on e_i. If $p_i \in e_i$ is verified as a legal vertex, e_i is an illegal edge (Lines 14-17). If Z_k is authenticated, the RkNN can be readily obtained (Line 22).

5   Experiments

5.1   Experimental Settings

In this section, we evaluate the performance of our methods experimentally. We implemented all methods in C++ and used cryptographic functions in the Crypto++ library[1]. Experiments are run on a PC with Intel Core2 Duo 3GHz CPU and 4 GB memory. We employ the SHA-256 as the hash function and 1024-bit RSA as the signature scheme. We evaluate our methods using two real-world datasets obtained from the U.S. Census Bureau[2]: 1) **NA** which contains 569k data points from North America, and 2) **LA** which consists of 1314k data points from Los Angeles. In order to evaluate bichromatic RkNN queries, we randomly divide these points into two sets of almost equal sizes. One of the set corresponds to the set of facilities and the other to the set of users. We vary k from 2 to 128 and the default value is 8. The page size is set to 4096 bytes.

VN-Auth in [4] is designed only for the monochromatic RkNN verification. It cannot be used to bichromatic RkNN verification directly. Now we describe a baseline method (**BAS**). Once the client obtains the $\mathcal{VO}$ from the LBS provider, BAS does the same preliminary process (Lines 1-8) as in Algorithm 2. Then, based on the data set D' (Line 6), BAS adopts the method in [1] to compute Z_k. The objects $o_i \in Z_k$ are the RkNN result.

Our approach proposed to verify both types of RkNN queries is based on Advance Influence Zone Authenticaion. We call the method designed for the client shown in Algorithm 2 **AIZ-Auth** . The approach designed for the server side depicted in Algorithm 1 is called **IZ-Veri**. Due to space limitation, only parts of experiment results are listed in following subsection.

---

[1] http://www.cryptopp.com
[2] http://www.census.gov/geo/www/tiger/

5.2   Various Testing

As shown in Fig. 9a, as the value of k increases, the size of $\mathcal{VO}$ in both algorithms grows larger. The reason is that, the cardinality of the result set for an RkNN query is not determined by the parameter k. but depends on the actual data distribution. Furthermore, some auxiliary objects used to verify query results are also contained in the $\mathcal{VO}$, while the auxiliary objects take a large part of the $\mathcal{VO}$.

Fig. 9b demonstrates the LBS time as a function of the query parameter k. In comparison to VN-Auth scheme at the server, our method IZ-Veri gains a better performance. This is mainly because i) Influence zone is the best known algorithm for RkNN retrieval, which is experimentally verified in literature [1] and ii) the computation of $\mathcal{VO}$ based on C_{v_i} of each vertex of influence zone is a high efficient method.

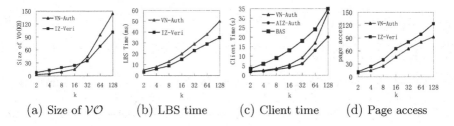

(a) Size of $\mathcal{VO}$ (b) LBS time   (c) Client time   (d) Page access

Fig. 9. Monochromatic RkNN, Varying k on NA

Fig. 9c depicts the client verification time as a function of the query parameter k. As the value of k increases, the number of candidate objects grow exponentially. AIZ-Auth gains better performance than VN-Auth. This is because, *AIZ-Auth* verifies each vertex and edge based on the C_{v_i}, which is more simple and efficient than *six equal partition method* adopted by VN-Auth.

Fig. 9d shows I/O cost *at server*. VN-Auth results in less I/O accesses our method. However, the efficient I/O cost of VN-Auth is based on expensive data transformation process at data owner (DO). As studied in literature [4], DO first has to transforms each object by attaching neighborhood and authentication information, which can take about more than 50 minutes on NA dataset [4]. However, our method does not need this transformation step.

The results of experiment for monochromatic RkNN on LA dataset are similar to that in Fig. 9. Due to space limitation, we do not show them here. In order to evaluate bichromatic RkNN queries, we randomly divide **LA** data set into two sets of almost equal sizes. One of the set corresponds to the set of facilities and the other to the set of users. Because no literature studied the bichromatic RkNN queries, we only compare the *BAS* and *AIZ-Auth* at client verification, which are shown in Fig. 10. *AIZ-Auth* has a better performance than BAS. Furthermore, although LA dataset is much larger than NA dataset, comparing with Fig. 9, we find that AIZ-Auth still has a steady performance, which indicates AIZ-Auth is quite scalable with respect to data size.

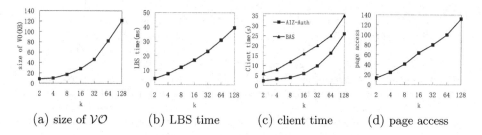

(a) size of $\mathcal{VO}$ (b) LBS time   (c) client time   (d) page access

Fig. 10. Bichromatic RkNN, Varying k on LA

6   Conclusion

In this paper, we present a framework for authenticating RkNN queries based on influence zone. The proposed methods can be used to authenticate both types of RkNN results. Experimental results show that our verification methods can gain a good performance. Our future work will study how to verify the continuous RkNN results.

Acknowledgments. This work is supported by the State Key Program of National Natural Science of China under Grant No. 61332001, National Natural Science Foundation of China under Grants Nos. 61173049, 61300045,61309002 and China Postdoctoral Science Foundation under Grant No. 2013M531696.

References

1. Cheema, M.A., Lin, X., Zhang, W., Zhang, Y.: Influence zone: efficiently processing reverse k nearest neighbors queries. In: 2011 IEEE 27th International Conference on Data Engineering (ICDE), pp. 577–588. IEEE (2011)
2. Cheema, M.A., Lin, X., Zhang, Y., Wang, W., Zhang, W.: Lazy updates: An efficient technique to continuously monitoring reverse knn. Proceedings of the VLDB Endowment **2**(1), 1138–1149 (2009)
3. Chen, Q., Hu, H., Xu, J.: Authenticating top-k queries in location-based services with confidentiality. Proc. of the VLDB Endowment **7**(1), 49–60 (2014)
4. Hu, L., Ku, W.S., Bakiras, S., Shahabi, C.: Spatial query integrity with voronoi neighbors. IEEE Transactions on Knowledge and Data Engineering **25**(4), 863–876 (2013)
5. Jing, Y., Hu, L., Ku, W.S., Shahabi, C.: Authentication of k nearest neighbor query on road networks. IEEE Transactions on Knowledge and Data Engineering **26**(6), 1494–1506 (2014)
6. Korn, F., Muthukrishnan, S.: Influence sets based on reverse nearest neighbor queries. In: ACM SIGMOD Record, vol. 29, pp. 201–212. ACM (2000)
7. Li, F., Hadjieleftheriou, M., Kollios, G., Reyzin, L.: Dynamic authenticated index structures for outsourced databases. In: Proceedings of the 2006 ACM SIGMOD International Conference on Management of Data, pp. 121–132. ACM (2006)

8. Lin, X., Xu, J., Hu, H., Lee, W.C.: Authenticating location-based skyline queries in arbitrary subspaces. IEEE Transactions on Knowledge and Data Engineering **26**(6), 1479–1493 (2014)

9. Merkle, R.C.: A certified digital signature. In: Brassard, G. (ed.) CRYPTO 1989. LNCS, vol. 435, pp. 218–238. Springer, Heidelberg (1990)

10. Mykletun, E., Narasimha, M., Tsudik, G.: Authentication and integrity in outsourced databases. ACM Transactions on Storage (TOS) **2**(2), 107–138 (2006)

11. Tao, Y., Papadias, D., Lian, X.: Reverse knn search in arbitrary dimensionality. In: Proceedings of the Thirtieth International Conference on Very Large Data Bases, vol. 30, pp. 744–755. VLDB Endowment (2004)

12. Wu, D., Choi, B., Xu, J., Jensen, C.: Authentication of Moving Top-k Spatial Keyword Queries (to appear) (2014)

13. Wu, W., Yang, F., Chan, C.Y., Tan, K.L.: Continuous reverse k-nearest-neighbor monitoring. In: 9th International Conference on Mobile Data Management, MDM 2008, pp. 132–139. IEEE (2008)

14. Wu, W., Yang, F., Chan, C.Y., Tan, K.L.: Finch: Evaluating reverse k-nearest-neighbor queries on location data. Proceedings of the VLDB Endowment **1**(1), 1056–1067 (2008)

15. Xie, M., Wang, H., Yin, J., Meng, X.: Integrity auditing of outsourced data. In: Proceedings of the 33rd International Conference on Very Large Data Bases, pp. 782–793. VLDB Endowment (2007)

16. Yang, Y., Papadopoulos, S., Papadias, D., Kollios, G.: Spatial outsourcing for location-based services. In: IEEE 24th International Conference on Data Engineering, ICDE 2008, pp. 1082–1091. IEEE (2008)

17. Yang, Y., Papadopoulos, S., Papadias, D., Kollios, G.: Authenticated indexing for outsourced spatial databases. The VLDB Journal, The International Journal on Very Large Data Bases **18**(3), 631–648 (2009)

18. Yiu, M.L., Lo, E., Yung, D.: Authentication of moving knn queries. In: 2011 IEEE 27th International Conference on Data Engineering (ICDE), pp. 565–576. IEEE (2011)

19. Yung, D., Lo, E., Yiu, M.L.: Authentication of moving range queries. In: Proceedings of the 21st ACM International Conference on Information and Knowledge Management, pp. 1372–1381. ACM (2012)

Author Index

rinted in the United States
y Bookmasters